Electricity in the 17th and 18th Centuries

ELECTRICITY IN THE 17TH AND 18TH CENTURIES

A STUDY OF EARLY MODERN PHYSICS

J. L. Heilbron

UNIVERSITY OF CALIFORNIA PRESS

Berkeley • Los Angeles • London

University of California Press
Berkeley and Los Angeles, California

University of California Press, Ltd.
London, England

Copyright © 1979 by
The Regents of the University of California

ISBN 0-520-03478-3
Library of Congress Catalog Card Number: 77-76185

1 2 3 4 5 6 7 8 9

FOR MY WIFE

Contents

Acknowledgments

This book began many years ago in a seminar conducted by T. S. Kuhn at the University of California, Berkeley. I am indebted to him and to other friends and colleagues, P. Forman, G. Freudenthal, R. Hahn, R. Home, A. Quinn, and S. Weart, for comments about the manuscript. Interpretation of texts has been much facilitated by the old histories of Gralath and Priestley, and by the more recent studies of Gliozzi, Torlais, Cohen, and Home. Permission to quote unpublished material has been granted by the Università Gregoriana (Rome), the Biblioteca Nazionale-Centrale (Florence), the Académie des Sciences (Paris), the Royal Society and the British Library (London), the Royal Observatory (Herstmonceux), the Staatsbibliothek (Berlin), the American Philosophical Society (Philadelphia), the Bancroft Library (Berkeley), the Yale University Library, and the Burndy Library (Smithsonian Institution, Washington).

Producing a work of this size and complexity offers endless opportunities for error. That I did not seize more of them is owing to the intelligence and vigilance of Judy Fox, Nancy Noennig, Carolyn Teragawa, and others at the Office for History of Science and Technology at the University of California, Berkeley.

A Note on the Notes

The notes give abbreviated titles of books and omit those of journal articles; full titles of both and other pertinent information will be found in the Bibliography. Arabic italics are used for volume numbers of journals, roman numerals for individuals of a multi-volume work or of a manuscript collection (excepting the Sloane Mss.). References within this book are given in the form '*infra* (or *supra*), XII.2,' meaning chapter XII, section 2. Small superscripts thus (^{2}) indicate the edition cited. 'x:y (1900)' signifies part or item y of volume x; if the volume number is not given, the form is '1900:y.' The following abbreviations are also used:

ADB	*Allgemeine deutsche Biographie.* 56 vols. Leipzig, 1875–1912.
AKSA	Kungl. Svenska Vetenskapsakademien, Stockholm. *Der Königl. schwedischen Akademie der Wissenschaften, Abhandlungen, aus der Naturlehre, Haushaltungskunst und Mechanik.* Tr. Abraham Gotthelf Kästner. Hamburg and Leipzig, 1749–65, Leipzig, 1766–83.
AS	Académie des Sciences, Paris.
BF	Benjamin Franklin.
BL	British Library.
BFP	Benjamin Franklin. *Papers.* Ed. Leonard W. Labaree et al. New Haven, 1959+.
BFP (Smyth)	Benjamin Franklin. *The Writings of Benjamin Franklin.* Ed. A. H. Smyth. 10 vols. New York, 1905–7.
CAS	Akademiiã nauk S.S.S.R., Leningrad. *Commentarii academiae scientiarum imperialis petropolitanae.*
CK	Carteggio Kircheriano. Mss. of Athanasius Kircher (mainly incoming letters). 13 vols. Università Gregoriana, Rome.
DBI	*Dizionario biografico degli italiani.* Rome, 1960+.
DM	William Gilbert. *De magnete.* London, 1600. *DM*(Mo) signifies the translation by P. F. Mottelay (New York, 1893); *DM*(Th) that of S. P. Thompson (London, 1900).
DNB	*Dictionary of National Biography.* 22 vols. Oxford, 1921–2.
DSB	*Dictionary of Scientific Biography.* New York, 1970+.
EO	Benjamin Franklin. *Experiments and Observations on Electricity.* EO without further qualification will refer to the annotated edition of EO^5 (London, 1774), published by I. B. Cohen as *Benjamin Franklin's Experiments* (Cambridge, 1941). Other editions, when required, will be plainly indicated, e.g., EO^1, the first English edi-

	tion (London, 1751–4); *EO* (Dal²), the second French edition of T. Dalibard (Paris, 1756); *EO* (Wilcke), the German version of J. C. Wilcke (Leipzig, 1758). For titles of these editions see under Franklin in the Bibliography.
FN	I. B. Cohen. *Franklin and Newton.* Philadelphia, 1956.
GGA	*Göttingische gelehrte Anzeigen.*
GM	*Gentleman's Magazine.*
HAS	Académie des Sciences, Paris. *Histoire.*
HAS/Ber	Akademie der Wissenschaften, Berlin. *Histoire.*
JB	Royal Society of London. Journal Book.
JHI	*Journal of the History of Ideas.*
JP	*Journal de physique.*
MAS	Académie des Sciences, Paris. *Mémoires.*
MAS/Ber	Akademie der Wissenschaften, Berlin. *Mémoires.*
Mss. Gal.	Manoscritti Galileiani. Biblioteca Nazionale-Centrale, Florence.
NCAS	Akademiia nauk S.S.S.R., Leningrad. *Novi commentarii academiae scientiarum imperialis petropolitanae.*
PT	Royal Society of London. *Philosophical Transactions.*
RHS	*Revue d'histoire des sciences et de leurs applications.*
RS	Royal Society of London.
RS Edin.	Royal Society of Edinburgh.
Sloane Mss.	Sloane Manuscripts. British Library, London.
VE	Alessandro Volta. *Epistolario.* 5 vols. Bologna, 1949–55.
VO	Alessandro Volta. *Le Opere.* 7 vols. Milan, 1918–29.

NOTE ON CONVERSIONS

The basic units of reference are the Paris livre of 1726 (silver) and the louis of 24 livres (gold). The value of any other currency is taken as the ratio of its precious metal content, as given by Martini, *Metrologia* (1883), to that of the livre or the louis. Some frequently used conversions:

Currency	*Symbol*	*Equivalent*
Livre	#	
British pound/guinea	£/gn.	25#
Dutch florin	f.	2#
Reichsthaler	RT	4#
Swedish daler (copper)	Dkmt	⅓#

Introduction

The Scientific Revolution of the seventeenth century did not affect the several branches of natural philosophy equally. Some sciences, like astronomy, mechanics and geometrical optics, already far advanced in antiquity, were then transformed into prototypes of modern, quantitative, instrumentalist physics. Other sciences, like chemistry, exchanged one set of unproductive concepts for another. Still others, like pneumatics and electricity, were created or recreated in the Scientific Revolution.[1] Some of these new sciences managed to join the first group, the transformed 'mixed mathematics,' by the end of the eighteenth century. Electricity led the way. Hence the propriety of treating its history as the exemplar of early modern physics.

This history had three stages. The first, which lasted until 1700, was one of narrow exploration and premature systematization, dominated by professional pedagogues and polymaths sympathetic to the natural philosophy of Aristotle. The second stage, which occupied the first half of the eighteenth century, brought wider exploration and the discovery of phenomena difficult to account for on the old philosophies. Here a new type, the unmathematical 'experimental philospher,' who followed Descartes or Newton as best he could and who earned some or all of his livelihood from his science, contributed most conspicuously. The third stage occurred during the last part of the eighteenth century when electricity was quantified, at least as regards the classic problems of electrostatics. The quantifying physicist accomplished his purpose by dropping scruples that had weighed with his predecessors and by adopting an instrumentalism that would have scandalized most of them.

Much of the historiography of the scientific revolution has centered on the development of terrestrial and celestial mechanics or on the spread of the 'corpuscular philosophy,' which many regard as the common denominator of the revolutionaries. Studies of eighteenth-century physics also tend to dwell on cosmological disputes, such as the squabbles between the sectaries of Descartes and Newton. These emphases have produced much excellent and essential work. They have also propagated serious distortions.

Our history shows, for example, that despite their disagreement over theory, in practice the Newtonian experimental philosopher thought in much the same terms as his Cartesian counterpart, aether being to the one what subtle matter

---

1. Cf. Kuhn, *J. Interd. Hist.*, 7 (1976), 1–31. Full citations are given under author's name in the Bibliography, abbreviations on pp. xiii–xiv.

was to the other;[2] that each side held experiment in high esteem; and that the achievement of quantification confounded the programs of both. It also shows that the 'Copernican Revolution' does not adequately represent the transition from medieval natural philosophy to classical physics. The bullish personality of Galileo, local jealousies, the post-Tridentine paranoia of the Roman Church, and the apparent bearing of scripture on questions of cosmic geometry combined to introduce into astronomy issues that divided men otherwise able to cooperate in the creation of a new science. Galileo's straw man Simplicio, the hack Aristotelian who opposes every novelty in the *Dialogue on the Two Chief Systems of the World,* still hoodwinks historians into believing that the peripatetics contributed nothing to the scientific revolution but unreasoning opposition.[3] Studies of the development of sciences that, like electricity, were theologically and cosmologically neutral, should help to provide a juster estimate of the contributions, the expectations, and the changing composition of the early modern physicists.

I begin with a sketch of the general conditions under which physics advanced during the seventeenth and eighteenth centuries. First comes an account of the general principles to which electrical theory at different times conformed, or that otherwise mediated its development: peripatetic philosophy, corpuscularism, Newton's attractions, Newtonian forces and fluids. Much of this ground has been tilled before; what is new is the extent of the coverage and an emphasis on application, rather than logical analysis, of the principles. Chapter I opens with indications of changes in the meaning and scope of 'physics,' and closes with examples of the successful mathematizing of its newer branches. This last section breaks new ground.

The second chapter describes the institutional frameworks in which physics was cultivated in the seventeenth and eighteenth centuries. I had two purposes in mind when preparing it. First, to show the opportunities offered and the constraints imposed by organized learning; second, to provide the beginnings of a demography of physicists, their numbers, salaries, and career goals. These factors conditioned the pace and extent of study of natural phenomena; at a level of support and urgency comparable to today's, early modern physicists no doubt could have brought knowledge of electrical phenomena from its very beginnings with Gilbert to the order given it by Dufay in a few years. It took a century and a third.

The single most important contributor to the support of the study of experimental physics in the seventeenth century was the Catholic Church, and, within it, the Society of Jesus. From about 1670 to about 1750, private lecturers played an important part in keeping up 'experimental philosophy;' while throughout most of the eighteenth century universities and academies dominated the investigation of physical, including electrical, phenomena. Chapter II

2. Cf. Martine, *Examination* (1740), 18: 'May not the *ignis* of some folks, the *aether* of others, the *materia subtilis* of others be at bottom (when thoroughly explained) one and the same thing. . . ?'

3. A recent example is Clavelin, *Philosophie* (1968), 391–5.

considers each group in turn, Jesuits, academicians, professors and private lecturers. Together Chapters I and II constitute an introduction to the history of early modern physics. A more comprehensive account would no doubt be desirable. None exists, however, whence the need, and, I hope, the utility of the first part of this book.

The malevolence of inanimate objects is nowhere better instanced than in the phenomena of frictional electricity. Their apparent caprice constantly frustrated the efforts of early theoreticians trying to reduce them to rule. Consider the effect of moisture on the surfaces of insulators and in the air surrounding them. The early electricians realized that contact with water enervated an otherwise vigorous electric, like amber, but they did not fully recognize the effect of humidity. On a sultry summer day, or in the presence of a sizable perspiring audience, experiments that had often succeeded might suddenly and inexplicably fail; while the operator himself, sweating at his task, helped to dissipate the charges he intended to collect. Nor did the early electricians recognize the inverse phenomenon, the promotion of electrical power by dessication, which may have extraordinary effects. In the winter of 1766–7, for example, after weeks of sunny, sub-zero weather, so much moisture had evaporated from the furnishings of the Royal Palace in St. Petersburg that much of the upholstery and even the floor, which generally conducted electricity, had become insulators. Charges appeared everywhere: by playing catch with a sable muff the crown prince could electrify himself enough to fling sparks from his fingers at the noses of his friends.[4] The early electrician secured many apparently inconsistent results by failing to appreciate that drying a substance might promote its electricity as much as wetting diminished it.

Unique triboelectric difficulties afflict glass. For one, the *texture* of the surface plays a role: substantial electrical charges may be developed by rubbing a smooth piece of glass against a rough one. The age and temperature of the surface are also consequential. Newly made glass often conducts passably and insulates only as its surface weathers, while glass abused in electrical experiments may briefly lose its triboelectric property if the rubbing exposes enough fresh surface. Conductivity also varies with temperature. A red-hot piece of glass may pass enough current to give a disagreeable shock to anyone in series with it and a charged Leyden jar. The electrical quirks of glass came to light only during the second half of the eighteenth century.[5]

Gems and precious stones, to which the electricians of the seventeenth century paid particular attention, are more capricious than glass. Robert Boyle owned two cornelians, only one of which he could electrify; and the gentlemen of the Accademia del Cimento, though they could usually excite diamonds,

4. Aepinus, RS Edinb., *Trans.*, 2 (1790), 234–44.

5. Priestley, *Hist.* (1775³), II, 178–82, 215–25, 230–1; Cavallo, *PT*, 67:2 (1777), 388; Achard, *Schriften* (1780), 264; Franklin to 'Dr. L.,' 18 March 1755, *EO*, 331. Cf. Harper, *Contact* (1967), *passim.*

invariably failed with stones cut or mounted in a certain way. Physicists now know that different cleavage planes of a crystal may have dissimilar triboelectric properties; strong charges can be produced by rubbing together surfaces corresponding to different cleavages of the same crystal. The old electricians, who did not know that the cut of a stone, its setting, and the manner of rubbing affect its electrical properties, found their data hopelessly conflicting.[6]

Making a virtue of necessity, the early electrician tended to emphasize the unruliness of triboelectric phenomena. A good example occurs in a message Newton sent the Royal Society of London on December 7, 1675:[7]

> I have sometimes laid upon a table a round piece of Glasse about 2 inches broad Sett in a brass ring, so that the glass might be ½ or ⅓ of an inch[8] from the table, & the Air between them inclosed on all sides by the ring, after the manner as if I had whelmed a little Sive upon the Table. And then rubbing a pretty while the Glass briskly with some ruff and rakeing stuffe, till some very little Fragments of very thin paper, laid on the Table under the glasse, began to be attracted and to move nimbly to and fro: after I had done rubbing the Glass the papers would continue a pretty while in various motions, sometimes leaping up to the glass & resting there a while, then leaping downe & resting there, then leaping up, & perhaps downe & up againe, & this sometimes in lines seeming perpendicular to the Table, Sometimes in oblique ones, Sometimes also they would leap up in one Arch & downe in another, divers times together, without Sensible resting between; Sometimes Skip in a bow from one part of the Glasse to another without touching the table, & Sometimes hang by a corner & turn often about very nimbly as if they had been carried about in the midst of a whirlwind, & be otherwise variously moved, every paper with a divers motion.

There were two novelties in Newton's message: glass replaced the usual generator, amber, and the paper bits flew to an untouched, not to a rubbed, surface. The intrigued Society undertook to repeat the demonstration; they succeeded once or twice and then failed altogether. Newton sent further instructions: the glass should be new, the paper bits triangles, the rubber rough, and the rubbing like the motion of a glass grinder. If the Fellows still could not budge the bits, they were to knock upon the glass with their finger tips; failing that, they should await Newton's next visit to London. They eventually succeeded by rubbing with stiff hog's bristles.[9]

The story has two points. First, it shows how little even informed savants, indeed an entire Royal Society, knew about electricity in the late seventeenth century. Their collective ignorance may serve as a benchmark for evaluating work one might have dismissed as trivial. The second point is that the regularities underlying electrostatic phenomena are very well hidden. Newton's

---

6. *Ibid.*, 86–9. A flat or 'table' diamond, if appropriately set and rubbed while worn, will charge like a condenser, and show very meager external signs. Cf. Magalotti, *Saggi* (1841³), 147, and *infra*, xvi. 3.

7. Newton, *Corresp.*, I, 364–5. The rationale of this experiment is considered *infra*, viii.1.

8. Cf. Newton, *Corresp.*, I, 392.

9. *Ibid.*, 392, 404, 407; Birch, *Hist.* (1756–7), III, 260–1, 271.

editor takes him to task for failing to find the principle of motion of the paper bits.[10] To have discerned the laws of plus and minus electricity, of superposed central forces, of insulators and conductors, in the whirling and skipping of the bits would have been a feat more remarkable than the invention of the gravitational theory.

The dominant phenomena had first to be identified. I have studied the detection of three of them in detail: repulsion, conduction and the sequence attraction-contact-repulsion (ACR). Since they are closely related, it is difficult to say when any one of them was first recognized. Several old electricians have been sponsored for the honor of discovering repulsion. None of them will do, however, if one requires the discoverer to have recognized repulsion as a distinct effect, coequal with attraction, and associated with conduction via ACR. Here we have a significant instance of the well-known problem of identifying a single individual as the discoverer of an effect the significance of which differs markedly according to the terms—to say nothing of the theory—used to describe and define it.[11]

The recognition of repulsion was completed by Dufay in the 1730s. Meanwhile conduction and the distinction between conductors and insulators had been discovered. Many other phenomena, including ACR, sparking and the two electricities came to light, and electricians, aware now of the basic electrostatic regularities, rushed to reduce them to system. They embroidered a theory developed in the seventeenth century: electric bodies shoot forth 'effluvia,' particles of electrical matter, which effect attraction and repulsion either by direct impact or by mobilizing the air. Many elaborate effluvial theories were invented in the 1740s, chiefly by continental physicists loyal to Cartesian principles. For several reasons I have been particular, perhaps overly so, in my accounts of these theories. First, because they illustrate the range of models and mechanical analogies exploited by physicists of the mid-eighteenth century. Second, because they show how mechanical principles were then used and understood by most physicists, who reasoned as if Newton and the Bernoullis had never existed. Third, because the details of qualitative theories, though often slighted, may, as in the case of electricity, determine the pace and direction in which physics moves.

The discovery of the Leyden jar in 1745 undermined all effluvial theories. It provided Franklin a proving ground for his theory of plus and minus electricity; in its later guises, as dissectible condenser, electrophore, condensatore and doubler, it accelerated the acceptance of a Newtonian approach to electrical theory. New apparatus played a capital role in altering ideas about electricity during the eighteenth century. I have paid special attention to the Leyden jar, the electrophore, and certain electrometers. To account adequately for the functioning of these devices, electricians invented the concepts of electrical ca-

---

10. Newton, *Corresp.*, I, 387, n. 6.
11. Kuhn, *Science*, 136 (1962), 760–4.

pacity, quantity of charge, and 'tension,' an anticipation of potential. These quantifiable quantities emerged slowly; once in hand they were immediately quantified, for example in Volta's equation (charge equals capacity times tension) and in the elaborate electrostatics of Coulomb, Cavendish and Poisson.

The concept of electric 'force' has an essential part in our story. Two usages must be kept distinct: force as observed macroscopic effect, e.g., moving force or kinetic energy, and *force* (italicized for identification) as an unobservable push or pull between elements of matter or charge. For reasons outlined in Part I, early Newtonians emphasized the first meaning; not until after 1750 did physicists become comfortable with the second. Now the triumph of Newton's theory of gravitation was to compute force by summing *forces*. It proved very difficult both conceptually and instrumentally to transfer his technique from celestial mechanics to terrestrial physics. The transfer was effected, after many false starts, by Robison, Cavendish and Coulomb. It was this move, and a similar simultaneous one in magnetic theory, that established a model, and presaged a future, for classical mathematical physics.

Electricity owed its lead to a circumstance peculiar to it: it attracted unusual and sustained attention. The demonstration of the identity of electricity and lightning, the discovery that electricity was one of nature's most powerful agents, made its theory a first priority. Physicists were conscious of the transformation Franklin's demonstration worked in their attitude towards electricity. Samuel Klingenstierna, professor of physics at the University of Uppsala, put the point in 1755, in an address on the latest discoveries in electricity: 'Forty years ago, when one knew nothing about electricity but its simplest effects, when it was regarded as an unimportant property of a few substances, who would have believed that it could have any connection with one of the greatest and most considerable phenomena in Nature, thunder and lightning?'[12] I have followed eighteenth-century discussions of lightning and lightning conductors as far as they concern the emergence of general electrostatic principles.

12. Klingenstierna, *Tal* (1755), 26.

PART ONE
Early Modern Physics and Its Cultivators

Now, to many a Royal Society, the Creation of a World is little more mysterious than the cooking of a Dumpling; concerning which last, indeed, there have been minds to whom the question, *How the Apples were got in*, presented difficulties.

—CARLYLE, *Sartor resartus*, I.1.

CHAPTER I

Physical Principles

1. THE SCOPE OF 'PHYSICS'

At the beginning of the seventeenth century 'physics' signified a qualitative, bookish science of natural bodies in general. It was at once wider and narrower than the subject that now has its name: wider in its coverage, which included organic and psychological as well as inorganic phenomena; and narrower in its methods, which recommended neither mathematics nor experiment. The width of coverage and the depreciation of mathematics derived from Aristotle; the indifference to experiment, as opposed to everyday experience, from the authors of peripatetic textbooks.

The *libri naturales,* or physical books of the Aristotelian corpus, begin with a treatise called *Physica,* which sets the categories of analysis of all natural bodies: form, matter, cause, chance, motion, time, place. After this *physica generalis* come the treatises of *physica specialis* or *particularis,* applications of the general principles to the heavens *(De caelo),* to inorganic nature *(De generatione et corruptione, Meteorologica),* to organic nature *(De partibus animalium),* and to man *(De anima).* The text books of the early seventeenth century offered epitomes of these ancient works, or rather epitomes of sixteenth-century handbooks, of which the most influential were the compendia of J. J. Scaliger and the enlightened commentaries of the Coimbra Jesuits.[1] Typical texts of the early seventeenth century are the *Idea philosophiae naturalis* (1622) of Frank Burgersdijck and the *Physiologia peripatetica* (1600) of Johannes Magirus, both long-lived, widely used and often reprinted, and now food only for the ultimate epitomizer, the historian.[2]

The authors of these texts were not physicists in the modern sense, but either professional philosophers or beginning physicians awaiting preferment in the practice of medicine. Above all the textbook writers were pedagogues, who aimed to supply not new material, but an improved arrangement of the old. One

1. Reif, *J. Hist. Ideas,* 30 (1969), 17–23; Ruestow, *Physics* (1973), 6, 17. The *Commentarii conimbricences* were widely used even in Protestant universities, e.g., Cambridge (H. F. Fletcher, *Milton,* II [1961], 169, 561), Oxford at least until 1678 (Reilly, *Line* [1969], 13), and many German schools. The Coimbra Jesuits define natural philosophy as the study of elements and of the bodies compounded of them. Coll. conim., *Comm. in Phys.,* I (1602), cols. 17–23.

2. Allen, *JHI,* 10 (1949), 225, 240, 243; Costello, *Schol. Curr.* (1958), 83–102; Ruestow, *Physics* (1973), 16–32; Thorndike, *Hist.,* VII (1958), 402–6; and H. F. Fletcher, *Milton,* II (1961), 168–75, including Magirus' table of contents.

went so far as to recommend false doctrines properly ordered over sound ones badly digested.[3] None suggested that physics might be advanced by experiment. The subtitle of Burgersdijck's *Idea* makes his objective clear enough: 'methodus definitionum et controversiarum physicarum,' a handbook of definitions and disputations for students wishing to wrangle over physics.[4] 'There are more things in Heaven and Earth, Horatio, than are dreamt of in your philosophy,' says Prince Hamlet. 'And more things in our compendia of physics,' answers a textbook writer, 'than can be found in earth or heaven.'[5]

For up-to-date general texts and reference works on physics, which describe experiments and the instruments used to perform them, one must look to books on natural magic, to J. C. Sturm's *Collegium curiosum* or the compendia of the Jesuit polymaths. Later we shall examine this literature, which kept the study of electricity alive during the seventeenth century. Now we need only confirm that, like early-modern physics, natural magic included all the sciences, physical and biological. According to Gaspar Schott, S. J., perhaps the best writer on the subject, 'magia universalis naturae et artis' covers vision, light, and everything pertaining to them; sound and hearing, with like accessories; white magic or applied mathematics; and the hidden, rare and uncommon things, the secrets of stones, plants and animals.[6]

The quantified portions of physical science fell not to physics in the seventeenth century but to 'mixed' or 'applied' mathematics, which customarily included astronomy, optics, statics, hydraulics, gnomonics, geography, horology, fortification, navigation and surveying. The association of mathematics with application gave philosophers who did not understand it a colorable cause to despise it; as John Wallis wrote of his experience at Cambridge in the 1630s, 'Mathematicks . . . were scarce looked upon as Academical Studies, but rather Mechanical, as the business of Traders, Seamen, Carpenters, Surveyors of Lands, or the like, and perhaps some Almanack-makers in London.' Neglect of mathematics in the English universities was doubtless linked to its odor of practicality, just as emphasis on it at Gresham College, London, reflected the concerns of City merchants and tradesmen.[7] 'Arithmetic,' says John Webster, in his well-known attack on the universities, '[is] transmitted over to the hands of Merchants and Mechanicks;' geometry is the province of 'Masons, Carpenters, Surveyors;' as for applied mathematics par excellence, 'in all the scholastick learning there is not found any piece . . . so rotten, ruinous, absurd and de-

3. Reif, *JHI*, 30 (1969), 29 quoting Bartholomaus Keckermann.

4. Ruestow, *Physics* (1973), 16.

5. Lichtenberg, *Aphorismen* (1902–8), III: 2, 37–8.

6. Schott, *Magia opt.* (1671), sig. 003v–004r. Cf. Heinrichs in Diemer, ed., *Wissenschaftsbegriff* (1970), 42. The Coimbra Jesuits take natural magic to be applied physics; Coll. conim., *Comm. in Phys.*, I (1602), cols. 18, 25.

7. Allen, *JHI*, 10 (1949), 226, 228, 231, 249; Johnson, *JHI*, 1 (1940), 430–4; Greaves, *Puritan Rev.* (1969), 65–7, 70; C. Hill, *Intell. Orig.* (1965), 122–3; Wallis in Scriba, RS, *Not. Rec.*, 25 (1970), 27.

formed' as Oxbridge astronomy.[8] No doubt Wallis and Webster exaggerated, but even those who defended the universities against the charge of neglect of mathematics conceded its tie to practical application.[9]

In the Jesuit schools, mathematics was taught, and taught well, but only in the vernacular, while the philosophy course spoke and wrote Latin. The mathematicians were so indulged because their technical terms, particularly those relating to fortification, could not be translated conveniently into the language of Cicero.[10] Perhaps the greatest mathematician trained by the Jesuits, Descartes, left school, he says, with the conviction that mathematics was 'useful only in the mechanical arts.'[11] Quantifying physics therefore implied a radical readjustment of the divisions of knowledge, including the downgrading of physics from philosophy to applied mathematics. It would be an uncomfortable process.[12]

'Physics' continued to be understood in its Aristotelian extent throughout the seventeenth century. Molière's bourgeois gentilhomme asks his philosophical tutor what physics is and receives in reply, '[the science] that explains the principles of natural things, and the properties of bodies; that discourses about the nature of the elements, metals, minerals, stones, plants and animals; and [that] teaches us the causes of all the meteors.' Molière's friend, the Cartesian physicist Jacques Rohault, says the same ('the science that teaches us the reasons and causes of all the effects that Nature produces') and he tries to give an account of everything, including human psychology, in a physics text that had a peculiarly long life.[13] John Harris' *Lexicon technicum* (1704) boils down Rohault's definition to 'the Speculative Knowledge of all Natural Bodies,' and adds angelology, on the authority of Locke. Then there is Sturm's important *Physica electiva*, which does not treat the organic world; not because the subject was foreign to physics, 'naturae seu naturalium rerum scientia,' but because Sturm died before he could reach it, only 2200 pages into his work.[14] Meanwhile nothing stopped or replaced Rohault's treatise, which reached a twelfth French edition in the 1720s, and frequently came forth in Latin, fur-

---

8. J. Webster, *Acad. exam.* (1654), 41–2.

9. Wilkins and Ward, *Vind. acad.* (1654), 36, 58–9; H. F. Fletcher, *Milton*, I (1956), 363–70, II (1961), 310, 314. Cf. Costello, *Schol. Curr.* (1958), 102–4.

10. Dainville, *XVIIe siècle*, no. 30 (1956), 62–8. What mathematics was taught at Oxbridge also used the vernacular, e.g., Blundeville's *Exercises;* H. F. Fletcher, *Milton*, II (1961), 311–21.

11. Descartes, *Discourse* [1637] (1965), 8. Cf. Boutroux, *Isis*, 4 (1921–2), 276–94, for the separation of mechanics (i.e., mathematical theory of simple machines) from physics in the 17th century.

12. See Ruestow, *Physics* (1973), 111–12, for the interesting case of B. de Volder.

13. Molière, *Bourgeois gentilhomme* (1667), Act 2, sc. 6; Rohault, *Traité* (1692⁶), I, 1. Cf. Fontenelle, 'Préface' to *HAS* (1699), vii: 'Ce qui regarde la conservation de la vie appartient particulièrement à la physique.'

14. J. Harris, *Lexicon technicum* (1704), s.v. 'Physicks'; Sturm, *Phys. electiva* (1697–1722). Cf. Schimank in Manegold, *Wissenschaft* (1969), 456–7; and Hartsoeker, *Conjectures physiques* (1706), who promises to complete his text with an account of biology.

nished with the notes of Samuel Clarke, which grew increasingly and belligerently Newtonian. Clarke's last version, translated into English by his brother John in 1723, was still used at Cambridge in the 1740s, long after its generous conception of the scope of physics—not to mention its Cartesian text and strange notes—was outmoded.[15]

Adoption of the modern meaning of 'physics,' like the developments in science it reflected, did not come abruptly. The word continued to be used in its older, broader sense even as it was being qualified and specialized. The lexicons naturally retained the oldest usage: Richelet (1706) gives the science of 'the causes of all natural effects,' and it is the same in the standard dictionaries in the chief European languages throughout the century. An exception is Johnson's *Dictionary* (1755 ff.), which has no entry for 'physics'; for 'physical' it offers a choice among 'relating to material or natural philosophy,' 'medicinal,' and, what some might prefer, 'not moral.'[16] Paulian's *Dictionnaire de physique* includes botany and physiology; that of Monge and his collaborators (1793) rejects them after showing their impropriety in entries for 'abeille' and 'abdomen.'[17] This subtle rejection scarcely ended the use of physics in the old inclusive sense. In the guide to 'Wissenschaftskunde,' as practiced in the Braunschweig gymnasium in 1792, we learn once again that physics is the science of 'all things that make up the Körperwelt,' and properly includes medical subjects as well as natural history.[18] The *Journal de physique*, founded in 1773, calls for papers in natural history; a leading German scientist recommends the study of agriculture as 'such an interesting branch of physics;' and the Paris Academy of Sciences in 1798 offers a prize in 'physics' for the best paper on 'the comparison of the nature, form and uses of the liver in the various classes of animals.'[19]

Yet the *Journal de physique* had, among its subclassifications, one for 'physique' in the modern sense, under which it published papers on mechanics, electricity, magnetism, and geophysics. Since these papers made up less than half the journal, most of the items in a periodical ostensibly devoted to physics were not classed as physics by its editors. Other examples of the simultaneous use of 'physics' in the ancient and modern senses may be found in the class designations of learned societies. Originally the Paris Academy had two

15. Hoskin, *Thomist*, 24 (1951), 353–63; Casini, *L'universo-macchina* (1969), 112–36; Hans, *New Trends* (1951), 51.

16. Richelet, *Dictionnaire françois*, 604; 'physics' had not yet appeared in the 1818 edition of Johnson's *Dictionary*, although *Encycl. Brit.* (1771), III, 478, allows it as a synonym for 'natural philosophy.' Cf. the *Vocabulario* of the Accademia della Crusca; J. C. Adelung, *Grammatisch-kritisches Wörterbuch* (1793²); P. C. V. Boiste, *Dictionnaire universel* (1823⁶).

17. Cf. Silliman, *Hist. Stud. Phys. Sci.*, 4 (1974), 140.

18. Eschenberg, *Lehrbuch* (1792), 169, 198–9, 217.

19. *JP*, 1 (1773), vii; Achard to Magellan, 6 Aug. 1784, in Carvalho, *Corresp.* (1952), 107; *MAS*, 1 (1798), ii–ix. Cf. d'Alembert's proposal of 1777 for a 'prix de physique' for questions in anatomy, botany and chemistry. Maindron, *Rev. sci.*, 18 (1880), 1107–17.

classes, one 'mathematical' (geometry, astronomy, mechanics), the other 'physical' (anatomy, biology, chemistry). In 1785 it added two new subclasses, experimental physics and natural history / mineralogy. Experimental 'physics' (new meaning) went into the class of mathematics, and natural history into that of 'physics' (old meaning). A similar juggle occurred in naming the divisions of the Koninklijke Maatschappij der Wetenschappen in 1807. The subgroup 'physics' then fell into the class of 'experimental and mathematical sciences' along with, and distinct from, anatomy, botany, chemistry, etc. In a draft of the organization, however, the class had been called 'physical and mathematical sciences' and the subgroup, 'experimental physics.'[20] The draft employs the old usage and the final version the new.

EXPERIMENTAL PHYSICS

The chief agent in changing the scope of physics was the demonstration experiment. The new instruments of the seventeenth century, and above all the air pump, having been invented, developed, and enjoyed outside the university, began to make their way slowly into the schools at the beginning of the eighteenth century. In discussing, say, the nature of the air, the up-to-date professor of physics not only talked but showed, extinguishing cats and candles *in vacuo* and weighing the atmosphere. Excellent pedagogues, they saw the advantage of similar illustrations of general concepts: the beating of pendula, the composition of forces, the conservation of 'motion' (momentum) in collisions, the principles of geometrical optics, the operation of the lodestone. Virtually the entire repertoire of experiments pertained to physics in the modern sense. There were three chief reasons for this narrowing. First, the biological sciences did not lend themselves readily to demonstration experiments. Second, the established instrument trade, which already made teaching apparatus like globes, telescopes, and surveying gear, could more easily supply the professor of experimental physics the closer his wants to those of his colleagues in applied mathematics. Third, Newton's first English and Dutch disciples, thinking to follow his experimental and mathematical way, radically restricted the purview of natural philosophy.

It is sometimes said that the adjectives in the title of Newton's major work, the *Mathematical Principles of Natural Philosophy* (1687), were intended to emphasize the distance between it and Descartes' *Principles of Philosophy* (1644), which had refashioned traditional physics in a qualitative manner. To Descartes' arrogance, breadth and imprecision Newton opposed caution, narrowness and exactitude: he confined himself to the application of mathematical laws of motion, said to be taken from experiment, to a few problems in mechanics and physical astronomy. Newton's limited mathematical principles

20. Maindron, *Académie* (1888), 50; R. Hahn, *Anatomy* (1971), 99–100; R. J. Forbes, *Marum*, III (1971), 6–7; Guerlac, *Hist. Stud. Phys. Sci.*, 7 (1976), 194–5n.

were immediately advertised as exhaustive in John Keill's *Introductio ad veram physicam* (1702), translated less presumptuously as *Introduction to Natural Philosophy* (1720), which does not pass beyond general mechanics. Keill was perhaps the first lecturer at Oxford to illustrate his course on natural philosophy with experiments; and, as will appear, one of his associates, J. T. Desaguliers, became the leading British exponent of the new experimental physics.[21]

The most influential of the narrowers of physics were the Dutch Newtonians, W. J. 'sGravesande and Pieter van Musschenbroek, whose teaching careers lasted from 1717 to 1761. Both drank in British natural philosophy at its source, 'sGravesande (who began his career as a lawyer) while on a diplomatic mission to London in 1715, Musschenbroek just after graduating M.D. at Leyden the same year. With the help of the Dutch ambassador to England, 'sGravesande, who had kept up a schoolboy interest in geometry, became professor of mathematics and astronomy at Leyden (1717). A few years later he published perhaps the first modern survey of physics, *Physices elementa mathematica experimentis confirmata, sive introductio ad philosophiam new-tonianam* (1720–1). It was incontinently translated into English, as *Mathematical Elements of Natural Philosophy,* in two competing editions, one made by Desaguliers, who reached print first by dictating to four copyists at a time, the other overseen by Keill, whose chief help was an old priest ignorant of natural philosophy. And these volumes were only hors d'oeuvres: 'sGravesande's book had two more Latin and four more English versions before he died in 1742.[22]

The French, after attacking 'sGravesande for preferring contrived experiments to 'simple, naive, and easy observations,' and for pretending that there was no physics but Newton's, tried to ignore him. Voltaire did not allow them to do so; he went to Leyden to ask the professor 'whose name begins with an apostrophe' for help in preparing his influential *Eléments de la philosophie de Newton* (1738).[23] When 'sGravesande's book did appear in French, in 1747, it bore the title *Eléments de physique,* etc., suggesting that, by then, 'physique' was understood to mean 'natural philosophy confirmed by experiments.' The inference is confirmed by the enthusiastic review in the *Journal de sçavans* for 1748, which extolled the *Eléments* for its 'very great quantity of curious experiments, which teach about everything now known in physics.' The same journal had earlier praised Musschenbroek's *Essai de physique* (1739) on the same ground.[24] Now both these books, deemed complete, omit the biological and geological sciences, and almost all of chemistry and meteorology.

By the middle of the eighteenth century the British and the French were

21. Schofield, *Mechanism* (1970), 25–8; *infra*, xii.1.

22. Brunet, *Physiciens* (1926), 41–2, 51, 75, 96; Allamand in 'sGravesande, *Oeuvres* (1774), II, x–xi, xxi; Torlais, *Rochelais* (1937), 19–20; Ruestow, *Physics* (1973), 117–19.

23. Knappert, *Janus*, 13 (1908), 249–57; 'sGravesande did not think very highly of Voltaire's popularization.

24. Brunet, *Physiciens* (1926), 104–5, 122, 128; Schofield, *Mechanism* (1970), 140–1; Schimank in Manegold, *Wissenschaft* (1969), 468.

composing texts in the Dutch style. Desaguliers wrote an elaborate *Course of Experimental Philosophy* (1734–44) in two volumes quarto that did not cover much more than mechanics.[25] J. A. Nollet issued six volumes of *Leçons de physique* beginning in 1743; except for a short digression on the nature of the senses, in connection with the question of the divisibility of matter, Nollet's lengthy text concerns only mechanics, hydrostatics and hydrodynamics, simple machines, pneumatics and sound, water and fire (from a physical point of view), light, electricity, magnetism, and elementary astronomy. The reviewers were impressed: 'Apart from a few general principles . . . the entire study of physics today reduces to the study of experimental physics.'[26] 'True physics is the science of the Newtons and the Boyles; one marches only with the baton of experiment in one's hands, true physics has become experimental physics.'[27]

In Germany the narrowing of physics was begun independently of the Dutch Newtonians by Christian von Wolff. His *Generally Useful Researches for Attaining to a more Exact Knowledge of Nature and the Arts*, completed in three volumes in 1720/1, describes demonstrations given in his lectures on physics, and every detail, 'to within a hair' as he says, needed to build the instruments to repeat them. 'We must spare no effort and no expense to permit nature to reveal to us what she usually hides from our eyes.' In the event Wolff left her some secrets; he restricted himself to gross mechanics, hydrostatics, pneumatics, meteorology, fire, light, color, sound and magnetism. Only two chapters of the work, some sixty of two thousand pages, concern biology and psychology; the one considers animals chiefly as subjects for investigation *in vacuo*, and the other treats sense organs as examples of optical and mechanical principles. Similarly a representative text of the next generation, J. G. Krüger's *Naturlehere* (1740), esteemed for its 'order, thoroughness and clarity,' gives up less than five percent of its space to plants and animals.[28]

The first important text explicitly to exclude 'the whole theory of plants, animals and man' from its domain was G. E. Hamberger's *Elementa physicae* (1735²), which drew its principles from Wolff's philosophy. Hamberger's book is particularly good evidence of a change in meaning of 'physics' since, as a physician, he might be expected to have advertised biological science where he could. The change in operational meaning was thus explained by the author of an excellent *Institutiones physicae* long used in Austria and Catholic Germany: the etymological meaning of 'physics,' the study of all natural things, 'physics in the largest sense,' is not a practicable subject. He confines himself to

25. Cf. *ibid.*, 463; Schofield, *Mechanism* (1970), 80–8.

26. Desfontaines, *Jugements sur quelques ouvrages nouveaux*, IV (1744), 49, quoted by Brunet, *Physiciens* (1926), 131n. Brunet emphasizes the Dutch ties of Nollet, who visited Leiden and London in 1736. Cf. *ibid.*, 108–9, 113–14, 117, 124–5, 151, and *infra*, xi.2.

27. Memorandum of 1762, quoted by Anthiaume, *Collège* (1905), I, 221.

28. Wolff, *Allerh. Nützl. Vers.* (1745–7³), I, Vorr., III, Vorr. and pp. 456–515; Börner, *Nachrichten*, I (1749), 75.

'physica stricte talis,' to general principles, astronomy, and the usual branches of experimental physics.[29]

The best German physics text of the eighteenth century, J.C.P. Erxleben's *Anfangsgründe der Naturlehre*, dates from 1772. It covers the material then standard: motion, gravity, elasticity, cohesion, hydrostatics, pneumatics, optics, heat, electricity, magnetism, elementary astronomy, geophysics. Its third edition (1784), brought up to date by Lichtenberg's incisive notes, sold out in eighteen months. More editions were called for, with still more notes; 'because of the fast trot of physics, much became old or useless while the book was in press.' It was translated into Danish; Volta toyed with an Italian version; while everyone, according to Lichtenberg, rushed to learn German 'for the admirable purpose of being able to read the best that is written in physics in Europe.'[30] There was also something passable in English.[31] None of these fine texts so much as hinted at the earlier intimacy between their subject and the biological sciences.

This liberation, or rather the demonstration experiment that effected it, had its dark side for serious savants. Demonstrations became too popular; people, even students, came to physics lectures expecting to be entertained. Kästner says that he gave up teaching from Erxleben's text because most of his students only 'wished to see physics, not to learn anything about it.'[32] A French school teacher at the turn of the century, Antoine Libes, scolds Nollet for serving up hasty, uncritical flim-flam, the 'plaything of childhood and the instrument of charletanism,' under the 'perfidious name of experimental physics.' 'Physique' had come to have a frivolous connotation. Daire, in his *Epithetes françoises* (1759), gives 'agréable' and 'curieuse' among its synonyms. The *Almanach dauphin* for 1777 names four Parisian practitioners under 'physicien.' One of them, Rabiqueau, operated a cabinet of curiosities filled with automata, 'which he makes play and move when asked [and paid] to'; another, Comus, 'known

29. Biwald, *Inst. phys.* (1779²), I, Prol., §§1–5. Similar moves are made in Maximus Imhof, *Grundriss der offentlichen Vorlesungen über die Experimental-Naturlehre* (1794–5), I, 1, 7, who defines physics in the old sense and lectures on physics in the new; in Beccaria's *Institutiones in physicam experimentalem*, for which see Tega, *Rev. crit. stor. fil.*, 24 (1969), 193 and in C. A. Guadagni, *Specimen experimentorum naturalium* (1779²), who defines physics inclusively and then restricts himself to a very narrow experimental physics, namely mechanics, hydrostatics, pneumatics, optics, for 'ad haec potissimum referri potest' (p. 5, 10–11).

30. Lichtenberg, *Briefe*, II, 220, 306; Herrmann, *NTM*, 6:1 (1969), 70–4, 80; see Brunet, *Physiciens* (1926), 93, for a complaint by Musschenbroek about the rapid outdating of physics texts. A measure of the value of Erxleben's text is the contemporaneous and yet very old-fashioned *Abrégé de physique* by the Berlin academician J. H. S. Formey.

31. By 1780 the British had several good texts to choose among, e.g., Adams or Nicholson, and no reason to consult Lichtenberg/Erxleben. For syllabi of lectures on 'experimental philosophy' at Cambridge toward the end of the century see F. S. Taylor in A. Ferguson, *Nat. Phil.* (1948), 152–3.

32. Kästner to J. E. Scheibel, 1 April 1799, in Kästner, *Briefe* (1912), 218. Cf. Kästner, *Selbstbiographie* [1909], 13, comparing the seriousness of the French officers who attended his courses during the Seven Years' War with the lightness of the German students.

for his extreme sleight of hand,' showed 'physical and magnetic recreations' that always amused the court; the other two, Brisson and Sigaud, were more serious physicists.[33]

Another force besides the demonstration experiment making for specialization of physics was applied mathematics. All our modernizing textbook writers advocated the use of mathematics in physics. 'sGravesande went so far as to place natural philosophy among the branches of mixed mathematics; for physics, he says, comes down to the comparison of motions, and motion is a quantity. 'In Physics then we are to discover the laws of Nature by the Phenomena, then by Induction prove them to be general Laws; all the rest is handled Mathematically.'[34] Musschenbroek and Desaguliers sound the same theme, and even Nollet, although he does without equations.[35] In fact the nature of their primary readership—university students with little mathematics and a general public with none—precluded elaborate proofs or geometrical deductions. Even the best texts do not use calculus; the experiments they serve up are designed not for quantitative analysis, but to assist, convince, and divert students who could not follow mathematical demonstrations.

Nevertheless the expectation that physics should be mathematical helped to redefine the traditional boundary between natural philosophy and mixed mathematics. Dutch and English Newtonians laid claim to optics, mechanics, hydrostatics, hydrodynamics, acoustics, and even planetary astronomy. By 1750 these subjects were recognized as constituting a special borderline, or, as we should say, interdisciplinary, group of 'physico-mathematical' sciences, or even 'mathematical physics.'[36] In each of these sciences, according to d'Alembert, one develops mathematically a single, simple generalization taken from experience as, for example, hydrostatics from the experimental proposition, 'which we would never have guessed,' that pressure within a liquid is independent of direction.[37] To be sure, there were few mathematical physicists—about one for every twenty pure mathematicians, according to an estimate of the early 1760s[38]—and they did not always pay court to experimentalists;

33. Libes, as quoted by Silliman, *Hist. Stud. Phys. Sci.*, 4 (1974), 143; *Almanach dauphin*, as quoted by A. Franklin, *Dict.* (1906), 570; cf. Fourier, *MAS*, 8 (1829), lxxvi; Torlais, *Hist. med.* (Feb. 1955), 13–25.

34. 'sGravesande, *Math. Elem.* (1731⁴), I, viii–ix, xii–xiii, xvi–xvii. Cf. Brunet, *Physiciens* (1926), 48–54; Ruestow, *Physics* (1973), 132.

35. *Ibid.*, 133–6; Desaguliers, *Course* (1763²), I, v; Nollet, 'Discours,' in *Leçons*, I (1764⁶), lviii, xci.

36. D'Alembert in *Encyclopédie* (1778³), XXV, 736, art. 'Physico-mathématiques;' *Prel. Disc.* [1751] (1963), 54–5, 152–5; Karsten, *Phys.-chem. Abh.* (1786), I, 151. Cf. the title of Grimaldi's masterpiece, *Physico-Mathesis de lumine* (1665).

37. D'Alembert in *Encyclopédie* (1778³), XIII, 613, art. 'Expérimental.' This example became a commonplace among mathematical physicists, e.g., Lagrange, *Mécanique analytique* (1788), in *Oeuvres* (1867), XI, 193; Bossut, *Traité* (an IV²), I, xix, xxiv, who stresses the difficulty of getting the initial generalization.

38. Beccaria to Boscovich, 31 May 1762 (Bancroft Library, U. California, Berkeley): 'Per venti puri matematici si stenti a trovare un fisico matematico.' Cf. Lambert to Karsten, 15 Sept. 1770

d'Alembert, for example, conveived that a subject once mathematized had no further need of the laboratory.[39] The essential point is not that few complete physicists could be found, but that the ideal had been recognized. The physicist, says Pierre Prevost, should be able to 'calculate, observe and compare.' But, as it is very difficult to excel at everything, and science, like all else, advances by division of labor, the physicist should emphasize either the experimental or the mathematical branch of his discipline.[40]

Since electricity was always considered a physical science, its place in the body of knowledge and its treatment varied with the fortunes of physics as a whole. It first received extended treatment in 1600, as a digression in a book about magnets. It found a place in scholastic compendia either near the magnet, as an example of attraction, or among complex 'minerals,' where the chief electric body, amber, was treated. The Dutch Newtonian texts consider electricity under 'fire,' a reclassification required by their dropping mineralogy and advised by the observation, early in the eighteenth century, of electric discharges in evacuated tubes. In the 1740s, owing to the discovery of spectacular phenomena easily reproduced, electricity became the leading branch of experimental physics, and the most popular source of diverting, and sometimes vapid, demonstrations. ('Electricity can sometimes become weak enough to kill a man.'[41]) As a serious study it commanded many monographs, and won its own extensive, independent section in the textbooks of natural philosophy. Soon it required its own texts, the best of which, Cavallo's *Complete Treatise on Electricity* (1777), spread into three volumes octavo in its fourth edition of 1795.

None of this was mathematical. Electricity, in common with other new experimental sciences of the seventeenth century, proved more difficult to quantify than the traditional subjects of 'physical mathematics.' In the mid-eighteenth century the great quantifier d'Alembert, despairing of yoking electricity to his favorite discipline, left it to the experimenters: 'That is mainly the method that must be followed with phenomena the cause of which reason cannot help us [find], and among which we see connections only very imperfectly, such as the phenomena of magnetism and of electricity.' About twenty years later, in 1776, Lichtenberg, professor of pure and applied mathematics at the University of Göttingen, allowed that the non-mathematical experimenter had done his share: electricity, he said, 'has more to expect from mathematicians than from apothecaries.' Another ten years and another quantifier and

---

(Lambert, *Deut. gelehrt. Briefw.*, IV:2 [1787], 277): 'Seit vielen Jahren brachte junge Leute von Universitäten kaum etwas mehr als die Mathesin puram mit.'

39. See d'Alembert's encyclopedia articles just cited; *infra*, i.5; Hankins, *D'Alembert* (1970), 94–6. For this attitude d'Alembert was criticized by Lalande (letter to Boscovich, 27 April 1767, in Varićak, Jug. akad. znan. i umjetn., *RAD*, 193 [1912], 239), and Euler by Clairaut (letter to Boscovich, *c.* 1764, *ibid.*, 222): 'Combien un able géomètre qui veut tout tirer de la théorie sans avoir recours aux expériences, peut s'écarter du vrai dans les sciences physico-mathématiques.'

40. Prevost, *Recherches* (1792), vii.

41. Lichtenberg in Erxleben, *Anfangsgründe* (1787⁴), 468, in reference to negative electricity. Cf. Musson and Robinson, *Science* (1969), 85.

organizer, W. C. G. Karsten, professor of mathematics and physics at the University of Halle, considered electricity a part of mathematical physics, although one 'not so entirely mathematical' as mechanics or optics.[42]

Karsten's classification corresponded to contemporary usage. The grouping of electricity (as experimental physics) in the class of mathematics by the Paris Academy in 1785 has been mentioned. A similar but subtler transformation occurred at the Petersburg Academy. From 1726 to 1746 its journal (*Commentarii*, later *Acta*) had two classes, mathematics and physics; the former included analytical and celestial mechanics, the latter everything from optics and hydraulics to botany and astronomy. Electricity was accordingly and, for the time, appropriately classed as physics. In 1747 a new class was added, 'physico-mathematics,' which took optics, hydraulics, heat, electricity, magnetism, and, increasingly, analytic dynamics; 'physics' retained, among the physical sciences, only meteorology, mineralogy, and chemistry.[43] The arrangement persisted until 1790, when 'mathematics' and 'physico-mathematics' united.[44] These moves corresponded to key conceptual innovations in the study of electricity, which it shall be our pleasure later to examine.

2. OCCULT AND OTHER CAUSES

The Aristotelian physicists concerned themselves with the true causes of things. Where the corpuscular or Newtonian philosopher saw few causes or none at all, the peripatetic could distinguish four general categories and several subspecies, one of which, termed 'occult,' became a password among the modernizing philosophers of the seventeenth and eighteenth centuries. To despise occult causes, to insist upon cleansing physics of them, was forward-looking; to accuse the enemy of advocating such bugaboos was always a good thrust in head-to-head philosophical combat. Cartesians and Newtonians flung the charge not only at peripatetics, but also at one another.[1] Much of our story turns upon the notion of occult cause.

ARISTOTELIAN NATURAL PHILOSOPHY

Aristotle's physics, as inherited by the Renaissance, was enriched or, as some said, polluted by the conflicting interpretations of scores of schools of

42. D'Alembert in *Encyclopédie* (1778³), XXV, 736, art. 'Physico-mathématiques'; P. Hahn, *Lichtenberg* (1927), 41; Karsten, *Phys.-chem. Abh.* (1785), 151.

43. Owing to this change, Kästner found himself reviewing 'physicomathematics' from St. Petersburg for the 'physics' section of the *Commentarii de rebus ad physicam et medicinam pertinentibus.* The Leipzig doctors who ran the *Commentarii* could not take physics with mathematics, 'a severiori enim mathesi medici nostri abhorrent.' Kästner to Heller, 1 Jan. 1752, in Kästner, *Briefe* (1912), 21.

44. Cf. the reorganization of the Royal Society of Science, Göttingen, in 1777–8: the omnibus class 'physics and mathematics' was divided into two, 'physics' getting chemistry and the biological sciences, 'mathematics' the usual mixed mathematics and electricity. Ak. Wiss., Gött., *Novi comm.*, 1 (1778), iii–iv, xvi–xx.

1. Cf. Genovese, 'Disputatio' (1745), quoted by Garin, *Physis*, 11 (1969), 220: '[Cartesianism] cessit Newtonianismo, iisque armis victus est quibus ille peripatetismum fugaverat.'

philosophy. One therefore cannot declare unambiguously the principles of six-teenth-century peripatetic philosophy. The school to which we shall subscribe is the Collegium conimbricense, the Coimbra Jesuits, who published commentaries on the Aristotelian texts in the first years of the seventeenth century. These commentaries recommend themselves on several grounds. They are authoritative and erudite; they stay close to the ancient texts; and they were very widely used. Descartes, among many others, learned his physics from them.[2]

The four scholastic categories of cause, among which we seek the occult, are the material, efficient, formal and final. They are more easily illustrated than defined. In Aristotle's own example of the making of a statue, the material cause is the bronze; the efficient, the sculptor's art; the formal, the statue's final figure; and the final, earning the sculptor a living, honoring the party sculpted, edifying the public.[3] The statue is an affair of art. In most cases of interest to the physicist, however, in natural processes, the number of causes reduces to three, the formal and final coinciding, or even to two, when the efficient cause is the nature or form of the body undergoing change.[4]

In Aristotle's philosophy, each individual is what it is in virtue of its 'form,' its defining principle, the sum of its 'actual' properties. 'Actual,' *actu*, signifies properties currently realized or activated as distinguished from potential ones; an animal now has the tendency to grow old, potentially of being old. Although each individual has but one form, Aristotle separates the characteristics it embraces into two groups, the 'substantial' or 'essential,' and the 'accidental.' Essential characteristics are those by which an individual belongs to a species; they explain why the world contains *kinds* of things—dogs, stars, marble, men. Accidental characteristics differentiate individuals of the same species one from another; they make it possible to distinguish between this dog and that, or between Plato and Socrates. Size, shape, color and 'attitude,' for example, are usually accidents, so that an individual six feet tall, thin, black and silent is no less a man than chubby, white, chattering Socrates. The form of an individual is the sum of its actual properties; the form is *not* the individual, however, and indeed has no separate existence except in the mind of the philosopher.[5]

A second principle, 'prime matter,' likewise incapable of independent existence, is necessary to bodies. Prime matter is the principle of materiality and potentiality; it reifies a given form to constitute an 'actual' body; and it readily exchanges one form for another to bring about change.[6] Just how one form

2. Descartes, *Corresp.* (1936), III, 185; Gilson, *Index* (1912), iv; *infra*, ii.1; *supra*, i.1, n. 1.

3. *Phys.*, II.3, 194^{b}16–195^a; *Metaphys.*, Bk. Δ, 1013^{a}25–1014^{a}15. Although these texts do not specify final or formal causes in the case of the statue, they imply those given here. Cf. Coll. conim., *Comm. in Phys.*, I (1602), cols. 327–9, 396–8.

4. *Phys.*, II. 7, 198^{a}25–28.

5. The Intelligences that regulate the motions of the heavens, the pure Form or Unmoved Mover of the world (*Metaphys.*, Bk. Λ, 1073^a–1074^{b}35), and the angels of the schoolmen all excepted. Cf. Coll. conim., *Comm. de Caelo* (1603^2), 267–8.

6. *Phys.*, I.7, 190^b 15–30, I.9, 192^{a}25–30, II.1, 193^{b}20; *Metaphys.*, Bk. Z, 1029^{a}20–30; *Gen. and Corr.*, I.3, 318^{b}18; *Cat.*, Chapt. 4, 1^{b}25–4^{b}20. Cf. Coll. conim., *Comm. in univ. dial.*, I (1607), 336–42. The degree of potentiality of matter has been differently interpreted according as one

succeeds another became a tough knot for the peripatetics. Some sixteenth-century philosophers, holding tight to the Aristotelian definition, continued to ascribe a single form to each individual, and referred the introduction of new forms to the stars or to God. Others, departing far from their original, in effect resolved an individual into a collection of independently-existing forms contained in an independently-existing piece of matter, like so many marbles in a box. The replacement of one of these 'substantial forms' by another amounted to no more than a change of place.[7] In this debased condition, with reified individual qualities, the theory could give an easy, empty, explanation of everything.

In certain sorts of change, called 'natural' by peripatetics, form can play the part of efficient as well as of formal and final cause. A standard example is the growth of plant or animal. An acorn—or better, this acorn—has, at this instant, in consequence of its form, the power to develop into an oak; a power that will become the efficient cause of development whenever artificial impediments to its action—being out of ground, being deprived of nutrients—disappear. The formal cause of growth is likewise form, understood not as the form of this acorn, but as the form of the oak to which it tends. This final form, when interpreted as the goal of growth, is also the final cause. Note that the acorn, or any plant or animal, has its power to change, or, to use the school term, its 'mover,' within it. Note also that this power, which is different for each natural species, *is not further analyzable*.

Precisely the same account can be given of the fall of a rock or the ascent of fire. Among their essential properties earth and fire possess, respectively, the qualities gravity and levity; when unconstrained, earth moves to the center of the world and fire towards its circumference. The form of a rock separated from the main body of the earth has among its accidental characteristics actually 'being on this shelf' and potentially being in any number of other places. The rock's form does not regard these possibilities indifferently: when the accidental constraint disappears, when the rock falls from the shelf, the element of gravity in its form moves it directly towards the center of the world. This is not a case of action at a distance, which Aristotle would not allow in the material world.[8] The center of the universe does not draw the stone: the stone 'knows' from the relevant element in its form, 'being on the shelf,' that it is separated, and it moves itself towards full actuality, it propels itself towards the ground, whenever possible.[9]

---

takes the texts of the *Physics* or the *Metaphysics* as fundamental. The difference is perhaps consequential for Renaissance scholastic physics; cf. Coll. conim., *Comm. in Phys.*, I (1602), cols. 205–6, 228–9, and de Vries, *Schol.*, 32 (1957), 161–85.

7. Dijksterhuis, *Mechanization* (1961), 281–4; Reif, *JHI*, 30 (1969), 26–7.

8. *Phys.*, VII.2, 243^a1-5, 244^b1-245^a20. Cf. Thomas Aquinas, *Commentary* (1963), 436–9; van Laer, *Phil.-Sci. Prob.* (1953), 80–94; Coll. conim., *Comm. in Phys.*, II (1609), col. 311.

9. The rock's 'knowledge' ultimately comes from the place it occupies; Aristotle's sublunary space is, as it were, full of sign posts directing rocks downward and fire upward. Cf. *On the Heavens*, IV.3, $310^a14-311^a15$; *Phys.*, II.1, 192^b12-15.

Opposed to the natural motions of organic growth and free fall are 'violent motions' that carry an object against the tendency of its form: flinging a javelin, compressing air, killing an animal, brainwashing a philosopher. In such cases the efficient cause necessarily lies outside the object moved. The same is true of another class of motions, which may be called 'indifferent,' motions to which the essential form offers neither encouragement nor resistance: displacing a rock horizontally, heating or cooling water, moistening or drying mud.

The last two qualities, the heating power (hotness) and the moistening (wetness) are, with their contraries coldness and dryness, the chief agents of change in the sublunary world. Aristotle considered them unique in combining the two attributes he deemed necessary for such agents: they are *tangible* and hence notify the philosopher of their working, and they come in contrary pairs each member of which can *act* upon the other. (Aristotle arbitrarily makes hotness and wetness active, and coldness and dryness passive.) The last criterion is of capital importance. Gravity and levity, for example, do not constitute an agent-patient pair. If one places a hot body in contact with a cold one, or a wet in touch with a dry, the first pair become lukewarm and the second damp. A rock, however, does not share its gravity with the shelf supporting it; however long they remain in contact the rock will sink and the shelf float.

The bodies constructed by the union of the fundamental qualities with prime matter are the 'elements' of the inorganic world. Aristotle accepted the view, already ancient in his time, that precisely four such elements existed, air, earth, fire and water; and he associated them with the fundamental qualities in such a way that, as observation showed, any two elemental bodies could interact. This condition required that each element be associated with a pair of fundamental qualities. The affiliations chosen by Aristotle, fire (hot, dry), air (hot, moist), water (moist, cold), earth (cold, dry), remained standard. This account does not, however, exhaust the essences of elemental bodies, for fire has levity as well as hotness and dryness, and earth has gravity as well as coldness and dryness. Gravity and levity, although invariably associated with earth and fire, cannot be derived from the four active qualities: all six are irreducible, singular powers.[10]

There is another capital distinction to be drawn between gravity / levity on the one hand and the active qualities on the other. The qualities—and 'secondary qualities' like hard / soft, rough / smooth, and brittle / malleable, which Aristotle supposes compounded of them—immediately identify themselves to the sense of touch. Gravity / levity do not. To be sure, a rock held in the hand gives one a sense of its gravity. But this sense records the force exerted to prevent the rock's natural motion; it in no way differs from felt resistance offered to any other push, and it vanishes when the rock rests on the ground. Our sense of touch alone cannot inform us of the gravity of the largest boulder.

10. *Gen. and Corr.*, II.1, 329ᵃ25–331ᵃ5; *Meteor.*, I.3, 341ᵃ25; *Cat.*, Chapt. 8, 9ᵃ30–9ᵇ10. Cf. Coll. Conim., *Comm. in Phys.*, I (1602), cols. 392–4; *Comm. de Caelo* (1603²), 424–9.

Similarly we have no sense of our own gravity, or of electricity. The scholastic philosophers distinguished clearly between these last qualities and the active ones. In their terminology, the active qualities and their compounds are manifest, and the gravitational *hidden* or, to say the worst, *occult*.

Philosophers admitted an occult gravitational quality to account for the apparent directed self-movement of heavy bodies. Magnetism presented a similar problem. Iron flies to the lodestone in roughly the same manner as a rock moves to the center of the world, driven by its peculiar self-actualizing form. Sublunary place, the 'sphere of influence,' as it were, of the world's center, directs the self-motion of the rock, while the magnet's characteristic quality, diffused through space, confers on the iron, or actuates, a power of self-motion, and guides it to union with the lodestone. Or so magnets operate according to Aristotelian commentators from Averroes to the Coimbra Jesuits. St. Thomas in particular took pains to explain the induction of self motion in iron, and to distinguish it in detail from free fall: gravity acts towards a point, from any distance; magnetism moves towards a body, and can be induced only over short distances.[11] Other scholastics, attending to the Aristotelian principles that the mover must be conjoined to the mobile and, except for souls and heavenly intelligences, can move only by being moved,[12] tried to explain how the lodestone could 'diffuse' its power to the iron without appearing to affect ('move') the intervening medium. 'They say [it is the testimony of Jean Buridan] that the magnet alters the air or water that it touches and propagates to the iron a quality which, because of a natural affinity between the iron and the magnet, attracts the iron but nothing else.' The magnet works just like that peculiar fish of St. Albert's, which numbs the hands of fishermen by doing something to the water.[13]

The account transmitted by Buridan is an adaptation of the medieval theory of 'multiplication of species,' according to which all bodies in the universe impress their peculiar qualities and powers (species) upon, and diffuse (multiply) them through, the surrounding medium. The multiplied species affect a body according to its nature. Consider the exemplar of multiplication of species, the propagation of light. An incandescent source (lux) imprints its species (lumen) on any transparent medium which, however, does not itself therefore become incandescent or colored; the lumen becomes manifest only at the surfaces of opaque bodies, or within translucent ones. Celestial influences operate in the same manner as light, but with greater discretion: they act preferentially upon certain special materials, which thereby may be made into

11. Thomas Aquinas, *Commentary* (1963), 433; Daujat, *Origines* (1945), I, 49–78; Urbanitsky, *Elektrizität* (1887), 10, 103–4; T. H. Martin, Acc. Pont. nuovi Lincei, *Atti*, 18 (1865), 99, 105.

12. *Phys.*, 241^{b}25–242^{a}25.

13. Buridan, *Quaestiones super VIII libros Physicorum*, VII, 4, quoted in Daujat, *Origines* (1945), I, 72. That St. Albert's fish, the torpedo, stuns by electricity was a discovery of the 18th century (*infra*, xix. 3).

medicines or talismans. The magnet is still more exclusive, for its species work visibly only upon iron, steel, and other lodestones.[14]

The medieval account of magnetism remained standard until the middle of the seventeenth century. One finds it, for example, in Nathanael Carpenter's influential *Geography Delineated*; in the widely used monograph of Vincent Léotaud, who followed St. Thomas' model despite his claim to expound a 'new magnetic philosophy;' and—an excellent testimonial to its persistence—it has left a trace in the anti-peripatetic atomistic compendium of Gassendi.[15] One also finds it constantly put forward as the paradigm of attractions, as the cleanest case of local motions effected by occult qualities: 'in the magnet God has offered to the eyes of mortals for observation qualities which in other objects he has left for discovery to the subtler research of the mind.'[16]

A convenient and representative baroque exposition of the results of this 'subtler research of the mind' may be found in a *New philosophy and medicine concerning occult qualities* published in Lisbon in 1650. Its author, Duarte Madeira Arrais, was a distinguished physician trained at the University of Coimbra.[17] His book therefore has a double authority: first, because it is informed by the commentaries of the Coimbra Jesuits; second, because its subject, occult qualities, figured prominently in medical theory in connection with growth, nutrition and the efficacy of poisons and purgatives.[18]

Madeira assumes that all philosophers admit the existence of the four elementary active qualities, of the secondary tactile qualities compounded from them, of 'manifest super-elemental qualities' like light, sound and impetus, and of the vital powers of animals. In addition, he says, there is a class of 'occult super-elemental qualities' like the virtues whereby the magnet draws iron, the remora stays ships and purgatives expel foul humors. These virtues are called 'occult' because, 'though manifest to the intellect, they are not apparent to the senses,' *sub humanos sensus non cadunt*. They must be super-elemental because 'remarkable effects' like the attractions of remoras and magnets, or the shock inflicted by electric eels, cannot arise from elemental qualities. Not only are hotness, dryness, coldness and moistness in any degree incompetent to produce magnetic qualities: they also act less rapidly and efficiently than, say, the virtues of magnets or scorpions.[19] As for the details of occult action, Madeira

14. R. Bacon, *Opus maius* (1900), II, 407ff.; Crombie, *Grosseteste* (1962²), 211–12; Coll. conim., *Comm. in Phys.*, I (1602), col. 394, II (1609), cols. 309–12; *Comm. de Caelo* (1603²), 196–9.

15. Carpenter, *Geographie* (1635²), 54–5; Léotaud, *Magnetologia* (1668), 31–3; Gassendi, *Opera* (1658), I, 347.

16. Dee, *Prop. aph.* (1568²), §xxiv.

17. Barbosa Machado, *Biblioteca* (1930²), I, 715–16.

18. Morhof, *Polyhistor* (1747⁴), II, 305: 'Nulla autem fecundior his qualitatibus occultis magis est, quam ars medica.'

19. Madeira Arrais, *Novae phil.* (1650), 1–19. The Coimbra Jesuits also say that the facts compel the philosopher to introduce occult qualities: 'Nec enim semper effecta ad quattuor primas qualitates, ut falso quidam opinantur, referri queunt.' Coll. conim., *Comm. in Phys.*, II (1609), col. 318.

generally follows the lead of St. Thomas. The occult quality, diffused about the substance that bears it, awakens a self-acting potency in appropriate neighboring bodies, which move themselves as their forms require.[20] Activated iron flies to the seat of its occult trigger, the magnet; a purgative stimulates a 'directive quality' in the affected humor, which guides it into the intestines.

Often the carriers of the stimulating power bear an external analogy or similitude to the substances they excite; but this similarity is not necessary, and where it does exist it is a formal, not an efficient cause.[21] Madeira does not believe in the doctrine of signatures, the notion that the outward appearance of herbs and stones, rightly read, reveals their powers and purposes. He sets few constraints against multiplication of occult qualities. He himself is circumspect. Others, however, had long since undermined the explanatory value of occult qualities by invoking them to resolve all the difficult phenomena, real or imaginary, treated in natural philosophy.

SYMPATHIES AND ANTIPATHIES

In the occult as elsewhere familiarity breeds contempt. It may be useful to identify a magnetic virtue the possession of which distinguishes a closed group of interacting substances; but when one ascribes several irreducible special qualities to every stone or plant or drug, one has a science of words, not of things.[22] 'The learned doctor asks me the cause and reason why opium puts people to sleep. A quoi respondeo / quia est in eo / virtus dormitiva / cuius est natura / sensus assoupire.' (Opium is a soporific because it contains a dormative virtue.) Thus Molière's candidate in medicine, answering his examiner in empty fractured Latin, to the great applause of the faculty. The same point was made by several sober physicists for whom Francesco Lana, a Jesuit obliged to teach the philosophy of Aristotle, may be allowed to speak. Ask most natural philosophers, says Lana, the cause of any natural phenomenon; 'they can only reply that it happens by an occult cause, that such is the nature of that substance.' And if you persist, and ask, say, how the occult cause whereby the magnet draws iron and not straw differs from that whereby amber draws straw and not iron? 'They reply, this is the nature of amber, and that the nature of the magnet.' Such people, according to Lana, bring disgrace and ruin to natural philosophy.[23]

The most extravagant occult qualities were the sympathies and antipathies

---

20. Cf. Cabeo, *Meteor.* (1646), I, 31–2; and, on the definition of occult qualities ('spectant ad facultates incognitas, causasque habeant incompertas'), Gassendi, *Opera* (1658), I, 449.

21. Madeira Arrais, *Novae phil.* (1650), 335, 352, 405–19.

22. To use the conceit in Fontenelle, 'Éloge' of du Hamel, *Oeuvres* (1764), V, 80: 'des idées anciennes et des nouvelles, de la philosophie des mots & de celle des choses, de l'École & de l'Académie.' Cf. Sprat, *Hist.* (1967), 113, 336; Flourens, *Fontenelle* (1847), 53–4. For an example of peripatetic evasion at its worst see Middleton, *Br. J. Hist. Sci.*, 8 (1975), 148.

23. Molière, *Le malade imaginaire* (1673), 3e intermède; Lana Terzi, *Prodromo* (1670), Proem., 3–4. Cf. Boyle, 'Occult Qualities,' unpublished Ms. quoted by M. B. Hall, *Boyle on Nat. Phil.* (1965), 59.

invoked especially by the hermetic philosophers and physicians of the sixteenth and seventeenth centuries. Gaspar Schott, S. J., an authority on sympathetic action, explains it in these words: 'Sympathetic effects arise from a friendly affection, or coordination and innate relation, of one thing to another . . ., so that if one is acting, or reacting, or only just present, the other also acts or is acted upon.' The operative quality or affection is not further analyzable. 'It originates directly from the particular temperament of each thing, being nothing but a certain natural inclination of one thing towards another.' An antipathy works in the same way, with aversion in place of inclination, just like—here Schott offers the inevitable analogy—magnetic attraction and repulsion.[24]

Sympathies and their opposites explained the fabulous as easily as they elucidated the natural. To Madeira the ship-staying power of the remora and the iron-pulling virtue of the magnet are equally acceptable, and he no more doubts the 'virtues of the tree of life of the Garden of Eden' than he does the efficacy of scammony or cinchona. Less critical writers speak readily and knowledgeably of the glance of the basilisk, the attraction of the weasel for the toad, the generation of minerals by celestial influence, fetal imprints produced by maternal appetites, and the curative power of the powder of sympathy.[25] This last extravagance represents the nadir of the doctrine of active qualities.

The seventeenth century credited Paracelsus with the invention of a powder or salve which could heal at a distance. No doubt this salve, made among other things of skull moss, mummy, and the fat and blood of a dead man, would be most beneficial when used as directed, viz., smeared upon a stick or napkin previously dipped in the wound and kept far away from the patient. According to the Paracelsians, the virtue of the salve, drawn out and fortified by celestial influences, flies to the injury by 'magnetic sympathy,' say between the separated particles of blood or, according to the great magus Robert Fludd, between the necrotic ingredients of the salve and the living flesh (opposites attract).[26] Many physicians and alchemists endorsed or improved the salve and its theory. One influential promoter was Sir Kenelm Digby, whom some contemporaries considered a respectable philosopher.[27]

Digby was an English Catholic who learned his philosophy from a Jesuit named Thomas White, alias Blacklow, and spent his early manhood serving the Stuarts in diplomatic missions on the Continent.[28] In 1623, at the age of

24. Schott, *Thaumaturgus* (1659), 368–70; *Phys. cur.* (1662), 1285ff. Cf. Kircher, *Magnes* (1641), 644.

25. Mousnerius (Fabri) lists these curious items, which he says are commonly alleged in favor of sympathetic action, in *Metaphysica* (1648), 283–5. Fabri rejects them all, including the remora, which he calls a 'mere fable.'

26. Debus, *J. Hist. Med.*, 19 (1964), 390–2, 404–11, quoting Fludd's *Philosophia moysaica* (1638). Cf. Thorndike, *History*, VII, 503–6.

27. Digby knew and was esteemed by Descartes, Mersenne, Fermat, John Wallis and Athanasius Kircher. See Gabrieli, *Digby* (1957), 197, 230–2; Petersson, *Digby* (1956), 120–8; Dobbs, *Ambix*, 18 (1971), 1–25.

28. Digby, *Two Treat.* (1645²), 180: 'To him [Thomas White] I owe that little which I know; and what I have, and shall set down in this discourse, is but a few sparks kindled by me at his great fire.'

twenty, he met a monk in Italy who gave him a secret recipe for the powder, which he was to dissolve in water and blood from the wound and set aside in a basin exposed to sunlight. Digby made the secret public in 1657, in a strange address to a 'solemn assembly' in Montpellier, where he had gone for the waters.[29] The disclosure included a theory of the cure, and, by way of illustration, several bizarre examples of natural sympathetic phenomena.

All bodies, Digby says, perspire when bombarded by light. The perspiration, or effluvium, which is characteristic of the emitter, tends to diffuse towards kindred substances, which preferentially absorb it. 'The reason hereof is the resemblance, and sympathy, they have one with the other.'[30] Everyone knows, for example, that a greater stench will attract a lesser. 'Tis an ordinary remedy, though a nasty one, that they who have ill breath, hold their mouths open at the mouth of a privy, as long as they can, and by the reiteration of this remedy, they find themselves cured at last, the greater stink of the privy drawing unto it and carrying away the lesser, which is that of the mouth.'[31] Similarly, in Digby's cure, the blood particles fly sympathetically to the wound whence they came, bringing with them the subtlest atoms of the healing powder of sympathy.[32]

Digby's tract was often reprinted, translated, and glossed.[33] The powder of sympathy and the similar technique of 'transplantation'—the transfer of poison to a sympathetic imbiber, for example, to a hair of the dog that bit you —also found favorable or agnostic treatment in eighteenth-century encyclopedias.[34] Such survivals occur almost exclusively in a medical context. By 1700 natural philosophy had largely freed itself from the animistic sympathies, from the innumerable occult qualities that had posed, according to the literary executor of the seventeenth century, 'the most vexed question of the age.'[35] Peripatetic physics, found guilty, among other reasons, by association, also fell victim to the purge.

The implication of guilt by association was commonly employed by corpuscular philosophers. Robert Boyle, for example, often found it advantageous to conflate hylomorphism and hermetic animism.[36] Up-to-date peripatetics met this gambit by denouncing hermeticism as loudly as the corpuscularians did,

29. Petersson, *Digby* (1956), 265–6.

30. Digby, *Late Disc.* (1658), 5, 11–12, 68. Digby says (*ibid.*, 68–75) that these sympathies derive from similarities in density and particle shape; but in practice they are non-mechanical occult qualities, *pace* Dobbs, *Ambix*, 18 (1971), 11, 25.

31. Digby, *Late Disc.* (1658), 76–110.

32. *Ibid.*, 133–41. Union of the atoms with sunbeams improves their efficacy.

33. At least 25 editions of *Late Disc.* were published by 1700. Physicians disputed the efficacy of the cure throughout the century. Petersson, *Digby* (1956), 272–4, 326; Schreiber, *Gesch.*, II:2 (1860), 416.

34. For Chamber's *Cyclopedia* see Shorr, *Science* (1932), 29–31, and A. Hughes, *Ann. Sci.*, 7 (1951), 354–6; for Zedler's *Universal Lexicon* (1732–50), Shorr, *op. cit.*, 60–8, 71–2; for Diderot's *Encyclopédie* (1751ff.), Thorndike, *Isis*, 6 (1924), 379–82.

35. Morhof, *Polyhistor* (1747⁴), II, 303.

36. Boyle, *infra*, v.2.

and by disavowing the miracle mongers, the hacks and novelty hunters, the dupes of Arab commentators, all who had muddied the pure Aristotelian water. Some reforming peripatetics adopted the tactics of the enemy. Niccolò Cabeo, S. J., cracked down on occult qualities and endorsed an eclectic experimentalism; his colleague, Honoré Fabri, put forth as 'purified' Aristotle a physics stained with the thought of Descartes.[37] But many physicists brought up on, and sympathetic to, Aristotelian principles, judged themselves unequal to purging the contaminated peripateticism of their day, and reluctantly embraced the radical alternative of corpuscularism.

An outstanding example of the frustrated Aristotelian reformer is the Minim monk, Marin Mersenne, the confidant of Descartes, the correspondent, guide and goad of much of learned Europe in the 1630s and 1640s. Mersenne, educated like Descartes in the Jesuit college at La Flèche, began his career hoping to rid traditional natural philosophy of hermetic mumbo-jumbo, of astrological influences, of bits and pieces borrowed from cabalists and magicians. A 'new Aristotle' would answer the 'grunting of the German beast' (Paracelsus), subdue the *cacomagus* (Fludd), secure religion against the animists and the pantheists, and, above all, distinguish the natural from the supernatural, blunt superstition and confound scepticism.[38] Eventually, to meet the threat, Mersenne gave up altogether the search for physical causes in the Aristotelian sense: and from the mid-1630s, under the inspiration of Galileo and others, he taught that 'true physics' could only be a descriptive science of motions. The rejection of essences, or rather of the claim to know them, sunk Mersenne's old program for reform along with the abuses he aimed to correct. He avoided shipwreck by jumping to the good ship 'Mécanisme,' which, although unknown and perhaps unknowable in its inner workings, at least sailed according to discoverable laws.[39]

Mersenne was not alone in identifying the Fludds and not the school philosophers as the greatest threat to true physics in the seventeenth century.[40] The same contempt for hermeticism spices the reply of the Oxford mathematicians, Seth Ward and John Wilkins, to the attack on the universities made by the Fluddist John Webster in 1654. Webster blasted the schools for despising the 'noble and almost divine Science of natural Magick,' which he understood to rest upon the doctrine of signatures, and for sticking to Aristotle. Wilkins and Ward replied that Aristotle was 'one of the greatest wits, and most useful that ever the world enjoyed,' and recommended his books, 'the best of any

37. Cabeo, *Meteor. comm.* (1646), I, 254; Fabri, *infra,* ii.1.

38. Lenoble, *Mersenne* (1943), 9–10, 29, 95, 147.

39. *Ibid.*, 361. "Les hommes ont introduit la sympathie et l'antipathie, et les qualitez occultes dans les arts et dans les sciences pour en couvrir les deffauts, et pour excuser leur ignorance . . . : car lors que l'on connoist les raisons de ces effets [magnetism, electricity] la sympathie s'évanoüit avec l'ignorance, comme ie demonstre dans le tremblement des chordes qui sont à l'unisson.' Mersenne, *Harmonie universelle,* quoted by Lenoble, *ibid.*, 371–2.

40. E.g., Gassendi, *Opera* (1658), III, 236–7, 251. Cf. the sudden conversion of Walter Charleton from hermeticist to corpuscularian in the early 1650s; Gelbart, *Ambix,* 18 (1971), 149–68.

Philosophick writings,' as correctives to the nonsense of the disciples of Hermes and Pythagoras.[41] Not that Wilkins and Ward were peripatetics; indeed, they were enthusiastic moderns, 'Copernicans of the Elliptical family.' That did not, however, blind them to their kinship to Aristotle and Ptolemy, or to their distance from true revolutionaries like John Webster.[42]

The role here assigned to hermeticism is much more modest than that claimed for it in some recent writing on Renaissance science. The hermetic magus, it is said, must necessarily have learned the sympathies of things by experience; moreover, the need to control and manipulate astrological influences, special qualities of bodies, and the harmonies of the world, directed him to the laboratory. 'In this way Paracelsian mysticism acted as a powerful stimulus towards the new observational approach to science.'[43] 'It is the Renaissance magus, I believe, who exemplifies the changed attitude of man to the cosmos which was the necessary preliminary to the rise of science. . . . The Renaissance magus was the immediate ancestor of the seventeenth century scientist.'[44] Although the claim may hold for certain traditions of alchemy and medicine that became something like chemistry and physiology in the seventeenth century, it fails for most of physics and mixed mathematics.

Perhaps the most powerful support for controlled and careful experiment in physics towards the end of the sixteenth century was the example of mixed mathematics, the requirements and achievements of architecture, fortification, navigation, Tychonian astronomy, optics. Galileo taught these subjects, which he had studied at a Florentine trade school set up specially to teach applied mathematics.[45] William Gilbert, the first important electrician, also taught mathematics; and the artisans whose methods he may have followed in his fundamental work on electricity and magnetism likewise had an interest in, and urgent need for, practical mathematics.[46] The number magic of cabalists and hermeticists did not contribute significantly to surveying, cartography, architecture, or exact astronomy; the *De occulta philosophia* (1533) of the celebrated magus H. Cornelius Agrippa did not encourage 'within its purview the growth of those mathematical and mechanical sciences which were to triumph in the 17th century.'[47] Numerology, like hermetic animism, was antithetical to the application of mathematics to practical problems.[48] Agrippa himself spurned the grubby calculations of astronomers: 'I omit [he says] their vain disputes about Eccentricks, Concentricks, Epicycles, Retrogradations, Trepidations, accessus, recessus, swift motions and circles of motion, as being the

41. J. Webster, *Acad. exam.* (1654), 68, 76–7; Wilkins and Ward, *Vind. acad.* (1654), 5, 46, and 22–3, where they spoof hermetic jargon.
42. *Ibid.*, 29; Debus, *Science* (1970), 37, 42, 48, 57–60.
43. Debus, *J. Hist. Med.*, 19 (1964), 391.
44. Yates in Singleton, *Art* (1968), 255, 258.
45. Geymonat, *Galileo* (1965), 7–10.
46. Cf. Zilsel, *J. Hist. Ideas*, 2 (1941), 1–32.
47. Yates in Singleton, *Art* (1968), 259.
48. Cf. Strong, *Procedures* (1936).

works neither of God nor Nature, but the Fiddle-Faddles and Trifles of Mathematicians.'[49]

An apparent exception to this antithesis is John Dee, an Elizabethan magus, alchemist, and navigational authority on whom recent hermetic champions heavily rely.[50] Dee wrote a mysterious alchemical book, the *Monas hieroglyphica,* and spent his declining years interviewing angels; but it was not his occultism that inspired his few—and, for their content, unimportant—contributions to applied mathematics, but his study early in life with the great geometers, cartographers, and instrument makers of Louvain, Gemma Frisius and Gerard Mercator.[51]

3. CORPUSCULAR PHYSICS

From a logical point of view corpuscularism was an extreme form of Aristotelian natural philosophy that recognized very few real qualities and—in its purest form—but one means of action, namely pushing, in the corporeal world. This radical parsimony proved immensely fruitful. The qualities supposed primary, like extension, motion, figure, impenetrability and inertia, which everyone understands intuitively, provided the basis for a 'comprehensible' and quantifiable physics, while most of the Aristotelian real characteristics were declared 'secondary,' creations of the perceiving mind, the business of the psychologist. Making do only with inert, sub-microscopic corpuscles and their motions, the revolutionary philosopher of the seventeenth century proposed to describe all the workings of lifeless matter and much of the operation of plants and animals.

The cynosure of the corpuscularians was Descartes, who tried to anchor his physics on unshakeable foundations. Everyone knows how he strove to doubt everything, but could not bring himself to doubt the existence of himself doubting; how the 'clarity and distinction' of his apprehension of the existence of his doubting mind became the touchstone for the truth of other propositions; and how, as the first of the propositions so secured, he gave the existence of an omnipotent God incapable of deceiving him about the truth of propositions he perceived clearly and distinctly. From this heady metaphysical journey Descartes brought back the clear and distinct principle of the equivalence of matter and extension (wherefore a void space is a contradiction in terms) and several

49. Agrippa, *Vanity* [1530] (1676), 86. The hermeticist distaste for exact computation and applied mathematics is emphasized by Debus, *Ambix,* 15 (1968), 15–25, who quotes van Helmont (p. 24): 'The Rules of Mathematicks, or Learning by Demonstration, do ill square to Nature. For man doth not measure Nature; but she him.'

50. Yates in Singleton, *Art* (1968), 262, 264; French, *Dee* (1972), 160–87.

51. Heilbron in Shumaker and Heilbron, *John Dee* (1978), 34–49. It is gratifying that Paolo Rossi, who has stressed magical elements in the thought of modernizing seventeenth-century philosophers like Bacon, now considers Yates' teachings to be wrong-headed and even mischievous. Rossi in *Reason* (1975), 259–64.

rules of motion, true perhaps of the world in which he found them, but mainly false in ours.[1]

In 1644 Descartes published his substitute for Aristotle, the *Principia philosophiae,* which begin with the metaphysical underpinnings just sketched, exhibit the rules of motion, and then apply—or appear to apply—them to the problems of physics in the widest sense. The freshness and ingenuity of the book made a great impression. To be sure it had a few passages not altogether intelligible, but the misunderstanding was more likely to be the fault of the reader than of the writer. Had Descartes not gone far beyond the ancients in mathematics, and said so? Was it not presumptuous to expect to be able to understand him in everything, even when, as in the *Principia,* he used no mathematics? Descartes anticipated his readers' difficulties. He advised them to read his book straight through, again and again, like a novel: 'on taking it up for the third time I dare say that you will find the solution of most of the difficulties previously noticed; and if a few remain, you can clear them up by rereading yet again.'[2] The quaintness of Descartes' formulations must not be allowed to obscure the influence of his physics nor the fact that its form— applications of firmly grounded rules of motion—is precisely that of Newton's.

Among the most puzzling and difficult problems faced by reductionists like Descartes was the explanation of attraction, particularly that of the magnet.[3] His ingenious solution, which formed the basis of magnetic theory for over a century,[4] exploits most of the characteristic features of his system. Cartesian lodestones owe their efficacy to a system of channels or pores which provide one-way passage for particles of appropriate shape. The pores may be likened to threaded gun barrels fitted with diaphragms to insure unidirectional flow. A magnet possesses two sets of pores aligned with its polar axis, each set admitting an opposite flow of minuscule, twisting, screw-like particles of precisely the size and pitch to wriggle through them. These helical or 'channelled' particles are necessary by-products of the creation of the universe, as appears from the following considerations.

At the beginning, we may imagine, the universe was an undifferentiated continuous bloc of matter: it was extended, and nothing more. God divided this matter into equal microscopic cubes, which he gave a powerful tendency to rotate about their centers. Since no void can exist—extension *is* matter— realization of the God-given rotational urge can occur only if the edges of contiguous cubes are ground to rubble, 'freeing' the potential spherulae they

---

1. *Principia* (1644), in *Oeuvres,* IX:2, 27–38.

2. *Ibid.,* 11–12.

3. Cf. Mersenne's continued struggle with it; Lenoble, *Mersenne* (1943), 366–7.

4. The most striking evidence being the winning of the Paris Academy's three prize competitions of the 1740s for the best 'explanation of the attraction of the magnet' by three embroiderers of Descartes' theory; L. Euler, AS, *Pièces,* 5 (1748), 3–4; Dutour, *ibid.,* 51–114; D. and J. Bernoulli, *ibid.,* 117–44.

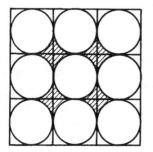

1.1 Sphericles of the second matter
within their parent cubes;
the shaded portions
are cross-sections of future
channeled particles.

circumscribe (fig. 1.1). The rubble and the little balls Descartes calls 'first' and 'second' matter, respectively; they consist of the same material, differing only in size, shape, and—an added quality—mobility.[5]

We are coming to the channelled particles. We have assumed that the spherulae remain where freed, so that their centers form a cubic lattice. Another, closer-packed arrangement is possible (fig. 1.2) and, according to Descartes, is favored by the mobility of the first matter. The close arrangement makes channelled particles. The rubble filling the interstices has, as its narrowest cross-section, the shaded lozenge of fig. 1.1. Imagine that a few layers of balls are exactly superposed on those of fig. 1.1: pores with lozenge-shaped cross-sections result. Suppose that because of their shapes and relative rest the

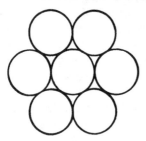

1.2 Close-packed arrangement
of the sphericles.

first matter in the pore forms a stable entity, which will resemble a grooved bar (fig. 1.3);[6] and suppose further that the movement of neighboring globules tends to twist the layers which molded the bar. The grooves become threads, right- or left-handed according to the direction of the twist. The headless cylindrical screws are channelled particles.[7]

The most general cosmic processes produce magnetism. The scrapings of mobile first matter, displaced during the packing of the second, combine to form an agitated spherical body, or sun, held together by the surrounding spherulae. The latter, together with the remaining interstitial first matter, pre-

---

5. *Principia*, in *Oeuvres*, IX:2, 126–9. In practice Descartes also endows his matter with inertia and, via impenetrability, with something like repulsion. Cf. Carteron, *Rev. Phil.*, 47 (1922), 261, 491–3.

6. Characteristically Descartes ignores the transverse pores; his purpose is to show how channelled particles might be formed, not to demonstrate that, in every case, they are so formed.

7. *Principia*, in *Oeuvres*, IX:2, 153–6.

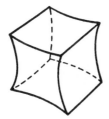

1.3 A channelled particle.

serve their motion by circulation in a vortex about the sun, a vortical or re-entrant curvilinear motion being the only type possible in Descartes' plenary universe. The channelled particles, adopted to potential pores in the vortical medium, or sky, move easily, tending towards the sun along the axis of the vortex and towards the circumference in the plane of its equator, where the centrifugal tendency is a maximum. The constant axial bombardment eventually opens two sets of threaded solar pores.

Meanwhile, channelled particles unable to penetrate the sun tend to settle on its surface. A completely covered sun, no longer able to maintain the surrounding whirlpool, might be pulled into a neighboring sun's swirl, where, if captured, it would play the part of a planet. Such a system might itself be captured, the old sun becoming a planet, and the old planet a moon. This was the origin of our earth and her satellite, which is retained by a little eddy in the vast vortex carrying the solar system. Our local eddy also causes the tides, the fall of heavy bodies, and, through the circulation of the channelled particles, magnetism.[8]

The threaded axial pores, which date from the earth's sunny past, are identical with the magnet's. The channelled particles traverse the earth as pictured (fig. 1.4a), emerge to be deflected by the air, whose pores are unfit to admit them, and, by bombardment, tend to orient magnets to receive them. Their course determines dip and declination, and their duality, or double-handedness, explains polarity.[9] To understand magnetic attraction and repulsion, note that the channelled particles tend to congregate where they find ready passage, namely about a lodestone, which they enclose in a mini-vortex. When two magnets, with contrary poles opposing, come so close together that the channelled particles issuing from the one just manage to reach the other before the air deflects them, they continue along the path of least resistance, enter the pores before them, and unify the magnetic mini-vortices (fig. 1.4b). The resulting vortical flow drives the air from the gap between the magnets; and the displaced air, circling to their back sides, pushes them to union. Magnetic 'attraction' is nothing but mechanical impulse.

A magnet draws iron nails in much the same way: its vortex threads them,

8. *Ibid.*, 157–74, 194–200, 209–10, 225–31. Cf. Scott, *Sci. Work* [1952], 167–94; Aiton, *Vortex Th.* (1972), 30–64.

9. Only one set of pores (right- or left-handed) in fact is needed, but it might be difficult to explain why it alone was produced.

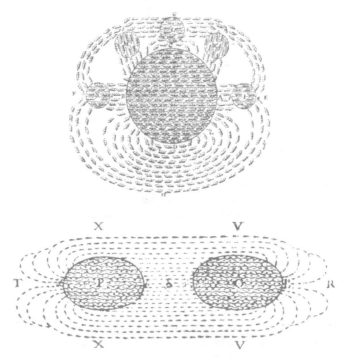

1.4 *Cartesian magnetism, from Descartes,* Principia *(1644): (a) the earth's magnetic vortices ABFGE and BAEHF, with oriented lodestones at I, K, L, M, and N (b) the joint vortex TXVRVX of lodestones P and Q with opposite poles facing.*

orients the diaphragms in their pores, which have the same shape as a lodestone's, and renders them magnetic in a sense contrary to its own. As for the repulsion between like poles, it occurs because the channelled particles issuing from them, being unable to enter the oppositely threaded pores of the opposing magnet, require space to execute their gyrations. And thus, by mobilizing the history and present economy of the universe, does Descartes rescue the archetype of actions at a distance from the grip of occult qualities.[10]

SPREAD AND REFINEMENT OF CARTESIAN PHYSICS

Interpreted on its own terms, as a secure metaphysical system, Cartesian natural philosophy menaced not only the received physics but also established religion. The eradication of most substantial forms and the elimination of all actions save pushes were not the only, nor the chief, irritants. In his account of the origin of our planet Descartes implied that there are many earths and many suns, and that the universe has no bounds. Moreover, despite a clever equivocation, his story of the

10. *Principia,* in *Oeuvres,* IX:2, 271–305.

capture of the moon by the earth, and of both by the sun, strongly endorsed, if it did not prove, the theories of Copernicus.[11] All this opposed both Aristotle and the Bible. Then there were delicate questions like the miracle of the eucharist, about which peripatetics could mumble apparently meaningful words ('Transsubstantiation saves the accidents while changing the essences of the bread and wine'), while Descartes, who had no such distinctions, struggled to show that, in this particular at least, his philosophy was compatible with mystery.

Above all Cartesian philosophy, with its methodical doubting, was subversive in spirit, the rallying point of novelty hunters, of modernists, of philosophical revolutionaries of all kinds. Alarmed authority responded first in the Netherlands, in the 1640s, when Descartes' teachings began to enter the universities, and especially the medical faculties, at Utrecht and Leyden. Their academic senates and curators, stirred up by their theologians, forbade the teaching of 'absurd, paradoxical, or novel doctrines,' of 'dangerous, new, or anti-Aristotelian theses,' that is, of Descartes.[12] In making, and then ignoring, these condemnations the Dutch were in advance of their time. In the 1660s prohibitions and injunctions against Cartesian philosophy were enacted all over Europe, testifying at once to the rapid spread of 'opiniones novae' and to the slow realization that they could not be stamped out by edict.

In August, 1662, the faculty of arts at the University of Louvain, agitated by the papal legate, called upon its professors to advise its students that though Descartes' writings 'might be well-considered in many points regarding natural philosophy, they contain some opinions not sufficiently in conformity with the sound and ancient doctrine of the Faculty.' Shortly thereafter the Faculty of Theology condemned six Cartesian propositions put forward in a thesis for a doctorate in medicine.[13] These moves perhaps inspired that of the Congregation of the Index, which in 1663 damned and prohibited certain of Descartes' writings 'until corrected.'[14] Lutheran theologians were no less vigilant than the Roman Catholics and the Dutch Calvinists. In Sweden, where Descartes had died of the cold in 1650, they fought grimly against the new opinions, which they almost succeeded in prohibiting at the University of Uppsala in 1664.[15]

The French also began to persecute Cartesians in the 1660s. In 1667 the Crown ordered the Chancellor of the University of Paris not to deliver his scheduled oration in honor of the reburial of the remains of the philosopher, just arrived from Sweden. A straw in the wind. In 1671 the Court, speaking through

---

11. The equivocation: since a body can be said to move only when it changes position relative to an object of reference, choose the 'object' to be the surrounding subtle matter; the earth is then at rest, because it remains in contact with the same portion of its vortex. *Oeuvres*, IX:2, 109. Prominent Cartesians of the second generation, e.g., Rohault and Malebranche, came out unequivocally for Copernicus; cf. Robinet, *Malebranche* (1970), 235.

12. Ruestow, *Physics* (1973), 36, 43–4; Lindborg, *Descartes* (1965), 30, 38–9.

13. Monchamp, *Hist.* (1886), 347–8, 354–69.

14. *Ibid.*, 338, 389–91, with text of the condemnation.

15. Lindborg, *Descartes* (1965), 93–6, 115–16.

the Archbishop of Paris, commanded the theologians of the Sorbonne to see to the enforcement of a ridiculous edict pronounced by the Paris Parlement in 1624, which prohibited the teaching of any philosophy but Aristotle's. After consideration the Parlement decided not to repeat its former folly. The famous spoof of Boileau may also have helped.[16] 'Whereas for several years an unknown person named Reason has tried to enter the schools of the University by force and to evict one Aristotle, the ancient and peaceful tenant of the said schools, the court . . . maintains and preserves the said Aristotle in the full and peaceful possession of his rights in the said schools, orders that he will always be taught and followed by the regents, doctors, masters of arts and professors of the said University, who, however, are not obliged to read him, or to know his language or his opinions.' The King did in fact succeed in driving the upstart temporarily from the higher schools of his realm.[17]

In this act of preservation he was ably assisted by the Jesuits. Already in 1650, in response to complaints about the teaching of 'new opinions' from several of its provincial administrators, the ninth general congregation of the Society drew up a list of proscribed propositions, 96 in all, of which fifteen touched upon Cartesianism.[18] The twelfth congregation (1682) repeated the prohibition against inculcation of novelties; the fourteenth reaffirmed it (1696); so did the fifteenth (1706), which pointed out thirty obnoxious doctrines drawn primarily from Cartesian physics, and the sixteenth (1732), which boiled the thirty down to ten.[19] Meanwhile the provinces clamored for the vigorous enforcement of the proscriptions.[20]

The repetition of these complaints and injunctions testifies to their ineffectiveness. A Jesuit professor who persevered in flagrantly disobeying them might be transferred to a third-rate school, or relieved of teaching duties;[21] but many advocated Cartesian doctrines without suffering more than an occasional reprimand. Even the Jesuit father who may have written the notice of Descartes for the Index, Honoré Fabri, was suspected, and quite rightly, of harboring Cartesian sympathies.[22] And why not, if the new philosophy contains anything useful, and one takes precautions? 'Just as formerly God allowed the Hebrews to marry their captives after many purifications, to cleanse them of the traces of infidelity, so after having washed and purified the philosophy of Monsieur Descartes, I [it is Gaston Pardies, S. J., professor of mathematics at Louis-le-

16. Cousin, *Frag. phil.*, III (1866⁵), 300–17.

17. *Ibid.*, 318–22; Mouy, *Développement* (1934), 170–1 (Boileau's *arrêt*); Lamprecht in Columbia U., *Studies*, III (1935), 196n.

18. Monchamp, *Hist.* (1886), 204–9. Cf. L. C. Rosenfeld, *Rev. Rel.*, 22 (1957), 14–40.

19. Pachtler, *Rat. stud.* (1887–94), III, 122–7, gives all forty propositions.

20. *Ibid.*, 121–3; Sortais, *Cartésianisme* (1929), 20.

21. *Ibid.*, 42–4. Perhaps the most interesting case is Y. André, transferred from Louis-le-Grand to Hesdin (Artois), then, since he persisted, removed from teaching (1713), finally locked up in the Bastille (1721); he thereupon apologized and returned to teaching mathematics. *Ibid.*, 23, 28–32.

22. *Ibid.*, 49–50; *infra*, ii.1.

Grand, an ornament of his Society] could very well embrace his opinions.'[23]

This liberty, and perhaps more, was virtually conceded by the general congregation of 1706; for the last of its 'prohibitions' allowed 'the defense of the Cartesian system as an hypothesis whose postulates and principles are well integrated [recte cohaerent] with one another and with their consequences.' Descartes became more and more acceptable to the Society in the eighteenth century; he may have erred, but he was a veritable Loyola compared with Voltaire.[24] The Jesuits came to teach Descartes' physics, and to esteem his mind. 'Even when Descartes lets us see the weaknesses and limits of human nature, he shows a penetrating, luminous, broad, methodical and systematic mind . . ., almost compelling us to follow him or, at least, to admire him.'[25]

While the Jesuits were painfully assimilating Cartesian physics, much of the rest of learned Europe adopted it. At the Dutch universities, after the troubles of the 1640s and 1650s, Cartesian fellow travellers captured the teaching of physics and several chairs in the medical schools. They retained their hold until the advent of the Newtonians.[26] The Swedes fought to a standstill, and agreed that Cartesian physics could be taught, but Cartesian hermeneutics proscribed, an accomodation that perhaps prepared the way for the flowering of Swedish physics in the eighteenth century.[27] In Italy, after ferocious battles in the 1660s in Naples, where Cartesian physics had been introduced by lawyers and radical physicians,[28] Descartes made his way into the northern universities. In 1700 his followers Guido Grandi, an excellent mathematician,[29] and Michelangelo Fardella, a great friend of the French Cartesians,[30] captured chairs of philosophy at

23. Pardies, *Lettre d'un philosophe à un Cartésien* (1672), in Ziggelaar, *Pardies* (1971), 119. Like Fabri, Pardies defended himself against charges of Cartesianism by the unpersuasive argument that he rejected Descartes' laws of motion (as did everyone else). *Ibid.*, 80, 115.

24. Sortais, *Cartésianisme* (1929), 89, 92; Werner, Ak. Wiss., Vienna, Phil.-Hist. Cl., *Sitzb.*, 102 (1883), 681–2; Bouillier, *Hist.* (1854), I, 557–80.

25. Regnault, *Origine* (1734), II, 358–9. Cf. P. Baudory's speech at Louis-le-Grand in 1744: Descartes 'erravit aliquando, quod humanum est, at non erravit errantium auctoritate, quod turpe ac imbecillum est.' Schimberg, *Education* (1913), 520.

26. Ruestow, *Physics* (1973), 73–88; Monchamp, *Hist.* (1886), 345, 476–83.

27. Lindborg, *Descartes* (1965), 325–38, 348–9. Conversely, the intellectual backwardness of Spain, according to the eighteenth-century reformer B. J. de Feijóo, owed not a little to an irrational fear of Descartes; as late as 1772 the Royal and Supreme Council of Castile ordered the University of Salamanca to have nothing to do with him. Browning in Hughes and Williams, eds., *Var. Patt.* (1971), 358, 365.

28. Fisch in *Science* (1953), I, 529, 544–5, 553–4; Berthé de Besaucèle, *Cartésiens* (1920), 3–17.

29. *Ibid.*, 55–7. Tenca, Ist. lomb., Cl. sci. mat. nat., *Rend.*, 83 (1950), 494–7, and Pavia, Coll. Ghisl., *Studi*, 1 (1952), 21–4; Grandi estimated that his Jesuit teacher of mathematics, Saccheri, was 'seven-eighths Cartesian,' the remaining part being reserved for good points in Aristotle and Gassendi.

30. Cromaziano, *Restaurazione* (1785–89), II, 84–5; Berthé de Besaucèle, *Cartésiens* (1920), 65–85; Werner, Ak. Wiss., Vienna, Phil.-Hist. Cl., *Sitzb.*, 102 (1883), 77, 132–3.

Pisa and Padua, respectively. Others brought the cause to Bologna, Rome and Turin.

Cartesian physics was taken up in England in the 1640s by the Cambridge philosopher Henry More, a Platonist who, for a time, thought that nothing could be finer than the system of Descartes; 'all that have attempted anything in naturall Philosophy hitherto [he said] are mere shrimps and fumblers in comparison to him.'[31] More urged that Cartesian philosophy be taught in all public schools and universities, particularly Cambridge, where he had already made it influential, and where Newton encountered it—and profited from it—in the 1660s.[32] Joseph Glanville, the self-ordained extirpator of dogmatists, the exploder of the 'steril, unsatisfying verbosities' of the schools, had nothing but praise for our prince of systematizers, 'the grand secretary of nature, the marvelous Descartes.'[33] Other fierce anti-peripatetics, like Robert Boyle, endorsed Cartesian corpuscularism as a good antidote to Aristotelian physics. But Descartes' day in England was brief. Even before the big guns of Newton and Locke came into play, the British had shivered before the materialist consequences of Cartesianism and drawn back.[34] More himself in 1668 had rejected Descartes' metaphysics as 'erroneously and ridiculously false,' but retained the physics; in 1671 he had come to regard all parts of the system as 'most impious, most inept, and entirely false.'[35]

Descartes then replaced Aristotle as the foil against which British physics tested its metal. Hooke could praise the 'most incomparable Descartes,' that 'most acute and excellent Philosopher,' for providing the law of refraction, and yet prefer his own ideas about the nature of light. Newton started with Descartes, and soon went far beyond him in optics; it took longer to elude the spell of the great solar vortex, and to replace Cartesian mechanism with the theory and apologetics of universal gravitation.[36] An apt symbol of the relationship between Cartesian and English physics is the text of Rohault in the editions

31. More to S. Hartlib, 11 Dec. [1648], in C. Webster, *Br. J. Hist. Sci.*, 4 (1969), 365. Cf. *ibid.*, 371: 'If Mr. [William] Petty [who did not share More's enthusiasm] should have twice the Age of an ordinarie man, and spend all his dayes in experiencing, he will never bring an instance against Des Cartes his principles of Light, of the Lodestone, of the Rainbow, of the Flux and Reflux of the Sea, etc.'

32. Lamprecht in Columbia U., *Studies*, III (1935), 208–9; C. Webster, *Ambix*, 14 (1967), 153, 156, 167–8; Power, *Exper. Phil.* (1664), sig. c.1ʳ, on the 'ever-to-be-admired Des-Cartes.'

33. Glanville, *Vanity of Dogmatizing* (1661), quoted by Lamprecht in Columbia U., *Studies*, III (1935), 201.

34. Pacchi, *Cartesio* (1973), viii–ix, 231–4, 248–50; cf. Lamprecht in Columbia U., *Studies*, III (1935), 185–7, 199, 229, and Bouillier, *Hist.* (1854), II, 491–8.

35. Lamprecht, in Columbia U., *Studies* III (1935), 219–25. Yet More appears to have taught Cartesian physics as late as 1674 (*ibid.*, 195).

36. Hooke, *Micrographia* (1665), Pref., 46, 54, 57; Whiston, *Memoirs* (1749), 8ff; cf. Koyré, *Newt. Stud.* (1965), 53–114, and the preface to the *PT* for 1693 (*PT*, 17, 581–2): 'Real Knowledge is a nice thing; and as no Man can be said to be Master of that which he cannot teach to another, so neither can the Mind itself, at least as to Physical matters, be allowed to apprehend that, whereof it has not in some sense a Mechanical Conception.'

of Clarke: the theories of the disciple of Descartes provided the stimulus for the criticisms and rectifications of the disciple of Newton.

While Descartes' star sank in England it rose in France, at least outside the universities.[37] The Oratorians, a teaching order second in importance only to the Jesuits, boasted several prominent Cartesians, above all Malebranche. The *noblesse de robe,* from whom Descartes sprang, welcomed his teachings, and played the part in France that physicians did in the universities of the low countries. The abbreviated obsequies of 1667, welcoming home Descartes' mortal remains, and the banquet that followed were attended by members of the Conseil d'Etat and a crowd of prominent lawyers.[38] In the 1650s the group maintained by the Parisian lawyer Henri-Louis de Montmor—a forerunner of the Paris Academy of Sciences—was sufficiently Cartesian to harass members who were not, and secure enough to encourage public lectures on the new physics.

Some of the lectures, particularly Jacques Rohault's, had an immense success. No one needed the old physics any more. Supply followed demand. Publication of texts on scholastic physics almost ceased in France. 'Such texts as do appear [lamented Gabriel Daniel, S. J., in 1690] are treatises on physics that assume the principles of the new philosophy.'[39] Rohault did not gain admission to the Paris Academy, but he led the Cartesians to the gate. After some official persecution, they entered in the person of Rohault's Joshua, Pierre-Sylvain Régis. That was in 1699, the year the Academy reorganized with the Cartesian reformer Malebranche as an honorary member and the moderate Cartesian Fontenelle as its secretary.[40] It remained Cartesian for two generations.

Several important new features entered Cartesian physics with Malebranche, whose life was a model of progressive enlightenment. In 1664, after completing his theology at the Sorbonne without enthusiasm or distinction, he ran across Descartes' physiological fragment, the *Traité de l'homme.* It was the first work of the master that he had seen. He read it straight through, 'with such excitement,' says Fontenelle, 'that it gave him palpitations of the heart, which obliged him sometimes to interrupt his reading.' It pushed him from scholastic darkness into the light of mathematics and physics. Ten years later the fruit of his learning and meditation appeared as the *Recherche de la verité* (1674–5).[41]

37. Cartesian physics officially entered the colleges of the University of Paris in 1721. Lantoine, *Hist.* (1879), 138–9, 150–1.

38. Cousin, *Frag. phil.*, III (1866⁵), 302; Bouillier, *Hist.* (1854), I, 254–5. 'Cette philosophie a été incroyablement applaudie partout . . . surtout parmi les nobles'; Pardies, c. 1670, quoted by Ziggelaar, *Pardies* (1971), 79.

39. Mouy, *Développement* (1934), 98–113; R. Hahn, *Anatomy* (1971), 6–8; *infra,* ii.4; Daniel, *Voyage au monde de Descartes* (1690), quoted by Sortais, *Cartésianisme* (1929), 59.

40. Mouy, *Développement* (1934), 145–7, 166–79.

41. Fontenelle, 'Eloge' of Malebranche, in Malebranche, *Oeuvres*, XIX, 1000. Grandi too came to mathematics from Cartesian philosophy; Tenca, Ist. lomb., Cl. sci. mat. nat., *Rend.*, 83 (1950), 495.

During the 1690s Malebranche climbed still further by mastering the Leibnizian calculus, an instrument more effective in hunting out the truths of physics than even the geometry of Descartes. He was joined in these studies by a group of strong mathematicians, later his colleagues in the Academy, all also Cartesian in physics, of whom the best known are Pierre Varignon and the Marquis de l'Hospital. Fontenelle shows us each of these men traversing the same road to Enlightenment: from scholastic night through Cartesian day to the blazing light of the differential calculus.[42] Their example and their work were to influence the mathematical physicists of the mid-century, d'Alembert, Maupertuis, and even Euler.[43]

Among Malebranche's modifications of Cartesian physics were a continuing repair of the laws of motion[44] and the discovery of several orders of whirlpools that Descartes had overlooked. Malebranche announced his discovery at the first session of the renovated Academy, in an important paper on optics. Since, he said, light is a pressure transmitted through the globules of the second matter, those globules cannot be hard seeds, as Descartes taught, but elastic balls; otherwise rays of different colors—pressure waves of different frequencies, in Malebranche's model—could not cross without destroying one another.[45] Nothing could be easier than to make the globules elastic: merely imagine them to be so many minute vortices in exceedingly rapid motion, striving to expand owing to 'centrifugal force,' but restrained by the similar tendencies of their neighbors. Much can be explained by these mini-vortices. Malebranche accounts for all elasticity and for solidity or hardness, which had given Descartes much trouble;[46] later he worked the mini-vortices into the theory of planetary motions, and the explanation of refraction;[47] and some of his followers found in them the explanation of electricity.

To understand the hold of Cartesian natural philosophy on the European mind one must understand it not as an ontology, but as an epistemology. Descartes pointed the way not to apodictic truth, but to intelligible physics: 'In the true philosophy,' says Huygens, 'one considers the cause of all natural effects in terms of mechanical motions. This, in my opinion, we must certainly do, or else renounce all hopes of ever comprehending anything in physics.'[48] And that

42. Robinet, *RHS*, 14 (1961), 238–43, and *RHS*, 12 (1959), 5–8, 15–16.

43. Hankins, *J. Hist. Ideas*, 28 (1967), 194, 201–5, and *D'Alembert* (1970), 17, 21, 118–20.

44. Mouy, *Développement* (1934), 292–304.

45. Malebranche, *MAS* (1699), 22–32; Robinet, *Malebranche* (1970), 277–8, 285–94; Mouy, *Développement* (1934), 289–91, 305–10; Duhem, *Rev. met. mor.*, 23 (1916), 77–9, 89–91. The objection seems to be that a hard globule must move as a unit, with one motion at a time, while a compressible ball can transmit many different pushes simultaneously. Cf. Fontenelle, *HAS* (1799), 19.

46. Mouy, *Développement* (1934), 282–8; Malebranche, *Oeuvres*, II, 326, III, 272–3 (text of 1712). Cf. Carteron, *Rev. phil.*, 47 (1922), 491–2.

47. Malebranche, *Oeuvres*, III, 283–4, 296–8; Mouy, *Développement* (1934), 310–14.

48. Huygens, *Treatise* [1690] (1945), 3. Cf. Koyré, *Newt. Stud.* (1965), 118, and Huygens' objection to Newton's optical theory of 1672: why suppose seven or eight colors in the spectrum

is all one must do to be a Cartesian physicist: the details of Descartes' mechanical pictures, his fanciful laws of motion, his metaphysical underpinnings, none of this need be—or often can be—accepted. As Fontenelle put it: 'Il faut admirer toujours Descartes, et le suivre quelquefois.'[49]

The Cartesian physicists of the late seventeenth Century did not recommend push-pull physics for its truth. Most placed physical truth, the essence of body, the ultimate cause of change, beyond the reach of the human mind. Fontenelle, who liked to poke fun at metaphysics ('which is to most people like an alcohol flame, too subtle to burn wood'[50]), declared that the process of collision, the only allowable interaction between extended bodies, is at bottom unintelligible to us.[51] The weakness of our minds, the false witness of our senses, and our share in Adam's fall showed Boyle that the hypotheses even of the corpuscular philosophy could only be doubtful conjectures.[52] 'Physical demonstrations can beget but a physical certainty.'[53] *A fortiori,* one could never choose between corpuscular explanations equally comprehensive, between Descartes' and Boyle's explanations of the spring of the air, for example; the same clock, after all, can be driven by a spring or by a weight.[54] Glanville hunted out paradoxes to prove the impossibility of ever comprehending matter.[55] Régis, despite his love of system, conceded physics to be 'problematic' and 'uncertain.' So did Huygens.[56]

---

rather than two, of which the remainder can be composed? For then 'it will be much more easy to find an Hypothesis by motion.' Newton, *Papers* (1958), 136; Guerlac in Wasserman, *Aspects* (1965), 327; McDonald, *Ann. Sci.*, 28 (1972), 219–21.

49. Fontenelle, *HAS* (1735), 139. Cf. Fontenelle, 'Digression sur les anciens et les modernes' (1687), *Oeuvres,* IV, 121: '[Descartes'] new way of philosophizing [is] much more valuable than the philosophy itself, a good part of which is false, or very uncertain.'

50. 'Eloge' of Malebranche, in Malebranche, *Oeuvres,* XIX, 1002. Failure to distinguish between agreement with Descartes' system and acceptance of the general principles of his physics has resulted in a large and inconclusive literature (for which see Lissa, *Cartesianismo* [1971], 33–4) on the 'question' of Fontenelle's Cartesianism; Lissa, *ibid.*, 37, concludes judiciously that Fontenelle 'libera dall'architettonica construzione metafisica di Cartesio un dottile strumento di indagine [fisiche], il cui spirito è fondamentale antimetafisico.'

51. Fontenelle, 'Doutes sur le système physique des causes occasionnelles' (1686), in *Oeuvres* (1764), IX, 45–6. Cf., Carré, *Philosophie* (1932), 20–1; Marsak, Am. Phil. Soc., *Trans.,* 49:7 (1959), 20–1, 31–2; Robinet, *Rev. synt.,* 82 (1961), 82–3.

52. Boyle, *Reason and Religion* (1675), in *Works,* IV, 164–5; van Leeuwen, *Problem* (1970), 93; Guerlac in Wasserman, *Aspects* (1965), 321–3. Cf. McGuire, *J. Hist. Ideas,* 33 (1972), 523–42, who emphasizes the connection between Boyle's nominalism in science and voluntarism in theology; Burtt, *Met. Found.* (1955), 166–302, who notes (p. 185) Boyle's anticipation of Newton's positivism; and Westfall, *Ann. Sci.,* 12 (1956), 107–10, who points to some passages from Boyle admitting a realist interpretation.

53. Boyle, *Excellency Theol.* (1674), in *Works,* IV, 42.

54. Boyle, *Works,* I, 12, II, 45–6, V, 74–5, quoted by Mandelbaum, *Philosophy* (1964), 90–1. Boyle's attitude towards physical hypotheses may have influenced Locke's; Gibson, *Locke's Th.* (1917), 205–31, 260–5.

55. *Scepsis scientifica* (1665), quoted by Gibson, *Locke's Th.* (1917), 257.

56. Mouy, *Développement* (1934), 147, 153; *DSB,* VI, 608.

Malebranche arrived at a similar view by a straighter path. Having a mind, he said, as good as anyone's, and yet being unable clearly to conceive how one body can act upon another, he declared that they do not, that the very notion was a vulgar error. Created beings cannot act; what we, in our ignorance, believe to be the push of one object upon another, or the stimulation of our senses by a third, are not acts but 'occasions' for acts.[57] God directly causes rebounds appropriate to the pushes and sensations agreeable to the stimuli. Being free from caprice, He acts always—or almost always, to leave room for miracles—in the same manner. The consequences of collisions may be confidently predicted by the laws of mechanics. Such laws, mathematically formulated, are the only true knowledge: 'I do not believe that there is anything useful which men can know with exactitude that they cannot know by arithmetic and algebra.'[58] Since these laws are God's choices, we can not deduce them *a priori*, as Descartes tried to do. But we must always strive to reduce our physics to one of impulse; for although we can have no conception of the 'mechanics' at work, we understand immediately that something must happen in collisions, something 'certain and incontestable,' and we can calculate the result.[59]

This last sentiment suggests why the 'true' physics must be mechanical. Mechanical theories have the advantage over all others of clarity, precision, completeness, and naturalness.[60] They are also relatively intelligible. To be sure the inner nature of a collision is not comprehensible; but there are degrees of unintelligibility, and the less the better. 'It is certain [says Fontenelle] that if one wishes to understand what one says, there is nothing but impulse, and if one does not care to understand, there are attractions and whatever one pleases; but then nature is so incomprehensible to us that it is probably wiser to leave her alone.' Our idea of impenetrability tells us immediately that something must happen when moving balls collide; but nothing suggests that, when mutually at rest and widely separated, they must attract one another.[61]

Boyle extols the mechanical philosophy first for its 'intelligibleness or clearness' (as against 'intricate' disputes of the peripatetics, the 'darkness and ambiguity' of the spagyrists); second, for its economy (requiring only matter and motion); third and fourth, for its radical simplicity (matter and motion being the simplest 'primary' concepts); and last for its comprehensiveness. He dismisses forms and qualities not because they are wrong or 'self-repugnant,' but because

57. Malebranche, *Oeuvres*, III, 203–12, 217–18. For the theological connections of the doctrine see Rodis-Lewis, *Malebranche* (1963), 135–8, 296–300.

58. Malebranche, *Oeuvres*, II, 292g.

59. *Ibid.*, II, 403: 'Il n'y a aucune raison, ni aucune expérience, qui démonstre clairement le mouvement d'attraction.' Cf. Robinet, *Malebranche* (1970), 77; Mouy, *Développement* (1934), 316–18; *infra*, i.4–5.

60. De Volder, as quoted by Ruestow, *Physics* (1973), 94–5.

61. Fontenelle, 'Eloge' of R. de Montmort, in *Oeuvres*, VI, 26; cf. Marsak, Am. Phil. Soc., *Trans.*, 49:7 (1959), 13–14.

'we conceive not, how they operate to bring effects to pass.'[62] 'It is more certain to reason from mechanical and intelligible principles than to depend upon novelties which are not expressed in ideas familiar to the mind.'[63] We shall take this insistence on mechanical explanation, on invoking 'mechanism as immediate cause of the phenomena of nature,' as the distinctive mark of Cartesian physics, and shall regard it as an epistemological condition, a requirement for securing the simplest, the most intelligible, the most fruitful, and the most satisfying of natural philosophies.[64] Or so it was for those who held with Fontenelle that 'nature is never so admired as when she is understood.'[65]

PRAECEPTOR GERMANIAE

Although Dutch Cartesianism bubbled over from Holland into a few neighboring Calvinist universities like Duisburg, the philosophy of Descartes in the strict sense never prospered in Germany.[66] The Jesuit schools fought it bitterly. The Lutheran universities of the north opposed it as subversive of sound theology. At Marburg, in 1688, they smothered a Cartesian work by a colleague, a professor of medicine. In the 1690s the theologians at the avant-garde University of Halle harassed the professor of 'new philosophy and mathematics.'[67] Although resistance to Descartes weakened in the German Protestant schools after 1700, that did not establish his philosophy, at least under his name. It was squeezed out between the school philosophy and the teachings of Leibniz, as modified, expanded and systematized by the 'preceptor of Germany,' Christian von Wolff.

The system owed much to Descartes, whose promise of a mathematical philosophy, and its extension to theology, Wolff hoped to realize.[68] He went to the University of Jena expressly to study mathematics under G. A. Hamberger, and began his own teaching career at Halle in 1706 as professor of that 'unknown and unusual' subject.[69] Soon he added physics, then metaphysics, philosophy, theology, law, to the increasing annoyance of his colleagues. He wrote text books on all these subjects, and in his own language, which obliged

62. *Excellency Corp. Phil.* (1674), *Works*, IV, 72. Cf. van Leeuwen, *Problem* (1970²), 75, 106.

63. Nollet, *Leçons* (1747-8), II, 477; cf. Nollet to Bergman, 20 Sept. 1766, in Bergman, *For. Corresp.* (1965), 285: 'Des causes mécaniques, qui sont les seules capables d'étendre les progrès de la physique expérimentale.'

64. Cf. Mairan, 'Eloge' of Molières, *MAS* (1742), 200; Bouillier, *Hist.* (1854), II, 569, 573–5.

65. 'Préface sur l'utilité des mathématiques et de la physique' (1733), in *Oeuvres*, V, 11.

66. Bouillier, *Hist.* (1854), II, 404–5; Paulsen, *Gesch.*, I (1896), 519; Bartholmess, *Hist.* (1850), I, 101.

67. Hermelink, *U. Marburg* (1927), 306–16, 330–1; Förster, *Übersicht* (1799), 49–50, 54.

68. Vleeschauwer, *Rev. belg. phil. hist.*, II (1932), 659–63, 676–7; Wolff, 'Lebensbeschreibung,' in Wuttke, *Wolff* (1841), 114, 121.

69. *Ibid.*, 146; Gottsched, *Wolff* (1755), 9–13; Vleeschauwer, *op. cit.*, 666–7.

him to invent a German philosophical vocabulary.[70] The mathematics book, first published in Latin in 1713, covered the full range of pure and applied subjects; it remained standard for half a century, and is still useful for its annotated bibliography.[71] The physics books stressed experimental confirmation and described in detail the apparatus necessary to achieve it.[72] They also had a wide circulation; they played the same role in Protestant Germany in the eighteenth century that Melanchthon's had in the sixteenth and seventeenth.[73] By 1715 Wolff was recognized as an authority in his own right and as the intellectual heir of Leibniz. The universities of Wittenberg and Jena begged him to add his light to their faculties; the city of Bologna sought his advice on water control; and the Czar of the Russias consulted him about the educational backwardness of his people.

Wolff's rational theology did not please everybody. When he went so far as to say a kind word about Confucius his enemies cried atheism and brought the matter before the King of Prussia, Frederick William I. The King ordered Wolff to leave his dominions forthwith (1723); the philosopher immediately found refuge and a higher salary at the University of Marburg, whence he continued to issue texts, now in Latin, on all respectable subjects.[74] His reputation, enhanced by martyrdom in the cause of academic freedom, brought him offers from the Petersburg Academy of Sciences, the universities of Göttingen and Utrecht ('under such circumstances as no professor in Holland had'), and even Halle, to which Wolff returned in 1740.[75] During the Marburg exile Wolff's philosophy spread to the Petersburg Academy and the universities of Leipzig, Jena, Tübingen and Würzburg, to mention only those most seriously infected. Count von Manteuffel set up an influential group called the Alethophiles to promote Wolff in Berlin.[76] And, doubtless most gratifying of all, the Marquise du Chatelet, sometime mistress of the great puffer of English philosophy, Voltaire, announced that Leibniz' was the only metaphysics that satisfied her and hired a Wolffian to initiate her into its mysteries.[77] Wolff himself sought a correspondence with the lady, the instrument, he hoped, for the conversion of the French. She gave him more than he could have hoped, an *Institutions physiques* prefaced by an abridgment of his philosophy, a clear and, what was unusual, a concise account that won the approval of both Wolff and

70. Gottsched, *Wolff* (1755), 30, 35–6, 42, 48, 51, and Beylage, 9–14, 17.

71. One finds Wolff's mathematics recommended into the 1770s.

72. Wolff, *Ausf. Nachr.* (1733²), 474, 476, 479; *supra,* i.1.

73. Paulsen, *Gesch.*, I (1896), 527; cf. Bartholmess, *Hist.*, I (1850), 99–100, 103.

74. Gottsched, *Wolff* (1755), 57–72; Förster, *Übersicht* (1799), 95–6, 140.

75. Gottsched, *Wolff* (1755), 90–1, 100, 102, and Beylage, 50, 67; Wolff in Wuttke, *Wolff* (1841), 154–5, 165. Negotiations for the return were begun under Frederick Wilhelm I and completed under his successor, Frederick the Great.

76. Gottsched, *Wolff* (1755), 104, 120–1; *infra,* ii.2. Cf. Paulsen, *Gesch.*, I, 546, for the Wolffian foundations of the University of Erlangen (1743).

77. Letter to Frederick the Great, 25 April 1740, in Du Châtelet-Lomont, *Lettres* (1958), II, 13; Du Châtelet-Lomont, *Institutions* (1740), 13. Cf. Barber, *Leibniz* (1955), 127–40.

the Alethophiles. 'What is certain, what leaps to the eye,' wrote Manteuffel, 'is that she has given up all the chimeras of her friend Voltaire, whom she far surpasses in precision and clarity of ideas.'[78]

Wolff's severe rationalism rests on two principles, one logical, the law of contradiction, the other psychological, the law of sufficient reason. The latter plays the same part in his system that the principle of clarity and distinction of ideas plays in Descartes': a phenomenon is satisfactorily explained, its existence fully understood, when its 'sufficient reason' can be given. All contingent truths depend on this proposition. The Principle of Sufficient Reason, says the Marquise du Chatelet, is the only 'compass able to guide us through the shifting sands of [metaphysics],' the only 'thread that can lead us in these labyrinths of error.'[79] What suffices for one philosopher, however, seldom contents another, and, in practice, the touchstone of Wolffian truth was the satisfaction of Wolff's reason.

That powerful instrument disclosed that the world consists of distinct, unextended 'units' or 'elements,' unselfconscious Leibnizian monads. Unfortunately, one cannot fully reduce physical phenomena to these elements; in particular, as unphilosophical mathematicians like Euler liked to point out, one cannot understand how they might make up an extended body. Wolff dismissed such objectors as metaphysically illiterate ('Euler is a baby in everything but the integral calculus'[80]) and built his physics not upon his elements, but upon what he called the basic 'phenomena' of matter: extension, inertia, and moving force.[81] The fundamental objective of physics is to reduce other phenomena to the basic set: 'per extensionem, vim inertiae et vim activam omnes corporum mutationes explicari possunt.' Everything works mechanically; the physical world is nothing but a clock; physics is 'mechanical philosophy.'[82]

The physicist should have nothing to do with occult qualities, for which, by Wolff's definition, he can give no sufficient, that is, mechanical reason. He must reject action at a distance, attractions understood as a primitive force. One can assign no reason that A and B, separated in space, should act upon one another; matter can act only by impact, by immediate contact and mutual 'obstruction;' where there appears to be attraction, as in the cases of electricity and magnetism, we must assume the existence of an unseen, mediating, material emanation.[83] As Mme du Châtelet put it: attraction is 'inadmissible, since it

78. Wolff in Wuttke, *Wolff* (1841), 178; Wolff to Manteuffel, 7 June 1739, in Ostertag, *Phil. Geh.* (1910), 38, 40. Cf. Droysen, *Zs. franz. Spr. Lit.*, 35 (1910), 226–38.

79. Du Châtelet-Lomont, *Institutions* (1740), 13, 22, 25.

80. Wolff to Manteuffel, 4 Aug. 1748, in Ostertag, *Phil. Geh.* (1910), 147–8. Cf. *ibid.*, 75: Euler 'understands not the least little things in philosophy.'

81. Wolff, *Cosmologia* (1732), §§226, 296, 298; the problematic standing of these 'basic phenomena' is discussed by Campo, *Wolff* (1939), I, 223–5, 230, 250–1.

82. Wolff, *Cosmologia* (1737²), §§74, 75, 79, 117, 127, 138.

83. *Ibid.*, §§133, 320–5, 149: 'Qualitas occulta dicitur ea, quae sufficiente ratione destituitur, cur subjecto insit, vel saltem inesse possit.' Cf. Leibniz to Wolff, 23 Dec. 1709, in Leibniz,

offers nothing from which an intelligent being can understand why the velocity and direction—the determinations of the being under discussion [motion acquired under the suppositious distance force]—are such rather than otherwise. Not even God could say how a body acted upon at a distance would move.' Sufficient reasons are mechanical causes. Hence, said the marquise, sounding the call to the colors, all true physicists should rally to a search for a mechanical explanation of gravity.[84] Wolff himself would have done so, successfully we are told, had he been able to perfect his own general physics.[85] He did hint that one should follow up the Cartesian accounts of magnetism, and he repeatedly pointed to electrical experiments as evidence that apparent attractions are mediated by a 'subtle matter.'[86]

The fact that Wolff could not stop to carry out his reductionist program was perhaps more an encouragement than a disappointment to others. On the one hand one could do physics as he did, 'acquiescing,' as he often said, 'in the phenomena;' here he limited himself to a description in terms of physical concepts such as cohesion and elasticity, inadmissible as fundamental powers but very useful as 'proximate causes,' 'physical' as opposed to 'mechanical' principles.[87] On the other hand one could seek deeper, mechanical explanations, assured by Wolff himself that the goal was possible and worthy of attainment. The Wolffian physicist, the last heir of the mechanical philosophy of Descartes, long protected Germany from the irrational and slipshod methods of Sir Isaac Newton.

4. ATTRACTION IN NEWTON

The 'Preface' to Newton's masterpiece, *Principia mathematica philosophiae naturalis* (1687), formulates its mission as follows: 'The whole business of philosophy seems to consist in this—from the phenomena of motions to investigate the forces of nature, and then from these forces to demonstrate the other phenomena.' What this means is plain from the body of the book. 'To investigate the forces of nature' means to infer mathematical propositions about forces, somehow known to exist; 'to demonstrate the other phenomena' means to compare quantitative data with logical consequences of the propositions. If the procedure succeeds, the propositions, according to Newton, must be regarded as true; for '[they] are deduced from Phenomena & made general by

---

Briefw. (1860), 113: '[Actio in distans] pugnat tamen cum magno illo Principio Metaphysico . . . , quod nihil sine ratione sive causa fiat.'

84. Du Châtelet-Lomont, *Institutions* (1740), 47–9, 328–34. Precisely the same line of reasoning appears in D. and J. Bernoulli, AS, *Pièces*, 5 (1748), 119.

85. Gottsched, *Wolff* (1755), 151.

86. Wolff, *Cosmologia* (1737²), §320; Wolff to Manteuffel, 8 March and 4 Oct. 1744 ('zur Zeit noch keine warscheinlichere Erklärung [of electricity] als durch die vortices cartesianas gefunden waren') and 8 Nov. 1747, in Ostertag, *Phil. Geh.* (1910), 65–7, 137–8.

87. Wolff, *Cosmologia* (1737²), §§235–8, 241, 292; cf. Campo, *Wolff* (1939), I, 240, 244–5.

Induction: w[hi]ch is the highest evidence that a Proposition can have in this philosophy.'[1]

In the special case of gravity Newton infers the accelerations of planets towards the sun and of satellites towards their primaries by examining Kepler's empirical laws in the light of certain mathematical principles about centripetal forces. That is investigating the forces of nature. He then applies the gravitational law to deduce the motions of the planets, the comets, the moon and the sea, which he shows agree most beautifully with the best data available.[2] In keeping with his conception of natural philosophy, Newton believed that he had thereby demonstrated the existence of gravity, of a centripetal force acting between every pair of material particles in the universe and causing their mutual accessions, or acceleration, in accordance with a simple mathematical expression. As to the ultimate nature of this force, its seat and mechanism, Newton —or, rather, the author of the *Principia,* for there were several Newtons— declined to speculate: since the phenomena did not settle whether gravity was innate to material particles, or a property of the space between them, or the result of a pressure from an interplanetary medium, he refused overtly to frame —or feign—an hypothesis. 'To us it is enough that gravity does really exist, and act according to the laws which we have explained, and abundantly serves to account for the motions of the celestial bodies, and of our sea.'[3]

Newton's view of the theory of universal gravitation failed to convince his contemporaries. Indeed, his apologetics are not very plausible. It requires more than a mathematical midwife to deliver from the phenomena the proposition that all pairs of particles in the universe—or even in the solar system— mutually gravitate. One needs besides some laws of motion and, in particular, the principle of rectilinear inertia, assumed to hold true of every bit of heavenly and earthly matter. But neither this principle, nor the universal applicability of any such principle, is evident, or perhaps even probable. Galileo inclined toward circular inertia and the Aristotelians made qualitative distinctions between sublunary and celestial behavior the basis of their physics.[4] One might object to Newton that, perhaps, the planets do not move like separated terrestrial objects; perhaps they go naturally in circles, which unknown, non-mutual, non-central agencies distort into the postulated Kepler ellipses? Newton saw this loophole. To close it he laid down three rules, which obliged the philosopher to ascribe similar effects to similar causes and to regard as *universal* those qualities of matter found to belong to, and to be unalterable in, bodies accessible to experiment. These rules have proved very, though not invariably fruitful, but they do

1. Newton to Cotes, 1712, in Koyré, *Newt. Stud.* (1965), 275. Cf. Newton, *Math. Princ.* (1934), 400: 'In experimental philosophy we are to look upon propositions inferred by general induction from phenomena as accurately or very nearly true.' This is the fourth *regula philosophandi,* which first appeared in *Princ.* (1726³); Koyré, *Newt. Stud.* (1965), 268–71.

2. *Math. Princ.* (1934), xviii.

3. *Ibid.,* 547. This is from the General Scholium, which first appeared in *Princ.* (1713²).

4. Cf. M. B. Hesse, *Forces* (1961), 146–7.

not secure the *existence* of universal gravity.[5] All one can say is that bodies act *as if* they are endowed with Newton's gravitational force; and even that is not strictly correct, as the precession of Mercury's perihelion seems to show.

Newton claimed that his powerful and fruitful, but nonetheless hypothetical theory of gravitation was a direct induction from the facts, and free from speculation. He had made precisely the same claim about his first scientific work, the paper on optics of 1672. He said that he had proved that white light is physically a composition of 'rays' differently refrangible, whereas his experiments had demonstrated only that a different index of refraction characterized each spectral color obtained from white light passed through a prism. In the ensuing many-sided controversy, Hooke hit upon an unexceptionable alternative hypothesis, that white light consists of a complex aether pulse, unanalyzed until its passage through the prism. Newton never recognized that his representation of light, though unexceptionable as a mathematical description, became as hypothetical as Hooke's when interpreted physically. He still claimed in the last edition of the *Opticks* (1717–8), which maintained that white light is a physical mixture of colored rays, that his purpose was 'not to explain the Properties of Light by Hypotheses, but to propose and prove them by Reason and Experiments.'[6]

It is not strange that Newton's first readers, who understood correctly that he believed in his universal centripetal accelerations, did not interpret his procedures in the later standard instrumentalist sense. Moreover, the intensity of his belief predisposed them to disregard his occasional disclaimers and to conclude —wrongly this time—that the *Principia* advanced a particular view of the cause of gravity. Its frequent references to mutual and equal attractions, to bodies drawing one another across resistanceless spaces, to powers exercised in proportion to mass, to accelerative forces diminishing as the square of the distance, made natural the inference that Newton held gravity to be an innate property of bodies, and to act immediately at a distance.[7]

One example is Proposition VII of Book III. It argues that 'all the parts of any planet A gravitate towards any other planet B,' a formulation which, as Newton's editor Roger Cotes told him, seems to imply the hypothesis that the power of gravitating resides *in* the several parts of matter.[8] Newton's early

5. *Math. Princ.* (1934), 398–400. The first two rules, re the multiplication of causes, appeared as 'hypotheses' in *Princ.* (1687¹); the third first occurs in *Princ.* (1713²), where all three are called rules. Cf. Koyré, *Newt. Stud.* (1965), 261–8; Cohen, *FN*, 575–9; McGuire, *Cent.*, 12 (1968), 233–60; *St. Hist. Phil. Sci.*, 1 (1970), 3–58; and *Ambix*, 14 (1967), 69–95.

6. *Opticks* (1730⁴), 1; Sabra, *Theories* (1967), 233, 273–97. A concept of the constitution of light akin to Hooke's figures in the discovery of the diffraction of X rays. Heilbron, *Moseley* (1974), 66–7, 71.

7. Cf. Koyré, *Newt. Stud.* (1965), 137, 149, 153–4; McGuire, *Arch. Hist. Exact Sci.*, 3 (1966), 206–48; Guerlac, *Newton* (1963), 10, 25. Buchdahl, *Minn. Stud.*, 5 (1970), 216, points to apparently realist references to gravity in *Principia*.

8. *Math. Princ.* (1934), 414; Edleston, *Corresp.* (1850), 153, 155. Cf. Hollman, Ak. Wiss., Gött., *Comm.*, 4 (1754), 224–6, who points to the same proposition.

English followers, his 'rash disciples' as Aepinus called them,[9] either misunderstood him to have considered gravity an essential quality of matter or, having understood his hedges, nonetheless believed gravity to have 'as fair a claim to the Title' of essentiality as any other property. The rashest, Cotes, professor of astronomy at Cambridge, provoked Continental natural philosophers already sufficiently aroused by insisting, in the preface to his edition of the *Principia*, that Newton's gravity had the same ontological status as the irreducible properties of Cartesian matter: 'either gravity will have a place among the primary qualities of bodies, or Extension, Mobility and Impenetrability will not.'[10] The same sort of irritant also appeared in the less official writings by John Keill, John Freind and George Cheyne, who thought that 'all the Particles of Matter endeavor to *embrace* one another.'[11]

Huygens and Leibniz had already criticized the *Principia* interpreted as Newton wished, as a book about effects, not causes: excellent mathematics, they said, but no physics; an *asylum ignorantiae*, a refuge for those too lazy or too ignorant to work out a clear, proper, mechanical account of gravitation.[12] The queries of the Latin *Opticks*, and the still stronger assertions of the disciples, who raised gravity to a quality inherent in matter, called forth wider opposition. The Leibnizians protested with particular vigor; for at the same time that the rash disciples insisted upon the primitivity of attractions, they were trying to appropriate to Newton all credit for inventing the calculus.[13]

In the *Acta eruditorum* of 1710 an anonymous Leibnizian, none other than Wolff, blasted Freind for taking attraction as a primitive force; Freind replied; and Wolff blasted him again, from a carefully fortified position. An attractive force may be admitted, he said, as a *pis aller*, as a phenomenon needing explanation; but if, with Freind, one supposed it innate and non-mechanical, which is to say inexplicable and unintelligible, we return to 'occult qualities,' to the

9. *Tent.* (1759), 5–6; Hutton, *Dict.* (1815), and *Encycl. Brit.*[3], allow the charge in their articles 'Attraction.'

10. *Math. Princ.* (1934), xxvi; Koyré, *Newt. Stud.* (1965), 159, 273–82; Cotes to Clarke, 25 June 1713, in Edleston, *Corresp.* (1850), 158–9. A primary quality is inherent and universal, but not necessarily essential, such that matter could not exist without it; God might have been able to make matter extended and mobile but not gravitating, or gravitating according to a law other than Newton's. The rash disciples explicitly made gravity primary but not essential; Kant was to go all the way. Cf. Tonelli, *RHS*, 12 (1959), 225–41.

11. Cheyne, *Philosophical Principles* (1715–6), as quoted in Koyré, *Newt. Stud.* (1965), 156, 282 and Bowles, *Ambix*, 22 (1975), 21; Keill, *PT*, 26 (1708), 97–110; Schofield, *Mechanism* (1970), 41–5; Freind, *Prael.* (1710), 4: 'Datur vis attractrix, seu omnes materiae partes à se invicem trahuntur.'

12. Koyré, *Newt. Stud.* (1965), 117–23, 140, 264–5, 273–4; Alexander, *Leibniz-Clark Corresp.* (1958); Guerlac, *Newton* (1963), 7, and Guerlac in Wasserman, ed., *Aspects*, 329. Cf. the review of *Principia* in *J. des sçavans*, 16 (1688), 328: 'Mr. Newton n'a qu'à nous donner une Physique aussi exacte qu'est [sa] Mechanique. Il l'aura donné quand il aura substitué de vrais mouvemens en la place de ceux qu'il a supposez.'

13. Wolff's shock at the Latin *Opticks* and *Principia* appears from his letters to Leibniz of 17 Aug. 1710 and 11 Dec. 1712, in Leibniz, *Briefw.* (1860), 124–5, 154.

'cant of the schools,' to 'sounds without content.'[14] A literal attraction, an action at a distance, exceeds the power of creatures, which, according to Leibniz and Wolff, can produce local motion only by pushing. Hence it implies either a perpetual miracle or a spiritual, non-mechanical agency. But neither alternative is admissible. Good philosophers do not call upon God to save the phenomena, nor invoke invisible and intangible, or 'inexplicable, unintelligible, precarious, groundless and unexampled' means of communication. 'Nobody'—it is still Leibniz—'nobody would have ventured to publish such chimerical notions . . . in the time of Boyle.' Alas! sighed Wolff, Newton was never more than a beginner in philosophy, and too weak to oppose the shameful quirks and crotchets of his disciples.[15]

In France, the learned press, the *Journal des sçavans* and the *Journal de Trévoux,* rallied support for the true, or Cartesian, physics. In the Paris Academy the mathematician Joseph Saurin, who in 1703 had parried a thrust at the doctrine of vortices, led the cause. Have nothing to do, he warned, with English gravity, one of those scholastic occult qualities. 'We need not flatter ourselves that, in physics, we can ever surmount all difficulties; but let us always philosophize from the clear principles of mechanics; if we abandon them, we extinguish all the light available to us, and we sink back into the old peripatetic darkness, from which Heaven preserve us.'[16] Fontenelle, who could never distinguish between innate gravity and sympathies, horrors, and 'everything that made the old philosophy revolting,' fired away at Newtonian occultism even as he commemorated the death of its founder: 'The continual use of the word Attraction, supported by great authority, and perhaps too by the inclination which Sir Isaac is thought to have had for the thing itself, at least makes the Reader familiar with a notion exploded by the Cartesians, and whose condemnation has been ratified by all the rest of the Philosophers; and we must now be upon our guard, lest we imagine that there is any reality in it, and so expose ourselves to the danger of believing that we comprehend it.'[17]

Newton had not found an adequate defense: the rules of philosophizing of the second edition of the *Principia* (1713) and its general scholium, hinting at a possible aether mechanism for gravity and claiming to advance no hypotheses, scarcely altered its character, while the explicit admission into the *Opticks* of

14. [Wolff], *Acta erud.* (1713), 307–14, a response to Freind's answer (*PT*, 27 [1710], 330–42) to the *Acta*'s review (*Acta erud.* [1710], 412–16) of Freind's *Praelectiones*. Wolff's authorship of the reviews appears from his letters to Leibniz of 6 June and 16 July 1710 (Leibniz, *Briefw.* [1860], 119–22).

15. Leibniz (1716) in Alexander, *Leibniz-Clarke Corresp.* (1958), 92, 94; Wolff to Manteuffel, 19 April 1739 and 30 Oct. 1747, in Ostertag, *Phil. Geh.* (1910), 61, 133–4. Cf. Malebranche to Ber-Berrand, 1707, in Malebranche, *Oeuvres*, xix, 771–2: 'Quoique M. Newton ne soit point physicien, son livre [the *Opticks*] est très curieux et très utile à ceux qui ont de bons principes de physique.'

16. Saurin, *MAS* (1709), 148; cf. *ibid.*, 133–4, and Fontenelle, *HAS* (1737), 115–17; Brunet, *Maupertuis* (1929), I, 21, and *Intro.* (1931), 9, 28–9.

17. Fontenelle, *Elogium* (1728), 12 (Newton, *Papers* [1958], 454); Flourens, *Fontenelle* (1847), 130–5.

1704—and, in larger measure, into its Latin translation of 1706—of microscopic forces acting directly at a distance only worsened the situation.[18] And,
when he came to approach the accusation directly, Newton was unable to distinguish sharply between his qualities and those of the peripatetics. The latter,
he says, are 'supposed to lie hid in Bodies, and to be the unknown Causes of
manifest Effects;' his are 'manifest Qualities, and their Causes only are occult.'[19] These distinctions do not, and did not, persuade. The rash disciples saw
no point in them: 'If the true Causes are hid from us, why may we not call them
occult Qualities?'[20] As for the enemy, Fontenelle, who had his schooling from
the Jesuits, rightly observed that Newton's specification was precisely that of a
scholastic occult quality. The same point was made by the Jesuit Regnault, in a
bit of dialogue he arranged between the principals. It runs like this. Newton:
'Attraction is a cause that I do not know; but after all it is the cause of sensible
effects, of phenomena.' Descartes: 'There you are, back to occult qualities; for
occult qualities were just the unknown causes of manifest effects.'[21]

As for Newton's attempted evasion—only the cause of gravity is occult—
it scarcely reassured continental philosophers that the English would, or that
anyone successfully could, seek a mechanical cause of Newton's mutual gravitation. Newton may have written that gravity might be effected by impulse; but
'could Sir Isaac think that others could find out these *Occult causes* which he
could not discover? With what hopes for success can any other man search after
them?'[22] By beginning wrong, by starting from a convenient mathematical fiction rather than from intelligible first principles, Newton has ended wrong:
since, according to the Cartesians, no mechanical cause of gravitation was possible on his theory,[23] he perforce had introduced an occult one.

NEWTON OF THE *OPTICKS*

The tight, self-justifying, towering mathematician of the *Principia* seldom appears in the more open, accessible and even romantic author of the *Opticks*.
The book, whose main business is to elaborate Newton's unhypothetical doctrine of light, ends with a set of imaginative conjectures, of 'bold and eccentric
thoughts,' of hypotheses not permitted others, but allowable to Newton because
he calls them 'queries.'[24] The first edition (1704) poses sixteen, all dealing with
light; the Latin translation (1706) and the second English edition (1717–8)

18. Koyré, *Newt. Stud.* (1965), 156–60.
19. Newton, *Opticks* (1952), 401 (Query 31, first published in *Optice* [1706], as Query 23).
20. Keill, *Intro.* (1720), 3–4; cf. d'Alembert, *Prel. Disc.* (1963), 82.
21. Regnault, *Origine*, III (1734), 66–7. Cf. the 'Disputatio physicohistorica' in Musschenbroek, *Elementa physicae* (Naples, 1745), I, 59, 73.
22. Fontenelle, *Elogium* (1728), 21; Newton, *Opticks* (1952), 376. Cf. Koyré, *Newt. Stud.*
(1965), 147–8.
23. E.g., Huygens. *Oeuvres*, XXI, 471: 'Je crois voir clairement, que la cause d'une telle attraction n'est point explicable par aucun principe de Méchanique.'
24. Priestley, *Experiments and Observations on Different Kinds of Air* (1781³), I, 258, quoted
by Cohen, *FN*, 191; Koyré, *Arch. int. hist. sci.*, 13 (1960), 15–29.

contain 23 and 31, respectively, and touch upon all branches of chemistry and physics. In their final form the queries are notable for their advocacy—as far as the interrogative form would permit—of a sometimes contradictory world redundantly filled with several aethers and with particles that act upon one another at a distance.

Newton's last guesses at the structure of the world, guesses that profoundly influenced eighteenth-century physics, owed their complexity to a double difficulty he never quite resolved. The first related to God's providence: should one assume that He continually and *directly* preserves His creation, or that He assigned its general maintenance to appropriate secondary, physical agents? Secondly, regarding the latter possibility, what sorts of agents ought one to consider?[25] Newton's earliest public answer to this question, which he sent the Royal Society in 1675, but which was not published until 1744,[26] supposed much the same range of mechanisms as appeared in the last edition of the queries. We are to imagine several distinct aethers, each very subtle and elastic, and 'some secret principle of unsociableness' and the reverse, whereby particles, both of aether and of grosser bodies, selectively flee and approach one another.

A good model of aether action, Newton says, is the behavior of the electrical vapor condensed in his telescope lens; just as the vapor spreads from and returns to the glass, driving light bodies before it, may not a thin, tenacious and springy local aether, constantly imbibed by our earth, 'bear down' upon objects in its path and so cause terrestrial gravity? And, just as friction elicits the electrical effluvia, may not the imbibed gravitational aether, transformed in the 'bowels of the earth,' appear again as air, gradually ascending and rarefying until it 'vanishes into the aetherial spaces?' The sun, too, may fancy this aether, which, in its rush to serve as 'solar fewel,' might push against the planets and retain them in their courses. Still more spirits are required to move the planets and their secondaries: Newton, not yet (1675) free from Descartes, assigns the job to 'aethers in the vortices of the sun and planets,' mutually unsociable aethers, moreover, to prevent their mingling and reciprocal destruction. There is also the special optical medium, which stands rarer in optically denser bodies, and which reflects, refracts and diffracts light corpuscles according to the gradient of its density and the direction of its vibrations.[27]

Although Newton took the trouble to sharpen these ideas,[28] they did not long dominate his physics. His capital discovery—which surprised and perplexed him—that the inverse-square law held precisely for planetary motions,

25. Cf. McGuire, *Ambix*, 15 (1968), 154–208.

26. A table of dates of composition and publication of Newton's work on physics is given in Thackray, *Atoms* (1970), 12.

27. These speculations were first published by Birch, *Hist.* (1756–7), III, 249–60 (Newton, *Papers* [1958], 179–90); for Newton's electrical experiment, *supra*, Intro., and for his earlier, quasi-Cartesian astronomy, Whiteside, *Br. J. Hist. Sci.*, 2 (1964), 117–37.

28. E.g., Newton to Boyle, 28 Feb. 1678–9, in Newton, *Papers* (1958), 70–3.

showed him that the celestial spaces must be resistanceless, aetherless voids. Since he thought it philosophically absurd and, what is worse, conducive to atheism, to ascribe to bodies the capacity to attract one another directly over sensible distances, he inclined to scrap secondary gravitational agents altogether, and to make God the immediate and omnipresent cause of the mutual accessions of bodies.[29] Now the chief reason for rejecting the gravitational aether—to clear the interplanetary spaces of material obstacles—did not apply to the optical aether, which did not need to extend far beyond the bodies whose interaction with light it mediated. Nonetheless Newton temporarily dropped the optical medium of the 1670s and, in the queries to the *Opticks* of 1704, assigned its functions to short-range forces by which the particles of bodies acted at a distance upon light 'rays' or corpuscles. Moreover, in the *Optice* of 1706, he expanded his earlier hints about the social intercourse of particles, explaining chemical phenomena in terms of specific, elective, microscopic attractions and repulsions between ultimate corpuscles of bodies.[30] Whether he conceived these forces to be innate, or direct manifestations of God's continuing activity does not appear; in either case the *Optice* nicely complements the *Principia*, which together constitute the most consistent of the world pictures that Newton decided to make public.

Neither Newton's opponents, nor his own restless intellect, nor, indeed, the advancement of knowledge, allowed him to stop here. When, after 1706, experiments showed that electricity,[31] which Newton had earlier associated with a special aether, might figure in the production of light, he conceived that the electrical effluvia might be the backbone of the frame of the world, the hidden bond between the attraction of the *Principia* and the light of the *Opticks*. In the General Scholium of 1713, partly in response to the criticisms of Leibniz and Wolff, he cautiously returned to secondary causes, and hinted at a new insight into the operations of a 'certain most subtle spirit which pervades and lies hid in all gross bodies.' Except for gravity, which Newton did not mention explicitly, this spirit shouldered all the tasks performed by the multiple aethers of 1675. By its 'force and action . . . the particles of bodies . . . attract one another at near distances and cohere, if contiguous; and electric bodies operate to greater distances, as well repelling as attracting the neighboring corpuscles; and light is emitted, reflected, refracted, inflected and heats bodies; and all sensation is excited . . .' This advertisement of the 'electric and elastic spirit,' which sits incongruously on the last page of the revised *Principia*, was amplified in the

29. Newton seems to have held this position for about 20 years after the publication of the *Principia*. McGuire, *Ambix*, 15 (1968), 154–208.

30. Cf. Koyré, *Newt. Stud.* (1965), 137; Lohne, *Arch. Hist. Exact Sci.*, 1 (1961), 400–2; Rosenfeld, *ibid.*, 2 (1965), 365–86; McGuire, *ibid.*, 3 (1966), 231–2, and *Ambix*, 15 (1968), 155–7, 161–4; Guerlac, *Newton* (1963), 5–13, 27–35. The change from the agnosticism of the *Principia* to the advocacy of distance forces in the *Optice* was echoed in J. Harris, *Lex. techn.* (1704–10); Bowles, *Ambix*, 22 (1971), 23–9.

31. *Infra*, viii.2; Guerlac in Hughes and Williams, *Var. Patt.* (1971), 156–7.

new queries to the *Opticks* of 1717–8, where a multipurpose aether uneasily shares the universe with interparticulate forces acting directly at a distance.[32]

The 31 queries of the last redaction divide into four almost equal groups. The first seven, identical with those in the earlier editions, refer the reflection, refraction, diffraction and emission of light, and the production of heat, to direct short-range attractions and repulsions between the particles of bodies and the rays of light. Queries 8–16, again largely unchanged, mention neither forces nor aethers explicitly.

The next set, queries 17–24, those newly composed in 1717, reinstate a subtle, elastic, active medium, rarer in optically denser media, to whose density gradients Newton refers refraction, diffraction and—though with less conviction—gravity. The optical mechanisms plainly conflict with those of the first set of queries. Moreover, they cover less ground, as they make no provision for reflection or emission, which the earlier scheme plausibly attributed to particle-ray forces directed oppositely to those responsible for refraction and inflection. And, of course, they do not avoid action at a distance, for, as Newton hints, one 'may'—indeed, must—suppose the aether to 'contain Particles which endeavor to recede from one another.'[33] As for the gravitational mechanism, the argument from resistance still applies, and all the more so as the optical medium must become denser the rarer the spaces it occupies.

The final set of queries, 25–31, being slight reworkings of those first published in 1706, return us to the aetherless world of short-range interparticulate forces. They argue against luminiferous and gravitational media, reattribute optical phenomena to particle-ray forces, and admit a host of interparticulate attractions and repulsions to explain cohesion, capillarity, elasticity, selective chemical combination and the power whereby 'Flies walk upon the Water without wetting their Feet.'[34]

Newton's stated purpose in posing his occasionally contradictory queries was to encourage the 'inquisitive' to search farther, to find and ultimately to quantify the short-range forces—acting directly and reciprocally among particles of light, aether and matter—through which God vicariously operates the universe. His own positive achievements would serve as guide and goal in this search, the *Opticks* showing how to design apt experiments and to reason semi-quantitatively about them, and the unique *Principia* illustrating the final steps toward a mathematical physics. Perhaps no one before the mid-eighteenth

32. The qualifiers 'electric and elastic' first appeared in the English *Princ.* (1729); their authority is an autograph interlineation in Newton's copy of *Princ.* (1713²). See Hall and Hall, *Isis*, 50 (1959), 473–6, and Koyré and Cohen, *Isis*, 51 (1960), 337.

33. Heimann and McGuire, *Hist. St. Phys. Sci.*, 3 (1971), 242–3, observe that the aether presented no difficulty of action at a distance to Newton, for the mode of existence of the repulsive force was precisely to fill the space between aether particles; in several manuscripts (*ibid.*, 244), Newton expressly says that the aether is nonmaterial and nonmechanical. Cf. Laudan's comments in *Minn. St.*, 5 (1970), 234–8.

34. Cf. Koyré, *Arch. Hist. Exact Sci.*, 13 (1960), 15–29; Newton, *Opticks*, 339–406.

century tried to implement Newton's arduous program.[35] The queries served instead as a quarry of qualitative images in the style of Descartes' *Principia philosophiae*. The distinctions Newton had tried to draw between fact, theory, hypothesis and query meant little to worshipful, uncritical sectaries who— being ignorant of or indifferent to the difficulties that had bothered him—all too often thought that 'Sir *Isaac* was *infallible* in everything that he *proved* and *demonstrated,* that is to say, in all his Philosophy.'[36]

5. FORCE AMONG THE EARLY NEWTONIANS

'Sir Isaac Newton has advanced something new in the latest edition of his *Opticks* which has surprised his physical and theological disciples.' Thus the London *Newsletter* of December 19, 1717[1], pointed to the new aether queries, which made patently inconsistent the corpus the disciples had undertaken to defend. They escaped from their predicament by the simple expedient of ignoring the revived aether, which does not figure significantly, if at all, in the authoritative texts of Keill (1720), Pemberton (1728), Desaguliers (1734), and 'sGravesande (1720), or in the influential popularizations of Algarotti (1737) and Voltaire (1738).[2] Not until the 1740s did Newton's aether become important for physical theory.[3]

Even without the aether Newton's expositors had some tidying up to do: freeing the word 'attraction' from the rash interpretation of Keill and Freind and the ambiguous usage of Newton; answering the Cartesian insistence upon the epistemological superiority of 'mechanical' explanations; and clarifying the number and interrelations of the many short-distance attractions and repulsions mentioned by Newton from time to time. The tone for much of this work was set by 'sGravesande.

'sGravesande's definition of gravitation returns to the most positivistic of Newton's: gravity is not an occult cause but a manifest effect.[4] 'When we use the Words Gravity, Gravitation, or Attraction,' says Desaguliers,[5] 'we have a Regard not to the Cause, but to the Effect; namely to the Force, which Bodies have when they are carried towards one another.' Attraction signifies no more than that, if left to themselves, bodies would move toward one another, 'force'

35. An exception must be made for optical theory, as developed by Robert Smith; cf. Steffens, *Development* (1977), 28–48. Smith was Cotes' cousin and his successor at Cambridge.

36. B. Martin, *Suppl.* (1746), 26. Cf. Casini, *L'univ.-macch.* (1969), 10.

1. Quoted in Kargon, *Atomism* (1966), 138.

2. Cf. Thackray, *Atoms* (1970), 104–6; Guerlac in Hughes and Williams, *Var. Patt.* (1971), 158; Schofield, *Mechanism* (1970), 19, 24, 29–30, 36.

3. *Infra,* i.6.

4. 'sGravesande, *Math. El.* (1731⁴), II, 207; Keill, *Intro.* (1745⁴), 4, adopts the same formulation ('So likewise we may call the Endeavor of Bodies to approach one another Attraction, by which word we do not mean to determine the Cause of that Action'), as does, e.g., Boerhaave's translator, Shaw, in Boerhaave, *New Method* (1741), I, 156n ('[Attraction signifies] not the cause determining the bodies to approach . . . , but the effect, i.e., the approach').

5. *Course* (1763³), I, 6–7.

meaning to Desaguliers and the Dutch Newtonians either momentum (or kinetic energy) or the inertia that maintains motion *(vis insita),* not a physical cause acting between bodies.[6]

The distinction was important tactically, as a help in routing the 'Army of Goths and Vandals in the Philosophical World,'[7] the Cartesians who persisted in imagining that (to speak with 'sGravesande) 'because we do not give the cause of . . . Attraction and Repulsion, . . . they must be looked upon as Occult Qualities.'[8] A good example is James Jurin's parry of the charge of occultism hurled at Newtonians by Wolff's disciple G. B. Bilfinger. Jurin, a frequent invoker of attractions, appealed to Book I of the *Principia:* 'I [Newton] use the words attraction, impulse, or propensity of any sort towards a centre, promiscuously and indifferently, one for another, considering those forces not physically but mathematically: wherefore the reader is not to imagine that by these words I anywhere take upon me to define the kind, or manner of any action, the causes or the physical reason thereof . . .'[9]

This defense was not made without cost, for it turned out that the quantities of fundamental interest to physicists were not forces as represented by Desaguliers and 'sGravesande, i.e., as macroscopic effects, but forces as microscopic causes, such as the suppositious mutual pull between all pairs of particles of matter. To study these quantities one had to do as Newton did in the *Principia:* one had to compute the force-effects to be expected from assumed force-causes, and to confirm the latter by the former; inhibitions against supposing force-causes delayed the fruitful application of the scheme. Perhaps for this reason the first attempt of the Newtonians to obtain a 'law of force' for the interaction of two magnets failed.[10]

There were two steps in the answer to the Cartesian insistence on the priority of mechanical explanations. First, a Lockean element: we know nothing but our ideas, derived ultimately from our unreliable sense experience; these ideas at best correspond to, but do not reach, the ultimate nature of things; in particular, we have no notion of the essence of matter.[11] 'Nothing can be present to our Minds besides our Ideas, upon which our Reasonings immediately turn,' says

---

6. *Ibid.,* I, 45: 'Motion is that Force with which Bodies change their Place. . . . Force and Motion mean the same thing': cf. 'sGravesande, *Math. El.* (1731⁴), I, 20–1. See also the examples in Ruestow, *Physics* (1973), 126–7, who, however, misinterprets them as statements about causes. For Musschenbroek's views, which were less instrumentalist than 'sGravesande's, see Hollman, Ak. Wiss. Gött., *Comm.,* 4 (1754), 228–30; Crommelin, *Sudh. Arch.,* 28 (1935), 138–9.

7. Desaguliers, *Course* (1763³), I, 21.

8. 'sGravesande, *Math. El.* (1731⁴), I, 17–18. Cf. Desaguliers. *Course* (1763³), I, 21.

9. Jurin, *CAS,* 3 (1728), 282–3; Newton, *Math. Princ.* (1934), 5–6. Cf. Jurin, *PT,* 30 (1717–19), 743; Voltaire, *Lett. Phil.* (1964), II, 27, 39–41; Boss, *Newton* (1972), 112–15.

10. *Infra,* i.7.

11. E.g., Locke, *Essay* (1894), I, 391–2, 410–15; II, 191–225. This was of course a standard theme in the Enlightenment; cf. Hume, *Enquiry* (1894), 32–3, and *Treatise* (1888), 16; Cassirer, *Philosophy* (1951), 53–6, 74; Hankins, *D'Alembert* (1970), 79–80, 107–110; Heimann and McGuire, *Hist. St. Phys. Sci.,* 3 (1971), 267.

'sGravesande; and these ideas do not take us to the bottom of things. 'What substances are, is one of the Things hidden from us. We know . . .some of the Properties of Matter; but we are absolutely ignorant, what Subject they are inherent in.'[12] We must content ourselves with such general regularities, or 'laws of nature,' as we can uncover: for 'we are at a loss to know, whether they flow from the Essence of Matter, or whether they are deducible from Properties, given by God to the Bodies, the world consists of, but no way essential to Body; or whether finally those Effects, which pass for Laws of Nature, depend upon external Causes, which even our Ideas cannot attain to.'[13] This general line, called 'modest' by its friends and 'pyrrhonistic' by its enemies,[14] became standard in early Newtonian apologetics: it is found not only among the Dutch,[15] but also in Pemberton, Maclaurin and even Keill,[16] and in the first Newtonian texts of Italy and Germany.[17]

The second step in responding to the Cartesians was to make a virtue of necessity. We know nothing of the essence of matter? Then we no more know how motion is transferred in collisions than how it is caused in attraction. One may recall that Fontenelle had formerly met this objection with the aristocratic notion of degrees of unintelligibility. His opponents now responded with the democratic doctrine of equality of incomprehension. 'The Cartesians reproach the Newtonians for having no idea of attraction. They are right, but there is no basis for their judgement that impulse is any more intelligible.'[18] 'Impulsion is a principle at least as obscure as that of attraction.'[19] 'The action of one body on another by contact is as inexplicable as *actio in distans*.'[20] We are therefore

12. 'sGravesande, *Math. El.* (1731⁴), I, xiv, x–xi.

13. *Ibid.*, I, xii; cf. Strong, *JHI*, 18 (1957), 68–9.

14. Cf. Tega, *Riv. crit. stor. fil.*, 4 (1969), 195; Nollet, 'Discours,' in *Leçons*, I (1743), lxiv.

15. Cf. Brunet, *Physiciens* (1926), 42–3, 63–75, 89–92; Ruestow, *Physics* (1973), 122–31.

16. Pemberton, *View* (1728), 'Intro.'; Maclaurin, *Account* (1750²), 115–17; Keill, *Intro.* (1745⁴), 11: 'We shall not here give a Definition of Body, taken from its intimate Notion or Essence, wherewith we are not perfectly acquainted, and perhaps never shall be.' Cf. Strong, *JHI*, 18 (1957), 59, 66, 76; and, for an elegant later version of the argument, Nicholson, *Intro.* (1790³), I, 3–5.

17. E.g., the strongly positivistic notes ('Nullatenus rerum operationes intelligimus,' etc.) added by Giuseppe Orlandi to the first Neapolitan edition of Musschenbroek's *El. phys.* (1745), quoted by Garin, *Physis*, 11 (1969), 214n; the methodological portions of Beccaria's unpublished 'Institutiones in physicam experimentalem' (c. 1750), quoted by Tega, *Riv. crit. stor. fil.*, 24 (1969), 194–5; and Lichtenberg's conventionalist annotations to Erxleben's *Anfangsgründe*, for which see Herrmann, *NTM*, 6:1 (1969), 81.

18. Condillac, *Traité des systèmes* (1749), quoted by I. Knight, *Condillac* (1968), 73. Cf. Voltaire, *Lett. phil.* (1964), II, 27; 'Vous n'entendez pas plus le mot d'impulsion que celui d'Attraction, et si vous ne concevez pas pourquoi un corps tend vers le centre d'un autre corps, vous n'imaginez pas plus par quelle vertu un corps en peut pousser un autre.'

19. Paulian, *Dictionnaire*, I (1773²), 188–9, art. 'Attraction.' Cf. *Encyclopédie méthodique*, art. 'Attraction.'

20. Playfair to Robison, 28 June 1773, citing Hume, in Olson, *Isis*, 60 (1969), 95, a favorite argument among the Scots (Heimann and McGuire, *Hist. St. Phys. Sci.*, 3 [1971], 295). Cf. Bailly

well-advised to be agnostic about the cause of gravity, and to proceed to the true business of the philosopher, an enquiry into its laws.[21]

THE FRENCH MATHEMATICIANS

The enquiry was rapidly advanced, the Cartesians confounded, and the cause of gravity triumphant, though still unknown, when the mathematicians of the Paris Academy—particularly Maupertuis, Bouguer, Clairaut and, somewhat later, d'Alembert—allowed themselves to be seduced by unsettled problems in the mathematical philosophy of Newton.[22] Two such problems deserve notice. The earlier concerned the shape of the earth, said to be oblate (shorter in polar than equatorial diameter) by Newton, and oblong (longer in the poles than the equator) by those who believed in measurements made by Jacques Cassini, head of the Paris Observatory. In 1728 Maupertuis, like the Dutch Newtonians a decade earlier, made a tour of the natural philosophers of London, and returned enthusiastic for the principles of Newton.[23] Four years later, in perhaps the first open exposition of Newton's astronomy in France,[24] Maupertuis argued for flattened poles. To resolve the controversy the Academy sent an expedition to Peru in 1735 and to Lapland in 1736.[25] The northern expedition, led by Maupertuis, unambiguously confirmed the shortening of the earth's axis, 'simultaneously flattening the poles and the Cassini.'[26] The mathematical young Turks thereupon took the offensive, and openly urged the adoption of attraction, understood as a mathematical hypothesis, on the sole ground that it saved the phenomena.[27]

Maupertuis hoped to slip his pioneering Newtonian expositions into the Cartesian Academy by oiling them with the arguments of the Dutch Newtonians, who accordingly thought his work the best physics France had yet produced.[28] We know only a few properties of things, and nothing of the essence

to Le Sage, 1 April 1778 (Prévost, *Notice* [1805], 299–300): 'Le phénomène de la communication du mouvement, quoiqu'aussi incompréhensible que celui de la pésanteur, est de notre connoissance plus intime.'

21. 'sGravesande, *Math. El.* (1731⁴), II, 215–16; Maclaurin, *Account* (1750²), 156.

22. Cf. Fontenelle, *HAS* (1732), 112: 'invité sans doute par une occasion d'employer la plus subtile géométrie.'

23. Maupertuis met Clarke, Pemberton, and Desaguliers, and was elected FRS; Brunet, *Maupertuis* (1929), I, 15.

24. D'Alembert, *Prel. Disc.* (1963), 89; Brunet, *Maupertuis* (1929), I, 22. In fact Bouguer, *Entretiens* (1731), has priority, as he claimed (*Entretiens* [1748²], 4), but his work was not published until after Maupertuis', and, as a dialogue, lacks system.

25. Todhunter, *Hist.* (1873), I, 93–102, 231–48; Brunet, *Maupertuis* (1929), I, 33–58, and *Clairaut* (1952), 30–53.

26. Maupertuis had become the tutor and technical advisor of the first continental expositors of Newton, Voltaire, Algarotti, and Mme du Châtelet (Brunet, *Maupertuis* [1929], I, 23–7); Voltaire, *Corresp.*, II, 377–89, 392, 400, 405–6 (letters of 1732).

27. According to Brunet, *Maupertuis* (1929), I, 188, Maupertuis used to give a dinner for the young Newtonians on the days of meetings of the Academy of Sciences, to which they would repair full of 'good spirits, presumption, and strong arguments.'

28. Voltaire to Maupertuis, 1738, quoted in Brunet, *Maupertuis* (1929), I, 58, 61.

of matter; attraction is just such a property, a fact not a cause; it would be 'ridiculous' dogmatically to exclude it from consideration. As for being intelligible, it is not less so than impulsion, for we have no conception of the true cause of either.[29] Assuming attraction (here Maupertuis in practice means force-cause), we may calculate many things of interest; and the law of squares is really a very nice law, mathematically speaking.[30]

In all this there is an air of apology. Maupertuis later called attention to the 'circumspection with which I presented the principle [of gravity], the timidity with which I dared hardly compare it with impulse, the fear I had when giving the reasons that had led the English to reject Cartesianism.' Despite his caution his performance made him many enemies, or so he said.[31] Bouguer supplied a similar argument with similar care, and ended with a proposal for peaceful coexistence: since Cartesians and Newtonians agree in admitting gravity as a fact and ignorance of its cause, they might easily be brought to cooperate with one another. 'For they need only say (the Newtonians perhaps without much believing it and the Cartesians without much hope) that the words attraction and weight shall signify only a fact while they wait for the discovery of its cause.'[32]

Maupertuis' younger colleague Clairaut dispensed with epistemological balm. Having swallowed gravitational attraction in Lapland ('revolting,' he said, when taken literally, but worth exploring as an hypothesis), he came before the Academy in 1739 with an essay on refraction based upon Newton's short-range attractive force. 'I've no wish at all to establish attraction as an essential property [he said]; I've no opinion on a question that is beyond my powers. My sole purpose is to show you how Newton uses attraction when he tries to explain refraction . . . I demand that you do me the kindness of listening.'[33] And they listened, much to the surprise of the President of the Royal Society of London: 'I remember the time when whoever would have spoken of attraction, as is now [1740] done before the philosophes, would have been as little noticed as someone wishing to resolve all difficulties by occult causes.'[34] Fontenelle had also to admit and lament the success of the attractionists. In 1737, in his Eloge of Saurin, he recalled the late mathematician's prayer ('may

29. Maupertuis, *Discours sur les différentes figures des astres* (1732), in *Oeuvres* (1756), I, 90–104; Brunet, *Maupertuis* (1929), II, 44–5, 346–7, 350, 357; Brunet, *Intro.*, 210–14; Koyré, *Newt. St.*, 162–3.

30. The shift to force-cause is evident in Maupertuis, *MAS* (1732), 343–62, which treats attraction 'geometrically, i.e., as a quality, whatever it may be, the phenomena of which may be calculated when one assumes it spread uniformly throughout all parts of matter, and acting in proportion to its quantity.' The law of squares is nice because a gravitating sphere, the most regular of bodies, then acts, as a whole, according to the same law as its parts; *ibid.*, 347; Brunet, *Maupertuis* (1929), II, 365–7.

31. Quoted by Bouillier, *Hist.* (1854), II, 561–2.

32. Bouguer, *Entretiens* (1748²), 24–5, 30–1, 47–9. Cf. Nollet, *Leçons* (1743–8), II, 476, and Abat, *Amusemens* (1763), 415–24: Newtonians like 'sGravesande and Maclaurin, who admit the possibility of an impulse theory of gravity, are in much the same case as the Cartesians.

33. Clairaut, *MAS* (1739), 263; Brunet, *RHS*, 4 (1951), 138–9, and *Clairaut* (1952), 58–9.

34. Hans Sloane to the abbé Bignon, 16 Oct. 1740, in Jacquot, RS, *Not. Rec.*, 10 (1953), 95.

God preserve us from . . . peripatetic darkness,' i.e., Newtonian attractions),
and added: 'Would anyone have believed that it ever would be necessary to
pray Heaven to preserve the French from prejudice in favor of an incomprehensible system, and moreover a foreign one; the French, who so love clarity, and
who are so often accused of liking only what originates with themselves?'[35] But
by then the battle was lost.[36]

The second intriguing Newtonian problem was the computation of the motion
of the moon. In a curious case of simultaneous non-discovery, Clairaut,
d'Alembert and Euler, each following his own method of approximation, calculated a value for the precession of the moon's apogee just half that observed. At
a dramatic meeting of the Academy in November, 1747, Clairaut announced
that the law of gravity failed for the moon, and proposed to save the
phenomena by adding a little universal r^{-3} force to the canonical inverse
square.[37] Bouger thereupon suggested that different portions of planets might
attract according to different laws of distance, some by the square, others the
cube, etc., Newton's law being a fortuitous average. The Cartesians had long
regarded the possiblity of such fiddling as a good reason to reject Newton's
procedures.[38]

Newton found a defender in Buffon, who, without bothering to calculate,
held that nothing so simple as gravitational attraction could obey a compound
law: should an r^{-3} term be needed, it betrayed the existence of another kind of
force, perhaps magnetic, a possibility that momentarily intrigued d'Alembert.
As for Clairaut, he dismissed Buffon's arguments as ignorant and metaphysical.
In the end neither an alteration in Newton's law nor the introduction of an
ad-hoc force proved necessary. Clairaut, dAlembert and Euler had each made a
mistake, as Clairaut was the first to discover, in 1749; whatever gravity might
be, it operated according to the law of squares.[39]

The episode had instructive consequences. For the non-mathematical Buffon
it confirmed a realist, and even a romantic interpretation of the gravitational
force. He was to teach that there was but a single primitive force in nature, 'a
power emanating from the divine power,' the cause of organization from the
original chaos, the agent of all physical and chemical phenomena, viz., New-

35. Fontenelle, *HAS* (1737), 117; *supra*, i.4.

36. Brunet, *Intro.*, 338–41; Wolff to Manteuffel, 19 April 1739 ('in Paris they think there are
no other philosophies than the Cartesian and Newtonian') and 11 July 1748 ('the so-called Newtonian philosophy has been well-received in France, Italy and Holland') in Ostertag, *Phil. Geh.*
(1910), 61, 147. According to d'Alembert, *Traité de dynamique* (1743), the Cartesians were then
'much weakened'; in its second edition (1758), they 'hardly exist.' Guerlac in Wasserman, *Aspects*
(1965), 318.

37. Clairaut hoped that the new law would also clear up small discrepancies between his calculations of the earth's shape and geodesic measurements. Brunet, *Clairaut* (1952), 82–3.

38. Fontenelle, *HAS* (1732), 113, 116; Bouguer, *Entretiens* (1748²), 51–5, where Clairaut's
lunar force is linked to Newton's short-range (r^{-3}) cohesion.

39. Brunet, *Clairaut* (1952), 82–7; Hankins, *D'Alembert* (1970), 32–5; Clairaut to Cramer, 26
July 1749, in Speziali, *RHS*, 8 (1955), 227; Maheu, *RHS*, 19 (1966), 221–3.

ton's inverse-square attraction. Apparent deviations from r^{-2} must be referred to shapes of the particles of matter and to posterity, which has the job of inferring the true shapes from the apparent forces.[40] Among mathematicians, however, the interaction with the moon advanced an instrumentalist interpretation of Newton. It showed that, as a calculating tool, the hypothesis of universal gravity withstood the finest tests available; and it suggested that, in case of failure, mathematicians would not scruple to treat the law of squares as an approximate and amendable description. 'It is extraordinarily difficult, or perhaps entirely impossible, to demonstrate the truth or falsity of Newton's law of attraction from the motion of the moon.'[41]

Despite their success with Newton's gravitational theory, the Parisian mathematicians did not hurry to analyze the phenomena of experimental physics in terms of attractions. Their attitude at the mid-century may be found in repetitious prefaces and philosophical essays by d'Alembert, and especially in his 'Preliminary Discourse' to the *Encyclopédie* of Diderot. D'Alembert has no trouble with gravity interpreted as an effect the cause of which we not only do not know, but are under no obligation to seek; so interpreted, he says, Newton's 'Theory of the World (for I do not mean his System) is today so generally accepted that men are beginning to dispute [his] claim to the honor of inventing it.'[42] This acceptance, he admits, came reluctantly, forced by youthful (French!) geometers concerned to establish precise relationships and willing, for the purpose, to suppose forces not evidently reducible to impulse.[43]

Although d'Alembert taught that the goal of physics was just such relationships,[44] he expressly appealed to philosophers to resist transposing attractions from celestial bodies to those about us. His reasons: firstly, these imitations of the gravitational theory will be less precise than their model and consequently the need for admitting them less evident; secondly, their laws would differ from gravity's, 'and it is not natural to think that attraction, if a fundamental [*primitif*] principle, is not uniform and absolutely the same for all the parts of matter;' and, thirdly, the most obvious candidates for Newtonian treatment, the phenomena of electricity and magnetism, 'seem to arise from an invisible fluid, and must make us question whether a similar fluid is not also the cause of other attractions observed between terrestrial bodies.'[45]

40. Buffon, 'Sec. vue' (1765), in *Oeuvres* (1954), 37, 39; 'Tr. aim.' (1788), in *Oeuvres* (1836), III, 76. Cf. Thackray, *Atoms* (1970), 159–60, 205–20.

41. Mayer to Euler, 4 July 1751, in Kopelevich, *Ist.-astr. issl.*, 5 (1959), 281–2. Cf. Bouguer to Euler, 30 May 1751, in Lamontagne, *RHS*, 19 (1966), 227.

42. D'Alembert, *Prel. Disc.* (1963), 81; *Elémens de philosophie* (1759), in *Oeuvres* (1805), II, 379, 420–1.

43. *Prel. Disc.* (1963), 90, 21; cf. Hankins. *D'Alembert* (1970), 24–5, 165–6.

44. *Prel. Disc.* (1963), 22; Grimsley, *D'Alembert* (1963), 240–1, 257.

45. *Elémens* (1759), in *Oeuvres* (1805), II, 421, 426. For the same reason Buffon also specifically exempted electricity from his reductionist program (*Oeuvres* [1835–6], III, 82–3) and even Boscovich hesitated (Costabel in Conv. Bosc., *Atti* [1963], 215).

There are two points of great interest here. For one, the argument implied—
what turned out to be the case—that mathematicians would not introduce
actions at a distance into electrical theory until electricians had exploded the
effluvial system. For another, it showed that, despite their positivist talk, the
French mathematicians had not yet grasped the full power, or admitted the
legitimacy, of instrumentalism. D'Alembert repeats a thousand times that the
first principles of things are unknowable, that we have no distinct ideas of
matter or of anything else, that we are 'condemned to be ignorant of the essence and interior constitutions of bodies.'[46] Moreover, he asserts just as often
that exact relations among the manifest properties of bodies constitute 'nearly
always' the highest knowledge we can attain, the limits prescribed to our understanding, and consequently the only goal we should set ourselves in
physics.[47] These propositions, together with the success of the gravitational
theory and d'Alembert's insistence on the need to quantify all the physical
sciences,[48] would endorse the ascription of promiscuous attractions to matter.
But d'Alembert would not—or could not—bring these ingredients together.

He began to warn against multiplying forces just after 1750. Simultaneously
his momentary friends Diderot and Buffon, vexed by the unnatural abstractions
apparently necessary to quantify physics, started up their program for the
peaceful extermination of mathematicians ('in less than a hundred years there
will not be three geometers in Europe'[49]). It is tempting to construe d'Alembert's warnings as attempts at accommodation, as efforts to bridle the unwhole-
some analytics of the mathematical physicist: 'having driven out the spirit of
system [he wrote], the spirit of calculation may dominate a bit too much in its
turn.'[50] In fact d'Alembert agreed neither with Buffon that mathematical
physics would work only for a very small range of phenomena, nor with Di-

46. E.g., *ibid.*, 440; *Prel. Disc.* (1963), 9, 25, 87; *Encyclopédie*, art. 'Cause'; d'Alembert to
Voltaire, 29 Aug. 1769, in Voltaire, *Corresp.*, LXXII, 284; d'Alembert to Formey, 19 Sept. 1749,
in Formey, *Souvenirs* (1797²), II, 363: 'L'impénétrabilité, l'essence de la matière, la force d'iner-
tie, etc., sont pour tous les hommes des énigmes inéxplicables,' a proposition approved by Formey,
Abrégé, I (1770), 286: 'La manière dont les propriétés résident dans un sujet, est toujours incon-
cevable pour nous.' Cf. Grimsley, *D'Alembert* (1963), 224–6, 240; Hankins, *D'Alembert* (1970),
99, 129–30, 152–3, 158; Guerlac in Wasserman, *Aspects* (1965), 330–1.

47. E.g., *Prel. Disc.* (1963), 22; *Traité* (1744), iii.

48. Among other reasons, to isolate 'fugitive and hidden effects' perhaps undiscoverable by observation alone. *Ibid.*, iv–vi; *Recherches*, I (1754), iii–iv; *Elémens* (1759), in *Oeuvres* (1805), II, 407–8;
cf. Pemberton, *View* (1728), 17. D'Alembert anticipated that the process of quantification would take
several centuries at least (*Réflexions sur la cause générale des vents* [1746], in *Oeuvres*
[1805], XIV, 13); often one can only collect facts and hope, as in the case of electricity (*Prel.
Disc.* [1963], 23–4, 29). Cf. Hankins, *D'Alembert* (1970), 88, 95.

49. A. M. Wilson, *Diderot* (1957), 187–98. Cf. Lagrange to d'Alembert, 21 Sept. 1781, Lagrange, *Oeuvres*, XIII, 368: 'Physics and chemistry now offer riches more brilliant and easier to
exploit than the deep and depleted mine of mathematics (*ibid.*, 99) . . . ; it is not impossible that
the places for Geometry in the Academy will eventually go the same way as the chairs in Arabic at
the Universities today.'

50. D'Alembert, *Elémens* (1759), in *Oeuvres* (1805), II, 466–7; cf. Hankins, *D'Alembert*
(1970), 75–6, 89–92; Lagrange to d'Alembert, 15 July 1769, Lagrange, *Oeuvres*, XIII, 140–1.

derot that the mathematician could succeed only by crude oversimplification.[51]

D'Alembert's warnings concerned the nature of allowable mathematical reductions; like most of his colleagues he did not incline to violate sound philosophy by multiplying abstractions unnecessarily.[52] The spirit of Descartes continued strong in them, especially in d'Alembert, who conceived the laws of motion and collision to be necessary consequences of an essential impenetrability of matter.[53] He could not abandon hope for an impulse theory of gravitation.[54] The manifold special attractions and repulsions of the *Opticks*, apparently unrelated to one another or to universal gravity, ran counter to Cartesian austerity. 'Is this language [the force talk of the *Opticks*] really that of good physics? Should we not worry that by accustoming ourselves to it, and by using attractions and repulsions in all sorts of ways, we will neglect investigations necessary to the advancement of knowledge, and so lose the opportunity of making discoveries?'[55] Attractionists do not examine phenomena whose causes are unknown with due attention: 'Sibi adeo, aliisque, ad ulteriorem veritatis cognitionem viam hoc ipso occludant.'[56]

6. FORCES AND FLUIDS

The early expositors, Keill, Freind, 'sGravesande, and Pemberton, blunted the problem of the multiplicity of forces by working primarily with attractions. Keill and Freind considered only attractions in their accounts of Newtonian short-range forces;[1] Keill very sparingly used repulsions in his text on true physics; 'sGravesande mentioned a few, such as the forces between oil and water and between mercury and iron, but he did not dwell upon them;[2] and Pemberton ignored them altogether. Their resistance to repulsion may be ascribed in part to the want of a universal repellent case comparable to gravity: 'If the laws of attraction were not better demonstrated than those of repulsion, I would never have become a Newtonian.'[3] Again, the multiple 'evident' attrac-

51. Roger, *Diderot St.*, 4 (1963), 230–1; Guerlac in Rockwood, *Becker's Heav. City* (1958), 23–4, wrongly associates d'Alembert's warnings with a concern for increased experimentation.

52. Cf. Paulian, *Dictionnaire* (1773²), arts. 'Attraction' and 'Répulsion': before admitting forces one must show that no extrinsic mechanical cause will do; multiplying forces is bad philosophy; 'nous sommes fâchés que le grand Newton ait insinué cette manière de procéder en physique dans plusieurs endroits de son optique.'

53. *Prel. Disc.* (1963), 17–18, 21; *Elémens* (1759), in *Oeuvres* (1805), II, 462; cf. Hankins, *D'Alembert* (1970), 87, 165–6. No doubt this belief and the Cartesian positions ('Mechanics is the base of all natural philosophy,' etc.) that he takes in several articles in the *Encyclopédie* (Briggs, *Col. U. Stud.*, no. 3 [1964], 42), conflict with d'Alembert's occasional skepticism (e.g., Hankins, *D'Alembert* [1970], 129).

54. *Prel. Disc.* (1963), 83; *Elémens* (1759) in *Oeuvres* (1805), II, 420–1.

55. Nollet, *Leçons* (1743–8), II, 479–80. Nollet accepts Newtonian gravity ('rien n'est si beau'), but no more.

56. Hollmann, Ak. Wiss., Gött., *Comm.*, 4 (1754), 244.

1. Keill, *PT*, 26 (1708–9), 97–110; Freind, *Praelectiones* (1710), 'Praef.,' 2–5.

2. 'sGravesande, *Math. El.* (1731⁴), I, 15. Cf. Thackray, *Atoms* (1970), 104, 118; Schofield, *Mechanism* (1970), 29–30, 43–5.

3. Paulian, *Dictionnaire* (1773²), art. 'Répulsion.'

tions made trouble enough. How are we to understand, or even to represent, the fact that matter particles attract one another according to two or more distinct laws of force, such as those of cohesion and gravitation?[4]

To appreciate the niceness of this problem, consider Newton's theory of matter, to which his expositors tried to hold firm: 'God in the Beginning form'd Matter in solid, massy, hard, impenetrable, movable Particles,' differing in size and shape, but otherwise—as Newton consistently and arbitrarily supposed—homogeneous.[5] From these primitive particles larger ones are compounded, apparently by attraction,[6] so that the smallest particle of, say, gold contains many primitives arranged in an exceptionally stable configuration. The particles move about in obedience to certain 'active principles,' 'such as is that of Gravity, and that which causes Fermentation, and the Cohesion of Bodies.'[7] How does it happen that particular configurations of homogeneous particles are always associated with certain active powers? How does it happen that between the collocations of particles constituting gold cohesion dominates, while between them and the collocations characteristic of aqua regia there is always fermentation; while, again, the principle of gravity acts among them all?

One could always respond to these questions, as Newton did occasionally, by retreating to phenomenology, by making of each 'active principle' a distinct 'law of nature' or 'manifest quality;' all one would need to know about, say, the law of cohesion is that 'all the Parts [of a body] have an attractive Force' that acts strongly at contact and vanishes at the least sensible distance.[8] But most physicists wishing to unify or systematize their concepts of matter preferred to develop one of two other Newtonian representations of the interrelations of forces.

(1) One might say—always reserving the question of the ultimate nature and seat of force—that each primitive particle of Newtonian matter acts according to the same law of force, which changes from attractive to repulsive and back again as the distance increases. 'As in Algebra, where affirmative Quantities vanish and cease, there negative ones begin; so in Mechanicks, where Attraction ceases, there a repulsive Virtue ought to succeed.'[9] Similarly, 'sGravesande, after defining the law of cohesion phenomenologically, states that, beyond its assigned range, the cohesive changes into a 'repellent Force, by which the Particles fly from each other.'[10]

(2) Or one might choose the crude but intelligible alternative of associating the several forces with as many distinct kinds of matter: ponderable matter

4. 'As it is easier to raise most Bodies from the Ground than to break them in pieces; that Force by which their Parts cohere, is stronger than their Gravity.' Desaguliers, *Course* (1763), I, 10.
5. *Opticks*, 400; *Math. Princ.* (1934), Bk. III, Prop. VI, cor. 4, p. 414.
6. *Opticks*, 394.
7. *Ibid.*, 401.
8. *Ibid.*; 'sGravesande, *Math. El.* (1731[4]), I, 11–12.
9. *Opticks*, 395; cf. *ibid.*, 389.
10. 'sGravesande, *Math. El.* (1731[4]), I, 12; cf. Desaguliers, *Course* (1763), II, 366ff; Thackray, *Atoms* (1970), 103.

cohering and gravitating, the 'matter' of heat self-repellent, those of light and, later, of electricity and magnetism, attractive and repulsive according to circumstance. Despite its evident conflict with Newton's matter theory, the last alternative nonetheless could claim Newtonian precedent in the neglected aether of the last edition of the *Opticks*.

A decisive step toward the refinement of these alternatives and the re-revival of the aether was the publication, in 1727, of Hales' *Vegetable Statics*. The volume presents Hales' unprecedented measurements of the amount of 'air' combined in organic and inorganic substances. How is one to understand his result that an apple can yield a quantity of air 48 times its bulk? According to Newton, air consists of particles which repel one another ('how . . .I do not here consider') with a force inversely as the distance;[11] to retain elastic air in a space 1/48th of what it occupies in its normal state would require a pressure of 48 atmospheres. Such a pressure, as Hales observed, would certainly burst the apple. His solution is to distinguish two states of air, one 'elastic,' the other 'fixed,' brought about by the destruction of the original elasticity, which may be restored by heat or fermentation.[12] As for the agent of the fixing, Hales, following Newton, picks sulphurous particles, which have a strong attraction. This attraction does not fix air by *force majeure*: air particles, when bound, lose their repellency, and become attractive of one another.[13] It appears that elastic air repels not only its own particles, but also those of common matter: while sulphur particles try to fix them, they—in violation of Newton's third law of motion—flee the sulphur. In a word, air may be in either an attractive or a repulsive *state*, but not in both simultaneously. As Hales put it, air is 'amphibious.'[14]

THE FORCE OR HOMOGENEOUS SOLUTION

The fundamental importance of Hales' work was immediately recognized by Desaguliers, who abstracted it in the *Philosophical Transactions* and soon was employing the repulsive force it legitimized in an attempt to account for evaporation. Desaguliers distinguished between solids, elastic fluids (airs), and inelastic fluids (waters), characterized, respectively, by particles in a state of attraction, repulsion, and — this is the novelty — attraction and repulsion.[15] Repulsion thereby ascended to the rank of attraction, both now 'first principles

---

11. *Opticks*, 376, 395-6; *Math. Princ.* (1934), Bk. II, prop. XXIII, pp. 300-2. Cf. 'sGravesande, *Math. El.* (1731⁴), I, 17: '[Air's] Elasticity arises from the force whereby its Parts repel one another.'

12. Hales, *Veg. Stat.* (1727), 93, 119-20.

13. *Ibid.*, 103, 110, 167-71, 179; cf. *Opticks*, 384-5; *FN*, 270-6.

14. *Veg. Stat.* (1727), xxvii, 178-9; cf. Quinn, *Evaporation* (1970), 47-59, and Hamberger, *El. phys.* (1735²), 146, on the difficulty of conceiving simultaneous attractions and repulsions.

15. Desaguliers, *PT*, 34-5 (1727), 264-91, 323-31; *PT*, 36 (1727), 6-22. The theory of evaporation standard to the mid-century assimilated it to chemical solution (of water in air). See Polvani, *Volta* (1942), 213, and Beckman, *Lychnos* (1967-8), 208-10.

of nature;'[16] the chief justification for the promotion, according to Desaguliers, being the behavior of airs and vapors as reported by Hales, and of chemical dissociation as discussed by Newton.[17] Still, Desaguliers did not face up to the problem of reconciling the concept of states of force with the simultaneous exercise of attractions and repulsions implied by the phenomena and required by the principle of equality of action and reaction. It was not until 1739 that he came clear on these matters. He then gave a theory of elasticity of solids that supposed particles simultaneously to attract (the cause of cohesion) and repel (the cause of elasticity) one another.[18] He ended by picturing the particles of matter at the centers of alternating spheres of attractive and repulsive force. Matter thereby remained homogeneous; whether its particles approached or receded from one another was an accident of distance.[19]

The line adumbrated by Desaguliers—the first of the solutions to the problem of the relationship among forces mentioned earlier—received its classical formulation in the hands of Roger Boscovich, S. J. In 1745, at the end of a dissertation on *vis viva,* Boscovich observed that attractions and repulsions, and even collisions, all became intelligible or at least simpler[20] if one conceived each of Newton's primitive particles to be a point, and required any pair of points to interact according to the same spherically symmetric, multivalued law of force, $f(r)$. At vanishingly small r, f is infinitely repulsive, rather than (as with the early Newtonians) strongly attractive, and so plays the part of impenetrability in the usual theories of matter. At a certain distance r_1 this repulsion vanishes, to be followed by an attraction, which extends to r_2, where a second repulsion sets in; after several such oscillations, f settles down to the usual gravitational attraction. Stable configurations of the primitives constitute the smallest particles of chemical substances, as in Newton's scheme. The net force exerted by two such configurations upon one another determines whether, at various distances, they will cohere or flee, combine chemically or ferment, constitute an aeriform vapor or interact magnetically.[21]

The same net forces account for collisions, the Cartesian theory of which Boscovich held to be untenable, as it violated the high principle of continuity and was unintelligible into the bargain.[22] His theory provided continuity, and

16. Desaguliers, *Course* (1763), II, 36; *FN,* 255–6.

17. Desaguliers, *Course* (1763), I, 7, 16–17, and II, 311–12. Cf. Schofield, *Mechanism* (1970), 81–7; *FN,* 236–7.

18. Desaguliers, *PT,* 41 (1739), 175–85.

19. Desaguliers, *Course* (1763), II, 336–50. This account of Desaguliers relies on Quinn, *Evaporation* (1970), 60–110.

20. Boscovich, *Theory* (1961), §4.

21. Boscovich, *De vir. vivis* (1745), §§47, 49. Cf. Costabel, *Arch. int. hist. sci.,* 14 (1961), 3–12; Marković in Whyte, *Boscovich* (1961), 128–9 and 135, emphasizing originality of the thesis that f becomes repulsive as r goes to zero.

22. It was in puzzling over the violation of continuity in the standard account of collision that, he says (*Theory* [1961], §§16–18) he first hit upon his new approach. Cf. *ibid.,* §§73, 127–8; *De vir. vivis* (1745), §41; and Nedeljković, *Philosophie* (1922), 120, 136, 138, 154.

although ultimately perhaps no more intelligibility than one invoking impulse, it had (he says) the very great merit of explaining everything with the same 'felicity:' bodies exchange motions in collision as they do in any other process, through forces that begin to operate before, and ultimately prohibit, contact.[23]

Boscovich applied his theory to many phenomena, including those of electricity; eventually his explanations filled a large book, printed in 1758 and again in 1763, which its author did not deem successful.[24] In fact it had some influence in England, particularly among the Scottish common-sense philosophers[25] and upon Joseph Priestley, who inferred from it—much to Boscovich's horror—the identity of matter and spirit.[26] The theory also intrigued John Michell and William Nicholson, who adopted Boscovich's account of impenetrability and extension in his influential *Introduction to Natural Philosophy*.[27] On the Continent, outside the Jesuit order, Boscovich seems to have had few followers. Although one can find scattered appreciations of his theory,[28] most physicists did not find it useful, and applied mathematicians had no idea how to attach its central problem, finding the form of f.[29]

THE FLUID OR INHOMOGENEOUS SOLUTION

The solution to the problem of the number and interrelation of forces favored by most physicists of the later eighteenth century was reification: the introduction of weightless substances as carriers of the forces associated with heat,

23. *Theory* (1961), §§102–3. Boscovich discusses theories in terms of convenience, utility, fertility, elegance, etc., and the reverse, not in terms of truth, which—as a disciple of Locke—he held to be unattainable in natural philosophy; Nedeljković, *Philosophie* (1922), 13–18, 189. In particular, the point atom and the universal force were not to be taken as 'real' in themselves, but as means of representing the phenomena distinctly (*Theory* [1961], §137); should it prove impossible—which Boscovich very much doubted—to account for the phenomena with only one law of force, he was prepared to admit others (*ibid.*, §§92, 517). Cf. Nedeljković, *Philosophie* (1922), 167–73, 180–2; Marković in Whyte, *Boscovich* (1961), 137.

24. Boscovich to A. Vallisnieri, Jr., 25 Aug. 1772, complaining that his theory has remained 'quasi sepolto' since it runs counter to the common philosophy (Gliozzi in Conv. Bosc., *Atti* [1963], 115–16). Cf. Marković in Whyte, *Boscovich* (1961), 147, and F. M. Fontana to Boscovich, 30 Aug. 1764 (Bosc. Papers, Berkeley): 'Io fin ad ora sono stato contrario all'universalità d'una tal legge [di continuità], stante che il mio lettore di Filosofia l'aura impugnata.'

25. Olson, *Isis*, 60 (1969), 91–103; Heimann and McGuire, *Hist. St. Phys. Sci.*, 3 (1971), 293–5.

26. Boscovich to Priestley, 17 Oct. 1778, in Varićak, Jug. akad. znan. i umjetn., *RAD*, 193 (1912), 208–10; cf. Thackray, *Atoms*, 189–92, and Heimann and McGuire, *Hist. St. Phys. Sci.*, 3 (1971), 270–3.

27. Hardin, *Ann. Sci.*, 22 (1966), 44–5; Nicholson, *Intro.* (1796⁴), I, 7, 15–17. Cf., *Encycl. Brit.*³, art. 'Earth'; Heimann and McGuire, *Hist. St. Phys. Sci.*, 3 (1971), 275; Schofield, *Mechanism* (1970), 242–6.

28. E.g., Steiglehner, Ak. Wiss., Munich, *Neue phil. Abh.*, 2 (1780), §31, in connection with the repulsion of negatively charged bodies.

29. 'There are indeed certain things that relate to the law of forces of which we are altogether ignorant, such as the number and distances of the intersections of the curve [$f(r)$] with the axis [of r], the shape of the intervening arcs, and other things of that sort; these indeed far surpass human understanding. . . .' Boscovich, *Theory* (1961), §102; cf. Schofield, *Mechanism* (1970), 239–40.

light, fire, electricity, and magnetism.[30] By the end of the century physicists distinguished two electric and two magnetic fluids, light corpuscles, phlogiston, caloric, and perhaps an aether or two. This multiplication of species had several short-term advantages. It immediately explained the existence of a force by the presence of its carrier. It promised that the force could be studied by isolating or concentrating the carrier. And it had a bias toward quantification; at a minimum, the intensity of the force could be made proportional to the 'quantity' or 'intensity' of its fluid. These advantages might be regarded as dearly bought. Both Cartesians and Newtonians of earlier generations would have considered the incoherence of physics—not to mention the inhomogeneity of matter—implied by the representation of specific force carriers as unscientific and weak-minded. The immediate causes of this fall in standards, of this permissive prodigality, were the re-revival of Newton's aether, the general acceptance of a material theory of heat, and the problems and popularity of electricity.

Among Hales' arguments promoting repulsion was the observation that without elastic particles all the parts of matter, being 'endued' with attractions only, would 'immediately become one inactive cohering clump.' It was therefore necessary that the vast mass of attracting matter be everywhere leavened with a 'due proportion of strongly repelling elastick particles.' Hales' experiments identified this yeast with the particles of air, which he had found to be capable both of joining with, and forcing apart, common attractive matter; 'that thereby this beautiful frame of things might be maintained, in a continual round of the production and dissolution of . . .bodies.'[31] To Hales, air is the principle of separation, elasticity, pressure; it plays much the same part in his natural philosophy as the aether occasionally did in Newton's.[32] Since it alone can assume a repulsive state, it must be qualitatively distinct from common matter.[33] Similarly Newton's aether cannot be ordinary matter: being a cause of gravity, it cannot itself gravitate, lest another aether be required to effect its gravitation, and so on.[34]

The parallels between air and aether appear clearly from a letter from Newton to Boyle, published for the first time in 1744. Although written sixty-five years earlier, it turned out to be of immediate scientific interest. It describes an aether that lies in all bodies in amounts inversely proportional to their

30. The process is recognized explicitly by Wilcke, *AKSA*, 2 (1781), 154–5.

31. *Veg. Stat.* (1727), 178. Cf. 'sGravesande, *Math. El.* (1731⁴), I, 17–18, on the dissolution of salt in water: the cohesion between the particles of salt becomes a 'repulsion' in solution, which, in 'sGravesande's language, probably means nothing more than that salt dissolves in water.

32. *Veg. Stat.* (1727), 162. Note that since Hales did not accept a matter of heat, he could not make 'fire' the repulsive principle. Cf. Schofield, *Mechanism* (1970), 75–9.

33. For other dualistic systems of the time see Greene's 'truely English, Cantabridgian and Clarian' *Principles of Philosophy* (1727), and G. Knight, *Attempt* (1748). Cf. Thackray, *Atoms* (1970), 132, 148–9; Heimann and McGuire, *Hist. St. Phys. Sci.*, 3 (1971), 289, 297–301.

34. Newton makes the point explicitly in a draft quoted by McGuire, *Ambix*, 14 (1967), 72–3; cf. *Opticks*, 404, allowing primitives of different densities and forces.

densities. The action of this aether derives primarily from the gradients set up in it across the interfaces between bodies of different densities; for example, the aether just outside the surface of a piece of glass surrounded by air gradually increases from that appropriate to glass to that characteristic of air. When pushing two smooth plates of glass together, one feels a resistance (or repulsion!) from the aether squeezed aside; but once the plates lie flat, the pressure from the circumambient aether holds them firmly together. The aether therefore is the principle both of cohesion and separation; once dissolved in it the particles of vapors 'endeavor to recede as far from one another, as the pressure of the incumbent atmosphere will let them.'[35]

Although this ancient letter conflicted with much in Newton's public writings, including the *Opticks'* aether queries, and although it ended with the usual disclaimer ('I have so little fancy to things of this nature, that, had not your encouragement moved me to it, I should never, I think, have thus far set pen to paper about them'), British natural philosphers took it as evidence that Newton had always believed in, and had virtually demonstrated, the existence of an active, springy, non-material aether. These inferences were drawn by Bryan Robinson, M.D., professor of physics at Trinity College, Dublin, who had taught that Newton's aether operated the nerves and muscles of the body.[36] In 1743 Robinson published a pseudo-mathematical account of the attractive, repulsive, elastic, cohesive and miscellaneous activities of the aether, most of which violate the laws of motion; and in 1745 he issued an aetherial chrestomathy derived from the *Opticks*, the newly published letter to Boyle, and his own work on muscle action.[37] All this publicity had an effect. Beginning in 1745, all significant British electricians postulated a special electrical matter identical with, or similar to, the springy, subtle, universal Newtonian aether.[38] At least one of these electricians, Benjamin Wilson, drew his inspiration directly from Robinson.

Another important carrier of repulsive force was the suppositious 'matter of heat,' or 'elementary fire,' which, in the influential representation of Herman Boerhaave, combined the properties of Newton's aether and Hales' air: a fluid *sui generis*, weightless, universal, penetrating all bodies, expansive, the principle of dilution, fluidity, and fermentation.[39] In Boerhaave's version fire particles exist always and everywhere in the same quantity, and their 'agitation'

---

35. Boyle, *Works*, I, 70–3, reprinted in Newton, *Papers* (1958), 250–3.

36. Schofield, *Mechanism* (1970), 109–10.

37. *A Dissertation on the Aether of Sir Isaac Newton* (1743), for which see Schofield, *Mechanism* (1970), 110–14, and *infra*, xii.4; *Sir Isaac Newton's Account of the Aether* (1745), for which see FN, 418–19.

38. Schofield, *Mechanism* (1970), 110, holds that Robinson did for the aether what Keill and Freind had done for short-range forces. Cf. Thackray, *Atoms* (1970), 137–9, and Guerlac in Hughes and Williams, eds., *Var. Patt.* (1971), 160–1.

39. Boerhaave, *New Method* (1741), I, 246–7, 254–5, 287. Cf. *FN*, 226–31; Metzger, *Newton* (1930), 215–24.

gives rise to temperature.[40] This proposition, neither plausible in itself nor easily reconciled with the assumed expansivity of fire, illustrates that even the leading physicists of the early eighteenth century had difficulty thinking exactly and consistently about forces. Boerhaave's doctrine of the uniform distribution of fire was attacked on many sides, perhaps most fruitfully by Joseph Black, who may have marched directly from his criticism to the discovery of specific heats. For our purpose the most interesting objections were those, like Nollet's, that also betray serious confusion in the application of the concept of force.[41]

Boerhaave's doctrine of the materiality and expansivity of fire prospered even where his theory of its distribution failed. His views, first made public in his lectures on chemistry at Leyden, became generally accessible in a pirated edition in 1724, and in 'sGravesande's text of 1731. The official version of Boerhaave's *Elementa chemiae* (1732) appeared in one French and two English editions before 1750; it brought over to the material theory of heat British physicists raised on the kinetic representations of Bacon and Newton, and latter-day Cartesians reluctant to admit special kinds of matter. No doubt the revival of Newton's aether and the work of Hales assisted this reception. By the mid-century most natural philosophers understood heat in terms of Boerhaave's elastic fire-fluid.[42] By then, too, electricians had constructed several theories postulating an electrical matter similar in many respects to elementary fire. Soon, however, the dependence was reversed, and progress in electrical theory, including the handling of forces, guided improvements in the theory of heat.[43]

By 1750 repulsion had been reified in air, aether, fire, and electricity. In the next few decades physicists accepted a second electrical fluid and other force-carrying imponderables like phlogiston, caloric, and the agents of magnetism. Special carriers of attraction likewise multiplied. The number of fundamental fluids became an embarrassment. But none of the many attempts to reduce it by identifying fluids apparently distinct or by introducing other imponderables succeeded.[44] Physics ended the century richer in essences than it had begun, and

40. *Ibid.*, 224–7; Boerhaave, *New Method* (1741), I, 245–6, 249–55.

41. McKie and Heathcote, *Discovery* (1935), 12–13; Nollet, *Leçons* (1743–8), IV, 177–8 (fire acts on all things, but is not acted upon), 185–6 (against Boerhaave's distribution), 203–9 (suggesting how common matter may capture fire).

42. Gibbs, *Ambix*, 6 (1958), 118–19; Nollet, *Leçons* (1743–8), IV, 161–3, 173, 207; J. C. Fischer, *Gesch.* (1801–8), V, 61–9, VII, 523–9; McKie and Heathcote, *Discovery* (1935), 28–9, 93; A. Hughes, *Ann. Sci.*, 8 (1952), 354–7. An interesting example of mid-century heat theory is Wallerius' (*AKSA*, 9 [1747], 272–81): common-matter particles cohere at short distances, but may be driven beyond the 'sphere of activity' of cohesion by subtle, mutually repellent fire particles; they may then repel one another to form a vapor, as Hales showed. The Secretary of the Swedish Academy of Sciences, P. Elvius, pointed out the connection between Wallerius' theory and that of Query 31 of the *Opticks* (*AKSA*, 10 [1748], 8–9).

43. E.g., Wilcke, *AKSA*, 2 (1781), 160–2, an analogy so strict that it appears to require the repulsion of warm bodies; cf. Oseen, *Wilcke* (1939), 250. Nollet, *Leçons* (1743–8), IV, 183, had earlier argued for the universality of fire from the analogy to electricity; cf. A. Hughes, *Ann. Sci.*, 8 (1952), 360.

44. The effort to reduce the number of fluids left a trace in *Encycl. Brit.*[3], art. 'Motion'; see Hughes, *Ann. Sci.*, 7 (1951), 367, and 8 (1952), 338–40. Cf. Lichtenberg's lectures on fire, electric-

more conscious of their hypothetical character. 'The adequacy of a proposed substance to explain a number of natural phenomena can never prove the existence of such a substance.'[45]

Instrumentalism This agnosticism and its implied instrumentalism—by all means invoke imponderables, feign hypotheses, multiply forces, if it is necessary to save the phenomena conveniently—are characteristic of the Newtonianizing physicists of the second half of the eighteenth century. This was the science of men who grew up familiar with attractions and repulsions and the mathematics needed to treat them; who disposed of more and better data than their predecessors, and lacked their epistemological sensibilities. 'I suppose that there is [in 1772] no physicist whom the terms accessus or attraction, and recessus or repulsion, offend, since in so many parts of physics these forces [cohesion, capillarity, dissociation, Hales' experiments] have become familiar.' Thus J. N. de Herbert, born 1725, professor of physics at the university of Vienna and a one-time Jesuit, who set himself the task of showing that electricity agreed with the rest of Newtonian force physics.[46] We have the same, but stronger, from Volta, born 1745: 'The dominion of the principle of mutual forces in chemistry and physics is today [1778] extensive and, in particular, it is becoming continually more evident in the phenomena of electricity.' Attractions so diverse as electricity, gravity, cohesion and the like may momentarily 'terrify the mind,' but soon 'experience domesticates them.'[47] By admitting repulsion between the particles of electrical matter, electrical theory could be made as simple as that of the planets.[48] Where we see no collisions we may assume an attraction.[49]

An instructive contrast with the older generation, the generation of d'Alembert, is afforded by J. H. van Swinden (born 1746), a great admirer of Newton as a methodologist, who earned his doctorate in philosophy at Leyden in 1766 with a dissertation on attractions. Like d'Alembert, and everyone else, he takes attraction to be effect not cause, and to operate everywhere according to inverse squares; and, again like d'Alembert, he excludes electricity, 'which arises from a fluid that we can see, smell and touch.' But van Swinden does not, as had d'Alembert, argue from the supposed mechanism of electricity that other apparent attractions should be referred to impulse; he enthusiastically admits cohesion and capillarity besides gravity, all following r^{-2} and all 'impossible to conceive.' Still, that is no reason to reject them anymore than our inability to

ity and the magnet (Hermann, *NTM*, 6:1 [1969], 76); Berthollet to van Marum, 30 July 1795, on 'proof' that electricity contains caloric (Sandoun-Goupil, *RHS*, 25 [1972], 242); and the many examples from Achard, *MAS*/Ber (1779), 27–35, to Voigt, *Versuch einer neuen Theorie des Feuers* (1793), cited in *Encyclopédie méthodique*, 76 (1819), 68–70.

45. Van Marum, 'Lectiones physico-chemicae,' (1793), as quoted by Levere in Forbes, *Marum*, I (1969), 245.

46. Herbert, *Th. phen.* (1778²), Praef., 19.

47. *VO*, III, 236, and *VE*, II, 510–11, respectively.

48. *EO* (Wilcke), Vorrede, sig. ++2.

49. Hutton, *Dict.* (1815²), I, 188, art. 'Attraction.'

understand collisions justifies denying the communication of motion.[50] Philosophy suffers from too great a desire to explain everything, *nimia omnia explicandi cupiditas*. Some years later van Swinden admitted electricity among Newtonian forces.[51]

By the last third of the eighteenth century the better physics texts were teaching an open and unfettered instrumentalism. We shall never get to the bottom of things; we should renounce the search for first causes; 'all these things are beyond the reach of our senses, consequently beyond the sphere of our understanding.'[52] 'We can explain nothing in nature completely, we can only derive one phenomenon from another.' Hence we should drop the old program of seeking an intelligible account of gravity and its cogeners: 'As we have no clear conception, or adequate idea, of any mechanical process by which attraction may be caused, all our reasoning on the subject must be not only hypothetical, but visionary.'[53] Under the circumstances one must be content with a physics of 'as if.' 'In explaining electrical phenomena by attraction and repulsion we claim nothing except that the phenomena are such that they would be the same if God had thought it suitable in fact to give the electric fluid the attractive and repulsive force we attribute to it.'[54] These forces are mathematical abstractions, 'ideas in the mind, not in the real world.'[55] The objective is utility: having supposed these forces, one can calculate, one can predict; geometers can compute though they do not understand.[56]

In these epistemological profundities geometers outdistanced philosophers. In 1777 the class of speculative philosophy of the Berlin Academy of Sciences proposed an essay competition on the question, 'What is the *fundamentum virium?*' The geometers complained of the obscurity and uselessness of the question; the King of Prussia, alerted by d'Alembert, ordered his academicians instead to propose, 'Is it useful to a person to be deceived?[57]

What serious opposition there was to instrumentalism among physicists centered on the Eulers, who insisted upon a mechanical account of magnetism and electricity and hoped for one for gravity, and on G. L. Lesage, who found the

---

50. *Dissertatio* (1766), 9–10, 21–40, 55. Cf. Moll, *Ed. J. Sci.*, 1 (1824), 198, and van Swinden to Deluc, 17 Mar. 1780, Deluc Papers, Box 4 (Yale).

51. Van Swinden, *Oratio* (1767), 16; Ak. Wiss., Munich, *Neue phil. Abh.*, 2 (1780), 3. Compare the earlier argument of Klingenstierna, *Tal* (1755), 26–7: physicists had always supposed that their task was to 'enter into Nature's essential inner constitution': in fact, we should observe and establish rules, and drop Cartesian scruples; in particular, mechanistic electrical theories are mere hypotheses, and always fail before new facts.

52. Beccaria, *Treatise* (1776), 382; cf. Mayer, *Anfangsgründe* (1812³), 7–8. For the same point in earlier research reports, *EO* (Wilcke), Vorrede, sig. ⁺⁺2, and J. A. Euler, *MAS*/Ber. (1757), 130. Cf. Lichtenberg to Wolff, 30 Dec. 1784, *Briefe* (1901), II, 174: 'The worst times for physics have been those in which one believed one could decide things which lie beyond the senses.'

53. Respectively, Kästner (1800), in *Briefe* (1912), 224, and Nicholson, *Intro.* (1782), II, 380.

54. Jacquet, *Précis* (1775), 54–5; cf. Haüy, *Exposition* (1787), xvi–xviii.

55. Karsten, *Phys.-Chem. Abh.* (1786), I, 121, 128.

56. Respectively, Haüy, *Traité* (1803), I, viii; *Encyclopédie méthodique*, 76 (1819), 71.

57. Formey, *Souvenirs* (1797²), II, 366–71.

cause of gravity in a rain of penetrating 'supra mundane' particles. The Eulers maintained a strong front against distance forces, 'mentis deliria,' hallucinations, according to the younger, 'arbitrary' and 'occult' in the opinion of the elder.[58] Their attempt at electrical theory, which will be described in its place, had a little life in Germany.[59] But Leonhard Euler's antique Cartesianism, as expressed qualitatively in his *Lettres à une princesse d'Allemagne* (1768), merely amused the younger mathematicians; 'a great analyst,' they said, 'but a poor philosopher.'[60] As for Lesage, even Euler detested his theories, preferring, he said, to 'admit ignorance about the cause of gravity than to take up such strange hypotheses.'[61] No one did so, although a quantitative account of gravitation can be developed on Lesage's terms.[62] The reason for this neglect, as given by Lesage's sympathetic colleague, J. A. Deluc: the overwhelming prevalence of the idea that 'the essence of forces, or the true differences of things, are beyond [the reach of] human ability.'[63]

7. QUANTITATIVE PHYSICS

In 1750 only a few parts of physics had fallen under the yoke of mathematics: hydrostatics, geometrical optics, much of mechanics, a fragment of pneumatics and thermodynamics. By 1800 the quantification of electrostatics, magnetism and thermodynamics was far advanced, and physical optics would soon enjoy similar preferment. The timing of this quantification owed nothing to the progress of mathematics; not until the turn of the nineteenth century did the electrician or thermodynamicist begin to require mathematical techniques not fully available a hundred years earlier. The quantification of electricity and the simple phenomena of heat awaited, first, a rise in the standards for work in physics and, second, improvements in the power, exactness and reliability of instruments.

RISING STANDARDS

'The determination of the relative and mutual dependence of the facts in particular cases must be the goal of the physicist; and to that effect he requires an

58. J. A. Euler, 'Disquisitio' (1755), 3–4; L. Euler, AS, *Pièces*, 5 (1748), 7, and letter to Müller, 30 Dec./10 Jan. 1761, in *Berl. Petersb. Akad.*, I (1959), 166.

59. *Infra*, xvi.3; cf. Achard, *JP*, 21 (1782), 199. In L. Euler's prize-winning essay on magnetism (*supra*, i.3 n. 4) one reads: 'nunquam dubitari quin omnes naturae effectus à causis mechanicis proficiscantur.' AS, *Pièces*, 5 (1748), 4.

60. Lagrange, *Oeuvres*, XIII, 132, 135, 147. Cf. Sarton, Am. Phil. Soc., *Proc.*, 88 (1944), 477.

61. Euler to Lesage, 8 Sept. 1765, in Prévost, *Notice* (1805), 390; Euler to Lesage, 13 Oct. 1761 and 16 April 1763, *ibid.*, 381–4.

62. Lesage thought that his impulsion theory called for a law of gravity of the form $1/(r^2 - r)$; Lesage to Boscovich, 20 Sept. 1763, in Costabel in Conv. Bosc., *Atti* (1963), 209–10.

63. Deluc, *Précis* (1802), I, 320. Cf. Deluc, 'Première esquisse du système de M. Lesage' (1781–2), Deluc Papers, Box 73 (Yale): Lesage aims to substitute 'agens physiques' for 'qualités occultes,' viz., attraction and repulsion.

exact instrument that will perform in an exact and invariable manner in every place in the world . . . The history of physics demonstrates a truth now [1782] sufficiently recognized: the physicist who does not measure only plays, and differs from a child only in the nature of his game and the construction of his toys.' Thus F. K. Achard,[1] permanent member of the Berlin Academy of Sciences, sounded a note often heard from continental natural philosophers during the latter eighteenth century. 'Everyone now [1773] agrees that a physics lacking all connection with mathematics and tied to a simple collection of observations and experiments would only be an historical amusement, fitter for entertaining the idle than for occupying the mind of a philosopher.'[2] 'En négligeant le calcul, on fait les expériences sans choix et sans desseins,' said Lambert, on entering Achard's academy in 1765.[3] Academician Le Roy, and professors Kästner, Karsten, and Volta, insist that physics, and especially electrical theory, cannot be advanced further without exact measurement. Lichtenberg does them one better, and advises that all of physics should be reexamined, 'from the ground up, and with all imaginable accuracy, using today's [1784] more complete instruments.'[4]

These statements, which could be multiplied a hundredfold, in themselves bring nothing new: Fontenelle had written that 'physics has substance only in so far as it is founded on geometry;' Boerhaave, in a famous address given in 1715, had insisted that mathematics supplied the only route to useful generalizations in science; and 'sGravesande, as already mentioned, took physics to be a branch of mathematics.[5] But these earlier writers did not practice what they preached.[6] Their successors did. Achard spent days and nights in his laboratory, thirteen in a row once on optical experiments; he left us, among other more useful things, measures of surface tension to four figures and elaborate investigations of comparative electrical conductivities.[7]

Lambert is an interesting case. A self-taught polymath, he took as his main

1. *JP*, 21 (1782), 196.

2. Paulian, *Dictionnaire* (1773²), art. 'Physique.' Cf. the sentiment of the Munich Academy of Sciences (1784): 'Auch zu der Experimentalphysik ist Mathematik und viel Mathematik erforderlich, und diejenigen welche bey jemanden, der keine Physik versteht, Experimentalphysik zu sehen glauben, lernen nichts weiter, als was sie von einem Taschenspieler lernen würden.' Westernrieder, *Gesch.*, I (1784), 276.

3. Quoted in Schur, *Lambert* (1905), 9.

4. Le Roy, art. 'Electromètre,' *Encyclopédie;* Karsten, *Phys.-chem. Abh.* (1786), I, 137–8; Kästner, ed. note to Bergman, *AKSA*, 25 (1763), 344–52, on p. 352 ('Yet the theory of electricity will remain uncertain as long as mathematics, the only way to make our knowledge of nature certain, is not applied more fully to it'); Kästner, 'Verbindung' [1768] in *Verm. Schr.* (1783³), II, 359; Lichtenberg, *Briefe* (1901), II, 149; P. Hahn, *Lichtenberg* (1927), 13–15.

5. Fontenelle, 'Préface' (1733), in *Oeuvres* (1764), V, 1–14, and Flourens, *Fontenelle* (1847), 174, 187; Brunet, *Physiciens* (1926), 44–5, 48–9.

6. Cf. Segner's *De mut. aer.* (1733), perhaps the earliest attempt to apply Newtonian theory to atmospheric tides. 'Without mathematics nothing can be done with a difficult physical problem,' he says, and serves up a result off by a factor of 400.

7. Achard, *Chem-phys. Schr.* (1780), 354–67, and *Sammlung* (1784), 20–45, 141–53; Stieda, Ak. Wiss., Leipzig, Phil.-Hist. Kl., *Abh.*, 39:3 (1928), 11–12, 173–4.

line the application of mathematics to physics and even to metaphysics. As a philosopher he worked out an epistemology similar to Kant's; as a physicist he sought effects linked by simple, general, and above all mathematical laws: as an experimentalist he advanced the quantitative study of photometry, pyrometry, hygrometry, and magnetism.[8] He talked as an equal to Leonhard Euler and to Georg Brander, respectively the leading mathematician and the leading instrument maker in Germany.[9] In a word, he was the perfect mathematical physicist: the mathematicians considered him an experimentalist with a 'rare talent for applying calculation to experiments;' the experimentalists thought him a mathematician with an unusual understanding of the behavior of instruments.[10] All of which (we are told) he accomplished by working from five in the morning to twelve at night, with a two-hour break at noon; the common experiences of life were to him so many occasions for calculations, and conversations opportunities for extemporaneous dissertations.[11]

The reduction of experimental data to law, or the deduction of law from first principles, is usually considered the domain of mathematical or theoretical physics. Here the physicists of the second half of the eighteenth century advanced over their predecessors in only a few isolated cases. But in respect of exactness of measurement, which constitutes the basis of quantitative physics, a great change occurred during the latter eighteenth century. Here Achard's measurements of surface tension, made without reference to a mathematical theory of capillarity, are representative. Similar labors occupied much of the life of M. J. Brisson, who in 1787 gave tables of specific weights to several significant figures, 'never entering any result as exact until the results of repeated measurements either showed no differences, or differences small enough to be neglected.' Brisson's attention to quantitative detail, to precautions to be taken, to reliability of instruments, is itself a good measure of the distance between his generation of experimentalists and the preceding one; for Brisson (born 1723) had learned his physics from Nollet (born 1700), who exercised only so much care as produced results, and seldom measured anything.[12]

Our methodologist van Swinden likewise recommended sedulous attention to exact observation, and supplied a heroic example by measuring the magnetic variation every hour of every day for ten years.[13] Another new man was J. A. Deluc (born 1727), who drove himself and his associates to distraction in his attempt to build meteorological instruments that would give reliable, and comparable, quantitative results. Recognizing a kindred soul in Brisson, Deluc oc-

8. Berger, *Cent.*, 6 (1959), 190, 196, 218–19.

9. See Lambert's correspondence with Euler in Bopp, Ak. Wiss., Berlin, Phys.-math. Kl., *Abh.* (1924:2), and with Brander in Lambert, *Deut. gel. Briefw.* (1781), III.

10. The opinions of, respectively, Lagrange, letter to d'Alembert, 3 Oct. 1777, in Lagrange, *Oeuvres*, XIII, 333–4, and of Saussure, *Essais* (1783), ix.

11. Lichtenberg, in Steck, *Bibl. lamb.* (1970²), xii–xiii.

12. Brisson, *Pesanteur* (1787), ii–iii, xiv–xv; Merland, *Biog. vend.*, II (1883), 11, 21–3; *DSB*, II, 473–5.

13. Van Swinden, *Oratio* (1767), 38–9; Moll, *Ed. J. Sci.*, 1 (1824), 199.

cupied him for eight months calibrating a Deluc thermometer against the last of Réaumur's surviving instruments.[14] With a fellow Genevan, H. B. de Saussure, Deluc liked to dispute about the corrections to be applied to the readings of hygrometers in the third and fourth places of decimals; both shared their odd passion with the public in big books on the errors of barometers, hygrometers and thermometers.[15] Deluc, 'sagacissimo e accuratissimo,' was a byword for precision and dependability; '[he] handles everything like his barometer, precise (so to speak) to the point of error; the least inaccuracy would ruin everything.'[16] It is therefore noteworthy that Deluc often supported imprecise, qualitative and even retrogressive theories, such as those of his friend Lesage.

A taste for fuzzy theory was precisely what precise experiment was calculated to correct. As Desaguliers observed, we are liable to mistake the causes of things unless we 'measure the Quantity of the Effects' each putative cause may produce.[17] It is just the evasions and obscurities made possible by ignorance of geometry that encourage and shelter the concocters of aethers and vortices, of contorted pores, threaded passages, hook-and-eye atoms, of a thousand impossibilities. Or so d'Alembert thought, adding that explanations resting on such fictions are 'so incomplete, so loose, that if the phenomena were completely different, they could very often be explained just as well in the same way, and sometimes even better.'[18] And it is true, as Haüy observed, that a great distance separates the physical theories of the mid-century, like Nollet's picture of electrical action, 'independent of law and rigorous method,' and the new quantitative theories created by Haüy's generation, based on 'exact measurement' and capable of calculating 'the various effects with such precision that they may be predicted.'[19] Nollet lived long enough to witness the Academy's swing against his brand of physics. In the year before his death he wrote his closest collaborator, E. F. Dutour, who had sent him a new paper in the old style: 'It seems to me that you often call on the configurations of the ultimate parts of bodies, on the arrangement of their pores . . ., on an unknown matter to which

14. Varenne de Beost to Deluc, March, April and Oct. 1765, Deluc Papers (Yale); Middleton, *Hist.* (1966), 117–18. Later van Swinden recommended recalibrating all surviving Réaumur and Nollet barometers; letter to Deluc, 12 April 1782, Deluc Papers (Yale).

15. E.g., Saussure, *Essais* (1783), table facing p. 122; Deluc, *Recherches* (1772), I, table facing p. 184. Landriani pointed out to Deluc (letter of 20 Oct. 1788, Burndy Library) that the public was not ready for so much hygrometry, 'owing to the attention it requires and the length of the discussion.' For the aficionado, however, these fat books, especially Saussure's, were 'incomparable' (Volta to Magellan, 28 Oct. 1783, *VO*, VI, 322), a 'masterpiece' (Senebier, *Mémoire* [an IX], 86, who applauds [p. 188] Saussure's 'désir insatiable d'acquérir des connaissances plus exactes').

16. *VO*, IV, 58; Lichtenberg, letter of 1781, in *Briefe*, I, 384–5. Cf. Lesage to Boscovich, 8 May 1772, in Varićak, *RAD*, 193 (1912), 212: 'on peut parfaitment computer sur l'exactitude de ses observations les plus délicates et ses expériences les plus difficiles.'

17. *Course*, I, v; cf. van Swinden, *Oratio* (1767), 38.

18. *Mélanges*, IV, 231, quoted by Hankins, *D'Alembert* (1970), 81; cf. Keill as quoted by Strong, *J. Hist. Ideas*, 18 (1957), 56: 'All these errors [of Cartesians, of course] seem to spring from hence, that men ignorant of Geometry presume to philosophize, and to guess at the causes of Natural Things.'

19. Haüy, *Traité* (1803), I, 337.

you assign a large role, etc. I ought not hide from you that the Academy is getting more and more difficult about this way of philosophizing.'[20]

The emphasis on precise measurement in physics profited from and contributed to efforts to raise the standard of scientific work in the later eighteenth century. The editor of the *Journal de physique* announced that he would not cater to dilettantes or browsers: 'We will not offer lazy amateurs matter purely for amusement, nor give them the sweet illusion that they know something about sciences of which they are ignorant.' Rather, the new journal would meet a need long felt for faster, cheaper, more useful and less parochial communication than the proceedings of learned societies provided.[21] The success of the *Journal* encouraged the foundation of other professional periodicals for short original articles and abstracts of academic memoirs; like the *Journal,* the *Bulletin des sciences* of the Société philomathique and the *Journal of Natural Philosophy, Chemistry and the Arts* (Nicholson's Journal) advertised themselves as international, fast, useful, cheap, and even 'accurate,' at least in the abstracting of papers published by others.[22] The effect of rising standards may also be seen in the establishment of—and the quality of the memoirs published in—such specialized journals as the *Annales de chimie et de physique* (1789) and the *Journal der Physik* (1790).[23]

Simultaneously scientific societies became more particular about their memberships. In 1776 the Council of the Royal Society of London, alarmed at the number of obscure foreign members, declared a moratorium on further admissions.[24] When it was removed election became increasingly more difficult. 'Never before [as van Marum, who sought admission, learned in 1791] has the honor of becoming a member been the object of such aspiration as it is now . . . They are getting very strict about foreign Members.'[25] The Società Italiana delle scienze, established in 1782 to overcome the jealousies, infighting and poor communication responsible, according to its founders, for the decay of Italian science, offered membership to any countryman of proven ability, 'recognized for his published work.' Similarly the Hollandsche Maatschappij der Wetenschappen, which began by admitting almost everybody, restricted itself from about 1795 to 'professionals who are professors, or who have acquired their reputations by works which they have published or presented to the Soci-

20. Letter of 13 March 1769 (Burndy): 'Il m'a semblé que vous appelez souvent à votre secours la configuration des parties primordiales des corps, celle de leur porosité. . . . une matière inconnüe à qui vous faites jouir de grands rôles, etc. Je ne dois vous dissimuler que l'Académie devient de plus en plus difficile sur cette manière de philosopher.'

21. *JP*, 1 (1773), i–vii; cf. K. Baker, *RHS*, 20 (1967), 264–7.

22. Soc. phil., *Bull.*, 1 (1791), iii–iv; *J. Nat. Phil.*, 1 (1797), iii; Lilly, *Ann. Sci.*, 6 (1948), 94; Neave, *ibid.*, 417–19.

23. Cf. Kronik, Bibl. Soc. Am., *Papers*, 59 (1965), 28–44.

24. Planta to Cowper, 11 Dec. 1778 (*EO*, I, 312, re Volta's candidacy).

25. Ingenhousz to van Marum, 11 March 1791, in Levere, RS, *Not. Rec.*, 25 (1970), 117, and R. J. Forbes, *Marum*, III, 37; cf. Ingenhousz to Magellan, 2 April 1787, in Carvalho, *Corresp.* (1952), 147.

ety.'[26] Even in the smaller local societies one notes a new seriousness of purpose. The management of the Gesellschaft der naturforschenden Freunde (Berlin), exasperated by a great increase in their scientific correspondence, decided to drop the usual flowery salutations and compliments, and begged their correspondents to do the same, and to come quickly to the point.[27]

INSTRUMENTS

Improved scientific instruments were the material cause and plainest expression of the rising standards and improving accuracy of physics in the later eighteenth century. After 1780 both the quality and quantity of physical apparatus commercially available increased sharply.[28] To take one measure of quantity: the number of *new* British firms making mathematical, optical, and / or philosophical instruments founded per decade remained between 25 and 30 from 1720 to 1780; in the eighties and nineties it averaged 48.[29] The same phenomenon may be followed on a finer scale in Scotland, where, on the average, ten instrument makers were active from 1730 to 1770, as compared to sixteen in the last two decades of the century.[30] In Holland, too, the number of instrument makers in business in the years 1770 to 1800 (about thirty each decade) greatly exceeded the number active earlier in the century (about five each decade, 1700–30, and fifteen each decade, 1730–50).[31] But these numbers give only a pale impression of the growth of the trade. In the first half of the century instrument firms consisted of the owner and a very few assistants; beginning in the 1750s with the shop of George Adams, London establishments grew prodigiously, sometimes—as in the case of the best makers, Peter Dolland, Edward Nairne, and Jesse Ramsden—to a staff of as many as fifty trained artisans.[32] A similar enterprise, employing thirty workers, was set up in Delft in the 1790s.[33]

London manufacturers supplied not only the British, but also much of the world trade in good scientific instruments. To be sure, the Dutch had excellent

26. Soc. ital., *Mem.*, 1 (1782), v–vi, ix; van Marum to Parmenter, 13 June 1817, in Levere, RS, *Not. Rec.*, 25 (1970), 115, and in R. J. Forbes, *Marum*, III, 34, 37; *infra*. ii.2.

27. Ges. naturf. Freunde, Berlin, *Schr.*, 1 (1780), v–x. Cf. Nicholson, *Intro.* (1790³), I, xi–xiii, on 'the solidity of argument, and precision of expression' of the best English physicists.

28. Cf. Daumas in Crombie, *Sci. Change* (1963), 418–19, and in Singer, *Hist. Tech.*, IV (1958), 403.

29. Compiled from the biographies in E. G. R. Taylor, *Math. Pract.* (1966), 152–353.

30. Bryden, *Scott. Sci.* (1972), 26. From Bryden's table, p. 28, one computes an average life of a little over 20 years for Scottish instrument firms. Assuming a half-life of 10 years for British firms as a whole, and ignoring survivals of firms founded before 1720, one finds the numbers active in each decade from 1740 to 1800 were 49, 54, 53, 55, 74, 85.

31. Compiled from Rooseboom, *Bijdrage* (1950), omitting watchmakers and men known only for a single, non-physical instrument, such as a compass or a telescope. Rooseboom omits those who made only barometers, thermometers, balances, surgical instruments, clocks (*ibid.*, 134–5).

32. Daumas, *Instruments* (1953), 311–20; E. G. R. Taylor, *Math. Pract.* (1966), 43; Bernoulli, *Lettres* (1771), 126.

33. By J. H. Onderwijngaart Canzius, who brought in workers from outside Holland; Rooseboom, *Bijdrage* (1950), 20.

workmen: Jan van Musschenbroek, for example, and Jan Paauw, who made, and indeed made possible, the demonstration apparatus of 'sGravesande and P. van Musschenbroek; and the transplanted German Daniel Fahrenheit, who set up in Amsterdam in 1717. But the Dutch trade did not extend much outside the Netherlands, nor, as the case of Fahrenheit shows, did it always recruit its best makers domestically. In 1790 the leading manufacturer of scientific apparatus in Holland was the Englishman John Cuthbertson, the builder of, among much else, the Teylerian electrical machine.[34]

In France, Nollet had overseen the making of instruments that were perhaps the equal of those of Jan van Musschenbroek and Desaguliers. But Nollet had trouble procuring competent workmen and his successor, Sigaud de Lafond ('as inexact a maker or director of makers of scientific instruments as mediocre physicist'[35]), could not hope to compete with the Dollands and the Ramsdens. Guild restrictions inhibited the development of the French industry, which did not begin to pick up until the 1780s, when Nicholas Fortin began to make precision instruments for Lavoisier, and the Paris Academy set up a 'corps d'ingénieurs en instruments,' which included Fortin, to evade trade regulations and encourage promising artisans.[36] And even then French scientists visiting England could see 'informed artisans unknown [in Paris], and instruments entirely different from ours.'[37]

Except for a very few men such as Lambert's friend Brander, the Germanies had no instrument makers with more than a local clientele before the end of the century; for large pieces and precision work they patronized the English.[38] The experience of Prof. J. G. Stegmann of Marburg, who worked three or four artisans for fourteen years to produce optical instruments 'rather far' below British quality, may be representative.[39] The Italians had to buy almost everything abroad, and bought English when they could afford to. 'The machines from Paris are very mediocre and moreover have suffered greatly in shipment,' Volta wrote of apparatus he had purchased on a trip to France and England in 1781–2. 'Those from London are bellissima, elegant, and arrived in perfect condition.'[40]

We may distinguish three sorts of scientific instruments in the expanded trade

34. Daumas, *Instruments* (1953), 123, 138, 326–33; Crommelin, *Descr. Cat.* (1951), 11–13; Hackmann, *Cuthbertson* (1973), 39–40; Crommelin, *Sudh. Arch.*, 28 (1935), 136–9; Cohen and Cohen-de Meester, *Chem. Week.*, 33 (1936), 379, 391.

35. Lambertenghi to Volta, 8 Nov. 1779 (*EO*, I, 384); Nollet, *Leçons*, I (1743), lxxxvii.

36. Daumas, *Instruments* (1953), 130–7, 339–85. Cf. Bugge, *Science* (1969), 171, for French physics instruments in 1798–9: 'very nice . . . , although the metal work and polish cannot be compared with English work.'

37. Ch. Messier to Magellan, 2 Feb. 1788, in Carvalho, *Corresp.* (1952), 153.

38. Daumas, *Instruments* (1953), 333–6; Hermann, *Phys. Bl.*, 22 (1966), 388–96; Körber, Cong. int. hist. sci., XIIIᵉ (1971), *Actes*, 6, 274–5.

39. Bernoulli, *Lettres* (1771), 46–7; cf. p. 68, on the British monopoly on good optical glass.

40. *VE*, II, 89, 91–2; Bernoulli, *Lettres* (1771), 171–3; Daumas, *Instruments* (1953), 324–5. Cf. Lichtenberg, *Briefe* (1901), I, 386, 389, complaining about the amount of expensive brass on English instruments.

of the second half of the eighteenth century. First, demonstration apparatus, required in quantity to decorate the 'physical cabinets' of wealthy amateurs and to illustrate the lectures of teachers of natural philosophy. A few substantial collections of demonstration apparatus were assembled before 1750, notably by Mårten Triewald at Newcastle, by the Landgrave of Hesse at Kassel and by Voltaire at Cirey,[41] and by the professional lecturers 'sGravesande, Nollet, P. van Musschenbroek, Desaguliers, and perhaps seven or eight others.[42] The great demand came after 1750. In England George III and Lord Bute, in France Louis XVI, the Duc d'Orléans, and the Duc de Chaulnes, in Italy the Grand Duke of Tuscany, inspired both the makers of instruments and the apers of fashion; a great many small cabinets (25 to 75 items, as against 250 to 350 in large holdings) came into existence; almost seventy private collections have been identified, and there must have been many more.[43] At the same time schools, colleges, academies and universities began to establish or augment collections, or to subsidize their professors' purchase of instruments. In so far as these collections included the best contemporary work, they were much superior to those assembled earlier in the century. Here is the estimate of the physicist J. A. Charles, who owned the best and most extensive demonstration apparatus in France in the 1790s (some 330 items), of the instruments of his predecessor Nollet: 'One finds in them neither the elegance of form, nor the beautiful workmanship, still less that severe precision that characterizes the most modern machines.'[44]

The second type of instrument multiplied or improved in the late eighteenth century was the measurer. Here improvement in quality can itself be measured. Perhaps the best-known advance was the Ramsden ruling engine (1773), which could divide an arc accurately into ten-second intervals, as compared with ten minutes and five minutes, the standard divisions of the sectors of 1700 and 1750, respectively.[45] This increase in precision, which depended upon im-

41. Tandberg, Lund, U., *Årrsk.*, Avd. 2, 16:9 (1920), 4–5; Kirchvogel, *Phys. Bl.*, 9 (1953), 259–63; Daumas, *Instruments* (1953), 189; Voltaire to B. Moussinot, 5 and 18 May, 1738 (*Corresp.*, VII, 156, 177), regarding the payment of 9–10,000# to Nollet. Triewald's collection (miscellaneous makers) went to the University of Lund, the Landgrave's (largely by J. van Musschenbroek) ultimately to the Hessisches Landesmuseum, and Voltaire's apparently to oblivion.

42. Desaguliers estimated that there existed only 10 or 11 competent and fully furnished lecturers in natural philosophy (according to Daumas, *Instruments* [1953], who dates the estimate 'about 1750,' although Desaguliers died in 1744). For examples of their instruments see Gerland and Traumüller, *Gesch.* (1899), 294–312.

43. Daumas, *Instruments* (1953), 189–94; Torlais in Taton, *Enseignement* (1964), 640–1; Chaldecott, *Handbook* (1951); G. Turner, *Ann. Sci.*, 23 (1967), 213–42 (Lord Bute); *JP*, 9 (1777), 42 (Florence). Many are mentioned in J. Bernoulli, *Lettres* (1777–9).

44. Daumas, *Instruments* (1953), 186n, quoting a note of 1789 on the occasion of the transfer of Nollet's collection (which Brisson, his heir, had sold for 1200# in 1792) to the Conservatoire des Arts et Métiers. Cf. Torlais in Taton, *Enseignement* (1964), 633–4.

45. Daumas, *Instruments* (1953), 249–50, 264–7; Skempton and Brown, RS, *Not. Rec.*, 27 (1973), 240.

provements in lathes, glass-making and metal-working,[46] extended to the measurement of physical quantities: the second half of the eighteenth century enjoyed significantly improved barometers and magnetic needles, standardized thermometers and hygrometers, and a choice of design of a new and characteristic instrument, the electrometer.

Take the case of the barometer. The mathematician Saurin considered it useless to correct the barometers of his day (1720s) for changes in temperature on the ground that the error fell within the limits of accuracy of even the best instruments, about ⅓ of a line (0.7 mm). Perhaps the prime cause of unreliability, air absorbed in the glass or dissolved in the mercury, was eliminated by boiling the mercury before closing the tube; thereafter corrections for temperature (which Deluc found to be about 0.06 lines per degree centigrade), as well as for capillarity and for several subtler effects, became significant. In about 1770 verniers were added; a Ramsden instrument of that date can be read to 0.1 lines. And about 1775 Ramsden introduced an index, which eliminated the effect of parallax and reduced the error in sighting the meniscus by perhaps an order of magnitude.[47] In 1777 the most advanced instruments could be read to a few thousandths of an inch, and the most advanced readers could scoff at the 'gross approximations' that had satisfied their predecessors.[48]

The thermometer had a similar history. In 1731 the usually meticulous Réaumur rejected the suggestion that he use brass instead of paper for his thermometer scales; that, he said, would be pushing accuracy to ridiculous lengths. Not until the 1740s did calibration between fixed points begin to be common, and even then differences in technique, especially in determining the setting for boiling water, created instruments literally incomparable. As late as 1777 the Royal Society found a variation of as much as 3.25 degrees Fahrenheit in the location of the boiling point on their instruments. At about that time, however, the better thermometers were accurately marked to a fifth or a tenth of a degree, and soon good thermometers with comparable fixed points—such as those made by Fortin and Mossy for Lavoisier, which could be read to one hundredth of a degree—became available.[49] In 1787 Charles le géomètre published elaborate formulae for correcting thermometers for dilation of the glass, in order to make their readings, as well as their fixed points, comparable. The same year Saussure climbed Mont Blanc with a perfected thermometer that he liked to read to 1/1000 of a degree and with which he determined the boiling point of water at the summit to an accuracy of 0.1 per cent.[50]

46. Daumas in Crombie, *Sci. Change* (1963), 421–3, and in Singer, *Hist.*, IV (1958), 382–4.

47. Middleton, *Hist. Barom.* (1964), 178–9, 188, 197, 243–5; Daumas, *Instruments* (1953), 273.

48. Shuckburgh-Evelyn, *PT,* 67:2 (1777), 524, 557n; [Fontana], *JP,* 9 (1777), 105.

49. Middleton, *Hist. Therm.* (1966), 80, 119, 127–8, 133; Daumas, *Instruments* (1953), 280–1; Cavendish, et al., *PT,* 67:2 (1777), 831. Cf. G. Turner in Forbes, *Marum,* IV, 257.

50. Crommelin, Ned. aardr. gen., *Tijds.*, 66 (1949), 327–31; Charles le géomètre, *MAS* (1787), 574–82; cf. Fontana's thermometer for measuring the heat of moonlight, *JP,* 9 (1777), 107.

The electrometer, about which there will be much to say later, came into existence about 1750, crudely made, without standards or standardization, and without much agreement on the part of its makers about what it measured. The progress of theory, the improvement of technique, and, above all, the need to standardize measurement—of the ratings of machines for trade,[51] of the shocks given in medical treatments, of the leakage of charge into the atmosphere—produced a strong demand for reliable instruments. It was a job for a Deluc, or so van Swinden told him, encouraging him to do for electricity what he had done for the atmosphere. 'Although electricity has been treated by a great many physicists, it has not yet been considered with the precision that physicists who care about mathematical precision could wish.'[52] In the event others developed the instruments, which existed by the mid-eighties. These in turn reacted upon the theory, simultaneously embodying and confirming the important relationship $Q = CT$ between the charge, capacity and 'tension' of an electrified conductor.[53]

The third of our three types of instrument is represented by the air pump and the electrical machine and their many accoutrements. These instruments, found in every respectable cabinet of the late eighteenth century, could be used for research as well as for demonstration. The power of both instruments increased dramatically after 1750. The usual pump of the mid-century, built according to the designs of Hauksbee, 'sGravesande and Musschenbroek, probably reached 1/40 or at best 1/50 atmosphere. About the same time Smeaton obtained an exhaustion of perhaps 1/80 at. by soaking the leather fittings of his pump in a mixture of alcohol and water.[54] The common pump of the 1770s, still considered an 'excellent machine' in France a decade later, attained 1/165 at. In the same period Nairne advertised an improvement of Smeaton's pump that could provide a vacuum of from 1/300 to 1/600 at. in six minutes' working.[55] These improvements enabled physicists to investigate, among other matters, the vexed question whether vacuum conducts or insulates.

As for the electrical machine, it was capable of generating about 10,000 volt when first introduced in the 1740s. Van Marum's white elephant of 1785, which employed a glass plate 65 inches in diameter, probably gave over 100,000 volt. The earlier machines when used to charge the favorite capacitor of the period, a boy or a gun barrel hung from the ceiling by insulating cords

51. The effect of competition appears clearly in Cuthbertson's search for better measures of electrical output; Hackmann, *Cuthbertson* (1973), 33, and in Forbes, *Marum*, III, 349–51.

52. Van Swinden to Deluc, 16 May 1783 and 23 April 1784, Deluc Papers (Yale): 'L'électricité, quoiqu'elle ait été traitée par un si grand nombre de physiciens, n'a pas encore été considérée avec cette précision que des physiciens, qui font quelque cas de la précision mathématique, pouvoient désirer.'

53. *Infra*, xix.1; cf. Daumas in Crombie, *Sci. Change* (1963), 428–30; Kühn, *Neu. Ent.* (1796), 218–83.

54. Smeaton, *PT*, 47 (1751–2), 420, estimating 10^{-3} at. with a pear gauge later shown to be inaccurate; Nairne, *PT*, 67:2 (1777), 619, 622–5, 634.

55. *Ibid.*, 635–6; *JP*, 30 (1787), 434; Daumas, *Instruments* (1953), 287.

(fig. 8.3), could accumulate an electrical energy of some 0.0008 joule. Van Marum's engine could supply his battery of 100 Leyden jars, put in service in 1790, with perhaps 3000 joule. The increasing power of these instruments will appear from Table 1.1.

TABLE 1.1

Energy Available in 18th-Century Sparks[a]

Date	Instrument	Spark length (inches)	Voltage (volt × 10^4)	Capacitance (farad × 10^{-9})	Energy (joule)
1747	Glass rod (Franklin's)	1	0.5	0.0005	.000006
1750	Globe machine (Franklin's)	2	1		
	with gun barrel			0.015	.0008
	with Leyden jar			2.0	0.1
1773	Cylinder machine (Nairne's)	14	3		
	with prime conductor			0.05	0.2
	with Leyden jar			2.0	0.9
	with 64 jars			130	58
1785/90	Plate machine (v. Marum's)	24	8		
	with prime conductor			0.2	0.6
	with Leyden jar			5.6	20
	with 100 jars			560	2000
	Storm cloud				5000000

[a] Spark lengths and capacitances from Finn, *Brit. J. Hist. Sci.*, 5 (1971), 290; potentials from Prinz, *Museosci.*, 11:5 (1971), 26, and 12:5 (1972), 14.

With the large installations of the end of the century one could electrocute small animals, melt several meters of wire, electrolyze water, magnetize needles, and study the chemical effects of electricity in motion.

SOME EIGHTEENTH-CENTURY QUANTITATIVE PHYSICS

The availability of good measuring instruments helped supply pressure for the establishment of quantitative relations between physical parameters. Numerical tables, some worked out to crowds of illusory decimals, were duly filled up by experimentalists. To obtain significant relations, however, a theory was required, and often an instrumentalist one. Examples from theories of heat and magnetism will illustrate the fitting together of improved measures, refined theory, instrumentalism and quantification. Each example shares features with the more difficult case we shall examine at large, the quantification of electrostatics.

Heat At the beginning of the eighteenth century there was no standard and reliable measure of any heat phenomenon, and no agreement about the nature of heat. There did exist a few quantitative rules, such as Newton's law of cooling and an unconfirmed theorem, an old saw among medical writers, about the final degree of heat of a mixture of bodies of unequal heats. The medical man assimilated the problem to that of the average price of a number of goods at different prices, whence $H = \Sigma m_i h_i / \Sigma m_i$, H being the final heat, m_i and h_i the masses and initial heats of the several bodies.[56] The question reopened when good thermometers became available and physicists, without much justification, took temperature as a measure of heat. In fact over a large range the quantity of free heat in a body is closely proportional to its temperature as given by a mercury thermometer, for mercury expands nearly linearly with heat. There seems to have been little persuasive evidence for this proposition before the experiments of Black and Deluc.[57]

In 1744, G. W. Krafft, professor of mathematics and physics at the Petersburg Academy of Sciences, attacked the old problem of mixing with thermometers probably obtained directly from Fahrenheit's successor Prins. With them he got for the final temperature T of the mixture of two masses of water m_1 and m_2 at initial temperatures t_1 and t_2, $T = (11m_1t_1 + 8m_2t_2) / (11m_1 + 8m_2)$. Recent experiments have shown that this peculiar formula is precisely what one obtains by performing as crudely as possible, ignoring heat absorbed by the mixing vessel and thermometer, or lost to the atmosphere. Krafft apparently regarded his task as a search for the numerical relation holding in a specific mixing operation; he omitted from consideration complications that theory might suggest, and he did not aim at—or even suggest the desirability of —a general law correctible according to circumstance.[58]

Not so Krafft's former student and, in 1745, his successor, G. W. Richmann. Richmann had an excellent physical intuition, which assured him of the linearity of Fahrenheit's thermometer and of the parochiality of Krafft's numbers. Richmann conceived his task to be the discovery of general quantitative relations, which, in the case before him, could not fail to be the old medical average, $T = (m_1t_1 + m_2t_2) / (m_1 + m_2)$. Careful measurements, taking into account what we would call the water equivalent of the instruments, confirmed Richmann's universal equation. 'The whole business,' he then wrote, 'shows clearly that physics should avoid mathematical abstractions with all diligence whenever possible, and attend to every circumstance in individual cases.'[59] In fact it more nearly shows the opposite: it was Richmann who began with an abstraction.[60]

56. Zubov in *Mélanges* (1964), I, 654–61, and Ak. nauk, Inst. ist. est. tek., *Trudy*, 5 (1955), 69–93; McKie and Heathcote, *Discovery* (1935), 54–9.

57. Deluc, *Recherches* (1772), §§418m–422rr; McKie and Heathcote, *Discovery* (1935), 125, citing Black's experiments of 1760; cf. Polvani, *Volta* (1942), 208–9. There had been earlier trials, e.g., Brook Taylor's on the linseed-oil thermometer (*PT*, 32 [1723], 291), and Richmann's on Fahrenheit's (*NCAS*, 4 [1752–3], 277–300). Cf. Middleton, *Hist. Therm.* (1966), 109, 124–6.

58. Krafft, *CAS*, 14 (1744–6), 218–39; McKie and Heathcote, *Discovery* (1935), 55–63.

59. Richmann, *NCAS*, 1 (1747–8), 152–67, 168–73; McKie and Heathcote, *Discovery* (1935), 65–76; *DSB*, XI, 432–4.

60. As to the nature of heat, Richmann, *NCAS*, 4 (1752–3), 278, held it to be 'a certain motion of certain corporeal particles,' a view Krafft probably shared.

Richmann's law answered an old problem in new language, and provided the opportunity for a splendid discovery through recognition of its limitations. As Richmann observed, 'only if we have accurately determined the properties of bodies can we legitimately infer other truths with certainty.'[61] In 1769 J. C. Wilcke, mathematician and experimental physicist at the Swedish Academy of Sciences, made the 'paradoxical' observation that water cooled below 0°C warms on freezing. With this paradox in the back of his mind he immediately grasped the significance of an observation made early in 1772.[62] Wishing to wash snow from a courtyard, he was surprised to find that hot water did not melt nearly so much as it should according to Richmann's law. He thereupon sought a new rule, by a new method. Having mixed hot water at temperature t_1 with melting snow and measured the resultant temperature T, he computed the difference between T and R, the final temperature to be expected from Richmann's law if water at 0°C had been used in place of snow. In the simplest case, all masses being equal, $R - T = 36$ and 3/28 degrees. Hence, as Wilcke concluded, it required somewhat more than 72° of heat to melt unit mass of snow at 0°C. He observed that these 72° must disappear, or as we would say become latent, in liquefying the ice, and that liquefaction occurs without change of temperature.[63]

Unknown to Wilcke, some ten years earlier (c. 1761) his discovery had been made by Joseph Black, then professor of chemistry at the University of Glasgow.[64] Black measured latent heat, as he called it, by exposing diverse mixtures of ice and water in a lecture room maintained at the comfortable Scots level of 47°F. He measured the time, and so the relative quantity of heat, that each mixture took to climb from its initial temperature of 32°F to that of the ambient air. Subtracting out the water equivalent of the containers, he calculated that the amount of heat required to melt unit mass of ice would heat the same mass of water from 32° to 173°F, or by some 78° on Wilcke's centigrade scale.[65] It is noteworthy that Black's line of thought was probably much the same as Wilcke's: he began with Fahrenheit's observation, as recorded by Boerhaave, that supercooled water warms on freezing, and inferred that, in solidifying, ice-cold water gives up heat without change of temperature.[66] In his lectures he represented this conclusion as obvious and commonsensical; for if, he said, ice immediately becomes fluid when the temperature rises above 32°F, we should have great spring torrents and floods, which would 'tear up and sweep away everything, and that so suddenly that mankind should have great difficulty to escape from their ravages.'[67] Whatever the merits of this

61. *NCAS*, 4 (1752–3), 241–2.

62. Wilcke, *AKSA*, 31 (1769), 87–108; Oseen, *Wilcke* (1939), 156, 174–8.

63. Wilcke, *AKSA*, 34 (1772), 93–116. Cf. McKie and Heathcote, *Discovery* (1935), 78–94.

64. Black's doctrine dates from 1757–8, his experiments from 1761, but he published nothing about either. McKie and Heathcote, *Discovery* (1935), 35; Black to Watt, 15 March 1780, in Robinson and McKie, *Partners* (1970), 83–4.

65. McKie and Heathcote, *Discovery* (1935), 17–20.

66. Guerlac, *DSB*, II, 177.

67. Quoted by McKie and Heathcote, *Discovery* (1935), 16.

argument, it was after the fact: both Black and Wilcke came to discover latent heat not by meditating about geophysical catastrophes, but by following up a quantitative discrepancy detected through the promiscuous use of the mercury thermometer. It may also be pertinent that, in contrast to the kineticists Krafft and Richmann, Black and Wilcke took heat to be an expansive substance capable of combining with matter, and thereby altering its state.[68]

From latent heats Black moved to specific ones, again by following up quantitative data from Fahrenheit and Boerhaave, who had satisfied themselves of what amounts to Richmann's law in the case of a mixture of equal volumes of the same substance.[69] (They worked with volumes rather than masses owing to Boerhaave's theory of the uniform distribution of heat.) In the case of mercury and water, however, the temperature of the mixture always fell out closer to the initial temperature of the water than the law allowed; to save the law it was necessary to mix three parts of mercury to two of water. Now Boerhaave held that this experiment confirmed his theory of heat distribution against the more common view that bodies hold heat in proportion to their mass; and indeed Fahrenheit's finding (2:3) was more favorable to Boerhaave than to his opponents, who should have expected about 1:13.[70] But Black refused to consider 2/3 equal to one; and when he compared Fahrenheit's result with measurements by the physician George Martine (1740), which showed that mercury both heated and cooled more quickly than an equal bulk of water, he concluded that bodies have different capacities for heat.[71] That solution may be regarded as instrumentalist. It saved a quantitative discrepancy at the price of implying unspecified connections between the matter of heat and the internal arrangements or chemical composition of bodies.[72]

Once again the parallel to Wilcke is striking. Boerhaave's theory of heat distribution appears in Musschenbroek's *Elementa physicae*, which came out in Swedish in 1747, together with notes by the translator, Samuel Klingenstierna, the first professor of physics at the University of Uppsala. Klingenstierna criticized the distribution law; Wilcke, who had been Klingenstierna's student,

68. Wilcke, *AKSA*, 34 (1772), 101, 107, and 2 (1781), 53–4. Black thought the fluid theory more probable than the kinetic because, e.g., it easily assimilated latent heat to chemical combination, which Lavoisier and Laplace (*MAS* [1780], 359) also took to be its strongest feature. One suspects that Black obtained more guidance from the material theory of heat than he allowed; cf. his famous researches on 'fixed air' (1753), a springy fluid of Hales' type, which alters the state of the bodies with which it combines, which can be recovered by heating, etc.

69. Boerhaave appears to say (*New Meth.* [1741], I, 290) that the final temperature T is $(t_1 - t_2)/2$; he obviously means that T lies $(t_1 - t_2)/2$ above t_2, not above zero. But cf. McKie and Heathcote, *Discovery* (1935), 76–7.

70. Boerhaave, *New Meth.* (1741), I, 290–1. The ratio should have been about 0.45 rather than 0.67.

71. McKie and Heathcote, *Discovery* (1935), 12–15; Guerlac, *DSB*, II, 178–9.

72. Richmann, *NCAS*, 3 (1750–1), 309, 323, 332–3, reports the rapid heating and cooling of mercury, points out the conflict with accepted theory, and tries to relate heat capacity (his term) to the configuration of particles and pores. But it is plain from *NCAS*, 4 (1752–3), 241–2, that he has not fully grasped the concept of specific heat.

probably picked up the subject from him; and in the 1770s Wilcke demonstrated that bodies hold heat in proportion neither to volume nor to mass. He hit upon the idea of specific heat capacity, though not the name, independently of Black, and found its measure in the following characteristic way.

Immerse a mass of metal at temperature t_1 in an equal mass of ice-cold water; record the temperature of the mixture T; calculate from Richmann's law the amount of water w at temperature t_1 which, when mixed with the same quantity of ice-cold water, will yield the same resultant T; then $w = T/(t_1-T)$ is the specific heat of the metal relative to that of water.[73] Wilcke probably had obtained w for gold and for lead before 1780, when he first learned of Black's work; he then measured it for ten other substances. Since, as in his measurements of the latent heat of fusion, he ignored the heat capacity of the calorimeter, his numerical results were not very good.[74] But the conceptual work had been done, and others more painstaking improved the measurements.[75]

These improvements in the theory of heat illustrate conceptual advance inspired by introduction of instruments and the consequent discovery of quantitative discrepancies between the results of measurement and the predictions of theory. Our second example, the quantification of magnetic force, offers improvement in measurement as a consequence of a previous clarification of concepts.

Magnetism Among pressing unfinished Newtonian business in the early eighteenth century was the establishment of the 'law' of magnetic force. Several important early disciples of Newton, particularly his assistants Francis Hauksbee and Brook Taylor, and the ever-inquisitive Musschenbroek, accordingly undertook to obtain by experiment a magnetic analog to the law of gravitation. Their procedure is instructive, and perhaps a little comforting to those who still fear physics; for it reveals that these Newtonian hierophants had failed to understand the foundations of their doctrine.

Hauksbee and Taylor placed a large lodestone so that its poles sat on an east-west line directed towards the center of a small compass needle; they measured the angle $\phi(d)$ between the needle and the magnetic meridian for several values of d, the distance between the needle's center and the closest pole of the lodestone; and they tried to find a value of n such that ϕ decreased as d^{-n}. They hoped that the procedure would give 'the Proportion of the Power of the Lode-

73. Oseen, *Wilcke* (1939), 232–4, 247–8.

74. Wilcke, *AKSA*, 2 (1781), 48–79; McKie and Heathcote, *Discovery* (1935), 95–108. Wilcke learned of Black in Magellan's *Nouvelle théorie du feu* (1780), which ascribes the discovery of latent heat to Wilcke on the ground of first publication. See the letters from Watt to Magellan in Robinson and McKie, *Partners* (1970), 80–1, 85–8.

75. E.g., Lavoisier and Laplace, *MAS* (1780), 373, who however obtained a poorer value for the latent heat of fusion with their famous ice calorimeter than had Black (75 as against 78); and Johan Gadolin, who got good values for the specific heats of some metals by scrupulously including the water equivalents of his apparatus and by reading temperatures to tenths of a degree. McKie and Heathcote, *Discovery* (1935), 108–15.

stone at different Distances' (Hauksbee) or 'the Law of Magnetical Attraction' (Taylor).[76] But it did not succeed: $\phi(d)$ went as d^{-2} at short distances and as d^{-3} further out.

And why should one expect any simple relationship between d and ϕ ? Why choose them and not, say, a trigonometric function of ϕ and the distance between centers, as variables? Compare Newton and the gravitational attraction between two bodies. Posit first an undetectable reciprocal r^{-2} attraction between every pair of *elementary* particles in the bodies. Derive thence the mathematical consequence that, should the bodies have spherically symmetric distributions of mass or be separated by distances very large compared to their diameters, each would experience a macroscopic force acting at its center, along the common line of centers, and decreasing as the square of the distance between centers. Hauksbee and Taylor did not start with an elementary magnetic force, and had no way to specify measurable variables of interest.

Taylor thought it his business to find, by experiment, 'what point within the Stone, and what point in the Needle, are the Centers of the Magnetical power.' In a word, he sought 'force' as an effect, and indeed as an effect of certain needles and lodestones, as contemporary Newtonian apologetics appeared to recommend. Had it succeeded, the hunt for centers would have allowed a formulation of the 'law of magnetism' that excluded or avoided postulating force-causes, or hypothetical interactions between microscopic entities. Unfortunately the best place for the lodestone's suppositious center fell outside its figure. 'From Whence it seems to appear, that the power of Magnetism does not alter according to any particular power of the distances.' Despite Taylor's failure, Daniel and Jean Bernoulli, in their prize-winning essay on magnetic theory, later advocated precisely the same search for 'what can be called in some sense centers of force.' They assumed that the law of distance was that of inverse squares, and proposed to find the two centers by experiment. The self-consciously Cartesian Bernoullis could propose an experiment identical in purpose to that of the Newtonian Taylor because both parties understood force to mean macroscopic effect.[77]

Consider next the flailings of Musschenbroek, who first took on magnetism excathedra in an elaborate dissertation published in 1729. He opens apologizing: 'You may well be amazed at an author who writes about a phenomenon [magnetic attraction] of whose cause he confesses himself ignorant.' But to do otherwise would be to run to hypotheses, which is to say fables, or mere opinion; like Newton, who rejected hypotheses, Musschenbroek will limit himself to careful description.[78]

The first property of the magnet needing attention is the 'proportion of the attractive forces at different distances.' On Christmas eve, 1724, Musschenbroek suspended one spherical magnet above another and measured the weight

76. Hauksbee, *PT*, 27 (1710–12), 506–11; B. Taylor, *PT*, 29 (1714–6), 294–5.
77. B. Taylor, *PT*, 31 (1720–1), 204–5; D. and J. Bernoulli, AS, *Pièces*, 5 (1748), 140.
78. *Dissertatio* (1754²), 7–8.

W required to counterbalance the attraction (fig. 1.5) as a function of the interval *d* between the magnets, not the distance between their centers. When collected into tables these numbers—as well as others obtained with other pairs of magnets—revealed nothing, or rather, as Taylor found, they showed that 'nulliam dari proportionem virium in diversis distantiis.' The most promising result, that the attractions seemed to diminish as the curvilinear volume between the balls, had no obvious interpretation. Krafft, after confirming Musschenbroek's measurements, threw up his hands: 'I cannot divine to what cause this proportion could owe its origin.' The possibility that the method might not meet the task did not occur to either of them. To be sure, Musschenbroek saw that the forces he measured were compounds of several attractions and repulsions, but he could not contrive a means to unscramble them. 'In such darkness it is best to suspend judgment, and to relate our observations [for the use of] a wiser and more serious age.'[79]

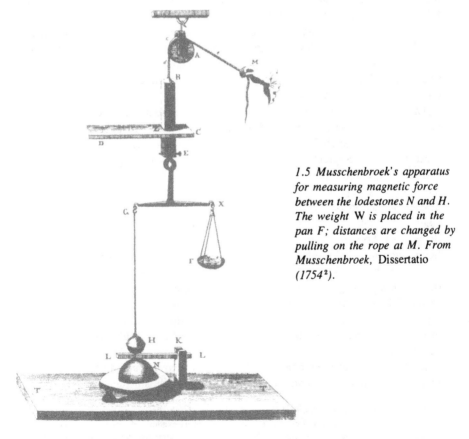

1.5 Musschenbroek's apparatus for measuring magnetic force between the lodestones N and H. The weight W *is placed in the pan F; distances are changed by pulling on the rope at M. From Musschenbroek,* Dissertatio *(1754²).*

79. *Ibid.*, 17, 26–7, 37–8; Musschenbroek, *PT*, 33 (1724–5), 372; Krafft to Wolff, 29 April 1740, in Wolff, *Briefe* (1860), 214. Cf. W. S. Harris, *Rud. Mag.* (1872²), 190–5.

Musschenbroek knew that Newton had found magnetic attraction to decrease roughly as the cube of the distance. 'Would that the experiments from which Newton gathered this result had been recorded! For perhaps that man of stupendous subtlety in mathematics found a way to segregate attractions and repulsions, the proportion of which he found to decrease as the third power of the distance.'[80] Even here Musschenbroek did not grasp the point; for he took Newton's achievement to have been the unscrambling of the *macroscopic* attraction exercised by one magnet on another from the simultaneous repulsion. He never did recommend the search for a law between magnetic elements, although Newton had pointed the way clearly enough.[81]

In his later texts Musschenbroek limited himself to reporting $W(d)$ for many pairs of magnets, and to working out spurious 'laws of attraction' for each arrangement. As for the macroscopic repulsive force, it proved more elusive than the attractive. Musschenbroek measured it by the weight $W'(d)$ required on the side G of his balance to counter the force between magnets with like poles facing. He found W' to be less than W and even negative (needed on side X of the balance) at very small d, when the repulsion sometimes became attraction.[82] It was of course hopeless to look for a 'law of repulsion' without distinguishing this last effect, a change in the strength of the magnetic elements (magnetic induction), from the property sought, the diminution with distance of the force of a magnet of fixed strength.

The cleansing of the subject was begun by continental mathematical physicists eager to master and to extend the methods of the *Principia*. Between 1739 and 1742 two French Minims, Thomas Le Seur and François Jacquier, both professors at the Sapientia in Rome, issued Newton's masterpiece enriched with many notes, some theirs, others contributed by J. L. Calendrini, professor of mathematics and philosophy at the Académie de Calvin (University of Geneva). The book has an odd pedigree: an anti-trinitarian author, Franciscan editors, Calvinist collaborators, and, as protector and dedicatee, the Anglican Royal Society of London; a miscegenation later compounded by the Presbyterian Scots, who reissued the book and attributed it to the Jesuits. The note on magnetism, probably Calendrini's, was intended to clear Newton's 'rough' result (the force of a magnet diminishes as r^{-3}) from the doubts raised by Musschenbroek.[83] Calendrini did not use a balance, but a needle (of length $2a$ = SN in fig. 1.6); and he developed his theory for the simple case that s = CM, the distance from the center of the magnet to that of the needle, is very large in comparison with a.

Let the 'force' of the magnet — the total force compounded of the actions of its poles — exerted upon an element dm of the needle a distance cM = r away

80. Musschenbroek, *Dissertatio* (1754²), 39. Newton, *Math. Princ.* (1934), 414.

81. *Ibid.*, 415: 'Magnetic and electric attractions afford us example of this [an integrated force, like gravity]; for all attraction towards the whole arises from the attractions towards the several parts.'

82. Musschenbroek, *Essai* (1739), I, 279–82; *Cours* (1769), I, 433–6.

83. Newton, *Phil. nat.*, III (1742), 39–42, tr. in Palter, *Isis*, 63 (1972), 552–8.

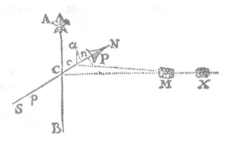

1.6 Calendrini's measurement. SN is the needle, M and X positions of the magnet, AB the magnetic meridian. From Palter, Isis, 63 (1972), 544–58.

be $Mf(r)dm$. If x is the distance Cc of dm from the center C, the turning moment of this force about C is $dT = Mf(r)dm \cdot x\sin\theta$, where $\theta = \angle$McN; and the total moment, $\int_{-a}^{a} dT$, must equal the total amount exerted by the earth's field H, $\int_{-a}^{a} Hdm \cdot x\sin\phi$, where $\phi = \angle$ACN is the departure of the needle from the magnetic meridian. In the extreme case $s \gg a$, $s \simeq r$ and $\theta \simeq \pi/2 - \phi$, whence $f(r)$, the quantity sought, is proportioned to $\tan\phi$. Calendrini found $f(r) \sim r^{-3}$, which he took to be the 'law' of magnetic action.[84] Note that he did not have the elementary law, but a compounded form appropriate only to a magnetic dipole at large distances. Calendrini apparently did not recognize these limitations.

Good steel bar magnets became available soon after Calendrini found the law of dipoles. These 'artificial magnets,' stronger and more uniform than lodestones, lent themselves to investigations of the elementary law of magnetic interaction. The first person to assert that magnetic poles interact according to the law of squares seems to have been John Michell, a Cambridge mathematician and one of the inventors of the method of making artificial magnets. Michell did not demonstrate how he deduced the law or that it saved the phenomena.[85] Nor did Tobias Mayer, professor of applied mathematics at the University of Göttingen, when, in 1760, ten years after Michell, he announced the same result.

The report of Mayer's work shows that he understood the error of the Musschenbroeks. Magnetism, he said, should be approached just as Newton did gravity: admit the 'force of a single part' of the magnet; do not worry about its cause, whether vortical 'or something worse;' measure its macroscopic effects, and secure laws valid for all magnetic bodies. Proceeding thus, we are told, Mayer accounted for all the phenomena 'with mathematical precision.'

84. Calendrini does not proceed quite so neatly, for his experimental arrangement admits values of $s \approx 3a$. He handles such cases by supposing that $dm = kxdx$ and by neglecting the change in θ with x, which allows him to define a magnetic center at P (CP = $2a/3$); the balancing of moments now gives $f(r) \sim \sin\phi/\sin\theta$, θ = angle MPN.

85. Michell, *Treatise* (1750), 17–19; *Traités* (1752), cxix–cxx. Cf. Hardin, *Ann. Sci.*, 22 (1966), 27–9.

He required only an inverse-square force between elements and the assumption that, in a bar magnet, the intensity of magnetism of each element is proportional to its distance from the geometrical center.[86]

Mayer's manuscripts on magnetism, once thought lost, have recently been published. They confirm the public announcement of his work and reveal its technique. Mayer cut through the difficulty of obtaining the elementary law of distance $f(r)$ and of magnetic strength $\nu(x)$ from composite measurements by postulating that $f \sim r^{-2}$ and that $\nu \sim x$, x being the distance from the center of an artificial bar magnet to any cross-section. On these assumptions he computed the angles $\theta(d)$ at which two magnets m and M, aligned as in fig. 1.7, would just part from one another. The measured values of θ serve as an indirect check of the conjectures behind the computations. Mayer found a good fit when he assumed that only 'symmetrical' parts of the magnets interacted, by which he meant parts distant from their respective centers by x_m and x_M, where $x_m/x_M = a_m/a_M$, a representing the length of a magnet.[87] His treatment was ruthlessly instrumentalist; his 'criterion of truth,' a successful mathematical theory. The phenomena of universal gravitation, he said, 'would not be better or more simply explained even if their ultimate causes were known.' The old Cartesian vortex hypothesis may or may not be false: what is clear is that it is 'useless and inept.' 'Nature would seem to have hidden these causes for the very reason that knowledge of them serves no useful purpose, thus reminding us once again of the truth of the dictum that Nature does nothing in vain.'[88]

The announcement of Mayer's 'law' called forth an instructive criticism from Aepinus, a gifted mathematical physicist who had just published a theory of electricity and magnetism of the first importance. He warmly endorsed Mayer's Newtonian goal, the search for quantitative laws. The search for causes had proved vain, he said; 'those who devote themselves to it, taking a cloud from Juno, concoct systems of dreams in the name of theories.' Furthermore, accord-

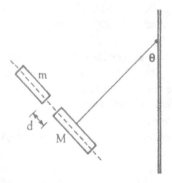

1.7 Mayer's technique for obtaining a law of magnetic force.

86. *GGA* (1760:1), 633–6; *GGA* (1762:1), 377–9; Kästner to Lambert, 13 Dec. 1769, in Bopp, Ak. Wiss., Heidelberg, *Sitzb.* (1928:18), 19, 21–3.
87. Mayer, *Unp. Writ.* (1972), III, 68–79, 83, 86.
88. *Ibid.*, 64–7.

ing to Aepinus, Mayer proceeded correctly in looking for laws of interaction between magnetic elements. But his laws were false: the magnetic center of a bar magnet does not always coincide with its geometrical middle; the intensity of magnetism of an element is not proportional to its distance from either center; and the force between elements cannot be of the gravitational form, which does not allow for induction.[89] Aepinus, who knew exactly what to look for, had no idea how to find it.

Not so Lambert, who admired Mayer as a 'genius of the first rank,' and who despised those who experimented at random, generating useless data, 'like most of that of a celebrated Leyden professor.'[90] Lambert, like Aepinus, advocated Newtonian instrumentalism: 'The example of gravity clearly shows that mathematical knowledge of the objects and processes of Nature depends only slightly on physical knowledge, and that the former can be advanced quickly and expanded wonderfully if the latter is always kept within narrow bounds.'[91] Magnetism has escaped Newtonianization because the macroscopic force between magnets is an integral of the elementary forces we seek but cannot isolate. We must therefore work with composites.

Lambert uses Calendrini's apparatus arranged so that the axis of the magnet need not be perpendicular to the magnetic meridian (fig. 1.8). He assumes that the 'force' of the magnet and of the earth on the needle are $Mf(r)g(\phi)$ and $Hg(\omega)$, respectively, M being the magnet's strength, H the earth's, f and g unknown functions, r an unspecified distance dependent upon SC, the separation of the south pole of the magnet from the center of the needle. Lambert's deduction that g = sine shows him at his best. He found ω and ϕ for two different positions of the magnet at the same distance SC (for example at d and L in fig. 1.8); since $Hg(\omega) = Mf(r)g(\phi)$, g could be deduced from $g(\phi_1)/g(\phi_2)$ = $g(\omega_1)/g(\omega_2)$, the subscripts indicating the two positions of equal r. Lambert recognized that the happy result was an average, since the obliquity of the magnet's action differed from point to point along the needle. As for f, Lambert, like the Bernoullis, *assumed* it to be inverse-square (surely a blemish in his method), whence r = const.$(\sin\phi/\sin\omega)^{\frac{1}{2}}$. But what to take for r? Lambert plotted the measured distances d from the south pole of the magnet to the north pole of the needle against $(\sin\phi/\sin\omega)^{\frac{1}{2}}$, and gathered that, to within 7 percent, r differed from d by more than the length of the needle! Hence the 'confirma-

89. Aepinus, *NCAS*, 12 (1766–7), 325–40. Mayer is defended by Hansteen, *Untersuchungen* (1819), 290–3, and, on the basis of the Mss., which show that Mayer realized the limited applicability of his work, by Forbes in Mayer, *Unp. Writ.* (1972), III, 9–11. For the problem of $v(x)$ see W. S. Harris, *Rud. Mag.* (1872²), 231–6.

90. Lambert to Mayer, 2 March 1772, in Lambert, *Deut. gel. Briefw.* (1781), II, 433; Lambert to Euler, 4 April 1760, in Bopp, Ak. Wiss. Berlin, Phys.-Math. Kl., *Abh.* (1924:2), 13. Cf. Lambert, *MAS*/Ber (1766), 26, and *Deut. gel. Briefw.*, II, 25, recording Musschenbroek's patronizing of Lambert.

91. Lambert, *Photometrie* (1760), §5. Cf. Lambert to Holland, 25 Sept. 1769, in Lambert, *Deut. gel. Briefw.* (1781), I, 325.

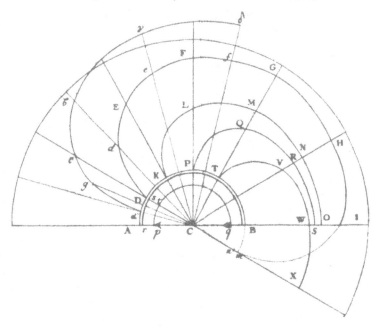

1.8 Lambert's method. The magnet is placed at points in the plane with its south pole S facing C and its axis on the line SC. Each curve is the locus of points S for which the needle takes a constant declination ω. For example, DdEdFfGHI shows the positions of the south pole of the magnet for which ω = 30°. Since the axis of the magnet lies along DC, dC, etc., one can read from the diagram the angular separation φ of the magnet's axis and the needle for the various configurations. From Lambert, MAS/Ber (1776), 22–48.

tion' of the inverse-square implied that the effective pole of the needle lay outside it.[92]

The 'force' Lambert sought, the integrated effect of the magnet on the needle, follows r^{-3}, not r^{-2}; and r is the distance between the centers CC' of the magnet and the needle, not a function of d. In fact his numbers satisfy $f = CC'^{-3}$ as well as his own law.[93] Ever on his guard against arbitrary assumptions, always developing his mathematical accounts in the widest generality, Lambert was nonetheless traduced by a vulgar physical theory. He preferred

---

92. Lambert, *MAS*/Ber (1766), 22–48. Note that the result, $f = f(d)$, undercuts the deduction of g, which depended on pairs of points at equal distances SP. Harris, *Rud. Mag.* (1872²), 196–209, gives a fair account of Lambert's work.

93. Hansteen, *Untersuchungen* (1819), 155–6, successfully obtains an integrated r^{-3} for Lambert's case using an elementary r^{-2}, and extravagantly praises his work ('Meisterstücke des Scharfsinns,' the only guide from the labyrinth [p. 294]), while admitting the handling of f is 'less satisfying' than that of g (p. 299).

Euler's plenum to Newton's void, attributed gravity to pressure, and conceived magnetic force to arise in the Cartesian manner, from a flow of magnetic aether.[94] He thought of the integrated magnetic force as a material current the density and power of which diminished, by geometrical necessity, as the square of the distance from its source.[95] As for a law for magnetic elements, Lambert rightly observed that it would be very difficult to deduce one from his integrated form.[96]

The foregoing will give a measure of the achievement of Coulomb, a Newtonian applied mathematician and an implacable foe of Cartesian explanations.[97] He had three advantages over his predecessors: a clear if crude representation of magnetism, a new method of measuring force, and, above all, long thin artificial magnets with well-defined poles. These magnets, two feet in length, of excellent steel, magnetized according to Michell's method as improved by Aepinus, acted as if their magnetism was concentrated about 5/6 inch from either extremity. Coulomb explained that their magnetic fluids, the cause or carrier of magnetic force, were confined to small regions; and he made the goal of his measurements the hypothetical force between elements of the fluids. The length and concentrated power of his magnets gave him a good approximation to isolated poles of strengths proportional to the quantity of the presumed magnetic fluid.

Coulomb suspended one such magnet horizontally in its magnetic meridian by a wire of known torsion and placed another vertically in the meridian with its north pole occupying the position from which it repelled that of the first (fig. 1.9). He had found that the force with which a twisted wire strives to unwind is proportional to its angle of twist, or torsion. In the case shown (fig. 1.10), the repulsion of the north poles balances the earth's field $H\sin\phi$ and the torsion $b\phi$, or, for small angles, a total force $\phi(H+b)$. If one now twists the dial clockwise, by, say, θ, the hanging magnet will rest at a new angle, ϕ, urged toward the meridian by a force $(H+b)\phi + b\theta$. Suppose the magnetic repulsive force to be c/r^2, where $r = 2a\sin\phi/2$ is the distance between the north poles. Then, for small ϕ,

$$c/a^2\phi^2 = (H + b)\phi + b\theta,$$

an equation Coulomb found to hold very nearly. The assumed law also gave the observed orientation of a compass needle exposed to the action of the long

94. Cf. *Photometrie* (1760), §18, promising to treat the Euler (wave) and Newtonian (particle) theories of light equally, and *MAS*/Ber (1766), 50–1, on the nature of lines of magnetic force.

95. Lambert to Euler, 15 Jan. 1760 and 6 Feb. 1761, in Bopp, Ak. Wiss., Berlin, Phys.-Math. Kl., *Abh.* (1924:2), 10, 18; *MAS*/Ber (1766), 32–5, 37, 54, 67. Cf. Daujat, *Origines* (1945), III, 482–7.

96. *MAS*/Ber (1766), 48; cf. *MAS*/Ber (1766), 73–7.

97. Coulomb, AS, *Mém. par div. sav.*, 9 (1780), §4 (Coulomb, *Mémoires* [1884], 9): 'Pour les [magnetic phenomena] expliquer, il faut nécessairement recourir à des forces attractives et répulsives de la nature de celles dont on est obligé de se servir pour expliquer la pesanteur des corps et la physique céleste.' Cf. Gillmor, *Coulomb* (1971), 176–9, 193–4.

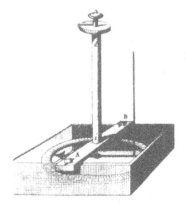

1.9 Coulomb's magnetic torsion balance. The wire of suspension, which runs in the housing di, can be twisted by the graduated knob; the vertical rod is the long thin magnet. From Coulomb, MAS (1785), 578–611.

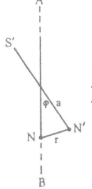

1.10 Coulomb's measurement. AB is the magnetic meridian, N the north pole of the vertical magnet, N'S' the suspended needle.

magnet by direct computation and summation of the *four* elementary interactions between pairs of poles.[98]

The results of the hunt for magnetic force are summarized in Table 1.2. Through most of the eighteenth century 'law of force' was an ambiguous expression, and the means of measuring any given force far from obvious. The complexities and false starts sketched for the magnetic case recur, redoubled, in the electrical; for it took longer to divine the appropriate conceptions in the apparently capricious phenomena of electricity.

98. Coulomb, *MAS* (1785), 589–90, 601–11 (*Mémoires* [1884], 138–45). Hansteen, *Untersuchungen* (1819), 306, criticizes Coulomb for not estimating the error incurred by ignoring the south poles; which, however, is easily done, and amounts to less than one percent in the least favorable case.

TABLE 1.2

Efforts to Obtain a 'Law of Magnetic Force,' 1710–1785

Date	Investigator	Method	'Law' as measured	Understood as integrated?	Sees need for elementary law?	Explicitly instrumentalist?	Uses artificial magnets?	Gets elementary r^{-2}?
1710/15	Hauksbee/ Taylor	needle	found none	no	no	no	no	no
1729	Musschenbroek	balance	found none	barely	no	no	no	no
1742	Calendrini	needle	r^{-3}	yes	no	no	no	no
1750	Michell	—	—	yes	yes	no	yes	as guess
1760	Mayer	gravity	r^{-2}	yes	yes	yes	yes	yes
1766	Lambert	needle	r^{-2} (wrong)	yes	yes	yes	no	no
1785	Coulomb	torsion	r^{-2} (right)	yes	yes	yes	yes	yes

CHAPTER II

The Physicists

We shall later examine the work of some 210 electricians active between 1600 and 1790. They make up two-thirds of all those then writing on electricity whose publications are noticed in the catalogs of the world's great collections.[1] They include everyone who made significant contributions to the understanding of electrical phenomena.

These early electricians may be divided into five groups according to their chief means of support: members of religious orders; paid academicians; professors; public lecturers; and 'others,' primarily artisans, practicers of professions (doctors, lawyers, ministers), and the independently wealthy. The results appear in Tables 2.1 and 2.2. Numbers in parentheses refer to our 210 electricians; the others, to our electricians plus additions from the catalogs. (Since people active in more than one time period are counted more than once, the total of the numbers in parentheses exceeds 210.) The sum S of the first four groups, A, B, C, D, is always greater than the number of 'others,' E: S/E ranges from 1.2 to 3.0, and, in each time period, is about the same whether calculated for all electricians or for our 210. We may therefore be confident that most early writers on electricity fall into groups A through D. Hence the rationale for taking these groups as 'the physicists:' they are identifiable, roughly homogenous, and they predominated, at least in the study of electricity.

This dominance is imperfectly indicated by the numbers. Except for certain unsalaried Fellows of the Royal Society of London, members of groups A through D were always the leaders, and those in E the followers. Paid academicians in particular made substantive contributions much out of proportion to their numbers. That the British did not conform to the general pattern is easily explained. Britain had neither Jesuits nor paid academicians, and interest in experimental physics declined at her universities during the middle of the eighteenth century, when study of electricity flourished at Continental institutions. A consequence and measure of this disparity are the strong showing of the British in Table 2.2.

The gross temporal variation in the numbers in the tables may be roughly accounted for as follows. Electricity did not claim the attention of many physi-

1. Compiled from Ekelöf, *Cat.* (1964–6), Frost, *Cat.* (1880), Gartrell, *Elect.* (1975), Rossetti and Cantoni, *Bibl.* (1881), and Weaver, *Cat.* (1909), omitting those who (a) wrote only on medical electricity or on the installation of lightning rods, or (b) are known for but a single article less than five pages in length, or (c) could not be identified. Members of this last group are represented in the catalogs by only one publication each.

TABLE 2.1

Electricians by Profession, 1600–1789

	1600–99	1700–39	1740–49	1750–59	1760–69	1770–79	1780–89
A. Jesuits	7(7)		6(2)	6(4)	6(2)	6(2)	1(1)
Univ. profs.	4(4)		4(1)	1(1)	1(1)	4(2)	
College profs.	1(1)		2(1)	4(2)	4(1)	2(0)	1(1)
B. Academicians[a]	3(3)	1(1)	4(4)	10(10)	4(4)	5(5)	10(7)
Big three[b]	1(1)		4(4)		3(3)	3(3)	5(4)
Others	2(2)	1(1)		10(10)	1(1)	2(2)	5(3)
C. Professors[c]	4(4)	10(10)	31(24)	22(16)	16(7)	29(15)	25(17)
Universities	4(4)	9(9)	24(20)	18(14)	11(5)	23(11)	15(12)
Colleges		1(1)	7(4)	4(2)	5(2)	6(4)	10(5)
D. Lecturers	3(3)	4(4)	4(4)	2(2)	2(2)	3(2)	3(3)
E. Others	10(10)	5(4)	43(32)	32(23)	19(17)	28(15)	40(26)
Britain	5(5)	5(4)	22(18)	14(13)	11(10)	15(12)	13(12)
Elsewhere	5(5)		21(14)	18(10)	8(7)	13(3)	27(14)
TOTALS	27(27)	20(19)	88(66)	72(55)	47(32)	71(39)	79(54)

[a] Salaried only, except for associés at AS.
[b] Paris, Berlin, Petersburg.
[c] Exclusive of Jesuits.

TABLE 2.2

Breakdown of 'Others' from Table 2.1

	1600–99	1700–39	1740–49	1750–59	1760–69	1770–79	1780–89
Artisans[a]		1(1)	12(9)	5(4)	4(3)	7(5)	7(4)
Britain[b]			10(8)	4(4)		6(5)	3(3)
Elsewhere		1(1)	2(1)	1(0)	4(3)	1(0)	4(1)
Law, Govt. or Military Service	1(1)		9(7)	9(4)		5(2)	9(4)
Practicing M.D.	3(3)	1(1)	8(5)	4(4)	3(2)	6(2)	8(5)
Regular Clergy[c]		1(1)	6(4)	5(3)	5(5)	2(1)	5(4)
Britain[b]			4(3)	1(1)	3(3)	1(1)	2(2)
Elsewhere		1(1)	2(1)	4(2)	2(2)	1(0)	3(2)
Independent	4(4)	1(0)	6(5)	8(7)	6(6)	5(5)	4(4)
Britain[b]	2(2)	1(0)	3(2)	6(6)	4(4)	4(4)	2(2)
Elsewhere	2(2)		3(3)	2(1)	1(1)	1(1)	2(2)
Others	2(2)	1(1)	2(2)	1(1)	1(1)	3(0)	7(5)
TOTALS	10(10)	5(4)	43(32)	32(23)	19(17)	28(15)	40(26)

[a]Includes instrument makers, apothecaries, engineers.
[b]Includes American colonies.
[c]Includes dissenting clergy.

cists until the invention of the electrical flare and the Leyden jar in the 1740s: these novelties brought five times as many people to study or play with electricity in the decade 1740–9 as had done so in the preceding thirty years. A decline in numbers occurred in the 1750s as the novelty wore off; that they fell by only 20 percent is owing to the restimulation of interest by the demonstration of the identity of lightning and electricity. The great drop in the 1760s corresponds to a general slump in the cultivation of the arts and sciences, caused in part by the Seven Years' War. The revival in the 1770s and 1780s owed much to the inventions of Volta.

1. JESUITS

Knowledge about electricity was kept alive during the seventeenth century by Jesuit polymaths. They also enriched the subject with valuable observations. In the eighteenth century the relative significance of the direct contributions of the Society sharply declined. It continued to be important indirectly, however, as Schoolmaster to Catholic Europe. One indication of its effectiveness is that it educated at least 20 percent of the 195 members of the Paris Academy of Sciences honored by éloges in its *Histoire* during the Ancien Régime.[2]

THEIR MAGISTERIUM

The reputation of the Jesuits as teachers dates from the foundation of their order. In the 1550s they received more invitations to establish schools than they could handle.[3] 'It is better for a town to found a college of Jesuits than to build highways or harbors,' the French proverb runs. These colleges taught poor children along with rich, opposed the spread of heresy, attracted and produced civilized and educated people, and—not to be ignored—brought in fresh money in the form of student expenditures for food and lodging.[4]

The Society's pedagogical principles, as set forth in its *Ratio studiorum* of 1599, guided the organization and conduct of all Jesuit schools in the seventeenth and eighteenth centuries. The system had many strengths: a steady curriculum, proof against educational faddists but adaptable to local needs; well-trained and educated teachers; dependability, regularity, punctuality and, above all, experience.[5] Both friends and foes of the order praised its schools. Ranke: 'Students learned more from the Jesuits in six months than from others in two

2. Information from Charles Paul, who is preparing a study of the éloges.

3. One example, that of the province of Austria (including Bohemia) will illustrate the multiplication of Jesuit colleges. The first foundation was at Vienna in 1551; 100 years later (1663) Austria had 15 colleges; another century (1767) and it had 38. Paulsen, *Gesch.*, I (1896), 403. There were 113 colleges in France in 1762 (Morney, *Origines* [1947²], 171), and 107 in Austria and Germany in 1750 (Pachtler, *Ratio*, I, ix–xx).

4. Chossat, *Jésuites* (1896), I, 103, 112–13.

5. H. Weber, *Gesch.* (1879), 87–91.

years.' Bacon: 'As for what pertains to pedagogy, it will be most briefly said: consult the schools of the Jesuits, nothing in use is better . . . If only they were ours.'[6]

The entire course of liberal studies took eight years, of which the first five were devoted to ancient languages and literature. Then came three years of 'philosophy:' logic and divisions of science in the first year; 'physics' (in the large Aristotelian sense) and mathematics in the second; and metaphysics, ethics and psychology in the third.[7] The scheme varied from place to place. At some of the larger colleges, for example at La Flèche beginning in 1626, metaphysics was studied with physics and mathematics became the main subject of the third year.[8] At other, smaller colleges, the philosophy course was reduced to two years, 'to save time,' or, as became necessary in German schools in the eighteenth century, to drop outmoded material, particularly metaphysics, from the curriculum.[9]

The 'physicists' (the second-year philosophy students) read Aristotle's *Physica*, the first book of *Generation and Corruption*, the *Meteorologica*, and, very briefly, *De Caelo*, treating only 'a few questions about the elements, and about the heavens only regarding its substance and influence.' The closer study of astronomy was the province of the professor of mathematics, who taught Euclid and 'something of geography or of the sphere or other matters that students like to listen to.'[10] The *Ratio* further provided that private instruction in 'mathematics,' which could include much of physical science, should be available to students with talent for the subject.

This important and forward-looking provision came at the urging of Christopher Clavius, the distinguished mathematician at the central Jesuit university, the Gregorian College in Rome.[11] It took almost a century to implement. The order had first to seek outside help[12] while Jesuit novices tried to develop a taste for mathematics. A crash program, planned at Rome, had Clavius giving a three-year course to ten students, who would go forth, teach, and populate the earth, or at least Jesuit colleges, with mathematicians.[13] The larger Jesuit

6. Ranke, *Popes* (1847³), 379; Bacon, *Advancement* (1915), 17; *Works*, I (1863), 445, 709. Cf. Schimberg, *Education* (1913), 521.

7. Farrell, *Jes. Code* (1938), 233; Fitzpatrick, *St. Ignatius* (1933), 131–4, 169–71. The philosophy course was not available at the smallest colleges, those with a staff of 30 and an income of 10,000# (according to regulations of 1603); philosophy first entered in 'middle' colleges, those with at least 60 religious and 15,000#. Rochemonteix, *Collège* (1889), I, 89–94. In fact the middle colleges were frequently smaller than the plan required, e.g., Montpellier in 1668 (26 religious, 7400#). Faucillon, *Collège* (1857), 59–61.

8. Rochemonteix, *Collège* (1889), IV, 27–50. At Louis-le-Grand, mathematics and metaphysics occupied the third year, physics and ethics the second; Dupont-Ferrier, *Vie*, I (1921), 181.

9. Chossat, *Jésuites* (1896), 234–6, 248–9; Ziggelaar, *Pardies* (1971), 49, 69.

10. Fitzpatrick, *St. Ignatius* (1933), 170–1, 175.

11. *Ibid.*, 130; Consentino, *Physis*, 13 (1971), 207–8.

12. E.g., Maurolico, who taught at the Jesuit college in Messina; in 1573–4 a group of Jesuits, including Clavius, was deputed to assist Maurolico to put his papers in order, and to produce a text needed by the Society. Saduto, *Arch. hist. Soc. Jesu*, 18 (1949), 133–40.

13. Pachtler, *Ratio*, II, 142–3, quoting a document of 1586.

schools of Italy and France made provision for the teaching of mathematics;[14] Germany remained behind, especially after the Thirty Years' War. Towards the end of the century, despite repeated requests from Rome for improvement, the German provinces 'scarcely had one professor who could teach the subject creditably at a large university.'[15]

The Jesuits had several reasons for emphasizing mathematics. According to a draft of the *Ratio* of 1586, it was necessary for other studies, including poetry (astronomical allusions), history (geography), politics (military technology), and ecclesiastical law (exact chronology), but above all for astronomy, navigation, architecture, and surveying.[16] These last subjects were important in the training both of Jesuits bound for foreign missions and of young aristocrats destined for high military or government service. Most 'noble youths,' says Schott, are interested in practical applications of mathematics, and come great distances to study with such Jesuit masters as himself and Athanasius Kircher.[17]

This instruction required instruments, of which the larger colleges often had a good stock: instruments for astronomy, geodesy, dialing, drawing, and, perhaps, for the demonstration or investigation of physical principles.[18] Since this apparatus catered to the interests of an influential portion of the student body, it was not considered a frill; when Schott took up a chair at the college at Würzburg he went far out of his way expressly to trade for mathematical books and instruments to replace those the college had lost during the Thirty Years' War, and to bring it in line with its better-stocked sister institution at Mainz.[19] The English knew how to appreciate what they lacked in this respect. 'This I must always affirm for the honor of my mother the University of Oxford, that if her children had the good utensils, which adorn the colleges of the Jesuits abroad, the world would not long want good proof of their ingenuity.'[20] Description of these instruments, some built to original designs, was a staple in the technical books that the Jesuits produced in profusion in the seventeenth century.[21] They were frequently asked for advice about the procurement of

14. Cf. Chossat, *Jésuites* (1896), 439; Ziggelaar, *Pardies* (1971), 76–7. Note the letter of the General to the Paris Provincial, 11 Sept. 1656, in Dainville, *RHS,* 7 (1954), 15: 'It is not without incredible sorrow that I have learned that mathematics and Hebrew are so neglected by our [priests] that there would be almost no one to teach them if we had to replace the older professors.'

15. Duhr, *Gesch.* (1921), III, 413. Cf. the efforts of the General, Tamburini, in 1724; Dainville in Taton, *Enseignement* (1964), 31–2.

16. Pachtler, *Ratio,* II, 141–2.

17. Schott, *Magia* (1671), sig.⁺⁺001v; *Pantometrum* (1660), sig. 0003r. Cf. the insistence of the city of Avignon that the Jesuits bring a mathematician, who turned out to be Kircher, into their college there in 1628; Chossat, *Jésuites* (1896), 234–6.

18. Dainville, *RHS,* 7 (1954), 6–21, 109–23. Note the special room for physics in the new college at Montpellier; Faucillon, *Collège* (1857), 88.

19. Schott to Kircher, 12 Sept. 1655, in CK, XIII, f. 49, quoted *infra,* iii.2. Cf. Duhr, *Gesch.* (1921), III, 414, for the apparatus at Ingolstadt, a tourist attraction in 1675.

20. E. Bernard to J. Collins, 3 April 1671, in Rigaud, *Corresp.* (1846), I, 159.

21. E.g., Schott, *Pantometrum* (1660); Zucchi, *Nova* (1649). Cf. the résumé of Schott's gadgets in Mercier, *Notice* (1785), 5–14.

instruments, and sometimes acted as intermediaries between purchasers and makers.[22]

A second reason for Jesuit emphasis on mathematics was that it offered a strong ground for confronting, or rather avoiding, the new physics. The mathematician, according to a dodge used by the ancient astronomers, aimed not at discovering the true principles of things, but at 'saving the phenomena,' at concocting adequate quantitative descriptions in the easiest possible way.[23] The Jesuits found this subterfuge comfortable, and taught optics, mechanics and astronomy without troubling about truth.[24] In gnomonics, or dialing, a subject they emphasized, they could reasonably adopt a geocentric point of view. When considering the solar system, they might use Copernicus' theory, treated as an hypothesis;[25] in the last quarter of the seventeenth century both the Gregorian College (Rome) and Louis-le-Grand (Paris) used texts that represented heliocentrism as a useful mathematical fiction.[26] The strength of this ploy decreased in time. In the eighteenth century, in the face of Cartesian and Newtonian physics, it no longer sufficed to protect the natural philosophy of Aristotle.[27]

These organizational generalities apply to Jesuit universities as well as to Jesuit colleges. In most cases the 'university' undergraduate or arts course was nothing other than the three years of philosophy prescribed by the *Ratio*. Humanities were taught at associated Jesuit grammar schools. The separate Jesuit university occurred primarily in Germany and to a lesser extent in Italy, where the Society took over existing institutions;[28] in France, the larger Jesuit colleges—such as Bordeaux, La Flèche, Lyon—were both grammar schools and provincial faculties of arts. In general, the Society could exercise stricter control over colleges than over universities; while the colleges were usually endowed, the universities had at least a part of their operating expenses from states or municipalities, which thereby retained the power to dabble in educational reform. The forced modernization during the eighteenth century of the outmoded curricula at the conservative Jesuit universities under Austrian control is a dramatic instance of government intervention.[29]

22. J. Doddington to Kircher, 17 and 31 Jan., 21 March and 24 Oct., 1671, in CK, V, ff. 21, 50, 97, 104; and the correspondence of Kircher with Leibniz, 1673 (Friedländer, Pont. acc. rom. arch., *Rend.*, 13 [1937], 229–47), and with G. A. Kinn, 1672 (J. E. Fletcher, *Janus*, 56 [1969], 267).

23. Duhem, *To Save the Phen.* (1969), *passim*.

24. Consentino, *Physis*, 13 (1971), 205. Cf. Descartes' annoyance at the superficiality of his teachers' use of mathematics; Descartes, *Discourse* (1965), 8.

25. E.g., Schott, *Cursus mathematicus* (1661), and Dechales, *Cursus mathematicus* (1674). Note the titles.

26. Eschinardi, *Cursus physico-mathematicus* (1687); Dechales, *Cursus seu mundus mathematicus* (1690³), who puts the matter thus: 'si l'on ne se rapporte qu'aux arguments de la science, sans égard à l'autorité de l'Ecriture, il n'y a aucun qui tranche définitivement pour ou contre le système de Copernic.' See Ziggelaar, *Pardies* (1971), 150–6.

27. Cf. Consentino, *Physis*, 13 (1971), 206–7.

28. Paulsen, *Gesch.*, I (1897), 383–407.   29. *Infra*, ii.3.

The teachers of philosophy and mathematics could hold several sorts of positions within the Society. The most junior taught at the smaller schools, and took their students through all three years of philosophy. They had little time, and usually little talent, for independent scientific work; many, perhaps most, did not continue in teaching careers.[30] Those deemed sufficiently conscientious and able advanced to the bigger schools, where they could specialize in one or another branch of philosophy. There were fifty positions in 'physics' in France in 1700, and 62 at the expulsion of the Society in 1761.[31] These positions often functioned as academic chairs, with the peculiarity, however, that their less distinguished incumbents frequently moved from college to college.

The professor of mathematics followed a different path. He was usually chosen from among the few who had shown mathematical ability, and received special lessons, during study of philosophy. Since their specialty occupied only a part of the physics course, they usually taught something else as well, chiefly languages or physics, or kept books for the college. During the early sevententh century they changed colleges frequently, but after 1660 their positions stabilized, owing, perhaps, to an increasing demand and higher appreciation for their services.[32] The demand came not only from students, but also from the general public and especially from government, which often used Jesuit mathematicians as consulting engineers. In France the value of the order as a source of technicians was recognized by the creation of seven royal chairs of hydrography in Jesuit institutions in or near seaports. By 1700 the French Jesuits had 21 other chairs for mathematics at their several colleges.[33]

During term the conscientious Jesuit professor had little opportunity to advance, or even to keep up with his subject. Typically he lectured two hours a day, conducted reviews, arranged for and presided over disputations, wrote theses for his abler students, offered lectures to the general public;[34] in addition he said masses, heard confessions, and attended to his own religious exercises. Recognizing that its savants needed occasional relief from these burdens, the Society set up positions free from teaching for distinguished specialists. During their leisure, which might last from two to six years, these scriptors wrote monographs or texts, or coordinated astronomical and geographical data sent from missionary or provincial colleagues. Clavius was alternately professor and scriptor from 1565 to 1612.[35] The Collège Louis-le-Grand, which the order wished to make a showplace, 'where science was done as well as disseminated,' sheltered some 90 scriptors between 1606 and 1672, including physicists and mathematicians such as Jacques Grandami and Noel Regnault.[36]

30. Cf. Dainville, *RHS*, 7 (1954), 14.

31. Dainville in Taton, *Enseignement* (1964), 33–4, 40–1.

32. Dainville, *RHS*, 7 (1954), 15; Huter, *Fächer* (1971), 7.

33. Dainville in Taton, *Enseignement* (1964), 33–4; cf. Ziggelaar, *Pardies* (1971), 28.

34. Chossat, *Jésuites* (1896), 100–3; Guitton, *Jésuites* (1954), 48; Ziggelaar, *Pardies*, 125; Schott, *Pantometrum* (1660), proem.

35. Phillips, *Arch. hist. soc. Jesu*, 8 (1939), 191.

36. Dupont-Ferrier, *Vie*, I (1921), 60–2, 81, 121, 150–1, and III (1925), App. A.

The official scriptors made up only a portion of the personnel supported by the Society in research and writing. Frequent sabbatical could be arranged for distinguished mathematicians. An interesting case is Gregory de St. Vincent, who had not only leave but assistants, furnished by his General in the expectation that many fine texts for the improvement of novices, and many fine monographs for the glory of the Society, would result. Gregory spent much of his time trying to square the circle; when he announced success he was called to Rome to be examined, lest his 'proof' embarrass the order in precisely the field it had made its own. The Roman mathematicians could not certify Gregory's circle-squaring. They nonetheless thought his other work worthy of support, which he continued to enjoy for forty years.[37]

Probably the most important result of all this writing was the preparation of standard texts and reference works, which, especially in the seventeenth century, served those outside the order as well. One thinks, for example, of Fabri's *Synopsis geometrica,* an introductory manual used by Leibniz, by Flamsteed, and, perhaps, by some of Newton's students;[38] and of Pardies' *Elémens de géometrie,* 'the plainest, shortest, and yet easiest Geometry' ever published, according to the preface to the eighth English edition.[39] Texts on the design and application of simple instruments, like Schott's on the pantograph and Zucchi's on elementary machines, also circulated widely.[40] Among works of reference Riccioli's *Almagestum novum* played a unique role: although belligerently anti-Copernican, it became a standard source for astronomical data even among adherents of the new cosmology. The Jesuit literature on the magnet well served students of the lodestone, again without regard to doctrine. And so the elder Huygens, who was no friend of the Order, exhorted the reformer of modern philosophy to rely upon the data of the Jesuits: 'For these scribblers,' he wrote Descartes, 'can serve you in matters *quae facti sunt, non juris.* They have more leisure than you to provide themselves with experiments.'[41]

The Jesuits gathered and disseminated much of their information through correspondence. One knows the important part played by the Minim monk, Marin Mersenne, in effecting communication among physicists and mathematicians in the 1630s and 1640s. But Mersenne was an individual, the Jesuits an organized and disciplined society; their world-wide missions housed a network of informants able to identify natural and artificial novelties, and to observe astronomical phenomena invisible from Europe. Requests for and dissemination of this information, and for the supply of rara naturalia, constituted much of the

37. Bosmans, *Biog. nat. Belg.*, XXI, cols. 145–51, 156, 168.

38. Fellmann, *Physis,* 1 (1959), 6–25; Leibniz, *Phil. Schr.* (1875), IV, 245; Flamsteed to Collins, 20 May 1672, and Collins to Newton, 30 April 1672, in Rigaud, *Corresp.* (1846), II, 146, 320.

39. Pardies, *Elémens* (1671); *Short, but yet Plaine Elements of Geometry* (1746⁸); Ziggelaar, *Pardies* (1971), 64–8; Wieleitner, *Arch. Gesch. Naturw. Tech.*, 1 (1908–9), 438.

40. Cf. Borelli to Collins, 10 April 1671, in Rigaud, *Corresp.* (1846), I, 165.

41. Huygens to Descartes, 7 Jan. 1643, in Descartes, *Corresp.* (1926), 186; Riccioli, *Almagestum* (1651); Koyré, *Metaphysics* (1968), 89–117; Flamsteed to Collins, 29 Sept. 1673, in Rigaud, *Corresp.* (1846), II, 168. For Jesuit magnetism see Daujat, *Origines* (1945), II, *passim.*

learned correspondence of the order. Kircher, for example, channeled foreign observations of eclipses of the moon to his colleague Riccioli, who rewarded him by naming a lunar crater in his honor.[42] Schott exchanged letters with much of learned Germany, including Otto von Guericke, the Lutheran mayor of Magdeburg, who found it convenient to announce his discoveries in Schott's encyclopedic works. The fine museum of the Settalas in Milan was filled with specimens furnished by Jesuit missionaries.[43] The order's connections also assisted the circulation of books at a time when no regular international trade in them existed. Jean Bertet, 'the true Mersennus of France,' a Jesuit mathematician, a student of Fabri's and friend of Pardies', labored for years to procure continental books on his subject for the use of English philosophers.[44]

The very reputation of the successful Jesuit savant might constitute a subtle but important aspect of his magisterium: since touring intellectuals tended to seek him out, he often functioned as the nucleus of a tiny international congress, an ever changing polyglot academy with shared interests, similar educations, and a common learned language. No study trip to Rome was complete without a visit to Fabri and Kircher; Arriaga, according to an old slogan, was a chief attraction of the capital of Bohemia, 'Pragam videre, Arriagam audire;' while the Minims Mersenne and Maignan drew even royal visitors to their cells in Paris and Toulouse.[45]

The powerful Jesuit educational system, its celebrated pedagogues and part-time researchers, made the Society the leading patron of physical and mathematical sciences during the seventeenth century. According to Leibniz, all the chief scientific men in Italy in the 1670s were Jesuits; the discipline and cooperation of the Society's savants served as the model, and perhaps the inspiration, for Leibniz' plans for a German academy of sciences.[46] Perhaps Bacon too took hints from them: 'partly in themselves [he observed] and partly by the emulation and the provocation of their example, they have much quickened and strengthened the state of learning.'[47]

Jesuit work provided starting points for investigations in many branches of physics and mathematics, especially optics, mechanics, magnetism and electricity; the Royal Society of London routinely reviewed Jesuit books and sought to

42. J. E. Fletcher, *Manuscr.*, 13:3 (1969), 157–8. Kircher was consulted on all respectable subjects (J. E. Fletcher, *Janus*, 56 [1969], 259–77) and many questionable ones; in a single year he examined and rejected ten designs for perpetual-motion machines (Gutmann, *Kircher* [1938], 14).

43. Fogolari, *Arch. stor. lomb.*, 14 (1900), 59, 113, 119; Rota Ghibaudi, *Ricerche* (1959), 46–7.

44. Rigaud, *Corresp.* (1846), I, 139–40, 151, 162, and II, 22–3, 530.

45. Oldenburg, *Corresp.*, V, 294, 423, 439; Eschweiler, *Sp. Forsch.*, 3 (1931), 253–85; Ceñal, *Rev. est. pol.*, 46 (1952), 111–49. Monconys, *Voyages* (1665–6), gives a lively picture of the international intellectual life of the day, when it was possible for one man to know 'most of ye Intelligent and Curious men in the world'; Oldenburg to Boyle, 24 Dec. 1667, re Carcavi, in Oldenburg, *Corresp.*, V, 79.

46. Friedländer, Pont. acc. rom. arch., *Rend.*, 13 (1937), 242–5. Cf. Grosier, *Mémoires* (1792), I, xxvi–xxviii.

47. Bacon, *Advancement* (1915), 41; *Works,* I (1863), 469. Cf. Reilly, *Line* (1969), 75.

open correspondence with the leading members of the Society.[48] As the national academies of science grew in size, resources and importance, and science returned to the secular universities, Jesuit patronage of mathematics and physics became less significant and appropriate. Moreover, the Society lost ground to the moderns by first opposing and then belatedly assimilating Cartesian philosophy. In Germany especially their defense of the old ways dropped their universities behind the leading Protestant schools during the eighteenth century. The German provinces took with peculiar literalness the Society's reaffirmation, at its general congregation of 1730–1, of the fundamental principles of Aristotelian philosophy.[49] But the Jesuits of France and Italy continued to train productive scientists until their expulsion and the dissolution of the Society (1773). 'It was the most beautiful work of man,' lamented a former student, J. J. de Lalande, astronomer, pensionary of the Paris Academy, professor at the Collège Royal, etc.; 'no human establishment will ever approach it.'[50]

THEIR ECLECTICISM

Although St. Ignatius had counselled his Society to follow St. Thomas in theology and Aristotle in philosophy, it did not long remain subservient to either. On the theological side the Jesuits produced their own doctor, Suarez, who did not hesitate to disagree with Aquinas.[51] Suarez' *Disputationes metaphysicae* (1597), freer and more orderly than the traditional Aristotelian commentary, have a vigorous, eclectic air;[52] they were used both in Jesuit colleges and in the Protestant universities of Holland and Germany.[53] During his student years Leibniz eagerly read the *Disputationes* and Descartes, it is said, carried them with him in his travels.[54]

The free, eclectic style of Suarez' popular metaphysics exercised an important though subtle influence on physics. Suarez' colleagues at Coimbra wrote their authoritative commentaries on Aristotle's natural philosophy in a style

48. Duhem, *Rev. met. mor.*, 23 (1916), 58–9; Reilly, *Arch. hist. Soc. Jesu.* 27 (1958), 340–4, and *Line* (1969), 89–90, 93–6; Oldenburg, *Corresp.*, VI, 119, 274, 317, and V, 315, 564; Boyle to Oldenburg, 3 April 1668 (*ibid.*, IV, 299): 'I am glad you are like to settle a correspondence with *Rome*, that being the chief center of intelligence.' In a quaint and ambitious reciprocation, the Jesuit college at Liège announced in its *Prospectus* of 1685 that it had perfected all the Royal Society's discoveries; Reilly, *Line* (1969), 14.

49. Hammermayer, *Gründ.-Frügesch.* (1959), 237–8; Prandtl, *Gesch.* (1872), I, 539; Paulsen, *Gesch.*, I, 425, and II, 103–4.

50. Quoted by Delattre, *Etablissements*, II, 1554n. Cf. Costa, *Rev. stor. it.*, 79 (1967), 849 f.; Varićak, Jug. akad. znan. umjet., *RAD*, 193 (1912), 248; Grosier, *Mémoires* (1792), I, vi–vii.

51. Mahieu, *Suarez* (1921), 522. Other commentators have tried to minimize the disparity; see Riedl in Smith, *Jes. Think.* (1939), 18n. A useful guide to the technical differences between Suarez and St. Thomas is Werner, *Suarez* (1889²).

52. Grabmann, *Mitt. Geist.*, I, 528–35; Jansen, *Phil. Jahrb.*, 50 (1937), 418–23.

53. Eschweiler, *Sp. Forsch.*, 1 (1928), 251–325, esp. 288–91; cf. Petersen, *Gesch.* (1921), 283–338.

54. Riedl in Smith, *Jes. Think.* (1939), 5; Eschweiler, *Sp. Forsch.*, 1 (1928), 254, 259; Gilen, *Schol.*, 32 (1957) 47.

between the old commentary and the new, freer synthesis; they did not fear to leave doubtful points unsettled, or to add essays that made possible 'a somewhat freer consideration of the topics than their original context provided.'[55] A symbiosis similar to that of Suarez and the Coimbra commentators established itself later at Prague, where Arriaga, an adept in Suarez' methods, presided over a group of physicists that included Gregory de St. Vincent, Juan Caramuel, Marcus Marci, and Balthasar Conrad.[56] It cannot be mere coincidence that several of those who advanced the study of electricity in the seventeenth century also played a part in the dissemination of Suarez' work.

Jesuit writings on natural philosophy contained much that conflicted with the Aristotelian doctrines that they purported to transmit, and that their authors were obliged to teach.[57] In the last third of the sixteenth century the Society hoped to establish a standard interpretation of peripatetic physics for use in its academies. It began by turning from the superficial summulae of the late Renaissance to Aristotle's texts, as in the products of Coimbra, which gave the original Greek, a Latin translation, and a running commentary, as well as the free expositions. The exigencies of the philosophical course also required shorter compendia or *cursus*; and these, despite the Society's best intentions, could never be made uniform. The general councils of the 1590s recommended the adoption of official textbooks, the characteristics of which are set forth in the early drafts of the *Ratio studiorum*; but the practicing pedagogues consulted, and especially the Spanish school, urged that the official doctrine be kept to a minimum.[58] 'In scientiis,' St. Thomas had said, 'auctoritas minime valet.' Early in the seventeenth century the Society formally abandoned the project for an official textbook of natural philosophy.[59] Such a text, by freezing the curriculum just as the new science began to develop, would seriously have handicapped Jesuit efforts to educate Catholic Europe for survival in the modern world.

As novelties unknown to Aristotle accumulated, the books of conscientious Jesuit natural philosophers departed more and more from their ancient model.

55. Coll. conim., *Comm. de Caelo* (1603²), 487. Among doubtful points: whether stars act by 'influence' or only by light and heat, *ibid.*, 200; whether qualities alone among accidents can act ('non videtur nobis absolutè pronunciandum'), *Comm. in phys.*, I (1602), col. 393; the number of categories ('suas habeat difficultates'), *Comm. in univ. dial.*, I (1607), 340.

56. Caramuel (of whom more below) was a Cistercian; St. Vincent and Conrad, Jesuits; Marcus Marci, a physician educated by the Jesuits, was only prevented by ill-health from becoming one himself. Eschweiler, *Sp. Forsch.*, 3 (1931), 253–85; Werner, *Suarez* (1889²), II, *passim;* Marek, *RHS*, 21 (1968), 109–30, and *Bohemia,* 16 (1975), 98–109.

57. 'In matters of any consequence let him [the professor of philosophy] not depart from Aristotle unless something occurs which is foreign to the doctrine that academies everywhere approve of.' *Ratio* (1599), in Fitzpatrick, *St. Ignatius* (1933), 168. Cf. *ibid.*, 151.

58. The Spanish Jesuits also opposed a definitive catalog of forbidden propositions; the first drafts of the *Ratio* had about 600; that of 1590, 200, that of 1599, none. Fichter, *Suarez* (1940), 137–44.

59. Jansen, *Phil. Jahrb.*, 51 (1938), 187; *Ratio* (1599), in Fitzpatrick, ed., *St. Ignatius* (1933), 119–21, 170–1; Hilgers, *Index* (1904), 194–206.

An early stage is represented by Niccolò Cabeo's commentary on Aristotle's *Meteorologica* (1646). Despite its title, it is not a classical commentary, but almost an independent *philosophia universalis,* a label under which it reappeared, unaltered, in 1686.[60] Cabeo no doubt chose the *Meteorologica* as his vehicle because, as the most practical and specialized of Aristotle's treatises on the inorganic world, it gave the 'commentator' the best purchase for an exposition of terrestrial physics.[61] Not that Cabeo professed to follow Aristotle closely: the true task of the peripatetic commentator, he holds, is first to clarify the writings, and then to judge their truth, 'not on his authority, because *ipse dixit,* but by plain reasoning.' 'If you never question Aristotle's doctrines your commentary will not be that of a philosopher but that of a grammarian.' And Cabeo does take shocking liberties, like interpreting the form of volatile substances as a material emanation.[62]

Jesuits soon dropped the commentary form in favor of comprehensive treatises, such as the *Physica* of Honoré Fabri. Though Fabri maintained the pious fraud of holding to hylomorphism 'most religiously,' and pretended to deduce all of physics from six peripatetic principles, viz., heat, impetus and the four elements, he in fact fashioned an adroit compromise between Aristotle and the corpuscularians.[63] Where he thought Aristotle erred, as in the theories of natural motion and celestial composition, Fabri did not pretend to follow: on the contrary, he admitted all the gains of early modern physics but the Copernican geometry. Professing to eschew hidden virtues, antiperistasis, sympathies, antipathies and atoms, he nonetheless invoked an occult cause (in the guise of 'magnetic particles') and exploited corpuscles so freely that contemporaries classified him as Cartesian or Gassendist. Moderns understandably find him eclectic.[64] Fabri always denied that he subscribed to corpuscular philosophies; his eclecticism, he insisted, was only in the eye of the beholder, misled by the commentaries of 'impious Arabs' and of 'modern scholastics, who have never understood Aristotle's thought.'[65]

More overtly eclectic is Fabri's younger colleague, Francesco Lana, a student of Kircher's. 'It is a most vulgar error,' he says, 'to believe that our mental images of the truth are unique, as if many likenesses cannot be made of

60. The title of this edition (Rome, 1686) is given by Ferretti-Torricelli, At. di Brescia, *Comm.* (1931), 384, but does not appear in Sommervogel. Lana Terzi, *Prodromo* (1670), 13, singles out Cabeo and Gassendi as the best modern writers on physics.

61. Descartes saw as much when he offered his *Météores,* which amounts to a free commentary on the *Meterologica,* as an example of the application of his new method to physics. Descartes, *Oeuvres,* IX:2, 15.

62. Cabeo, *Met. Ar. comm.* (1646), I, 'Ad lect.,' and p. 253.

63. Fabri, *Physica,* I (1669), 'Auctor-Lect.,' vii–viii, xxxviii; *Epistolae* (1674), 34–5; Morhof, *Polyhistor* (1747⁴), II, 267, 334–5; Cromaziano, *Restaurazione,* I (1785), 131.

64. Daujat, *Origines* (1945), II, 350–75 (Fabri's magnetism); Lasswitz, *Gesch.* (1890), II, 460, 486, 490 (Fabri's eclectic corpuscularism); Morhof, *Polyhistor* (1747⁴), II, 218; Pernetti, *Recherches* (1757), II, 122.

65. Fabri, *Epistolae* (1674), 56.

the same statue. Accordingly I will use many models [*imagines*] to express clearly the truth to which the science of natural things can attain: the more representations of an object are present to the intellect, the more intimately and clearly the object is known; what one model lacks, another will supply.' Lana supplements Aristotle's hylomorphic and elemental doctrines with the *tria prima* of the Paracelsians; with purposeful Democritean atoms; with sympathies, antipathies, powers, principles, the fixed, the volatile, the acid, the alkaline.[66] The primary obstacles to the advance of science, according to Lana, are excessive attachment to antiquity and, beyond that, to pet subjects, as Aristotle to logic, Plato to theology, Proclus to mathematics, Gilbert to magnetism, and chemists to fire.[67]

The knowledgeable natural philosopher changes his principles to suit his problems; when Aristotle falters he must take another guide. This program, which might be called 'moderate peripateticism,' was not peculiar to Lana or to the Jesuits. Juan Caramuel, for example, the friend of Arriaga, correspondent of Kircher's, mathematician, papal envoy and bishop, had earlier endorsed the same method. 'In the dead of winter, in a failing light, you come to a difficult crossing you do not dare to attempt; Peter approaches from the opposite direction, carrying a bright lantern. You cross, he crosses: but you do not follow him because you have used his light; you continue your journey in your original direction. It often happens thus in our philosophizing. We come to a difficult and obscure place; we hale Zeno, Plato, Aristotle; their doctrines are a torch that dissipates the darkness. We cross in the greatest security; and yet we do not therefore follow them, but proceed as we had begun.'[68]

One of the most original directions was that chosen by Emmanuel Maignan. Schooled by the Jesuits of Toulouse, Maignan determined to follow the religious life among the neighboring Minims when his masters awarded a classmate a literary prize he thought he had earned. Such intellectual sensitivity was then not characteristic of the *minimi minimorum,* an austere order of Franciscans, some of whom had not troubled to learn to read; but the few savants the order did produce and encourage in the seventeenth century were aggressive and independent thinkers, untied to medieval doctors, opposed to frilly argumentation, and interested in matters accessible to experiment. These qualities appear in Maignan as they do in his colleagues Mersenne and J. F. Niceron.[69]

Maignan's chief work is a *Cursus philosophicus* (1653), which, despite its global and pedestrian title, is an original treatise primarily devoted to natural

66. Lana Terzi, *Magisterium*, I (1684), 'Auctor-Lect.' Cf. Morhof, *Polyhistor* (1747⁴), II, 266, who holds that the eclectic method is the soundest for physics, 'ut non omnia omnes videant.'

67. Lana Terzi, *Prodromo* (1670), 5–6; cf. the Idols in Bacon's *Novum organum.*

68. Caramuel, *Rat. real. phil.* (1642), 62. Cf. Ceñal, *Rev. fil.,* 12 (1953), 101–47; Fernández Diéguez, *Rev. mat. hisp.-amer.,* 1 (1919), 121–7, 178–89, 203–12; Glick, *Isis,* 62 (1971), 279–81.

69. Bayle, *Dictionnaire* (1740⁵), III, 280–3; Rochot, *Corresp.* (1966), 7; Whitmore, *Order* (1967), 112–19; Ceñal, *Rev. est. pol.,* 46 (1952), 111–49; Nicéron, ed., *Mémoires,* XXXI (1735), 346–53.

philosophy.[70] He has no patience with received authority in science. 'I would think myself exceedingly stupid,' he writes, 'were I to deny the truth of what I see and touch in my experiments, and credulously to embrace what someone else has concluded from some abstract and fanciful little arguments.' The Aristotelians are his main targets. He rejects hylomorphism, 'the solemn daily teaching of the peripatetic schools;' the *horror vacui;* and the distinction between potency and act, which he calls 'voces sine re,' or claptrap. Yet he is no corpuscularian. He conceives that there are many kinds of matter, the magnetic and the non-magnetic, for example, whose different properties cannot be reduced to differences in the sizes, shapes and motions of constituent particles. Maignan is a qualitative atomist: he holds that bodies act upon one another only during contact, but denies that the act is purely mechanical. Among the non-mechanical qualities his corpuscles possess are principles of self-movement and sympathetic powers competent to activate them: a magnet, for example, emits streams of characteristic particles that trigger and direct the self-movements of neighboring pieces of iron.[71]

Maignan's qualitative atomism, with its sympathies and principles, was a fair compromise between radical corpuscularism and the traditional philosophy.[72] His *Cursus* well illustrates the breadth of options open to moderate peripatetics of the seventeenth century, to those who wished to develop the received philosophy according to their own lights, and to come to grips with the new physics of Galileo and Descartes.

During the eighteenth century the Jesuits stopped struggling to reconcile physics and mathematics with the doctrines of Aristotle and Ptolemy. As noticed earlier, they were teaching Copernicus 'hypothetically' before the end of the seventeenth century. Later Descartes, and still later Newton, entered their textbooks. These liberties were acknowledged after the fact, in an edict of 1751, which freed Jesuit physicists of Aristotle except for 'the first principles of natural body. . . , namely of matter and form in the peripatetic sense, of prodduction *de novo,* and of the existence of some absolute accidents.'[73] There is no difficulty in clothing any physical theory in the language of hylomorphism.

THEIR CONSTRAINTS

To balance our rude portrait of Jesuit physicists we should notice two obstacles they met with in their work. The first was the problem of support. The gluttonous intellect of a Kircher consumed an amount of treasure little consistent with the vow of poverty, or with the funds made available by superiors often hard

70. The second edition (1672), 742 pages in-folio, devotes 75 pages to logic, 50 to metaphysics and the rest to 'philosophia naturae, seu physica.' Cf. Ceñal, *Rev. fil.*, 13 (1954), 15–68.

71. Maignan, *Cursus* (1673²), 127, 143, 195, 226, 590–6, 608; *infra*, vi.2; Ceñal, *Rev. fil.*, 13 (1954), 15–68; Lasswitz, *Gesch.* (1890), II, 492–3.

72. Cf. Jansen, *Phil. Jahrb.*, 50 (1937), 433–6; Sander, *Auffassungen* (1934), esp. 9–10.

73. Pachtler, *Ratio*, III, 435f. Cf. Specht, *Gesch.* (1902), 199n; Dainville in Taton, *Enseignement* (1964), 49n; Bednarski, *Arch. hist. Soc. Jesu*, 2 (1933), 213.

pressed to make ends meet. Kircher found his Maecenas among Catholic and Protestant princes, who sometimes contributed so handsomely to his treasury that he was able to assist the work of others. The celebrated museum grew from a private bequest.[74] Occasionally proceeds from the sale of books might be made available for scholarly purposes: Schott secured special permission to handle money in connection with his publications, and any surplus probably ended in new instruments.[75] To do original work in the sciences the moderate peripatetic probably needed outside support and certainly required the cooperation of his colleagues and superiors, the use of the order's instruments and facilities, and free time. These advantages were not always easy to procure. Would-be physicists assigned to remote, impoverished or understaffed colleges, where the customary teaching apparatus might be incomplete, experienced considerable, and perhaps insurmountable difficulties.

Opposition from colleagues posed another obstacle. Simplicios did exist, and exercised their influence both unofficially, through local cabals, and institutionally, through the censorship of the press. The irregular practices were probably the more effective: conservative rectors or prefects of studies might burden their modernizing professors with routine assignments or secure their transfer to non-academic positions. A cabal of conservatives at the college of Lyon, alarmed at Fabri's aggressive taste for novelties, is said to have engineered his reassignment from France to Italy, and from teaching to the papal bureaucracy. It is difficult to estimate the harm done the eclectics and moderate peripatetics by such maneuvers. In Fabri's case it backfired; he found Rome congenial, enjoyed the support of the Jesuits' General, and became more productive, innovative and combative than before. Similarly Pardies, under attack from his colleagues at Bordeaux for his weakness for Cartesian physics, was transferred by the General to Louis-le-Grand, where he could do even greater mischief.[76]

The practical working of the censorship is also difficult to determine from its formal acts. Few scientific works ever earned a place on the *Index of Prohibited Books*. The Church directed its vigilance against errors in faith, morals and canon law;[77] its sortie against the Copernicans carried it into territory it had previously avoided and could not hope to hold. After the blunder over Galileo (1633), the Church again avoided indexing physics books,[78] and sought to sway their authors during review. The Roman censorship was particularly officious. Schott, who knew that its dilatory, frightened and often ignorant functionaries had annoyed even Kircher, refused an invitation to return from Germany to the Collegio Romano in 1664 primarily because he wished to

74. Brischar, *Kircher* (1877), 43–59; Villoslada, *Storia* (1954), 183–4; Schott to Kircher, 16 June 1657, in CK, XIII, 45: 'Ringratio molto a Vª Rª per li 6 scudi offertimi di nuovo per li miei studij.' Maignan, *Cursus* (1673²), Praef., acknowledges the financial support of Cardinal Spada.

75. Duhr, *Gesch.* (1921), III, 590–3.

76. Vregille, *Bull. Soc. Gorini*, 9 (1906), 5–15; Ziggelaar, *Pardies* (1971), 69–78.

77. Putnam, *Censorship* (1906), I, 182–93, and II, 127–9.

78. Galileo first appeared upon the *Index* in 1664; Mendham, *Lit. Pol.* (1830²), 175–7.

write, and at Würzburg the censors gave him little trouble.[79] But even in Italy few if any scientific books were suppressed. On balance, the censorship seems rather to have harassed than guided; a physicist willing to suffer delay and indignity would probably see his book approved much as he had written it, provided, of course, that he did not openly espouse heliocentrism. To pick a late example, the first volume of the proceedings of the Bologna Academy of Sciences was delayed for several years because the unenlightened authorities— Bologna then (1730) being a Papal State—demanded that whenever the proceedings mentioned Copernicus, or discussed the moderns 'respectfully and charitably,' the editor characterize the novelties as ingenious hypotheses. He refused, saying that it would make the Academy ridiculous. That was in 1729. Two years later, when the enlightened Lambertini, later Benedict XIV, became Archbishop of Bologna, the proceedings appeared as originally planned.[80]

The differential workings of the censorship, as applied to doctrinal and scientific matters, may be illustrated by the experiences of Maignan and Fabri, each of whom has managed to secure a permanent place on the *Index*. Maignan had vigorously attacked the natural philosophy of the schools, and ended closer to Descartes than to Aristotle. He was indexed not for assaulting peripatetics, however, but for trying to justify the practice of usury.[81] Fabri published a *Physica* in 1670 that strayed close to Descartes, praised Galileo, and took a soft line towards Copernicus. It was not for this that he then found himself in the jails of the Inquisition, but for an untimely attack upon the Jansenists, and an excessive defense of the slippery doctrine of probabilism.[82] 'As for his books on philosophy, don't worry about them,' writes the General, Oliva, to the Provincial of Lyon. 'One can indulge a man of such parts.'[83] The point is important, as Fabri's contretemps has given rise to the idea that his brethren sacrificed him because he had 'busied himself with science.'[84] In fact, Fabri's scientific connections helped extricate him from the clutches of the Inquisition. Cardinal Leopold dei Medici, who had known Fabri as a correspondent of the Florentine Accademia del Cimento, intervened on his behalf and helped restore him to a position of honor and influence.[85] We may take it that the Congregation of the Index had no interest in electricity.

79. Schott to Kircher, 21 Aug. 1664, in CK, VIII, 110: 'Hactenus non multum desideravi redire Romam, quoniam existimo me hic habere meliorem occasionem scribendi et imprimendi libros meos, quam ibi.'

80. Bortolotti, *Storia* (1947), 157–8.

81. Maignan, *De usu licito pecuniae* (1673), prohibited 24 Oct. 1674; *Index* (1948), 293.

82. Fabri, *Apologeticus doctrinae moralis societatis Jesu* (1670), prohibited 27 Jan. and 22 March 1672; *Index* (1948), 168. For Fabri's favorable view of Copernicus see Thorndike, *Hist.*, VII, 665–70.

83. Guitton, *Jésuites* (1954), 55–6. The Jesuits were never very enthusiastic about the *Index;* cf. Hilgers, *Index* (1904), 205.

84. Middleton, *Experimenters* (1971), 324; Boffito, *Bibl.*, 44 (1942), 176–84; cf. Middleton, *Br. J. Hist. Sci.*, 8 (1975), 144–6, 154, acknowledging that during the last third of the seventeenth century churchmen in Rome were free to study and write about the new physics.

85. Boffito, *Bibl.*, 44 (1942), 176–84. Ceyssens, *Franz. St.*, 35 (1953), 401–11, does not support his claim that, after 1670, Fabri had trouble gaining official approbation for his books.

2. ACADEMICIANS

Most eighteenth-century natural philosophers managed to enter one or more of the learned academies whose rapid multiplication was a characteristic of the age. The proliferation of these institutions is suggested by Table 2.3. The more important societies are listed separately; all, including those lumped together as 'others,' published memoirs and / or offered prizes for papers on scientific subjects. The list could be extended indefinitely by admitting unproductive societies and intellectual drinking clubs. Cross-national comparisons on the basis of the table should be hedged, since, owing to the state of scholarship, coverage for France is better than that for other countries, especially the Germanies.

Leaving aside differences in quality, we may divide these organizations into two natural classes in two different ways. The first division opposes the special to the general: the scientific academies, like London, Paris, Bologna, Stockholm, Haarlem, to the all-purpose society of 'sciences, arts, et belles lettres,' such as Berlin, Petersburg, Brussels, Edinburgh, and most of the academies of France. Only the activities of the scientific classes of these general societies will concern us. The second, and more important division opposes open academies, with large, coopting, self-supporting memberships, to closed institutions, a few men supported in whole or part by princes who established their salaries, appointed their colleagues, and fixed their number.

CLOSED SOCIETIES

The model of closed institutions was the Académie Royale des Sciences, Paris, set up as a part of the bureaucracy of Louis XIV. In return for instruments, quarters, and salaries that ranged from 1500# to 2000# for ordinary members and 6000# to 9000# for foreign heavyweights like Huygens, the academicians were to advance science and to apply it as government consultants on technological problems, patent applications, and other technical matters. The Academy's procedures were first codified at its reorganization in 1699; after a few further refinements promulgated in 1716, its structure and functioning remained essentially unaltered until 1785.

According to the constitution of 1716, the Academy had 44 regular scientific members, distributed horizontally into six subject classes, three 'mathematical' (geometry, astronomy, mechanics) and three 'physical' (chemistry, anatomy, botany), and vertically into three levels, adjuncts, associates, and pensionaries. Two adjuncts, two associates and three pensionaries constituted a class; the perpetual secretary and treasurer, both pensionaries, completed the company. No one living outside Paris, or 'attached to any religious order,' was eligible for regular membership. There were also several categories of irregular members: occasional supernumerary regulars; twelve *honoraires,* usually unscientific; twelve *associés libres,* unconnected with any class and ineligible for promotion; six (later eight) foreigners, always distinguished scientists; and an indefinite number of 'correspondents,' provincials or foreigners given the right—

TABLE 2.3

Numbers of Academies Founded 1660–1800

	1660/1724	*1725/49*	*1750/74*	*1775/99*
Britain				
RS, London	1662			
RS, Edin.				1783
Manch. Lit. and Phil.				1781
France[a]				
AS, Paris	1660			
AS, Bordeaux	1712			
AS, Lyon	1724			
AS, Montpellier	1706			
AS, Toulouse		1729		
Others	2	6	5+12[b]	2
Germanies				
AS, Berlin	1700	[1744]		
SW, Göttingen			1751	
AW, Munich			1758	
Others [c]		1	4	2
Italy				
AS, Bologna	1714			
AS, Turin			[1757]	1783
Netherlands[d]				
HM, Haarlem			1752	
Teylers Stichting				1778
Others			2+ 1[b]	11
Elsewhere				
AS, St. Petersburg		1725		
AS, Stockholm		1739		

[a]Compiled primarily from Delandine, *Couronnes* (1787); purely literary societies, even those that toward the end of the century occasionally ventured into agriculture or commerce, have been discarded.

[b]'Sociétés d'agriculture.'

[c]Compiled from McClellam, *Int. Org.* (1975), 249–56, 499–502, an admittedly incomplete sample.

[d]Compiled from Rooseboom, *Bijdrage* (1950); Bierens de Haan, *Holl. Maat.* (1752), 35–6; Muntendam in Forbes, *Marum,* I (1969), 5–6.

and the honor—to communicate scientific news and the results of their own researches to a specified regular member of the Academy.[1]

The duties of the academician remained as before: to work at his science and to advise when consulted. To maintain the pace of work, the Academy met twice a week when not on one of its statutory vacations, which amounted to

1. Maindron, *Académie* (1888), 19, 23, 48.

fourteen weeks a year. At each meeting, according to the regulations of 1716, one pensionary and either an associate or an adjunct were to read an original paper; if enforced, the provision would have produced 152 papers a year.[2] A great many excellent papers were written, and published in the Academy's *Mémoires*, which began to appear regularly from 1699, or rather from 1702, when the volume for 1699 was issued. These *Mémoires* and their accompanying *Histoire* (chiefly summaries of the memoirs and stylized éloges of deceased members) set the eighteenth century its highest standards of scientific work and — through the éloges — its ideal of an honorable, dedicated, selfless career in science.

The Parisian academician needed dedication. In principle only the pensionaries of each class received salaries. After 1775 they had 3000#, 1800# and 1200#, in order of seniority; salaries had probably averaged the same (2000#) since 1716. Occasionally an associate might receive up to 500#, and very rarely, as in the case of d'Alembert, an adjunct might receive something too. In addition, one earned a small sum for attendance at meetings.[3] Already in 1716 the academicians complained that they could not live in Paris on 1500# a year; 3000# was probably nearer the minimum needed during most of the century.[4] Consequently only the senior pensionaries had anything approaching a sufficiency from their academic employment, and then usually only after a long wait; promotion generally followed seniority, although the King reserved the right to intervene and the company itself sometimes perpetrated an irregularity.[5]

Nor were provisions for research generous. The Academy's regular budget for general expenses appears to have been about 12,000#. Additional grants might be made for special projects, for example the 12,000# given Réaumur annually to do experiments and to prepare the *Descriptions des arts et métiers*. The fate of this grant is instructive. The government agreed that on Réaumur's death it would go to the Academy. When he and his authority died in 1751 the government punctually honored its commitment: Réaumur's 12,000# went to the Academy, and the Academy's 12,000# went back to the treasury.[6]

To work and to eat the Paris academician had to have a private income, a pension from outside the Academy, or a job. The prestige of his position and his access to influence usually brought him a post, if he needed one, associated with his specialty: teaching at the Collège Royal or at the state military, naval, or technical schools; editing the Academy's almanac, the *Connaissance des temps,* which brought 800# toward the end of the century; special consulting at the mint, mines, or the government porcelain works; drawing maps for the

2. Chapin, *Fr. Hist. St.,* 5 (1968), 385–93.

3. Maindron, *Académie* (1888), 100–1; Chapin, *Fr. Hist. St.,* 5 (1968), 384–6; Hankins, *D'Alembert* (1970), 43.

4. Bioche, *Rev. deux mondes,* 107:2 (1937), 181; R. Hahn, *Min.,* 13 (1975), 501–13. The total for salaries was 50,000# in the 1770s, of which 42,000# went to pensionaries (the secretary and treasurer each had 3000#) and 8000# for doles ('petites pensions').

5. E.g., in the election of Dufay to associate in 1724; cf. Chapin, *Fr. Hist. St.,* 5 (1968), 387, and Hankins, *D'Alembert* (1970), 137.

6. K. Baker, *RHS,* 20 (1967), 248.

Navy; writing for encyclopedias; inspecting industries; serving, semi-retired, in the military.[7] This moonlighting might make ends meet; it also reduced ouput and, perhaps, deflected some from an academic career. 'What kind of work can one expect of savants forced to spend their days on the pavements of Paris instead of in their studies? Is a man who comes home tired and distraught ready for work that demands his full powers? Will he spend his evenings doing experiments? . . . A gifted young man who wishes to follow his scientific bent will find himself opposed by family and friends who do not want to see him enmeshed in studies which, while they might bring him some glory, will certainly lead him to starvation.'[8]

The Academy of Berlin, established in 1700 by the King of Prussia, tried to follow the lead of Paris, including class divisions, salaries, corresponding members, practical applications and instruments furnished by the government. Its income, however, which derived from a monopoly on the sale of almanacs, could not initially maintain more than a president, a secretary, two astronomers and household help. Only a small fraction of the seventy regular members it engaged between 1700 and 1740, when Frederick the Great came to the throne, had had anything but token support; in 1721 only its anatomist received a pension.[9] Moreover the Academy had little international importance; its journal, rightly called *Miscellanea,* appeared at long intervals filled with inhomogeneous and often stale material. These blemishes caused the livelier academicians themselves to crave reform. In 1743, after joining with new spirits drawn to the capital by Frederick's promise, they set up a new, and newly Frenchified, academy.[10]

According to its revised statutes (1746), the Académie Royale de Berlin was a general learned society, consisting of four classes (experimental philosophy, mathematics, speculative philosophy, literature) each made up of three pensionaries and three associates. In addition, there were a top-heavy administration (a president, a secretary, a director for each class), correspondents, and emeriti or 'veterans.' Financial support came from the almanac income, which had grown considerably with the acquisition of Silesia, and which Frederick allowed to reach the Academy intact, a policy his father had not often followed.[11]

Frederick wished his academy to have the best brains in Europe. Shortly after his accession he tried for the most advertised brain in Germany, Wolff's. Wolff preferred to remain a professor and to return in triumph to Halle. As head of his

7. R. Hahn, *Min.*, 13 (1975), 501–13; cf., regarding Clairaut's income, Doublet, *Bull. sci. math.*, 38:1 (1914), 95, 189–90.

8. Maindron, *Académie* (1888), 106, quoting an anonymous Ms. of c. 1720, apparently the work of leading academicians.

9. Harnack, *Gesch.* (1900), I:1, 75, 158–9, 230–1, 240–4; Dunken, *Deut. Ak.* (1960). In 1737 the Academy's mathematical class had six members and its medical-physical class eight, mostly physicians. Hagelgans, *Orbis* (1737), [Pt. ii], 9.

10. Harnack, *Gesch.* (1900), I:1, 248–92.

11. *Ibid.*, 230–1, 299–302; Gottsched, *Hist. Lobs.* (1755), 108–9.

mathematical class Frederick captured Euler from the Petersburg academy. It was as well that Wolff did not come, for Euler detested the monad, which he took in a physical sense and held to violate both mathematics and the Christian faith.[12] The other chief scientific members of the new academy were Newtonians, an orientation urged by Voltaire and Algarotti, who happened to be in Berlin in the early 1740s. For president Frederick recruited Maupertuis, who had the double and unusual advantage of being both attractionist and French. For director of the physics class he chose his physician, J. T. Eller, who had studied in Leyden and even in London. He was a 'passionate Newtonian,' according to Wolff, 'who hotly rejects everything that does not agree with his pitiful principles, which he incorrectly thinks are Newton's.' Frederick had also hoped to have a Dutch Newtonian, but none came.[13]

Although Wolff himself had declined to fight in Berlin, his epigoni J. G. Sulzer and Samuel Formey, members of the Academy's non-scientific classes, were prepared to do battle. The first rounds went to the anti-Wolffians. But after the deaths of Maupertuis (1759) and Eller (1760), the Wolffians gained ground, supported by extreme representatives of the French Enlightenment, whom Frederick delighted to honor. This odd alliance was celebrated in 1766 in a speech on the 'Reconciliation of the Philosophies of Leibniz and Newton' given in the Academy on the occasion of the King's birthday. Later that year Euler sought refuge in Petersburg from rampant rationalism and what he took to be Frederick's disfavor. In the twenty years of his directorship the Berlin Academy had come to rival that of Paris. The old *Miscellanea* were replaced by regular *Mémoires,* printed in French; as Maupertuis explained, his native tongue was perfect, and also rich in words for describing the advances of science.[14] To be entirely Parisian the *Mémoires* wanted only an *Histoire,* which, according to the Academy's secretary Formey, Maupertuis refused to supply out of antagonism to the inventor of the genre, Fontenelle.[15] This imperfection was corrected in 1770.

Salaries at Berlin ran a little higher than at Paris. The average seems to have been about the same, namely 500 RT; but the better emoluments, like the 1600 RT (some 6400#) paid Euler or the 12,000# given Maupertuis, greatly exceeded even the best Paris pensions.[16] Moreover Berlin was cheaper than Paris.[17] It also appears that Frederick wished to raise the average salary of his

12. Winter, *Registres* (1957), 16, 32, 37–8.

13. Brunet, *Maupertuis* (1929), I, 77–8; Stieda, Ak. Wiss., Leipzig, Phil.-Hist. Kl., *Abh.*, 83:3 (1931), 7, 13; Wolff to Manteuffel, 29 Aug. 1747, in Ostertag, *Phil. Geh.* (1910), 117, and Manteuffel's answer, *ibid.,* 118. For Eller, *DSB*, IV, 352–3; *MAS*/Ber (1761), 498–510.

14. Winter, *Registres* (1957), 45–8, 52–6, 68, 84, 89, 91; Brunet, *Maupertuis* (1929), I, 127; Stieda, Ak. Wiss., Leipzig, Phil.-Hist. Kl., *Abh.*, 83:3 (1931), 30–7; Calinger, *Ann. Sci.*, 24 (1968), 239–49.

15. Formey, *Souvenirs* (1797²), II, 254.

16. Harnack, *Gesch.* (1900), II:1, 487–91; Bartholmess, *Hist.*, I (1850), 179; Winter, *Registres* (1957), 14, 17, 19; Clairaut to Cramer, 27 April 1744, in Speziali, *RHS*, 8 (1955), 221.

17. Von Freyberg, *Lehmann* (1955), 52.

academicians, for in trying to woo Mayer from Göttingen in 1758 Euler offered 550 RT (later 700 RT) and the assurance of a raise at the next vacancy, 'for the king intends to reduce the number of members and contrariwise to increase their salaries.'[18] As at Paris, at Berlin academicians had access to further technical jobs, teaching or advising; an unusually successful example is F. K. Achard, once the Academy's leading electrician and director of its physics class, who received an additional pension of 500 RT for assistance to the tobacco industry.[19]

The third major European foundation on the Paris model was the Academy of Sciences of St. Petersburg, consisting of three classes (mathematical, physical, rhetorical) of salaried members, whose number fluctuated according to supply and the policy of the regime. The Academy was the last project of Peter the Great, who conceived that Russia might reach the eighteenth century within a generation or two if the natives could be trained, and the government advised, by Western experts. His academicians were to advance and popularize their sciences, train talented Russians in an 'academic university,' and direct the translation of Western texts to reach those enticed by the popularizations. At the head of this pedagogical pyramid was to sit the inevitable Wolff, with a salary of 2000 rubles (2500 RT). But in 1724, after protracted negotiations, Wolff rejected the offer and stayed in Marburg, where he had just found refuge from the theologians of Halle.

The following year Peter died, and although his successors maintained his academy, they did not implement his general instructional scheme.[20] This failure by no means inconvenienced the academicians. They could do their own work, as long as their relatively high salaries were paid (from 600 to 1000 rubles [750 to 1250 RT] or more[21]). And, if they could endure the weather, the constant changes of government, the dictatorship of the bureaucrats placed over them, the jealousy of the Russian members, and their own bickering, they might find St. Petersburg very 'agreeable.' Or so Euler said in transmitting an offer of 1000 rubles plus moving expenses to Mayer, adding that no one had yet complained of the life and forgetting the reasons that had prompted him to accept a call to Berlin.[22]

Many of the first academicians were Germans or Swiss Germans recruited by

18. Euler to Mayer, 15 May 1753, in Forbes and Kopelevich, *Ist.-astr. issl.*, 10 (1969), 404. In fact, however, salaries were rarely increased, 'however good or bad they are'; Lagrange, *Oeuvres*, XIII, 258.

19. Harnack, *Gesch.* (1900), I:2, 481, 512; Stieda, Ak. Wiss., Leipzig, Phil.-Hist. Kl., *Abh.* 39:3 (1928), 16.

20. Kunik in Wolff, *Briefe* (1860), xiv–xv, xxii–xxv, xxix–xxxiv; Blumentrost to Wolff, 23 May and 27 Dec. 1723, *ibid.*, 167, 171.

21. *Ibid.*, 167, offering between 700 and 800 roubles; cf. *ibid.*, 177–82, and Lipski, *Isis*, 44 (1953), 349–54. The academy's income derived from customs dues and a monopoly on calendars; Kunik in Wolff, *Briefe* (1860), xxix; Vucinich, *Science* (1963), I, 71, 88, 96.

22. Euler to Mayer, 11 June 1754, in Kopelevich, *Ist.-astr. issl.*, 5 (1959), 391; Vucinich, *Science* (1963), I, 72–82.

Wolff, whom the Russians retained at 300 RT a year to suggest candidates and negotiate contracts.[23] Not unnaturally these pioneers inclined toward Wolffian philosophy. The contract of Christian Martini, professor of natural philosophy, bound him to 'expound physics according to Wolff's principles.' His successor, G. B. Bilfinger, 'who [as Lambert put it] in many respects rendered greater service to Wolff's philosophy than did Wolff himself,' had already infected the University of Tübingen with his master's teachings. Euler and Daniel Bernouilli, although they despised Wolff's metaphysics, did not reject his physics. They would have agreed with the spirit of Bilfinger's effort to save the theory of vortices, for which he won a prize from the Paris academicians in 1728: 'Nothing is simpler than the Cartesian vortices; hence I think everything should be tried before they are given up; and if they will not work properly, I should wish them to be changed as little as possible.'[24]

Mechanistic physics had a prosperous career in Russia. Euler dominated mathematics and physics in the Academy until 1741, and continued to influence their development from Berlin. His colleagues Krafft and Richmann shared his views, as did the first important Russian member of the Academy, Lomonosov, who had been trained by Wolff at Marburg. The Academy's first Newtonian physicist, Aepinus, arrived in 1757. Aepinus had moved into the government bureaucracy by the time Euler returned in 1766 to reinvigorate the tradition of Cartesian physics, which, in the easily digested form of his *Letters to a German Princess,* enjoyed a vogue in Russia in the late 1760s and 1770s.[25]

The Kings of France and Prussia, and the Czar of Russia, could afford to spend upwards of 15,000 RT a year on their academies. Lesser potentates contented themselves with lesser academies and fewer academicians. George II of England, as Elector of Hanover, refused to spend more than 450 RT on his Societät der Wissenschaften in Göttingen; it opened in 1751 with five members whose academic salaries summed to 400 RT. They came cheaply because they were professors, and hence already on the royal payroll. The Society improved its income and science by publishing an almanac, a few volumes of memoirs and an important literary review, the *Göttingische Zeitungen* (later the *Göttingische Gelehrte Anzeigen*), which played a role in the history of electricity. But learned publications are not a dependable source of revenue, and the Society remained financially, intellectually, and, by the coopting of professors, personally tied to the university.[26]

23. Blumentrost to Wolff, 19 March 1725, Wolff, *Briefe* (1860), 185; cf. *ibid.*, 43–6.

24. Boss, *Newton* (1972), 105, 110n; Winter, *Registres* (1957), 8–9. For the bitter squabble between Bilfinger and Bernoulli see Ak. nauk, *Materialy,* I, 501–67. Lambert, *Neues Organon,* §632, quoted by Wahl, *Zs. Phil. phil. Kritik,* 85 (1884), 68.

25. Boss, *Newton* (1972), 139, 146–51, 162, 169, 211, 215; *infra,* xvi.3. The *Letters* were written in 1760–1 but not published until 1768. On the negotiations for Euler's return to Petersburg (final settlement: his salary, 3000 roubles; his son's, 1000; and moving expenses for 14 people), see Stieda, Ak. Wiss., Leipzig, Phil.-Hist. Kl., *Abh.*, 83:3 (1931), 26–35.

26. Joachim, *Anfänge* (1936), 29, 36–7, 68–9; Gött., Soc. Wiss., *Sekul.*, 40–53. The membership increased to a maximum of about 15 in the 1770s.

A similar arrangement had existed since 1714 in Bologna. The Accademia delle Scienze dell'Istituto di Bologna had an endowment, library and instruments from a wealthy and well-travelled townsman, Luigi Ferdinando Marsigli, and a building and operating expenses from the city, with the approval of the Pope. Professors were appointed to the Institute to teach and to study subjects outside the university curriculum. Although it was not, as Fontenelle fancied, 'Chancellor Bacon's Atlantis in actuality, and the dream of a savant come true,' the Bologna Academy, like that at Göttingen, was a valuable adjunct to its university.[27]

Among the smaller proprietary academies not tied to universities those of Turin and Munich deserve notice. In 1783 Victor Amadeus III, King of Sardinia, granted a royal charter to a group that had existed in Turin since 1757; it thereby became the Académie Royale des Sciences, at a cost to the king of a building and a 'generous allotment' for its upkeep, for instruments, prizes and medals.[28] The original group, although small, had 'provide[d] models in everything,' according to Lambert, and not least in electricity. The larger Academy strove to retain its quality by admitting only those who had already acquired a reputation by their published works.[29]

The Akademie der Wissenschaften of Munich, established in 1758, received its charter, meeting place, instruments and operating expenses from the Elector of Bavaria and from the usual source of academic funds, the calendar monopoly. In the 1760s it had an income of over 8000 guldens (21,000#) annually, of which about 3000 went to pay its secretary, the directors of its two classes (historical, philosophical), and a few 'professors,' whose emoluments ranged between 600 and 800 guldens (1600 and 2100#) plus lodging, fuel and light. Two-thirds of the members elected between 1759 and 1769 were bureaucrats or priests, in equal measure, and consequently already salaried by state or church. Lay professors made up less than a tenth of the company.[30] In contrast to the Protestant Göttingen Society, the Catholic Munich Academy was set up in opposition to the local university, run by the Jesuits at Ingolstadt. One of the Academy's founders, who wished to make it a 'lodge of Wolffians,' urged the exclusion of Jesuits, 'because they are scholastics and Jesuits,' and supported the appointment of Lambert, a Protestant, to a paid professorship. 'What has orthodoxy to do with mathematics, physics, chronology and calendar making?'[31] Several resourceful experimental physicists were to hold the position

27. Bortolotti, *Storia* (1947), 149–53.

28. Anon., Acc. Sci., Turin, *Mém.*, 1 (1784–5), ii–iii, xiii, xix, xxxii–xxxiii.

29. *Ibid.*, xix; Lambert to Euler, in Bopp, Ak. Wiss., Berlin, Phys.-Math. Kl., *Abh.* (1924:2), 29: 'cette société donne en tout des modèles.'

30. Doeberl, *Entwicklungsgeschichte*, II (1928), 321–2; Hammermayer, *Gründ.-Frühgesch.* (1959), 106, 164, 193, 298, 368.

31. *Ibid.*, 239–40, 248; Lambert thought the pension of 600 guldens (1600#), for which he obliged himself to some administrative work and the provision of three original memoirs a year, sufficient. Lambert to Kästner, 24 March 1761, in Bopp, Ak. Wiss., Heidelberg, *Sitzb.* (1928:18), 14–15.

briefly occupied by Lambert, and one or two made contributions to the study of electricity.[32]

OPEN SOCIETIES

The home of the Newtonians, the Royal Society of London, had obtained its charter, a silver mace, and a property worth £1300 from Charles II in the 1660s; its operating expenses at first came from the pockets of its Fellows, who agreed to pay two pounds for admission and a shilling a week thereafter.[33] New members were first elected upon the proposal of a Fellow, and two-thirds of the votes of those present; the recruit entered on the same footing as the other regular members, there being none of those 'disagreeable distinctions' (as Voltaire called them) that ordered the Paris Academy. 'The Royal Society of London lacks two things most necessary to mankind,' he says, 'rewards [pensions] and rules. It is worth a small fortune in Paris for a geometer or a chemist to be a member of the Academy; it costs one in London to be a Fellow of the Royal Society.'[34] The Fellowship was free and poor, and long poorer than it needed to be; its unpaid dues mounted steadily, to almost £2000 in 1673. Newton, who had been excused his dues, moved energetically against malingerers on assuming the Society's presidency in 1703.[35] Further tough measures and gifts of land and stock secured the Society's finances without raising the dues. After 1752 the admission fee was five guineas, the increase owing to the Society's assumption of financial responsibility for the *Philosophical Transactions* in that year. Most new Fellows preferred a composition fee or life membership, which rose from 20 guineas to 26 after 1752.[36] Although by no means a small sum, it probably worked no hardship on those otherwise eligible for election.

At first recruitment proceeded apace, and the Society numbered 199 in 1671. It declined to about 150 in 1700, and picked up again with Newton's presidency. The annual recruitment almost doubled, from about nine a year in 1700 to fifteen anually between 1701 and 1720; for the rest of the century it averaged 23. The membership reached 303 in 1741, 545 in 1800. In the same period its income rose from £232 to £1652.[37] More members meant more money; for a time admission came very easily, particularly for foreign candidates. Voltaire had the idea that anyone who declared his love of science, and deposited his fee, was immediately received a member, and d'Alembert is said to have boasted that he could arrange the election of any traveller bound for England,

32. Ildefons Kennedy (Benedictine), F. X. Epp (ex-Jesuit), M. Imhof (Augustinian), Coelestin Steiglehner (Benedictine). Cf. Westenrieder, *Gesch.*, II (1807), 110–11, 390, 415.

33. RS, *Record* (1940⁴), 9, 11, 24, 93; Weld, *Hist.* (1848), I, 100, 145.

34. RS, *Record* (1940⁴), 10–11; Voltaire, *Lettres* [1734] (1964), II, 170–1.

35. RS, *Record* (1940⁴), 44–5; Weld, *Hist.* (1848), I, 231, 250; Hunter, RS, *Not. Rec.* 31(1976), 17–18, 24–5, 49–57.

36. RS, *Record* (1940⁴), 36, 94–5; Weld, *Hist.* (1848), I, 181–2, 523, II, 43.

37. RS, *Record* (1940⁴), 49; Weld, *Hist.* (1848), I, 473, II, 51, 227–8; Hunter, RS, *Not. Rec.*, 31(1976), 26–32.

'should he think it an honor.'[38] The society countered such censure by requiring each nomination to be put forward, in writing, by at least three Fellows, and by admitting no more than two foreigners a year, 'till [their] number be reduced to eighty.'[39]

Since the ordinary income of the Society seldom exceeded £1000 before 1790, and in its first years did not reach half that, it could not afford to engage much help. Its earliest paid staff, exclusive of household help, consisted of the principal Secretary and a Curator. The former, Oldenburg, was to have £40 a year, the latter, Hooke, £30 and an apartment in the Society's rooms. But it appears that Oldenburg received no regular salary until 1669, and that Hooke agreed in 1662 to 'furnish the Society every day they meet [once a week], with three or four considerable experiments, expecting no recompense until the Society get a stock enabling them to give it.'[40] That happened in 1664. Thereafter special curators of experiments were engaged from time to time, the last of whom, J. T. Desaguliers, held the post from 1714 to 1743. His pay varied from about £10 to £40, depending upon the number of experiments he furnished. After Desaguliers' time the post lapsed, as the Society's statutes of 1775 explain, because the Fellows themselves had become 'so well acquainted with the mode of making experiments, that such accomplished curators have not been found necessary.'[41] Meanwhile the burden of the Secretary had been divided, and a clerk-librarian engaged. In 1780 the two chief secretaries, both members of the Society, received £70.5, a little more than junior pensionaries of the Paris Academy, and the Clerk, a full-time employee ineligible simultaneously to be a Fellow, had £220. These emoluments were raised to £105 and £280 respectively, in 1800, in acknowledgment of the inflation at the end of the century.[42]

Salaries and household expenses left little for research. The Society gradually built up a good library and collection of curiosities, mainly by donation, but it could not treat members to the services of mechanics or instrument makers or commission special apparatus. Very rarely it paid for a research project, as when it gave a man a guinea for permission to transfuse twelve ounces of sheep blood into him. (Why sheep's blood? Because, according to the transfusee, who survived the trial, 'the blood of a lamb has a certain symbolic relation to the blood of Chirst, since Christ is the agnus Dei.'[43]) A worthier project was measuring the gravity of a mountain, a work of Newtonian piety that cost the

38. Voltaire, *Lettres* (1964), II, 171; d'Alembert as quoted in Weld, *Hist.* (1848), II, 152.

39. These measures were introduced in 1728–30 and 1761, respectively. RS, *Record* (1940[4]), 49–51; Weld., *Hist.* (1848), I, 459–61.

40. *Ibid.*, 137–8, 173, 204, 360; RS, *Record* (1940[4]), 11. According to Magalotti, who attended meetings in the 1660s, Hooke performed experiments chosen by the Secretary from among those suggested by the Society. Crino, *Fatti* (1957), 159.

41. RS, *Record* (1940[4]), 30; Weld, *Hist.* (1848), I, 286–7, II, 87–8.

42. *Ibid.*, I, 302–3, II, 228–9. Halley resigned his fellowship in 1686 to become the Society's first Clerk, at a salary of £50. F. Hauksbee, Jr. received about the same (£47.35) in 1763. Henderson, *Ferguson* (1867), 273–4.

43. Weld, *Hist.* (1848), I, 220–1.

Society almost £600 in the early 1770s. More consistent with its means was appeal to the government for grants for meritorious projects, such as expeditions to observe the transits of Venus in the 1760s. The budget of one observer, the same man who measured the attraction of the mountain, Neville Maskelyne, may be of interest: estimated expenses for eighteen months came to £290, of which £140 were needed for 'liquors.'[44]

Since the Fellows undertook almost all their researches at their own charge and initiative, and since their scientific attainments varied greatly, their activities ranged from inexpensive observations of two-headed cows to elaborate calculations in the style of Newton. The *Philosophical Transactions* were of unequal quality. 'It is not astonishing that the memoirs of our Academy are superior to theirs,' says Voltaire; 'well disciplined and well trained soldiers must in the end overpower volunteers.'[45] But volunteers, who outnumber regulars ten or twenty to one, can afford frequent misfires, provided some among them are clever or lucky enough occasionally to hit the mark. Even during the presidency of Martin Folkes, when, according to the Society's nineteenth-century historian, the *Philosophical Transactions* had more than their usual quantity of puerile and trifling papers,[46] they also carried reports of important and original experiments, particularly on electricity.

The Royal Society had a few domestic descendants late in the eighteenth century, notably its namesake at Edinburgh (1783) and the Manchester Literary and Philosophical Society (1781), the latter organized as a corrective to indolence: 'Science, like fire, is put in motion by collision,' its founders said, confidently looking forward to productive bumps among its members.[47] Both societies supported themselves by dues, a guinea per member per year; neither could afford even part-time support for a research or teaching position. The wealthiest and, for our subject, the most important of the learned academies inspired by English example was created not in Britain, but in Sweden, as a result of the pushing of, among others, Mårtin Triewald, who had lived for many years in England, knew the leading Newtonians there, and was himself a member of the Royal Society.[48]

True to its model, the Swedish Academy of Sciences at Stockholm at first received nothing from the crown but the right to call itself royal. It had its operating expenses from its members, a third of whom were aristocrats or high civil servants, and a fifth professors. Their contributions, from a ducat (about 0.5 guinea) to 300 Dkmt (100#) could scarcely have caused them hardship.[49] Help soon came, however: bequests that amounted to well over 70,000# in

44. *Ibid.*, II, 79–82, 14–15.
45. Voltaire, *Lettres* (1964), II, 171–2.
46. See Weld, *Hist.* (1848), I, 483–5.
47. Manch. Lit. Phil. Soc., *Mem.*, 1 (1781–3), vi; RS, Edinb., *Trans.*, 1 (1788), 1–15. For the elaborate bumping behind the RSE see Shapin, *Br. J. Hist. Sci.*, 7 (1974), 1–41; for bibliography of provincial English academies, Schofield, *Hist. Sci.*, 2 (1963), 76–7.
48. Lindroth, *Hist.* (1967), I, 2–4, 14; B. Hildebrand, *Förhistoria* (1939), 140–2, 257, 281–2.
49. Lindroth, *Hist.* (1967), I, 28–32, 102–3; the membership reached 64 in 1742, and then rose to, and remained about, 100 (*ibid.*, 12, 15).

35 years, and, most important, control and profit of the almanac business, acquired in 1747. By 1765 the almanacs brought between 15,000# and 20,000#; in the sixties and seventies the total income of the Stockholm Academy— including income on its endowments, often lent to members at 6 percent— was about half that of the Paris Academy.[50] Most of this money went to support research or travel. Only a little over a fourth of the total went to salaries. The most important office of the Society, the secretaryship, worth only 1800 Dkmt (600#) in the 1740s, brought a good salary, 3000#, in 1771. The emolument for the only other significant paid post, a lectureship (later professorship) in experimental physics, also rose from below subsistence level (2000 Dkmt when founded in 1759) to adequate (6000 Dkmt or about 2000# in 1776).[51]

The lecturer in physics had to show experiments to general audiences and to report his researches for publication in the Academy's *Handlingar* or *Transactions*. Accordingly the Academy made available substantial funds for the purchase of instruments. The first full-time lecturer, Johan Carl Wilcke, who held the post for almost forty years, made capital contributions to the study of electricity. After accepting his position, he published his work, as anticipated, in the *Handlingar*. Fortunately their value was so great, particularly for technology and physics, that it proved financially profitable to translate them into German in their entirety. The indefatigable Kästner saw the opportunity and seized it: he taught himself Swedish and issued 53 volumes of faithful translations between 1740 and 1790.[52] These translations, a vigorous correspondence, a capable membership and an excellent income made the Stockholm Academy one of the leading learned societies of the Enlightenment.

A less distinguished form of the Academy of the London type bred promiscuously in the provinces of France. To be sure these societies often aped the Paris Academy. They divided themselves into classes (mathematical, physical, historical, literary) and orders (regulars, associates, correspondents, honoraries), fixed the number of regulars (from eight, as at Agen, to forty, as in the final specifications for Lyon and Bordeaux), and set up requirements of residence and attendance.[53] But they followed the Royal Society of London in most important respects: meagre resources; miscellaneous memberships, drawn mainly from nobles, lawyers, and high clerics;[54] and the freedom to study what they pleased, so far as their poverty and competence allowed. Usually they

50. *Ibid.*, I, 104–9, 147, 150, 153.

51. *Ibid.*, 45–6, 51–2, 462–70.

52. *Ibid.*, 185–98, 208. On the practical orientation of the *Handlingar, ibid.*, 114–15.

53. Delandine, *Couronnes* (1787), *passim;* J. Bernoulli, *Letters,* II (1777), 131–2. The most Parisian in organization was Montpellier, which enjoyed the privilege of submitting one memoir a year for publication by the Paris Academy. Dulieu, *RHS,* 11 (1958), 232–3.

54. A good example is the general academy (science, literature, technology) of Besançon, established in 1752, with 40 members, almost all 'gens d'Eglise, de robe et d'épée'; the 'scientific' members were a physician, an engineer, and a surgeon (Cousin, *RHS,* 12 [1959], 327). Cf. the legal-financial domination at Pau (Desplat, *Milieu* [1971], 41), Bordeaux, Dijon, and Châlons-sur-Marne (Roche, *Livre* [1965], 176).

chose to study the natural and human history of their region, its curiosities, agriculture, commerce and industry. Often enough they did nothing at all. 'They make many promises, but their ardor soon cools; few are willing to force themselves to compose thorough and thoughtful works.'[55] The pretentions, incompetence, and jealousies of provincial academicians often made them ridiculous. 'Ci git qui ne fut rien pas même academicien,' reads a gravestone in Dijon.[56] Members of the Paris Academy, particularly Condorcet, tried unsuccessfully to improve provincial performance by encouraging the association of smaller societies with larger ones,[57] such as Bordeaux, Lyon, Montpellier or Toulouse, which contributed their mites to science, and provided a few opportunities for scientific careers.

The history of the Bordeaux Academy, almost the earliest and certainly the most active of these societies, will illustrate their vicissitudes and opportunities. Founded in 1712 primarily by the parliamentary aristocracy under the protection of a local magnate, the Duc de la Force, the Academy at first struggled to meet its housekeeping expenses. Most of its income came from dues, which were set so high (300#) that few paid them regularly; they were dropped altogether towards 1732 when income from 60,000# given by de la Force became available. Other wealthy citizens made gifts or bequests: the Academy soon had its own buildings, a library, a natural-history collection, a cabinet de physique, and an average income, from 1739 to 1771, of about 3000# a year. It began to overextend itself. In the 1740s it bought 4500# of physical instruments; it hired a curator and a librarian, and sponsored public lectures; by 1761 it could no longer meet its debts or pay its employees. Further gifts and better administration renewed its income, which averaged 7000# a year from 1772 to 1780.[58] The same financial difficulties, and, though less often, the same accomplishments, are met with elsewhere.[59]

Bordeaux was an academy of 'sciences, literature and arts.' At first it emphasized physics, and, in the 1740s, experimental physics. After the midcentury it moved towards technology, agriculture and commerce. These subjects then were fashionable; a dozen provincial agricultural societies were created in the 1760s. Even academies that had been entirely literary, such as Arras, Caen, Grenoble, Nîmes and Pau, began to cultivate agronomy.[60] 'In our

55. Sequier (Nîmes) to Condorcet, 1774; quoted by Barrière, *Académie* (1951), 349.

56. 'Here lies one who was nothing, not even an academician'; the gravestone is exhibited in the Musée des Beaux Arts, Dijon. Cf. Roche, *Livre* (1965), 105.

57. Baker, *RHS*, 20 (1967), 258, 262, 266–77. Cf. Priestley, *Hist.* (1775³), I, xviii–xx, recommending progress by subdivision of labor, by funnelling money from large academies to small.

58. Barrière, *Académie* (1951), 20, 25–7, 31–3, 39, 97–8.

59. For example Montpellier, too poor to publish its memoirs regularly, nonetheless supported a lecturer in physics from 1780, with the help of the Estates of Languedoc (Dulieu, *RHS,* 11 [1958], 234–7). There was a similar position at La Rochelle from 1785 (Torlais, *RHS,* 12 [1959], 111–25) and one in chemistry at Dijon from 1783 (Delandine, *Couronnes* [1787], I, 259–63).

60. Barrière, *Académie* (1951), 350–1, 354–5; Delandine *Couronnes* (1787), I, 190–1, 244–5, 276, II, 54–7, 64–9. Cf. Roche, *Livre* (1965), 163–8.

time,' wrote a correspondent of Voltaire's 'all the women [he might have said provincial academicians] had their beau esprit, then their geometer, then their abbé Nollet; nowadays [1760s] it is said that they all have their statesman, their politician, their duc de Sully.'[61] It was in the first half of the century that the provincial academies contributed to the study of electricity. Thereafter they occasionally helped indirectly, as the Montpellier Academy encouraged Coulomb, but the subject no longer answered the interests, as it eluded the competence, of most provincial academicians.[62]

The establishment of scientific societies came late to Holland, among other reasons because the universities and independent lecturers satisfied much of the interest in experimental philosophy. In 1756, however, domestic pride and foreign example encouraged the formation of the Hollandsche Maatschappij der Wetenschappen at Haarlem. The organization of this body, unique at the time, anticipated that of the Kaiser Wilhelm Gesellschaft in our century: funds were supplied by 'directors,' who paid an admission fee of 60 florins (130#) and dues of $f.50$ (later $f.100$); the work was done by 'scientific members,' who had access to the society's books, instruments and collections, but no salary. It began with 23 members, all local, including several professors from Leyden; foreign members were elected beginning in 1758 and, as we have seen, so promiscuously as to bring discredit on the Academy, and measures to decrease their number.[63]

To minister to its collections and correspondence, the Academy employed a secretary at $f.700$ (from 1777 $f.1000$) and, from 1777, a curator at $f.300$ and fringe benefits that included housing among the specimens. This curator, who held the job throughout the century, was Martinus van Marum, the leading Dutch electrician of his time.[64] His work on electricity was supported by Teyler's Tweede Genootschap, a small general-purpose learned society set up at Haarlem under the will of a Peter Teyler in 1778. As a member of this society and, from 1784, the director of its library and collections, van Marum had $f.1500$ a year and the confidence of its officers, who supplied the money for his great electrical machine.[65]

ACADEMIC FUNCTIONS

The purpose of the academies was the advancement of science and technology. The Royal Society's charter of 1663 specifies its goal as 'promoting Naturall Knowledge,' which the members took to mean useful information (as opposed to scholastic speculation) about the physical world (as opposed to the super-

---

61. Pierres de Bernis to Voltaire, 26 July 1762, in Voltaire, *Corresp.*, XLIX, 139–40. Nollet stands for experimental physics; Sully for finance, commerce, transportation.

62. Cf. the prize questions, *infra*, ii.2.

63. Bierens de Haan, *Holl. Maat.* (1752), 3, 8–11, 273–4; *supra*, i.7.

64. Bierens de Haan, *Holl. Maat.* (1752), 43–4; Muntendam in Forbes, *Marum*, I, (1971), 17–18, 41. Van Marum became secretary of the Society in 1794 while retaining the curatorship; the Society then could not afford to pay its secretary anything (*ibid.*, 33–4).

65. *Ibid.*, 5, 20–1; *infra*, xviii.

natural).[66] The same concept recurs in the names of many later foundations: the Hollandsche Maatschappij tot Voortsetting en Aanmoediging van Nuttigge Konsten en Wetenschappen, the American Philosophical Society for Promoting Useful Knowledge, the Kungl. Svenska Vetenskapsakademie [for improving] Vetenskaper och Konster 'som tiena til en almän nytta.'[67] The program of the New Atlantis, of perfecting the arts by perfecting the sciences, of simultaneously seeking fruit and light, or the useful and the true, also informed the plans made by Leibniz for the Berlin Academy, and the hopes of the provincial academicians of France.

The balance between the two parts of the program was struck in different ways by different institutions. Stockholm excepted, the more prestigious the society the closer it stood to pure science; in 1798 Frederick William II of Prussia reprimanded his academicians for having moved too far into 'speculative investigations' at the expense of 'works of general utility.'[68] There is no doubt that the academies' implied promise to improve man's estate, as well as the services they rendered as technological consultants and patent officers, gave rationale for much of their support. To us, however, the chief feature of the academies is not their promise to be useful but their explicit dedication to the 'cultivation,' 'advancement,' or 'promotion' of their sciences, or, in a word, to research. 'Ein Academiste muss erfinden und verbessern oder seine Blösse unvermeidlich verrahten.'[69]

It was commitment to research that distinguished the academician. Not until the end of the eighteenth century, and then only in the leading universities, did the cultivation of science—as opposed to its preservation and dissemination —begin to be a responsibility of the professoriate. In theory the Academy complemented the University; the one taught the known, the other explored the unknown. The putative duties of the Petersburg academicians, to 'cultivate their sciences and give a course of lectures once a year,' implied a new sort of institution, 'not a complete university, not an academy of sciences, but rather a combination of both.'[70] This oddity exacerbated the problems of recruitment: should one seek able and ambitious men, eager to join an academy where they might win reputations as savants, or should one settle for a more common and docile type, 'those who aspire only to be professors?'[71] As it happened the brilliance of the first recruits and the failure of the 'academic university' reduced the planned hybrid institution to a first-class academy.

The best example of the complementary character of academy and university

66. Weld, *Hist.* (1848), I, 126, 138.

67. The elucidation of the purpose of the Stockholm Academy comes from the introduction to the first number of its *Handlingar;* B. Hildebrand, *Förhistoria* (1939), 374.

68. Quoted by Westenrieder, *Gesch.,* II (1807), Vor.

69. Haller (1751), quoted by Joachim, *Anfänge* (1936), 52.

70. Blumentrost to Wolff, 23 May 1723 and Feb. 1724, in Wolff, *Briefe* (1860), 167, 173. Cf. *Encyclopédie,* art. 'Académie': 'Une académie n'est point destinée à enseigner ou professer aucun art, quel qu'il soit, mais à en procurer la perfection.'

71. Memo by A. Golovkin, 1724, in Wolff, *Briefe* (1860), 181.

is Göttingen. The founders of the university wished to create it and its research arm, the Societät der Wissenschaften, together. The Society would provide a supplementary salary to the ablest professors to enable them to reduce their large teaching loads; on their 'released time,' as we might say, they were to meet once a week, 'improve and elucidate' their sciences, and produce an annual volume of memoirs.[72] The regime decided to create the university first. The Society followed fifteen years later, with the modest numbers and stipends already mentioned. Its ambitions were anything but modest: 'to increase the realm of knowledge with new and important discoveries, encourage professors both to write solid works and to apply themselves to their lectures, and to spur students to praiseworthy zeal for science and good morals.'[73] If we disregard Bologna, the arrangements at Göttingen were unique in Europe, and proved useful in argument against those who objected when 'learned university professors or those capable of making discoveries become academicians.'[74]

As this argument suggests, universities did not always welcome the establishment of academies in their neighborhoods. The academic spirit of free enquiry opposed the professoriate's commitment to established learning. The conflict was most serious in Southern Germany and Austria; the Jesuits there helped to defeat plans for establishing an academy in Vienna in 1749–50, and the one in Munich was set up against their protests and, as we have seen, to their exclusion.[75] Another ground for professorial opposition to academies was jealousy, expressed as a fear that the new institutions would diminish the lustre of the old. An example is the campaign mounted by the Senate of the University of Leyden against the Hollandsche Maatschappij. 'The lustre of the University [they said] does not derive entirely from the merit and importance of its professors, but also from its authority [which would be greatly impaired] if a second society of letters were established in the province.' Look at France and England. Oxford and Cambridge have declined since the establishment of the Royal Society, and the University of Paris, formerly so famous, 'has scarcely been heard from since the Royal Academy has been made to flourish there under the particular protection of the King.' This interested argument could not arrest the progress of the Maatschappij, but it did result in introducing two restrictions in its charter: the new society should not sponsor public lectures and —to hide its light as much as possible—its publications were to be entirely in Dutch.[76]

72. Joachim, *Anfänge* (1936), 5–8; Smend, Ak. Wiss., Gött, *Festschrift* (1951), vi.

73. *Ibid.*, 15–16.

74. F. A. Wolff, 'Berliner Universitätsdenkschrift' (1807), quoted by Smend in Ak. Wiss., Gött., *Festschrift* (1951), vii.

75. Huter, *Fächer* (1971), 6; Huber, *Parn. Boic.* (1868), 3, 14; Westenrieder, *Gesch.*, II (1807), 93. The Jesuits were excluded 'non par les loix mais par voie de fait': Lambert to Kästner, 24 Jan. 1764, in Bopp, Ak. Wiss., Heidelberg, *Sitzb.* (1928:18), 16. Universities and neighboring academies sometimes worked together; besides Göttingen, Uppsala and Stockholm (Lindroth, *Hist.* [1967], I, 33), and Halle and Berlin.

76. Bierens de Haan, *Holl. Maat.* (1752), 32–5; the victorious society later joined with the University in unsuccessfully opposing the establishment of a rival academy in Rotterdam, the

The argument of the Leyden Senate had some merit; at least as regards natural philosophy, the University suffered the decline it anticipated. The academies did attract men who, had they not had such opportunities, might have increased the stock of research-oriented members of the universities. Or, to put the point the other way, the academies very quickly lost their importance as scientific institutions, though perhaps not as pressure groups, once research became an expected, and supported, professorial activity.

The eighteenth-century academy promoted science in three ways. First, intramurally, by the mutual encouragement of its members at its frequent meetings, which, as a rough rule, took place weekly for the most distinguished societies, fortnightly for the less, and monthly for the least. To this must be added salaries and instruments that supported the work of members of academies rich enough to provide them. A second service was the publication of the results of the researches of members and associates. The memoirs of the chief academies of the Paris type carried only the work of its members; those of more open societies, such as the *Philosophical Transactions* of the Royal Society, printed papers by unaffiliated people when submitted through a Fellow. The majority of scientific work published in the eighteenth century appeared in the periodicals of learned societies; both the big book and the independent scholarly journal, like the *Journal des sçavans* and the *Acta eruditorum*, became rapidly outmoded as outlets for research results after 1700. The connection of the decline of the *Acta* with the multiplication of academies was recognized at the time.[77]

There is balance in all things. The proliferation of academies created quantities of publications often difficult to procure[78] and impossible to survey. Moreover, they did not always appear regularly, and even those that did came out a year or more after the memoirs they contain were first presented. The delay of the *Philosophical Transactions* averaged eighteen months; of the *Mémoires* of the Berlin and Paris academies, two and three and a half years, respectively; of the Petersburg *Commentarii*, almost five years. Hence the dates printed on the memoirs and cited in our notes should not be interpreted as the time at which they became generally available. Taking into account the delay caused by difficulties of travel and the inefficiencies of the book trade, one should allow on the average a minimum of three years from the date of presentation of a memoir (usually also the date of the volume containing it) to the time of its arrival, printed and bound, on the library shelves. To improve the system review journals were started, such as the *Commentarii de rebus in scientia naturali et medicina gestis* (Leipzig, 1752–98), and, more valuable yet, periodicals that both excerpted academic publications and provided quick and

---

Bataafsch Genootschap der Proefondervindelijke Wijsgebeerte (*ibid.*, 37–9), which countered by pointing out the number of competing academies in France.

77. Cf. Kästner to Maupertuis, 15 April 1750, in Kästner, *Briefe* (1912), 8–9.

78. Much of the learned correspondence of the time concerns the procurement of books and journals. Even Wolff at Marburg had trouble obtaining the *Commentarii* of the Petersburg Academy. Wolff to Schumaker, 1727, in Wolff, *Briefe* (1860), 97.

accessible publication of research results. Of these the most important in the early history of electricity was the *Journal de physique* (Paris, from 1773).[79]

The third and characteristic promotional activity of the academies was the prize competition. A society too poor to award an occasional prize for an essay on a subject of its choosing scarcely qualified as an academy.[80] The major institutions offered one and sometimes more prizes a year; and they were prizes worth having, ranging from the 2500# Rouille prize at Paris through the 50 or 60 ducats (550# to 660#) of Munich and Berlin down to about 300#, as at Bordeaux, Dijon, Marseille, Montpellier and Rouen.[81] The poverty-stricken society at Pau could manage only a little over 100#. Since a respectable award amounted to a sizeable fraction of the income of the smaller academies,[82] they could offer them only when a donor could be found. That happened surprisingly often. Delandine, writing in 1787, lists 1037 competitions proposed in France alone since the foundation of the Paris Academy, and warns that the number of jousts has begun to exceed the supply of knights: 'Already there are no longer enough men of letters and of science to compete successfully for the number of prizes annually proposed.'[83] Many of the competitions, as he says, were slight and parochial, and their subjects and winners soon forgotten; but those of the national academies often brought intense competition, advanced the careers of the victors, and influenced the course of science.

In theory academies chose prize questions for their timeliness, and judges did not know who had written the essays presented in competition. In practice distinguished savants might be invited to compete and the question set to attract them,[84] while the judges could frequently identify the authors of the anonymous papers before rendering their verdict. This foreknowledge came most easily in mathematical tourneys, in which the same men were alternately the judges and the judged, and knew one another's handwriting. One complained that Parisian academicians favored contenders from Berlin over those from St. Petersburg; that d'Alembert could not win a competition of which Euler was a judge, or Daniel Bernouilli one over which Maupertuis had influence.[85] The ethical level of the business may be gauged from Lagrange's remark that, had

79. For the chief rationale of the *Journal*, the proliferation of journals, see *JP*, 1 (1773), iii–iv; K. Baker, *RHS*, 20 (1967), 262–7; *supra*, i.1.

80. Cf., Roche, *Livre*, 157–8, and Lambert's proposal for instant glory for the Munich Academy: offer a big mathematics prize to attract Euler and friends. Lambert to Kästner, 24 Jan. 1764, in Bopp, Ak. Wiss., Heidelberg, *Sitzb.* (1928:18), 16.

81. Maindron, *Académie* (1888), 13–14; Delandine, *Couronnes* (1787), *passim;* Dulieu, *RHS*, 11 (1958), 235; Barrière, *Académie* (1951), 351; Roche, *Livre*, 160.

82. Desplat, *Milieu*, 73–5, 123.

83. Delandine, *Couronnes* (1787), I, vii. Consequently prizes were withheld more and more frequently in France after 1750; Desplat, *Milieu*, 72; Roche, *Livre* (1965), 161–2.

84. Cf. Bouguer to Euler, 8 April 1754, in Lamontagne, *RHS*, 19 (1966), 238.

85. Frisi's complaints in Costa, *Riv. stor. ital.*, 79 (1967), 873–4; d'Alembert to Lagrange, 26 April 1776, in Lagrange, *Oeuvres*, XIII, 316; Lamontagne, *RHS*, 19 (1966), 229; Hankins, *D'Alembert* (1970), 45–6, 49, 59; Winter, *Registres* (1957), 71.

he known that Condorcet was the author of an essay for a certain competition, he 'would have made an effort to have the prize awarded to him.'[86]

The choice and phrasing of the question, and the informed adjudication of the prizes, gave many opportunities to influence the development of a field. The Paris Academy, for example, liked to reward theoreticians of the vortex, as in Bilfinger's victory in 1725 and the essays on magnetism of the 1740s; no doubt the hope of winning 1000# or 2000# strengthened the resolve of wavering Cartesians. The same influence was exerted by the two most important competitions on electricity, the Berlin of 1745 and the Petersburg of 1755; both in the statement of the question and in the awards the academicians approved and confirmed a Cartesian approach. Other prize questions directed attention to the relation between electricity and magnetism (Lyon, 1747), and between electricity and lightning (Bordeaux, 1748).[87] Perhaps the most important contribution of the prize competitions to the study of electricity was Coulomb, who came to his measurements of electrical force by following up work for his winning entry for the Paris prize on magnetism of 1777.

The continental organized prize competition fit neither the permissiveness nor the purse of the Royal Society of London. Rather than stimulate research on a specified subject, the Society preferred to reward, chiefly with nonnegotiable honors, the author of any discovery or invention that its administrators deemed worthy. Two such distinctions encouraged eighteenth-century physicists, the Copley medal, formally established in 1736,[88] and the Bakerian lectureship, initiated in 1775. The endowment of each was £100, making the income, and hence the value of medal or lectureship, about £5 (125#). Many received awards for electricity. Among Copley medallists were Gray, Desaguliers, Watson, Canton, Franklin, Wilson, Priestley, Volta; among Bakerian lecturers, Ingenhousz and Cavallo.[89] The Manchester Literary and Philosophical Society likewise rewarded the best work done, rather than the best answer to a set question.[90] So did the Stockholm Academy until 1760; thereafter it gave prizes in the continental manner, but—perhaps to illustrate its most frequent subject, economy—usually of comparatively little value.[91]

---

86. Lagrange to d'Alembert, 1 Oct. 1774, in Lagrange, *Oeuvres*, XIII, 292.

87. Delandine lists 11 French prizes on electricity, which divided into two groups: 4 prizes, 1747–9, on physical properties; 7 prizes, 1760–83, on applications to plants and animals (3 prizes not awarded). There were at least 18 competitions on electricity throughout Europe in the Ancien Régime; Barrière, *Académie* (1951), 136.

88. The income from the Copley bequest, received in 1709, initially paid part of the salary of the curator, Desaguliers; the first award of a medal was made in 1731, to an electrician. RS, *Record* (1940⁴), 112, 345; Weld, *Hist.* (1848), I, 385; Wightman, *Physis*, 3 (1961), 346–8.

89. RS, *Record* (1940⁴), 345–6, 364–5.

90. Manch. Lit. Phil. Soc., *Mem.*, 1 (1781–3), xvi, offering a silver medal of about two guineas' value annually to 'encourage the exertions of young men.'

91. From about 50# to 300#; Lindroth, *Hist.* (1967), I, 143, 150; Nordin-Pettersson, Sv. Vet., *Årsbok* (1959), 435–516.

3. PROFESSORS

Between one-quarter and one-half of the electricians active at any time during our period were 'professors' in a university or secondary school (Table 2.4). The rationale for grouping all professors together is that the level of instruction in the university's philosophical faculty, where the physicist usually held forth, did not differ much from what prevailed in the better secondary schools. It was not unusual for a professor to exchange a university for a college post or to hold both simultaneously. Some secondary schools had stronger philosophy courses than some universities; the Jesuit college at Lyon, the Académie de Calvin at Geneva, the Scuole palatine in Milan, the gymnasium in Nuremberg, the dissenting Academy at Warrington, were more distinguished than the 'dwarf universities' (to borrow Eulenburg's phase) of Germany and Italy.[1]

Although in principle universities differed from colleges in possessing schools of law, medicine and theology, they often lacked some, and sometimes all, of these 'higher faculties.' The only reliable mark of a university was a legal one, its right to grant degrees. Our profile of the eighteenth-century professoriate will be drawn primarily from information about universities so defined. For reasons already given, however, it will apply to instructors in senior classes in secondary schools as well.

Almost every European—as opposed to British—university in 1700 had a professor responsible for instruction in physics. There were perhaps 75 such men, and almost all professed a literary, all-inclusive physics. Their number did not increase much during the century. Consequently, to account for the entries in Table 2.5, we must suppose that, by the 1740s, a good fraction of physicists were doing experiments.[2] This new activity, when expressed in alluring demonstrations, brought a wider audience than literary physics could command, and, by its requirements of space and equipment, made the professor a more expensive—and consequently a more valuable—member of the faculty. His prestige and value also rose outside the university, at least among those concerned to modernize instruction; for experimental physics, like economics, history, vernacular instruction and Cameralwissenschaft, breathed the spirit of Enlightenment.

THE SETTING

The professor of experimental physics was not commonly an experimental physicist, or expected to be one. The proposition we have already met—the professor teaches, the academician researches—was emphasized as strongly by the eighteenth-century professoriate as by the spokesmen for learned societies.

1. Eulenburg, *Frequenz* (1804), 2–3. See Borgeaud, *Hist.*, I (1900), 498–9, for the discussion of a defeated proposal to seek university status for the Geneva Academy in 1708.

2. Table 2.5 includes contributions from professors other than physicists, and omits experimental work other than electricity done by the physics professoriate. If the former were subtracted and the latter added, the numbers in the table would doubtless be larger.

TABLE 2.4

Ratio of Professorial to All Electricians

	1600–99	1700–39	1740–49	1750–59	1760–69	1770–79	1780–89
Professorial	9(9)						
Table II.1, row C	4(4)	10(10)	31(24)	22(16)	16(7)	29(15)	25(17)
Ibid., row A	5(5)		6(2)	5(3)	5(2)	6(2)	1(1)
		37(26)	27(19)	21(9)	35(17)	26(17)	
Ratio	.33(.33)	.50(.53)	.42(.39)	.38(.35)	.45(.28)	.49(.44)	.33(.31)

TABLE 2.5

Breakdown of 'Professors'[a] from Table 2.1

	1600–99	1700–39	1740–49	1750–59	1760–69	1770–79	1780–89
Catholic Universities[b]	2(2)	2(2)	3(3)	7(5)	4(3)	13(6)	7(5)
Austria					1(0)	1(0)	1(0)
France		2(2)			1(1)		
Germany			3(3)			3(2)	2(2)
Italy[c]	2(2)			7(5)	2(2)	9(4)	4(3)
Protestant Universities	2(2)	7(7)	21(17)	11(9)	7(4)	10(5)	8(7)
Britain	1(0)		1(1)		2(1)	1(1)	1(1)
Germany	1(0)	7(5)	13(10)	6(5)	1(0)	5(3)	5(4)
Netherlands		2(2)	2(2)	2(1)	2(1)	2(0)	1(1)
Scandinavia			4(3)	2(2)	2(1)	1(0)	
Switzerland			1(1)	1(1)	2(2)	1(1)	1(1)
Secondary Schools		1(1)	7(4)	4(2)	5(2)	6(4)	10(5)
Protestant		1(1)	5(3)	3(1)	2(2)	2(1)	3(0)
Catholic			2(1)	1(1)	3(0)	4(3)	7(5)
TOTALS	4(4)	10(10)	31(24)	22(16)	16(9)	29(15)	25(17)

[a] The author (or respondent) of a doctoral thesis is counted as one-half; all fractions are rounded *up*.
[b] Exclusive of Jesuits, for whom see Table 2.1, row A.
[c] Lombardy is counted as Italy.

One such affirmation occurs in Johann Christian Förster's centennial history of the University of Halle, a context that gives it a special authority, for Förster was a progressive professor of philosophy, an economist, the supervisor of the University's botanical gardens, and Halle was one of Germany's most advanced higher schools. According to him, 'a professor by no means needs to discover new truths or to advance his science. Should he do so, he is in fact more than an academic teacher, he has done opera supererogationis,' he has worked beyond the call of duty.[3] The prescient pro-rector and academic senate of the University of Marburg sounded the same theme in 1786 in opposing the establishment of a Hessian academy of sciences that might tempt and turn professors from their Hauptwerk, teaching.[4] The minister most responsible for the foundation of the University of Göttingen, G. A. von Münchausen, wanted to require the faculty to improve their sciences as well as their students. They soon set him right. 'Whoever justly considers the various duties and offices of professors,' said Albrecht von Haller, speaking at the opening of the Göttingen Society of Sciences, 'will easily see that so great a burden falls upon them that it is entirely unfair to ask them to do any special scientific work, *peculiares singularium inquisitionum labores.*' 'To do more than teach,' says his colleague Michaelis, 'is to do the work of societies of science.'[5]

And yet there is no doubt that by 1790, at the leading universities, calls and promotions came most easily to those who contributed their bits to science. As early as 1750 Tobias Mayer's appointment to Göttingen specified not only that he teach applied mathematics, but also that he devote himself to 'Forschungsarbeiten,' and later, in countering an offer from the Berlin Academy, the Hanoverian government acceded to Mayer's wishes for research facilities as well as for an increase in salary.[6] Even when their contracts did not explicitly require it, Göttingen professors were expected to write, and did so, 'as if the entire empire of letters acknowledged their academic scepter.'[7] Those who did not measure up, who neither wrote nor researched, were as out of place among Göttingen professors as (to quote their colleague Kästner) 'mouse turds among pepper grains.'[8]

Perceiving that research and writing made the Göttingen faculty glorious, the weak University of Vienna thought to improve itself by ordering each of its professors to publish two papers every year.[9] There were other straws in the wind. At Pavia, Volta extracted many improvements for himself and his labora-

3. Förster, *Übersicht* (1799), 2–3; Meusel, *Gel. Teut.* (1786⁵), II, 381. Cf. Turner in Stone, *University* (1974), II, 505–28.

4. Hermelink and Kähler, *Phil.-Univ.* (1927), 458–9; cf. Paulsen, *Gesch.* (1896–7), II, 133–4.

5. Haller (1751) and Michaelis (1768), quoted by Joachim, *Anfänge* (1936), 2.

6. Quoted by E. G. Forbes, *Jahrb. Gesch. oberd. Reichs.*, 16 (1970), 149, 155; Mayer to Euler, 6 Oct. 1754, in Kopelevich, *Ist.-asst. issl.*, 5 (1959), 414. Cf. Schimank, *Rete*, 2 (1974), 207ff.

7. Bose (Wittenberg) to Formey (Berlin), 4 Aug. 1754 (Formey Papers, Deut. Staatsbibl., Berlin).

8. Kästner, *Briefe* (1912), 215–18; Müller, *Abh. Gesch. math. Wiss.*, 18 (1904), 135–6.

9. Kink, *Gesch.* (1854), I, 594.

tory from the Austrian government of Lombardy on the strength of his reputation as a discoverer. His justifications of these benefits show how matters stood. More room, he said, would allow '[me] to busy myself in research and to give private courses on it to capable students;' more money (a raise of 600 lire or 440#) would bring 'all my talents [to bear] on advancing the science I profess, and the instruction of students of it.'[10] Note the order in which he put his obligations.

Just after the turn of the century a Hanoverian minister responsible for the affairs of the University of Göttingen made the new concept of the university explicit. He allowed that faculty had a two-fold obligation: on the one hand, 'to preserve, propagate, and, where possible, to increase the sum of knowledge;' and, on the other, to teach, guide, and inspire. 'Without research, there is a great fear that we shall concentrate only upon the useful, to the ruin of science and, eventually, of teaching itself.'[11]

Now the University of Göttingen had been established and was maintained as Hanover's counterweight to the strong Prussian institution at Halle, itself founded in 1694 as a competitor to the then leading universities, Leipzig (Saxony) and Jena (Weimar).[12] The University of Pavia had been brought from the decadence characteristic of Italian higher schools in an effort to show the benefits of the enlightened despotism of Maria Theresa, who turned her attention to university reform in the 1750s. The happy results of this 'particular attention' greatly impressed the census-taker of Europe's intellectual riches, Jean III Bernoulli, when he visited Pavia in 1775.[13]

Most of the universities of Europe were behind Göttingen and Pavia in 1790. Among the institutions that might stand comparison with them in physics were, in Germany, Halle and perhaps Leipzig;[14] in Italy, Turin and perhaps Bologna; in Switzerland, Geneva. The Dutch universities had by then lost the ascendency in physics that they enjoyed earlier in the century. Neither Utrecht, which Musschenbroek left in 1740 for Leyden, nor Leyden, where he died in 1761, was able to replace him; the lead in research in physics passed to Van Marum, who had no university post, and to Van Swinden, who left Franeker for a post in Amsterdam in 1785.[15] At Paris virtually all the productive professors were also academicians. Great Britain defies generalizations.

The eighteenth century seems not to have been a prosperous time for universities. They were pinched for money by war and inflation and they seldom had

10. Volta to Wilzeck, 15 Jan. 1785 and 3 Feb. 1786, *VE*, III, 283, 330.

11. Brandes, *Betrachtungen* (1808), 16–18, 31; *ADB*, III, 241–2.

12. Selle, *U. Gött.* (1937), 4, 12; Förster, *Übersicht* (1799), 14, 32.

13. Bernoulli, *Lettres*, III (1779), 56–68, 63, 66.

14. Fester, *Gedike* (1905), 13, 21, 78, 87–8; Hermelink and Kähler, *Phil.-Univ.* (1927), 390.

15. Ruestow, *Physics* (1973), 153; Kernkamp, *Utrech. U.* (1936), I, 211; *DSB*, XIII, 183–4; Spiess, *Basel* (1936), 146, giving a student's view of the relative strengths of Leyden and Utrecht in 1760. Cf. Cuthberston, *Prin. Elect.* (1807), v, on the low state of experimental physics in Holland in 1769, and Z. Volta, Ist. lomb., *Rend.*, 15 (1882), 32, on Volta's estimate of Utrecht in 1782 ('the instruments of physics are nothing much').

first claim on the resources of their controlling prince or municipality. Enrollments stagnated or fell. The total annual matriculations in the German universities averaged 4200 from 1700 to 1750, and then declined almost linearly to about 2900 in 1800. Oxford and Cambridge fell from about 300 each in 1700 to some 200 in 1750; Cambridge then stagnated while Oxford recovered half its losses, to about 250.[16]

At its largest the eighteenth-century university was not very large. The biggest in Germany, Halle, never had a faculty greater than forty; its student body fluctuated between about 680 and 1500, and averaged 1000. The next in size, Jena and Leipzig, averaged 930 and 740 between 1700 and 1790. The smallest, Herborn and Duisburg, sixty and eighty, respectively.[17] The range in the Scottish universities was similar: at the end of the century St. Andrews had about 100 and Edinburgh, one of the few older foundations to grow during the period, 1000, up from 200 in 1720.[18] Oxford ranged between 1000 and 1700, at the outside.[19] The Italian universities and the Jesuit colleges doubtless stayed below 1000.[20]

These figures do not imply that professors of physics lacked students. In schools where the old order of learning still held, where philosophy preceded professionalization, he would teach all students who survived into their second year. Such programs characterized the Jesuit universities and colleges. At Würzburg, for example, an average of about forty students annually were 'physicists;' there, and at Dillingen, Freiburg and Fulda, an average of 65 percent of the student body was enrolled in the philosophy faculties.[21] There were also Protestant schools with fixed and frequented curricula in arts, particularly the Scottish universities, which required 'natural philosophy' in the fourth and final year.[22]

In the German Protestant universities, however, enrollment in the philosophy faculty was usually very small, often less than ten. Students went directly into the higher faculties; in a sample of six universities—Duisburg, Erlangen, Göttingen, Halle, Kiel, Strassburg—43 percent of the student body matriculated in theology, 38 percent in law, and 11 percent in medicine.[23] Often these

---

16. Eulenburg, *Frequenz* (1904), 132; Stone in Stone, ed., *University* (1974), I, 6.

17. Eulenburg, *Frequenz* (1904), 146, 153, 164–5, 319. Eulenburg's figures differ from Gedike's contemporary survey, which makes Leipzig 1200–1300 and Jena and Göttingen each between 800 and 900 in 1790 (against Eulenburg's 670, 780 and 810, respectively). Fester, *Gedike* (1905), 33, 78, 87.

18. Great Britain, *Sess. Papers*, 37 (1837), 248; Dalzel, *Hist.* (1862), II, 307–25.

19. These figures come from the matriculations via the multiplier 5.6 (Stone in Stone, *University* [1974], I, 87), which is probably too high; the table (*ibid.*, 95) suggests one between 3 and 4.

20. E.g., 400 at Parma in 1775 (Bernoulli, *Letters,* III [1779], 184); 'a few hundreds' at Bologna (Simeoni, *Storia* [1940], 89–90); still fewer at Modena (Pietro, *Studio* [1970], 146) and Ferrara (Visconti, *Storia* [1950], 91–2).

21. Computed from Eulenburg, *Frequenz* (1904), 207–9, 310, 312.

22. Morgan, *Scot. U.* (1933), 72–4; Rait, *U. Aberdeen* (1895), 202, 300; Morrell, *Isis,* 62 (1971), 160–1, 207.

23. Eulenburg, *Frequenz* (1904), 207.

students cared only for professional training, for 'jus, jus et nihil plus,' as the lawyers said.[24] Under these circumstances it took a good man to draw audiences to courses in experimental physics; and audiences he needed, for it was their approbation and fees that made possible the purchase of the necessary apparatus.

The successful professor of experimental physics had to be a showman. In Vienna, where he had a captive audience, he was nonetheless directed by statute to strive for '[die] nöthigen Popularität.'[25] Playing to the gallery did not improve the morale of serious savants. We already know that Kästner gave up teaching the standard lecture course because his students came only to be entertained. The expression of the dilemma of his class of pedagogue is best left to one of them, John Robison, professor of natural philosophy at the University of Edinburgh, who had rights to fees from fourth-year students, if he could entice them to stay. 'As I endeavor to conduct my lessons in such a manner that [some students may learn something], I render them less pleasing to the generality of my hearers, who aim at nothing but getting a superficial Knowledge, or, more properly speaking, whose only aim is a frivolous amusement. This renders me a very unpopular teacher, and as I cannot think of becoming a showman, I do not expect to grow rich in the profession.'[26]

Despite the need to perform, despite stagnant enrollments and rising costs, the professors of experimental physics managed to establish their subject firmly in universities and colleges during the eighteenth century. And, despite the consensus that the professor need do nothing more than teach, a few performed supererogatory works, and inspired students to do the same.

THE FIRST COURSES IN EXPERIMENTAL PHYSICS

A very few professors were illustrating their lectures on physics with occasional demonstrations by or just after 1700. Those whose performances had some influence may be counted upon the fingers of one had: Burchard de Volder, a moderate Cartesian at Leyden, the inventor of an improved air pump; J. C. Sturm, who had studied at Leyden and who developed demonstrations based upon the experiments of the Accademia del Cimento, at Altdorf;[27] G. A. Hamberger, a disciple of Sturm's, at Jena; and Pierre Varignon, the follower of Malebranche, at the Collège Mazarin.[28] The introduction of experimental physics into the universities had therefore begun before the advent of Newtonian experimental philosophy. For a time, it proceeded on the continent under

24. Paulsen, *Gesch.* (1896–7), I, 531–2, II, 127.

25. Meister, Ak. Wiss., Vienna, Phil.-Hist. Kl., *Sb.*, 232:2 (1958), 95 (an edict of 1774).

26. Robison to Watt, 22 Oct. 1783, in Robinson and McKie, *Partners* (1970), 130. Cf. Grant, *Story* (1884), I, 241–2; Kästner, 'Verbindung' (1768), in *Verm. Schr.* (1783³), II, 363.

27. Klee, *Gesch.* (1908), 28–30; Crommelin, *Sudh. Arch.*, 28 (1935), 131; Günther, Ver. Gesch. Stadt Nürn., *Mitt.*, 3 (1881), 18; Will, *Nürn. Gel.-Lex.*, III (1757), 800–9; *ADB*, XXXVIII, 39–40.

28. Steinmetz, *Gesch.* (1958), I, 130-2, 206, and Jöcher, *Allg. Gel.-Lex.* (1750-1), II, 1338 (Hamberger); *infra*, ii.4 (Varignon).

Cartesian fellow travellers opposed to Newton's methods and bemused by his apologetics.

One indigenous European pattern, common to the Protestant universities of Germany and Scandinavia, may be illustrated by the career of Christian Wolff. His precocious interest in Cartesian method led him to the study of mathematics and physics, which he chose to pursue at Jena under Hamberger. He went to Halle in 1706 as professor of mathematics. A professor of medicine then taught physics as a second field; Halle's founders had tried to pry Sturm from Altdorf, but he had declined to move. Wolff was soon offering a course in physics based on Sturm. Later he formally took responsibility for instruction in the subject.[29] The course succeeded. Wolff's modernized Cartesian physics, demonstrated in the style of Sturm and Hamberger, spread to many universities of Central Europe between 1720 and 1750, replacing either biological physics taught by a member of the medical faculty or literary physics taught by a professor of philosophy.[30]

Among institutions that followed this pattern were Marburg, to which Wolff himself brought it in 1723;[31] Kiel, where a Wolffian physicist, J. C. Hennings, arrived in 1738, in succession to a Cartesian physician;[32] Leipzig, where in 1750 J. H. Winkler, a Wolffian electrician, took a chair just released by a philosopher and poet, who had had it from a physician;[33] Uppsala, where the first professorship of physics, established in 1750 under external pressure, went to a former student of Wolff's, Samuel Klingenstierna.[34] In these later cases, the new physics came with Newtonian admixtures; for example, Klingenstierna was directed to teach the usual range of mechanics, 'aerometry' (Wolff's specialty), and 'the discoveries of Newton.' Nonetheless the native contribution remained in evidence: the most popular physics texts in Protestant Germany in the 1730s and 1740s were probably Wolff's *Nützliche Versuche* and the *Elementa physices* of G. E. Hamberger,[35] who succeeded his father at Jena in the 1720s.

The little stream of Sturm, Hamberger and Wolff is easily overlooked against the flood of Anglo-Dutch Newtonianism. Just after 1700 Newtonian epigoni, installed in Oxbridge chairs, began to offer courses of experimental philosophy. At Cambridge Newton's successor as Lucasian professor of mathematics,

29. Förster, *Übersicht* (1794), 27, 29, 55, 95

30. Paulsen in Lexis, *Deut. U.* (1893), I, 28–32, and *Gesch.* (1896–7), I, 532; Nauck, *Beit. freib. Wiss.-Univ.gesch.*, no. 4 (1954), 10–11; Helbig, *U. Leipzig* (1961), 55.

31. Hermelink and Kähler, *Phil.-Univ.* (1927), 348, 384n.

32. Schmidt-Schönbeck, *300 Jahre* (1965), 30–5. It appears that Hennings lacked some of the apparatus for the lectures he announced.

33. Leipzig, U., *Festschrift*, IV:2 (1909), 26–9.

34. Anon. in Uppsala, U., *Aarsk.* (1910), 35–42; cf. Hilderbrandsson, *Klingenstierna* (1919), 12–17.

35. *ADB*, X, 470–1; Börner, *Nachrichten*, I (1749), 52–71. For Hamberger's text, Stieda, *Erf. Univ.* (1934), 27.

William Whiston, and Roger Cotes, named the first Plumian professor of astronomy and natural philosophy in 1706, collaborated on such a course. Their association came to an unnatural end in 1710, when Whiston was unseated for unorthodoxy. Cotes carried on alone until he died in 1716, leaving his professorship and his lectures to his cousin, Robert Smith.[36] At Oxford the rash John Keill, then deputy Sedleian professor of natural philosophy, offered a course in experimental physics with the assistance of Desaguliers; it lasted until 1712, when Keill became Savillian professor of astronomy. The Keeper of the Ashmolean Museum moved into the void, from which he was chased by James Bradley, who succeeded to Keill's professorship in 1721.[37] These professorial lectures represented only a portion of the interest in experimental physics at Oxbridge in the first quarter of the eighteenth century. Many tutors assigned reading in Newton and Keill as well as in Rohault, and some offered experimental demonstrations.[38] The decline of physics at the ancient English universities set in after the death of the first generation of Newtonian professors.

The work of Keill, Cotes and Desaguliers inspired the authoritative texts of 'sGravesande, whose excellent order, convenient size, and copper plates did much to spread the cause of experimental physics on the continent.[39] Among the first institutions to teach physics in the Dutch style were the universities of Duisburg and Utrecht, which shared the services of 'sGravesande's agent Musschenbroek in the 1720s. Musschenbroek was trained as a physician, but taught as a philosopher, within the philosophy faculty.[40] Giessen probably also belongs in this group; in 1729 it set up a chair of 'physica naturalis et experimentalis' for a professor of medicine who had studied and travelled in Holland.[41]

The Catholic universities of Germany and Austria were slower to take up the new physics. The Jesuits who dominated the philosophical faculties there held to the *Ratio studiorum* and the regental system, and stayed suspicious of Descartes. Efforts to introduce Cartesian physics and Sturm's demonstrations into Ingolstadt failed early in the century. Towards 1730 the medical faculty renewed the attempt, and succeeded by threatening to teach experimental physics itself.[42] Responsibility and initiative for developing the subject fell to Joseph Mangold, S. J., a Cartesian of Euler's type. The course of development at Dillingen was similar. In its case the local bishop led the attack against tradi-

36. Hans, *New Trends* (1951), 49–50; *DSB*, III, 430–3.

37. Hans, *New Trends* (1951), 47–8; *DNB*, II, 1074–9; *DSB*, VII, 275–7.

38. Mayor, *Cambridge* (1911), 55, 457; Wordsworth, *Scholae* (1877), 68; Rouse Ball, *Hist.* (1889), 94–5; Gunther, *Early Sci.*, I (1920), 196; Frank, *Hist. Sci.*, 11 (1973), 253–5.

39. Brunet, *Physiciens* (1926), 40–2, 48–54, 61; *supra*, i.1.

40. Kernkamp, *Utrech. U.* (1936), I, 305. Duisburg was closely associated with the Dutch schools; Ring, *Gesch.* (1949²), 179.

41. Jöcher, *Allg. Gel.-Lex.* (1750–1), IV, 1523–4; *ADB*, XXXIX, 615–16; Lorey, Giess. Hochsch., *Nach.*, 14 (1940), 23–31, 15 (1941), 83–7.

42. Schaff, *Gesch.* (1912), 6, 154–6; Prandtl, *Gesch.* (1872), I, 541, 610; Hofmann, *Math.* (1954), 14–15.

tion; in 1745 he recommended the texts of the 'acatholic' Wolff. By the 1750s the Dillingen Jesuits were giving experimental demonstrations and lecturing on physics in German.[43]

The acceptance of experimental physics at the German Catholic universities in the 1750s owed not a little to Maria Theresa's reforms of the Jesuit-run higher schools of Austria: Freiburg i./B., Graz, Innsbruck, Prague and Vienna. A government commission during the reign of her father, Charles VI, had already criticized the philosophical faculty for teaching 'empty subtleties;' its insistence that Descartes' physics be *introduced* into the schools tells plainly enough what was considered up-to-date in Vienna in 1735.[44] The Austrian universities, particularly Vienna and Innsbruck, did acknowledge experimental physics in the 1740s, but Aristotle remained their official guide until 1752, when the Empress insisted on reducing the philosophical course to two years, on using German in instruction, and on eliminating metaphysics, ethics, and 'everything useless.' The old curriculum, she said, had nothing to do with the common concerns of men and states; 'the ungrounded theory (which can not be confirmed by experience) of peripatetic form and matter is henceforth entirely forbidden;' class time 'shall be devoted to true physica experimentalis.'[45]

The progress of the institutionalization of experimental physics accelerated after the Seven Years' War. In the German Protestant universities true Fach-Physiker began to appear, men trained by the experimental physicists who had established themselves in the 1740s and 1750s. At Kiel, for example, the first physics course thoroughly illustrated by experiment was initiated by a Wolffian professor of medicine; J. F. Ackermann, who from 1763 also held a chair in the philosophy faculty; his successor, C. H. Pfaff, M.D., who had finished his studies under Lichtenberg at Göttingen, made his career as a professor of physics.[46] Other Göttingen graduates garnered physics chairs at Giessen (G. G. Schmidt) and Altdorf (J. T. Mayer).[47]

The suppression of the Jesuits in 1773 resulted in a short-run improvement in the former Jesuit universities of Germany and Austria even though—or rather because—instruction continued in the hands of ex-Jesuits. As an Austrian commission charged to consider their replacement reported in 1774: 'We do not have their equal in the mathematical sciences and they are cheaper to maintain than lay professors.'[48] The suppression allowed the concentration of resources and the substitution of modern, specialized, vernacular texts for the Jesuit compendia. For example, Ingolstadt adopted Erxleben's *Anfangsgründe*; Freiburg,

43. Specht, *Gesch.* (1902), 197–200, 317–18; Sommervogel, *Bibliothèque*, V, 481.

44. Paulsen, *Gesch.* (1896–97), II, 107.

45. Schreiber, *Gesch.*, II:3 (1860), 7–11; Fester, *Gedike* (1905), 44; Meister, Ak. Wiss., Vienna, Phil.-Hist. Kl., *Sitzb.*, 232:2 (1958), 89–91. For the reforms see Kink, *Fesch.* (1854), I, 424–590.

46. Schmidt-Schönbeck, *300 Jahre* (1965), 30–8, 59.

47. Lorey, Giess. Hochsch., *Nach.*, 15 (1941), 83–7. Mayer soon left for Erlangen, which was well regarded in 1790; Fester, *Gedike* (1905), 70, 74; Kästner, *Briefe* (1912), 244.

48. Halberzettl, *Stell.* (1973), 141, 167, 189; Schreiber, *Gesch.*, II:3 (1860), 50–1. Frederick the Great also kept them on in Silesia; Paulsen, *Gesch.* (1896–7), II, 100–1.

Sigaud's *Anweisungen zur Experimentalphysik*; and Vienna, still reforming, chose Biwald's *Institutiones physicae,* up-to-date in all but language.[49]

In France the Jesuits could not hope to retain their control of education if they opposed novelty as strongly as their German brethren. They had always emphasized applied mathematics; and in the late seventeenth century, as we have seen, they accepted a government charge to provide for instruction in navigation and associated sciences, 'to pray and teach hydrography.'[50] The earliest recorded 'exercises de physique expérimentale' in the French Jesuit system occurred either in colleges associated with hydrographers or in ones where physics was customarily taught by men trained in mathematics: at Aix in 1716, Lyon in 1725, Louis-le-Grand in 1731, Pont-à-Mousson in 1740, Marseille in 1742.[51] Meanwhile the père physicien had identified more closely with his subject: in 1700 30 of the 80 colleges teaching physics used the regental system; in 1761 the figures were 23 and 85.[52]

The higher schools of Paris, the University and the Collège Royal, institionalized the new physics relatively late, partly because the many public teachers of mathematics and physics in the capital allowed them to shirk responsibility. The university's first professor of experimental physics, J. A. Nollet, had been an unusually successful public lecturer; his chair, founded in 1753 at the Collège de Navarre, was a gift of Louis XV.[53] (Other colleges, especially Harcourt, Louis-le-Grand, and Mazarin, which had strong traditions in mathematics, occasionally offered instruction in experimental physics either by the professor of mathematics or by a private lecturer hired for the purpose.[54]) Nollet further broke with tradition by lecturing in French. Although few in the university followed his example—physics in most colleges continued Latin and literary until the Revolution, and the Faculty of Arts did not endorse a vernacular textbook until 1790[55]—opportunities for teaching and learning experimental physics continually improved. For example, in the 1760s Sigaud de Lafond, later of the Academy of Sciences, and his nephew Rouland, instrument makers and public lecturers, advertised themselves as 'demonstrators in experimental physics' at the University, which probably meant that for a fee they brought their own equipment to demonstrate in the class of a professor of physics or philosophy.[56] The arrangement satisfied d'Alembert: 'The University of Paris furnishes convincing proof of the progress of philosophy among us.

49. Schaff, *Fesch.* (1912), 181; Zentgraf, *Gesch.* (1957), 13; Meister, Ak. Wiss., Vienna, Phil.-Hist. Kl., *Sb.,* 232:2 (1958), 93; Paulsen, *Gesch.* (1896–7), II, 110.

50. Dainville, *Géographie* (1940), 435–9; Dainville in Taton, *Enseignement* (1964), 28.

51. *Ibid.,* 40–1; Schimberg, *Education* (1913), 518–19n; Lallamand, *Hist.* (1888), 259.

52. Dainville in Taton, *Enseignement* (1964), 29.

53. Torlais in Taton, *Enseignement* (1964), 627; Jourdain, *Hist.* (1888), II, 274–6, 382–6. For private lecturers, *infra,* ii.4.

54. Lacoarret and Ter-Menassian in Taton, *Enseignement* (1964), 141–5; A. Franklin, *Hist.* (1901²), 199, and *Recherches* (1862), 162–75, 109–10; Guerlac, *Isis,* 47 (1956), 212.

55. Lantoine, *Hist.* (1874), 152; Anthiaume, *Collège* (1905), I, 221.

56. Torlais in Taton, *Enseignement* (1964), 633, 637.

Geometry and experimental physics are successfully cultivated there . . .Young masters train truly educated students who leave their [course of] philosophy initiated into the true principles of all the physico-mathematical sciences.'[57]

Outside Paris Nollet's course, and his earlier public lectures, also had significant results. They probably helped the case of experimental physics among the Jesuits, as he claimed that they did among the Oratorians. After the expulsion of the Jesuits, they brought him into demand as consultant to new professors and administrators wishing to set up or to continue instruction in experimental physics.[58] Caen in 1762, Bordeaux in 1763, Pau, Strasbourg, Draguignan, Amiens, all sought his advice.[59]

The Collège Royal, founded in the sixteenth century, had always been more modern than its medieval sister. An example of its precocity was the slow transformation of one of its old chairs of Greek and Latin philosophy into a professorship of physics, a change made permanent with the appointment of Varignon (1694) and his successor Privat de Molières (1722). The chair nonetheless retained its title until 1769. It was then converted into a chair of physics, 'His Majesty having recognized that the two chairs of Greek and Latin philosophy have had little audience since physics has been enhanced by the discoveries of the moderns.' Four years later the Crown combined the chairs of Hebrew and Syriac and established a chair of 'mechanics' (in 1786 changed to 'experimental physics') on the income of the suppressed orientalist.[60]

In Italy the institutionalization of experimental physics had not proceeded very far by mid-century; instruction in philosophy remained literary, and almost exclusively in the hands of clerics, who offered mixtures of Aristotle and Descartes. An exception is Padua, where the Cartesianism of Fardella and the Galilean tradition of applied mathematics met to produce Giovanni Poleni, engineer, philosopher, mathematician and, from 1739, tenant of a new chair 'ad mathematicam et philosophiam experimentalem.'[61] Bologna also made provision for modern instruction in physics and mathematics. In 1737, in association with the moribund University, the Institute established by Marsigli in 1714 set up a professorship of experimental physics. It was later held by Giuseppe Veratti, M.D., among other electricians.[62]

The tempo increased in the 1740s. Benedict XIV encouraged the reform of the Sapienza at Rome 1744–6, which brought a chair of 'rational and ex-

57. D'Alembert in 'Elémens' (1760), *Oeuvres* (1805), II, 460–1.

58. Nollet, *Leçons*, I (1754⁶), xii–xiv; Jourdain, *Hist.* (1888), II, 175–6, 274–6; Nollet to Dutour 7 Sept. 1768 (Burndy): 'Les nouveaux collèges depuis l'expulsion des Jésuites veulent tout faire de la physique expérimentale.'

59. Torlais in Taton, *Enseignement* (1964), 628; Mornet, *Origines* (1947²), 181; Irsay, *Hist.*, II (1935), 115.

60. Anon., *Rev. int. ens.*, 5 (1883), 406–7; Sédillot, *Bull. bib. stor. sci. mat. fis.*, 2 (1869), 499–510, 3 (1870), 165–6.

61. Vidari, *Educazione* (1930), 132–4, 180–97; Favaro, *U. Padova* (1922), 67–8, 142.

62. Bortolotti, *Storia* (1947), 147–58; Simeoni, *Storia*, II (1944), 116. The institute provided for a professor of experimental physics from the beginning (*ibid.*, 126).

perimental philosophy' (later 'experimental physics') held first by the Minim mathematician François Jacquier. Pisa set up a professorship of experimental physics in 1746, for C. A. Guadagni; in 1748 Turin freed its physics chair from philosophers and entrusted it to G. B. Beccaria, a major figure in the early history of electricity; and similar attempts, apparently not entirely successful, were made at Pavia and Naples.[63] In 1760, if not before, professors of experimental physics existed at Perugia and Modena; in 1764–5, owing to reforms introduced by the King of Sardinia, Charles Emanuel III, the impoverished universities of Cagliari and Sassari had them too.[64] In the 1770s the Austrians made Pavia a showplace; Modena added a second chair of physics; and the pint-sized enlightened despot, Duke Ferdinand I of Parma, enriched his university with the confiscated wealth of the Jesuits and built for it a lecture hall for physics that Volta took as the pattern for his own.[65]

In English universities the cause of experimental physics had several champions at mid-century. Smith and Bradley lived through the 1750s, and found an unlikely colleague in Thomas Rutherford, Regius Professor of Divinity at Cambridge, who wrote an important textbook on natural philosophy. Rutherford's career instances a difficulty in determining the level of academic activity at eighteenth-century Oxbridge: although Rutherford never lectured as university professor of divinity, he taught experimental physics regularly as a member of his college. Similarly, Bradley's successor, singled out by Adam Smith to represent those who had 'given up altogether even the pretence of teaching,' often lectured privately on experimental physics.[66]

There were no doubt many dons and professors sunk in ignorance and indolence, such as Charles Beattie, appointed Sedleian Professor at Oxford in 1720, 'not on account of any skill (for he hath none) in Natural Philosophy, but because he is much in debt to [his] college, occasioned by his Negligence as Bursar.'[67] Yet there was the counterweight of professors who taught well in their or someone else's statutory field, and the dons, particularly at Cambridge, who kept science alive in their colleges. One should not make much of the well-known estimate that the fraction of active English 'scientists' who had been educated at Oxbridge fell from two-thirds in 1650 to one-fifth in 1750.[68] Everything depends upon the definition of scientist. Among those with mathematical training, like the electrician Henry Cavendish, a good fraction

63. Italy, *Monografie,* I (1911), 299–300, 547–55; Scolopio, *Storia,* I (1877), 58; *infra,* xv.4.

64. Italy, *Monografie,* I (1911), 78, 166, 429–39, 442.

65. Mor, *Storia* (1953), 93, 175–8; Pietro, *Studio* (1970), 42–95; Italy, *Monografie,* I, (1911), 246–7, 270; Volta to Wilzeck, 4 March 1785, *VE,* III, 295. Bernoulli, *Lettres,* III (1779), 181–3, comments on the excellence of the new facilities for science at Parma.

66. *DNB,* XVII, 499, IX, 1267; Wordsworth, *Scholae* (1877), 72. Cf. the case of Isaac Milner, Jacksonian and later Lucasian professor at Cambridge; *DNB,* XIII, 456–9; Winstanley, *Unr. Camb.* (1935), 131.

67. Godley, *Oxford,* (1908), 82–91; Mallet, *Hist.,* III (1927), 124–6; Winstanley, *Unr. Camb.* (1935), 129–32, 151–2, 179–82.

68. Hans, *New Trends* (1951), 34.

attended Cambridge, where the mathematical tripos called for close study of the *Principia* and treatises on mechanics, optics, and hydrostatics.[69]

Experimental physics came into the Scottish universities when they abolished the regental system and established chairs of natural philosophy. That occurred at Edinburgh in 1708, Glasgow in 1727, St. Andrews in 1747, and Aberdeen in 1753. The fixing of the chairs coincided with large purchases of demonstration apparatus at Edinburgh (1709) and Glasgow (1726); smaller acquisitions by the other schools about 1715 suggest that even before they stabilized their chairs they made some provision for the new science.[70]

In 1703 the Rector of Calvin's Academy at Geneva, praising the 'brief, yet substantial, incomparable, royal,' method of Descartes, insisted on the need for mathematics and experiment.[71] Mathematics prospered first; a chair was set up in 1724 by the municipal authorities against the opposition of the Church, and awarded jointly to two able young men, Calendrini (the collaborator of Jacquier and Le Seur) and Gabriel Cramer. Both then urged the appointment of a professor of experimental physics. That partly came to pass in 1737, when Jean Jallabert received an 'honorary' (unsalaried) post. Jallabert went off to England, France and Holland, procured apparatus, met Musschenbroek and Nollet, became a Fellow of the Royal Society, and returned to lecture to applause. He subsequently followed Cramer, who had followed Calendrini, into one of the Academy's two chairs of philosophy. Their chair passed to Saussure, an excellent physicist and electrician, in 1762; and it became a chair of experimental physics in all but name under him and his hand-picked successor, M. A. Pictet.[72] The rise of Calvin's Academy to a leading center of physics owed not a little to the fact that Calendrini, Cramer, Jallabert, Saussure and Pictet were wealthy members of Geneva's governing class.[73] They could afford to accept half-chairs and honorary posts and, after election to professorships, to do what they pleased.

BUILDING UP THE CABINET

The professor and his institution shared responsibility for the upkeep and increase of instrument collections according to their relative power and poverty. At Protestant universities, except in Scotland, the professor of experimental physics was expected to furnish some if not all of his equipment; large private collections resulted, such as those of 'sGravesande and Musschenbroek at Leyden, Winkler at Leipzig, Lichtenberg at Göttingen, Bose at Wittenberg, Jallabert and Saussure at Geneva.[74] The extent of the collection, and hence of

69. Rouse Ball, *Hist.* (1889), 191, quoting a source of 1772.

70. Dalzel, *Hist.* (1862), II, 304; Gr. Br., *Sess. Pap.*, 38 (1837), 303; Murray, *Memories* (1927), 110–11; Coutts, *Hist.* (1909), 195; Cant, *College* (1950), 83.

71. Borgeaud, *Hist.*, I (1900), 458–8; cf. Montandon, *Développement* (1975), 45–9.

72. Borgeaud, *Hist.*, I (1900), 483–4, 503–4, 569, 573–7.

73. Montandon, *Développement* (1975), 51–3, 59.

74. Crommelin, *Desc. Cat.* (1951), 13, 21; Lichtenberg, *Briefe* (1901), II, 136, 259–60; Leipzig, U., *Festschrift* (1909), IV:2, 29; Bernoulli, *Lettres*, III (1777), 6; Borgeaud, *Hist.*, I (1900), 569.

instruction, was proportional to the depth of the professor's pocket, 'there being no public subsidies for the improvement of knowledge.'[75] At Kiel, for example, demonstrations sometimes lapsed during the tenure of the Wolffian Hennings, for want of means rather than will; his successor, J. F. Ackermann, had made money at doctoring and provided what was needed.[76] Bose's complaint, that he had to pay 'ready money for all my instruments, without exception, from my air pump to my funnel,' suggests that the professor / demonstrator was not regarded as a good credit risk.[77]

The situation provided an opportunity for wealthy faculty to poach on the preserve of the professor of physics. At Jena, for example, J. C. Stock, M.D., tried to win the physics chair by investing 100 RT (400#) in demonstration apparatus. At Lund Daniel Menlös succeeded in buying his way into a professorship over a man agreed to be his better by promising to acquire for the university the Triewald collection, then (1732) unrivalled in Sweden. At Halle a professor of law, Gottfried Sellius, briefly monopolized instruction in experimental physics. He had married a rich wife, with whose dowry he bought excellent instruments, some made of silver, and all very elegant; but he lived beyond his means and had to flee his creditors, leaving physics to the less ambitious pedagogues from whom he had snatched it.[78]

Despite noteworthy exceptions, such as the annual subvention of 200 RT for 'expensive instruments, books, etc.,' in physics and mathematics given Wolff at Marburg from 1724, the remarkable generosity of the curators of Utrecht during the tenure of Musschenbroek,[79] and the one-time purchase for 6000 Dkmt (2000#) of a 'complete' apparatus to Musschenbroek's specifications for Klingenstierna at Uppsala in 1740,[80] continental Protestant universities did not begin to acquire substantial collections until their professors, who had made the initial investments and suffered the depreciation, started to die off. Then the institution might purchase a working apparatus at a good price, leaving responsibility for further acquisitions to the new professor. Leyden bought 'sGravesande's collection for $f.3931$ (8400#), and several pieces from Musschenbroek's, which fetched $f.8864$ altogether when auctioned in 1761.

The perennially poor University of Duisburg bought half of J. J. Schilling's

75. Kästner to Haller, 1 Aug. 1755, Kästner, *Briefe* (1912), 35.

76. Schmidt-Schönbeck, *300 Jahre* (1965), 35–8.

77. Bose to Formey, 26 July 1750 (Formey Papers, Staatsbibl., Berlin): 'Il me faut payer argent comptant tous mes instrumens, sans en excepter aucun, depuis ma pompe pneumatique jusqu' à l'entonnier, pourtant je suis professeur en physique. A plus forte raison mes téléscopes, la pendule, les micromètres, etc., ne sont acquis qu'à mes propre[s] dépens, n'étant que des opera supererogationis, où je travaille pour ainsi dire par pur[e] magnanimité.'

78. Steinmetz, *Gesch.* (1958), I, 205; Förster, *Übersicht* (1799), 100–1; Leide, *Fys. inst.* (1968), 29–33.

79. Gottsched, *Hist. Lobs.* (1755), Beylage, 34; Kernkamp, *Utrecht. U.* (1936), I, 209–10. The curators bought many instruments when Musschenbroek came in 1723, and again in 1732–3 (almost *f.* 2000 [4300#] worth) when he was pondering accepting a call from Copenhagen.

80. Anon., Uppsala, U., *Aarsk.* (1910), 32–4. One should perhaps also except Lund, which acquired the Menlös-Triewald collection in the 1730s, and Greifswald, which began to build a collection just after 1750. Dähnert, *Sammlung,* II (1767), 828, 889, 985, 999–1000.

collection at his death in 1779.[81] Göttingen bought Lichtenberg's in 1787–8, having meanwhile added nothing to the physical instruments in use there in the 1750s.[82] Leipzig got Winkler's in 1785 for 1064 RT, and brought it up to date in 1808 by acquiring the instruments of his successor, K. F. Hindenberg, for 1000 RT; Altdorf did the same, building upon Sturm's old collection and adding to it, in 1780, that of another of its professors, M. Adelbulner; Marburg and Giessen acted similarly at the turn of the century.[83] Occasionally a philanthropic professor donated his instruments, as Kratzenstein did at Copenhagen (1795), adding an endowment the income of which was still an important part of the economy of the physics institute in 1900.[84]

The case of Halle, that 'garden of free arts and sciences,'[85] is particularly interesting. During his first tenure there, Wolff received a little dole for instruments, which was not continued to his successor in the mathematics chair, J. J. Lange, who gradually built up his own collection. When Wolff died in 1754, leaving Halle for the second time, Segner accepted his post on condition that the University provide adequate teaching apparatus. One thought to buy Wolff's. The heirs wanted a just price. Lange, however, was willing to sell cheaply; and the University could meet its commitment to Segner at little expense by adding a few new machines to those it bought from him. Not until the end of the century did Halle make adequate provision for acquiring physical instruments.[86]

The acquisition of instruments brought with it responsibility for maintenance and modernization. Duisburg, despite its poverty, spent 15 or 20 RT a year on its 'new' physical collection,[87] and Schmidt eventually obtained help from Giessen. A 'mechanic' might be engaged to keep the instruments in order and to help with the demonstrations: Lund employed such a person from the mid-1730s, Leyden from 1752, Utrecht from 1768. All also had to make provision for storage, sometimes, as at Lund, with great difficulty.[88] In 1795 Göttingen obtained a mechanic and other universities did so soon after the turn of the century. No doubt the best known of the tribe is James Watt, who in 1757 took the post of instrument maker set up at Glasgow in 1730.[89]

Catholic universities could not follow the Protestant pattern of acquisition

81. Rooseboom, *Bijdrage* (1950), 15, 107; W. Hesse, *Beiträge* (1875), 90–1.

82. Pütter, *Versuch*, I, 242, II, 267, III, 489; L. Euler to Fred. II, 7 Oct. 1752, in Stieda, Ak. Wiss., Leipzig, Phil.-Hist. Kl., *Ber.*, 83:3 (1931), 54.

83. Leipzig, U., *Festschrift* (1909), IV:2, 28–31; Günther, Ver. Gesch. Stadt Nürn., *Mitt.*, 3 (1881), 9, 30; Lorey, *Nach.*, 15 (1941), 86–7; Hermelink and Kähler, *Phil.-U. Marb.* (1927), 756–7. Cf. Bernoulli, *Lettres* (1771), I, 7n, regarding Frankfurt/Oder.

84. Snorrason, *Kratzenstein* (1967), 55–6; Copenhagen, *Poly. Laere* (1910), 27. The collection was worth at least 20,000# (4000 Rigsdaler).

85. Börner, *Nachrichten* (1749–54), I, 72.

86. Förster, *Übersicht* (1799), 222, 233; Schrader, *Gesch.* (1894), I, 570, 578; Euler to Fred. II, 2 and 20 Nov. 1754, in Stieda, Ak. Wiss., Leipzig, Phil.-Hist. Kl., *Ber.*, 83:3 (1931), 54–6.

87. W. Hesse, *Beiträge* (1875), 91.

88. Rooseboom, *Bijdrage* (1950), 15–18, 32; Kernkamp, *Utrecht. U.* (1936), I, 212–13; Leide, *Fys. inst.* (1968), 35–44.

89. Mackie, *U. Glasgow* (1954), 218.

since many of their professors, as members of religious orders, had very small incomes, and as a rule could not charge fees for courses. Consequently from the beginning instruments were usually provided by the university or by gift. In 1757 Dillingen set up a 'mathematical-physical museum' with the help of 1000 gulden (2600#) from its chancellor;[90] by 1787 it had a very small annual sum, 25 gulden, for instruments. At the same time (1754), Ingolstadt started a 'physical-chemical cabinet' for 1200 gulden.[91]

Both the course and nature of the acquisitions may be illustrated by the purchases of the Jesuit college at Bamberg, of which a full account survives. The first large acquisition was an air pump, the next, in 1747, an electrical machine. Buying increased between 1749 and 1753, and then declined until 1771. In 1789 the college was again looking for a better air pump and an improved electrical machine, for each of which it could pay 650#. By the end of the century it had 125 instruments for general physics, 45 for electricity, 60 for optics, 8 to demonstrate the theory of heat, and 6 for magnetism. These treasures created the usual storage problem; a special cabinet, built in 1755, was much enlarged in the 1790s.[92]

A similar pattern and similar timing occurred in Italy. During the pontificate of Benedict XIV (1740–58), the university of Rome got a 'theater' for experimental physics on the top floor of the Sapienza, a small budget, and instruments, the first six, including an air pump, gifts of Benedict himself. Benedict also enriched the Bologna Institute with a set of instruments made from 'sGravesande's designs. Nollet judged it to be 'assez ample' when he saw it in 1749, but missing a few things, about which he spoke to the Pope, 'who listened favorably,' during an audience in Rome.[93] Nollet also saw and approved the 'rather complete' collection bought for Poleni by the University of Padua in the 1740s; by 1764 it contained 392 items.[94]

Pisa had a cabinet by mid-century, stocked by the private collection of Dutch instruments made by its first professor of experimental physics, C. A. Guadagni; the university put up a small annual sum for apparatus, and, in 1778, partly compensated Guadagni for his total personal outlay with a gift of 100 zecchini (about 1200#). By the end of the century the professor of physics had 500# for the gabinetto and 200# for experiments.[95] Modena's gabinetto was established in 1760, on much the same basis as Pisa's: as nucleus it took the instruments of its professor, the Minim Mariano Moreni, whom it repaid in 1772 with a tiny annuity; from 1777–8 it made available 200# annually for

90. Specht, *Gesch.* (1902), 199–200, 530; Schmid, *Erinnerungen* (1953), 94–5, says that Weber expanded the collection 'modestly,' little money being available.

91. Günther, Ver. Gesch. Stadt Nürn., *Mitt.*, 3 (1881), 10; cf. Schaff, *Gesch.* (1912), 155–6, for earlier purchases.

92. H. Weber, *Gesch.* (1879–82), 338–42.

93. Spano, *Univ.* (1935), 50–1, 253; Nollet, 'Journal' (1749), f. 153v.

94. *Ibid.*, f. 91–2; Favaro, *Univ.* (1922), 142.

95. Italy, *Monografie*, I (1911), 259; Occhialini, *Notizie* (1914), 3, 17–18; Scolopio, *Storia* (1877), I, 79; Nollet, 'Journal' (1749), f. 127v.

improving the collections, and for two Capuchins to look after them.[96] Parma, under its modernizing duke, established a gabinetto in 1770 and furnished it with instruments in the style of Nollet with the help of money and material confiscated from the Jesuits.[97]

Turin acquired the nucleus of its collection in 1739, with the purchase of the many instruments that Nollet had brought to teach physics to the crown prince and with the hiring of a mechanic to keep them in order.[98] The most notable case of government benefaction to Italian physics was Pavia, where Volta established a direct channel to the Governor of Lombardy. He obtained thousands of lire for instruments, grants for foreign travel to select them, salary for a mechanic, a laboratory, storerooms, and a 'teatro fisico,' a large lecture hall where he entertained and instructed the large audiences which, with his reputation as a discoverer, supported his claim upon the treasury.[99]

In France a few Paris colleges—Navarre, Royal, and, to a lesser degree, Louis-le-Grand—had substantial collections by 1790. Navarre's, for example, numbered 235 pieces. Many of the former Jesuit colleges had important instruments, for instance, Dijon, Poitiers, Puy; even the tiny college at Epinal had two electrical machines, an air pump, and devices to illustrate the principles of mechanics, hydraulics and optics.[100] Many of these items found their way into the institutions that divided up the educational empire of the Jesuits in the 1760s.

The Jesuit universities in Austria were furnished with instruments in consequence of the Theresian reforms. Those of 1752 established a cabinet at the University of Vienna; those of 1774, an assistant to set out the apparatus before lecture. The collection excelled in models of machines, which aroused the admiration of Volta.[101] Freiburg had important pieces of apparatus, including an air pump and an electrical machine, by mid-century. It made further acquisitions through a special fee for degrees, introduced in 1752; set up a special cabinet sometime before 1756; and, as at Bamberg, made substantial additions in the 1780s.[102] Innsbruck had built up an impressive 'physical-mathematical cabinet' by 1761, for which a mechanic was engaged in 1774; meanwhile its professor of physics, Ignatius Weinhart, S. J., had assembled a good private

96. Modena, U., *Annuario* (1899–1900), 177; Pietro, *Studio* (1970), 39; Mor, *Storia* (1953), 260–2.

97. Italy, *Monografie*, I (1911), 251; Bernoulli, *Lettres*, III (1779), 181–3.

98. Torlais, *Physicien* (1954), 53–4; *DSB*, X, 145; Nollet, 'Journal' (1749), f. 11v.

99. Correspondence between Volta and Wilzeck, 1785–6, in *VE*, III, 283–4, 295, 311, 401. Volta had a regular budget of 725 lire (560#) annually for instruments around 1780; *VE*, I, 409–10.

100. Torlais in Taton, *Enseignement* (1964), 633; Dainville, *ibid.*, 40–1; Delfour, *Jésuites* (1902), 269.

101. Haberzettl, *Stellung* (1973), 141; *VE*, II, 246; Böhm, *Wiener U.* (1952), 62; Meister, Ak. Wiss., Vienna, Phil.-Hist. Kl., *Sb.*, 232:2 (1958), 36, 95.

102. [Kangro], *Beitr. Freib. Wiss. Univ. Gesch.*, 18 (1957), 10–11; Schreiber, *Gesch.*, II:3 (1860), 109. Bamberg also instituted a fee of from 0.5 to 1 RT in 1785 from each student taking physics to defray the cost of experiments; H. Weber, *Gesch.* (1879–82), 340.

collection, which he wished to be preserved for his order, in whose imminent resurrection he trusted.[103] Good collections were also made at Graz and Tyrnau.[104]

In Scotland the acquisition of instruments depended upon the generosity of friends of the universities. Money for the purchases at Edinburgh and Glasgow previously mentioned came from the town councils; Aberdeen received gifts for instruments from its graduating classes from 1721 to 1756, and from the Society for the Encouragement of Manufactures in Scotland between 1781 and 1785; Edinburgh raised £600 for apparatus during Robison's tenure of its chair of natural philosophy (1773–1797).[105] Oxbridge followed the continental practice. Cotes' instruments seem to have been passed down, probably by purchase, and were still in use in 1776; the collection of his successor once removed, Anthony Shepherd, had a reputation for excellence. Similarly Bradley had bought his apparatus from a previous lecturer, Whiteside the Ashmolean Keeper, for the large sum of £400. No doubt the instruments were sold once again at his death.[106]

PROFESSORIAL FINANCES

Perhaps the chief reason that eighteenth-century professors were not expected to do original work is that they seldom had time for it. Their academic salaries barely answered their needs; to live comfortably they taught more than their contracts required, consulted if they could, wrote textbooks, ran boarding houses, or, if clerics, assumed a share of the chores of parish or monastery.

Taking the simplest case first, the University of Paris established a fixed hierarchy of salaries in 1719, the year in which it abolished fees for courses: 1000# for professors of philosophy, 600# or 800# for regents in the lower forms.[107] The more popular professors suffered from the change, and even those whose income increased had little to celebrate, for 1000# did not support life in Paris. No doubt the fact that most of the professors were clerics made the system work. Salaries climbed slowly by award of supplements; just after the expulsion of the Jesuits the philosophy professors had supplements about equal to their salaries (2000# in all), and in 1783 they reached 2400#, still a very modest income. It is not surprising that Paris professors who were not also academicians contributed little to the progress of science. In the provinces the

103. Huter, *Fächer* (1971), 59–61.

104. According to Bernoulli, *Lettres*, I (1777), 50–4.

105. P. J. Anderson, *Studies* (1906), 151–2; Gr. Br., *Sess. Pap.*, 35 (1837), 132, 169–70; 38 (1837), 303–4; Dalzel, *Hist.* (1862), II, 449–50; *DSB*, XI, 495–7.

106. Winstanley, *Unr. Camb.* (1935), 151; Bernoulli, *Lettres* (1771), 117–18; Hans, *New Trends* (1951), 52; Gunther, *Early Sci.*, I (1920), 200–1.

107. Targe, *Professeurs* (1902), 188, 196–7; Jourdain, *Hist.*, (1888), II, 162–8; a livre of 1719 had 0.83 the silver content of our standard, the livre of 1726.

Jesuit monopoly kept salaries down until the 1760s, when they rose to 1200# in the bigger institutions.[108]

In Italy also, clerical monopoly of the lower faculties—as well as a lower standard of living—kept down salaries throughout the century. At Naples, for example, the physicist got about 900# in 1740, exactly twice the salary of the janitor. At Pavia he recieved some 500#, as against about 800# for the professor of law. There is evidence of improvement in the 1770s and 1780s: Naples was paying about 1300# in 1777; Catania gave about 500# in 1779 and 800# in 1787.[109] Volta's salary reached almost 4000# in 1795, and the senior physics professor at Pisa had about 2700#. These should be regarded as minimum amounts: often a living allowance or perhaps a house was added, and fees for degrees might bring something. At Modena, for example, professors participating in the examinations for the laurea got about 15# each from successful candidates; even the janitors had a share; and since two-thirds of the fee was returned to unsuccessful candidates, the entire university had a stake in preventing failures. The improvement in facilities and salaries helped upgrade Italian physics to the point that, in 1784, Lichtenberg could try to obtain a grant for travel to the peninsula on the ground that 'Italy is now, perhaps more than Britain, the home of true physics, der Sitz der wahren Naturlehre.'[110]

The differences between Oxbridge and the Scottish universities are no better illustrated than in professorial emoluments. The Oxbridge professor had a salary, fixed by statute, in return for which he lectured publicly, that is gratis, on a specified subject. Salaries varied according to the wishes and wealth of the founder of the chair, the prestige of the subject, and the date of foundation. The Lady Margaret Chair of Divinity at Cambridge brought some £1,000 at the end of the eighteenth century, when the Regius professorships in languages, law and medicine still yielded the £40 fixed for them by Henry VIII. The chairs for science, as relatively late foundations, usually carried adequate emoluments, from £100 (the Lucasian) to £300 (the Lowndean); college fellowships paid less, sometimes less than £40, and often about £60. Tutoring brought a pound a student a term. With £100 (2500#) a don might be comfortable, with £200 well off, and with £40 'almost destitute.'[111]

The salaries came whether the professor had few students or many, or indeed whether he lectured or not. Consequently his interest, as Adam Smith remarked, was 'directly in opposition to his duty.' No doubt Smith correctly

108. Targe, *Professeurs* (1902), 308; Lacoarret and Ter-Menassian in Taton, *Enseignement* (1964), 135; Gaullier, *Collège* (1874), 508; Montzey, *Hist.* (1877), II, 158.

109. Amodeo, *Vita*, I (1905), 13, 61, 155; Pavia, U., *Contributi* (1925), 120; Catania, U., *Storia* (1934), 258.

110. Volpati, *Volta* (1927), 42–8; Scolopio, *Storia* (1877), 79; Mor, *Storia* (1953), 110; Schaff, *Gesch.* (1912), 180–1; Lichtenberg to Schernhagen, 30 Sept. 1784, in Lichtenberg, *Briefe* (1901–4), II, 147.

111. Winstanley, *Unr. Camb.* (1935), 97, 101–2, 121, 129, 151–2, 171–3; Mallet, *Hist.*, III (1927), 124; Godley, *Oxford* (1908), 83; Frank, *Hist. Sci.*, 11 (1973), 256.

associated the sinecurism of the Oxbridge professoriate of his time with its inability to exact fees for statutory lectures. It could, however, charge for additional services: Bradley, as Savillian professor of astronomy, asked 3 gns for his course on experimental physics, and drew an average attendance of 57. The Plumian professor at Cambridge also charged for experimental physics; as late as 1802 the incumbent, Samuel Vince, still advertised lectures for the conventional 3 gns.[112]

The exception in England was the rule in Scotland, where professors collected fees of 2 or 3 gns a student. The total incomes of physicists and mathematicians ranged from about £150 to a little over £300. At the turn of the century fees accounted for about half the total at Aberdeen, for a fourth or less at St. Andrew's, for two-thirds to five-sixths at Edinburgh. Robison, for example, had a salary of £52 and an average of £260 a year from students in the late 1790s.[113] The Scottish natural philosopher worked harder to earn more than his English counterpart. No doubt economic incentive helped to make Edinburgh the leading British university in the late eighteenth century.

But the lands of academic opportunity were Holland and Protestant Germany. A man with a reputation might exact a large salary as a price for accepting or refusing a call; he could negotiate important fringe benefits, such as a free dwelling, firewood, bread and beer; and he could complete his happiness by attracting crowds of students to courses for which he was entitled to charge fees. Wolff was particularly successful at this game. He had 200 RT (800#) when he began at Halle in 1706; after calls to Leipzig, Jena and Petersburg, he got 600 RT.[114] Marburg hired him in 1723 at 500 RT with perhaps as much again in fringes; which, as he said, was 'nothing trifling in Germany,' although only half the income of the Italian singers in the Hessian opera.[115] Prussia brought him back to Halle in 1740 at the price of 1000 RT. All the while he took in substantial fees. His 'private' lectures at Marburg had as many as 100 auditors; his total income from university sources—salary plus fringes plus fees—exceeded 2000 RT annually by 1724. He had in addition royalties from his books and rent from student lodgers (1 or 1.5 RT a week).[116] Taking lodg-

112. Smith, *Wealth of Nations* (1880), II, 345–6; *DNB*, II, 1074–9, XX, 355–6; Rouse Ball, *Hist.* (1889), 104. Cf. Wordsworth, *Scholae* (1877), 255n.

113. Gr. Br., *Sess. Pap.*, 35 (1837), 51–64, 130 (tabulated by Morrell, *Isis*, 62 [1971], 165); 37 (1837), 247–8. Cf. Dalzel, *Hist.* (1862), II, 324; Grant, *Story* (1884), II, 298.

114. Wolff to Blumentrost, 24 April 1723, in Wolff, *Briefe* (1860), 14. Other results of well-played calls: G. E. Hamberger's extra professorships and the dignity of Hofrath for refusing invitations to Altdorf, Göttingen and Halle (Börner, *Nachrichten*, I [1749], 63–5); Musschenbroek's instruments and raises (to 3000#) for remaining at Utrecht (Kernkamp, *Utrecht. U.* [1936], I, 132–3).

115. Wolff, *Briefe* (1860), 23 (7 May 1724); Hermelink and Kähler, *Phil.-U. Marb.* (1927), 347n; Gottsched, *Hist. Lob.* (1755), Beylage, 34. The fringes: 115 bushels of corn, 90 of barley, 55 of oats, 5 of peas; 10 sheep, 2 pigs, 167 pounds of fish; 164 gallons of wine; free housing in the observatory.

116. Wolff, *Briefe* (1860), 14, 25, 36, 103, 114; cf. Wuttke, *Wolff* (1841), 68.

ers was very common. Professors had large houses because their private courses, and very often their public ones as well, had to be taught in their homes. It was frequently necessary to realize some income from the extra rooms in their mansions.[117]

Wolff was rewarded for more than his physics and mathematics. A more representative entrepreneur is Kästner, who began teaching mathematics at Leipzig for 200 RT, 'which would perhaps be enough for me if I were as abstemious in pleasures of the mind as I am in those of the body.' To satisfy his lust he wrote reviews for learned journals, which often paid him in books; attendance at his courses, about twenty students a year, all beginners, brought little in fees. Vigorous effort raised his total income to between 400 and 500 RT. In 1755 he was called to Göttingen to replace Segner; Göttingen gave him 'more than I could ever hope to have had at Leipzig,' and opportunities for offering advanced courses at unusually high fees.

Salaries were not particularly high at Göttingen, and did not inflate so quickly as at other universities; Lichtenberg remarked in 1784 that Saxony (meaning Leipzig), Weimar (Jena), and Mainz paid more, while Gedike was surprised to find in 1789 that many Göttingen professors had between 300 and 400 RT.[118] (Gedike's poorest universities, Altdorf, Erlangen, and Giessen, paid their professors of philosophy 50 to 150 RT; his best, Jena, Leipzig and Wittenberg, 400 to 600, or more.[119]) But Göttingen excelled in the size of its fees. By teaching between four and five hours a day, which probably did not much exceed the average, Kästner brought his income to over 1000 RT a year.[120]

Fees charged for instruction—as opposed to premiums for degrees—became more and more important in the finances of the German professor at precisely the time that experimental physics was entering the university curriculum. In principle one gave public lectures in courses necessary for degrees in return for one's salary, and offered private instruction, for a fee, in specialized or advanced subjects. During the eighteenth century the public lectures rapidly declined, and the private courses, especially in the leading universities, came to fill most of the curriculum, even in the philosophical faculty; by the beginning

117. Paulsen, *Gesch.* (1896–7), I, 536, II, 13.

118. Lichtenberg, *Briefe* (1901–4), II, 137–8, 141; Fester, *Gedike* (1905), 17. The Catholic University of Mainz does not belong in the group because its high salaries—Lichtenberg pointed to an offer of 1800 RT—compensated for its professors' inability to charge fees (*ibid.*, 47).

119. Cf. W. Hesse, *Beiträge* (1875), 66, giving salaries of two philosophers at Duisburg in 1775; the physicist Schilling, aged 72, 270 RT after 47 years' service; J. A. Melchior, aged 54, a 20-year veteran, 128 RT. Pfaff had only 300 RT when he began at Kiel in 1793 (*ADB*, XXV, 582–3). When Duisburg professors complained about their poverty, they were turned away with the agreeable information that other Prussian universities, Königsberg and Frankfurt/Oder, suffered equally. Ring, *Gesch.* (1949²), 179–80.

120. Kästner, *Briefe* (1912), 10, 30, 36–7, 59, 65, 123, 213–14; Müller, *Abh. Gesch. math. Wiss.*, 18 (1904), 103n. Paulsen, *Gesch.* (1896–7), II, 142, estimates the average teaching at 20 to 24 hours a week, and observes that Kant once offered 34.

of the nineteenth century, fees were charged in the main courses leading to degrees. Among the causes of this remarkable evolution were the economic pressures of the secular inflation of the eighteenth century and the advance of knowledge, the creation of new subjects and new approaches that could be construed as material beyond the purview—and hence beyond the responsibility—of the public lecture.[121] Fees for a course *privatim* in the philosophy faculty ranged from 1 to 6 RT, depending on the university and the reputation of the professor; according to Gedike's numbers, Göttingen's 4 to 6 RT was about twice the average charge at German Protestant universities in 1789. A course *privatissima* came still dearer, at between 15 and 20 RT, or perhaps even twice that, at Göttingen.[122]

The right to charge fees meant little unless students enrolled in sufficient numbers. At the larger universities one could take as much or more in fees as in salary. The case of Kästner has been mentioned. Similarly Lichtenberg took 80 louis d'or (about 450 RT) from 112 auditors in 1784, a considerable improvement over the 40 students with which he started in 1777 / 8. Wolff averaged over 100 at Marburg. Gilbert had 40 in physics and 12 in mathematics at Halle in 1801–2; Kratzenstein had between 30 and 40 at Copenhagen.[123] At the smaller schools little could be got. Andreas Nunn, for example, had no takers for a course *privatim* in experimental physics at Erfurt in 1755. At Duisburg, where matriculations averaged less than 50 a year, not much could be hoped for. And even in the bigger schools difficult or specialized subjects might not pay. C. A. Hausen, professor of mathematics at Leipzig, lectured publicly, 'for no one would give money to hear about conic sections,' and even then he had few auditors.[124]

To complete the picture of professorial income we must add in premiums for degrees (which cost from 43 RT for an MA to 132 RT for a doctorate in theology at Göttingen in 1768), fees for preparing theses (30 RT at Halle), royalties, payment for collaborating in learned journals or reviews, gifts from dedicatees, and so on.[125] It is very difficult to estimate the income from these sources, which could be considerable for a well-placed man. On the whole, an able and energetic professor of physics could do better at a leading Protestant German university than at any other at the end of the eighteenth century.

121. Paulsen, *Pr. Jahrb.*, 87 (1897), 138–41; *Gesch.* (1896–7), II, 128–9.

122. Fester, *Gedike* (1905), *passim;* Müller, *Abh. Gesch. math. Wiss.*, 18 (1904), 82; Pütter, *Versuch*, I (1765), 319, reporting the higher figure—indeed 30 to 100 RT—for 1765. Other indications of fees: 2–6 RT for 5 hours' *privatim* at Halle (Schrader, *Gesch.* [1894], I, 108–9); 3.5 RT for two months', probably eight hours', *privatim* at Basle in 1760 (Spiess, *Basel* [1936], 113); 4 RT for a semester's *privatim*, 50–100 *privatissima*, at Marburg (Hermelink and Kähler, *Phil.-U. Marb.* [1927], 385).

123. Lichtenberg, *Briefe* (1901–4), II, 127, 228, 335; Hermelink and Kähler, *Phil.-U. Marb.* (1927), 391n; Schrader, *Gesch.* (1894), I, 635; Snorrason, *Kratzenstein* (1967), 53–4.

124. Stieda, *Erf. U.* (1934), 19, 27; W. Hesse, *Beiträge* (1875), 48; Kästner, *Selbstbiographie* [1909], 6–7.

125. Pütter, *Versuch*, I (1765), 320; Schrader, *Gesch.* (1894), I, 108–9.

It remains to compare the price of the standard instruments to professorial incomes. A benchmark is the gift to Harvard of a good apparatus in the style of Hauksbee, bought new in Britain in 1727 for about 3000#, or the estimate of 200 gns (5000#) for a full outfit of books and instruments made by the University of Aberdeen in 1726, or the value of Jallabert's excellent collection (4500#) assembled in the 1740s.[126] (The so-called 'complete' apparatus in the style of Musschenbroek, purchased in London by the University of Uppsala in 1740 for about 2000#, must have lacked something.) From these data it appears that the cost of a full set of instruments in the 1730s was about equal to the annual income of a well-paid professor of physics.

Fifty years later, Volta drew up a list of instruments 'needed' to bring his cabinet up to the mark. It ran to 9300# for purchases in France and England, plus an unestimated charge for items to be made locally. And that did not include apparatus for electricity, 'for which at the moment there is nothing much good in the gabinetto.' Such an expenditure was beyond Volta's means, and well beyond what was absolutely required; he wished an apparatus that would not only instruct his students but impress his many foreign visitors, who 'will view the physics cabinet with the same satisfaction and surprise—and will talk about it everywhere—as they already see, praise, and admire the botanical garden, the chemical laboratory, and the museum of natural history.'[127] We may take it that a meagre but serviceable demonstration apparatus could be bought at less than 5000# at the mid-century, and that a full and fancy one cost upward of 10,000# in the 1780s.[128]

The items in the cabinet of greatest interest to us—air pumps and, above all, electrical machines—were among the most expensive. The average cost of Volta's desiderata was 150#. Jan van Musschenbroek's double-barreled air pump sold for $f.300$ (650#) in 1736; Martin's best pump cost 35 gns (900#) in 1765; Nairne's standard pump cost as much in the 1780s, and considerably more with accessories.[129] Nairne's standard electrical machine, a serviceable but not elaborate model with a six-inch globe, cost 170# in 1765, when Martin's best large machine brought 490#. Nairne's big cylinder machine could be bought for 480# in 1779; Cuthbertson wanted over 2000# for his completely furnished three-foot plate machine in 1782.[130] Up-to-date electrical machines and air pumps went beyond the reach of most professors in the 1780s.

126. Cohen, *Tools* (1950), 133; Rait, *Universities* (1895), 295–6; Borgeaud, *Hist.*, I (1900), 571. The Scots did not succeed in raising the money.

127. Volta to Firmian, 13 March 1780, *VE*, III, 455–67.

128. This estimate agrees with the cost of the second complete Harvard apparatus, acquired in 1765/6 for about £400 (9800#), for it contained duplicates of several expensive items. Millburn, *Martin* (1967), 131–5, 142–3.

129. Crommelin, *Desc. Cat.* (1951), 33; Lichtenberg, *Briefe* (1901), II, 6; *VE*, II, 146; cf. Henderson, *Life* (1867), 216. The Harvard air pump, bought in 1727, also cost about 650#; Cohen, *Tools* (1950), 141.

130. *BFP*, XII, 259; Hackmann, *Cuthbertson* (1973), 52; Lichtenberg, *Briefe*, I (1901), 277, 338, and *GGA* (1786:2), 2012–13; Millburn, *Martin* (1976), 131, 219. An early plate machine, of

4. INDEPENDENT LECTURERS

Electricity figured prominently in the repertoire of independent lecturers on experimental physics. They attracted many to its study, and occasionally made advances in it themselves. Their chief goal was popularization and entertainment, the reduction of the latest discoveries to the level of 'the meanest capacities' able to afford the service. One offered to explain everything 'in such a plain, easy and familiar Manner, as may be understood by those who have neither seen or read anything of the like Nature before.'[1]

The mean capacities could choose from a wide range of purveyors. At the top were the public lecturers associated with learned societies. We already know the permanent lectureships at Stockholm and Munich. Other academies, such as Bordeaux, colleges of the University of Paris, and secondary schools in France and Britain also occasionally engaged a 'physicist' to instruct and amuse them. A second class of lecturer consisted of members of learned societies who set up independently of their institutions. A notch lower, perhaps, came the unaffiliated entrepreneurs, who taught in rented rooms, and the itinerant lecturers, who performed in public houses. At the bottom of the heap were the hawkers of curiosities, the street entertainers, and the jugglers who held forth at the fairs of Saint Laurent and Saint Germain.[2]

LECTURERS OF THE BETTER CLASS

In France, public lectures of quality on the mechanical philosophy go back to the middle of the seventeenth century, when Jacques Rohault began his 'Wednesdays' dedicated to experimental illustrations of the physics of Descartes. Educated by the Jesuits in Amiens, Rohault took readily to mathematics, mechanics, and Descartes, set up in Paris as a tutor in geometry, visited artisans 'for the pleasure of seeing them work,' and won himself a fortune and a wife above his station, a lady sacrificed by her Cartesian father 'for the sake of the philosophy of Descartes.' Although he became chief of the Cartesian physicists, Rohault by no means slavishly followed Descartes; like all the successful public lecturers, he had to care more for the phenomena than for the system; not metaphysics but clarity, eloquence and manipulative skill brought in paying auditors 'of all ages, sexes and professions.'[3] Even the physicists regarded him favorably.[4]

---

five-foot diameter, was made for the Duc de Chaulnes in 1777 for 800#; Ingenhousz, *PT*, 69 (1779), 670.

1. Ferguson (1764), quoted in Harding, *Hist. Ed.*, 1 (1972), 149.

2. Cf. A. Franklin, *Dictionnaire* (1906), 570; Kästner, 'Verbindung,' in *Verm. Schr.* (1783³), II, 364; and Pujoulx, *Paris* (1801), 33: 'Hé comment les sciences ne feraient-elles pas des progrès rapides! Les savans courent les rues, et nos boulevards sont devenus des écoles de physique.'

3. Savérien, *Hist.*, VI (1768), 5–20; Mouy, *Développement* (1934), 108ff.; Pacaut, Ac. sci., Amiens, *Mém.*, 8 (1881), 5, 9. Mouy, *Développement* (1934), 112, is doubtless correct in rejecting the canard (Savérien, *Hist.*, VI [1768], 22) that the pendant Pancrace in Molière's *Marriage forcé* (1664) is based on Rohault.

4. Mouy, *Développement* (1934), 187n.

Among Rohault's emulators Pierre-Sylvain Régis and the physician Pierre Polinière were most conspicuous. Régis, a student of the Jesuits at Cahors, went to Paris for theology but gave it up on hearing Rohault. Admitted to discipleship, he was sent in 1665 to the provinces to lecture publicly on Cartesian physics. He returned to Paris after Rohault's death and lectured to great applause until the Archbishop of Paris shut him down, 'in deference to the old philosophy.'[5] Régis thereupon offered 'private' courses to the mighty, among them the archbishop, who is said to have become the most enthusiastic of his auditors. Régis entered the Paris Academy at its reorganization in 1699; although he was too old and ill for academic work, 'his name [as Fontenelle gracefully put it] served to ornament a list on which the public would have been surprised not to find it.'[6]

Polinière also developed against a Cartesian background. He was educated by the Jesuits at Caen and at the University of Paris, where in 1695 he initiated a course of experimental physics under the auspices of the professor of philosophy at the Collège d'Harcourt. He perhaps took as his inspiration and model his professor of mathematics, Varignon, who occasionally used experiments to illustrate his lectures at the Collège Mazarin. Polinière's course was clear, intelligent, 'a mortal blow [we are told] to the physics of Aristotle.'[7] He worked hard at improving standard demonstrations, which he interpreted in undogmatic Cartesian terms. His enterprise was sometimes rewarded by the discovery of new phenomena, like electroluminescence, which might be shown to advantage. He succeeded before general audiences, in guest lectureships at Parisian colleges, and before the Regent, young Louis XV, and Fontenelle. Yet, like Rohault, Polinière remained outside the Academy of Sciences, which first admitted such an entrepreneur in the person of Nollet, whose lecturing began just after Polinière's death.[8]

With the help of academicians whose assistant he had become, Nollet went to London to seek the advice of Desaguliers, and to Leyden to inspect the instruments of 'sGravesande. He found the apparatus so expensive that he could finance it only by building and selling duplicates: before mounting the podium he had first to enter the workshop. 'I wielded the file and scissors myself [he wrote of that time]; I trained and hired workmen; I aroused the curiosity of several gentlemen who placed my products in their studies; I levelled a kind of voluntary tribute; in a word (I will not hide it) I have often made two or three instruments of the same kind in order to keep one for myself.'[9] By 1738, when

5. *Ibid.*, 146, 166–7.

6. 'Eloge de Régis,' *Oeuvres*, V, 92.

7. Savérien, *Hist.*, VI (1768), 167–73; Hanna in Gay, ed., *Eight. Cent. St.*, 16–18. Fontenelle, *Oeuvres*, VI, 261, traces college lectureships in experimental physics to the example of the semi-private lectures arranged by the apothecary M.F. Geoffroy for the benefit of his son, Etienne-François, c. 1690.

8. Polinière, *Expériences* (1718[2]), Préf.; Corson, *Isis*, 59 (1968), 402–13; Brunet, *Physiciens* (1926), 101–2; Savérien, *Hist.*, VI (1768), 185–7.

9. Nollet, *Programme* (1738), xviii–xix; Grandjean de Fouchy, *HAS* (1770), 121–37; Lecot, *Nollet* (1856), 1–13; Torlais, *Physicien* (1954), 1–40.

Nollet provided his course with a formal syllabus, his business could handle an order from Voltaire for instruments costing over 10,000#.[10]

Nollet's *Cours de physique,* which incorporated phenomena he had discovered, was perhaps the most popular exhibition of its kind ever given. In 1760 he drew 500 paying customers.[11] He aimed to be useful and agreeable, to entertain his auditors as he disabused them of their 'vulgar errors, extravagant fears, and faith in the marvellous.'[12] People of all conditions flocked to hear him, including duchesses, whose carriages piled up before his doors, and princes of the blood, who 'honored the master with their close attention, and brought away the kind of knowledge that is always an ornament to the mind, and confers luster on the most distinguished birth.' In 1739 Nollet entered the Academy as adjunct mechanic and in 1757 he became a pensionary.[13]

Several others managed to follow Nollet's example. His protégé Mathurin-Jacques Brisson marched at his heels: assistant to Réamur, public lecturer, academician, professor.[14] Similarly Sigaud de la Fond, member of the academies of Montpellier and Angers and, in 1796, also that of Paris, amused the Parisian *grand monde* with experimental physics—'Nollet improved,' he said—from about 1767.[15] The best of these later public lecturers was J. A. C. Charles, who turned to experimental physics at the age of 35, when an economizing ministry abolished his petty bureaucratic post. 'He was left with what happily suffices for those who are to excel in the arts, the free disposition of his time and talents.'[16] In 1781, after eighteen months of study, he began to lecture. His skill, plus the advertisement of a journey in his hydrogen-filled balloon, brought him a large audience. With their fees he built up what in 1795 was judged to be the most complete collection of demonstration apparatus in Northern Europe, all fashioned in the style of Nollet, Brisson, and Sigaud de la Fond.[17]

The first public physics course in London was inaugurated in 1704 by an important electrician, Francis Hauksbee, who began as an instrument maker and gave the public material like Polinière's, but interpreted on Newtonian

10. Voltaire, *Corresp.,* VI, 191, VII, 156, 176, 261: 'C'est un philosophe [Nollet], c'est un homme d'un vray mérite qui seul peut me fournir mon cabinet de physique, et il est beaucoup plus aisé de trouver de l'argent qu'un homme comme luy.'

11. Ferrner, *Resa* (1956), xliii; Tolnai, ed., *Cour* (1943), 64.

12. The same sentiment appears in Polinière, *Expériences* (1718²), Préf.; cf. Savérien, *Hist.,* VI (1768), 190.

13. Nollet, *Programme* (1738), xxxv–xxxvi; Marquis du Châtelet to Francesco Algarotti, 20 April 1736, in Du Châtelet-Lomont, *Lettres* (1958), I, 112. Cf. *ibid.,* 93; Lecot, *Nollet* (1856), 19; Torlais, *Physicien* (1954), 41–63, 203–4.

14. Torlais, *Physicien* (1954), 234–6.

15. *Ibid.,* 232–3; Torlais in Taton, *Enseignement* (1964), 630–1.

16. Fourier, *MAS,* 8 (1829), lxxiv.

17. *Ibid.,* lxxvi; Bugge, *Science* (1969), 154, 166–8. For the situation in the provinces, where demand picked up briskly in the '70s and '80s, see Mornet, *Origines* (1947²), 316; Torlais in Taton, *Enseignement* (1964), 634; G. Martin, *RHS,* 11 (1958), 214, reporting difficulty in obtaining subscribers in 1747.

principles.[18] The success of his lectures may be inferred from the eagerness of two separate parties to continue them after his death in 1713, namely his nephew, Francis Hauksbee the Younger, and Desaguliers, who perhaps left Oxford for the purpose. Desaguliers' efforts proved the more attractive to the public, who consumed, on the average, some six cycles of his lectures every year.[19]

Desaguliers also succeeded to another post of Hauksbee's, that of occasional curator of experiments to the Royal Society of London. The position, which had fallen into desuetude by the end of the seventeenth century, was apparently revived for Hauksbee when Newton became president of the Society in 1703. As we know, its incumbent had to prepare and exhibit experiments to the Fellows at their weekly meetings, an onerous task if, as in the case of Hauksbee and Desaguliers, at least some of the demonstrations rested upon original work, and reimbursement for out-of-pocket expenses was not always prompt. (When expenses were slight, payment might be immediate, as when Desaguliers 'made a present of a Worm vomit'd by a Cat, for which he had thanks.'[20]) Desaguliers' burden appears from a memorandum addressed to the Society to justify an expenditure of about £10 for the construction of four new machines. 'Before I bring anything to the Society I spend many Days about it at Home to try the Experiments before Hand; and adjust the Machines; so that the time expended and accidental Charge that Way, is often more than double the Cost of the Machines, especially because it often happens that the whole Instrument is thrown by, when I find it is not worth the Society's Notice.'[21] Doubtless apparatus tested at the Society's expense found its way into his lecture room.

Hauksbee and Desaguliers had close ties with leading applied mathematicians and physicists. Hauksbee began his lectures with James Hodgson, an assistant of the Astronomer Royal, John Flamsteed,[22] and in his curatorial capacity Hauksbee often worked with Newton. The younger Hauksbee teamed up with William Whiston, whom we have met as a Cambridge professor of mathematics.[23] Desaguliers had worked with the belligerent Newtonian, John Keill, one-time professor of astronomy at Oxford.[24] As Hauksbee's successor, Desaguliers became the most active of Newton's agents in the Royal Society.

Desaguliers had no successor of equivalent stature: unlike Paris, London proved unable to generate or to support such men in the second half of the

18. Hauksbee, *Phys. Mech. Exp.* (1709).

19. Guerlac in *Aventure* (1964), I, 228–53; *infra*, viii.1; Desaguliers, *Course* (1763³), I, ix, says that he completed his 121st lecture cycle in 1734.

20. JB, XIV, 279 (RS).

21. *DNB*, V, 850–1; Torlais, *Rochelais* (1937), 1–14; Desaguliers to RS, 29 Oct. 1733, in Misc. Corresp. D:2, ff. 71–5 (RS).

22. Rowbottom, Cong. int. hist. sci., XIᵉ (1965), *Actes*, IV (1968), 198–9; Hug. Soc. Lond., *Proc.*, 21 (1968), 191–206.

23. Hans, *Trends* (1951), 49–50, 137, 142–3; Whiston, *Memoirs* (1749), 235–6; *DNB*, VI, 175–6 (Hauksbee Jr.)

24. *Supra*, ii.3.

eighteenth century. From time to time amateurs offered to show the public the latest scientific discoveries; such as one Rackstrow, who '[took] impressions from life, [and made] them up in plaster,' and also demonstrated electricity, about which he wrote a pamphlet at the urging of his friends, 'not being proof against flattery.'[25] There is evidence that in the late 1750s London could not maintain a single distinguished independent public lecturer. S. C. T. Demainbray, a disciple of Desaguliers', a successful lecturer in both France and England in the 1740s and early 1750s, the owner of what Franklin judged to be the best demonstration apparatus in the world, could in 1758 'hardly make up an audience in this great City [London] to attend one course a winter.'[26] In the same year James Ferguson, a painter who had been trying for a decade to support himself by his excellent courses on astronomy and experimental physics, contemplated leaving the capital: 'There are at present more than double the number [of hopeful demonstrators] which might serve the place, people's taste lying but very little that way; so that unless something unforseen happens, I believe my wisest course will be to leave London soon.'[27]

Ferguson managed to make a living, and a good one, by going on tour: Bristol and especially Bath, where, like Martin before him, he was always warmly received by the fashionable and unoccupied water-takers. Reading, Gloucester, Salisbury, Liverpool, Newcastle, and the growing industrial towns of the Midlands, which lacked facilities for adult education and recreation, supported many itinerant lecturers; at Manchester the public subscribed to at least one course a year from 1760 to 1800.[28] Naturally these courses varied in quality, from the authoritative lectures and original demonstrations of a Ferguson to the entertainments of Gustavus Katterfelto, a German who worked the Midlands with 'electricity, and a few other tricks of physics, and a little of the art of conjuring.'[29] The best of the itinerant lecturers—Adam Walker, Henry Moyes, and others—included London in their circuit, but few besides Ferguson maintained their headquarters there. Some numbers will illustrate the extent to which British public lecturers in natural philosophy shifted their attention to the provinces after 1740. Taking as a sample the fifty or so individuals about whom something is known, one finds seven public lecturers active in London in each decade from 1710 to 1740, but only four in the forties and fewer thereafter; in the case of itinerant lecturers, two were active per decade before 1740 and ten on the average from then until the end of the century.[30]

25. Rackstrow, *Misc. Obs.* (1748), i–ii. For reasons set out *infra*, x.2–3, public demonstrations of electricity reached a peak in the late 1740s. As Henry Baker sneered in 1747, many then earned 'great deal of money shewing a course of Electrical Experiments at a Shilling for each Person'; G. Turner, RS, *Not. Rec.*, 29 (1974), 64.

26. Franklin to Kinnersley, 28 July 1759, *BFP*, VIII, 416; *DNB*, V, 780–1. For colonial itinerant lecturers, see Stearns, *Science* (1970), 510–11.

27. Ferguson to A. Irvine, 17 Jan. 1758, in Henderson, *Ferguson* (1867), 225.

28. Musson and Robinson, *Science* (1969), 102n; Gibbs, *Ambix*, 8 (1961), 111; Millburn, *Martin* (1976), 38, 49–51.

29. Musson and Robinson, *Science* (1969), 101–2n.

30. Compiled from *ibid.*; Gibbs, *Ambix*, 8 (1961), 111–17; Mumford, *Manchester* (1919);

The Netherlands also supported several distinguished independent lecturers. Fahrenheit was perhaps the first; he supplemented the revenues from his thermometers by giving public instruction in experimental physics in Amsterdam from 1718 to 1729. Shortly after he retired, Desaguliers made a triumphant tour in Holland, billed by Musschenbroek as 'one of the most famous philosophers of the age.'[31] Several lesser intellects, impressed by the spectacle and the profits, set up in Amsterdam and elsewhere. After 1750 lecturers became attached to the newly-founded scientific societies, despite the prohibition against the sponsorship of public courses forced on the academies of Haarlem and Rotterdam by the University of Leyden. The most important of these men, who perhaps numbered a dozen in all, was the ubiquitous van Marum, who lectured first without sponsorship and then moved under the wing of the Teyler Genootschap.[32] A similar development occurred in Sweden, where the public lectures of Triewald, conceived in the style of Desaguliers and 'sGravesande, were later replaced by those of the Stockholm academician Wilcke.[33]

FINANCES

The London lecturers in the 1720s asked two to three guineas for a course; the itinerants of the 1760s one guinea for subscribers, or a half crown for a single lecture. Since the latter customarily gave twelve lectures and the former twice that, the average cost per session remained between one and two shillings throughout the century. Subscriptions were payable in advance, and arranged before the lecturer came to town; usually he required a guaranteed minimum, twenty or thirty paid-up clients locally, or forty or more if he had to travel, before he would agree to perform. A good lecturer could make something in this way. In four months in 1763 Ferguson grossed £139 in Bristol and Bath, and the same area yielded over twice as much in 1774. Adam Walker made 600 guineas in Manchester and Liverpool in 1792; James Bradley and Desaguliers averaged 420 and perhaps 300 guineas, respectively, in Oxford and London, in the 1720s and 1730s.[34]

Special events might bring special emoluments, such as the £120 that Whiston had from a 'numerous and noble audience' for a lecture on an upcoming solar eclipse. This was good money, and above average expectation, as we

Turner in Forbes, *Marum*, IV, 1–38; Fawcett, *Hist. Today*, 22 (1972), 590–5. The total for the 1730s agrees well with the figure (eleven or twelve) given by Desaguliers, *Course*, I (1734), Pref.

31. Torlais, *Rochelais* (1937), 22; Rooseboom, *Bijdrage* (1950), 21; Cohen and Cohen-de Meester, *Chem. Week.*, 33 (1936), 379–83.

32. Muntendam in Forbes, *Marum*, I, 16–17; Brunet, *Physiciens* (1926), 98; Dekker, *Geloof weten.*, 53 (1955), 173–6; Hackmann, *Cuthbertson* (1973), 15.

33. Beckman, *Lychnos* (1967–8), 187–93; Tandberg in Lund., U., *Årssk.*, avd. 2, 16:9 (1920), 4–5; *supra*, ii.2.

34. Hans, *Trends* (1951), 47–8, 139, 142–3, 147–8; McKie, *Endeavor*, 10 (1951), 48–9; E. Robinson, *Ann. Sci.*, 19 (1963), 31; Musson and Robinson, *Science* (1969), 104, 145, 164–5; Henderson, *Ferguson* (1867), 272, 340–1, 348, 376, 408; Fawcett, *Hist. Today*, 22 (1972), 590–5; Millburn, *Martin* (1976), 61–2.

learn from a letter from Smeaton to Benjamin Wilson, then (1746) contemplating an itinerant lectureship: 'I don't take ye shewing ye wonders of Electricity for money is much more considerable than ye shewing any other strange . . . sight for ye same end, however if £200 could be got by a worthy employment in y^t way I don't see where is ye harm as there is no fraud or Dishonesty in it.'[35] One needed a capital of good will, a little information, and an apparatus costing about £300.[36]

The price of subscription shows that English lecturers did not aim at the common man. John Roebuck, acting as advance man for Henry Moyes, 'procured him the Countenance and favor'—that is the subscriptions—'of some principal gentlemen' in the neighborhood.[37] Erasmus Darwin recommended that young ladies improve themselves 'by attending the lectures in experimental philosophy, which are occasionally exhibited by itinerant philosophers.'[38] Benjamin Martin, writing of the lecture circuit of the 1740s and 1750s, permits us no doubt about his clientele. 'There are many places I have been so barbarously ignorant, that they have taken me for a *Magician*; yea, some have threaten'd my life, for raising Storms and Hurricanes: Nor would I show my face in some Towns, but in company with the Clergy or the Gentry, who were of the Course.'[39] And when gentlemen lost interest in natural philosophy, most of the audience of the independent lecturer disappeared. One cause of the depressed market for physics in London around 1760 was a diversion of interest to current events, local intrigue and the Seven Years' War. 'Some Notice may be taken abroad, of what is new and ingenious in Matters of Natural Philosophy; but here we think of nothing but Politicks, Money and Pleasure.'[40]

The Parisian purveyor of natural philosophy also suffered from changes in fashion, for his prosperity depended in large measure on pleasing the ladies. In the 1740s they flocked to Nollet, in such numbers as to drive away the gentlemen: 'It seems that among the fashionable only women are still [1749] able to meddle publicly with physics.'[41] Interest appears to have declined in the 1760s, only to rise to a new pitch in the 1780s. Paris then supported many independent lecturers. The 'lycée' of Pilatre de Rozier, 'la vogue de Paris,' had 700 sub-

35. 24 Sept. 1746, Wilson Papers, f. 22 (RS); Whiston, *Memoirs* (1749), 204–5.

36. For a good collection such as Ferguson's (Henderson, *Ferguson* [1867], 453) or John Whiteside's (Turner in Forbes, *Marum*, IV, 18); one could get by with £100, as did Caleb Rotheram (Musson and Robinson, *Science* [1969], 90). Demainbray's expenditure, estimated by Franklin at £2000 (*BFP*, VIII, 416), was probably largely for fine furniture; cf. the cost of university cabinets (*supra*, ii.3).

37. Roebuck to Watt, 14 Aug. 1777, Musson and Robinson, *Science* (1969), 145n. Mathew Bolton performed the same service for John Warltire in 1776 and 1779; McKie, *Endeavor*, 10 (1951), 48–9.

38. *Plan for the Conduct of Female Education in Boarding Schools* (1797), quoted by E. Robinson, *Ann. Sci.*, 19 (1963), 30n; cf. Millburn, *Martin* (1976), 73: 'It is now [1755] growing into a fashion for the ladies to study philosophy.'

39. B. Martin, *Supplement* (1746), 28–9n.

40. Symmer to Mitchell, 30 Jan. 1761, Add. Ms., 6839, f. 309 (BL).

41. J. B. Le Roy to the comte de Tressan, 26 Aug. 1749, in Tressan, *Souvenirs* (1897), 6.

scribers in 1785, mostly women, and Swiss guards at its doors. That year its professor of physics, Antoine de Parcieux, successor of Nollet at the Collège Navarre, offered two complete courses on natural philosophy.[42] He faced very strong competition: to mention only those of the highest quality, Brisson advertised two courses, Charles four, and Sigaud's successor, Rouland, no fewer than ten.

Descending a level, we find Jacques Bianchi, rue St. Honoré, puffing several series in electricity and offering to provide, at a cost of 55 louis (1320#) and within six months of order, a complete outfit for demonstrating the truths of experimental physics. He did not exhaust the opportunities of either buyer or seller. A competitor of Bianchi's, one Bienvenue, prepared an apparatus for his auditors to take on summer holidays, to forestall ennui; while several characters at the Palais Royal showed electricity, automatons, funny mirrors, and 'amusing experiments.'[43] These last gentlemen, from the Bianchi's down, contributed nothing conceptually or instrumentally to our subject: '[Their] cabinets all contain the same items, sold by the same shops . . . A hundred such collections would not furnish the apparatus for a coherent course of instruction.'[44]

The cost of these lectures appears to have been independent of their quality. Charles and Brisson got one louis (24#) per month for courses of two or three months' duration, meeting probably three hours a week (a cost of about two livres an hour); Rouland's standard offering was 12 lessons for 24 livres; and the others asked between 1.2 (Bienvenue) and 3 (Bianchi) livres per 'séance.'[45] Public lectures in France were therefore about as expensive as those in England.[46] Who patronized them besides the *grand monde* and the 'foreigners, women and savants' known to have frequented Charles'?[47] Students, perhaps, but of what? Or maybe the highest class of artisan, of whom we occasionally find traces in the lecture halls, or rather public houses, used by itinerant lecturers in England?[48]

Most of the public lecturers in France and England—and *a fortiori* in other countries where demand was less—supplemented their incomes with other related work, particularly designing, improving, or making apparatus, teaching in

42. Torlais in Taton, *Enseignement* (1964), 634; R. Hahn, 'Sci. Lect.'; Mornet, *Origines* (1947²), 284–6.

43. R. Hahn, 'Sci. Lect.,' compiled from advertisements in the *Journal de Paris;* Daumas, *Instruments* (1953), 195–6.

44. Charles, 1794, as quoted in Daumas, *ibid.*, 196.

45. Compiled from R. Hahn, 'Sci. Lect.,' and, for Charles, from Bugge, *Science* (1969), 167. Fahrenheit had asked a little more, about 40# (*f.* 18.7) for 16 lessons; Rooseboom, *Bijdrage* (1950), 21.

46. A louis d'or had a value slightly less than a guinea; Lichtenberg, *Briefe*, II, 71.

47. France, *Elvire* (1893), 13; Fourier, *MAS*, 8 (1829), lxxvi.

48. Musson and Robinson, *Science* (1969), 108–9, 113–15, 132. If we are to credit Benjamin Donne's advertisement in the *Bath Chronicle* for 29 Dec. 1774 (Robinson, *Ann. Sci.*, 19 [1963], 28), 'few schools or academies' then offered 'courses of Lectures in Experimental Philosophy upon a proper apparatus, at only one guinea per annum additional expense.' Cf. *ibid.*, 32.

secondary schools, tutoring,[49] surveying,[50] and writing books. It is difficult to estimate what these activities might bring. One knows that the instrument business of Nollet, and, for a time, of Martin, were profitable, and a successful text, based upon tested lectures, might be rewarding as well as influential. Ferguson's books earned something: he sold the copyright to his *Astronomy* for £300, which helped to set him up as a lecturer.[51] Several English itinerant lecturers taught in secondary or vocational schools, and a few London lecturers were closely associated with the Little Tower Street and Soho Academies.[52] In the 1720s James Stirling, who owned an interest in Little Tower Street, had an annual income from his public and private lecturing of some £200, which he found to exceed his needs.[53]

We shall meet many of these frugal and able men again.

49. E.g., McKie, *Endeavor*, 10 (1951), 49; Henderson, *Ferguson* (1867), 251.
50. Robinson, *Ann. Sci.*, 19 (1963), 31, 35; *Ann. Sci.*, 18 (1962), 197, 205.
51. Henderson, *Ferguson* (1867), 52–3.
52. Musson and Robinson, *Science* (1969), 41, 119; Hans, *Trends* (1951), 82–93.
53. James Stirling to his brother, 22 July 1729, in Tweedie, *Stirling* (1922), 14.

PART TWO

Electricity in the Seventeenth Century

When I use the term attractions I do so with the qualification that I do not mean it in any way whatever.
—MRS. WILFER in Dickens, *Our Mutual Friend*

Chapter III

William Gilbert and the Amber Effect

Although the ancients knew amber's peculiar ability to draw light objects, and medieval philosophers had occasionally exercised their speculative powers upon it, the 'amber effect' did not become the basis of an independent branch of knowledge until the seventeenth century.[1] This improvement followed the discoveries that many substances besides amber could be made to exhibit the effect, and that certain characteristics seemed to distinguish it sharply from other processes previously lumped together as 'attractions.' In a word, the amber effect became 'electricity,' and those substances which displayed it, 'electrics.' We owe this term, derived from the Greek word for amber, to William Gilbert, whose experimental work gave the first intimation of the catholicity of electrical phenomena.[2]

1. GILBERT'S STYLE

Gilbert intrigues all students of the Scientific Revolution. His chief work, *On the Magnet,* published in London in 1600, is one of the earliest monographs devoted to a particular branch of terrestrial physics and one of the first published reports of an extensive series of linked, reconfirmed experiments. In these items *De magnete* anticipated and even inspired some of the best scientific work of the seventeenth century. It may also seem to mark the break with scholastic philosophy. Gilbert was a sedulous advocate of his own originality, of the novelty of his 'new style of philosophizing,' of his 'unheard-of doctrines, opposed to everyone's opinions'; he wrote, so he says, not for common schoolmen, not for 'smatterers, learned idiots, grammatists, sophists, wranglers and perverse little folk,' but for independent spirits, for 'true and honest philosophizers' who could appreciate the improvements of the age.[3] Some have taken this Renaissance bombast at face value, and accepted Gilbert as a revolutionary hero.[4] That is a mistake. Beneath his rant and puffery Gilbert is a moderate peripatetic and not above plagiarizing those he criticizes.

1. For ancient and medieval knowledge and opinion about electricity see Benjamin, *Hist.* (1898), Chaps. i–iv; Daujat, *Origines,* I (1945), *passim;* Urbanitzky, *Elektricität* (1887), 67–110; T. H. Martin, Pont. acc. sci., Rome, *Atti,* 18 (1865), 97–123.

2. 'Electricity' (electricitas) originally meant the property of attracting like amber, a sense it retained into the eighteenth century. Cf. Heathcote, *Ann. Sci.,* 23 (1967), 261–75.

3. *DM* (Th), *ij–iij; cf. *DM* (Mo). xlviii–xlix.

4. Cf. M. B. Hesse, *Brit. J. Phil. Sci.,* 11 (1960), 1–10, 130–42.

Gilbert spent twelve years at St. John's, Cambridge, where he matriculated in 1558. As an undergraduate he would at first have followed the curriculum prescribed in the statutes of 1549, which had been designed with an eye to the improvement of navigation: mathematics and astronomy in the freshman year, dialectics in the second, and philosophy, the usual Aristotelian physics, metaphysics and ethics—in the third and fourth. In the year after he came up, however, Elizabeth directed that rhetoric and dialectics receive greater emphasis, at the expense of mathematics and philosophy. When he graduated in 1560, his primary achievement, if he had learned his lessons well, would have been a capacity to wrangle facilely in Latin syllogisms over themes taken from, and with weapons provided by, Aristotle.[5]

On graduation Gilbert probably became a 'physik fellow,' a scholar supported in the study of medicine without the usual fellowship qualification regarding holy orders. Medical studies, like the undergraduate curriculum, were still largely literary and medieval; Galen and his commentators, the only authorities recognized by the Royal College of Physicians, remained the standard texts. Gilbert's private reading during his long preparation for the M.D. doubtless ranged far beyond the prescribed curriculum. He apparently kept up the technical studies cultivated under the favorable regimen of Edward VI, for he became the college's examiner in mathematics in 1565.[6] And, perhaps, before taking his doctorate in 1569 he began to feel the attraction of the lodestone.

The new physician broke from the small-town environment in which he had spent most of his life; he may have traveled to the Continent, and certainly visited London where, by the mid-1570s, he had established a prosperous practice. He rose in his profession, attended the aristocracy in what, alas, often proved their last illnesses, mixed easily with the intellectuals in the busy capital, and became an influential member of the stodgy and conservative College of Physicians.[7] As he rose he continued his private studies, 'with many pains, and vigils and expenses.'[8] Two books resulted, which he was in no rush to publish, perhaps from fear they might injure his practice. The first, *De magnete,* appeared in the same year, 1600, that Gilbert attained the pinnacle of his profession, an appointment as doctor to Queen Elizabeth and the presidency of the Royal College of Physicians.[9] The second book, *De mundo nostro sublunari*

5. Roller, *The De Mag.* (1959), 66; Simon, *Education* (1966), 252–3, 271–2. Cf. Costello, *Schol. Curr.* (1958), 41 and *passim;* and, for a more favorable view, Curtis, *Isis,* 51 (1960), 112–13, and *Oxford* (1959).

6. Roller, *loc. cit.;* Clark, *Hist.* (1964), I, 165; Langdon-Brown, *Some Chapts.* (1946), 22; Curtis, *Oxford* (1959), 152–4; Allen, *J. Hist. Med.,* 1 (1946), 117–18.

7. Roller, *The De Mag.* (1959), 67–83; Clark, *Hist.* (1964), 127, 167. See Kargon, *Atomism* (1966), 5–12, and Johnson, *Astr. Thought* (1967), *passim,* for some of Gilbert's associates.

8. *DM* (Th), *ij. Langdon-Brown, *Some Chapts.* (1946), 23, citing Harvey, whose father was a colleague of Gilbert's, reports that Gilbert spent £5,000 on his researches.

9. Gilbert's friend E. Wright disclosed (*DM* [Mo], xliv) that *De magnete* was withheld for almost eighteen years and insinuated (*ibid.,* xxxvii) that Gilbert's concern for his career contributed to the delay: 'who shall deem these studies trifling and in no wise sufficiently worthy of a man consecrated to the graver study of medicine?' The coincidence of Gilbert's elevation and the ap-

philosophia nova, on which Gilbert worked until the year of his death (1603), was first published in 1651, apparently from a manuscript once owned by Bacon.[10] Despite their disparate histories and characters, these volumes are companion pieces: magnetism, the noblest of terrestrial virtues, whose accidents form the subject of the painstaking and original investigations of the high-handed *De magnete,* serves as the guide to a reformulation of sublunary physics in the plodding, traditional, and no less supercilious *De mundo.*

On the Magnet everywhere bears the stamp of its author's character, profession and experience. Most evident, perhaps, is the skeptical, skillful physician equally at home testing the reputed medical virtues of powdered lodestones and traditional accounts of magnetic motions. We also meet the wealthy, urbane hobbyist, rushing to buy 'varieties of specimens and foreign products never before seen,' merchandise 'lately brought from distant regions by traders and mariners,' and hurrying to test their magnetism.[11] And we encounter the would-be technologist, an expert on terrellas, or spherical lodestones, tiny models of the earth, who hopes to improve the art of navigation: reading travel books and interviewing master mariners have made him an expert on the working of 'that wonderful director in sea-voyages, that finger of God,' the compass needle, which he hopes to adapt to finding position at sea.[12] Far more, however, than the work of the observant physician, the hobbyist or the technologist, *On the Magnet* is the artifact of a philosopher. The true purpose of magnetism, and the ultimate justification for its study, is its fundamental role in the operations of nature.

Beneath its messy and 'excrementitious' rind, the earth, according to Gilbert, has a pure, continuous magnetic core, which orients our globe in the heavens just as it swings a compass needle to the north. The magnetic virtue, moreover, provides the ground or condition of the life we know. With the help of celestial powers it causes the diurnal rotation of the earth, whereby terrestrial creatures imbibe the sun's animating rays, and retire before they are fried; and it slowly twists the earth's axis, resulting in the precession of the equinoxes, which by continually altering the astrological aspects, insures vitality and diversity here below.[13] Magnetism is the most noble, characteristic, necessary and competent of sublunary qualities. Gilbert can liken it only to a soul, and indeed the human

---

pearance of *De magnete* is striking. Other English physicians have worried about the harm the publication of their non-medical writings might do their careers; Thomas Browne put off printing *Religio medici* several years on this account, and Thomas Young let some of his contributions to optics and Egyptology appear anonymously, lest his frivolous interests alarm his patients. Finch, *Browne* (1950), 95–8; Wood and Oldham, *Young* (1954), 76, 197.

10. Kelly, *The De Mundo* (1965), 15–16, 20; Jarrell, *Isis,* 63 (1972), 94–5.

11. *DM* (Th), 9; cf. *DM* (Mo), 17.

12. *DM* (Mo), 223, 181, 253–4, 297–300, xliii–xliv; *De mundo,* 45: 'Pertissimi naucleri Hispani, Angli, Belgae, cum quibus nos saepius sermonem habuimus . . .'; *DM* (Th), 4: 'It is plain that no invention of man's device has ever done more for mankind than the compass.' Cf. Roller, *The De Mag.* (1959), 154, 161–2; Zilsel, *J. Hist. Ideas,* 2 (1941), 1–32.

13. *DM* (Mo), 18–24, 61–8, 180, 333–5. This description is not necessarily Copernican, as it omits the annual motion; Gilbert never committed himself to heliocentrism although he evidently

soul, shut up as it is in our degenerate, infirm, nonmagnetic bodies, suffers by his comparison: 'Confined, as it were, by prison bars [our souls] send not forth their effused immaterial forms beyond the limits of the body.'[14]

Gilbert's large view of the nobility and competence of magnetic souls was original with him, and a surprise to others; recall Bacon's sneer, 'Gilbert, our countryman, hath made a philosophy out of the observations of a lodestone.'[15] But the *mode* of operation of the magnetic soul offered nothing novel. Gilbert says that it 'effuses' its power about it, 'informs' the mass it dominates, and seeks and is sought by any separated magnetic spirit within its sphere of virtue. All this suggests the medieval mechanics of the multiplication of species. The 'effused [magnetic] forms reached out and are projected in a sphere all round, and have their own bounds;' they operate by 'immaterial act,' by 'incorporeal going forth;' and they pull iron by activating its form and inspiring it to self-motion.[16]

The scheme differs from multiplication of species chiefly in the medium of propagation, which in the old representation was material, and in Gilbert's appears to be the space or 'incorporeal aether' surrounding the earth.[17] Consider his account of lumen, the exemplar of medieval species: lumen is the 'act' proceeding from lux, the 'effusio lucis,' an immaterial something manifest only when it meets resistance in diaphanous substances; it cannot be said to be 'in' the vacuous interplanetary spaces, which are incapable of 'conceiving' or 'arresting' it, and which it traverses instantaneously. Travel through the diaphanous and resistant takes time. Perhaps Gilbert assimilated magnetic propagation with that of lumen; he calls both 'formae effusae,' 'entia corpora apprehendentia sine materia.'[18] Neither is so unscholastic as to act at a distance.[19]

Gilbert also remains peripatetic in spirit when he explores those cosmological consequences of *De magnete* that patently violate the letter of Aristotle's teaching. We pass over the celestial consequences—the fixity of the heavens, the extraterrestrial void that makes possible the earth's frictionless rotation, the scattering of the stars in space—since neither Gilbert nor Aristotle cared particularly for the physics of things lying beyond the moon. The doctrine of the earth's rotation, however, violated the principles of sublunary motions, the foundation of peripatetic dynamics and a major support of the four-element theory, itself the basis of peripatetic chemistry and meteorology. How is the

---

inclined towards it. Cf. Roller, *The De Mag.* (1959), 173, and Johnson, *Astr. Thought* (1967), 216. Even peripatetics might ascribe the precession of the equinoxes to a motion of the earth, e.g., Cesalpino, *Quaest. perip.* (1593³), 65–7, 70–1.

14. *DM* (Mo), 312; ibid., 36–8, 107–9, 304–11.

15. *Advancement of Learning* (1605), in *Works*, I, 461, III, 293.

16. *DM* (Mo), 112–13, 308–9; *supra*, i.2.

17. *DM* (Mo), 326; *De mundo*, 30, 64–8; Lasswitz, *Gesch.* (1890), I, 318–19.

18. *De mundo*, 50–2, 210, 214, 215; *supra*, i.2. Cf. the praise that Gilbert, who seldom has a good word for anyone, bestows upon 'the godlike Thomas' Aquinas for his ideas about lodestones; *DM* (Mo), 5, 103–4.

19. For a contrary view see Hesse, *Br. J. Phil. Sci.*, 11 (1960), 133–4, 138; Krafft, *Sudh. Arch.*, 54 (1970), 114, 130–1. Cf. Debus, *J. Hist. Med.*, 19 (1964), 389–417.

rotation to be understood? According to Gilbert, it is a magnetic phenomenon quite *natural* to our globe. Earth is not merely cold, dry and heavy; it has, in addition, 'some principal, efficient, predominant potencies that give to it firmness, direction and movement.'[20] Now Aristotle had constructed the qualities of the other elements from those of earth and their opposites, levity, hotness and moistness. Gilbert had to abandon this elegant procedure, since the prepotent powers of his earth lacked evident contraries. His solution, which further enhanced the cosmic role of magnetism, was to declare earth unique, the only *bona fide* element in the sublunary world.

This is an uneasy position for one reputed to be the father of experimental science. It is neither plausible nor useful to consider air and water, not to mention animate creatures, nothing but lodestones. Gilbert saved the phenomena by tacitly admitting a second element or principle, corruption. Everything we see about us, he says, is 'defective, spoilt and irregular;' even lodestones and iron ores contain degenerate stony material. Supralunary influences, solar light and astral radiations, cause this corruption, which, happily, is only skin deep. A degeneration gradient extends from the outer reaches of the atmosphere, where the denaturing celestial powers luxuriate, through water, nonferrous rocks, magnetic sands and clays, iron ore and lodestones down to the earth's pure, self-moving magnetic nucleus.[21]

No conscientious philosopher brought up on the teachings of the 'master of those who know' could contentedly refer all common physical phenomena to the corruptive virtues of astral radiations. When, in *De mundo,* Gilbert came to rewrite Aristotle's *De Caelo* and *Meteorologica,* he found he could not do without positive sublunary principles, which, in practice, readmitted much of the peripatetic theory of the elements.

These principles are fire and a certain *succus,* or juice, both innate to the earth's magnetic core. The *succus,* warmed by the fire, tends to rise to the crust, which it penetrates with the help of the corruptive cosmic rays. Within the crust the warmed *succus,* called humor, may coagulate to form salts, minerals and metallic ores; outside the crust the irradiated humor may appear as water or, when more fully degenerated and dissipated, as air. Fire is here the principle of levity and the chief cause of sublunary change, just as the peripatetics taught; and water and air, the 'effluvia of the earth,' are transmuted from it by solar heat, much in the Aristotelian manner.[22] As for gravity, it is the natural tendency of a body for union with similar material; it implies an 'inclinatio ad unitatem,' a predilection for a like mass and not, as Aristotle taught, for a particular spot. This predilection is by no means magnetic, as the middling density of iron evinces, nor is it entirely *sui generis.* It resembles the mechanism of the amber effect.[23]

20. *DM* (Th), 68 ff., 179, 338.   21. *Ibid.,* 34–5, 59–68, 184.

22. *De mundo,* 37–41, 47–8, 107–11; cf. Kelly, *The De Mundo* (1965), 25–44.

23. *De mundo,* 47–8, 59–60, 105–6; *DM* (Th), 109–10; *DM* (Mo), 170: 'Inanimate natural bodies do not attract and are not attracted by others on the earth, excepting magnetically or electrically.'

2. GILBERT'S ELECTRICITY

Gilbert devoted a long chapter of *De magnete* to amber, in order, he says, 'to show the nature of the attachment of bodies to it, and to point out the vast difference between this and the magnetic actions.' He had to distinguish sharply the pure bond of sympathy uniting iron and lodestones from the promiscuous behavior of amber lest the scribblers and pettifoggers raise damaging but irrelevant objections to his magnetic philosophy.

The distinction was crucial. The ancients had associated the attractions of amber and the lodestone, and explained both in the same general way.[24] Gilbert recognized a difference in kind. Neither phenomenon, according to him, is a true attraction. On the authority of the chief Aristotelian commentator, Averroes, he lays it down that true attraction implies violence and coupled motions, as when a horse pulls a cart. Electrical action is violent, but not attractive, since, in Gilbert's opinion, the drawing electric does not move; he prefers to call it 'incitation,' although he often uses 'attraction' for convenience. Magnetism, on the other hand, is both mutual and nonviolent; it forms an independent class of activity, a coming together, a 'coition.'[25]

Although Gilbert may have invented this terminology, he was not the first to insist upon the distinction. A catalog of differences between electric and magnetic phenomena had been drawn up in the mid-sixteenth century by another physician, Girolamo Cardano, and buried in his great book *De subtilitate*. Gilbert doubtless knew this catalog, for Cardano followed it with an explanation of electrical attraction for which he is roundly cudgelled in *De magnete*.[26]

Cardano recognized five distinctions: (1) amber draws many kinds of bodies, the lodestone only iron (2) the attraction between lodestone and iron is mutual, the amber effect is not (3) the lodestone, unlike amber, acts across interposed objects (4) the magnet pulls only towards its poles, amber everywhere (5) amber's force is improved by gentle heat and friction, which do not affect the magnet's. These observations are acute, and, except for the second, correct. Gilbert accepted them all, and added three of his own: the magnet pulls heavier weights than amber can; surface or atmospheric moisture inhibits electrical but not magnetic action; and the attractive property of amber, unlike that of the lodestone, belongs to a wide variety of substances.[27]

This last observation was Gilbert's most important contribution to the study of electricity. It may have resulted from his desire to strengthen his list of

24. *DM* (Mo), 75; Daujat, *Origines* (1945), I, 26; Roller and Roller, in Conant, ed., *Harvard* (1957), II, 543–639. Cf. Benjamin, *Hist.* (1898), 294–9, and Roller, *The De Mag.* (1959), 96–7, 127.

25. *DM* (Mo), 74, 97–8; *De mundo*, 104–5.

26. Cardano, *De subt.* (1550), 222–3; cf. *DM* (Mo), 80, and Benjamin, *Hist.* (1898), 246–50. Cf. Regnault, *Origine* (1734), I, 145: 'L'on dit assez ordinairement que le jour ne suffisoit pas pour les bisareries de Cardan.'

27. *DM* (Mo), 86, 97.

electromagnetic *differentiae*. As separated bits of true earth, lodestones should differ sharply from amber, a concretion of terrene *succus*; but why should amber have powers without analogy to those of other substances similarly concreted? The amber effect or, to use Gilbert's new word, 'electricity,' ought to be a common property of hard, compacted bodies! From the writings of another Italian physician, Girolamo Fracastoro, he knew that diamond attracted chaff when rubbed; therefore, perhaps, he began his investigation with precious stones and gems. To assist the search he lifted another item from Fracastoro, this time without acknowledgment, a small pivoted metallic needle that Fracastoro had used to show that amber draws silver as well as chaff.[28] This 'versorium,' as Gilbert called it, is a better detector of weak electricity than bits of paper or straw, since the effect of induction is stronger and free from complications of charges induced in the surface bearing the chaff. Using the versorium Gilbert added at least twenty-three items to the inventory of electrics, including sulphur, glass and sealing wax, and a persuasive new item to Cardano's list of distinctions.

Historians have recognized in Gilbert's separation of the amber effect from magnetism the essential first step in the history of electricity, as well as an exemplar of proper scientific method.[29] The methodological soundness of Gilbert's procedure was not so obvious, however, to a younger contemporary, a man whose opinions ought not to be summarily dismissed: Francis Bacon.[30] In Bacon's view not enough magnetic information had been accumulated to permit the nice distinctions Gilbert wanted to draw. Bacon thought Gilbert's procedure—separating magnetism immediately from other phenomena, constructing a magnetic theory, and subsequently using it to explain many of those same phenomena—an illustration and warning of the evils worked by the Idol of the Cave, by attachment to a pet theory; he would have begun by compiling a natural history, say of 'attractions,' and only then proceed to analyze, divide and distinguish.[31]

Bacon's objection may appear to be but another example of his characteristic inability to perceive what was valuable in the work of his contemporaries. Although Gilbert's rigid separation of electricity and magnetism has not survived, it made possible the advances that caused its rejection. That is the way of science: fruitful distinctions made at one level eventually reveal hidden connec-

28. Benjamin, *Hist.* (1898), 295–7; Fracastoro, *De symp.* (1550), 69. Gilbert knew Fracastoro's device, since he cites the chapter describing it (*DM* [Mo], 82). See Gliozzi, *Per. di mat.*, 13 (1933), 1–14, for Fracastoro's text and Thompson, *Notes* (1901), 38–41, for identification of Gilbert's electrics.

29. E.g., Benjamin, *Hist.* (1898), 299; Roller, *The De Mag.* (1959), 92, 127; Hoppe, *Gesch.* (1884), 3; Roller and Roller, in Conant, ed., *Harvard* (1957), II, 548.

30. Bacon knew both *De mundo* and *De magnete*, the latter probably only superficially. M. B. Hall, *J. Hist. Ideas*, 12 (1951), 466–7; Kelly, *Physis*, 5 (1963), 249–58, and *The De Mundo*, 75–96.

31. Bacon, *Novum organum* (1620), in *Works*, VIII, 77, 93. Cf. Benjamin, *Hist.* (1898), 315–31; M. B. Hesse, *Br. J. Phil. Sci.*, 11 (1960), 141–2, and Roller, *Isis*, 44 (1953), 10–13.

tions at a deeper one. And yet there is something in Bacon's objection. Gilbert's methodological distinctions may appear profound and fruitful when seen against the history of electricity from his time to Einstein's; considered with respect to the seventeenth century, they look less productive, and in some respects damaging. The need then felt to distinguish between electrical and magnetic phenomena not only helped to separate two fields of study: as Bacon perhaps divined, it also affected their content, and the approach and expectations of their cultivators. For example, the supposed one-sidedness of electrical attraction and two other fanciful items later introduced made it difficult for electricians of the seventeenth century to recognize mutuality in electrical interactions, and practically impossible for them to discover conduction and electrostatic repulsion.

Equally important consequences followed from distinctions (1), (3) and (5) above, and from Gilbert's discovery that bodies as diverse as common glass, sulphur, resin, precious stones and sealing wax possess electricity. Almost everyone who proposed a model of electrical interaction in the seventeenth century expressly avoided occult qualities: they admitted only manifest qualities, both the purely mechanical, like extension and motion, and the elemental, like hotness and moistness, all of which act by material contact. Even men like Gilbert and the Jesuit polymaths, who referred magnetism to an immaterial quality or innate sympathy, insisted that electrical attraction implied a material bond between interacting bodies. But if magnetism is a sympathy, why not electricity? Because, says Gilbert, electricity can be excited in many dissimilar substances, while magnetism is a property peculiar to the lodestone.[32] Furthermore, electrics draw light objects of every description, including smoke, save only fire, flame, and the thinnest air, while magnets act only on iron and on one another. And they must be rubbed before they will draw, while magnets need no preparation at all.

In sum, electricity cannot be an innate sympathy: not innate, for it does not arise spontaneously; not a sympathy, because the diversity of electrics and the heterogeneity of the objects drawn preclude a common occult quality. *Quid clarius?* Gilbert allows himself to strut and sneer at the 'over-inquisitive theologians' and 'light-headed metaphysicians,' who, ignorant of the discoveries he has only just announced, treat electricity 'esoterically, miracle-mongeringly, abstrusely, reconditely, mystically.'[33]

The argument that electricity cannot be occult has two interesting aspects. First, it shows that Gilbert rejected innate virtues or sympathies in the case of electricity not because he objected to them in principle, but because electrical phenomena were incompatible with them. Second, his reasoning relates to obvious differences in the physical properties of bodies interacting electrically; nothing could have been easier than to invent an occult quality common to all electrics and attracted objects. Fracastoro had imagined that hair, leaves, twigs, and other chaff drawn by amber contained a common buried principle. Gilbert

32. *DM* (Mo), 86.   33. *Ibid.*, 74–5, 77.

thought the idea too silly to require refutation.[34] Two centuries after the publication of *De magnete*, however, such universal, hidden principles were the foundation of electrical theory.

Since Gilbert has ruled out electrical sympathy, electricity must arise from direct action of matter. One can go further. In accordance with the principle, 'no action can be performed by matter save by contact,'[35] a principle admitted by corpuscularians and scholastics alike, Gilbert infers that a material bond must exist between the excited electric and the drawn chaff. He deduces the precise nature of the bond by confronting a fact, the friction necessary to awaken an electric, with a fancy, his abbreviated peripatetic chemistry.

As is evident to everyone, he says, the earth's crust and all things within it are compounded of two kinds of humor, one fluid and humid, the other firm and dry. The watery humor, related to the mercurial principle of the alchemists and to Aristotle's moist terrestrial exhalation, mediates electrical attraction. Friction causes bodies concreted from the humor to release a subtle vapor, a light perspiration of fine humoral particles. These effluvia proceed abroad whence, encountering light objects, they effect that union commonly (but improperly) known as electrical attraction.[36]

How does the thin bridge of effluvia bring a light object to an electric body? Gilbert economically finds the mechanism in the watery nature of the effluvia, in a manifest elemental active quality. All bodies, he says, are 'cemented together by moisture;' wet bodies on the surface of water attract one another when sufficiently close, and drops of water on a dry surface unite when contiguous. The 'attraction' (via surface tension) of small sticks floating side by side affords a complete analogy to the operation of the effluvia, which 'lay hold of the bodies with which they unite, enfold them, as it were, in their arms, and bring them into union with the electrics.'[37]

Gilbert's theory easily accommodated the few additional facts known to him, amber's failure to draw flame and its reluctance to act across fine cloth screens or humid air: flame consumes the effluvia, cloth intercepts them and atmospheric moisture, by clogging the pores of electrics, prevents their emission.[38] Less persuasively, he makes the humid effluvia the agent—or, rather, the model—of his gravitational mechanism. He says that rocks fall 'electrically,' and that 'the matter of the earth's globe is brought together and held together by itself electrically;' by which he means that the earth's proper emanation, the air, mediates gravitational attraction as the subtle humid effluvia do the electri-

34. *Ibid.*, 82.

35. *Ibid.*, 92. This does not conflict with Gilbert's explanation of magnetism via immaterial agencies.

36. *Ibid.*, 83–4. The chapter on electricity is entitled, 'Of the attraction exerted by amber, or more properly the attachment of bodies to amber.'

37. *DM* (Mo), 92–5.

38. *Ibid.*, 85–6, 91, 95–6. An implicit difficulty arose here, as Gilbert had found that some fluids—oils—did *not* impair an electric's action when spread on its surface; the diverse effects of coating liquids became a great puzzle for electricians (*infra*, v.2).

cal, and that fallen objects are retained by the watery humor, the cement of the earth's degenerate crust.[39] This analogy is evidently incompatible with Gilbert's earlier distinction between the violence of electricity and gravity's 'natural inclination towards union.'[40]

Gilbert's watery effluvia provided a universal theory of electricity accounting qualitatively for all relevant phenomena recognized in 1600. Indeed, in one respect at least, the theory was better than he realized: the mutuality of electrical action seems an obvious inference from the analogy to the swimming sticks. Nonetheless Gilbert emphatically denied the possibility of reciprocity. The supposed lack of mutuality constituted an important distinction between electrical and magnetic phenomena. The desire to maintain it and, perhaps, a limited confidence in the applicability of mechanical analogies, caused him to miss this interesting, and easily testable consequence of his model.[41]

He did not overlook another consequence. The watery effluvia bring light objects to the electric by 'enfolding' or 'holding' them; they grasp, but do not push. Repulsion consequently *cannot be an electrical effect*. One might expect that, in their outward course, the effluvia should push small objects aside; but Gilbert expressly states that because of their great subtlety, the effluvia cannot repel: 'Were this effluvium as dense as air, or the winds, or the fumes of burning saltpeter, or as the thick, foul effluvia emitted with much force from other bodies . . . it would repel everything, and not attract.'[42]

Here Gilbert denies the possibility of repulsion before attraction, whereas the reality, as we know, is attraction, then repulsion. He speaks to this point as well. The ancients had a theory of the amber effect, preserved by Plutarch, which traced the motion of the attracted chaff to an inrush of air agitated by exhalations from the rubbed electric.[43] A persuasive theory, and mechanically more orthodox than Gilbert's; he recognized its challenge, quite rightly as events proved, and fired upon it with whatever he could muster. He remarked, for example, that hot bodies, which pull air currents towards themselves, are not attractive; and that tiny diamonds should not be able to mobilize enough air to push forward long thin hairs or leaves. The strongest of his arguments bears

39. *DM* (Mo), 94, 97; *De mundo*, 106: 'Corpora ad tellurem mediantibus effluviis feruntur; corpuscula ad electrica per tenuissima effluvia propria, a communis telluris effluviis diversa.' Cf. the analogy between gravity and magnetism in Gassendi, *Opera* (1658), I, 346–8.

40. Roller (*The De Mag.*, [1959], 110) has rendered a critical passage in a way that suggests that bodies drawn electrically move themselves. The original runs, 'electrica corpora alliciunt tantum, allectum non immutatur insita vi, sed materiae ratione sponte appulsum incumbit,' and Roller's translation, '. . . electric bodies merely allure. The enticed object is not altered by the incorporated force, but exerts itself [!] by its own reason of substance to an approach.' The sense is rather, 'electrics only allure [i.e., they do not run together like magnets]; the drawn body is not altered by an innate force, but freely accedes, so far as its mass permits, to the [external] force driving it towards [the electric].' The rendering in *DM* (Th), 60, is unintelligible; that in *DM* (Mo), 97, correct as to sense, although it mistranslates 'allectum' as the electric, and not the body drawn.

41. *Infra*, v.2.

42. *DM* (Mo), 94–5.

43. Daujat, *Origines* (1945), I, 26; Roller and Roller, in Conant, ed., *Harvard* (1957), II, 545.

on the question of repulsion. If the air impels the attracted objects then, he says, they ought to rebound from a broad flat piece of amber. But no such thing occurs: 'In addition to attracting bodies, electrics hold them for some time.'[44] (Leaves and paper bits *do* stick to amber, there being little exchange of charge.) These arguments recur throughout the century.

Despite the poverty of its subject matter, Gilbert's theory of electricity had many subtle implications. Its tacit components and limitations, no less than its explicit doctrine, both advanced and inhibited later investigations. Although some of Gilbert's successors defended the rival theory of the impelling air, and all rejected his peculiar watery effluvia, it was always *De magnete* that defined the subject of electricity, identified its problems, and provided the framework for their resolution.

44. *DM* (Mo), 88–9, 176: 'All electrics attract objects of every kind; they never repel or propel.'

CHAPTER IV

The Jesuit School

1. NICCOLÒ CABEO

Cabeo was educated in the Jesuit college of Ferrara, where, according to his friend and biographer, Antonio Libanori, ancient languages, Tuscan literature, philosophy, Euclid and theology came as easily as games to him. His teachers thought him a prodigy, and invited him to join their company. He entered his novitiate in 1602, and subsequently taught mathematics, astronomy, and natural and moral philosophy in several Italian cities. After 1622 he followed the occupations of itinerant preacher, consulting engineer, experimental physicist,[1] writer, and stoic philosopher, 'humble in all his dealings, modest in dress, and blameless in character.'[2] Cabeo wrote two important books, the original *Commentary* on Aristotle's *Meteorologica* noticed earlier, and a *Philosophia magnetica,* published in Ferrara in 1629.The *Magnetic Philosophy* attracted wide attention. Although modeled on *De magnete,* it contained many new and clever experiments, and challenged Gilbert's views on the natures of magnetism—the earth, according to Cabeo, is not an animated lodestone—and of electricity.[3]

Cabeo opens his account of electricity with a list of new electrics found by systematic exploration with a versorium, thereby showing himself to be an independent authority, and not the sort of armchair philosopher lampooned by Gilbert. He proceeds to discriminate between electrical and magnetic phenomena, repeating the distinctions drawn by Cardano and Gilbert,[4] and adding two unhappy items of his own: (1) the magnet transmits its power to the iron attached to it, but an electric cannot 'communicate its faculty' to other bodies by contact; (2) under certain conditions magnets repel and flee from one

1. His work as roving physicist included the measurement of acceleration in free fall, which he undertook at the request of Baliani in collaboration with his confrère and former schoolmate Riccioli. Moscovici, *Expérience* (1967), 49–58.

2. Libanori, *Ferrara* (1665–74), I, 145–6, II, 213–14; Gimma, *Idea* (1723), II, 548–9; Borsetti, *Hist.* (1735), II, 349; *DBI,* XV, 686–8.

3. The rejection of the earth-magnet had larger implications, for Gilbert had associated the magnetic soul and the diurnal rotation (*DM* [Mo], 328–33; *De mundo,* 159–60), which Cabeo perforce denied. Galileo's school therefore tended to depreciate *Philosophica magnetica* (e.g., Galileo to Cesare Marsili, 21 April 1629, in Galileo, *Opere* [1890], XIV, 36). Cf. Descartes to Mersenne, 25 November 1630, in Descartes, *Corresp.* (1926), II, 562; and Magrini, *Arch. stor. sci.,* 8 (1927), 22–39.

4. These distinctions had meanwhile been widely accepted and occasionally exploited. A curious example is Rudolf Goclenius, *Synarthrosis magnetica* (1617), a defense of the weapon salve against the attacks of the Jesuit Jean Roberti; the relevant passage is quoted by Debus, *J. Hist. Med.,* 19 (1964), n. 75.

another, while electrics flee nothing, but draw and join themselves to all substances indiscriminately.[5] Neither distinction has held up. The first denies in principle the possibility of electrical conduction. The second would appear no less categorically to rule out electrical repulsion; nonetheless, as will appear, Cabeo has been recognized as its discoverer.

Cabeo next turns to earlier opinions about the cause of electrical attraction. Only Gilbert's, he finds, merits examination, and it is unsatisfactory. Whatever attraction is, it does not involve a humor that cements all things: 'These are words introduced for eloquence, not for explaining the cause and method of attraction.' Experiment shows that some things concreted of humor—metals and certain gems, for example—do not attract; and that others, which contain no more humor—like glass and other gems—do. In any case, Gilbert's watery effluvia can not act as advertised. Fluids adhere in proportion to their viscosity: since cohering glass plates separate more easily in air than under water, they should part readily in a subtle humor.[6]

What then causes the attraction between floating sticks that Gilbert saw as the prototype of electrical interactions? Cabeo distinguishes four species of attractions. He notes the abused attraction *ob simultudinem naturae*, the standard sympathy, which he accepts as the cause of magnetic action. Then comes attraction by gravity or levity, through which a body tends towards its place by a native quality. The third attraction operates when bodies move to fill a vacated place, or rarefied air condenses. Finally there is attraction proper, *ad-tractio*, by which one body draws a second through others conjoined, as in pulling a boat with a rope. The third attraction is implicated in electrical phenomena, and the second in Gilbert's experiments with floating bodies.[7]

According to Archimedes, water in a container strives to arrange itself so that its surface lies on a sphere concentric with the earth; according to Aristotle, dryness and moistness are opposed active principles. Cabeo deduces that a dry floating object depresses the water supporting it: the water poises where its Archimedean tendency to level balances its Aristotelian antipathy to contact with dryness (Fig. 4.1). Wet the body and the water rises, its gravity now equilibrated by adhesion. Each of Gilbert's wet twigs carries 'swellings' of water violently prevented from fulfilling the call of their gravity. The amount of disturbed liquid can be reduced, however, and the adhesive force still satisfied, if the sticks come together: when their swellings touch, the water reduces its unnatural displacement and pushes them together. The explanation of Gilbert's experiment, says Cabeo, is the gravity of water, different *toto caelo* from electric attraction; for—to use a standard argument—while gravity is a tendency of a thing toward a certain place, bodies move toward electrics regardless of place, and from every direction.[8]

5. Cabeo, *Phil. magn.* (1629), 180, 182.
6. *Ibid.*, 183–5.
7. *Ibid.*, 192. Cf. Thorndike, *Hist.*, VII, 424, for action at a distance in Cabeo's *Commentarii*.
8. *Phil. magn.* (1629), 190–2.

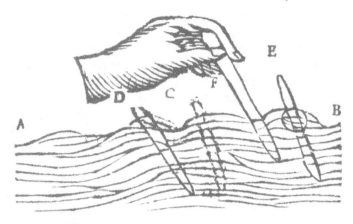

4.1 Cabeo's account of electrical attraction. The dry stick F depresses the water, the wet ones D,C, E raise it. From Cabeo, Phil. magn. (1639), 190–2, after Gilbert, DM (Th), 58.

Nor can electricity be a sympathy, as Cabeo proves by elaborating the argument we have met in Gilbert. Electrics attract all bodies, but otherwise have nothing in common; to refer their action to an occult electrical quality we should have to assume that all light objects desire all electrics, and vice versa, as goods to complete their natures. But this is opposed both to reason and to experience: to reason, because a body with certain manifest qualities ought not desire the same good as a body with opposite qualities; to experience, because if such a good existed we should observe a concursus, a running together or mutual attraction between electrics and drawn bodies. (Note that to establish a distinction, Cabeo, like Gilbert, denied the mutuality of electrical attraction.[9]) Electricity being neither gravity nor sympathy must act like a vacuum or like pulling on a rope. Cabeo tacitly dismissed the rope. He was left with the vacuum, the scheme of Plutarch. Friction opens an electric's pores to streams of subtle effluvia, which beat back and rarefy the surrounding air; the air returns to re-establish its former density, driving light bodies before it.[10]

Cabeo marshalled imposing supporting evidence. He observed that chaff attracted by a large flat electric tends to go to the edges (where the field is greatest), a consequence, he says, of the concentration there of effluvia projected from the center of the electric. In other experiments attracted sawdust particles adhered to one another. Cabeo thought that the wild fluctuations of the far ends of these threads, from which particles occasionally flew off, were an ocular demonstration of the suppositious aerial motions.[11] Thirdly, while checking on Gilbert Cabeo observed that a strong electric sometimes drew

9. *Ibid.*, 191.
10. Galileo also inclined towards Plutarch's theory; *Opere*, III:1, 13–14, 399.
11. *Phil. magn.*, 192–4.

scraps of iron or wood with such force that they rebounded (*resilio*) to a distance of three or four inches. On the strength of this remark he has been sponsored as the discoverer of electrostatic repulsion.[12] He did not consider that he had found anything novel; on the contrary, he saw in this 'repulsion' confirmation of his mechanism of attraction, and even an *experimentum crucis*. Were sticky effluvia the agent of electricity, drawn bodies could never rebound, but once arrived must remain attached as Gilbert wrongly said they did. With Plutarch's mechanism, on the other hand, nothing could be plainer: 'repulsion' is mechanical rebound.

Cabeo's work on electricity deserves admiration. He added valuable observations to his subject, and strove to relate them to his stock of old-fashioned explanatory categories. He recognized the complexity of an experimental science like electricity (a 'res difficillisima'), and he professed himself prepared to modify his theory should experiments require it.[13] Much of the creative work on electricity during the seventeenth century did center on the testing of his theory, which the Jesuits at the Collegio Romano adapted and extended.

2. KIRCHER'S CIRCLE

The chief of this Jesuit science was the 'Oedipus of his age,' the 'glory of Germany,' the all-round polyhistor Athanasius Kircher, who in 1634 became Professor of Mathematics, Physics and Oriental Languages at the Gregorian College in Rome.[14] He was born in 1601, the son of a lawyer who had taught theology at a Benedictine convent and subsequently entered the service of the Abbot of Fulda. When Protestants destroyed the abbot's political position, the elder Kircher retired; 'one ounce of freedom applied to improving the mind [he declared] is worth a thousand pounds of princely titles.' Athanasius must have been a disappointment to his erudite parent. Sent to the Jesuits at Mainz, he so failed to distinguish himself that the Rector of the college at first declined to accept him as a postulant. Prayer and hard study cleared his mind; he was admitted to the novitiate in 1618, at the outbreak of the Thirty Years' War. After many vicissitudes including gangrened legs healed by the Virgin and a fall in the freezing Rhine, Kircher completed his philosophy at Cologne, and began to teach humanities. In 1624 he transferred to Heiligenstadt in Saxony; on the way he had his customary adventures, fell in with bandits, and was very nearly hanged.

12. E.g., Magrini, *Arch. stor. sci.*, 8 (1927), 37; Dibner, *Early Elect.* (1957), 14; *DBI*, XV, 687.

13. Cabeo refers to electricity in his *Commentarii* (IV, 411–12) primarily to buttress the theory that odoriferous bodies emit effluvia. This argument, which recurs in Boyle (*Works* [1772²], III, 279–83), suggests that Cabeo thought his readers readier to admit electrical than odoriferous effluvia, at least in the case of hard, compacted bodies.

14. See Kircher's autobiography (1684), excerpted in Brischar, *Kircher* (1877) and Reilly, *Studies*, 44 (1955), 457–68. The honorific titles come from Kircher's unpublished correspondence: 'edipo de nostri tempi,' 'Oedipus huius seculi' from F. Traviginus (CK, VIII, 81, 144); 'germaniae meae decus gloriosum' from Q. Kuhlmann (*ibid.*, IX, 45).

In Heiligenstadt Kircher's rise began. A common amusement among the educated then was the collection of natural curiosities and the construction— and exhibition—of 'magical' machines: devices for creating optical illusions, for communicating at a distance, for moving or suspending weights without evident means, for turning water into wine.[15] This 'artificial magic' served the seventeenth century as exhibitions of 'experimental philosophy' served the eighteenth. Kircher made mechanical and pyrotechnical magic at Heiligenstadt to divert representatives of the Elector of Mainz; he soon found himself at the Electoral Court, assigned to invent mechanical amusements and to survey parishes recently returned to Catholicism. He also tried to puzzle out the hieroglyphics he encountered in a book describing a great feat of 'mathematical magic,' the raising of the obelisk of Sixtus V.[16] In 1629 he became Professor of Mathematics at the Jesuit college in Würzburg, where he completed the first of several books on magnetism and had the satisfaction of at least one promising student, Gaspar Schott.

Two years later the Swedes drove the Jesuits from Würzburg, and Kircher from Germany. He reached the College at Avignon and the home of Nicolas de Peiresc in Aix-en-Provence. That sage, astronomer and orientalist delighted in Kircher, whose ingenuousness and intelligence he thought equal to the hieroglyphics: 'his face bears witness to his great piety and innocence, and to the nobility and acuteness of mind that have allowed him to advance so far in the discovery of many secrets both of nature and antiquity, and [in the mastery] of all the languages of Christiandom.'[17] Peiresc's support proved more beneficial for physics than for philology. When, in 1633, Kircher was appointed to teach mathematics in Vienna as successor to Kepler,[18] his powerful friend intervened: Kircher must go to Rome, where important Egyptiana awaited his attention. Peiresc wrote the Jesuit General and the Pope himself. Kircher's orders were changed.

Kircher's peculiar genius luxuriated in the intellectual climate of Rome. He produced one polymathic volume after another. With a gift left the Gregorian College in 1652 he established a museum, part collection and part laboratory, where he practiced natural magic on visitors.[19] His fame spread. Materials and queries, specimens and paradoxes arrived from all corners of Europe, and from every barbarous place the zeal and curiosity of the Jesuits took them. Kircher handled this information with an odd mixture of gullibility and scepticism; a bit credulous, perhaps, regarding wonders of nature, such as the poisonous winged

15. Cf. Thorndike, *Hist.*, VII, 596–621, VIII, 204; Houghton, *J. Hist. Ideas*, 3 (1943), 190–205.

16. Accomplished by 140 horses and 900 men working levers and pulleys, precisely the sort of Archimedean feat, or 'engineering extravagance,' touted in John Wilkins' *Mathematicall Magic* (1648). See Shapiro, *Wilkins* (1969), 43–6; Dibner, *Moving* (1952), 26–40.

17. Peiresc to F. Barberino, 10 Sept. 1633, in Rizza, *Peiresc* (1965), 89–90.

18. Reilly, *Line* (1969), 29.

19. Villoslada, *Storia* (1954), 183–4; Sommervogel, *Bibl.*, art. 'Kircher,' *ad fin.*: Bedini, *Techn. Cult.*, 6 (1965), 15.

dragon of Rhodes or the German who vomited a six-foot snake, but a fierce debunker of extravagant wonders of art, such as the pretensions of the alchemists, the weapon-salvers, and the authors of perpetual-motion machines.[20] He and Schott, and the Jesuits in general, felt an obligation to combat superstition; despite being gulled occasionally themselves, they rescued many from more serious infatuations. The hermeticists still claimed a following, and even the Royal Society of London worried about—and somehow managed to test—whether tarantulas could escape from a ring of powdered unicorn's horn.[21]

Among Kircher's heterogeneous interests, magnetism, at once a natural wonder and an agent of artificial magic, figured prominently; and along with magnetism, in the manner of Gilbert and Cabeo, he touched upon electricity. In his first account of the subject (1641), Kircher followed Cabeo closely, including the new differentiae and the implied rejection of the possibility of electrical conduction.[22] He had no new electrics to report, and little first-hand experience. Twenty-five years later he was an authority, or rather, an impresario, for electrical demonstrations, some of his own devising, had become showpieces at his museum. He required large, catchy effects, visible at a distance: the dancing of a smoking candle bombarded by the effluvia from amber;[23] the drawing of large suspended objects (fig. 4.2), like the gloves of a visiting prince; the curtseying, 'to the great surprise of all,' of bright, sunny flowers to a bit of black coal. Natural magical parlor games like these did much to keep up interest in electricity; *homo ludens* appears frequently in our history. Kircher the impresario anticipated Hauksbee the demonstrator, whose professional entertainments at the Royal Society of London rejuvenated the study of electricity early in the eighteenth century.[24]

MAIGNAN'S ATTACK

Kircher's audience at the Collegio Romano included Gaspar Schott, Francesco Lana, and perhaps Cabeo, who contributed to discussions on electricity and magnetism there.[25] These discussions intensified towards the mid-century,

20. Thorndike, *Hist.*, VII, 505, 568–75, examples mainly from Kircher's *Mundus subt.*; Kestler, *Phys. kirch.* (1680), 36–7. Cf. Regnault, *Origine* (1734), I, 161.

21. Thorndike, *Hist.*, VII, 571, 596, 599, 603–7; Weld, *Hist.* (1848), I, 113. That Kircher's attacks against alchemists told appears from John Webster's complaint against *Mundus subt.*, 'Stuffed with lies and scandals against Paracelsus, Arnoldus, and Lully' (Thorndike, *Hist.*, VII, 568).

22. Kircher, *Magnes* (1641), 640–50; *Magnes* (1643²), 563–71.

23. Large pieces of amber, especially specimens filled with the corpses of insects, were prized in collections of rarities; Kircher's museum had several, as did the celebrated cabinet of Manfredo Settala. Fogolari, *Arch. stor. lomb.*, 14 (1900), 105.

24. *Mundus subt.* (1665), II, 76–7; Kestler, *Phys. kirch.* (1680), 35; Buonanni, *Rerum natur. hist.* (1773), 109–11; Schott, *Thaum. phys.* (1659), 377. Cf. J. Huizinga, *Homo ludens* (1955²), 105–18, 187.

25. B. Conrad, S. J., to Kircher, 13 Jan. 1646 (CK, III, 51, and V, 266): 'De tractione succini

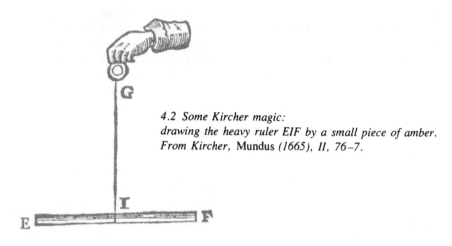

4.2 *Some Kircher magic:*
drawing the heavy ruler EIF by a small piece of amber.
From Kircher, Mundus *(1665), II, 76–7.*

when Emanuel Maignan and Tomaso Cornelio, both also of Kircher's circle, fell out over the air theory. Cornelio, a 'hot-headed Calabrian,' 'a follower of Descartes and a great favorer of things new,'[26] had his early education from the Jesuits. He then went to Galileo's disciples Torricelli and Cavalieri, to whom he was recommended by a friend of Kircher's, the mathematician Michelangelo Ricci. He became doctor of medicine, professor of mathematics at the University of Naples, and a founder and leading light of the novelty-hunting Neapolitan Accademia degli Investiganti. His chief academic employments were to defend Cartesian philosophy, which he may have introduced into Naples, against local scholastics, and to wrest physiology from the dead hand of Galen.[27]

To combat the prevailing view that digestion requires a sympathy between food and body, Cornelio undertook to show that attractions like electricity, magnetism, and the phenomena attributed to nature's distaste for voids, operated mechanically. In electricity he followed Cabeo, with one exception. As Cornelio wished to rescue nature from the horror of the vacuum, he could not place the dilation and contraction of air in a special category of attractions. He held, rather, that the greater gravity of the air beyond the range of the effluvia (whose bombardment 'thins' the atmosphere near the electric) caused Cabeo's

R. V. adhuc examinet, an subsistat ille modus explicandi pro quo habebit adiumentum ex illo experimento quod per accidens meis thesibus de horologiis, illis nuperimis, adieci. Puto enim, quod & P. Cabeus fatetur, sic non posse explicare.' Since the theses, *Propositiones gnomonicae de perfectione solarium horologiorum* (Prague, 1645), have not come to hand, 'ille modus explicandi' still wants explanation.

26. Sir John French to Prince Leopold, 24 Nov. 1663, in Fabroni, *Lettere* (1773–5), I, 266; Fisch, in Underwood, ed., *Science* (1953), I, 523–4, 527.

27. *Ibid.*, 553–4; Colangolo, *Storia*, III (1834), 216–20; Gerini, *Nuovo ris.*, 9 (1899), 429–30. Gurlt and Hirsch, *Biogr. Lex.* (1884–8), II, 81, credit Cornelio with recognizing irritability before Haller. He was well-regarded by Malpighi; Oldenberg, *Corresp.*, IV, 272 and V, 508.

vortex. According to Cornelio, gravity is a pressure in a subtle, ubiquitous medium. Our understanding of electricity would thus be complete, were it not that effluvia competent to drive back air might also be expected to repel small objects in their path. Cornelio raised and razed this objection: effluvia work by many blows individually inappreciable, of such a nature, therefore, as to act upon finely divisible air but not upon macroscopic bodies.[28]

Cornelio's views had the merit of being falsifiable. Maignan came forward, wrangling as follows.[29]

> If [1] the heavier, agitated air really drives light bodies to the amber, doubtless [2] it is also the air, by its gravity, which holds those bodies against the amber. But that is false, for the following reasons. The consequence of the major [1] is obvious, because if the bodies are not retained by the air, they must either hold themselves or be held by the amber. Now the first alternative is impossible, as the bodies possess a greater gravity than the air. Likewise the second, for the bodies would instead be driven away by the repeated, or rather by the continuous blows of the issuing effluvia. . . .
>
> As for the minor [2]: first if you move the amber with the bodies attached to it into a place where the air has been made thicker and therefore heavier, in the manner supposed, then those bodies should immediately fall off, for lack of a sustaining force. But this is false. Second, if while corpuscles adhere to a piece of amber A, you approach a second piece B, then B lightens the air, and hence [by hypothesis] the corpuscles' support; for its effluvia will bombard the air, making it as light as before, and incapable of sustaining the corpuscles. But they do not fall off, as is clear from experiment. . . .[30]
>
> Perhaps you will say that those corpuscles are carried away by the air, both by gravity and by impulse, just as by a wind. But this will not do either, for if so, once the impulse ceases (as it must) the corpuscles will fall from the amber to the ground, as when the wind throws little stones, straw, or dust against a wall.

This last paragraph was aimed at Cabeo and at Maignan's 'very good friend' Kircher.[31]

About the time Maignan published his criticism of Cornelio, Gaspar Schott, who like Kircher had been driven from Würzburg by the Swedes, joined his former teacher in Rome, after twenty years as professor in the Jesuit college at Palermo.[32] The 'serious controversy over the cause of electrical traction' stirred

28. Timaeus Locrensis (T. Cornelio), *Epist.* (1648), 8, 28–30, 55–60; reprinted in Cornelio, *Prog. phys.* (1663), because (p. 111), 'multi sunt inventi viri plane eruditi, qui nostrum libellum laudibus extulerunt.' Like Plato's *Timaeus* and Descartes' *Principia*, Cornelio's essay preaches without practicing mathematical physics. See the review in *PT*, 2 (1667), 576–9.

29. Maignan, *Cursus* (1673²), 399–400, unchanged from the first edition (1653).

30. The adherence of chaff to an electric is more noticeable than its repulsion. Little bits of metal, which acquire charges readily and do not stick so easily show the repulsion better, but Maignan and his colleagues did not employ them.

31. Middleton, *Hist.* (1964), 10–16. Maignan did not pursue his experiment with the two pieces of amber; had he suspended piece A, as was Kircher's practice, he might have made the paradoxical discovery that two rubbed pieces of amber repel one another.

32. Duhr, *Gesch.* (1921), III, 590–3.

up by Maignan bothered Schott, who excelled in collecting, classifying, and compromising the facts. He had no opportunity to defend Cabeo before returning to Mainz in 1655, 'a terrestrial paradise,' where, so he wrote Kircher, he 'felt healthier than elsewhere, happy and contented, and lacked nothing.' He soon transferred to the reopened college at Würzburg, stopping at Frankfurt to trade for items needed by the plundered school, which had 'very few volumes of mathematics and no instruments.'[33] His superiors left him time for mathematical studies, allowed him to handle money in connection with his books, and comforted his body, grown soft in Italy, with a private stove. Thus indulged and relatively free from censorial harassment, Schott declined invitations from Kircher to return to Italy.[34] He scribbled prodigiously, 9000 printed quarto pages in a decade, chiefly on mathematics and natural magic. Among the first outpourings from his tepid chamber was his influential improvement of Cabeo's theory.[35]

Schott perceived that the Jesuits had gone too quickly in rejecting Gilbert's 'incomprehensible verbiage.' Cabeo had proved that watery effluvia would not do, but something might yet be made of stickiness. Assume that rubbing an electric produces viscous particles that both agitate the air and adhere to small bodies, rebounding from solid objects supporting them in accordance (as Schott conjectured) with the law of spectral reflection. The disturbed air assists the return of the effluvia, which paste the chaff against the electric. Maignan's objections fell to the ground. The victory came at a cost, however: contrary to Cabeo's theory, which assumed nothing about the chemical constituents of elec-

33. Schott to Kircher, Maguntiae (Mainz), 12 Sept. 1655 (CK, XIII, 49): 'Sono stato in questo Collegio di Maguntiae gia 3. messi, con tanto mio gusto e con tanta quiete, che me pare di essere stato nel paradiso terrestre. Son stato sempre sano piu che in nessun'altra parte; allegro e contento e non me sia mancato niente. Vorrei restare sempre qui; ma preferisco la voluntà delli superiori alla mia. Vado ad Herbipoli [Würzburg] a leggere la Mathematica, e la Morale, per essere un maggior numero di scolari che qui. Piutanto ho fatto li miei essercitij spirituali, ad esso vado alla fiera di Francofortia, per vedere se possa cambiare le reliquie delli libri salviati da quel naufragio, del quale ho scritto altra volta con . . . qualche altro libro necessario, massimente perche in Herbipoli sono pochissimi libri di mathematica, e nesun'instrumento, et al contrario qui vi e abondanzia dell'uni e dell'altri.'

34. Schott to Kircher, Herbipoli, 23 Nov. 1663 (CK, IX, 157): 'Ringrazio da cuore a V^a R^a per l'affetto verso di me, e per il desiderio di vedermi costa. Ma mi pare che sono gia frappo vecchio (cio e di 56 anni fatto) et assai indebolito per li studij. Vorrei piu tosto qui essere libero da ogni altri occupazioni, fuora di leggere la Matematica, e di scrivere i libri.' Eventually Schott saw difficulties on the horizon at Würzburg also, as appears from a letter to Kircher of 21 Aug. 1664 (Ck, VIII, 110): 'Hactenus non multum desideravi redire Romam, quoniam existimo me hic habere meliorem occasionem scribendi et imprimendi libros meos, quam ibi. Si mathematica cathedra, aut aliud officium meis studijs commodum mihi definitur, non refutarem. Propter certam causam timeo ne hic impedimentum injiciatur studijs meis.' A century later opportunities had much improved in Rome, where Boscovich (also a Jesuit) had 'tutti i commodi che può avere un Religioso . . . senza avere il minimo impiego; colla totale libertà di applicarmi a modo mio.' R. Boscovich to P. Frisi, 6 Nov. 1764, in Costa, *Rev. stor. ital.*, 79 (1967), 835–6.

35. Sommervogel, *Biblioth.*, art. 'Schott'; Schott, *Thaum. phys.* (1659), 376–83; Thorndike, *Hist.*, VII, 596–607.

trics, Schott's compromise implied retrogradation from Gilbert's chief insight, a tacit limitation of electricity to 'viscous,' 'pinguid,' or 'bituminous' substances. Kircher at once adopted his colleague's suggestion, as did his student Francesco Lana, who brought it to its fullest development.

LANA'S MAGISTERIUM

While in Rome Schott had helped instruct Lana, who completed his theology at the Collegio Romano in the early fifties, when Kircher started to arrange the museum and Paolo Casati held the chair of mathematics there.[36] Lana did not fail to profit from so favorable a conjunction of luminaries. He studied independently with Kircher, and, with fellow student Daniel Bartoli, assisted in the experiments of Casati. He mastered natural philosophy, but without great satisfaction, for he found its branches to be differently, even contradictorily, treated. At the conclusion of his studies, if his later testimony be credited, he decided to try to establish a complete and consistent approach to the subject, firmly based on experiment.[37]

The fulfillment of this design, made doubly difficult by the demands of the ordinary itinerant professorship, required almost forty years. Lana began, as was usual, by teaching humanities, probably at the Jesuit College in Terni. From there he moved to his home town, Brescia, and to philosophy, which he taught for three years and improved with experiments on the barometer; and thence to Ferrara, to Cabeo's old chair of mathematics at the University. Some years later he was again in Brescia determining the declination, and then in Bologna assisting (as had Cabeo) in Riccioli's measurements of gravitational acceleration; and again in Ferrara, writing long letters, mainly on sound, to Bartoli. He retired to Brescia, where he founded a short-lived academy of mathematicians and natural philosophers called *Philo-exotici,* and devoted himself to completing his great work.[38] The first volume of *Magisterium naturae et artis* left the press in 1684; the third, which was not intended as the last, followed posthumously eight years later.

The *Magisterium* was announced in 1670 in a *Prodromo,* today remembered for its description of a balloon-like aircraft. It is a superb book on natural magic, which Lana took as the subject of his prologue because he felt it to be in worse case than other branches of natural philosophy. It suffered not only from those general causes noticed earlier by which he explained the unfinished condition of the arts and sciences in his time, but principally from the pervasive influence of della Porta and his followers, who worked off nonsense, supersti-

36. Mazzuchelli, *Nuova racc. op. sci. fil.,* 40 (1784), 1–132; [Rosa], *Civ. catt.,* 1 (1932), 211–22, 424–37; Villoslada, *Storia* (1954), 335; Lana Terzi, *Magisterium* (1684–92), I, 150–4, 385, 508, and II, 176.

37. *Ibid.,* I, Pref.

38. Bartoli, *Lettere* (1865), 86–7, 102, 108–9; [Lana Terzi], *Acta erud.* (1686), 556–65. The acts of the 'Philoexotics' commence 25 March 1686 and end with Lana's death in February 1687.

tions and trivia in the name of 'this most beautiful part of natural philosophy.'[39] Only a very few moderns had improved on these unnatural magicians: among the Jesuits Lana esteemed only Cabeo, Riccioli, Grimaldi, Zucchi, Casati and Bertoli, among extramural authorities, Boyle and the Accademia del Cimento, and perhaps Sir Kenelm Digby.[40] An interesting list, on which one misses Schott and Kircher, through whom Lana first encountered respectable natural magic, and learned to despise the della Portas.[41]

The *Prodromo* contains proposals ingenious and impracticable, like the airship; others only impracticable, like a method of transmitting messages where neither vocal nor visible signals will do; and others both practicable and valuable, like improved techniques for writing for the blind and hints for the manufacture of lenses. Similar devices and inventions enliven the *Magisterium,* where a typical section begins with experiments, follows with principles and causes, and ends with an account of diverting, useful and instructive contrivances. Some sections, including electricity, lack the applications, but all are illustrated by experiment; for the experiments, and not their explanations, constitute the uniform approach to natural philosophy that Lana strove to attain.

The design and completion of these experiments was a gigantic task. The effort involved in the *Prodromo* alone so fatigued and sickened its author that he set his great project aside. The *Physics* of Fabri, also intended as a uniform treatment, rekindled his ardor and ambition. Other duties, lack of assistance, and that peculiar clerical disorder, the reconciliation of the 'immodest expense' of the undertaking with a vow of religious poverty, conspired to delay his progress, and in the end prevented him from bringing to completion his *Great Guide to Nature and the Arts.*[42]

Lana's treatment of electricity is no discredit to his method. He announced new electrics, which, following the example of the Accademia del Cimento, he compared in strength to those already established. He tested and amended the results of all previous investigators; he described delicate new experiments; and he presented fairly the theories that divided electricians. In a word, he assembled an encyclopedia, the most extensive and valuable account of electricity published in the seventeenth century.[43]

Lana's original experimentation dealt not with large-scale, average effects,

39. Cf. Valentinus, *Adv. fall.* (1591), 1, condemning black arts, astrology, etc., but approving natural magic, 'natural and true, coming from the occult nature and virtue of things, an intimate and excellent part of Philosophy.'

40. Lana Terzi, *Prodromo* (1670), Pref.; *Magisterium* (1684–92), I, Pref.; Bartoli, *Lettere* (1865), 88.

41. Schott was reluctant to employ the word 'magia' in the titles of his books because, he says, the pretensions of Albertus Magnus, della Porta *et al.* had made it 'detestable': only Kircher's example decided him to use it, and then with careful qualification. Schott, *Mag. opt.* (1671), Sigs. OO2^r, OO4.

42. Lana Terzi, *Magisterium* (1684–92), I, Pref. The original plan called for twelve volumes.

43. *Ibid.,* III, 287–312. The chapter, cited hereafter by Lana's paragraph numbers (§), amounts to about 15,000 words, its nearest competitors to 6,250 (Boyle) and 2,650 (Accademia del Cimento).

but, following Cabeo, with the behavior of the individual attracted scrap. A bit of chaff responded best when placed on mercury, less well when on wood or paper, and least when swimming on water. A floating feather followed a piece of amber, but would not leave the liquid; green vegetables came with difficulty; and wet chaff resting on glass could not be budged at all. A paper bit could be urged over a greater distance when bent to its support than when lying flat.[44] Lana had no worthy successor in these experiments before Gray and Dufay.

Lana also called attention to the irregular trajectories of attracted bodies. The irregularity instanced a deep difference between electrical and magnetic action: 'Iron flies to the magnet by the shortest path, while chaff approaching amber is often seen to wander and turn about in the air, circling around the place to which it was first urged.' Once arrived, it prevents the electric from drawing other bits to the same place, and resists attempts to remove it. Eventually it drops off of its own weight.[45] Lana appreciated the irregular as well as the uniform features of electricity, and he did not recognize electrical repulsion.

Lana's discussion of theory proceeds like Cabeo's. First comes the now-standard 'proof' that electricity cannot be a sympathetic quality. Next, a string of syllogisms with the Aristotelian premises, 'all true action of bodies consists in local motion,' 'no body can move itself to local motion,' and 'the mover must be contiguous or immediately applied to the moved,' yields the electrical effluvium which, however, cannot be the whole story, were it as sticky as Gilbert pretended. No gluey emanation, according to Lana, could cart chaff back to the electric against the opposing force of fresh effluvia. Some other agency, to wit the atmosphere, must cooperate. Like Cabeo, Lana uses the supposed absence of 'repulsion'—his word—to infer the intervention of the air.[46]

Lana goes a step beyond Schott by associating the action of effluvia on the air with their adhesive property. He considers electrics to be 'igneo-sulphureous' bodies whose emanations expand the near air by heat; at any weakening or cooling of the stream of effluvia the denser air beyond must rush in. Lana's pedagogy requires a picture (fig. 4.3), the first of many attempts to illustrate the

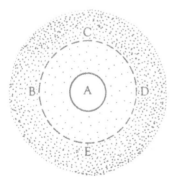

4.3 *Lana's map of the effluvial regions around an electric. After Lana,* Magisterium, *III (1692).*

44. §§27, 35–8.   45. §§15–17, 20.   46. Cf. Aristotle, *Phys.*, VII, 1–3.

course of effluvia: A is the electric, ABD the heated region, and BCED the volume of cold dense air. Once arrived at A the chaff remains entangled in the viscous, igneous matter on the surface until the returning air has lost its impulse. When the weight of the attracted corpuscle overcomes the weak adhesive, the electrical interaction ceases.

Lana's double-duty effluvia, which expand air and fix chaff, gave a complete and plausible account of electrical phenomena in keeping with received physics. The theory had a long life: it was primarily in Lana's persuasive form that Schott's compromise dominated German theory in the first decades of the eighteenth century.[47]

47. The accurate and approving review of Lana's electrical ideas in *Acta erud.*, 12 (1693), 145–50, is a harbinger of this dominance.

CHAPTER V

Testing the Air Theory

1. THE ENGLISH APPROACH

Not until the 1640s did Gilbert's countrymen begin to free his watery humor from the objections of Cabeo.[1] The chief actors then were Sir Kenelm Digby, Stuart diplomat, traveller, and miscellaneous philosopher; his Jesuit side-kick and former mentor, Thomas White, alias Blacklow; and his one-time literary opponent, the physician Sir Thomas Browne.

Digby took up electricity, along with everything else, in a peripatetic, animistic, corpuscularian hodge-podge written in Paris, where he enjoyed the enforced leisure of a Royalist refugee. In contrast to Gilbert, Digby took amber as the prototypical electric; he deduced that electrical effluvia have an unctuous, viscid, non-watery principle. When shot out, '[as] steams . . . issueth from sweating men or horses,' they resemble so many tensed lute cords; if they strike a foreign body they retreat, just like 'the little tender horns of snails used to shrink back if anything touched them, till they settle in little lumps upon their heads.'[2] The sticky 'strings of bituminous vapor' drag back any chaff they may encounter. The model eludes Cabeo's chief objections to Gilbert, but again at the cost of restricting electricity to 'bituminous' substances.[3]

There remains the experiment of the rebounding corpuscles, Cabeo's strongest argument against adhesive effluvia. Following Gilbert, Digby denied the phenomenon: a body driven by Cabeo's wind would indeed rebound from the electric, whereas they stick one to the other, 'turn them which way you will, as though they were glewed together.' But Digby did see rebounds. In trying to explain them away he noticed the reattraction of ricocheting corpuscles. Could that occur through any admissible gyration of the air? Digby said no. On the hypothesis of elastic, sticky, retentive threads, however, it follows immediately.[4]

1. Only one earlier mention of the watery effluvia in English literature has come to light: Carpenter, *Geographie* (1635²), I, 54, distinguishes electricity from magnetism, one of his main topics, by Gilbert's *differentiae,* including: '*Electricall* bodies, as *Gilbert* well confirms by experiments, draw other bodies to them by reason of a *moist* effluence of vapors, which hath a quality of joyning bodies together. . . . But the *magneticall* coition cannot be other than an act of the magneticall form.'

2. Boyle, *Electricity* (1675), in *Works,* IV, 345–6; Peterson, *Digby* (1956), *passim;* Descartes, *Corresp.,* II, 304, 364; Digby, *Two Treat.* (1644), Chap. xix.

3. Cf. Crosland, *Hist. Stud.* (1962), 109–16.

4. The corpuscles might have discharged themselves against a third object, lost their electricity to the air, or moved to a stronger part of the amber's field, setting up the 'Aepinus effect' (*infra,* xvii.2).

Before he left England in 1643 Digby published patronizing *Observations* on Browne's *Religio medici,* which ensured a wide sale for the obscure doctor's book. Thus encouraged, Browne wrote *Pseudodoxia epidemica,* a compendium of balanced judgments on controverted questions of natural philosophy.[5] Electricity receives a chapter, enriched by original experiments: 'a faithful and candid observer' (it is Boyle's judgment), Browne found seven new electrics, increasing by one-fourth the contemporary inventory, and reported an effect that some consider the discovery of electrostatic repulsion.[6]

The report appears in one of three arguments for the theory, 'granted by most,' that effluvia cause electrical attraction. With the first two we are familiar: electrics must be made 'perspicable' to act, and surface moisture or an interposed object destroys electricity by hindering the 'effluxion.' Here is the third:

> If also a powerfull and broad Electrick of wax or Anime [resin] be held over fine powder, the Atomes or small particles will ascend most numerously unto it; and if the electric be held unto the light, it may be observed that many thereof will flye, and be as it were discharged from the Electrick to the distance sometime of two or three inches, which motion is performed by the breath of the effluvium issuing with agility; for as the electric cooleth, the projection of the Atomes ceaseth.

For Browne, as for every seventeenth-century electrician but Huygens, electricity was by definition a faculty of attraction. The effect he described— electrical 'discharge,' to use his word—was a by-product, a side-effect, of attraction, useful perhaps as corroboration of the action of effluvia, but inconsequential for designing electrical theory. The 'discharge' occurs only for a short time, within a short distance of a powerful electric literally going at full steam. Because of their 'thinness' the outgoing steams in general do not drive objects *from* the electric; only when dilating on their return can they pull anything with them.

The model of Digby, Browne, and Thomas White, who published an epitome of Digby in 1646,[7] guided English thought about electricity until Hauksbee's time. A measure of the pervasiveness of the model is the doubly derivative treatment of electricity given in Walter Charleton's 'augmentation' of Gassendi's 'repair' of the atomic philosophy of Epicurus. Gassendi had discussed electricity among 'so-called occult qualities,' which he wished to banish from physics. Nothing, says he, can act at a distance: hence electrics must

<hr>

5. Browne, *Pseud. epid.* (1646), 79–82; Chalmers, *Osiris,* 2 (1936), 28; Finch, *Browne* (1950), 104–19.

6. E.g., M. B. Hesse, *Forces* (1961), 91.

7. White, *Inst. perip.* (1646); *Perip. Inst.* (1656), 77; Schott (*Th. phys.* [1659], 376–83) knew Digby via 'Thomas Anglicus,' i.e., White. White's writings, which strayed far from the *Ratio studiorum,* had some currency in England. He is a principal source for John Webster, who quotes this characteristic bit: 'The things that are feigned of Arthur and his Knights . . . are not more difficult to be persuaded of, than the composure of the heavens, which the preceding age us hath taught.' Webster, *Acad. exam.* (1654), 50.

possess a mechanism analogous to the tongue with which a chameleon 'attracts' a fly *in distans*. But what answers to the lizard's retractive muscles? The *linguae* of amber, according to Gassendi, operate either directly, like elastic cords stretched and released, or mediately, by stirring up the air. Charleton follows his master up to the alternatives, but he cannot bring himself to mention the hypothesis of the Jesuits.[8]

HONORÉ FABRI

The English theory found an independent and creative exponent inside the Society in the person of the unconventional Fabri. For a time Fabri taught philosophy at the important Jesuit college in Lyon. His argumentative exposition of new ideas, like the circulation of the blood, which he claimed to have discovered independently of Harvey, ended his teaching there under circumstances earlier related.[9] Reassigned to Rome in 1646, he became theologian to the Penitentiary, a curial office concerned with dispensations in difficult matters of conscience. He retained this position until 1680, when he retired to busy himself with history and apologetics.[10]

The Penitentiary left Fabri time to exercise his talent, curiosity and combativeness on natural philosophy. Shortly after his arrival in Rome he was crusading in favor of Jesuit's bark, cinchona, as a febrifuge. He associated himself with Kircher and with Ricci, through whom he became, in 1660, a correspondent of the Accademia del Cimento. He promptly irritated the academicians by a lengthy and losing dispute with another of its correspondents, Christiaan Huygens, over the existence of Saturn's rings. During the sixties he published a set of *Physical Dialogues,* the widely used *Geometrical Synopsis,* and the *Physics* that revived Lana. Sometimes his inventiveness got the better of his judgment, as in the wrangle with Huygens, and in the essay in apologetics that earned him two months in a papal prison in 1670.[11]

Although historians of science have not treated Fabri well,[12] his contemporaries thought him one of the first physicists of the age. Leibniz placed him with Galileo, Torricelli, Steno and Borelli for his work on elasticity and the theory of vibrations, and alone with Galileo for his efforts 'to rationalize experimental phoronomics [kinematics].' He is a paragon of Aristotelian commentators, a man 'more than erudite, of a frank and judicious liberty in philosophizing.' Mersenne rated him 'a veritable giant in science,' or so he wrote Descartes in the spring of 1646, announcing that 'the Jesuit of Lyon' had projected an encyclopedia of philosophy and mathematics in eighteen volumes.

8. Gassendi, *Opera* (1658), I, 540; Charleton, *Physiologia* (1654), 345–6.

9. *Supra,* ii.1.

10. Descartes, *Corresp.,* VI, 357; de Vregille, *Bull. soc. Gor.,* 9 (1906), 5–15; Delandine, *Manuscrits* (1812), I, *passim.*

11. *Catholic Encyclopedia* (1913–4), VIII, 372–4; Fermi, *Magalotti* (1903), 99–101; Heilbron, Cong. int. hist sci., XIIe (1968), *Actes,* 3B (1971), 45–9; *supra,* ii.1.

12. See Caverni, *Storia* (1891–8), I, 213; Ornstein, *Rôle* (1928), 81–2; Poggendorff, *Gesch.* (1879), 371–4; Fellmann, *Physis,* 1 (1959), 6–29, 73–102.

After two volumes had been published and acclaimed, Mersenne wrote that some preferred Fabri's work to the *Principia philosophiae*. Descartes procured a copy: '[Fabri] is a man of great wit and enthusiasm,' he agreed, and worth reading had he not misunderstood Cartesian mechanics. These lines might stand as Fabri's appraisal of Descartes, 'a man of singular genius and ingenuity,' but the author of an impossible philosophy, grounded on 'thoroughly fallacious' mechanical principles.[13]

Despite the contretemps with Huygens, the Accademia del Cimento esteemed Fabri, whose 'candor . . . makes his learning additionally agreeable.' In appreciation of his communications, Leopold gave him some of the delicate glass apparatus for which the Accademia was and is so celebrated.[14] One of the academicians, J. A. Borelli, a man most difficult to please, had no recourse but to admire the 'prolix and eccentric' Jesuit; for, as he observed regarding the causes of planetary motions, Fabri's ideas were often very similar to his own. 'I think the keenness of this priest's mind is truly admirable,' he wrote to Leopold, 'as well as the great learning, frankness, and conviction with which he treats innumerable difficult and recondite matters.'[15]

Fabri's interest in electricity, which extended over a half-century,[16] had its fullest expression in 1665, in a dialogue ostensibly devoted to capillary action. 'A most ingenious and dear friend' having suggested to one of the colloquists, Antimus (who represents Fabri), that the capillary force is electrical, he proposes an excursus on electricity. The proposal is welcomed by Augustinus and Chrysocomus, who play the parts of Sagredo and Simplicio.[17]

Antimus must first outline the state of knowledge. He brings nothing new to us except the fact that amber, which loses its electricity when moistened with water, retains it when covered with oil, 'as I was told some years ago by a man of great genius, my very good friend Lorenzo Magalotti.'[18] Antimus' rapid

13. Leibniz, *Hypothesis physica nova* (1671), in *Phil. Schr.* (1875–90), IV, 208, 216, 240; Fabri to Leibniz, 14 Nov. 1671, *ibid.*, 241–4; Leibniz to Fabri [1677?], *ibid.*, 245; Descartes, *Corresp.*, VII, 171, 210, 282–3; Fabri, *Epistolae* (1674), 5, 27 ff.

14. Ricci to Leopold, 26 July 1660, in Fabroni, *Lettere* (1773–5), II, 73–5. Cf. *ibid.*, I, 307; II, 75–8; and Fabri to Leopold, Idus Feb. 1661, in Mss. Gal., CCLXXVI, 99–100: 'Caeterum, pro illa supellectile vitrea, variis thermometris, nobili, aliisque organis, ad gravitatem liquorum definiendum cocinne et apposite fabricatis, immortales, et quam possum maximas gratias habeo.' Communications from Fabri appear *ibid.*, CCLXIII, 97, and CCLXXXIII, 60 ff. Other recipients of Leopold's beautiful apparatus were Manuel Settala (Fogolari, *Arch. stor. lomb.*, 14 [1900], 93) and the Accademia degli Investiganti, founded in 1663 (Fisch in Underwood, ed., *Science* [1953], I, 527).

15. Caverni, *Storia*, I (1891), 212–13; Borelli to Leopold, 27 Feb. 1665 and 29 Dec. 1666, in Fabroni, *Lettere* (1773–5), I, 124–6, 131–3.

16. Fabri, 'Synopsis recens inventorum' (Ms. at the Bibl. Naz., Florence), §9: 'Cuncta haec nova et amoena [phaenomena electrica] primus inveni iam ferè a 50. annis.' He first wrote of electricity in his *Philosophia universa per propositiones digesta* (Lyon, 1646), of which no copy has come to hand.

17. Fabri, *Dialogi* (1665), Pref., 176–82. Some of the quotations in the text are abridged paraphrases.

18. Magalotti was the secretary of the Accademia del Cimento, whose experiments he reported to Fabri (*infra*, v.2–3).

survey confuses Augustinus; he demands an explanation, an 'hypothesis,' a model to make the physics, the *veritas physica*, clear to him. Antimus obliges with the theory of viscuous threads.

The model seems entirely arbitrary to Chrysocomus. 'Who ever saw or felt these filaments? And who can't concoct [*fingo*] any number of similar hypotheses?' Antimus replies in two ways. First he says that Chrysocomus wants too much: who has ever seen the effluvia in odors or contagious emanations? For that we should require a 'divine microscope.' Then he lists the phenomena from which the existence of the filaments can be inferred: the great irregularity in the motions of the attracted chaff (as if it were hit by streams moving in many directions); its resistance to removal (as if caught in a net); the destruction of electricity by moistening, breathing upon, or heating the electric (which impedes or destroys the filaments); the stoppage of attraction by interposing a screen (which entangles the filaments); and the high-pitched ring produced when rubbed amber is brought up to a thin sonorous membrane held against the ear (the sound of filaments striking the membrane). 'Do you mean then that electricity cannot be referred to a fear of the vacuum?' 'Some people believe that is its cause. But I [Antimus] think that they are entirely wrong, among other reasons because many experiments show that no body by itself moves anywhere from fear of the vacuum.'

Antimus' opinions could be studied in Fabri's *Physics* as well as in the *Dialogue*, and in popular texts like Duhamel's *De corporum affectionibus* and Sturm's *Physica electiva*.[19] Plagiaries, which Fabri professed not to mind, also carried the theory, which can be traced up to the middle of the eighteenth century.[20]

2. MUTUALITY

Of the several phenomena tacitly excluded by theories of effluvia—repulsion, conduction, the universality of electricity, the reciprocity of electrical interactions—only the last forced itself upon the attention of seventeenth-century physicists. Its recognition, by Fabri and by Robert Boyle, has an interest beyond the detection of a new phenomenon. It was a consequence of taking a mechanical model literally. The physicists of Gilbert's age did not widely advocate mechanism; its effective explanations touched only a few disconnected phenomena. The virtuosity, inventiveness, and optimism of Descartes, however, and the example of the *Principia philosophiae* persuaded many that mechanical models offered the only hope for a precise and comprehensible physics. One's expectations rose. One demanded more from models, perhaps even a complete fit with phenomena, with little or no 'negative analogy,' as modern philosophers put it.[21]

19. Fabri, *Physica* (1669–71), IV, 212–15; Duhamel, *De corp.* (1670), 467–79; Sturm, *Physica*, II (1722), 1044–5, 1096, 1106–7.

20. Fabri, 'Synopsis recens inventorum,' *loc. cit.*: 'Haec [principia electrica] deinde aliqui ex me licet innominato, quod molestum non facit; illud autem de crepitu omiserunt, quia nondum editeram, 2. partem Dialog.' Pohl, *Tentamen* (1747), 35, gives the theory of the viscous threads.

21. Hesse, *Forces* (1961), 24, 100.

Boyle held that a 'good hypothesis' or model must be 'fit and sufficient' to explain the chief phenomena and 'at least consistent with the rest;' whereas an 'excellent hypothesis' must 'enable a skilful Naturalist to foretell future Phaenomena by their Congruity or Incongruity to it; and especially the events of such experiments as are aptly devis'd to examine it, as Things that ought or ought not, to be consequent with it.'[22] In this terminology, physicists early in the century expected only good hypotheses, while those of Boyle's age demanded excellent ones. As for Fabri's sanguine expectations regarding the fit of mechanical models, they are plain enough in the speeches of Antimus and in the *Epistolae* of 1674, where Fabri denied unconvincingly that he followed Descartes, Gassendi and Epicurus.

FABRI AND THE ACCADEMIA DEL CIMENTO

The short-lived Accademia del Cimento (1657–1667) consisted of some ten men supported by Duke Ferdinand and Prince Leopold dei Medici. Its best known and most productive members were G. A. Borelli, Vincenzo Viviani, and Francesco Redi. It appears that Viviani, who liked to style himself the last pupil of Galileo, designed most of the Academy's electrical experiments.[23] The printed record of the Academy's labor, their *Saggi* or *Essays* (1667), do not reveal the extent of their concern with electricity, give no rationale for it, and neither acknowledge Viviani's role in, nor explain the reasons for any of the experiments. All this is characteristic of the *Saggi,* which strive to conceal the thought and preparation behind the Academy's investigations, and avoid identifying individual contributions.[24]

It is said that Magalotti, the principal author of the *Saggi,* chose this impersonal style lest the Church persecute individual members for their opinions about the nature of electricity, magnetism, the vacuum, the hot and the cold, and similar matters fascinating to the Holy Office.[25] That is a serious misinterpretation: the Church no longer cared to martyr physicists, and in any case the opinions of the academicians were no more scandalous than those of Fabri or Lana. Magalotti chose his style for two reasons. First, the peculiar organization of the Academy precluded other forms of publication. The group was Leopold's creation and depended financially upon him and his brother. The Academy tried to act as a whole: individuals proposed experiments; the group decided their importance, feasibility and relevance. The usual, or at least the initial rou-

22. Boyle, 'The Requisites of a Good Hypothesis,' in M. B. Hall, *Boyle* (1965), 134–5. Cf. Boyle, *Excellency Mech. Phil.* (1674), in *Works*, IV, 68–70, 77–8; M. B. Hall, *Boyle* (1965), 189, 191, 195, 207. The changing expectations of the mechanical philosophers have received little attention; see Laudan, *Idea* (1966), 71–8, 89–93, 162–5, and *Ann. Sci.*, 22 (1966), 73–104.

23. Nelli, *Saggio* (1759), 110–111n; Middleton, *Experimenters* (1971), 39.

24. Antinori in Acc. del Cim., *Saggi* (1841³), 3–133; Targioni-Tozzetti, *Notizie* (1780), *passim;* Abetti in *Celebrazione* (1958), 3–10. For the lesser known members of the Academy, the del Buono brothers, Dati, Rinaldini, Marsili and Oliva, see Antinori in Acc. del Cim., *Saggi* (1841³), 70–80; Nelli, *Saggio* (1759); and Middleton, *Experimenters* (1971), 26–40.

25. *Ibid.*, 6, 54, 72, 333.

tine called for Borelli, Oliva and Rinaldini to set out the necessary instruments in the palace in the evening; the next day the Academy, together with visiting dignitaries and interested friends, if any, would jointly carry out the approved experiments.[26] This group participation reproduced on a grander scale the established procedure at Kircher's museum, and anticipated the early practices of the Paris Academy of Sciences.[27]

The communal approach did not work. Jealousy flourished, especially between Viviani and Borelli.[28] The Academy was inherently unstable. Though customarily accorded a ten-year life, in fact—excepting a last flourish in 1667 and observations of the comet of 1665—its work was completed by 1662, when Magalotti began to draft the *Saggi*. Nor were its early years vigorous: the surviving registers indicate that it operated, altogether, for less than two years of its nominal ten, and that it peaked in the late spring and summer of 1660. Some, believing incorrectly that the group met only at Leopold's command and in his presence, have laid these interruptions at his door; but it is likely that the jealous academicians could only work together when Leopold or his brother kept order. Magalotti would have found it impossible, had he considered it desirable, to identify individual contributions in the *Saggi*. Leopold, who financed the publication as he had the experiments, not unreasonably considered it *his* book: attributions could only have irritated the Prince and increased the jealousy of the academicians.[29]

If the impersonal style of the *Saggi* arose from causes peculiar to the

26. Antinori in Acc. del Cim., *Saggi* (1841³), 93–5.

27. For the contemporary and similar practices of the Accademia degli Investiganti and of the Museo Settala see Fich in Underwood, ed., *Science* (1953), I, 528, and Fogolari, *Arch. stor. lomb.*, 14 (1900), 98, 114–15, respectively.

28. Antinori in Acc. del Cim., *Saggi* (1841³), 70, 106–7; Middleton, *Experimenters* (1971), 28–9, 34, 310–16.

29. Targioni-Tozzetti, *Notizie* (1780), I, 412, 555; Antinori in Acc. del Cim., *Saggi* (1841³), 109; Fermi, *Magalotti* (1903), 78–85. It is often said (bibliography in Favaro, Ist. ven., *Atti*, 71:2 [1912], 1173–8, to which may be added Eccher in *Vita* [1903], 389, and Ornstein, *Rôle* [1928], 78) that the cause of the final dissolution of the Academy in 1667 was Leopold's elevation to the cardinalate, the relation being either that his new duties prevented his attendance at its meetings, or that the Pope exacted its destruction as a condition for the office. As to the first, the Academy did meet without Leopold, and he was in Rome but twice, for short periods, after his preferment; as for the second, it is typical of the imaginings of those who can see the seventeenth century only through the eyes of Galileo. The single qualification Leopold had or needed for the cardinalate was being a Medici prince; indeed, the Pope asked Ferdinand to choose which of his brothers should have the honor (Middleton, *Experimenters* [1971], 319–24). After his investiture Leopold continued to support experiments, and he might at first not have considered the interruption of 1667 more final than earlier ones. The suspension of that year became permanent because the leading academicians themselves dispersed; even so Magalotti urged Leopold to continue the sessions (Fabroni, *Lettere*, I [1773], 312). Cf. Fermi, *Magalotti* (1903), 83–5; Middleton, *Experimenters* (1971), 319–26; and Leopold to L. de' Vecchi, 15 Dec. 1667, in del Lungo, *Arch. stor. ital.*, 76 (1918), 110; 'If at the moment more experiments are not published and if the ardor with which we began these activities has cooled, the real reasons are the occupations of myself and of many academicians.'

Academy, Magalotti's refusal to theorize in print was a policy, a consequence of a scientific method similar to Bacon's. Before theories, before hypotheses even, must come experiments, natural histories, facts. The Academy offered accurate experiments, and no more. They would obtain accuracy by repetition, 'provando e riprovando;' they would avoid misconstruction by presenting results without explanation or hypothesis.

Magalotti knew that neither he nor the Academy had realized these ideals: despite group efforts, experimental uncertainties remained; despite editorial care, stray hypotheses might have survived. He declined responsibility for rash conclusions drawn from his lapses. 'If sometimes, as a Transition from one Experiment to another, or upon what occasion so ever, there shall be inserted any speculation, we Request they may be taken always for the thoughts, and particular sense of one of the Members, but not imputed to the whole Academy, whose sole Design is to make Experiments, and Relate them.'[30] Fortunately the Academy's register of experiments, or 'Diario,' has survived. It reveals the rationale of many of the *Saggi*'s experiments, including those on electricity.

Except for a few experiments done in July 1657, which erroneously identified steel as electric, the Diario records no work on electricity until May 1660.[31] On May 20 the Academy ended a twenty-month intermission with the investiture of Magalotti; the main reason for its revival may have been the electrical experiments that monopolized its time during the following week. These bore on two questions: the debilitating effects of coating liquids, temperature and screening on the activity of amber, and the identification of electrics.[32] The rationale of the first enquiry is not far to seek; supposing the effluvium to be a distinct chemical species, the academicians examined its response to heat, cold, and the surface contaminants through which they made it pass. They tried 63 liquids in all, which gives a measure of their interest in the

30. [Magalotti], *Essayes* (1684), Pref. Middleton, *Experimenters* (1971), 80–1, and *Br. J. Hist. Sci.*, 4 (1969), 283–6, observes that the *Saggi*'s title should be rendered 'Examples of Experiments in Natural Philosophy.' The Academy's motto appears to come from Dante, *Paradiso*, III, 1–3: 'Quel sol che . . . di bella verità m'avea scoverto, provando e riprovando, il dolce aspetto.'

31. There are three copies of the register: Mss. Gal., CCLX, written by Magalotti, his predecessor Alessandro Segni, Rinaldini, Oliva, and Viviani; CCLXI, which proceeds only to February 1658, and appears to be a contemporary copy; and CCLXII, dating from the late eighteenth century and marred by ignorant errors of transcription, e.g., 'Cesio' for 'Cabeo' (Cf. Middleton, *Experimenters* [1971], 42–5). There is no account of the electrical morning of July 3, 1657, in CCLXI; CCLX, 242^r and CCLXII, 12^v, have the following: 'Si provo se fosse vero cio che dell' ambra scrive il Gilberto, cioe bagnandola, doppo esser riscaldata, con acquavite non tira punto, e si trova fals[iss]imo, perche non solo tira, ma anche della med[esim]a distanza. [CCLXII adds 'Falsa.'] L'istess'ambra con l'interposizione d'un velo benche sottisissimo non tira. [CCLXII adds 'la paglia.'] Il diamante tira il foglio, la paglia, et ogni altro corpo e l'istesso fa l'acciaio. [CCLXII: 'Questo ultimo e falso.']' 'Acciaio' is not a copying error; the *Saggi* emphasize that, in demonstrating the nonelectricity of metals, rocks, coral, etc., one must guard against impaling the chaff on minute, sharp protuberances from the nonelectric's surface.

32. *Essayes* (1684), 128–32; *Saggi* (1841^3), 143–7; Middleton, *Experimenters* (1971), 230–3, 363.

matter and of its appropriateness for communal investigation.[33]

The motive for the investigation of screens was, the Diarist says, to discover 'the resistance sufficient to impede the attraction of amber.' The academicians began by interposing sheets of paper punctured first with a fine needle, then with the points of scissors, finally with a large nail. The holes grew; the attraction did not. They made gratings of silk threads, hairs, and paper strings, *none* of which interrupted the action; and windows of unravelled thread, lint, fine cloth, and leaves of precious metals, which shut it out.[34] All this was perplexing: Gilbert's idea, that screens debilitated electricity by entangling effluvia, did not do justice to the phenomena. Nor did the *Saggi,* which suppress all results incompatible with the mechanical explanation of the behavior of screens. Rediscovery of the caprices of interposed objects was to provide a persuasive argument against theories of effluvia.[35]

Among the reports of these indecisive experiments one finds the following surprising intelligence: 'It is commonly believed, That *Amber* attracts the little Bodies to itself; but the action is indeed *mutual,* not more properly belonging to the *Amber,* than to the Bodies moved, by which also itself is attracted.' This the academicians ascertained by suspending a piece of amber by a thread, or upon a pivot; the electric 'made a little stoop to those little Bodies, which likewise *proportionally* presented themselves thereto, and readily obeyed its call.' The Diario does not record this important demonstration. The reason? It was not invented by the Academy.

In October 1660 Magalotti was working on a tractate on electricity, now lost, that described an experiment 'praiseworthy for its novelty.'[36] An undated manuscript, probably also from 1660, suggests the nature and importance of the novelty: 'His views,' says Magalotti, speaking of Cabeo,. 'are refuted by experience, for the *ambra versoria* follows all bodies presented to it.'[37] The *ambra versoria* is the demonstration of mutuality recorded in the *Saggi.* Where did Magalotti, an uninventive physicist who contributed few if any research proposals to the Academy, obtain his information? Among the drafts of the *Saggi*'s sections on electricity is an unsigned note (fig. 5.1). The first of its

33. Mss. Gal., CCLX, 66ʳ–69ᵛ; Targioni-Tozzetti, *Notizie* (1780), II, 554–5; Abetti, ed., *Opere* (1942), 476–7.

34. Targioni-Tozzetti, *Notizie* (1780), II, 555; Mss. Gal., CCLX, 130ʳ; CCLXII, 102. The puncture experiments were done in August, 1660.

35. The chaff feels the field of the electric diminished by that of charges induced in the screen. The extent of the diminution depends upon the conductivity and inductive capacity of the screen, its shape and thickness, and the distances between the interacting objects. If the screen conducts well, sits close to the electric, and is grounded, one has an 'air electrophore,' which gives almost no field beyond the barrier (*infra,* xviii.3).

36. Letter of 26 Oct. 1660, in Magalotti, *Lettere familiari* (Florence, 1769), 1, 68, quoted by Fermi, *Magalotti* (1903), 105. The Marchese Paolo Ginori Venturi, who owns many of Magalotti's papers, has kindly informed me that his collection does not include the tractate; Fermi, who searched hard for it, doubted that it ever existed.

37. Targioni-Tozzetti, *Notizie* (1780), II, 555–6; cf. Magalotti, *Lettere* (1740²), 87.

numbered points reveals that 'a piece of sealing wax suspended freely and then rubbed approaches other bodies.'[38] This sentence, not to mention the hand, identifies the author, the only contemporary who regularly used sealing wax as an electric, Honoré Fabri; while a glance shows the whole to be a résumé of the speeches of Antimus in the *Dialogi physici* of 1665.

Fabri knew Magalotti, who had been educated at Rome, probably in the Collegio Romano, before advancing to the University of Pisa, where he attracted the attention of Viviani and Borelli.[39] Both Fabri and Magalotti became attached to the Academy in the late spring of 1660. By October Fabri was under pressure to contribute something useful, having irritated the academicians with tiresome repairs to the artificial hypothesis he had devised to defeat Huygens' Saturnian ring. The squabble over Saturn did not show Fabri at his best. As Ricci hinted to Leopold and Magalotti, the affair arose from Fabri's professional obligation to combat Copernicans. 'Otherwise,' he wrote, 'the father is not partisan, and will come straight to the point when you want his opinion, without your academicians having to respond if they do not wish to.'[40] The spare, informative memorandum of fig. 5.1 is perhaps the peace offering.

Fabri probably predicted mutuality from his model and looked for it for confirmation. Antimus used the *succinum vertibile* to complete the proof, sketched by Gilbert and repeated by the Jesuits, that electricity cannot be a sympathy because amber draws all bodies promiscuously. Fabri's spokesman could add that amber also approaches all bodies indiscriminately. 'If it reacted to only one kind, like magnets to iron and vice versa, there might be some analogy; but because it approaches all, it cannot do so through a native, intrinsic force, which is silly even to think, much less to say.'

ROBERT BOYLE

For a quarter-century following the publication of his first scienfic work (1660), Boyle was head propagandizer of the corpuscularian faith. The simplicity of his models (a pile of fleece to represent the 'spring of the air'), the extent and variety of his experiments, his sincerity, probity and undogmatic assertiveness, made him an effective advocate. His ignorance of scholastic philosophy, which enabled him to conflate Aristotle's teachings with alchemy, magic, and debased forms of peripateticism, also helped.[41] Among the happiest of his essays, the eleventh and last in an anti-peripatetic collection published in 1675, is a pious fraud purporting to rescue electrical theory from the grip of the occultists.[42]

38. Mss. Gal., CCLXII, 97ʳ.

39. Fermi, *Magalotti* (1903), 23–5; Fabroni, *Lettere*, I (1773), 307, and II (1775), 76.

40. Fabroni, *ibid.*, II, 101–4. Fabri later (*Dialogi* [1665], 89–91, 212) handsomely conceded defeat to Huygens. Cf. Huygens, *Oeuvres*, XV, 391–402.

41. P. P. Wiener, *Phil. Rev.*, 41 (1932), 596–7, 604; *supra*, i.3.

42. *Exper. Mech. Orig. Qual.* (1675–6), in *Works*, IV, 235, 345–54; cf. M. B. Hall, *Boyle* (1958), 90–1, and Power, *Exper. Phil.* (1664), 57–8, 155, who fancies that the sight of electric and

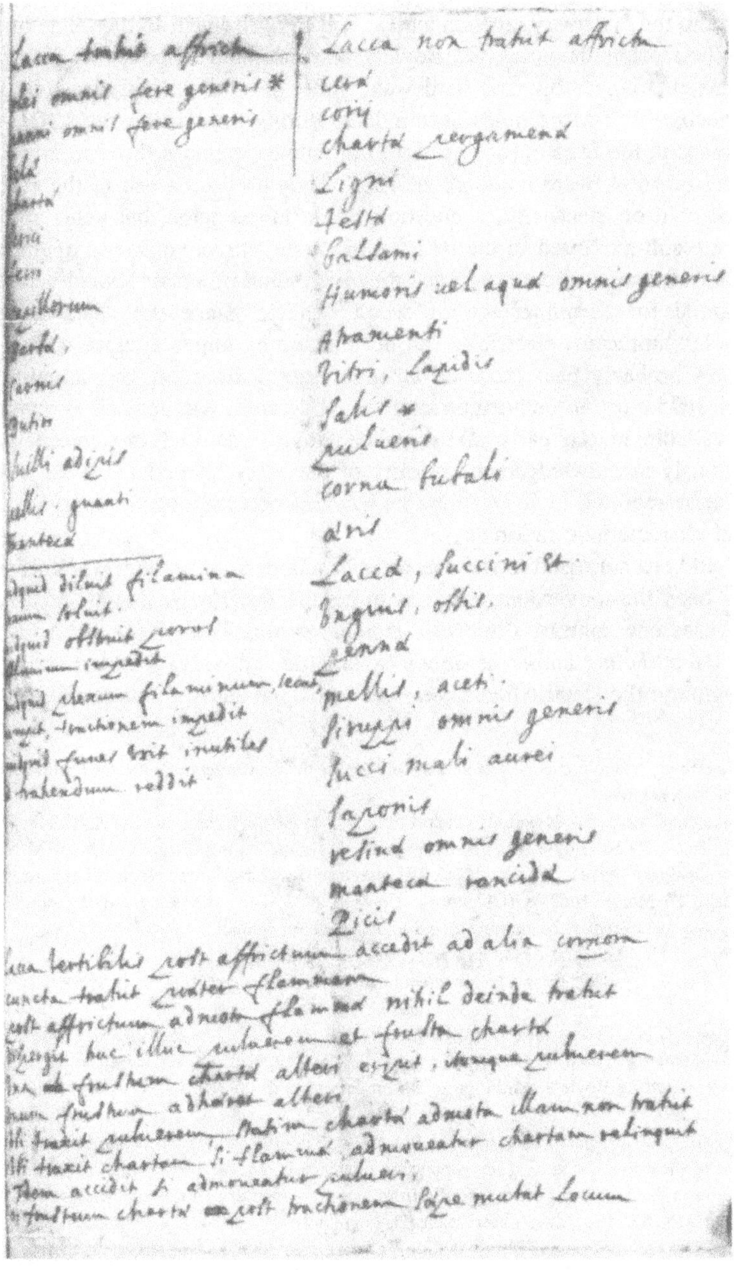
…acca trahit affricta
…li omnis fere generis *
…nni omnis fere generis
[illegible]

Lacca non trahit affricta
cera
corij
charta pergamena
Ligni
Testa
balsami
Humoris vel aquae omnis generis
Atramenti
Vitri, Lapidis
Salis *
pulueris
cornu bubali
auris
Lacca, succini et
Unguis, ossis
gemma
mellis, aceti
Sirupi omnis generis
Succi mali aurei
Saponis
retinae omnis generis
manteca rancida
Picis

…acca [illegible] post affrictum accedit ad alia corpora
cuncta trahit praeter flammam
post affrictum admota flammae nihil deinde trahit
[illegible] huc illuc pulverem et frustum chartae
[illegible] frustum chartae alteri [illegible], itaque pulverem
[illegible] frustum adhaeret alteri
…i trahit pulverem, statim charta admota illam non trahit
…i trahit chartam si flamma admoveatur chartam relinquit
Idem accidit si admoveatur pulveri
… frustum chartae post tractionem saepe mutat locum

5.1 *Fabri's minute on electricity. From Mss. Gal., CCLXII, f. 97.*

Boyle read Italian and knew of the *Saggi* almost as soon as they were printed.[43] Hence his account of mutuality, which first appeared in the tract on electricity of 1675, may have derived from them. Other evidence suggests that he came to the discovery independently. The advertisement to the tracts of 1675 traces their origin in notes for Boyle's discourse on saltpeter (1661), which mentions rubbing amber and hard wax, although not explicitly in connection with electricity.[44] More notes accumulated during the writing of 'A Diamond that Shines in the Dark' (1664).[45] Still more notes, perhaps those referred to in an essay of 1669, were made about 1665. They propose a test of the effect of extreme cold on electricity, a question Boyle investigated that year, and they report a result exploited in the tract of 1675: the attractive power of the caput mortuum of amber. (Boyle reasoned that distillation of amber should destroy its 'substantial form' and effective 'occult quality;' since the caput mortuum nonetheless attracted, electricity can not operate by innate sympathy.[46]) These old notes probably also recorded an independent discovery of mutuality: the tract of 1675 cites no authorities later than Gassendi, Kircher and Browne, the best available in the early sixties when Boyle was studying electricity; he scrupulously acknowledged the priority of the 'very learned Fabri' in a minor observation reported in 1673;[47] and he gave his detection of mutuality a plausible and characteristic rationale.

He had been surprised to see one piece of amber draw another, for amber had always been the mover and terminus in his electrical experiments. 'And as in many cases one contrary directs to another, so this Trial suggested a further.' Could the attracting amber be drawn by an attracted body? The test challenged his ingenuity; the electric had to be freely movable and yet able to return easily

---

magnetic effluvia in a microscope would 'sensibly decide the Controversie 'twixt the Peripatetick and Atomical Philosophies.'

43. *Works*, I, xxiv. Cf. Magalotti's report on a visit to Boyle; Crino, *Fatti* (1957), 158; Waller, *It. Stud.*, 1 (1937), 58. Boyle repeated some of the experiments, not always successfully; Middleton, *Experimenters* (1971), 283, 291, 296. His attention was drawn to the electrical experiments by Oldenburg, 17 March 1667-8 (Oldenburg, *Corresp.*, IV, 248): 'there is nothing new in [the *Saggi*], as to us, except it be perhaps some experiments of amber.'

44. *Works*, IV, 235, and *Phys. Essays* (1661), in *Works*, I, 365.

45. *Works*, I, 789–99; the notes are mentioned in *Phys. Essays* (1669²), in *Works*, I, 451–2, 456.

46. *Atmos. Cons. Bodies* (1669), in *Works*, III, 279, 281–3; M. B. Hall, *Boyle* (1965), 252–5; Boyle to Oldenburg, 1665, Oldenburg, *Corresp.*, III, 58, 77, 570. Cf. C. Merett to T. Browne, 8 May 1669, reporting Boyle's exhibition of the amber effect to the RS, in Browne, *Works* (1964²), IV, 361.

47. *Essays . . . of Effluviums* (1673), in *Works* III, 659–706, which teach that effluvia have the character of their sources; among the supporting evidence is the 'fact' that pools of quicksilver are often found in the heads of gilders who die from breathing mercury vapor. Cf. McKie, *Sci. Prog.*, 29 (1934), 253–65. The minor observation (Boyle, *Works*, III, 681–2): a suspended hair or other light body (apparently charged from handling) is sometimes repelled, not attracted, by the electric. Like Browne before him, Boyle judged the effect, caused by 'the briskly moving streams that issue out of the amber,' to confirm his theory of attraction.

to rest after excitation. A thick piece of amber, suspended from a silk string and rubbed horizontally, consented to follow the pin cushion used to excite it. The motion confirmed the model, for to Boyle *only* the amber was excited: it had the 'power of approaching the Cushion by virtue of the operation of its own streams.'

Boyle's conception of this mechanism appears clearly from the affair of the virtuosa's locks. He learned from a 'Fair Lady' that 'false locks of hair,' well combed and curled, often stuck to the cheeks of the beauties they adorned. When his informant indignantly rejected his first explanation—sticky rouge— he considered electricity. That only increased the oddity. He could suppose the locks to be electrified by combing, and, in keeping with his late discovery, to pull themselves to ladies' cheeks by their sticky elastic effluvia. A test with a chicken feather lying on an excited piece of amber confirmed the mechanism: the feather's branches attached themselves to his finger, because it and the tendrils were all engaged in the gluey embrace of the amber's effluvia. Unfortunately, the available force, which Boyle once planned to measure, seemed inadequate to the task. The most powerful electric known, amber, could barely budge the locks; how could they raise themselves? It did not occur to him that the cheek too was electrified.[48] The business remained a conundrum, which passed from history with the bloom of its discoverer.

3. ELECTRICITY *IN VACUO*

Having undermined Cabeo's theory by the demonstration of mutuality, the Florentine academicians attempted to destroy it by generating electricity in a space void of air. Knowing nothing in August, 1660, of the improved air pump Boyle and Hooke had just built in Oxford,[49] they had to make do with the 'Torricelli space' at the top of a barometer (fig. 5.2). The glass tube EDABC has openings at A, C, and E. Having tied a cover around C, an academician introduces his hand through DE, making the air seal tight by binding his arm with the bladder DEHI (fig. 5.2a). Another academician then fills the tube with mercury via the opening at A, throws in some bits of paper, seals the orifice, places the tube in the mercury-filled vessel as shown, and removes the bladder at C. The mercury descends in the tube below B, leaving a partial vacuum. The hand thereupon produces a piece of amber, rubs it against a cloth previously pasted at K, and tries to draw whatever bits of paper have not gone down the tube with the mercury. The experiment failed. If the ligature binding the blad-

48. The electrified lock induces an opposite charge in the lady's cheek, while dry hair, being a poor conductor, offers amber no such advantage. For Boyle's concern with the magnitude of electrical force, cf. M. B. Hall, *Boyle* (1965), 253–4.

49. The Academy probably first saw a copy of Boyle's *New Experiments . . . Touching the Spring of the Air* (1660), in 1662, in the Latin translation of 1661 (Fulton, *Bibliography* [1961²], nos. 13, 19); Middleton, *Experimenters* (1971), 263, 332. That the vacuum experiments were expected to disprove Cabeo's theory appears from Magalotti's note in Mss. Gal., CCLXVIII, 81 (Abetti, *Opere,* 411).

der to the arm excluded the air, it also stopped the flow of blood. The academicians replaced the arm with an amber-tipped stick (LM in fig. 5.2b); but they found that, however tight the ligature, it always admitted air during the rubbing. 'Still desiring some fruit of this experiment,' they arranged (fig. 5.2c) to twirl the stick against the cloth without any movement to and fro.

This arrangement gave reproducible results: the amber would never attract the straw or paper left on the glass when the mercury subsided. That did not content the academicians, however; the results, which, against their expectations, seemed to confirm Cabeo, were too regular; air could not be entirely excluded from the tube; *provando e riprovando*. Air was admitted through A, the amber rubbed and offered to the paper. Still no reaction. Suspecting, then, that particles of mercury or of the glue used to mount the cloth might have

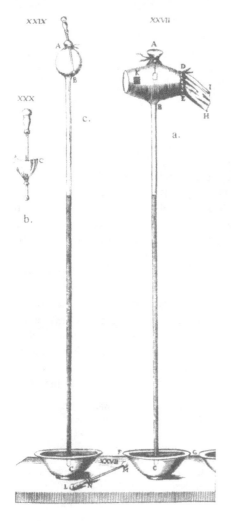

5.2 *The apparatus with which the Accademia del Cimento failed to detect electricity in vacuo: (a) the initial arrangement (b) the stick replacing the academician's arm (c) the final arrangement. From Acc. Cim.,* Saggi *(1841).*

choked the amber's pores, the academicians took countermeasures; but all their 'diligence was in vain, for whether the vessel were full or empty of Air, the Amber attracted not.' And that, alas, was all they could 'with truth report of an Experiment attempted so many ways unsuccessfully.'[50]

Boyle also tried to determine whether 'the motions excited by the air had a considerable Interest' in electrical attraction. A piece of amber that retained its virtue for over fifteen minutes in air was 'well chafed' and suspended in a glass receiver just over some chaff. A sturdy lad worked the pump; the air was withdrawn and the amber lowered to its prey. '[We] perceived, as we expected, that in some Trials, upon the least Contact, it would lift it up; and in others, for we repeated the Experiment, the Amber would raise it without touching it, that is, would attract it.'[51]

The followers of Cabeo were not annihilated by Boyle's *experimentum crucis*. They explained that his pump had failed to remove all the air; and they argued, probably correctly, that the Torricelli space—in which the Accademia del Cimento detected no electrical activity—was emptier than the *vacuum boyleanum*.[52] Here nature played a trick on electricians. The dielectric strength of air—its resistance to a disruptive spark discharge—decreases in proportion to the pressure down to about 1/1000 of an atmosphere, when it begins to grow rapidly. Boyle's suspended amber may be compared to one plate of a condenser, and the metal platform of the pump bearing the receiver to the other. The minimum sparking potential at a plate separation of 1 cm and an exhaustion of 1/30 at., about the best Boyle's pump could do, is roughly 2000 volt. His 'purposely turn'd and polish'd' amber, without points or protuberances, did not exert a field of 2000 volt / cm, and so retained its electricity. In the eighteenth century, owing to improvements made by Hauksbee and Smeaton, pumps could reach almost to 1/600 at., where a very small surface charge, enough to produce about 350 volt / cm, can dissipate itself by sparking.[53] By extrapolation it appeared that a 'vacuum' 'conducted' perfectly, and that the air played an essential role in electrical interactions. Not until the mid-nineteenth century did air pumps reach the region of increasing dielectric strength below 1/1000 at.

The Torricelli space can insulate well if the mercury has been carefully

50. *Essayes* (1684), 43–6; *Saggi* (1841³), 51–4; Middleton, *Experimenters* (1971), 143–5. Both Borelli (Mss. Gal., CCLXVII, 19ᵛ) and Rinaldini (*ibid.*, 24–40) proposed further refinements, apparently never implemented; cf. Targioni-Tozzetti, *Notizie* (1780), II, 606, 612–13. Rinaldini's objections to Magalotti's draft report of the vacuum experiments (Mss. Gal., CCLXVII, 37ʳ) perfectly exemplify the Academy's vigilance: 'Non solo il dato dovrebbe esser postato con più modestia ma dovrebbe esser ancora detto con disiunzione in questa forma. Esperienza proposta per conietturare, o se l'aria si richiegga per mez[z]o all'attrazione elettrica, o almeno se non essendocci tal mezzo si osservi varietà alcuna nell'attrazione suddᵃ fatta nel vuoto.' Cf. Abetti, *Opere*, 303–4, 331–2.

51. Boyle, *Electricity* (1675), in *Works*, IV, 34.

52. Lana Terzi, *Magisterium*, I (1684), 342, III (1692), 287–312.

53. Smeaton, *PT*, 47 (1751–2), 421–2; Loeb, *Fund. Proc.* (1939), 411; *supra*, i.1.

cleansed of air and water vapor. Such a vacuum was not attained before the 1780s. Its demonstration then did not convince many electricians, who preferred the evidence of their air pumps. The *vacuum torricellianum* employed by the Accademia del Cimento certainly did not reach 1/1000 at., and leakage around the bladder probably reduced it to above 1/100. Perhaps their experiment failed because the mercury, in sliding down the tube to open the Torricelli space, electrified the glass, which stayed charged in the relatively poor vacuum and attracted the paper. This effect will play an important part in our story as the celebrated 'mercurial phosphorus.'[54]

54. W. Morgan, *PT*, 75 (1785), 272–8; *BFP* (Smyth), VI, 221; *infra*, viii.1

Chapter VI

Immaterialists

Three seventeenth-century physicists, Madeira, Maignan and Guericke, employed principles other than material effluvia to explain the amber effect.[1] They formed no school and had no disciples. They are nonetheless individually interesting both for their work and as exceptions to the rule that the facts of electricity forced natural philosophers, whether radical or conservative, corpuscularian or peripatetic, to regard the amber effect as mechanical.

1. MADEIRA

'Electricity' is advertised on the title page of Madeira's book, *On Occult Qualities,* along with the powers of music, the bite of the tarantula, and the virtues of a tree from the Garden of Eden. Madeira was a competent physician, an authority on the treatment of syphilis and doctor to John IV of Portugal;[2] like many physicians, he attributed the metabolism of the body to hidden powers or virtues. In *De Qualitatibus occultis* (1650), he argued the existence of such powers chiefly on medical and pharmaceutical grounds, and confirmed it by appeal to the odd items among which he placed electricity.

To recapitulate Madeira's argument, on which we have touched before, the forms of bodies are characterized by the elemental qualities, hotness, coldness, moistness and dryness, by secondary qualities compounded of them, and by 'occult' higher powers or virtues different from either. He infers the existence of occult powers on three counts. First, the elemental qualities cannot explain such phenomena as magnetism, the detention of ships by the remora, and the stupefying sting of the torpedo, for no elements, and hence no mixture of elements, act like magnets, remoras, or torpedos. Second, some effects are brought about more quickly and economically than elemental qualities can manage: the spread of poison from a scorpion bite, the metabolizing of food, the dominion of the brain over the muscles. Third, many philosophers already admitted 'manifest super-elemental qualities' including sound, light and sometimes impulse, none of which seemed further reducible. Why, Madeira asks,

---

1. Bacon also has a place here since he includes electricity under 'virtues' that may act at a distance for the 'gain,' or to supply the 'want,' of the bodies that possess them. 'For electricity [of which Gilbert and others after him have devised such stories] is nothing else than the appetite of a body when excited by gentle friction—an appetite that does not well endure the air, but prefers some other tangible body.' *Novum Organum* (1620), in *Works,* IV, 218.

2. Barbosa Machado, *Biblioteca* (1930²), I, 715–16; *Grande enciclopedia portugesa e brasileira,* art. 'Arrais'; *supra,* i.2.

should one object to another order, more elevated still, competent to produce more recondite effects?[3]

Madeira admits electricity into the occult on the first count, which obliges him to prove that elemental qualities do not suffice. Against Cabeo's theory he throws up the classical objection that the air ought to propel the chaff outwards, not backwards; against other theories he can only bluster. 'Who believes,' he asks, 'that the very little heat generated in the rubbing can elicit vapors or effluvia from such extremely hard bodies as diamond or glass?'[4]

Having proved the incompetence of the effluvia, Madeira argues their irrelevance, by appealing to a 'cognate' phenomenon. He means magnetism, which his authorities had carefully distinguished from electricity. The magnet acts across screens of wood, stones or metal through which a material vapor could not penetrate; consequently, he says, it must operate by direct communication of qualities. With this consideration Madeira inverts the usual argument: 'Why turn to occult [!] effluvia in electrical attractions for no reason or evidence, in order to avoid an occult quality? Those who do so merely fall into Scylla, hoping to avoid Charybdis.'[5]

Freed from their scholastic and bombastic form, these considerations are not inappropriate. To accomplish their work material effluvia did require properties different from any displayed by familiar mechanical systems. Still, undisputed cases of action by sympathy or by occult qualities, like magnetism, fitted the accidents of electricity worse than the mechanical analogies. The heterogeneity of electrics and of the bodies they draw remained the crucial facts, ones that Madeira noticed but did not explain. For this omission other electricians held him up to ridicule. And that he richly deserved for his parting piece of bombast: 'Electrical attraction provides an optimal argument for occult virtues, an argument not weakened a jot by the silly figments of Gilbert, Cabeo and their followers.'

2. MAIGNAN

Emanuel Maignan, the son and grandson of professors, was much admired by his friends and, according to his biographers, preferred by his church for his independence of mind.[6] At seventeen, after schooling by the Jesuits, he entered the convent of Minims in Toulouse, where he studied philosophy under a worthy peripatetic who encouraged his objections to Aristotle's physics.

Maignan's intellectual strength determined his superiors to hasten his ordination and the commencement of his teaching career. In 1636 he was transferred

3. *Qual. occult.* (1650), 13–16.

4. *Ibid.*, 542–3.

5. *Ibid.*, 453.

6. Bayle, *Dictionnaire* (1740[5]), III, 280–3; Whitmore, *Order* (1967), 166–71; Ceñal, *Rev. pol.*, 46 (1952), 111–49; Nicéron, ed., *Mémoires*, XXXI (1735), 346–53; Louyat, Cong. Soc. Sas., 95(1971), Sec. Sci., *Competes rendus* (1974), I, 15–29.

to Rome, to the convent of Trinità del Monte, established in 1623 as a graduate school for Minims.[7] In Rome Maignan made a reputation as a philosopher and mathematician, met Mersenne, joined Kircher's circle, and wrote an important work on optics, *Perspectiva horaria* (1648), no longer recognized at its true value. In 1650, after 15 years in Rome, Maignan returned to Toulouse. He crowded his cell with a lathe and other machinery with which he built the apparatus for many of the experiments recorded in his major work, a *Cursus philosophicus* first published in 1653. His fame drew visitors like the lodestone iron; people of all ranks sought him out, from the artisans he instructed to Louis XIV, who spent some hours of his honeymoon in Maignan's monkish retreat. Louis invited him to Paris, but he declined, preferring, he said, to finish his days in the convent where his religious life began.[8] His hope was fulfilled in 1676.

Maignan speaks of electricity in a chapter of his *Cursus* concerned with 'natural body, heavy or light, with electric and magnetic motion, and with other similar matters.'[9] All instance stimulated self-motion, the awakening of intrinsic powers by external agents: just as contact with earthy effluvia causes bodies to gravitate, so magnets and electrics excite their prey to self-motion via a vapor or corporeal spirit. Consider two cognate bodies A and B, A lying within B's 'spiritual sphere' and receiving its effluvia. Because of their sympathy, B's spirit stimulates A to emit a vapor that in turn increases B's productivity. The greater the 'friendliness' (amicitia) of the bodies, the faster their spirits flow and the sooner their self-moving principles actuate.

Maignan's spiritual emetics also perform some of the services of the lines of force introduced into electrostatics in the nineteenth century:[10]

Not only do the spirits set the bodies in motion, they also in a certain sense carry them off or draw them forth: for such spirits are not like sand or little particles of other kinds, rather they cohere together by a natural connecting bond, like the rings in a chain. . . . Therefore since the spirits are linked together, and each is attached to its font, their emission does not thoroughly disperse them, and they cohere to their source as much as they can.

Now these two bodies, or fountains, together with their spirits are excited by the impulse of each other's effluvia, as I have said; and because of the linking just described, it is much easier for them (unless they are restrained by a greater force) to follow in the direction of the spirits proceeding from them, than to constrain the spirits' flow. Hence, I say, once the effluvia have been exchanged they play the part of goods or goals, enticing the fountains on as it were by loving hands. Whence it is that their motion is a mutual approach, if they have equal facility in acceding to the forces acting on them . . . and even if both are held back, they still tend equally toward one another.

7. Bonnard, *Hist.* (1933), 172–3.
8. Montucla, *Hist.* (an VII²), I, 730; Middleton, *Hist.* (1964), 10–16, 32–4, 58.
9. Maignan, *Cursus* (1673²), 358–62, 388–400; Oldenburg, *Corresp.*, V, 227.
10. Maignan, *Cursus* (1673²), 360.

Note that the chains of effluvia do not pull bodies together mechanically, as would Fabri's filaments, but provide the path and the incentive that prompt them to continue to move themselves. Antipathy or repulsion follows the same principle, the chains being *less* concentrated between the bodies than elsewhere. If Maignan had had to illustrate the distribution of his concatenated spirits in the space about pairs of sympathetic or antipathetic bodies, his diagrams would necessarily have resembled those of lines of force between unlike and like charges, respectively.

Maignan realized that his theory of electricity deviated from dominant views. We already know his objections to Cabeo's theory. As for the English approach, it failed to explain the return of the effluvia to the electric. Maignan dismissed the analogy to stretched cords; only an external force could reverse the flow of effluvia. Experiment ruled out the two forces that occurred to him: extrinsic cold, which might counteract heat acquired during rubbing; and reflection from chaff, 'as when,' he says, 'using a received analogy [Digby's], the horns of a snail strike against leaves and similar matter.'

Cold will not do because an excited electric continues to work when chilled. Nor can effluvia be returned by contact. Approach an excited electric to a slab of wood too large for it to move. The effluvia supposedly rebound from the wood, and yet the electric continues to draw chaff: therefore either the effluvia do not return, which violates the hypothesis, or new ones have been emitted. But we can repeat this experiment many times without renewing the friction, which implies that after each exposure to the wood the electric sends forth new effluvia. And this conflicts with the theory's own explanation of the ruin of electricity by flames: if the electric produces new effluvia after the wood has diverted the earlier flow, why should it not do as much when the older ones are consumed by fire?

Since material theories failed, Maignan reverted to sympathy: 'Electric motion is like magnetic, that is, it arises from an electric effluvium that solicits straw and other objects and determines them through an innate principle of motion to fly off to the amber, etc., as to bodies consonant and cognate with them.' There are differences, however, between electricity and magnetism. First, the 'sympathy' of electrics is 'more universal,' a touchy point that Maignan wisely does not enlarge. Second, electric spirits, being more 'corpulent' than magnetic, are difficult to excite and easy to interrupt. Third, since fat effluvia tend to stick in the pores of their sources, an external force, namely friction, must be applied to release them. (The difference regarding mutuality, an error incompatible with a sympathetic theory, necessarily does not appear.)

Maignan leaves other applications to his readers, as he must run to exercise his principles on the agitated problem of the weasel and the toad.[11] He pauses

11. The toad was thought to attract and then to slay the weasel, which raised a choice philosophical problem: since the weasel approaches uncoerced, its motion must be natural; but as the outcome is its own destruction, its motion must be violent. Maignan, *Cursus* (1673²), 400–1, makes their unhappy association 'violent-natural,' like the reputed attraction of iron-bearing ships

only to assure the faltering that they need not scruple to suppose that spirits can issue from substances as 'arid' as glass. There is no lack of evidence of such spirits, he says, serving up one oddity that stirs curiosity about the others. He holds that pills made from vitrified antimony will purge twenty asses, without suffering change in weight; a feat that, among less elevated thoughts, suggests that the effective matter in the pills is a spiritous substance, and not their grosser parts.

3. OTTO VON GUERICKE

Guericke ornamented a leading family of the free city of Magdeburg. During the days of his father and grandfather, each of whom became mayor, the prosperous Lutheran town had grown rich, strong and populous. Guericke was educated to help guide and preserve his little state. After preliminaries at the University of Leipzig and some law at Jena, he went to Leyden to complete his equipment with modern languages, surveying, fortification, architecture and military engineering. He returned home in about 1626 to marriage and his place among the aldermen of Magdeburg.[12]

The town was soon to need good counsel. In 1631 imperial armies under Tilly besieged it; the Swedish representative, insisting that his ever-victorious master, Gustavus Adolphus, would relieve them, urged the city fathers to stand firm. Eventually Tilly's uncontrollable troops overran the barricades. While they raped and murdered at will, an equally uncontrollable fire broke out that consumed the wooden town. Five of every six citizens perished, 25,000 in all; Guericke bought his freedom at a price that left him little more than his shirt. Set adrift as Kircher and Schott had been the year before, he naturally moved in the opposite direction; while the Jesuits, driven out by the Swedes, found their way to Rome, the Lutheran patrician left a homeland devastated by Catholic troops to seek service with the Scandinavians. He returned to Magdeburg in 1632, charged to rebuild and to refortify the ruins.[13]

Towards 1642 Guericke's surviving fellow townsmen made him their chief negotiator. He arranged for the removal of foreign garrisons, became a mayor, represented Magdeburg at the peace congress that closed the war (1648), and strove to gain imperial recognition of the ancient privileges of his city. This mission caused him to attend the Imperial Diets; and it was at one of these meetings, in Ratisbon in 1654, that he chose to reveal his invention of the air pump.

---

to magnetic cliffs near the poles. The weasel contains a poison attuned to the toad's, which moves the unwitting vermin 'naturally' toward its end. Fabri, *Metaphysica* (1648), 285, ascribes such interactions to fear, noise, or the breath of the predator.

12. Hoffmann, *Gesch.* (1850), I, 331–5; Schimank, *Guericke* (n.d.), most of which appears in *Beit. Gesch. Tech. Ind.*, 19 (1929), 13–30; Windgårdh, *Fra fysik. verd.*, 12 (1950), 119–39.

13. Guericke, *Belagerung* (1912²); C. V. Wedgwood, *Thirty Years' War* (1961²), 253–7; Schimank, *Guericke* (n.d.), 16–21.

He had chosen this form of advertisement deliberately: the influential delegates would appreciate the novelty and carry news of it throughout Germany. The Bishop of Würzburg proved the best herald. He bought Guericke's apparatus, carried it home, and showed it to the local authority on instruments, Gaspar Schott. Schott published the first account of the pump, with Guericke's permission, in his *Mechanica hydraulico-pneumatica* of 1657, which inspired the work of Hooke and Boyle. The Lutheran engineer found the Jesuit scribbler useful. The celebrated demonstration of the Magdeburg spheres and several lesser exercises, called collectively 'Experimenta nova magdeburgica,' were reported in Schott's book on natural magic published in 1664. Guericke chose the same title for his own account of his researches, which appeared at Amsterdam in 1672.[14]

Guericke's formal knowledge of natural philosophy never advanced far beyond what he had learned as a student in the early 1620s. If the citations in *Experimenta nova* are reliable, he then read carefully such modern monographs as Gilbert's *De magnete* and Galileo's *Letters on Sunspots,* and through them came to Copernican cosmology.[15] He was staggered at the immensity of the world, as implied by the absence of stellar parallax; at the size, number and distances of the stars; and at the great reaches of apparent emptiness. 'Most of all,' he says, 'I was intrigued and puzzled by that vast intermediate and endless space, and perpetually desired to plumb its nature.'[16]

The war at first precluded, then abetted the indulgence of this desire. As his finances improved and the pressure on him eased, he found the contemplation of nature a 'gate of tranquility,' 'the quiet of the most irenic truth,' a welcome relief from the heavy responsibilities of defense and reconstruction.[17] Now he came to his old problem as an engineer used to large installations; he determined to make interplanetary space here below. The 'vacuous space' manufactured by his air pump had precisely the properties required: it seemed not to resist the motions of ponderable bodies; it stifled sounds, which do not reach us from the stars, and passed light, which does. Assuming the identity of interplanetary space and his artificial vacuum, Guericke turned to design a mechanism adequate to run a Copernican world imbedded in a timeless, resistanceless void.[18] In the style of Gilbert he assigned most of the job to planetary

14. Schott, *Tech. cur.* (1664), 8–86; Schimank, *Guericke* (n.d.), *passim;* Guericke, *Exper. nova* (1672), Pref.; documents in Guericke, *Neue Vers.* (1968), (9)–(45).

15. Although Guericke cites several later monographs, the most recent being Huygens' *Systema saturni* (1659), he does so more for scholastic completeness than to record intellectual debt. For facts he relies on Jesuit compendia, particularly Kircher's, whom he respected. Cf. Schott, *Tech. cur.* (1664), 29; Schimank, *Organon,* 4 (1967), 27–37; and the notes of Schimank and Fritz Krafft in Guericke, *Neue Vers.* (1968), (232)–(331).

16. *Exper. nova,* (1672), **1, 53–4.

17. *Ibid.,* **1–**2. Guericke also liked to contrast 'peaceful' experimental philosophy with the 'bellicose' speculations of the schools, where 'tam acriter disputarunt & quilibet sententiam suam semel conceptam, mordicus ut miles arcem contra oppugnantem hostem, defendit.'

18. The central role of the problem of space in Guericke's research is emphasized by Kauffeldt, *Guericke* (1968), 14–19, 92, 141, and Cong. int. hist. sci., XI[e], 1965, *Actes,* 3 (1968), 364–8.

souls or powers. Following his own bent, he found a way to reproduce these 'mundane virtues' in the laboratory.[19]

ELECTRICITY AND THE MUNDANE VIRTUES

The planetary powers are neither substance nor accident, but 'effluences that have their seat in mundane bodies and flow forth from them.' These emanations may effuse through a vacuum and can be corporeal, like air, the effluvium of the earth, or incorporeal, like the penetrating magnetic virtue. In either case their powers decrease in strength from the neighborhood of their source, where they are 'denser, more compressed and stronger,' to the limits of their 'spheres of virtue.' When Guericke says that 'their nature is to act at a distance,' he means that their effects spread instantaneously through the 'medium' of space.[20] The terrestrial virtues include the *conservative,* whereby bodies gravitate and the earth retains whatever is necessary to its well-being; the *expulsive,* which expels harmful materials like fire and keeps the moon at a distance; the *directive,* which fixes the earth's axis during the annual revolution; the *gyrational* (vertens), responsible for the diurnal rotation; the *impulsive,* or inertia; and *virtutes lucens, soni* and *calefaciens,* the causes of light, sound and heat.[21]

The problem is to exhibit these powers in the laboratory, an undertaking no less formidable than the manufacture of space. Here is Guericke's solution: take a glass phial the size of an infant's head; fill it with sulphur finely ground; heat the phial, cool, break the glass and recover a beautiful, symmetrical sulphur globe; fit the ball with a handle and place it in a wooden frame, as in figure 6.1a; rub it and the terrestrial virtues, or some of them, will appear. The most evident is the conservative: the excited globe attracts all light objects and retains them when rotated. 'The terrestrial globe is, as it were, placed before your very eyes, for the earth attracts all things on its surface, and carries them with it in its diurnal motion.'[22] As for the expulsive virtue, the globe sometimes repels what it has attracted because of a conflict of natures, and sometimes on caprice, *pro lubitu.* The rejected bodies remain unacceptable until they correct their dispositions through contact with a third object. Guericke advises that this phenomenon is best observed with a feather, which will hover in the globe's

19. No documents fix the commencement or order of Guericke's researches. Since the vacuum experiments required time and great expense, perhaps 20,000 gulden in all (Kauffeldt, *Guericke* [1968], 134), they probably did not begin before Guericke's shift to diplomacy in the early 1640s; and, as he had ample opportunity to announce them at diets before Ratisbon, they probably did not succeed much before the early fifties. According to its preface, *Exper. nova* (1672) was nearly finished in 1663; it went to the publisher in 1670, delayed by other claims on its author's time and his rusty Latin. Cf. Krafft in Schmauderer, *Buch* (1969), 103–29.

20. *Exper. nova* (1672), 125–6. Cf. Kauffeldt, *Guericke* (1968), 72, 144.

21. *Exper. nova* (1672), 125–51; cf. Schimank's notes in Guericke, *Neue Vers.* (1968), (271)–(283).

22. The ball in fact has a stronger *virtus conservativa* than the earth's, for it carries objects against their gravity. Note that the purpose of the wooden frame was to facilitate this demonstration; it is not, as is often alleged, a prototype of later electrostatic machines. Cf. Rosenberger, *Abh. Gesch. Math.,* no. 8 (1890), 69–88.

sphere of activity and remain suspended above it when it is paraded about the room (fig. 6.1b).

Play with the feather can further illustrate the operation of the virtues. First, the feather sometimes extends its fibres when stuck on the globe or when swimming in air, and it attracts or is attracted by all objects in its vicinity. Second, the repelled feather always opposes the same face to the globe, as does the moon to the earth. Third, the feather will fly back and forth between the globe and a neighboring object. Fourth, a linen thread suspended so as almost to touch the excited globe flees the contact of an approaching finger. Fifth, a long linen thread will draw chaff at one end when the globe is excited at the other. Sixth, the suspended feather, when presented to a neighboring body, often retracts its branches because of 'antipathy'; similarly, if one tries to grasp the fibres, they draw back, curl up, and apply themselves to the globe.

a   b

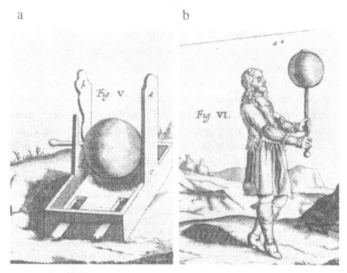

6.1 Guericke's demonstrations of the mundane virtues: (a) his so-called electrical machine (b) parading a feather supported by the expulsive virtue of the sulphur globe. From Guericke, Exp. nova *(1672).*

Excited in the usual way, the sulphur sphere also exhibits *v. califaciens* and *v. sonans* (it feels warm and crackles), and in a dark room it glows like pounded sugar (*v. lucens*). Dropped on the toe its *v. impulsiva* becomes only too evident. This, however, is as far as one can go. The globe possesses neither *v. vertens* nor *v. directiva*, the latter property belonging only to the lodestone.

There is little doubt that Guericke came to these experiments under the influence of Gilbert, accepting, on the one hand, his association of electricity and gravity, and rejecting, on the other, his notion that the earth is a magnet.[23]

---

23. Guericke benefited from Kircher's criticism in rejecting the earth-magnet; Krafft, *Sudh. Arch.*, 54 (1970), 135–6.

For Guericke a lodestone terrella did not adequately represent the earth; hence the sulphur ball. But why sulphur? If electricity were the only consideration, why bother with the ball when its glass mold was equally electrical, more durable, easier to handle, cheaper and free of the stench of brimstone?[24] The answer appears in the travel book of Monconys.

On a visit to Magdeburg in October, 1663, he was entertained by bad wine, worse beer, and a yellow globe about a foot and a half in diameter. Nine different minerals, according to Monconys, made up the ball, which imitated the attraction of the earth and the behavior of the moon before his eyes. Stories about this marvel spread through Germany, and eventually reached Leibniz. Guericke answered his request for information with a description of a globe made of 'diverse minerals together with sulphur poured into a two-foot sphere.' Here the minerals play a role so prominent that Leibniz could not miss them; thereafter he referred to Guericke's handiwork not as a sulphur sphere, but as a *globus mineralis*.[25] The sulphur supplied a matrix in which to combine the materials Guericke thought the most important constituents of the *mundus subterraneus*; the brimstone globe could reproduce most of the virtues of our planet because it was, and was designed to be, a literal simulacrum of the earth.[26] Of course the sulphur, being electric, also guaranteed that the ball would have at least one of the telluric virtues, the conservative, which Guericke knew how to awaken. Imagine his delight and excitement when he found that the thing worked, that the rubbing indeed called forth other mundane virtues!

Those who credit Guericke with the discovery of repulsion or of the electric light because his *virtutes expulsiva* and *lucens* manifest electricity do him an injustice.[27] His world was richer in powers than ours, and he made a point of separating them. Only the conservative virtue was to him electrical: 'The attraction of the sulphur is the same as electricity, and is either identical with, or arises from, the *virtus conservativa*.'[28] This association, and perhaps a straightforward analogy to the communication of *virtus impulsiva,* led Guericke to the only one of his discoveries that he would have regarded as electrical. In order to save his incorporeal conservative virtue from possible objections

24. Most recently Smolka, *Acta hist. rer. nat. necnon tech.*, special issue, 2 (1966), 43–56, has raised this old problem without solving it.

25. Monconys, *Voyages* (1695²), Pt. III, 75–80, 481; Guericke, *Exper. nova* (1672), 150, citing the first edition (1665–6); Guericke to Leibniz, 6/16 June 1671, in Leibniz, *Phil. Briefw.* (1926), 119–20. The passages from Monconys and the correspondence with Leibniz are also given in Guericke, *Neue Vers.* (1968), (46)–(49), (81)–(95).

26. Had Guericke embedded a lodestone in the ball it would have added the 'directive' to its other virtues. Cf. Guericke to Leibniz, 29 Aug./8 Sept. 1671, in Leibniz, *Phil. Briefw.* (1926), 151.

27. E.g., Benjamin, *Hist.* (1898), 403; Wolf, *Hist.* (1952²), I, 304–5; Hoppe, *Gesch.* (1884), 4–5. All follow Dufay, *MAS* (1733), 25–35; cf. Cohen, *Ann. Sci.*, 7 (1951), 207–9. Kauffeldt, *Guericke* (1968), 77, 147, accepts Guericke as the discoverer of repulsion, sparking and conduction.

28. Cf. Rosenberger, *Abh. Gesch. Math.*, no. 8 (1890), 89–112; Heathcote, *Ann. Sci.*, 6 (1948–50), 293–305.

brought from Cabeo's theory, he invented the fifth of the experiments noticed earlier, which demonstrated the transmission, or communication of electricity. 'We cannot concede,' he wrote, 'that the attraction occurs by the intervention of the air, because we can see by experiment that the sulphur globe, when excited by friction, can also exercise its *virtus* [*conservativa*] through a linen thread an ell or more in length, and draw something or other at the far end.' This important fact had to be rediscovered before it permanently entered the body of knowledge.[29]

REACTIONS TO *EXPERIMENTA NOVA*

The masterpiece of the amateur septuagenarian philosopher disappointed its readers. His qualitative arguments in favor of Copernicus were passé; his description of the vacuum experiments added nothing to Schott's; his apparatus was inferior to Boyle's. The old-fashioned program of mundane virtues, moreover, violated the sensibilities of the younger generation. 'Nowadays,' Leibniz warned the old man, 'one despises everything said about virtues and qualities and wants to explain everything in terms of size, shape and local motion.'[30] By common consent the only novelties in *Experimenta nova* were the manipulations with the sulphur ball.

The experiments were known in England as early as 1663, when Monconys reported them to Sir Robert Moray, a leading light of the Royal Society. Moray thought Guericke had used some trick, and deplored his refusal to 'communicate the secret of this little business.' The Society also knew of the experiments from Leibniz, who described them to Oldenburg in 1671.[31] A year later, when Hooke brought a copy of *Experimenta nova* to the Society and recommended that the globe experiments be tried, John Locke announced that he had already done so, and promised to exhibit his sulphur ball. By the next meeting, November 13, 1672, he had forgotten the matter, nor did he remember to bring his sphere the following week. Boyle lost patience at this procrastination. On November 27 he produced a ball of sulphur melted into a glass globe, which, 'like electrical bodies, attracted several light substances.' In addition, and of great interest, Boyle showed that 'feathers being first attracted by this sulphur ball would leave this electrical body and pass to one not electrical, untouched, as to a glass phial.'[32]

We seem at last to be on the brink of the discovery of electrical repulsion. But there is no mention of the globe or of electricity in the records of the

29. *Infra*, viii.3. Note that the inventor of the air pump did not try to disprove Cabeo by exhibiting electrical action *in vacuo;* his technique was more elegant, easier and less objectionable than Boyle's or that of the Accademia del Cimento.

30. Leibniz to Guericke, 17/27 Aug. 1671, in Leibniz, *Phil. Briefw.* (1926), 145–6.

31. Monconys, *Voyages* (1695²), Pt. II, Suppl., 137–8; Leibniz, *Phil. Briefw.* (1926), 168; Guericke, *Neue Vers.* (1968), (96)–(112).

32. Hooke, *Diary* (1935), 10, 12; Birch, *Hist.* (1756–7), III, 59, 61, 63. Cf. *PT*, 7 (1672), 5103–5.

Society's meetings in December, nor in January, nor indeed throughout the entire year 1673. Locke never brought his sulphur ball and no one again asked him for it. Guericke's influence on Boyle's continuing electrical experiments appears only in the occasional use of feathers. Nowhere does Boyle associate the expulsive or the *lucens* virtue with electricity, although he knew the work of Browne and Cabeo and had himself noted a glow in rubbed diamonds.[33]

Guericke's work fared little better on the Continent despite energetic advertisements there. Unable to invoke his usual outlet—Schott died shortly after receiving a description of the globe experiments—Guericke published an account of the conservative and expulsive virtues in the *Miscellanea curiosa* for 1671. He also made an agent of Leibniz, who mentioned the virtues to his correspondents and sent copies of *Experimenta nova* to his friends.[34] Still, except for Huygens, no one seems to have followed up Guericke's revelations. The only account of experiments with a sulphur globe made in the late seventeenth century that have come to light mentions neither Guericke nor electricity.[35]

The infertility of the globe experiments is not surprising. They are difficult to reproduce; even an electrician as practiced as Dufay never managed them.[36] Guericke recognized the problem and tried to reduce it by distributing globes and giving lessons in their manufacture. Nonetheless success depended on the skill and even on the chemical composition of the operator. As Guericke advised Leibniz, people with soft, moist hands—like his professorial son, and probably most academics—invariably failed. He traced his own aptitude to his former employment as a mechanic, but even his tough hide was not infallible.[37]

Secondly, the globe experiments were anachronistic. The purpose for which Guericke had invented them was not congenial to the moderns, who did not care to perfect experiments to prove the existence of sympathetic qualities or virtues. Also the special, almost magical character of the sulphur globe put them off. Gilbert's recognition of the generality of the amber effect had been the electrical discovery of the century. Guericke's globe, on the other hand, was a unique object, and he never pointed to any other material, except the earth itself, that enjoyed its peculiar properties. His discoveries had all to be made again.

33. Boyle in *Experiments* (1664), 413–23; Harvey, *Hist.* (1957), 380.

34. *Misc. cur.*, 2 (1671), 461–70; Leibniz, *Phil. Briefw.* (1926), 221–2.

35. AS, *Histoire depuis 1666 jusqu'à* . . . *1699*, II (1733), 233–4, which notes a crackling noise and the attraction of feathers.

36. Dufay, *MAS* (1733), 458.

37. Guericke to Leibniz, 13/23 Oct. 1671, in Leibniz, *Phil. Briefw.* (1926), 158–9. This is one of the subtler forms of interaction between science and technology.

Chapter VII

The Cartesians

Descartes came to speak of electricity toward the end of the fourth part of his *Principia philosophiae,* following his brilliantly imaginative discourse on the magnet. He apologized for opening the subject, for he had wished to limit himself to prominent natural phenomena; electricity, although common to several substances, seemed too parochial to invite attention on its own account. Moreover, the empirical evidence did not satisfy him; before he could know certainly how electrics work, he would have to try many tiresome experiments. Nonetheless he felt obliged to offer an explanation. Just before discussing the magnet he had revealed the nature of fire, and the constitution of its product, glass; and this constitution conflicted with what he took to be the common view of electrical attraction, the theory of Digby and White.[1]

Cartesian glass owes its properties to the fact that fire makes its constituent or third-matter particles smooth and plane. When, on cooling, the fire-matter recedes, the little balls of second matter and the requisite filler of the first replace it; glass owes its firmness to the plane contact of its third-matter particles, its transparency to its complement of the second, and its elasticity to the resistance to distortion of its pores. Its electricity also derives from the smooth joining of its larger corpuscles. Amber and other oily and sulphurous substances, unlike glass, have interlinked branching parts, and it is just this branching which gives them their unctuous character.[2]

Since glass contains no branching third matter, its electricity—and by implication amber's also—must result from something else. Descartes spied the cause in the shape of the pores. In figure 7.1, ABCDEFG is a cross-section of pieces of third matter abutting one another so as to outline the raviola-like pore abcdef. The pore itself extends from one external surface of the glass to another. It can accommodate only one sphericle of the second matter, H, in each frustum whose height equals the diameter of the ball; first matter fills the rest of the volume. Descartes supposes that particles confined in these channels tend to form long narrow fillets by attaching themselves to one another. The fillets move rapidly, since they are composed of flighty first matter, writhing and twisting within the glass in a vain effort at escape. They are confined by the surrounding air, whose passages suit them less than their home pores to which, during the original cooling of the glass, they had adjusted their shapes.[3]

1. *Principia* (1644) in *Oeuvres,* (1964²), IX:2, 305–6.
2. *Ibid.,* 266–71, 235, 241; cf. *supra* i.3.   3. *Ibid.,* 306–7.

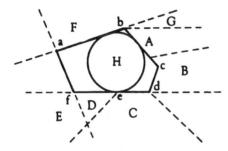

7.1 The structure of electrics according to Descartes.

The friction needed to awaken electricity agitates the fillets enough to enable them to penetrate air passages and the pores of small neighboring bodies. The foreign environment quickly overwhelms them, however, and they withdraw with the tidbits in whose pores they have become engaged. Now what is true of glass in this respect should apply to all or at least to most electric bodies: they possess passageways contrived to encourage the formation of fillets, and to prevent their casual loss to the atmosphere.[4] An electric is characterized by the shape and arrangement of its elementary particles and not, as the English school held, by its possession of a particular chemical principle.

The primitiveness of Descartes' machinery may obscure the importance of his contribution. He has attempted to do without secondary qualities entirely— to explain electricity without the help of fattiness and glueyness. Other seventeenth-century corpuscularians like Boyle hoped ultimately to explain those qualities by the shapes and arrangements of the particles constituting fatty and gluey bodies. They did not proceed to the reduction, however, and so helped perpetuate the unproductive conundrum, how substances with such disparate properties as glass, gems, resins and bitumins could possess the same active chemical principle. As late as 1730 the chemist Caspar Neumann felt obliged to chide people who thought that both glass and amber could owe their electrical power to the emission of *partes oleosae*.[5] But by Neumann's time most electricians in France, England and Holland had left sticky effluvia for mechanism. This is not to say they were Cartesians of strict observance, but that the uncompromising corpuscularism of the latter seventeenth century, in whose creation Descartes had played the leading role, had colored their thinking.

1. EPIGONI

Descartes' early disciples followed him closely in electrical theory. Rohault's *Traité de physique* (1671) treats electricity just where Descartes did, im-

---

4. *Ibid.*, 307–8.
5. Neumann, *Lect. publ.* (1730), 61, 64: 'Wer demnach die Elektricität beim Bernstein expliciren will, der bleibe vorher erst das weisse Glass zu considieren und vorher all hier die Elektricität auch statt habe.'

mediately after magnetism, and repeats everything, the narrow pores, the fillets, the resistance of the air, the errors of the English.[6] Precisely the same account was given, without acknowledgement, by Rohault's competitor, Jean Baptiste Denis, sometime physician to Louis XIV and to Charles II. Neither monarch, happily, left his person solely to Denis, a faddist who advocated blood transfusion as the grand elixir, or general remedy, for organic infirmities. He did indeed cure with its help all the ills of two of his countrymen, whose deaths precipitated a decree outlawing the practice. Meanwhile, beginning in about 1664, Denis ran an informal Academy at his home, offering lectures and perhaps demonstrations in mathematics and natural philosophy.

A lecture on amber gave Denis occasion to speak of electricity.[7] He opens by flogging that dead horse, the sympathetic explanation of electrical attraction, and then proffers Descartes' alternative as his own. He stresses the superiority of 'his' explanation to theories that suppose a fatty effluvium; what fat is there in glass? Denis is not an uncritical plagiarist. Why, he asks, does the electrical matter not push away in its outward journey the small objects it pulls back on its return? The answer: the fillets leave the electric in straight lines, and pass directly through bodies without striking their 'solid parts'; after frustration and disarrangement in the air, the returning fillets snag in the pores they had traversed freely before. Note that Denis, a professional lecturer on natural philosophy, knew nothing of Cabeo's 'rebounds' or Browne's 'discharges'; not the slightest hint of electrostatic repulsion had penetrated to him.

Malebranche also accepted attraction by entanglement as the cause of electricity. The theory proved useful. In 1689 Jacques Aymer, a peasant of Dauphiné, began to find all sorts of things—underground water, buried treasure, murderers—with the aid of a divining rod. The local clergy were concerned. Could the effect be natural? A Cartesian theory of subtle particles was invoked. It did not satisfy everyone. Malebranche was consulted, and pronounced the business supernatural. 'Bodies can only act upon one another by impulse,' he said; apparent violations of the principle, like electricity and magnetism, arise from subtle matter in motion. Dowsing cannot work similarly, since the strength of a material emanation decreases with distance from the source, while the rod responds better to buried objects than to ones close to hand. In short, from the Cartesian physics of electricity and magnetism Malebranche inferred, correctly so far as we know, that dowsing is the work of the devil.[8]

The second generation of Cartesians refined this physics in the manner most natural to them. Rohault's successor as chief expositor of Descartes' natural

6. Rohault, *Traité* (1692⁶), II, 214–15; *Physica* (1718⁴), 408; *Roh. Syst.* (1735³), II, 186–7; Thorndike, *Hist.*, VII, 689.

7. Denis, *J. sçav.*, 3 (1672–4), 222–33; Oldenburg, *Corresp.*, IV, *passim;* Hoff and Guillemin, *J. Hist. Med.*, 18 (1963), 103–24.

8. Malebranche, letters of 1689, in *Oeuvres*, XVIII, 503–5, 510–19. Cf. Thorndike, *Hist.*, VIII, 495–500; Barrett and Besterman, *Div.-rod* (1926), 17, 27–31; and Lorraine de Vallemont, *Phys. occ.* (1709²), 54–5, 81–5, who offers a Cartesian theory of dowsing and electricity.

philosophy, Pierre Sylvain Régis, perceived that in returning to the electric the fillets must twist aside, 'in such a way that they make a little vortex around the amber similar to that formed around the lodestone by the magnetic matter.'[9] It was clear enough that the electrical matter must participate in a vortex; for, as that excellent pedagogue and mongrel Cartesian Wolferdus Senguerdius reminds us in his textbook discussion of electricity, 'a circle is required to complete every motion.'[10] It remains to explain how the vortex does the trick. Régis' solution was to compromise with Cabeo: the whirlpool of fillets knocks aside the near air; that beyond the vortex, 'acting by its weight,' pushes the straw towards the amber. Another, more orthodox possibility was suggested by a student of Maignan's, François Bayle, Professor of Medicine at the University of Toulouse. In a text devoted to a critical comparison of scholastic and Cartesian physics, Bayle explained that the old doctrine ('which refers the obscure to the more obscure') could teach nothing about electricity, whereas the new brings a flood of light.[11] The fillets necessarily set up a vortex, which drives chaff to an electric exactly as the terrestrial vortex causes gravity. Since the Cartesians had not yet puzzled out the workings of the *grand tourbillon*, it would not be fair to hold Bayle to details.

The electrical vortex was the standard explanation of the amber effect in French physics until the discoveries of the 1730s. The subject was considered closed. In the popular *Entretiens physiques* of the Jesuit Noel Régnault, for example, the wise old physicist Eudoxe, having intitiated his young disciple Ariste into the mysteries of Cartesian magnetism, tests the indoctrination by asking for an explanation of electricity. The conversation is short.[12]

Ariste: A violent agitation, a kind of heat caused by friction, . . . makes a sort of vortex emerge from precious stones, glass, etc., as in the case of the magnet. The vortex passes through chaff more easily than through air, which it compresses and drives aside; but soon the air dilates and, returning victorious, pushes the chaff it encounters . . .
Eudoxe: Voilà bien des mystères philosophiques assez heureusement dévoilés!

2. HUYGENS

Huygens' father, a great friend of Descartes, bred his son on Cartesian physics. 'The novelty of the shapes of his [Descartes'] little particles and of his vortices [was] most agreeable,' Huygens later wrote of his adolescent reading of the *Principia philosophiae*. 'But having since found them to contain some things obviously false and others most improbable, I freed myself from my infatua-

9. Régis, *Système* (1691), III, 510–13; Mouy, *Développement* (1934), 145–67. The Jesuit Pardies seems to have taught a similar theory about 1670. Ziggelaar, *Pardies* (1971), 74.

10. *Phil. nat.* (1685²), 361; the first edition (1681) has no electricity. For Senguerdius' compromise between traditional and Cartesian physics see Ruestow, *Physics* (1973), 78–84.

11. Bayle, *Inst. phys.* (1700), II, 314–17. Cf. Whitmore, *Order* (1967), 184.

12. *Entr. phys.* (1732²), I, 276–7; cf. Regnault, *Origine* (1734), II, 144, 251. Regnault's *Entretiens* was a popular introductory work; cf. Borgeaud, *Histoire*, I (1900), 564–5.

tion.'[13] And yet Huygens' physics never relaxed the fundamental requirement of Descartes', that all phenomena must be referred to the pushes of material particles. As he wrote in 1679: 'Nearly all contemporary philosophers agree that it is only the motion and shape of the corpuscles of which everything is made that produce all the admirable effects that we see in nature.'[14] The task of the Cartesian physicist is to save the usual perplexing phenomena—gravity, electricity, magnetism, etc.—by the mechanics of subtle matter. Such was Huygens' goal from at least as early as 1660–1, when, during his first visit to Paris, he watched with the keenest interest Rohault's operations on the lodestone.[15]

In a memorandum of 1667 Huygens lists gravity, magnetism and elasticity as the most pressing and promising subjects for the vorticist;[16] electricity may have been added in 1672, under the inspiration of Guericke's work. Certainly the sulphur globe fascinated Huygens. He and Jacob Spener, who had been instructed by Guericke himself, spent much of April 27, 1672, making the ball, taking care to melt the sulphur over low heat, for boiled sulphur, according to Spener, has no 'attractive virtue.' All precautions, and the recommended addition of sal ammoniac, were useless: 'The ball was nothing much, and attracted only at very short range.'[17] Huygens did not return seriously to the subject for almost twenty years, when or just after revising his old notes on gravity for the press.[18] Perhaps contemplating the terrestrial vortex led him to the electrical, and to those 'hypotheses for explaining this admirable attraction and its various phenomena' that he hinted at in a letter to Leibniz in November, 1690. In December, still vigorously pursuing subtle matter at the age of 61, Huygens asked Leibniz for additional information about electrics that Guericke might have left. 'I find the effects of amber even more difficult to explain than those of the magnet, especially some new phenomena that I found not long ago by experiment.'[19]

Huygens replaced the effete sulphur globe with a small amber sphere; as detecters he used feathers, chicory seeds, and little flocks of wool.[20] Their behavior was capricious. Sometimes the wool when dropped upon the sphere stuck there; sometimes it was driven away as quickly as it came, to a height of six or eight inches; sometimes it fled the ball before contact. 'This diversity is most noteworthy.' Could it be reduced to rule? Many trials yielded an extraordinary regularity: wool let fall from dry, cold fingers almost invariably adhered

13. Huygens, *Oeuvres*, X, 403.
14. *Ibid.*, XXII, 710.
15. *Ibid.*, III, 210, 215, 252; Mouy, *Développement* (1934), 187.
16. Huygens, *Oeuvres*, XIX, 553; cf. *ibid.*, 632; XVI, 327; XXII, 512, 641–2.
17. *Ibid.*, XIX, 611; Huygens to Leibniz, 23 Feb. 1691, in *Oeuvres*, X, 22; cf. IX, 496.
18. See Mouy, *Développement* (1934), 187–92.
19. Huygens to Leibniz, 19 Dec. and 18 Nov. 1690, in *Oeuvres*, IX, 572, 539. Cf. *ibid.*, XXII, 408–10, 649, 653–4, 756.
20. The following account is drawn from notes dated 1692–3 (*Oeuvres*, XIX, 612–16); there does not seem to be a record of the experiments of 1690.

to the sphere. To test the evident corollary, that the sphere will generally drive away moist bodies it has attracted, Huygens brought up his amber to a silk string hung like a catenary; the string stuck (after attraction) when dry and fled when wet, but always adhered when dipped in oil. A little paper ball, suspended from a silk string and lightly dampened, likewise was repelled after being drawn to the excited sphere.[21] 'Whence this hydrophobia? Guericke did not notice it.' The answer, he thought, involved evaporation, which endows watery surfaces with atmospheres able to abet the *communication of electrical vortices.*

Rubbing an electric causes a 'trepidation' of its surface, which in turn agitates the subtler parts of the surrounding subtle matter. These parts include the luminiferous aether, about which Huygens had just published his classic work; that is plain from Guericke's *virtus lucens,* and from the glow of rubbed diamonds as described by Boyle. The subtlest portion of the aether is not involved, however, for electrical effects, unlike magnetic, can be screened. There is no outpouring of electrical matter, nothing leaves the electric's pores; for if one side of a thin slice of amber is rubbed, the other does not attract. The famous 'electrical vortex' is then an agitated aether set in motion by a suitably vibrating surface. The explanation of Guericke's feather experiments and their variants lies ready to hand. 'Evidently a certain vortex of invisible matter surrounds the flock of wool, which vortex originates in, and is transmitted by that set up about the sphere by the rubbing. The two vortices impede and prohibit the flock's motion towards the sphere without difficulty.'[22] Consequently two flocks of wool, each having acquired a vortex from the sphere, will repel one another without preliminary attraction. As for the accession of a repelled flock or feather to the finger, it occurs because the finger, having no vortex, tends to acquire that of the light object, with which it eventually rests at the center of a common whirlpool. Confirmation of the mechanism may be found in the behavior of water towards which the amber ball is moved. As the ball approaches, the water rises to meet it, perhaps, Huygens conjectured, because of a drop in air pressure between them; but at closer distances the water falls, pushed down by the pressure of the amber's electrical vortex.[23]

Huygens' acute application of the ideas of Descartes to the discoveries of Guericke is unquestionably the high point of seventeenth-century studies of electricity. Guided by the vortex, Huygens recognized the existence of electrostatic repulsion, which had eluded many shrewd investigators armed, or rather disarmed, with the ordinary theory of ejected effluvia. In his picture, repulsion is coordinate with attraction; he grasped the relation between attraction, electrification by communication, and repulsion. Furthermore, he knew that mois-

---

21. This is what happens. Nonmetals will not easily acquire a charge by brushing against an electrified amber sphere unless surface moisture or another contaminant promotes the conduction. Cf. *supra,* iv, n. 30.

22. *Oeuvres,* XIX, 615.

23. The water swells under the usual attraction and falls when a spark passes.

ture promoted communication and that two objects electrified by a third repel one another. Moreover, he associated luminosity and electricity; in his view Guericke's *v. lucens,* as well as *v. expulsiva,* are electrical in nature, like *v. conservativa.* The rapid development of the study of electricity in the 1730s and 1740s started where Huygens stopped. Since he published not a word about his discoveries, they, like Guericke's that had inspired them, had first to be made again.

PART THREE

The Great Discoveries

Toutes les apparences sont que l'électricité et le magnétisme agissent par les écoulements de matière. Voilà
ce qui est dans le royaume de l'impulsion: mais l'empire de l'attraction *non est hinc.*
> —Voltaire to Dortous de Mairan, 11
> Sept. 1738, Voltaire, *Corr.*, VII, 368.

Kunftige Jahrhunderte werden dies, worin wir leben,
bewundern und immer eingestehen, dass keines der
verflossenen an grossen und natürlichen Entdeckungen
so fruchtbar gewesen ist.
> —Ingenhousz, in Wiener, *J. Ingenhousz*
> (1905), 190.

Il est fort ordinaire en Physique, que la théorie ne soit
confirmée par l'expérience.
> —Deluc, *Modifications* (1772), II, 140.

Chapter VIII

Hauksbee and Gray

1. HAUKSBEE'S EXPERIMENTS BEFORE THE ROYAL SOCIETY

On December 5, 1703, the Royal Society enjoyed two maiden performances, the chairmanship of Newton, its new president, and the demonstrations of Francis Hauksbee, who was to become its chief experimentalist. Their simultaneous initiations were related. Newton, who wished to revive the earlier practice of weekly experimental demonstrations in order to recall the Society, grown comfortable and a trifle frivolous, to its proper activities, required a talented operator, a good Newtonian with the ingenuity of a Hooke. He appears to have picked Hauksbee for the post. How the obscure demonstrator, who lamented his lack of a 'Learned Education,' had prepared himself, and how he came to Newton's attention, are unknown.[1]

Since Hauksbee's major task was to prepare demonstrations that a sizable and discriminating audience could easily observe and appreciate, he needed large-scale, striking effects, sturdy equipment, reproducible and dependable results. These constraints were real: when he did not perform to the Society's satisfaction, his earnings probably suffered.[2] Consequently electricity, which he knew to be weak, capricious, and shy in company, did not recommend itself to him on its own account; he came to it by a circuitous route that affected his subsequent researches.

Hauksbee's first performances, commonplace exercises with an improved air pump, ceased in February 1705, when at the Society's request he examined the luminosity of phosphorus.[3] Proceeding in a way natural to him and reminiscent of Boyle, he tried the effect of varying the pressure; his subsequent discovery, that the luminosity increased with the exhaustion, probably inspired his first original experiments, an investigation of the 'mercurial' or 'barometric' phosphorus.

1. Hauksbee, *Phys.-Mech. Exp.* (1719²), 'Epistle'; Guerlac, *Arch. int. hist. sci.*, 16 (1963), 113–28, and in Cohen and Taton, *Aventure* (1964), I, 228–53. Guerlac conjectures that Hauksbee served his apprenticeship under Boyle; by 1704 he was an established instrument maker. Rowbottom, Hug. Soc. Lond., *Proc.*, 21 (1968), 199; E. G. R. Taylor, *Math. Pract. Tudor Eng.* (1954), 285, 296–7.

2. Cf. Hauksbee to RS, 19 March 1706–7, in RS, Sci. Pap., XVIII:1, 106: 'Notwithstanding I have not Made soe many Experiments before ye Society this Yeare as ye Last, Yet my Diligence & Application in ye pursuit of them has been Altogether as Great; For in Experiments of this Nature One must Expect to meet with more Disappointments, than Discoveries. Yet (I thinke) some that I have made this Yeare are not Contemptable ones.'

3. Hauksbee, *PT*, 24 (1704–5), 1865–6.

The glow sometimes visible in the Torricelli space of a jostled mercury barometer had perplexed physicists since its discovery in 1675. When Hauksbee took up its study, Jean Bernoulli's Cartesian explanation—according to which the jostling produced light from the first matter *contained in the mercury*—held the field. The business remained obscure, however, because neither Bernoulli nor anyone else knew how to ensure the appearance of the glow.[4] Hauksbee's first contribution was to render the phenomenon reproducible by eliminating the barometer. He put a pound and a half of quicksilver in a glass basin within an evacuated vessel, opened a stopcock, and watched the inpouring air break the mercury into gleaming droplets. He dribbled mercury onto a glass cylinder placed in a vacuum; it ran down, tearing at the glass in a splendid shower of silver and fire. It was just this brisk, tearing motion, Hauksbee found, that ensured the phosphorus; droplets that did not roll did not glow. The vacuum was not essential. The fire fall ceased only after half the air had been readmitted, and even in the open air shaken mercury might yield detached pinpoints of light. Here were demonstrations worthy of the Society! Hauksbee showed them on at least eight occasions between April and November of 1705.[5]

Having dispensed with the high vacuum and the barometer, Hauksbee proceeded to eliminate the mercury. From November 21, 1705 through the following February, he used an elaborate machine (fig. 8.1) to show the Society that a glimmer always appeared when two objects were chafed together *in vacuo*.[6] The culmination of these experiments, and the prelude to Hauksbee's study of electricity, was a brilliant inversion of his usual arrangement. Heretofore he had applied friction to objects in the medium—vacuum, partial vacuum, or open air—where the light was to appear; now, using the machine depicted in figure 8.2, he rubbed the external surface of an evacuated globe and obtained within it a light intense enough to illuminate a wall ten feet distant and to make legible letters in a nearby book. As the air returned, the glow became ragged and branched until, at atmospheric pressure, it vanished altogether, to appear as tiny sparks outside the globe, on the fingers of anyone who advanced his hand to within an inch of the glass.[7]

This experiment underlined the initial problem of the mercurial phosphorus, namely the provenance of the light; for the globe's glow appeared to originate in the glass, while the phosphorus and the shower of fire seemed to come from the quicksilver, as Bernoulli's theory required. The problem intrigued Hauksbee, who probably understood luminescence in Newtonian terms, as the ejection—by rapid, heat-inspired, internal vibrations of a body—of light corpuscles previously trapped in it.[8] He recognized that his experiments did not

4. Cf. Harvey, *Hist.* (1957), 271–6.
5. JB, X, 102, 105–7, 112–15; Hauksbee, *PT*, 24 (1704–5), 2129–35.
6. JB, X, 115, 117–23; Hauksbee, *PT*, 24 (1704–5), 2165–75.
7. Hauksbee, *PT*, 25 (1706–7), 2277–82.
8. Newton, *Opticks* (1952), Qu. 8–11; Hauksbee, *PT*, 25 (1706), 2277–82.

8.1 Hauksbee's chafing machine: the crank NN and pulley CC turn the small wheel m and thereby the object hh within the cylinder ff resting on the plate of an air pump EE. From Hauksbee, Phys.-Mech. Exp. *(1709).*

indicate unambiguously whether the mercury or the glass donated the corpuscles. (The light in fact comes from molecules of the residual air.) The Society discussed the question on October 30, 1706, and suggested to Hauksbee that he try to obtain a flicker from glass drops broken *in vacuo.* The following week he read his report of the glow in the empty globe, which he still declined to accept as proof that quicksilver contributes no light to the mercurial phosphorus. Even Newton found this caution excessive: the glowing globe 'evinc'd,' he said, 'that Light proceeded from the subtle effluvium of the glass, and not from the gross body' employed in the rubbing.[9] Although unconvinced by Newton, Hauksbee abruptly altered the direction of his researches.

HAUKSBEE'S ELECTRICITY

The emission of light corpuscles *within* an evacuated globe on rubbing it *without* may have reminded Newton of his old demonstration that chafing the top of

---

9. JB, X, 139, 148; cf. RS, Sci. Pap., XVIII:1, 100. For other slips of Hauksbee from Newton's line see Schofield, *Mechanism* (1970), 64–8.

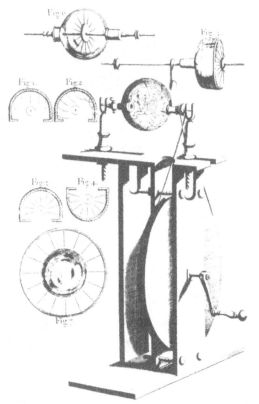

8.2 Hauksbee's prototypical electrical machine together with several deployments of 'Hauksbee's threads.' From Hauksbee, Phys.- Mech. Exp. *(1709).*

a telescope lens caused electrical vapor to pour from its bottom. Perhaps New-ton directed Hauksbee's attention to the parallel and to the unpublished descrip-tion of the experiment. At the meeting of November 13, 1706, Hauksbee pre-sented himself before the Society armed with a hollow tube of fine flint glass, a club thirty inches long and an inch in external diameter, with which to demon-strate the force of electricity.[10]

This weapon, the invention of an exhibitioner concerned to secure grand effects, became the standard electrical generator, more efficient, regular, sturdy and powerful than the amber bits, precious stones and wax sticks earlier in use. Its effects were further enhanced by the leaf-brass with which Hauksbee re-placed the straw, chaff and paper scraps used by his predecessors. When ex-posed to the tube's effluvium the brass bits did a complex dance that Hauksbee described much as Newton had, though with an important difference in lan-guage. He occasionally saw pieces of brass violently driven off the perspiring glass, 'repelled,' as he put it, 'to distances of four or five inches.'[11] Sometimes

10. JB, X, 149, under 'the magnetic [!] quality of a glass cylinder.'

11. Hauksbee, *PT*, 25 (1706–7), 2327–31, probably performed on 13 Nov. and 11 Dec. 1706 (JB, X, 146, 149). Although Hauksbee never acknowledged earlier electricians, he consulted them, for he spoke familiarly of electrical effluvia and gave the usual explanation of their inhibition by

the repulsion exceeded the attraction; replacing leaf-brass with lampblack, he found that particles unable by their own gravity to make a noise when dropped on paper might sound when returned by the tube. Did Hauksbee's strong electric and sensitive detectors at last bring into focus the repulsion so many had looked at and none but Huygens had seen? Certainly motions from the tube figure more conspicuously in Hauksbee's account than in those of Cabeo or Browne.[12] Yet, as will appear, Hauksbee discovered repulsion only in a Pickwickian sense.

For the rest, Hauksbee's initial essay into electricity reports the puzzling results of a characteristic enquiry, an examination of the tube's luminosity and electricity as functions of its *internal* pressure. Towards light, he found, it behaved as the machine-spun globe, glowing brightly within when exhausted and flickering weakly without when filled. But its electricity—and here was the puzzle—acted oppositely to its luminosity; the tube, which worked powerfully enough when filled, proved impotent when void of air. (Hauksbee's 'vacuum' furnished enough ions effectively to neutralize the triboelectric charge on the tube's outer surface.)

These experiments having pleased the Society, Hauksbee continued them in a direction suggested by the 'electric spiderweb,' the suppositious source of the tickling experienced on presenting the forehead or the back of the hand to the tube.[13] (The tube's *pull* on the experimenter's own hairs causes the sensation.) Hauksbee inferred that the effluvia, which his new generator had enabled him to touch, stood or flowed fixedly in a manner he might hope to determine. He made a glass cylinder seven inches in height and diameter and surrounded it with a wooden hoop supporting threads at equal intervals; when he spun and rubbed the cylinder, the threads extended themselves radially towards an axial point determined by the location of his hand (fig. 8.2). He did not miss the evident analogy to the grand fact of the Newtonian universe: the threads 'seem to Gravitate,' he wrote, 'or were attracted in a direct line to the Center,' a static symmetry more instructive than the confused motions of Newton's paper bits.

At one point, however, the analogy failed utterly. When the 'Attracting or Electrical Effluvia' (an awkward phrase for those who think Hauksbee discovered electrical repulsion) held the threads and he approached his finger to their pointing ends, they *shrank from its touch*.[14] Seeking further information, he placed the threads within the glass, supposing that friction would agitate the cylinder's electrical vapor as it had the globe's light corpuscles. The internal threads arranged themselves like the spokes of a wheel, as he had perhaps

moisture and screens. He may not have known the literature well, however, as he reported spurious results relating to the action of electricity on smoke, flame and liquids, which reference to the *Saggi* would have corrected.

12. Indeed, some later writers took Hauksbee's noisy lampblack to show that repulsion is stronger than attraction. Taglini, *Lettere* (1747), 142.

13. Hauksbee, *Phys.-Mech. Exp.* (1719²), 58.

14. Hauksbee, *PT*, 25 (1706–7), 2332–5; JB, X, 151–3 (Feb. and March, 1706–7).

anticipated; but they reacted to the motion of his hand even more oddly than the circumscribed ones:

> Now when this Experiment was made by hanging the Threads near the outside, it was very odd (as before related) to see them fly the Approach of a Finger; yet how much more surprizing is it to see the same perform'd even when a Body so solid as Glass interposes; which shews the subtility of the Effluvia, the Body *from which* it [!] is produced seeming to be no Impediment to its motion; Besides it seems very much to resemble or emulate a Solid, since Motion may be given to a Body, by pushing the Effluvia at some distance from it: But what is still more strange is, That this Body (I presume to call it so) altho' so subtil as seemingly to perviate Glass, will not . . . affect a light Body thro' a piece of Muslin: Now whether the Muslin absorbs the Effluvium, or what other Laws it may be subject to, I cannot tell, but sure I am 'tis very amazing.[15]

Hauksbee's explorations yielded a novel and eccentric picture: the effluvia do not rush about in a wind or vortex, but stand in continuous, stiff, glass-piercing chains, rigid enough to support light objects and, as the threads' flights indicated, to remain intact when pushed aside by the finger.[16] He intended this stiffness literally, and troubled to confirm it. He mounted two globes, one inside the other, on a special spinning machine. The inner vessel, which he had evacuated, glowed when he turned and rubbed the outer; an ingenious experiment, which he took to show that the effluvia from the external globe could 'act the Part of a Solid Body' in generating light by friction. (One can obtain a similar effect by rubbing a hollow tube enclosing a fluorescent one.) Hauksbee exhibited this striking phenomenon several times, lastly in April 1707, and then put the subject aside.[17]

GRAY'S INTERVENTION

On January 3, 1708, Stephen Gray, a dyer by profession and an occasional contributor of astronomical and microscopical notes to the *Philosophical Transactions*, sent the Society's secretary, Sir Hans Sloane, an account of novel experiments on light and electricity.[18] Gray's interest was awakened by the 'Strongness of the phenomena together with the facility of operation' of the portentous glass tube; his work began where Hauksbee's had stopped, with mapping the course of the 'Luminous and Electric Effluvium.' Lacking Hauks-

15. *PT*, 25 (1706–7), 2374. The rubbed glass induces charges of the same sign on the finger and the threads, which therefore repel one another.

16. Coincidentally Hooke (*Post. Works*, [1705], 183) also referred electricity to an 'Aether . . . resembled to a solid.' Hooke defines an electric as a body whose internal parts possess a peculiar facility of vibration, and offers the following 'mechanical analogy' to its action: a hammer (the vibrating internal parts) strikes a horizontal, spring-supported rod (the aether) bearing a sliding weight (the attracted body), which is driven toward the source of the disturbance. Hooke also gave a fluid analogy; both were perhaps exhibited as early as 1671. Cf. Waller, 'Life of Dr. Hooke,' *ibid.*, xiv-xv.

17. Hauksbee, *PT*, 25 (1706–7), 2413–15; JB, X, 156, under April 30.

18. Sloane Mss. 4041, f. 83; full text in Chipman, *Isis*, 45 (1954), 33–40.

bee's expensive apparatus, Gray explored the sensible range of electrical action using a down feather. With it he rediscovered the effects of Guericke's expulsive virtue: once drawn to the tube the feather would go to any neighboring solid body, and thence return to the tube, vibrating ten or fifteen times before coming to rest; or, after its first 'reflection' from the glass, it might hover in mid-air, or climb and drop in time to the rubbing of the tube. Effluvia flowed in paths more intricate than Hauksbee's radially directed threads suggested.

Gray initially ascribed the feather's behavior to a mixture of flows, one direct from the tube, the other reflected from neighboring objects. The adherence of a hovering feather to a large object placed between it and the tube caused him to reconsider. (The object cut off the flow, which, by hypothesis, should ricochet from more remote bodies and drive the feather downwards.) A second hypothesis made all objects surrounding the tube themselves *emitters* of effluvia: 'As all bodies Emitt soe they Receive part of the Effluvia of all other bodies that Inviron them . . . the attraction [being] made according to the current of these Effluvia.' Although, as Gray observed, it is difficult to conceive how rubbing a glass stick can electrify distant objects, his picture, which in effect replaces Cabeo's air current with effluvia, has much to recommend it, as it neatly and even quantifiably represents induction effects.[19]

Gray discovered an arc discharge that confirmed the hypothetical return flow. Hauksbee had occasionally spoken as if the light appearing on a finger approached to a rubbed, rotating globe *began* at the hand: Gray now found the glow to be conically shaped, with its vertex at the finger. The obvious inference, that the light proceeded from the hand, agreed with the sound accompanying the discharge, which Hauksbee also had noticed: according to Gray, the noise originated in 'the strikeing of the Glass by the Effluvia in the Rapit motion from the finger.' He did not distinguish between luminous and electrical effluvia. He thought Hauksbee had identified the two in the case of glass, and he discovered that light was 'inherent in the Effluvia' of other electrics, specifically in those of sulphur, amber and sealing wax.

The Society directed Hauksbee to try what was new in Gray's experiments. Three weeks later, on January 28, he read a paper, apparently not extant, 'giving his Thoughts of Mr. Gray's Letter and Experiments,' and on February 4 he showed the Society the experiments it had not yet seen. He also reaffirmed the theory of abraded glass effluvia with 'stiff and continu'd Parts' and supported it with an account of the antics of the hunted feather.[20] The affair does not display Hauksbee to advantage. The experiment was not his and did not confirm his principles, for the stiff and continued effluvia did not explain the feather's oscillations.

The story has another round. On June 2, 1708, Gray was permitted to attend

19. Whittaker, *Hist.* (1951²), I, 243, 279–80.

20. Hauksbee, *PT*, 26 (1708–9), 82–6; JB, X, 175, and RS, Sci. Pap., XVIII:1, 113, under 4 Feb. 1708–9. The first confirmation showed that a sand-filled tube did not draw as well as a normal one, suggesting that external attraction required continuity of the effluvia across the glass.

a meeting of the Society at which Dr. Samuel Wall, a physician of doubtful pedigree, a mail-order druggist, a 'promoter of the spagyric art,' showed electrical experiments.[21] He had been much interested in an artificial phosphorus made from excrement. In his fastidious chemistry, an 'oleosum' common to urine and dung made the glow; and since he took amber to be a 'mineral oleosum,' he inferred that it too might phosphoresce. He obtained a glow from polished amber rubbed in the dark, and crackly sounds, light and something sensible to his finger from an amber taper drawn briskly across a piece of wool. All of which resembled thunder and lightning, and succeeded best when the sun was 18° below the horizon. Wall moved on to diamond, 'from its being electrical as well as the [amber],' and thence to gum-lac; both proved luminescent, 'well assur[ing]' him that it was just the light complement of electrics that rendered them electrical. 'Coy Nature which disdaines to bee embrac'd / By ev'ry love pretending Swaine / You have in all her dark recesses trac'd / And did her by industrious courtship gaine.'[22]

Neither Wall nor Gray was a Fellow of the Royal Society. Neither's paper need have been printed. In its wisdom—or bias—the Society chose to publish the derivative piece of the questionable doctor rather than the prior and original contribution of the country tradesman. It did not admit Gray until 1730. As for Gray's paper, Hauksbee appropriated what he found valuable in it and suppressed or glossed over what conflicted with his own ideas. He not only stole and misrepresented the experiment with the feather, but his last work on electricity of importance (June 1708)—testing the luminous and electrical qualities of the effluvia of wax, sulphur, and resin—exploited without acknowledgement hints first given by Gray.[23]

Here justice intervened. In investigating these heterogeneous effluvia Hauksbee ran upon a fact so disconcerting that it ended his investigation of electricity. He brought a piece of rubbed sealing wax to the outside of an unexcited globe containing some strings and saw them move. 'Very amasing,' he says, 'very astonishing.' Apparently glass passed not only its own effluvia, but also those of sealing wax. The special argument with which he had explained the differential opacity of glass and muslin to vitreous effluvia fell to the ground. And if the subtle emanation from wax could pass glass, why did it fail to traverse the thinnest cloth barriers? Here Hauksbee abandoned the unequal struggle.[24]

21. JB, X, 189; Sloane Mss. 1731A, ff. 21–2, 27–8, 47; 'fautor spagyricae artis' (f. 37). A birth chart (*ibid.*, f. 26) dated 9 Jan. 1652/3, may be Wall's; he was dealing in healing powders in 1680 (*ibid.*, f. 47).

22. Wall, *PT*, 26 (1708–9), 69–76; John Whitehall, 'To Dr. Wall,' Sloane Mss. 1731A, f. 154.

23. Hauksbee, *PT*, 26 (1708–9), 87–92; JB, X, 190, 192. Regarding the RS and tradesmen see Millburn, *Martin* (1976), 35–8, 108. Wall was proposed FRS (JB, X, 192) but died before election.

24. Hauksbee invented one more electrical experiment. He coated the *inner* surface of a glass globe with wax, resin or sulphur, leaving a portion free to permit observation, and saw the outlines of his hand in the *concave* surface of the wax, etc., when he excited the globe. The effect being mysterious and entertaining he repeated it at least three times between June, 1708 and the following

2. EARLY NEWTONIAN ELECTRICITY

In 1709 Hauksbee brought out a book that appears to be a reprinting of his papers, in their original order, with all repetitions and inconsistencies. In fact, although the *Physico-Mechanical Experiments* could not have cost its author many pains, it does differ from his earlier accounts in two significant respects: by many verbal changes that conflict with the unaltered text, and by an Appendix that flatly contradicts the theory developed in the book. These additions instance the practical difficulties of maintaining Newton's methodological distinctions and they help answer the apparently frivolous question whether, or in what sense, Hauksbee discovered that elusive oxymoron, electric repulsion.

HAUKSBEE AND REPULSION

In Newtonian parlance, 'force' can be meant either descriptively, as in the accelerations of the *Principia*, or physically, as in the innate sociabilities of the *Opticks*. In revising for the *Physico-Mechanical Experiments*, Hauksbee worked descriptive forces into accounts that otherwise used the language of effluvia. No contradiction need have resulted; an alternative to effluvial theory would be one involving physical, not descriptive forces. In practice, however, one sees the phenomena in accordance with the language used. In his initial descriptions, Hauksbee emphasized the irregularity of the motions of the test bodies: they flew about fitfully, 'promiscuously Ascending and Descending,' 'repell'd' to be sure, but also 'thrown,' 'return'd,' 'suspended,' or fixed to the glass, whipped by an aether wind; while in the revised version, though the motions still afford variety, 'the *Attractive* and *Repulsive* Forces (whatever they are)' now exert themselves 'as it were by turns, the one drawing up, the other beating down the light Bodies,' which 'often repeat this alternate *Rising* and *Falling*.' The hovering feather is now 'repell'd' by neighboring solid bodies, rather than 'return'd'; and the thread experiments provide 'a plain Instance of a *Repulsive* and *Attractive* Force,' indeed of centrifugal and centripetal forces, 'so that in these smaller Orbs of Matter, we have some little resemblances of the Grand Phaenomena of the Universe.'[25]

Elsewhere the effluvia go their way as before. They are attractive or electric, never repellent; electricity is an 'attractive power' and 'repulsion' an occasional consequence of vigorously projected effluvia. They remain stiff and continuous, like '*Physical Lines* or *Rays*,' whose parts 'adhere and joyn to one-another, in such manner, that when any of 'em are push'd, all in the same Line are affected by that Impulse given to others.' They still produce light by attrition, not as before because of their stiffness, but because of their momentum; Hauksbee likens them to particles of a subtle fluid, probably to bring them into line with

---

March (RS, Sci. Pap., XVIII:1, 115, 123–5). Accounts appeared in *PT*, 26 (1708–9), 219, 391, 435, and *Phys.-Mech. Exp.* (1719²), 268–78, without explanation.

25. Hauksbee, *Phys.-Mech. Exp.* (1719²), 54–5, 67, 74–5, 143, 154–5.

the theory given in the Appendix.[26] When he employs descriptive forces he sees repulsion as coequal with attraction, and both as regular causes of irregular effects; when he talks of effluvia, the uniformity and equality vanish and the aether wind blows as it lists.

The Appendix drops forces and makes the 'disorderly, fluctuating and irregular' motion of effluvia the ultimate agent of electricity. But not the sole agent. Hauksbee now revives Cabeo's theory, without acknowledgment. Since the hollow tube will not draw when evacuated, he infers it needs air inside it; and since he cannot excite electricity *in vacuo,* he concludes that external contact with the atmosphere also is required. (Hauksbee's vacuum apparently conducted better than Boyle's.) The tube needs air within either to assist in launching the effluvia or to prevent them from retreating; external air is required 'to carry the little Bodies (which we say are attracted) to the Tube.' The atmosphere advances when layers of heavier air, remote from the tube, overcome nearer ones, which the heat of the rubbing has rarefied. The mechanism explains every apparent caprice. Irregularities in the rubbing produce disordered effluvia. These heat and beat the air fitfully, which 'may be sufficient to account for the various uncertain motions of the little Bodies carried towards the Tube.'[27]

What then causes the regular orientation in the thread experiments? In the external case, the uniform rotation could produce a continuous flow of radially coursing air; a similar explanation could not hold for the internal case, however, for the source of denser air, the middle regions of the cylinder, would soon be exhausted. The same objection applies to the attraction of bodies across the walls of sealed vessels. Still, Hauksbee's form of the air theory had its uses. It would seem a natural representation of his force description of electrical phenomena: the issuing effluvia cause repulsion, the returning air attraction.[28] He did not exploit this possibility, choosing rather to emphasize the irregularity of the motions and of the wind that drove them. Did he recognize repulsion?

From the outset Hauksbee had doubted the identity of luminous and electrical effluvia, partly because of reservations about the emitter of the mercurial phosphorus and partly because the tube would glow but not attract when exhausted. The new doctrine severed all connection: light might appear when either the internal or the external surface bounded a vacuum; Hauksbee's air theory required atmospheric exposure of both to produce electricity. He ended his im-

26. *Ibid.,* 59, 64, 81, 87, 141–2, 145, 152.

27. *Ibid.,* ?41–7. This is not good Newtonian physics. If heat expanded the contiguous air, a convection current would set in; rising hot air would carry chaff *from* the tube, an answering breeze would bring new bits to it.

28. This would approximate Gray's theory with the puzzling secondary flow transformed into an air current. Gray's ideas influenced Hauksbee's Appendix, which opens (*ibid.,* 238–40) with a seemingly gratuitous attack against the notion that the effluvia come not from the electric, but from surrounding bodies. Cf. R. Home, *Arch. Hist. Ex. Sci.,* 4 (1967–8), 203–17, and *Effl. Theory* (1967), 30, 37–8.

posing work on electricity by rejecting the conjecture that had prompted him to undertake it.[29]

NEWTON AND THE AETHER

Newton continued to think that Hauksbee's experiments had uncovered a fundamental connection between the two grand classes of Newtonian phenomena, the optical and the attractive. At least two ways of developing the connection stood open to him: he could adapt the traditional corpuscular approach to electricity, or he could try to recast electrical along the lines of gravitational theory. The second alternative, which proved the more fruitful, fit with the world picture sketched in the queries added to the *Optice* (1706), where physical, short-range attractive and repulsive forces direct the traffic between particles of matter and of light. Nonetheless Newton chose the first approach: he saw in the connection of light and electricity a powerful argument for reviving and unifying the mechanisms he had flirted with thirty years earlier. This old attachment, objections to his occult qualities, traditional electrical theory and, not least, Hauksbee's electrical spider web, which made effluvia palpable, all helped drive him to this retrograde step.[30]

The first public intimation of his new viewpoint, the General Scholium of the revised *Principia* (1713), introduces a subtle, all-pervasive spirit, by whose 'force and action' material particles 'attract one another at near distances,' light is 'emitted, reflected, refracted [and] inflected,' animals sense and move and electrics operate, 'as well repelling as attracting the neighboring corpuscles.' No hint of the nature or mechanism of this spirit is intended; Newton does not mean to endow it with physical forces by which it attracts *and* repels light bodies.

Several drafts of the Scholium have survived, which show that he then identified the spirit with the electric vapor or effluvium. One draft asserts that 'attraction' between particles, the force of cohesion and capillarity, is 'of the electric kind,' and of very short range, as 'attraction without friction extends only to small distances and at greater distances particles repel one another.' 'Attraction' and 'repulsion' here again are meant descriptively, not physically; if the interparticulate forces are electric, their agent must be an effluvium or aether, not forces innate to the corpuscles.[31] The draft then describes the 'elec-

29. Cohen, *FN*, 35, suggests that Hauksbee could have resolved the question of the electrical nature of the mercurial phosphorus 'once and for all . . . [by] a simple *experimentum crucis*.' In the proposed experiment threads suspended near the top of a shaken barometer approach it when the glow appears. But such an experiment shows merely that in one case (and Hauksbee supplied others) electricity and light can be excited simultaneously, not that their natures are the same. One could as well identify the sound and heat developed in driving a nail.

30. Cf. Newton, *Opticks* (1952), 341; Guerlac, RS, *Not. Rec.*, 22 (1967), 46; Hawes, *ibid.*, 23 (1968), 200–12; Dukov, Akad. nauk, Inst. ist. est. i tekh., *Vop.*, 7 (1959), 120–7.

31. Hall and Hall, *Unpub. Sci. Pap.* (1962), 336–7. Hawes, *Ann. Sci.*, 24 (1968), 121–30, fails to make this distinction and so arrives at an 'electric' innate short-range particulate attraction unintelligibly distinct from the 'spirit' of electricity.

tric spirit,' a 'most subtle medium,' highly penetrating, active and vibratory, the vehicle of animal sensation and nutrition, the agent of chemical combination and dissociation, and the cause of the emission, reflection, refraction and inflection of light. There is still no indication how the spirit performs its duties and no reference to the usual phenomena of electricity.

A second draft provides a clue: 'As the system of the sun, planets and comets is set in motion by the forces of gravity and its parts persist in their motions, so smaller systems of bodies seem to be driven by other forces and their particles to be moved diversely, and especially by the electric force. For the particles of most bodies seem furnished with an electric force, and to act upon one another at small distances even without friction, and those which are most electric through friction emit to great distances a certain spirit by which they sometimes attract, sometimes repel, and sometimes agitate light bodies diversely.'[32] It appears that the electric spirit when quiescent mediates interparticulate attractions and, when driven from electrics by friction, causes electrical motions by an aether blast.

A sketch for an unpublished query originally intended for the *Opticks* of 1717–8 confirms this interpretation. Here, to make an electric cohesive force plausible, Newton argues that the 'virtue' of electrics might not be generated, but only expanded by rubbing, 'for the particles of all bodies may abound [!] with an electric spirit wch reaches not to any sensible distance from the particles unless agitated by friction.' (Newton usually thought of particles as composites, which may help to explain how they can 'abound' with spirit.) Another draft query asks, 'Do not all bodies therefore abound with a very subtile, but active, potent, electric spirit by wch light is emitted, refracted, & reflected, electric attractions & fugations are performed, & the small particles of bodies cohere when contiguous, agitate one another at small distances & regulate almost all their motions amongst themselves?'[33]

The operations of the electrical spirit are clear enough. At close distances it causes cohesion; at small distances it may 'agitate,' drive off or repel other particles, perhaps by its vibrations, but not through repulsive force; at large distances it blows light corpuscles about, as when leaf gold exposed to a 'cilynder of glass newly rubbed' is 'agitated and carried about by the Etherial Effluvium as with a wind.'[34] No more than Hauksbee did Newton fully recognize electrical repulsion. In the end his view of electricity remained what it had been in 1675.

The printed *Opticks* of 1717–8 shows little trace of Newton's speculations about electricity.[35] What survives is the optical-gravitational aether, which

32. Hall and Hall, *Unpub. Sci. Pap.* (1962), 350–4; the translation in the text differs somewhat from the Halls'.

33. Quoted by McGuire, *Ambix*, 15 (1968), 175–6.

34. A draft for the *Opticks* of 1717 / 8, quoted by Guerlac, RS, *Not. Rec.*, 22 (1967), 48.

35. The most indelible are in *Opticks* (1952), 341 (Qu. 8), where the electric vapor rushes from Hauksbee's globe, strikes a piece of paper with great force, and agitates itself enough to emit light

works by pressure and is nowhere identified with the electrical effluvium. Indeed the media are incompatible. As the agent of cohesion, the electrical spirit should be present in bodies in proportion to their densities, while the optical-gravitational aether must stand rarer in denser bodies. Even the task of representing short-range chemical repulsions and cohesive attractions by the same mechanism must ultimately have appeared hopeless to Newton. Electricity itself offered no guide to the hidden operations of the electric spirit; electrical phenomena did not even illustrate them, for the operations required a mechanism, and the electrical motions only a wind. No wonder Newton decided not to electrify the *Opticks*.[36] The book nonetheless has played a part in the history of electricity. While some continental physicists owed it their first acquaintance with Hauksbee's globe, certain English philosophers, approaching their master more closely than they realized, found in its aether a misleading guide to the hidden mechanics of electricity.

EARLY TEXTS

All four English versions of 'sGravesande's text have the same account of electricity, placed at the head of a section on fire, for 'sGravesande associated fire and light, and connected both with electricity via Hauksbee's observations.[37] He interpreted the thread experiments and the impotence of evacuated tubes in accordance with Hauksbee's latest ideas. But in explaining the motions of the leaf-brass under the tube, 'sGravesande introduced a new and influential concept, a modification of the effluvial wind, namely 'a certain Atmosphere,' contained in and about glass, set into vibration by friction, and maintained by the elastic response of the tube's parts to the shock of attrition.

Although a vibrating atmosphere implies greater regularity than the facts warrant, it has the advantage of placing motions away from the tube on a par with those toward it; for 'sGravesande, following and oversimplifying Hauksbee, repulsion must occur as often as attraction. Accordingly, he breaks with the seventeenth century by amplifying the definition of electricity to include repulsion: 'Electricity is that Property of Bodies by which (when they are heated by Attrition) they attract, and repel lighter Bodies at a sensible distance.'[38] The transformation is noteworthy. Like Hauksbee, 'sGravesande uses

---

(from the paper?); and *ibid.*, 376 (Qu. 31), where, as an aside, Newton conjectures that a short-range unexcited electrical attraction may exist.

36. Hawes, *Ann. Sci.*, 24 (1968), 124–30, suggests, doubtless correctly, that Newton would not have approved the addition of 'electrical and elastic' to the General Scholium's 'certain spirit' in the English *Principia* (1729); not, however, because he never seriously inclined towards such a spirit, but because he had discarded it in 1718.

37. *Math. Elem.* (1721), II, 2–13; 'sGravesande's account remained adequate through the fourth edition (1731), but was antiquated by the fifth (1737).

38. Cf. 'Various Experiments concerning Electrical Attraction, and Repulsion' in the *Exper. Course* (1726) of Isaac Greenwood, one-time student and lodger of Desaguliers, who became the first professor of natural philosophy at Harvard. Cohen, *Tools* (1950), 31–6; Lemay, *Kinnersley* (1964), 48–9.

'attraction' and 'repulsion' descriptively, the physical agent being the pulsating atmosphere. Unlike Hauksbee, however, he replaces the aether blast with a constant throb, which, though inadequate to save the phenomena, yet brings him closer to recognizing the underlying electrostatic regularities.

'sGravesande's model also posed implicitly what became a severe difficulty for effluvialists. The older theories ascribed a progressive motion to effluvia; Hauksbee also employed static ones, rigidly interconnected and firmly attached to the glass tube. All these pictures invoked common macroscopic mechanical effects. Not so 'sGravesande, whose pulsating atmospheres have no evident reason for remaining in contact with their fountains. Does the resistance of the air prevent their evaporation? Or do they adhere by macroscopic attractions at a distance? 'sGravesande does not confront the choice. Later electricians found neither alternative viable.[39]

Musschenbroek's earliest account of electricity (1731), the estimable notes to his Latin translation of the *Saggi,* is fuller, more circumspect and less coherent than 'sGravesande's.[40] The notes update the *Saggi* with the results of Boyle and Hauksbee, without offering a theory; if Musschenbroek inclined to one it was probably 'sGravesande's, which he later entertained favorably.[41] Two curious points emerge from the notes. To resolve the discrepancy between the results of Boyle and Hauksbee, who respectively did and did not succeed in generating electricity *in vacuo,* Musschenbroek hit on an unprecedented and prescient idea; remarking that Boyle used amber and Hauksbee glass, he concluded that these substances have different electricities. (This is not the discovery of the two electricities, but an arbitrary solution to an isolated difficulty.) The second point concerns repulsion. Musschenbroek records without emphasizing Hauksbee's observation that light bodies are sometimes repelled farther than they are attracted; he squeezes it into a note that also gives an excellent prescription for rubbing the tube—always end with your hands together—and a warning about atmospheric humidity. He does not recognize that the *Saggi* require updating about repulsion, and concentrates on the perplexing behavior of screens, the opacity of muslin and the transparency of glass.[42]

Neither Hauksbee nor his Dutch interpreters succeeded in modifying traditional theory to accommodate his results. Such an accommodation would in any event have had a short life. It could not have survived the work of Gray, who, in 1729, made the most important electrical find since Gilbert's.

GRAY'S EARLY CAREER

Stephen Gray, son of a dyer and brother to a dyer, carpenter and grocer, was born in Canterbury in 1666. Although he followed his father's trade, he had a

---

39. Cf. Home, *Br. J. Hist. Sci.*, 6 (1972), 136–40.

40. Musschenbroek, *Tentamina* (1731), I, 70; II, 82, 86–7, 91. The Italian editions of the *Saggi* had then become scarce and expensive. Targioni-Tozzetti, *Notizie* (1780), I, 459.

41. *Elements* (1744), I, 187, 196–7, emphasizing the 'wonderful fits' of the effluvia to and from the electrics. 'Fit' is Newtonian jargon for periodic behavior; see *Opticks* (1952), 281–3, 347–8.

42. Musschenbroek, *Tentamina,* in *Coll. acad.*, I, 30, 154–6.

fair education, including enough Latin to puzzle out Scheiner's *Rosa ursina* (1630) when he began to observe sunspots. He may have studied in London or Greenwich for a time, perhaps under his 'Learned Friend,' the Astronomer Royal John Flamsteed.[43] Gray was a devoted amateur; he invested 'the far Greatest Part of my time that the avocations for a Subsistence would permitt me' in pursuit of science and pseudo-science, optics, astronomy, electricity, and the Canterbury ghost.[44] From the beginning Gray's scientific work was characterized by simple experiments, 'for the most part Naturall, being ushered in with very little assistance of Art;'[45] by alertness to unanticipated effects; by cautious explanation of anomalies; and, especially in his later astronomical observations, by a quest for accuracy that won him the regard of the scrupulous Flamsteed.

In 1707, on the strength of his observations, Gray was brought to Cambridge by Roger Cotes, who planned to set up an observatory at Trinity College. Cotes might have expected Gray also to help with demonstrations for his lectures on hydrostatics and pneumatics, which were given at the observatory in collaboration with Whiston. Electricity and the mercurial phosphorus, as displayed by Hauksbee, figured among 'the more hidden properties of the air.'[46] Cotes may have drawn Gray's attention to Hauksbee, or Gray his; it was at Cambridge that Gray began to study and demonstrate electricity.[47] Otherwise he neither liked nor profited from the association with Cotes. The professor, supported by Newton, hoped to make a catalogue of the fixed stars, a grandiose project Gray thought both silly and insulting, for Flamsteed had devoted much of his life to such a catalogue and the Cambridge observatory as yet had few of the necessary instruments.[48] Our innocent from Canterbury also found his employers too 'Mercenary.' He resigned after little more than a year and was back in Canterbury by September, 1708. But his old trade had become too strenuous for him. In July, 1711 he petitioned Sloane to intercede for him with the Governors of Sutton's Hospital.

43. Cowper, *Roll* (1903), 39; *DSB*, V, 515–17. Besides the association with Flamsteed, Gray's friendship with Henry Hunt of Gresham College, a minor functionary of the Royal Society who sent Gray the *Phil. Trans.* (Gray to Hunt, RS, Guard Book G.1, f. 50), suggests a stay in London. If this is the Henry Hunt made free of the Dyers' Company of London on July 8, 1687 (Dyers' Co., Register, Guildhall, London), Gray's trade connections may have brought him to science. Cf. Chipman, *Isis*, 49 (1958), 429.

44. A Mrs. Veal, no whit inconvenienced by her death, dropped in on her old friend Mrs. Bargrave, who advertised the apparition (1705); it aroused great interest and inspired a story by Defoe. For Gray's part see Higenbottam, *Arch. Cant.*, 73 (1959), 154–66.

45. Gray to Sloane, 31 July 1711 and 4 May 1700 (Chipman, *op. cit.*, 419, and RS, Guard Book G.1, f. 54, resp.).

46. Cotes, *Hydr. Pneum. Lect.* (1775³), 'Heads of the Course.' It is not known when electricity first entered the lectures. Cotes thought the subject too obscure even for speculation (*ibid.*, 153).

47. See William Stukeley's reminiscences of Gray's electrical demonstrations at Cambridge quoted by Cohen, *Isis*, 45 (1954), 43, 46.

48. The Observatory did not succeed. During Cotes' time it lacked money; after being completed with College fees derived from doctor's degrees (some £20 each) it fell into desuetude, and was pulled down in 1797. Edleston, *Corresp.* (1850), 198–201; Price, *Ann. Sci.*, 8 (1952), 2–3.

This foundation, known as the Charterhouse because it occupies the site of a Carthusian monastery, was established in 1611 as a day school for poor boys and a home for eighty gentlemen pensioners. Many of these 'brothers' were distinguished men. Gray felt he had a claim on admission, for he told Sloane that he had undertaken his astronomical studies 'not altogether without hopes that some greater advantage might at one time or another attend [them] then barely the satisfaction of my inclinations.' As an inmate of the Charterhouse he would have time to pursue his 'inquieries Relating to Astronomy and navigation and might hapyly find out something that might be of use.' The improvement of navigation then being a matter of national policy, the Governors could consult both charity and patriotism in admitting Gray, which they did in June 1719, on the nomination of the Prince of Wales.[49]

Shortly after his admission to the Charterhouse Gray showed the Royal Society that hair, feathers, silk, paper, and gilded ox guts are electric. The paper describing these revelations[50]—which were not without interest, as implying a new class of non-rigid or semi-solid electrics—and a letter of 1706 forwarded by Flamsteed were the only communications by Gray printed during Newton's presidency of the Royal Society. The coincidence of Gray's silence and Newton's tenure has prompted the suggestion that Newton, who disliked Flamsteed, deliberately muzzled his protégé.[51] In fact, of the six letters supposedly stifled, Hauksbee pilfered one, two contain no scientific news, and the others, which concern sunspots, did not appear because the earliest of them complained about Sloane's editorial liberties. 'I was not Extraordinarily well Pleased to see [in the *Transactions*] my account of spots in the sun observed by me in june and july last, that being but a mean Peace of work.'[52] Gray lacked the modern savant's eagerness for print and Sloane did not repeat his indiscretion.

Perhaps Gray's sojourn at Cambridge resulted in some coldness between himself and Newton's clique; yet he attended several meetings of the Royal Society during Newton's time,[53] and he chose its secretary to assist his entry into the Charterhouse. The Newtonians in the Society tried to help him. 'I am very much obliged to you,' writes Sloane's successor, Brook Taylor, to Keill, 'for the great readiness, you are pleased to shew, to assist Stephen Gray upon my account. He is a very fit person for the service of the R:S: wherefore I thought to have recommended him very heartily; but the poor man is so very bashful, that I can by no means prevail upon him to think of that business now it seems to be so near by the death of Hunt, he has such dreadful apprehensions of the presence of so many virtuosos.'[54] Gray could not bring himself to con-

49. Chipman, *Isis*, 49 (1958), 419, prints Gray's appeal to Sloane; cf. Courtney, *Not. Qu.*, 6 (1906), 161–3, 354.
50. Gray, *PT*, 31 (1720–1), 104–7.
51. Chipman, *Isis*, 49 (1958), 428–9.
52. Gray to Sloane, 3 April 1704, in RS, Guard Book G.1, f. 56. Sloane had not published all of Gray's earlier letters, for example two of seven between May 1700 and January 1702.
53. JB, X, 189; XII, 63–4.
54. Taylor to Keill, 3 July 1713, in RS, Corr., LXXXII, 5.

front the wits and remained in Canterbury, perhaps until his admission to the Charterhouse in 1719.[55]

No conspiracy foiled Gray. There was no reason to think that he would accomplish anything extraordinary, certainly not in electricity; in his own estimate, his best chance lay in astronomy and navigation, studies in which he had collaborated with the Astronomer Royal. His first publication on electricity, the routine piece on feathers and ox guts, gave no cause to doubt his assessment.

3. ELECTRIFICATION BY COMMUNICATION

In February, 1729, having conceded defeat in an attempt to electrify metals by heat, friction and percussion, Gray thought to awaken them by exposure to the effluvia of a glass tube. He reasoned that just as the tube 'communicated a Light' to bodies, it might also, under suitable circumstances, 'at the same Time communicate an Electricity.'[56] This was the last and least step in his thinking. The idea that the tube's emanation might stimulate a flow of effluvia from bodies it struck was a residue of Gray's theory of 1708. He had already fruitfully exploited the idea, for he had discovered the electricity of feathers by noticing that, if a plume tied to a stick were presented to the tube and then withdrawn, its fibres would adhere to the wood, 'as if there had been some Electricity communicated to the Stick or Feather.'[57] (The communication could be understood by supposing Hauksbee's stiff effluvia capable of weakly electrifying wood or feathers by friction.) For many years before his frustration with recalcitrant metals provoked him to investigate it, Gray believed in the possibility of weak communicated electricity.

A shrewd observation made during a preliminary trial led him to the discovery of a communicated electricity far different from what he had anticipated. He used a flint-glass tube corked at both ends to keep out the dust; a wise precaution, as Hauksbee had shown, for contaminants within the tube reduce its electricity. And a fortunate precaution. Gray wondered whether the stoppers themselves altered the tube's power. They did not. While gleaning this information he found that a down feather released near the end of the tube went not to the glass, but to the cork. 'I then held the Feather over against the flat End of the Cork, which attracted and repelled many Times together; at which I was much surprized, and concluded that there was certainly an attractive Vertue communicated to the Cork by the excited Tube.'[58] It is a classic example of the chance that favors the prepared mind. Pursuing a detour before looking for an effect he expected to be very weak, Gray found precisely what was to be the object of careful research. One not alert to the possibility of communicated

55. According to Stukeley, Gray lived with and assisted Desaguliers in London before entering the Charterhouse. Cohen, *Isis*, 45 (1954), 46. Cf. J.B., XI, under 4 June 1719.
56. Gray, *PT*, 37 (1731-2), 18–44.
57. Gray, *PT*, 31 (1720-1), 104–7.
58. *PT*, 37 (1731-2), 18–44.

electricity might well have missed the significance of the feather's preference for cork.

Gray exploited his discovery by trying how far he could communicate the 'attractive Vertue.' He fixed an ivory ball on a stick thrust into the stopper; it received a stronger virtue than did the cork in the first exploratory experiment. He then varied the nature of the 'line' (heretofore the stick) and of the 'receiving body' (the ball). For the former he substituted iron or brass wire, and packthread;[59] for the latter, a shilling, a tea kettle, a silver pint pot, stones, bricks, tiles and vegetables. All proved satisfactory. The metals were 'strongly Electrical, attracting the Leaf-Brass to the Height of several Inches;' and so Gray succeeded at last in awakening their hidden electricity.

Note the distinction that Gray drew between the line and receiving bodies. Priestly severely takes him to task for it: 'Mr. Grey [!],' he says, 'had not properly considered the line of communication, and the body electrified by it, as one and the same thing, in an electrical view, differing only in form, as they were both alike conductors of electricity.'[60] This is good physics but bad history. Gray, as we know, had hoped to stimulate the emission of effluvia across the air; his discovery seemed to him to be just the effect he sought, made unexpectedly powerful with the line in place of air. He confirmed that the communicated electricity went primarily to the ball by showing that the wire line attracted most weakly nearest the tube. (Gray's experiment works by *induction*: the tube end of the wire exhibits little electricity because the negative charge induced in the cork neutralizes the positive charge on the glass.) All Gray's experiments and explanations make excellent sense if we do *not* interpret them as Priestley recommended.

Having succeeded with the metals Gray returned to the range of the tube's influence. He replaced the stick with a fishing rod, the rod with a pole, the pole with a length of packthread, until, on the morning of May 31, 1729, he reached 52 feet, having suspended a thread of 34 feet—the greatest vertical drop available to him—from an eighteen-foot pole inserted in the tube. He then had literally reached the end of his rope. No other convenient free drops presented themselves, and horizontal transmission seemed impossible. This unpleasant fact emerged when Gray failed to excite the ivory ball through a thread suspended from the ceiling by short lengths of the same material.

At this point, June 30, 1729, Gray gathered up his tubes and thread for a visit to his friend Granville Wheler, a wealthy, able scientific amateur with a large country house admirably suited to the new experiments.[61] Wheler thought Gray's dismissal of horizontal transmission premature. Perhaps silk supporters would work? Gray thought it possible, since the 'Smallness' (narrowness) of

59. Packthread was stout cord used for wrapping packages.

60. Priestley, *Hist.* (1775³), I, 53.

61. Wheler (1701–70), a fellow of Christ's College, Cambridge (1722–4), and an Anglican cleric, became FRS in 1728. He made his home in Kent; it is not known how he met Gray, a man twice his age and a pensioner in the London Charterhouse.

the silk might prevent the loss of the 'Vertue.' They managed thereby to transmit electricity a few hundred feet, until the silk parted under the strain. (The silk cords must support the jerks of the electrifying friction as well as the weight of the thread.) They replaced the silk with brass wire, supposing its narrow gauge might make it suitable, but without avail. Their endeavor to increase the mere distance, or quantity, of the transmission thus culminated in a capital discovery, the recognition of a qualitative distinction of the utmost importance. 'The Success we had before, depended upon the Lines that supported the Line of Communication, being Silk, and not upon their being small.'[62]

This eventful discovery occurred on July 3, 1729. For the next thirty months Gray and his associates, Wheler and John Godfrey,[63] enjoyed a monopoly in the study of communicated electricity. They devoted themselves 'separately and jointly' to two main enquiries, 'communicating . . . our Thoughts as we went along & Mr. Grey [!] giving [them to the] World.'[64] Despite Gray's repetitious reports, which might give the impression of confusion of purpose,[65] their investigations were sensible and coherent. First they sought other substances which, like silk, might serve as supporters or which, like ivory, metal and vegetables, might play the part of receivers. To the former category they added hair, glass and resin;[66] in the latter they found soap bubbles, water, a map of the world, an umbrella, and an unfortunate 47-pound child, no doubt a Charterhouse charity boy (fig. 8.3).

8.3 *Gray's charity boy as pictured in Doppelmayr's* Neu-ent. Phaen. *(1744).*

62. Gray, *PT*, 37 (1731–2), 18–44.

63. Gray knew Godfrey, a cousin of Flamsteed's, from about 1715, the year Godfrey became FRS; Gray to Flamsteed, 4 Jan. 1715/6, in Flamsteed Papers, Roy. Obs. Like Wheler, Godfrey was a country gentleman who made science, especially astronomy, his hobby; he lived to a great age if the obituary in *GM* (1787:1), 454 is his. Cf. Chipman, *Isis*, 49 (1958), 417.

64. Wheler to B. Wilson, 7 Sept. 1748, Add. Ms., 30094, f. 71 (BL).

65. *FN*, 370; cf. Cohen's more favorable view in *EO*, 37.

66. Gray exploits but does not announce these discoveries; cf. Priestley, *Hist.* (1775³), I, 40.

Second, Gray's group aimed to elucidate the mechanics of communication. They retained the idea that had guided Gray, that effluvia from one body could electrify another by driving secondary flow from it. They never specified the mechanics of stimulation, any more than they explained the original emission from the tube, but they undertook to trace the course of the stimulating effluvia. In their first trial, a lead weight fastened to the rafters by a hair cord hung over leaf-brass placed on a wooden stand.[67] When Gray brought the tube near the weight without touching it or the line, the weight attracted the brass, which, however, lay dormant when the tube was placed near the ceiling. This was precisely the test for communicated electricity that Gray had projected at the outset: in the first case effluvia can cross the air in enough strength to excite the electricity of the weight; in the second, the air gap exceeds their range, and they cannot flow down the supporting hair cord. What if a transmitting line or pole offered an easier path? Gray suspended two corks at the same level by hair lines and connected them by a horizontal packthread. Both drew the leaf-brass when the tube approached the connecting line, the cork further from it pulling more strongly.[68]

They tried many variations of this last experiment, suspending hoops and poles from hair cords, bringing the tube to one end of the apparatus and drawing brass at the other (fig. 8.4), all showing that 'the *Electric Effluvia* will be carried in a *Circle,* and be communicated from one *Circle* to another.' The effluvia jumped air gaps to and through threads several hundred feet long, and flew to wooden hoops from a line running axially through their centers. For an encore, the effluvia leapt from the tube into the feet of a dangling boy, ran along a pole he held and into the left arm of a lad standing on glass 'cakes,' thence across his body and out his right hand, to fall upon and agitate the inevitable leaf-brass.[69]

Two more investigations complete Gray's study of effluvia. He tried the electricity of a glass ball *in vacuo,* and found that it pulled over the same distance as in air. Priestley intimates that had Gray known Boyle's work he might have saved himself the trouble;[70] but Gray's misleading experiment was not superfluous in view of Hauksbee's revival of the air theory and his failure to generate electricity in evacuated vessels. The second investigation examined whether solid objects received a stronger communicated electricity than hollow ones. Two oak cubes, one solid and one hollow, were suspended and joined by a packthread; when the tube stood over the center of the connecting line, both attracted equally. This curious and important experiment seems to have been an unsuccessful test of stimulated emission. Gray expected that communicated

67. Gray, *PT,* 37 (1731–2), 42, had noticed that the leaf-brass ascended further from a narrow, paper-covered stand than from the floor or a table; cf. *supra,* iv.3, and *infra,* ix.2.

68. *PT,* 37 (1731–2), 34. The tube is positive, both corks negative by induction; the tube's field tends to neutralize that of the cork to which it is closer.

69. *Ibid.,* 36–42, 399–404. These experiments conflate inductive and conductive effects.

70. *Ibid.,* 285–91; Priestley, *Hist.* (1775³), I, 50.

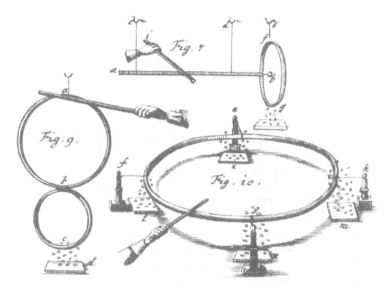

8.4 *Gray's arrangements to map the course of effluvia: from center to circumference, from hoop to hoop, from one part of a circle to another. From Doppelmayr,* Neu-ent. Phaen. *(1744).*

'electric attraction' would be 'proportional to the Quantity of Matter in Bodies,' and not, as is the case, to their surface areas.[71]

Gray's experiments proved revolutionary; no one ever managed to square his discoveries with effluvial theory. The apparent coherence of his own view had been achieved by treating communication apart from the mechanism of attraction. Gray's effluvia ran easily through threads, with difficulty through air, and not at all through silk, glass or resin, while the usual, or attractive effluvia penetrate glass but not muslin. Trouble might have been glimpsed as early as 1731, when Gray, after recommending glass bricks as supporters, mentioned that Wheler had drawn leaf-brass across five superposed layers of glass![72]

71. Gray, *PT*, 37 (1731–2), 35; the negative result did not disconfirm stimulated emission.
72. Gray, *PT*, 37 (1731–2), 399, 405–6.

CHAPTER IX

Dufay

Old Gray's rambling report of his collaboration with Wheler and Godfrey caught the interest of Charles François de Cisternay Dufay, a young man as different from Gray in temperament, class, education and cast of mind as in age and nationality. In 1733 Dufay was 35, energetic, brilliant, thorough and orderly, already a leading member of the Paris Academy of Science, independently wealthy, equally at home among academicians, ministers and high society, open and good-humored, with a taste for Italian burlesques and the satires of Swift.[1] He proved the ideal successor to Gray and Hauksbee. In one stroke he ordered their scattered insights and discoveries, disentangled what he called 'simple rules' (the dominant electrostatic regularities!), and demonstrated that electricity was a general property of material bodies.

Dufay came from a family that had followed military careers uninterruptedly for over a century. He himself entered the Regiment of Picardy as a lieutenant in 1712, at the age of fourteen, fought in Spain, and retained his commission until he joined the Academy as adjoint chemist in 1723. This last step was not taken without family influence and precedent. Dufay's father had been educated by the Jesuits at Louis-le-Grand, where he contracted a bibliophilia that his military adventures only enhanced. 'The muses,' he said, 'assuaged the wounds of Mars,' who, in 1695, put an end to his soldiering by carrying off his leg. He returned to Paris and devoted himself to raising a family and accumulating a library. Charles François grew up 'like an ancient Roman, raised equally for arms and for letters.'[2] No fuzziness of thought, no intellectual pretense or superficiality did the retired warrior allow in his domain. 'He used words sparingly, and never blabbered, but spoke as if he had thought out everything beforehand; so great a lover of truth was he, and so utterly free from deceit, that if he heard anyone speak insincerely, he immediately became dumb, and reproached the speaker's want of probity by his very silence.' This aggressive discipline which, with his peg-leg, earned the elder Dufay the nickname 'the wooden devil,' reappears softened and redirected in the son; the high standards of the old soldier became the rigorous, thorough, painstaking experimental method of the academician.[3]

We must also look to the father to understand how Dufay, a 25-year-old

1. Fontenelle, 'Eloge de Dufay,' *Oeuvres*, VI, 372–83. Dufay's tastes are inferred from the books he retained from his father's library; Du Fay, *Bibl. Fay.* (1725), 'Bibl. Lect.' and p. 108*, nos. 1090, 1898, 1991–2, 2055–6. Of the 198 items kept, about 70% dealt with the Gallican controversy.

2. Fontenelle, *Oeuvres*, VI, 373.

3. Du Fay, *Bibl. Fay.* (1725), Pref.; Marais, *Journal* (1864), II, 490.

infantry officer who had written nothing, jumped directly from the army to the academy. Despite his prickliness the elder Dufay had easy access to high places. The most distinguished and closest of his powerful friends was his old schoolmate the Cardinal de Rohan, a great noble and the chief churchman of the realm, whose relatives happened to command Charles François' regiment. The Cardinal knew young Dufay, and in 1721 took him on an expedition to Rome; shortly after their return to France Dufay found himself a candidate for adjoint chemist at the Academy.[4] Doubtless Rohan had put him forward. The Academy's leading scientist, Réaumur, managed the candidacy, which, with the support of its titular head, the abbé Bignon, culminated successfully on May 14, 1723. Dufay recognized his debt to Réaumur. 'It is you alone who, through pure goodness [!], have been kind enough to notice me just to raise me to a degree of honor that my want of merit placed so far from me. I can, however, assure you that so signal a favor will encourage me to try to supply by hard work what I otherwise lack and that I will spare no effort to justify your choice to the Academy.'[5]

Dufay quickly redeemed his promise. His first academic paper (1723), on the mecurial phosphorus, already displayed the characteristics that distinguished his work: full command of earlier writings, clear prescriptions for producing the phenomena under study, general rules or regularities of their action, thorough investigations of possible complications or exceptions, and cautious mechanical explanations of a Cartesian type. He never became doctrinaire, however, and even such staunch Newtonians as Voltaire and the Marquise du Châtelet considered him an ally against the vorticists.

The investigation of the phosphor, which Dufay did not connect with electricity, did not provide a continuing line of research. For several years he flitted from one subject to another—the heat of slaked lime, the solubility of glass, geometry, optics, magnetism—until returning, in 1730, to phosphorescence, and another memoir important for the development of his method. Chemists had long been acquainted with a few minerals like the Bologna stone (BaS) and Balduin's hermetic phosphor (plain CaS) that glowed after exposure to light. Great mystery surrounded these expensive and supposedly rare substances. Dufay detested mysteries and held as a guiding principle that a given physical property, however bizarre, must be assumed characteristic of a large class of bodies, not of isolated species. He found that almost everything except metals and very hard gems could be made phosphorescent; he depressed the phosphor market by describing his procedure; and he became sensitive to the endless small variations in the physical properties of bodies. 'How different things behave that seemed so similar, and how many varieties there are in effects that seemed identical!'[6]

4. Du Fay, *Bibl. Fay.* (1725), Pref.; C. Desmaze, *Infanterie* (1888), 85, 113.

5. *Corr. hist. arch.*, 5 (1898), 306–9; cf. Torlais, *Esprit* (1961²), 29–30.

6. Dufay, *MAS* (1723), 295–306; *MAS* (1730), 524–35; Mme du Châtelet to Dufay, 18 Sept. 1738, in *Lettres* (1958), I, 261. Cf. Brunet, *Pet. non.*, 3:2 (1940), 1–19, and Becquerel in Paris, Mus. nat. hist. nat., *Centenaire* (1893), 163–85.

In 1732 Dufay was made Intendant of the Jardin du Roi in the hope that his sympathy, energy and knowledge of the world would enable him to repair the damage done by his predecessor. Shortly thereafter he learned about Gray's work, perhaps through Bignon, himself alerted by Sloane: 'These experiments on electricity [are] surprising,' Sloane had written, 'no one here is offering to explain them.'[7] During the remainder of his short life Dufay responded brilliantly to both challenges, the king's garden and the pensioner's electricity; and we may say of his electrical work as of his botanical, that he transformed a collection of miscellaneous weeds into the first garden of Europe.[8]

1. THE UBIQUITY OF ELECTRICITY

Dufay begins with two methodological innovations: a short history, made necessary, he says, because previous writers had ignored earlier work, and a list of the questions he proposes to investigate.[9]

1. Can all bodies be made electric by rubbing, and is electricity a common property of matter?
2. Can all bodies receive the electric virtue, either by contact or by close approach of an excited electric?
3. What bodies stop, and which facilitate the transmission of the virtue; and what bodies are most strongly attracted by an excited electric?
4. What is the relation between the repulsive and attractive virtues; are they connected or totally independent?
5. Is the strength of electricity augmented or diminished by a void, compressed air, elevated temperature, etc.?
6. What is the relation between electricity and the faculty of producing light, which is common to most electrical bodies, and what can be inferred from this relation?

Dufay's answers to the first two queries put an end to the listing of individual new electrics. Every body can acquire electrification by friction—or 'par lui-même' as he termed it—except metals and substances too soft or fluid to rub.[10] One need only dry and warm the body before rubbing it. Dufay examined everything; and, when he was done, electricity, the power to draw when rubbed, had become an almost universal property of matter.

Communicated electricity proved still more ecumenical. All bodies save flame displayed the virtue after *contact* with an excited electric, the metals now being more powerfully affected than other substances. These experiments did not succeed without precaution: the body to be electrified had to rest upon an

7. Sloane to Bignon, 29 March 1733, quoted in Jacquot, *Naturaliste* (1953), 22.

8. *DSB*, III, 214–17; Jusieu, in Paris, Mus. nat. hist. nat., *Ann.*, 4 (1804), 1–19; Bidal in Paris, Mus. nat. hist. nat., *Arch.*, 11 (1934), nos. 34, 152–5, 157 (suggesting that Dufay put half his salary back into the Jardin); Laissus in Taton, *Enseignement* (1964), 287–341.

9. *MAS* (1733), 23–35, 73–4. Dufay found that he could not adequately credit his predecessors as their want of system made it difficult to find anything in their writings; *MAS* (1734), 235–6.

10. Not until the late 1770s was it shown conclusively that insulated metals electrified by friction. Hemmer, *JP* (1780), 50–2.

electric support of sufficient thickness, or, as we would say, had to be properly insulated. This important principle, already implicit in the work of Gray, became known as the 'Rule of Dufay;' though it was not quite correct as Dufay and his successors understood it, it guided them successfully for over a decade, until an apparent violation of it forced a revolution in electrical theory.

The next point was to know which bodies electrified by the mere approach of the tube or conducting cord. The answer, all bodies whatsoever, did not surprise Dufay; like Gray, he considered near approach and contact to operate by the same mechanism, the conveyance of electrical matter that electrified by stimulating the emission of effluvia. He therefore could make no distinction between conduction and induction. Momentarily, however, he did not require it, for the test he used for electrification by communication did work by conduction. He passed the tube among the test objects and then, having withdrawn it, investigated their electricity. Some—those to which, unknown to Dufay, a charge had jumped—attracted threads presented to them. There was, however, something peculiar about these experiments, something 'one could not foresee,' an effect that vexed Dufay until he had reduced it to rule: the outcome of the experiments depended on the nature of the stand supporting the test objects. If of glass or wax, then metals, books, wood, etc., placed upon it acquired the virtue powerfully, while good electrics did so only weakly; whereas if wooden or metallic, the electrics again acted insipidly but the nonelectrics did not respond at all.[11] Although Dufay offered no explanation, one followed naturally from Gray's theory: the effluvia conveyed to metals, etc., and any stimulated from them should run easily through the wooden supports and vanish into the ground.

A consequential application of these results was Dufay's prescription for electrifying water, a feat Gray had already accomplished by passing the tube over and under a water-filled wooden bowl screwed into a small stand *resting on glass*. 'The only way to succeed in electrifying [liquids] by the approach of the tube is to put them in a little glass, porcelain or faience jar, and to place the jar on a wax or a glass support. One would try in vain,' Dufay warned, 'using a platform made of wood or metal.' The same is true if one chooses to electrify water by contact, say by dipping the end of the conducting line into it: the dish or bowl containing the water must rest on a support of dry glass.[12] No one who accepted the Rule of Dufay—as all practiced electricians did—could intentionally have invented the Leyden jar.

In the fall of 1733 Dufay took on a collaborator, J. A. Nollet, who was to

11. Any charges placed on the metal, etc., by the breakdown of the insulating air will be lost through the grounded conducting supports when the tube is withdrawn. Gray, *PT*, 37 (1731–2), 42, had mentioned that the nature of the stand supporting the brass affected the distance over which the tube could draw it. This effect differs from Dufay's, however, as it depends on charges induced in the surface of the supporter. Cf. Lana's experiments, *supra*, iv.3.

12. Gray, *PT*, 37 (1731–2), 227–30; Dufay, *MAS* (1733), 34, 84. If the glass or porcelain vessel containing the water has a perfectly dry exterior, the glass support is unnecessary.

succeed him as the dean of French electricians. They extended the range of conduction to 1,256 feet of packthread. They found that glass, wax, and other electrics, as well as silk, were suitable supporters; that metals were better 'receivers' than ivory balls; and that moistening the line improved the communication. They confirmed that electricity could jump air gaps up to a foot across, and that 'the electric matter runs freely in the air, without being fixed in any body.'[13] Everything seemed to conform to Gray's model. Except for a few troubling observations.

At the end of his report of the discoveries made by himself, Godfrey and Wheler, Gray had hinted that the strength of electrical attraction depended on the color of the drawn body. Dufay felt himself obliged to investigate, under point three of his program. He procured nine silk ribbons, one black, one white, the others of the colors assumed primitive in Newton's *Opticks,* and hung them from a horizontal wooden stick like socks from a clothesline. He approached the tube parallel to the stick: the black and white ribbons were drawn most strongly, the red least of all, the others to an intermediate degree. 'I did not doubt then,' Dufay wrote, 'that, as M. Gray said, color counts for much in electrical experiments.'[14] He sought further evidence by fixing differently colored cloths to a wooden hoop one inch deep and six in diameter, placed above leaf-brass resting on a table. The tube drew across the Newtonian colors, but not across the black or the white. The result agreed perfectly with the earlier: one need only assume that black or white drinks up 'electrical matter' more vigorously than red or green. (The assumption explains the screening; it seems to require an auxiliary hypothesis, that attraction is stronger when both parties are electrified, to account for the first set of experiments.)

Dufay went on to other questions, always taking care to vary the colors of his apparatus, which greatly added to the tedium of his researches. Gradually, for reasons he could not identify, he came to doubt the effect that had seemed so certain. He tried whether the tube could act upon light, and whether different colors of light, thrown onto a white ribbon, would affect its screening. All negative. He then thought to warm the screens before use, and discovered the secret of Gray's effect. When warmed, all the cloths acted like the red one; when moistened, all behaved like the white. The perplexing electricities of the colored ribbons derived from moisture retained by the dyes and mordants used in their manufacture. Effluvia apparently could not penetrate moist bodies.

But wetting conducting lines promotes the flow of the electrical matter! A similar difficulty occurs in Gray's theory. It is not yet explicit; it begins to surface only in Dufay's later experiments. He places a sheet of metal, then a pane of dry glass on his wooden hoop; the virtue penetrates the metal but not the glass. Wet glass also cuts off the attraction. If, however, he uses a hoop of perfectly dry glass, both the pane and the metal sheet pass the virtue. Why?

13. Dufay, *MAS* (1733), 233–54.

14. Dufay, *MAS* (1733), 235; Gray, *PT,* 37 (1731–2), 44. Cohen (*FN,* 373) wrongly makes the attraction depend on the color of the *attractor.*

Because, he says, moist bodies and metals grasp the effluvia and so arrest their passage;[15] while they freely penetrate glass, as one can see by moving the tube outside a sealed bottle containing a feather. In the case of the glass hoop and pane, the effluvia fly from the tube through the air, across the pane, and through the air again to the leaf-brass. But in conduction experiments the electrical matter passes freely through metals and moist bodies *only* if they rest on a substance, like glass, that can prevent loss of effluvia to ground.

2. REPULSION AND THE TWO ELECTRICITIES

Point four of Dufay's program required him to dispel the fog in which Hauksbee and 'sGravesande had left electrical repulsion. He first thought there was no such thing; that—as Gray had believed—light bodies apparently driven from the tube were drawn away by surrounding objects made electrical by communication. Réamur suggested an *experimentum crucis*: place gun powder near the edge of a card and try whether the tube drove back any grains beyond the card, where neighboring bodies could not attract them. It did. Dufay, now persuaded of the existence of a 'real repulsion,' cast about for another plausible connection with attraction. 'Finally,' he says, 'having reflected that bodies the least electric by themselves [like metals] were more vigorously attracted than others, I imagined that an electric body might attract all those that were not so, and repel all those that became electric by its approach, and by the communication of its virtue.'[16]

To test this strained analogy, Dufay dropped a gold leaf onto the tube, as in Hauksbee's experiment, and chased it about the room. The leaf was attracted only when unelectrified, and, when repelled, was electrified; first attraction, next communication of electricity, then repulsion. Dufay delighted in this simple rule, which we shall designate the regularity 'ACR.' It explained many 'bizarre' and 'incomprehensible' phenomena. He rightly considered it a great discovery, for it constitutes the full recognition of electrostatic repulsion.

That this recognition depended closely upon, and even was a corollary to, the discovery of electrification by communication, also appears from the unpublished notes of Huygens noticed earlier, and from the work of Gray's collaborator Wheler. Wheler later wrote that he had recognized ACR in the autumn of 1732, before Dufay began to study electricity, and the following summer he showed Gray confirmatory experiments. Wheler recognized their importance, but, 'unwilling to be an Author,' he was anticipated by Dufay. He published his work in 1738, Gray's death having freed him from any scruple about appearing apart from his collaborator.[17]

Figure 9.1 schematizes a representative demonstration. Wheler rubs the tube

---

15. For further confirmation of the holding property of moisture Dufay (*MAS* [1733], 243–4) adduces experiments of Hauksbee's type in which the tube is moistened or filled with sand; electrical action without is impaired by the retention of effluvia within the tube.

16. Dufay, *MAS* (1733), 458.

17. Wheler, *PT*, 41:1 (1739–40), 98–117.

T; the conducting thread A is attracted and repelled several times in succession. He joins T and A by the thread xz; A is immediately and continuously repelled, without preliminary attraction. Finally, he grounds xz via the thread yG; A flies to T and remains there. A second demonstration, where a silk string B lies along the thread A, shows how the nature of the drawn object, or the rate at which it imbibes effluvia, affects its repulsion from the tube. Wheler understood that these experiments, and the regularities they illustrated, required fundamental revision of earlier theory. Henceforth, he says, 'a mere Vibration of the Parts of the Tube' and of the associated atmosphere 'is not sufficient to account for the Electrical *Phaenomena.*' Nor is a vague, undifferentiated Newtonian force: 'Electricity is not so properly called an attractive and repulsive Virtue, as a Virtue attractive of those Bodies that are not attractive themselves, and repulsive of those that are.'[18]

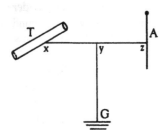

9.1 Diagram of Wheler's demonstration of ACR; T is the tube, A, xz, and yG, conducting threads.

Wheler did not suggest a mechanism to account for ACR. Dufay tried his hand at a Cartesian one. The electrified tube, like all electrified bodies, supports a vortex whose extent is indicated by the height at which the gold leaf floats. When unelectrified, the leaf is drawn to the tube by the vortex; on arrival it develops its own vortex, which drives it away from the tube's; when suspended, it goes to neighboring unelectrified bodies by the mechanism of vortical attraction, surrenders its vortex and recommences the cycle. The vortices and the rule that two electrified bodies repel one another helped Dufay to explain such difficult phenomena as the dance of feathers and leaves stirred up by excited electrics. Everything follows from the rule and the rider that apparent exceptions to it arise from entanglements of the drawn bodies with protuberances on the surface of electrics.

The internal thread experiment proved more troublesome. A thread evidently flees the finger because the excited globe has electrified both. But then why does the glass itself not repel the threads? According to Dufay, the 'electric virtue' is so evenly distributed within the globe that nowhere does it possess enough vigor to push the threads away; the finger 'unifies' the dispersed vortex surrounding the glass and concentrates the power necessary to displace the threads.[19] The Hauksbee experiments always remained beyond the reach of theories of effluvia.

18. Cf. Dufay, *MAS* (1733), 459–60, 463–4, 474–5.   19. *MAS* (1733), 464–5.

The direct evidence for the rule ACR was not yet complete. Dufay had shown that the tube repelled bodies it electrified. Would such bodies repel one another? Would one tube repel bodies electrified by a second? Does the electricity of different bodies differ only in degree? Dufay expected that experiment would answer all three questions affirmatively. He suspended two gold leaves above the tube; as he had foreseen, the leaves repelled by the tube repelled one another. He then brought a rod of gum copal up to one of them. He was 'prodigiously disconcerted,' flabbergasted, at the result: the leaf, in violation of his new-found rule, flew to the rod and stuck there:

> I confess I expected an entirely different effect, because according to my reasoning the copal, being electric, ought to have repelled the leaf, which was likewise electric. I repeated the experiment many times, believing I had not presented the rubbed part of the rod to the leaf, which accordingly came to the copal as it would to any [unelectrified] body; but, having satisfied myself completely on that score, I was entirely convinced that the copal would attract the leaf which the tube repelled.

Further exploration showed that resin and wax rods behaved like copal, and that rock crystal acted like glass. Dufay jumped to the 'bold hypothesis'[20] of the two electricities: all substances that exhibit the amber effect behave either vitreously, like glass, or resinously, like copal; a body electrified by communication receives the type of electricity the communicator possesses. Objects with dissimilar electrifications attract, those with like electrifications repel one another.[21]

What determines which electricity is excited by friction? Does it depend on the chemical character of the rubbed body? Does the nature of the rubber count? Dufay knew that the obvious association—all 'resinous' substances possess one electricity and all 'vitreous' substances the other—did not hold. Everything had to be tested. But only once: the electricity of a substance, he thought, was characteristic of it, and independent of the nature of the rubber. In a rare lapse he had satisfied himself of this convenient principle with a cursory examination: silk rubbed against silk and against his hand in both cases electrified resinously; wool and feathers prepared in the same two ways were always vitreous. The supposed inherence of the different electrifications suggested the existence of two electrical matters with different mechanical properties. Fortenelle volunteered a rough analogy: mixing vortices of the same kind resembles the addition of so many volumes of water; mixing dissimilar vortices is like combining water with 'pulverized matter' that dissolves in it without increasing its volume.[22]

Dufay's discovery of the two electricities—of the two opposed types of electrification—was a triumph of care and method, of professional training and standards. He was more energetic, more thorough, more obligated than Wheler.

---

20. *HAS* (1733), 3.   21. Dufay, *MAS* (1733), 467–9.
22. *Ibid.*, 470–3; Fontenelle, *HAS* (1733), 12. Home, *Effl. Th.* (1967), 58–9, observes that Dufay never explicitly supposes two qualitatively distinct electrical matters.

After discovering ACR, Wheler too had perceived the need to verify that objects electrified by the tube or by friction repel one another.[23] But he did not vary sufficiently the materials he employed; having shown a repulsion between tubes of glass, he discontinued his experiments. Had he substituted a piece of wax, resin or amber for one of the tubes, he too would have discovered the two electricities.

3. SPARKS, SHOCKS, AND STIMULATED EMISSION

Dufay's method required that he replace Gray's favorite apparatus, the dangling charity boy, with a trained observer. He had himself suspended: the leaf-brass flew to his face and hands when the tube was held to his toes, just as Gray had described. If, however, he took the plate supporting the brass in one hand, the metal shavings refused the solicitations of the other, though they darted to the hand of a gentleman who stood upon the floor. This significant effect, to which we shall return, enjoyed Dufay's attention only momentarily, for he immediately encountered another that appeared to him utterly astonishing, or as astonishing as anything could be 'in a subject where one meets the marvelous at every turn.'[24]

A gold leaf, used as a detector, had settled on his leg. An observant and frugal bystander stretched out his hand to recover the leaf. He and the hovering experimenter earned an unexpected and unpleasant shock, accompanied by a snapping noise. When done in the dark, the business gave a spark as well as a snap and a shock.[25] It was the pain that Dufay thought most remarkable. Why did the tube not strike so fiercesomely when applied directly? How was the spark related to the shock? And why, as Dufay found, did he see the one and feel the other when a piece of metal, or another person, but not when an *electrique par lui-même* approached him?

He began to answer these questions toward the end of 1734.[26] A review and repetition of the principal experiments on the production of light by friction in precious stones and evacuated globes showed that their luminosity and their electricity could not proceed from the same cause.[27] Dufay distinguished an 'electrical matter' from a 'matter of the electrical light,' itself distinct from common light, as Hauksbee's experiment with the wax-lined globe evinced. All bodies, whether electrified or not, are surrounded by 'atmospheres' composed

23. Wheler, *PT*, 41:1 (1739–40), 98–117.

24. Dufay, *MAS* (1733), 251.

25. Nollet, who assisted at this demonstration, later recorded the surprise the eerie phenomenon had caused him and his mentor. Priestley, *Hist.* (1775³), I, 46; Sigaud de Lafond, *Précis* (1781), 153.

26. In his sixth memoir. Memoir IV, on repulsion, announced ACR and the two electricities; Memoir V (read 21 July 1734) corroborated (1) inhibition by moisture and (2) the action of electricity *in vacuo*. Dufay thereby sided with Gray and Boyle against Hauksbee.

27. *MAS* (1734), 512–18. Dufay remarks that rubbing can produce a strong light and a weak electricity, a strong electricity and a weak light, or one and not the other, according to circumstance. He thinks the mercurial phosphorus luminous but not electrical.

of their own substance vaporized. The matter of the electric light, which glows softly when traversing the gossamer atmosphere of glass, encounters in the atmospheres of living and metallic bodies something that transforms it into 'real and sensible fire.' A spark is this light-matter inflamed, a shock its burning of the skin; 'one might contrive therewith to ignite dry combustible materials wrapped about a living body.'[28]

On December 27, 1733, Dufay sent the Duke of Richmond a résumé of his first four memoirs for publication in the *Philosophical Transactions*.[29] There Gray saw it. Neither ACR nor the two electricities caught his attention. He admired instead the 'luminiferous experiments,' the sparks and shocks of dangling boys, which he and Wheler undertook to expand. They inferred that a metal object could replace the boy, and found it to be so, with qualifications. Blunt objects gave a single snap and a shock, like animal bodies. A bar pointed at both ends, however, gave conical glows, which appeared at each approach of the tube though no ponderable body stood opposite the points.[30] Gray continued these researches at the Charterhouse during the autumn of 1734. The lights coaxed forth by the ancient philosopher excited much wonder. A frequent visitor to the Charterhouse, the poet Anna Williams, daughter of a colleague of Gray's, was one of the first to see the display, the memory of which drew from her some verses on the occasion of Gray's death.[31]

Gray made two further important observations about the lights. He noticed that after he had removed the tube from the pointed bar he could still elicit glows at either end by bringing his finger up to the point.[32] Next, substituting a shallow dish of water for the bar, he saw the liquid rise under the tube and suddenly collapse, with a snap and a spark. The collapse suggested an important experiment. A boy, strung up and and electrified, extends his finger towards the hand of a gentleman standing on wax cakes. A spark jumps, the shock is felt; the boy loses and the gentleman gains electricity, as shown by the excursions of a 'pendulous Thread'—a rough electroscope introduced by Gray—applied to each. The process can be repeated three or four times, the boy's loss being the gentleman's gain.[33] These experiments seriously challenged the theory of stimulated emission that had guided Gray to the discovery

28. *MAS* (1734), 520, 522.

29. *PT*, 38 (1733–4), 258–66. The Duke of Richmond, grandson of Charles II, an aristocratic ornament of the Royal Society, was the sort of person to whom Dufay would have access.

30. Gray, *PT*, 39 (1735–6), 16–24.

31. They run as revised by Samuel Johnson: 'No more shall Art thy dext'rous hand require / To break the sleep of elemental fire / To rouse the pow'rs that actuate Nature's frame / The momentaneous shock, th'electrick flame / . . . Now hoary sage, pursue thy happy flight / With swifter motion haste to purer light / Where Bacon waits with Newton and with Boyle / To hail thy genius, and applaud thy toil.' Williams, *Miscellanies* (1766), 42–3; Philip, RS, *Not. Rec.*, 29 (1975), 198.

32. The bar, though electrified by sparking from the tube, cannot break down the air when the tube is withdrawn; the charged induced in the finger augments the field near the point enough to cause breakdown again.

33. Gray, *PT*, 37 (1731–2), 227–30; *PT*, 39 (1735–6), 166–70.

of communication. The boy evidently communicated electricity to the gentleman only by parting with his own.

After repeating these experiments, Dufay concluded that the passage of a spark between two persons conveys all or part of the vortex of the electrified partner to the other, who, if not grounded, will appear electrified by communication.[34] Transfer of electricity and sparking depend on the bonding between the natural atmospheres of bodies and their acquired electrical vortex. The atmosphere of an electrified iron poker, for example, strongly retains its vortex. Thrust your finger into it; *your* atmosphere, which also fancies electrical matter, pulls violently on the poker's vortex: the electrical matter noisily changes hosts, with a display of sparks owing to its associated 'matter of the electric light.'[35] This picture and the mechanism it supposed—the accumulation and communication of electrical vortices or haloes—were to guide electricians for a quarter of a century.

Dufay's substantive discoveries—ACR, the two electricities, shocks and sparking—are but one aspect, and perhaps not the most significant, of his achievement. His insistence on the importance of the subject, on the universal character of electricity, on the necessity of organizing, digesting and regularizing known facts before grasping for new ones, all helped to introduce order and professional standards into the study of electricity at precisely the moment when the accumulation of data began to require them. He found the subject a record of often capricious, disconnected phenomena, the domain of polymaths, textbook writers, and professional lecturers, and left it a body of knowledge that invited and rewarded prolonged scrutiny from serious physicists.

34. The interdependence of the work on either side of the Channel gave rise to the rumor, doubtless spread by people jealous of Dufay's accomplishments, that he had stolen from Gray; Mme du Châtelet was entirely correct in indignantly rejecting the charge (letter to Thieriot, 22 Dec. 1738, in Voltaire, *Corresp.*, VIII, 89; cf. *ibid.*, 118). 'J'aime passionnément Mr. du Fay et la verité'; Voltaire to ?, 10 Dec. 738, *Corresp.*, VII, 41–2.

35. Dufay, *MAS* (1737), 95–7.

CHAPTER X

Electricity Beyond the Rhine

Despite the discoveries of Gray and Dufay there was little general interest in electricity in 1740. Five years later nothing was more fashionable. 'Persons of quality' traveled about to see the electricity of famous professors; the Berlin Academy, in royal session, chose electrical experiments to entertain its guests; a Mr. Smith offered 'all lovers and judges of experimental philosophy' at Bath the sight of his 'electrical phenomenon' from ten in the morning to eight at night; at a certain inn in Amsterdam a lecturer stood ready six hours a day, five days a week. For a time electricity, as Haller put it, took the place of quadrille.[1] The origin of this revolution was as peculiar as the event itself. Rather than starting where Hauksbee, Gray and Dufay had labored, the craze began beyond the Rhine, in a land long barren of electricians.

1. THE FIRST GERMAN ELECTRICIANS

The barrenness may be illustrated by a dissertation on the electrical properties of amber, defended at the University of Königsberg in 1714 by a student of its author, Heinrich von Sanden, Professor of Physics, Rector of the Medical School, member of the Berlin Society of Sciences. Sanden rehearses the hypotheses of the seventeenth century, selects Lana's version of the air theory, and demolishes the others as best he can. He argues arbitrarily and scholastically: effluvia cannot have the properties Digby or Browne or the Cartesians assign them because . . . well, because it is not the nature of effluvia to have such properties. The action of the air, on the other hand, appears from the incoherent approaches of the chaff. As for Boyle's supposed exhibition of electricity *in vacuo*, it argues rather the inferiority of his air pump than the incapacity of Lana's theory.[2]

A similar approach and the same conclusion may be found in Sturm's *Kurzer Begriff der Physic* (1713). Sturm, who learned his physics in the 1650s and 1660s, describes the sticky effluvia and the impelled air, and chooses the latter. Although the founder of the Collegium curiosum (sive experimentale!) and, as

---

1. Gralath, Nat. Ges., Danz., *Vers. Abh.,* 1 (1747), 278–9, and 2 (1754), 399; *HAS I* Ber (1745), 11–12; [Haller], *GM,* 15 (1745), 194; Heilbron, *RHS,* 19 (1966), 133–42; Millburn, *Martin* (1976), 5.

2. Von Sanden, *Dissertatio* (1714), 5, 13, 17, 24, 29–33; Jöcher, *Allg. Gel.-Lex.,* IV, 117–18. Von Sanden appears to have kept himself current by abstracting articles for *Acta eruditorum;* the review of Hauksbee in *Acta erud.* (1709), 237–9, which repeats Hauksbee's definition ('vis attractiva seu electricitas vitri'), is said to be his.

we have seen, perhaps the first professor in Germany to illustrate his physics by experiment, Sturm was an armchair philosopher about electricity.[3] His contribution is an analogy between electrical action and the recoil of artillery pieces: the recoil is caused by air rushing toward the gun, the rush by a rarefaction of the air in front of the muzzle, and the rarefaction by steams issuing from the firing.[4] Wolff too advertised the sufficiency of the air theory, and gave no hint that experiments might prove useful in the study of electricity. It was not in any case a subject on which he wished to dwell: it smacked too much of those detestable attractions. Only in the late 1740s, after the English had revived Newton's aether, did Wolff begin to find the subject 'useful,' and its experiments worth recommending as so many confutations of 'Keill's whimsies about attractive force.'[5] In 1730, however, it was represented as a closed and uninteresting branch of natural philosophy.[6]

It was reopened in 1734 by Johann Jacob Schilling, professor of mathematics and philosophy at the University of Duisburg, a fellow traveller with the Dutch Newtonians. His predecessor at Duisburg was Musschenbroek, who perhaps influenced his appointment and the subject of his inaugural lecture, 'The Laws of Newtonian Nature;' and he began his own investigation of electricity with the ideas and information in 'sGravesande's text.[7] Schilling rejected the air theory in favor of pulsating atmospheres, whose existence he demonstrated as follows: grasp a portion of the tube tightly enough to 'compress' its vibrating 'fibres' and so prevent them from driving the atmosphere; remove your hand, and the tube will draw only opposite its 'uncompressed' parts. (You discharge the tube wherever you touch it.)

The atmospheres alone could not satisfy Schilling. His Newtonian world required active unsociabilities among matter particles; he held that moisture destroys electricity not by clogging pores, as in the old theories, but by the same repulsive force that prevents the mixture of oil and water. To confirm the existence of this force he adduced a novel demonstration, a glass bauble swimming

3. There is no mention of electricity in the thick volumes recording the experiments of the Collegium.

4. Sturm, *Kurz. Beg.* (1713), 560–3, begins with the traditional argument against sympathies from the promiscuity of electrical interactions. Cf. Sturm, *Phys. el.*, II (1722), 1096, 1106–7, which endorses Schott's version of the air theory.

5. Wolff, *All. nützl. Vers.* (1745–7), 150. Cf. Wolff, *Gedanken* (1716), 22–3; *Kl. Schr.* (1755), 77–82; Wolff to Manteuffel, 30 Oct. and 8 Nov. 1747, in Ostertag, *Phil. Gehalt* (1910), 133–4, 137–8; and the curious letter to Gralath quoted by Schimank, *Zs. f. techn. Phys.*, 16 (1935), 247: 'Nachdem man in England die attractionem universalem einführen wollen, so hat man auch angefangen, die experimenta de electricitate vor die Hand zunehmen. Da ich nun von derselben eben kein Freund bin, so habe zu der Zeit, da ich die Experimente geschrieben, auch nicht darauf Acht gehabt, indem ich von den wenigen, welche dazumal vorhanden waren, keinen besonderen Nutzen zu zeigen wusste.'

6. Gralath, Nat. Ges., Danz., *Vers. Abh.*, 1 (1747), 263; Fisher, Br. Soc. Hist. Sci., *Bull.*, 2 (1956), 49.

7. Schilling, *Misc. ber.*, 4 (1734), 334–43. Schilling does not acknowledge his evident debt to 'sGravesande.

about in a dish of water in obedience to the waving of the tube. On the side of the ball nearest the tube the repulsive force annihilates the power of the electrical atmosphere, leaving unopposed the vapor pressure on the far side to push the ball towards the tube! Frightful physics, no doubt, for a Newtonian and a professor, but a very pretty experiment. It caught the attention of an imaginative young instructor at the University of Leipzig, G. M. Bose, who was to lead the first generation of German electricians.

Meanwhile the textbook writers continued as before. Hamberger's popular *Elementa physices* copies and Germanizes 'sGravesande by making the air necessary to the motion and maintenance of the atmosphere. Börner and Krüger take much the same tack.[8] In the early 1740s Northern literature on electricity began to improve, not from the experiments of its authors but from their belated discovery of the investigations of the English and the French. Krüger had excerpted Gray and Dufay (exclusive of the two electricities) for his *Naturlehre* of 1740, and in the same year a student of Klingenstierna's named Mårtensson reviewed the achievements of Dufay, 'cui extra controversiam palma deferenda est.' Krüger later described the novelties more fully, as did Mårtensson, who underscored the open and incomplete state of contemporary theory.[9]

The first German book devoted solely to electricity was J. G. Doppelmayr's scholarly *Neu-entdeckte Phaenomena* (1744). Doppelmayr, who had studied in Altdorf under Sturm and in England and Holland, taught mathematics at the gymnasium in Nuremberg and devoted his energies to acquainting his countrymen with the scientific advances of the West. He won election to the Berlin Academy and the Royal Society of London for his trouble. *Newly Discovered Phaenomena* ordered the data reported by 'sGravesande, von Sanden, Gray, Dufay, and Musschenbroek, who had himself recently updated his account of electricity. Although Doppelmayr favored Hauksbee's air theory, he did not seek to establish it as a system; his contribution, as Bose put it, was to 'command electricity to speak the language of the fatherland.' He brought the literary tradition up to date and prepared his countrymen to appreciate the first significant, original German contributions to the study of electricity, the contemporaneous monographs of Bose and C. A. Hausen.[10]

2. BOSE, HAUSEN, AND THE ELECTRICAL MACHINE

Bose was born in Leipzig in 1710, the son of a local judge. At the age of seventeen he lectured publicly on physics and mathematics while studying medicine. His enterprise, not to mention his 'fiery intellect, ready memory,

---

8. Hamberger, *El. phys.* (1735²), 473–8, (1741³), 451–6 (where the mercurial phosphorus is 'proved' nonelectrical); Börner, *Physica* (1742), 416; Krüger, *Naturlehre*, 1 (1740), 521–8. A later edition of the *Naturlehre* (1771⁵), 554–72, presents electricity without reference to Franklin.

9. Mårtensson, *Diss. phys. el.* (1740), esp. 5, 16–21, and *Diss. phys. grad.* (1742), esp. 35–52; Krüger, *Zuschrift* (1743, 1745²). Contrary to the usual practice in the Swedish universities, Mårtensson wrote his own dissertation; Heyman, *Symbola* (1927), 192.

10. Musschenbroek, *Essai* (1739), I, 254–72; Bose, *Tent. el.*, I:iii (1744), 95–6; *DSB*, IV, 166–7. A German translation of Dufay's papers appeared at Erfurt in 1745.

penetrating judgment, tireless industry and insatiable curiosity' earned him the post of Assessor (Beysizer), or junior lecturer, to the Philosophical Faculty at the University of Leipzig.[11] In 1738 he was called to Wittenberg as Professor of Natural Philosophy. There his contribution to science was more inspirational than substantive; 'generally one can say of him that he wrote loftily of the wonderful and the sublime.'[12] He also had a genius for inventing striking demonstrations and a knack for asking stimulating questions. To complete the picture we must add a great ambition to be recognized by those he considered illustrious: the royal societies of France and England, eminent foreign scientists, the Grand Mufti, lesser aristocrats in any form and, strangely for one who lectured *ad ceneres Lutheri,* the Roman Pontiff. In this last vanity he overreached himself. The Theological Faculty at Wittenberg was not pleased when Bose advertised his slight relations with Benedict XIV; years of feuding followed; and it required Frederick the Great to put an end to the affair.[13]

While still at Leipzig Bose ran across Schilling's paper in the Berlin *Miscellany,* which, as he later understated it, 'almost drove me mad with delight.' 'I wondered, I doubted, and I sought the cause' of the antics of the swimming glass balls.[14] A frantic search among the glassmakers of Leipzig failed to produce the tube essential to the experiment. While lamenting his luck, he spied a large alembic just removed from the fire. In an instant, 'with the vigor of youth,' he ripped the beak from the vessel, rubbed it eagerly, and presented it to some bits of paper. Nothing happened; the day was warm and the professor moist. He persisted and succeeded. He then regularly demonstrated Schilling's experiment to his students, who thought him a magician.

The poet-physicist was not long content with his small knowledge. He turned to Gilbert, to Musschenbroek's edition of the *Saggi,* to Boyle and to 'sGravesande. From the Dutch books he learned of Hauksbee's thread experiments, which he repeated, having fashioned what remained of his alembic into the globe of an electrical machine. He also drew sparks from the vessel, to the great delight of his students, who witnessed the 'little miracle' two by two, lest their combined perspiration destroy it. The great size of the rotating globe kept him, so he says, in constant fear; it did indeed explode one day, as we shall see.

Not until September 1737, shortly before his departure for Wittenberg, did Bose learn of Dufay's first papers, which had then been in print for over two years.[15] He inhaled the French memoirs 'in one breath, at one sitting.' He

11. Börner, *Nachrichten,* 1 (1749), 789–91; *DSB,* II, 324–5.

12. Waitz. *Abhandlung* (1745), Vorb.

13. Mercati, Pont. accad. sci., *Acta,* 15 (1952), 57–70; Bose to Formey, 21 Jan. 1750 (Formey Papers, Deutsche Staatsbibl., Berlin).

14. 'Obstupui, dubitoque diu, causam requiro.' Almost Ovid, *Metam.,* XIII, 940 (Obstupui, dubiusque diu, quae causa, requiro); Bose, *Tent.,* I:i (1738).

15. Judging from *Misc. ber.,* 5 (1737), 109–12, Schilling also knew nothing of Gray and Dufay in 1736/7. German booksellers doubtless deserve some of the credit for his ignorance; cf. Kronick, Bibl. Soc. Am., *Papers,* 59 (1965), 41–2.

repeated everything, 'with incredible patience,' and found it all correct. He also found it tedious. It occurred to him that the machine he had constructed for Hauksbee's experiments would also serve for Dufay's: he needed only an arrangement for communicating the electricity from the whirling alembic to an insulated nonelectric. Once this latter body, the 'prime conductor,' was electrified, one could run through the repertoire of conduction and sparking experiments swiftly and conveniently. At first Bose mounted the prime conductor—a sword, iron bar, or cannon barrel—on glass caps, but soon he began to suspend it by silk cords, which became the standard arrangement.[16] The design of the machine (figure 10.1) remained basically the same throughout the 1740s, although some electricians preferred cylinders to globes, and some employed several vessels at once (figure 10.2). In the only noteworthy improvement, Andreas Gordon, a Scottish Benedictine who taught an anti-scholastic, or 'useful and pleasant,' philosophy in the order's cloister at Erfurt, employed a spring-mounted leather cushion to do the rubbing, thereby freeing one pair of hands and probably increasing the available voltage.[17]

The application of Hauksbee's machine to the experiments of Gray and Dufay more than rewarded Bose's enterprise. Not only did the experiments succeed more quickly and easily than with the tube, the effects were also more powerful, dependable, and consistent. Bose subsequently celebrated these advantages, together with himself and Dufay, in a few verses deservedly forgotten:

> Immortal Cisternay, vous m'avez prévenu.
> Voici mon hommage. Tout près, je vous suivis.
> La verité le veut. La verité des sages.
> Elle triomphe enfin malgrès l'envie en rage.
> Jamais ton souvenir chez moi perira
> Tant que je suis en vie. Bose te louera.
> Mais tout ce que tu fis, fut par des tubes caves,
> Qui sont bons, mais qui font l'essay tardif et grave.
> Je pris le premier, quelle commodité!
> La sphère de Hauksbej en de temps tant soit peu
> Tout devient plus fort, surpassant la creance,
> Le tube étoit plus tard, foible, à non chalance.[18]

16. *Tentamina*, I:ii (1743), 53 ff.

17. Cf. Saxtorph, *Elect.-Laere*, I (1802), 63–212; Hoppe, *Gesch.* (1884), 13–15; Dibner, *Early El. Mach.* (1957); Priestley, *Hist.* (1775³), I, 88–91; Espenschied, *El. Eng.*, 74 (1955), 392–7; *ADB*, XLIX, 461–2.

18. Bose, *Electricité* (1754), 41–2. 'Immortal Cisternay, you who preceded me / here is my hommage. I followed you closely. / Truth wishes it, the truth of philosophers / which triumphs in the end over all jealousy. / Your memory shall be with me / so long as I live, Bose will praise you. / But everything you did was done with hollow tubes / which are good, but make slow and heavy work. / I was the first to take—how convenient! / the sphere of Hauksbee. In a time as short as you please / everything becomes stronger, surpassing all belief. / The tube was slower, feeble, insipid.'

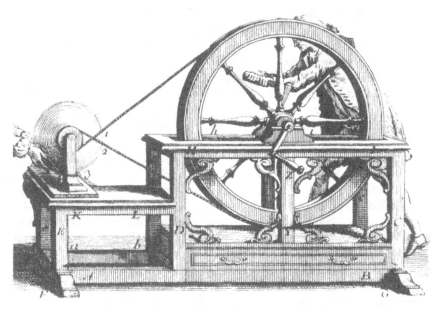

10.1 The classic electrical machine, as pictured in Nollet, Essai *(1750²).*

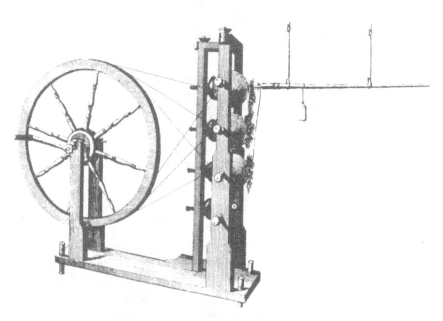

10.2 A multiple-globe machine with prime conductor. From Priestly, Hist. *(1775³).*

Bose did not limit himself to repeating earlier work. With the new power at his disposal he created novel demonstrations, some edifying, some amusing, and all calculated to draw attention to himself and to electricity.

Here are a few. Insulate a large dinner table and one of its chairs, which you will occupy; run a wire from a concealed prime conductor to within reach of the special seat; set the machine in motion, grasp the wire, touch the table, and watch your guests jump when the sparks fly from their forks. Or try the dollar trick, forcing a friend holding a coin in his teeth to relinquish it against his will when you discharge the prime conductor through it (and him). If the friend is female, she can serve as the Venus electrificata:

> Une seule fois, quelle temerité,
> à Venus sur poix je donnai un baiser.
> La peine vint de près. Les levres me tremblerent,
> La bouche se tourna, presque les dents brisent.[19]

If something more edifying is required, beatify someone, electrify him while he sits insulated and wearing a pointed metal cap (fig. 10.3), a pretty effect of which Bose long maintained a monopoly, as he neglected to mention the special headwear of his saints.[20] These demonstrations, known directly or through Bose's *Tentamina* of 1744, made a great impression. One said of them, as Pliny had of the books of the ancient philosophers, 'tam varia sunt quam ipsa natura.'[21]

We have what purports to be an eyewitness report of one of Bose's exhibitions in a book published anonymously in Venice in 1746. In the winter of 1739, the story goes, an Austrian count detained in Venice and bored with the usual amusements fell in with a virtuoso who took him to the home of a philosophical contessa.[22] The hostess welcomed the stranger, hoping to learn something about the electricity of the Paris Academy of which—and this a realistic touch—he was ignorant. But he could describe a lecture by Bose. It began with soothing words delivered as assistants arranged wax cakes and other paraphernalia. Bose asked several people, including the count's fiancée, to stand upon the cakes and join hands, forming a conductor that stretched across

19. *Tentamina*, I:ii (1743), 58, 61, 65–9; *Electricité* (1754), 54. 'Once only, what temerity! / I kissed Venus, standing on pitch. / It pained me to the quick. My lips trembled / my mouth quivered, my teeth almost broke.' Anyone scandalized by the experiment is advised to throw himself into the ocean.

20. *Tentamina*, I:ii (1743), 61–2; I:iii (1744), 79–80; II:i (1745), 16–17; Priestley, *Hist.* (1775³), I, 188–9. Gordon, *Versuch* (1746), 59–60, placed two points on the cap and transformed the beatification into 'Moses Strahlen,' after the horns of the patriarch. A 'natural beatification,' or St. Elmo's fire, is said to have blazed about Jallabert's gold-trimmed hat during an Alpine excursion. Freshfield, *Saussure* (1920), 87–8.

21. Börner, *Nachrichten*, I (1749), 791; cf. the long and laudatory review in *Nye tidender om laerde og curieuse sager*, quoted by Snorrason, *Kratzenstein* (1967), 22–8.

22. [Squario], *Dell'elett.* (1746), 1–48; *Hamburgisches Magazin*, 1 (1747), 154–71, pans our 'novella filosofica e galante.'

10.3 The beatification as practiced and revealed by Rackstrow, Misc. Obs. (1748). Electricity runs from a prime conductor to the plate above the crown, which is evacuated to enhance the effect.

the room. The first person grasped the gun barrel of the electrical machine, which was furiously spun by an assistant, while the last held his hand above a plate filled with gold leaves, which leapt into the air, to the delight of the company. Then the candles were removed and the rest of the audience approached the human conductor, which spat sparks and prickles at everyone who tried to touch it. The party continued in this fashion until interrupted by a noise like a clap of thunder. When the professor recovered himself enough to order the candles relit, he discovered that the globe of his electrical machine had exploded into a thousand pieces, a result, the count fancied, of the frantic pace at which it had been rotated.[23]

One of Bose's apparently frivolous displays was to inspire a pivotal event in our history, the invention of the Leyden jar. Bose knew from Dufay that water could play the part of a nonelectric body in drawing sparks from an electrified object. Always a lover of paradox, he proposed to reverse the phenomenon, to try whether the electrical fire could be obtained from water as well as from metals. He electrified water in a drinking glass, drew sparks from it with his finger or at sword-point, and, as usual, made the most of his wares:

> Mais, dites vous, est cela ou flamme ou étincelle?
> Que nous voyons si clair dans l'eau, et la plus belle.
> Plutot notre rien Titan attirera,
> Et le lion fuit la chevre d'Angora,
> Le soleil en Juin sera en sagittaire,
> Le diamant sera coupé par notre verre,
> Jamais le feu pourra de ses ondes sortir,
> Jamais le même feu en pourra rejaillir
> Avec tant de vigueur, qu'un fond de la bouteille
> En est illuminé. Comblé de la marveille.[24]

We shall return to these fireworks.

As a theorist, Bose shone primarily as an enemy of action at a distance. 'Who would have expected the English to endorse such an aberration,' he says, 'the English to whom experimental philosophy owes more than to any other nation?' Physicists who believe in such a thing, 'who worship either God and Newton, or no God at all' and who do not like Bose's poetry, might just as well admit sympathies, antipathies, antiperistasis, and the vires expultrix, fermentat-

---

23. *Dell'elett.* (1746), 42–7. Bose speaks of such an explosion in *Tentamina* I:ii (1743), 59, and *Recherches* (1745), xxxvi. They were not uncommon. See Abat, *Amusemens* (1763), 352–8, and Nollet, Journal, f. 99, who reports that a bursting globe cost the physician Pivati (*infra*, xi.3) an eye.

24. Bose, *Electricité* (1754) 47–8. 'But, you say, is it flame or spark / that we see so clearly and beautifully in the water? / As like our small might will attract a Titan / the lion flee the Angoran goat / the sun be in Sagittarius in June / or our glass scratch a diamond. / Never can fire jump forth from the water's waves / never can it spring back so vigorously / as to illuminate the bottom of a glass. / [And yet it does.] Wonder of wonders!' Cf. *Tentamina*, I:ii (1743), 64, and Heilbron, *Isis*, 57 (1966), 264–7, for a more prosaic description.

rix and formatrix.[25] This fustian did not impede the anglicization of electricity, which followed upon developments to which Bose's electrical machine and promotional propaganda largely contributed.

HAUSEN

Bose first formally described his machine in an address delivered at Wittenberg on November 3, 1743. On October 3 preceding, Frau Prof. C. A. Hausen signed the preface to a treatise by her late husband, the professor of mathematics at the University of Leipzig, in which the application of Hauksbee's machine to the experiments of Gray and Dufay is fully set forth, with no acknowledgment to Bose.[26] Historians have therefore hedged their accounts of the innovation. Priestley credits Bose only with the introduction of the prime conductor; Hoppe agrees, but awards the innovation to a student of Hausen's named Litzendorf; Dibner safely sets Bose and Hausen experimenting 'concurrently.'[27]

It is unlikely that Hausen and Bose hit upon applying Hauksbee's machine to the new experiments entirely independently. Hausen completed his studies at the University of Leipzig in 1714, became Ordinarius in mathematics there in 1726, and subsequently often served as Dean of the Philosophical Faculty. Since Bose studied at Leipzig in the 1720s and taught there until 1738, he unquestionably knew Hausen well, and was perhaps his assistant; for Hausen, although not interested in the usual practical exercises associated with his chair, did show experiments in his private lectures on physics, devoted primarily to demolishing Hamberger's *Elementa physices*.[28] Bose could not have approved his superior's views. Hausen inclined toward English ideas, with which he had been infected during a *Gelehrtenreise* to the West, when he had met Halley, Desaguliers, Keill, Clarke and Newton himself.[29]

Hausen's colleague, friend and editor,[30] the Wolffian philosopher and litterateur J. C. Gottsched, wrote that Hausen's interest in electricity and the machine was awakened by the brief reference to Hauksbee in Newton's *Optice*. Hausen himself said that the machine 'suggested itself' to him as he repeated the Anglo-French experiments, implying that his example inspired Bose's work. The evidence, however, favors the reverse line of influence. Bose used a globe in 1737, and Hausen's experiments date mainly from 1743. Bose always,

25. *Tentamina*, I:i (1738), 8–11; Bose to Formey, 2 Nov. 1756 (Formey Papers, Deutsche Staatsbibl., Berlin): 'Les attractionnaires de profession et les critiques par habitude seront peut-être choqués par ma poésie.'

26. Hausen, *Novi prof.* (1743).

27. Priestley, *Hist.* (1775³), I, 87–8; Hoppe, *Gesch.* (1884), 13, giving no authority; Schimank, *Zs. f. techn. Phys.*, 16 (1935), 248, giving 'Litzendorf' and citing 'a reliable report'; Dibner, *Early Elect. Mach.* (1957), 24.

28. Kästner, *Selbstbiographie* [1909], 7–8; *ADB*, XV, 440–1.

29. Hausen, *Novi prof.* (1743), Praef.

30. According to Gralath, Nat. Ges., Danz., *Vers. Abh.*, 1 (1747), 269. Cf. Fontenelle to Gottsched, 16 Oct. 1732, in Danzel, *Gottsched* (1855), 342.

and Hausen never, claimed credit for the application of Hauksbee's globe to the discoveries of Gray and Dufay. Who can doubt the queer story of the cannibalized alembic?[31]

Hausen's experiments were popularized under the aegis of Wolff's champion Manteuffel.[32] With the endorsement of the Count, a former Saxon cabinet minister, everyone in Leipzig, or at least all proper society, crowded to see Hausen perform. Reports of his wonders reached the Royal Electoral Prince in Dresden who, with an appropriate entourage, repaired to Leipzig to witness them. He was too late for Hausen. The professor's successor performed well enough, however, to induce one of the party, the Chancellor of Poland, to procure an electrical machine, and to carry it in triumph across the Vistula;[33] and that, sneered the *Gentleman's Magazine,* was indeed a wonder, for those regions were not then known for their 'polite tastes.'[34]

Manteuffel brought together Hausen and J. H. Winkler, also a friend of Gottsched and fellow Wolffian, who was to become the most distinguished of the Leipzig electricians. When he met Hausen, in 1742, Winkler had just risen to a professorship in classical languages, and knew nothing about electricity. Hausen used to tell of his progress at Manteuffel's Sunday dinners for the Leipzig learned; Winkler, who also attended, was so intrigued that he resolved to repeat all the electrical experiments described by Musschenbroek.[35] He soon became adept enough to attract two princes of Saxony, a cabinet minister, the Russian Ambassador, the Major-General of the Polish Army, and Maria Anna, Archduchess of Austria and Governess of the Spanish Netherlands.[36]

Hausen's book is noteworthy not only for its full description of the electrical machine, but also for its electrical theory, the prototype of those ingenious, rococo, tiresome expositions of effluvial mechanics that became the principal business of the electrician of the mid-century. It is a fancy, muddy version of Dufay. Electrification is vortification; friction elicits the electrical matter, a universal fluid, from the pores of appropriate bodies; air resistance deflects the rapidly issuing effluvia into circular or spiral paths. They remain to clothe their source because an attractive force—which the Newtonianizing Hausen freely

31. Hausen, *Novi prof.* (1743), xi–xii, 1; Gralath, Nat. Ges., Danz., *Vers. Abh.,* 1 (1747), 278–9; Bose, *Electricité* ?1754), 26; Wolff, *Kl. Schr.* (1755), 77–82. Cf. Nollet to Dutour, 27 April 1745 (*infra,* xi.3).

32. Formey, *Souvenirs* (1797²), II, 39–44; Danzel, *Gottsched* (1855), 7–70. On Hausen's attitude toward Wolff, Kästner, *Briefe* (1912), 212.

33. Bose, *Electricité* (1754), 26; Winkler, *Gedanken* (1744), Vorr.; Gralath, Nat. Ges., Danz., *Vers. Abh.,* 1 (1747), 279; Wolff, *Kl. Schr.* (1755), 80.

34. *GM,* 15 (1746), 194.

35. Reicke, *Neues* (1923), 79–80, 173; Gottsched, *Hist. Lobschr.* (1755), 83–4; *ADB,* XLIII, 376. Note the interview of Gottsched and Winkler with Frederick the Great, who knew Winkler's work on electricity; Reicke, *op. cit.,* 90.

36. Winkler, *Gedanken* (1744), Vorr. Cf. the four-hour performance before the Danish court on 18 Dec. 1744 by Christian Hee, later professor of mathematics at the University of Copenhagen. Everyone liked the sparks from the King's beard. Snorrason, *Kratzenstein* (1967), 40–1.

admits—exists between them and the particles of common matter. Bodies endowed with vortices of 'dissimilar strengths' draw together; after an exchange of electrical matter, leading to equalization, they push themselves (or the air pulls them) apart.[37] Because they have a stronger attachment to their effluvia, resinous bodies yield weaker vortices than vitreous ones, a disparity responsible for the phenomena of the two electricities. Sparks and glows are explosions of the vortices caused by bodies, like metals, that act strongly on the electrical matter. Hausen was much concerned with luminous effects, especially sparks drawn from liquids, interests his immediate successors shared.[38]

Hausen's system was retrograde in re-establishing the air as a principal agent in electrification and in obscuring, with its clumsy vortices, the regularities Dufay had uncovered. It was not well-received, though one distinguished electrician, the author of an improved system in the same style, conceived that Hausen would have accomplished wonders, had he lived.[39]

3. THE BERLIN PRIZE AND THE GRAND EFFLUVIAL SYSTEMS

From Wittenberg and Leipzig the electrical furor proceeded to Berlin, where, despite the distractions of the first Silesian wars, Frederick's new Academy was cultivating *Wissenschaft* with unusual vigor. Its public inauguration on January 23, 1744, was a brilliant occasion. After official speeches the class of physics entertained the company—which included all princes of the royal house, ministers Prussian and foreign, and 'a crowd of persons of distinction'—with experiments on electricity.[40]

Most of the experiments were standard. The last one tried, however, was quite new, arresting, important and appropriate: an ordinary member of the new academy, a physician named C. F. Ludolff, the author of a paper on electrical barometers, accomplished what Dufay had not been able to do, the ignition of combustibles by an electric spark.[41] He succeeded by using warmed 'spiritus

37. Hausen, *Novi prof.* (1743), 2, 10, 27–9, 34–42. Hausen 'justifies' these assertions by an impotent appeal to 'well-known theorems on vortical action.' Let x indicate radial distance from the axis of the tube, $y(x)$ centripetal acceleration, $z(x)$ density of the vortical matter. Hausen correctly infers that the centrifugal force / unit volume is yz. He then sets $y \sim x^n$, $z \sim x^{-s}$, fiddles and befuddles, and persuades himself that the *centripetal* force at x on a unit volume of a body of density D is zy/D, whereas it is $y(z - D)$; the body is only drawn into the whirlpool if it is specifically lighter than the vortical fluid. Hausen was a good geometer but evidently no algebraist; he avoided the calculus (Kästner, *Selbstbiographie* [1909], 9).

38. Hausen, *Novi prof.* (1743) 7, 11–12, 32–3, 41–5. Since in these experiments Hausen used a prime conductor, an iron bar five feet along, one wonders why Priestly and Hoppe credit Bose with its introduction.

39. Waitz, *Abhandlung* (1745), Vorb. Cf. Gralath, Nat. Ges., Danz., *Vers. Abh.*, 1 (1747), 275–6; Winkler, *Gedanken* (1744), Vorr.; and Bose, *Tentamina*, I:ii (1743), 71, who makes good game of the vorticists.

40. Harnack, *Gesch.* (1900), I:1, 248–92; *HAS*/Ber (1745), 9.

41. *Ibid.*, 12–13; *MAS*/Ber (1745), 3–7; Bose, *Tentamina* I:ii (1743), 26–7, had asked, 'Potestne hoc [electrico] igne, corpus, quod facile eum concipiat, ardente flamma accendi,' a query doubtless inspired by Dufay. After Ludolff's success, Bose (*Tentamina*, I:iii, 76–7) claimed prior-

Frobenii,' a combination of alcohol and sulphuric acid, which he presented to the prime conductor in a grounded metal spoon.[42]

One soon found that alcohol, turpentine, and gun powder could replace Frobenius' spirit; and that the human body, a drop of water, or a piece of ice could substitute for the prime conductor. People flocked to see electricians draw fire from ice and throw thunderbolts from their fingers. The effect also had some practical value. 'What's the use of all this *philosophische Narrenpossen*?' an unsympathetic observer asked Andreas Gordon. 'To improve the sense of smell.' The skeptic demanded proof. Gordon climbed on wax, grasped the prime conductor in one hand and a spoonful of brandy in the other, and brought the utensil, 'which I could hardly hold from laughter,' up to the heckler's nose.[43]

One of the Parisian provisions of the Berlin Academy required its members annually to select 'an important and useful matter from science or literature' as subject of an essay competition. For the first contest, announced in May, 1744, they asked for 'the causes of electricity.' The Hessian Landgrave's Finance Minister, the mining engineer J. S. von Waitz, won the prize.[44] His essay and those of the runners-up follow the lead of Dufay, Hausen, and Winkler, whose first book on electricity appeared early enough for the contestants to consult. A comparison of the schemes of Winkler and Waitz will illustrate the range of questions and solutions characteristic of electrical theory in the middle of the Age of Reason.

First one must decide the nature of the electrical matter and the manner of its excitation. Our authors agree about its universality, and its strong appetite for common matter; Winkler thinks it a fluid *sui generis,* akin to common fire; Waitz identifies it with the matter of light and heat. Taking a hint from Newton, Winkler conceives the electrical matter to stand about all objects, even when unexcited, in the form of atmospheres, whose density, contrary to that of the optical aether, decreases outwards; friction sets the atmospheres into radial or vibratory, not vortical motion, and operates either by reducing the 'quasi-gravity' between electrical and common matter, or by mechanical compression, which causes atmospheres to vibrate like springs. Waitz does without these

---

ity, having himself—so he says—already ignited alcohol and gunpowder when he had posed the query. But it seems most unlikely, considering the showman he was, that Bose would not have advertised his success, particularly in view of Dufay's failure. Cf. Gralath, Nat. Ges., Danz., *Vers. Abh.*, 1 (1747), 285–9.

42. *HAS*/Ber (1755), 11–12; Waitz, *Abhandlung* (1745), Vorb. This arrangement, perhaps because it differed completely from Dufay's in which the tinder rested on the body providing the spark, did not easily suggest itself. Gralath, who heard of Ludolff's success before learning the experimental details, could not repeat it; neither, he suggests, could Bose (*ibid.,* 288–9). Gralath, Nat. Ges., Danz., *Vers. Abh.,* 1 (1747), 287–9.

43. Gordon, *Versuch* (1746), 66–85, describes many ways of kindling spirits. Cf. Bose, *Tentamina,* I:iii (1744), 78; Gralath, Nat. Ges., Danz., *Vers. Abh.,* 1 (1747), 287–8; Watson, *PT,* 43 (1744–5), 481–7.

44. *HAS*/Ber (1777), 48–54; Waitz, *Abhandlung* (1745), Vorb.

envelopes. In his opinion rubbing electrifies by despoiling bodies of their normal fire, as appears from the spark thrown from the finger into a freshly chafed glass surface. Neither electrician mentions the two electricities.[45]

Next one must confront Dufay's dominant regularity, ACR. Winkler holds that the agitated atmospheres 'cling' to the chaff, reacquire their quasi-gravity, and bring their prey to their electrics; proximity stimulates the natural envelopes of the chaff, and repulsion ensues mechanically. Waitz prefers a hydrodynamical model, which the Eulers later exploited. The depletion of fire in the excited electric causes an influx of the 'elastic' electrical matter from environing bodies. A small, mobile, emitting object moves by jet propulsion to the electric, which draws out its remaining fire; the object then absorbs electrical fire from the environment, and is pulled away by an unspecified linkage between fire particles. To Waitz there is no attraction between bodies evacuated to different degrees, no force between bodies equally despoiled, and no true repulsion. As for electrification by communication, Winkler refers it to the stimulation of the pre-existing atmospheres, without transfer of electrical matter, while Waitz conceives the large channels of electrics to pull the fire from the smaller pores of contiguous metals.[46] Both authorities recommend the Rule of Dufay. Finally, to each the ubiquitous sparks and glows arise directly from the fire in or associated with electrical matter.

The other prize essays differ from Waitz' only in mechanical details and style of presentation. The brief second-place paper is the work of J. F. Unger, Mayor of Einbeck;[47] the prolix writer of the anonymous third paper was probably Johann Gottfried Teske, Professor of Physics at the University of Königsberg; the author of the last paper, which is in French, has not been identified. None introduces new experimental data of any consequence. They divide as follows over the most pressing questions concerning electricity.

The electrical matter they hold to be universal and *sui generis*, different from common fire, attracted to ponderable bodies in proportion to their densities. Two authors think it elastic owing to Newtonian repulsive forces or to Malebranche's sub-vortices. Unger thinks it a cohesive medium perpetually seeking equilibrium. Friction causes elastic bodies—electrics—to expel their electrical matter, which surrounds them as atmospheres either quiescent, eruptive or vortical.[48]

ACR. The quiescent and cohesive atmosphere draws by adherence of the particles of the electric matter to one another and to the immersed object; the vortical, by a density gradient; the eruptive, by 'the law of centrifugal forces.'

45. Winkler, *Gedanken* (1744), in *Recueil* (1748), 69–73, 76–8, 88–112, 145; Waitz, *Abhandlung* (1745), 22–3, 49–64; Gralath, Nat. Ges., Danz., *Vers. Abh.*, 2 (1754), 380–92.

46. Winkler in *Recueil*, 98, 101, 146–51, and *Eigenschaften* (1745), 114, 131–5; Waitz, *Abhandlung* (1745), 31–6, 67–70.

47. As identified by Gralath, Naturf. Ges., Danzig, *Vers. Abh.*, 2 (1747), 392–3.

48. Waitz et al., *Abhandlungen* (1745), 75–7, 82–8; 98–9, 112–20; 199, 237, 287. Here semicolons distinguish the essays, which occupy pp. 71–87 (Unger's), 90–192 (the prolix electrician's), 193–237 (the French essayist's).

Contact in all instances confers an atmosphere. Repulsion follows in the first case from the action of the air, which pulls on the conferred cohesive atmosphere; and in the others from the elasticity of the electrical matter.[49]

The two electricities. Unger omits Dufay's triumph entirely and the French essayist rehearses Fontenelle's image. The prolix writer says that the effect cannot be understood solely in terms of relative vortical strengths; one must add that resinous atmospheres have more sulphur particles than vitreous ones.[50]

Electrification by communication. Glass is semi-permeable to the electrical matter, a fact one should note in applying the Rule of Dufay. One essayist thinks the communicated atmosphere is composed of electrical matter from the communicating body; the others derive it at least partly from the natural complement of the receiving body.[51]

Sparks and glows. Fire and sulphur particles present in the atmospheres glow when put into motion by the passage of electrical matter from one atmosphere to another. The appearances depend on the strengths of the interacting atmospheres and on their mixtures of fire, sulphur and electrical matter; metallic and animal bodies have sulphurous envelopes; vigor of sparks is proportional to difference in vortical strengths.[52]

Hauksbee's threads. Unger suggests that the electrical matter attracting the threads runs into the finger; the French writer follows Dufay; the prolix individual conjectures that the finger pushes aside some of the effluvia from the rubbed external surface, which would otherwise penetrate the glass and support the threads.[53]

These rough, amateurish, superficially plausible systems, the work of a classics professor, a mining engineer, a small-town mayor, a physics professor and an unknown, are typical of qualitative baroque physics. *Post factum,* with less predictive value than the phenomenological rules they inexactly re-express, they confronted no real constraints, no standards of precision that might decide between one set of mechanical images and another. They are casual, passing conjectures on a subject of current interest. That the self-conscious new academicians of Berlin distinguished Waitz's and three similar essays says much about the state of the nonmathematical physical sciences in the mid-eighteenth century. Contemporaries thought the Berlin prize appropriately bestowed, though an anonymous critic complained that Waitz had not attained to causes, and Gordon objected that understanding of the relevant physics of light, fire, attraction and repulsion was too meager to justify the competition at all.[54]

49. *Ibid.,* 76–7; 126–32; 199, 210.   50. *Ibid.,* 183–91; 210–11.

51. *Ibid.,* 84, 88, 100–8, 124–6; 200–3.   52. *Ibid.,* 75–6; 149–70; 220–7.

53. *Ibid.,* 85, 191–2; 214.

54. Piderit, *Diss.* (1746), 15–18, citing Gordon and the anonymous author, perhaps Hollmann, of a 'Sendschreiben' of 1745 to the Berlin Academy.

CHAPTER XI

Electricity in France after Dufay

1. CARTESIAN HOPES

Fontenelle seized on the 'prodigious' and 'practically incredible' new phenomena of electricity as plain and timely proof of the beleaguered system of vortices. The mere need for preliminary heating and rubbing, he wrote in résumé of Dufay's first memoir, shows that electrified bodies are surrounded by a vortex of very subtle and very mobile matter. 'There is nothing conjectural about these vortices any more,' he continued. Electrification by communication is the sharing of a vortex.[1] Collisions between the conferred and the conferring vortices bring about repulsion and suspension. As we know, the two electricities, though an 'unforeseen paradox,' presented Fontenelle with no difficulty of principle. The course of subsequent researches only strengthened his conviction that electricity provided an ocular demonstration of Cartesian principles. 'The vortices indicated by the attraction and repulsion of electrified bodies,' so he wrote in review of Dufay's last memoirs, 'are confirmed every day.'[2]

Yet the theory of vortices does not agree with the approximately linear trajectories of drawn and driven bodies and the possibility of suspension.[3] The consequence, that electrical atmospheres are not awhirl, was drawn before the Paris Academy in July 1734 by Joseph Privat de Molières, successor to Varignon at the Collège Royal and devout Malebranchiste. Molières had left the Congregation of the Oratory, in whose provincial colleges he had taught humanities, to sit at the feet of Malebranche; this pedigree and geometry brought him into the Academy as adjoint mechanician in 1721.[4] Molières' chief purpose was to save the great solar vortex by referring the impossibilities alleged by Newtonians— how can comets move against vortices powerful enough to carry planets? how can planets submerged in a whirlpool obey Kepler's laws?—to an improved mechanics of mini-vortices. 'The difficulty of the enterprise, the danger of its execution, the merit and reputation of his adversaries could not stop him.'[5] The program was carried out in four volumes ostensibly devoted to the 'queer busi-

---

1. *HAS* (1733), 6–13. Fontenelle's mechanism, an exact echo of Huygens' (*supra*, vii.2), goes beyond anything Dufay had said explicitly.
2. *HAS* (1737), 3.
3. This became a common ground for rejecting electrical vortices.
4. Savérien, *Hist.*, VI (1768), 221–2; Mairan, *HAS* (1742), 195–205; AS, Proc. verb., LIII (1734), f. 216; *DSB*, XI, 157–8.
5. Mairan, *HAS* (1742), 202. Molières differed from Malebranche in supposing a hard core to his mini-vortices.

ness,' in the view of the Marquise du Châtelet, of conciliating the systems of Newton and Descartes. In fact the only fundamental Cartesian teaching Molières altered was the doctrine of centrifugal forces, about which he admitted Descartes knew nothing; otherwise it was Newton who had 'lost the track' by 'sequestering his great riches under the somber veil of attraction and the void.'[6]

Molières' electricity works by a subtle matter composed of mini-vortices and confined in the pores of bodies by the tiny whirlpools that constitute the globules of the second matter. Friction causes most objects to expel their mini-vortices, which expand in the pores of the air until equilibrated by the globules. The expelled material forms an elastic atmosphere, whose spring derives from the vortical balance. The atmospheres can stimulate the emission of mini-vortices (or electrify by communication); all bodies yield to the stimulus, even metals, whose hold on electrical matter friction cannot overcome. The air succumbs easily, a point of theoretical importance: as the mini-vortices flow from their confining pores, an answering current sets in from the air and surrounding objects, a current that both preserves the Cartesian plenum and, by its collisions with the initial outward flow, gives rise to sparks and glows.[7]

Molières' explanation of ACR may be clear from fig. 11.1. The glass globe A possesses a symmetric electrical atmosphere whose density decreases with distance from its center; XMY in fig. 11.1a represents one of its isobars. Each element of the surface of the gold leaf B emits electrical matter in proportion to the density of the contiguous portion of A's atmosphere. Molières supposes that a secondary atmosphere results, with elliptical isobars like MN, B lying near the focus furthest from A. The vortex-particles of the second matter with which the electrical strives for equilibrium now endeavor to place B in the center of its atmosphere, and push it towards A. Since B is thereby accelerated, it will have moved towards the near focus by the time of union (fig. 11.1b); the elastic

a.   b.

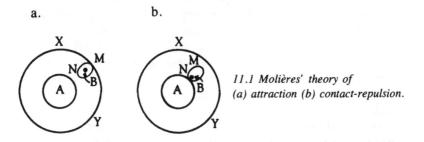

11.1 Molières' theory of (a) attraction (b) contact-repulsion.

6. Molières, *Leçons*, I (1734), iii–x; Molières to Sloane, 20 April 1734, in Sloane Mss., 4053, f. 202; Marquise du Châtelet to Maupertuis, 19 Nov. 1738, in Du Châtelet-Lomont, *Lettres* (1958), I, 271. The Marquise referred to another compromiser, the abbé Gamaches, who deplored the games most electricians played: 'Ils ont beau assembler leurs amis et s'électriser tous ensemble avec tous les meubles . . . on ne voit pas que cela les avance vers la cause de l'électricité ' E. S. de Gamaches, *L'Hypothèse des petits tourbillons* (1761), 326, quoted in Rodis-Lewis, *Malebranche* (1963), 337.

7. Molières, *Leçons*, III, 429–37; cf. Brunet, *Introduction* (1931), 320–4.

mechanism of the equilibrating matter then pushes the leaf away until equilibrium (suspension) is achieved. The hovering leaf may now be drawn by a wax rod because the vitreous atmosphere, having 'dissolved' the resinous, extends itself elliptically, and sets in train the mechanism of attraction.

Molières' extravagant Cartesianism was losing ground when the third volume of his *Leçons* appeared in 1737. 'It is a house collapsing into ruins, propped up on every side,' the Marquise du Châtelet wrote Dufay. 'I think it would be prudent to leave.' But the residents of the crazy tenement declined to move. Molières fought desperately and died, it is said, of a cold caught when leaving the Academy in a red-hot rage.[8] His disciples de Launay and E. S. de Gamaches carried on the struggle, not without the sympathy of senior academicians. Much to the annoyance of the Marquise, Réaumur had written Voltaire about 'the great debt that physics owed to P. Malebranche for having acquainted it with so many different orders of vortices.' And Dortous de Mairan, Fontenelle's successor as Secretary of the Academy, not only approved Molières' program, but found support for it in the writings of the Antipope: Newton must have intended a system like the mini-vortices, according to Mairan, for otherwise he would have had to ascribe the spring of his aethers to actions at a distance, a peripatetic perversion of which we cannot suppose him capable.[9]

Others came forward to improve Molières' electricity. Already the Malebranchiste Mazières, in his *Traité des petits tourbillons* (1727), had guessed that electrical attraction and repulsion originated in a double current of mini-vortices. This possibility was exploited in the 1740s by Laurent Béraud, S. J., professor of mathematics and director of the observatory of the Jesuit college at Lyon. He was a painstaking and exact astronomer—it took him ten years to establish the longitude of Lyon—and an ingenious if superficial physicist.[10] In 1748 his ingenuity brought him the prize offered by the Academy of Sciences of Bordeaux for the best answer to the question whether electricity and magnetism are related. Yes, according to Béraud, both operate by subtle-matter vortices; electrics are characterized by a gradient in their aether, which is denser within than in the immediately contiguous air. Rubbing expands the imprisoned mini-vortices, which rush out until reaching air with a normal content of aether; they are driven back, setting up a dual current that accounts for attraction and repulsion. 'The mechanism of electricity seems simple to me.' Béraud tried it out again in 1757, and received honorable mention in a prize competition on the 'true cause of electricity' held by the Petersburg Academy. One reason for the longevity of this tired theory was its superficial agreement with the received effluvial system of the time. 'I am the more attached to my

8. Savérien, *Hist.*, VI (1768), 234–5. For the dispute see Brunet, *Isis*, 20 (1934), 367–95, and Aiton, *Ann. Sci.*, 14 (1958), 141–5.

9. Mme du Châtelet to Maupertuis, 7 Sept. 1738, in *Lettres* (1958), I, 243 (Réamur); *ibid.*, I, 261 (Mairan); *HAS* (1749), 53–8.

10. Delattre, *Etablissements* (1939–57), II, 1553.

opinion [Béraud had written in 1748] because it is not opposed to the system of effluent and affluent matter of M. l'abbé Nollet, who has written so well on the subject; I pride myself on profiting from his views and his experiments.'[11]

2. NOLLET'S APPRENTICESHIP

Jean Antoine Nollet was a peasant from Pimprez, a village about sixty miles north of Paris. His father, who could scarcely write his name, agreed reluctantly to allow him to be educated—for the priesthood of course—when the local curé perceived the boy's powers. At eighteen, after humanities at the provincial college of Clermont, Nollet went to Paris to prepare for his ministry. He supported himself as a private tutor while working his way through the prescribed curriculum, whence he emerged, in 1724, a Master of Theology. But he did not become a priest. As recreation from theology he had amused himself with glass blowing and enameling, at which he became adept; he discovered that he preferred the manual arts to preaching, and experimental physics to theology. He resolved to end his clerical vocation, to retire with a deaconship and the equivocal title 'abbé,' and to seek a livelihood in the unpromising borderland between art and science.[12]

In 1728, at about the time he became deacon, he was invited to join the Société des Arts, a small group with a grandiose scheme for promoting technological innovations in France and abroad.[13] The Society, which included Polinière and Nollet's future colleagues in the Paris Academy, Grandjean de Fouchy, Clairaut and La Condamine, expired in two years, partly from the selfish opposition of the Academy and partly from the disparity between its ambitions and its means. Its financial resources derived from the Comte de Clermont, whose ample income, drawn from the richest abbeys in France, did not suffice for his women and his savants.[14] The short-lived Society probably decided Nollet's future, however, since it was doubtless through contacts made there that he became Dufay's assistant in 1731 or 1732.

After a year or two with Dufay, Nollet became chief of Réaumur's laboratory, where he fertilized frogs and improved his new superior's new thermometer.[15] Dufay, continuing to help him, arranged those trips to England and Holland that brought acquaintance with Desaguliers, 'sGravesande and the brothers

11. Béraud, *Dissertation* (1748), 24; Béraud in J. A. Euler et al., *Dissertationes* (1757), 136–9, 167–9, 183–5. For the Petersburg competition, *infra*, xvi.3; for the close association of Béraud and Nollet, Nollet, 'Journal,' ff. 219–20.

12. Lecot, *Nollet* (1856), 1–10; Quignon, *Nollet* (1905), 3–4; Torlais, *Physicien* (1954), *passim*.

13. Cf. the attempt to enlist Cramer as a foreign member; Clairaut to Cramer, 8 Jan. 1730, in Speziali, *RHS*, 8 (1955), 203.

14. Lecot, *Nollet* (1856), 13; Quignon, *Nollet* (1905), 5; Cousin, *Clermont* (1867), I, 1–10, 108; R. Hahn, *Anatomy* (1971), 108–10. Cf. Condorcet, *HAS* (1788), 37–49; Grandjean de Fouchy, *HAS* (1770), 121–37, and *HAS* (1765), 144–59; d'Alembert, *Oeuvres* (1805), XI, 405–25.

15. Torlais, *Physicien* (1954), 29 ff.; *Esprit* (1961²), 81 ff.

Musschenbroek. After the Dutch visit of 1735 Nollet set up as a private lecturer. In 1739 the Academy admitted him as adjoint mechanician, and the King of Sardinia invited him to instruct the heir apparent at Turin; in 1741 the Bordeaux Academy sponsored his lectures and three years later he entertained the Dauphin and the Queen at Versailles (fig. 11.2).[16] Eventually he collected a chair of physics at the Collège de Navarre (1756), an annual lectureship at the military schools of La Fère and Mézières, the succession to Réaumur as pensionary in the class of mechanics (1757), appointment as 'Preceptor to the Children of France' (1758), and the glamor of 'a middle-sized star in the heaven of the learned.'[17]

One can see from the six volumes of his *Leçons de physique* that Nollet's presentations were clear, ordered, comprehensive and timely.[18] He took his material from astronomy and experimental physics, particularly mechanics, optics, pneumatics, magnetism and electricity. Electricity figured in his lectures from the very beginning. His first syllabus, published in 1738, promised performances of the Anglo-French experiments and demonstrations of Dufay's rules, including those of the two electricities. The electrical performances, played on three dozen instruments (about 10% of those required for the entire course), aroused great interest, especially at the Bordeaux Academy, which was inspired to set electricity as the subject of its essay competition of 1742.[19] Until 1745 these demonstrations remained rehashes of the work of Hauksbee, Gray and Dufay. In February of that year word reached Nollet of the antics of Bose and ignition by sparks. They interested him both as lecturer and as physicist; he threw himself into their study, from which he emerged, three months later, with the elements of the theory of simultaneous effluence and affluence.

3. THE SYSTÈME NOLLET

'About three months ago,' Nollet wrote E. F. Dutour in April 1745, 'I learned that after Mr Hauzen and other German scientists Mr Bose, Professor of Physics at Wittenberg, used to excite electricity with glass globes, rather than with tubes; that he did not spin these globes with a wheel of small diameter, as we do, but with a large wheel, like a cutler's; that rubbing in this way, which,

16. The course was much talked about at Versailles, and earned Nollet several handsome presents. Cf. Luynes, *Mémoires* (1860–5), V, 452–3, VI, 3, 479.

17. Torlais, *Physicien* (1954), 51–63, 203–4; AS, *Index biog.* (1968), 409; J. Teleki, re Nollet's public lectures in 1760: 'N . . . gehört, wie ich glaube, am Gelehrtenhimmel unter die Sterne mittlerer Grösse' (Spiess, *Basel* [1936], 99). Cf. Gottsched to Ledermüller, 22 Feb. 1756, in Reicke, *Neues* (1923), 43: 'Ich schätze unser Vaterland glücklich, dass wir auch wirklich noch solche geschickte und eifrige Naturforscher haben, die das Reich der Wissenschaften, allen Reamürs und Nollets zu Trotze erweitern können.'

18. The *Leçons* were often reprinted and widely used. Cf. J. Black to J. Watt, 19 Feb. 1768, in Robinson and McKie, *Partners* (1970), 10–11.

19. Nollet, *Programme* (1738), 99–104; *DSB*, X, 145–8. The instruments employed by Nollet at Turin passed to the University there and thence, ironically, into the hands of the arch-Franklinist G. B. Beccaria. Nollet, *Lettres* (1760), 149 and Journal, f. 11; *infra*, xv.4.

11.2 Nollet entertaining at Versailles. From Nollet, Essai (1750²).

perhaps, he had only intended for convenience, he electrified much more strongly than we, to the point of igniting spirit of wine, etc., something we never could do; that the hair of a man so electrified became luminous, which he jokingly calls *beatifying electricity*; that sparks from his fingers killed flies; that drops of his blood looked like drops of fire in the dark; and that water dripped more quickly from a leaky vessel when the vessel was electrified. As you can imagine, these facts greatly intrigued me, and I didn't sleep until I had had a great wheel built, fully equipped with globes. . . . '[20]

Using this machine Nollet invented two dozen demonstrations and a new theory. He wrote Dutour: 'In the light of all these experiments and observations and of many others done before mine, I've made myself a system, of which I'll try to give you a rough idea; I would not take the risk of presenting it without all the necessary experimental evidence to anyone but yourself, who are fully conversant with the subject.'[21] The theory sketched for Dutour was the substance of 'Conjectures' on the cause of electricity that Nollet read to the Academy two days later, and of his book, *Essai sur l'électricité des corps*, first published in 1746.[22]

The 'Conjectures' affirm the existence of an electrical atmosphere, a halo of electrical-matter-in-motion surrounding electrified bodies. Nollet appeals to the testimony of the usual four senses, to the sparks, the pricklings, the hissings and snappings, and the peculiar odor—which distressed his dog and his parrot—surrounding a working electrical machine.[23] The electrical matter, which Nollet supposes to be a combination of elementary fire and grosser mate-

20. Nollet to Dutour (a corresponding member of the Academy who became Nollet's staunchest supporter against the Franklinists), 27 April 1745 (Burndy):'Il y a environ trois mois que jay appris que Mr Bose professeur de physique à Wittenberg, d'après Mr Hauzen et quelques autres scavants d'Allemagne et premièrement (?) de Berlin, au lieu de tubes electrisoit des globes de verre qu'il faisoit tourner non avec des roues de petits diamètres, comme nous, mais avec de grandes roues de couteliers; qu'au moyen de ce frotement, où il n'auroit cherché peutêtre que l'aisance, il electrisoit beaucoup plus fort que nous, au point d'enflammer l'esprit de vin etc ce que nous n'avons jamais pu faire; qu'un homme électrisé de cette manière devenoit lumineux par sa chevalure, ce qu'il appele par plaisanterie, l'électricité béatificante; que les étincelles qui sortoient de ses doits tuoient les mouches; que les goutes de sang d'un homme électrisé ainsi, parroissoient des goutes de feu dans l'obscurité; qu'un vaisseau qui s'écoule, étant électrisé, avoit un écoulement accéléré. Ces faits comme vous le pouvez croire, me misent aux [anges] et je ne dormis point que je neusse fait faire une grande roue toute équipée des globes. . . .'

21. *Ibid.*: 'Apres toutes ces expériences, ces observations et beaucoup d'autres qui avoient déjà été faites avant les miennes je me suis formé un système, dont je vais tacher de vous traire une légère idée; à tout autre que vous je ne hazarderois pas de le donner, pour extrait, car, ce sont les preuves d'expériences qui peuvent seules le faire valoir, mais vous êtes au fait de la matière.'

22. Nollet, *MAS* (1745), 107–51. The *Essai* appeared in Spanish and Italian in 1747, and again in French in 1750; the German, English, Dutch and Latin editions mentioned in *HAS* (1746), 30, are probably fictitious.

23. *MAS* (1745), 110–12; *Essai* (1750²), 67. 'Car une substance qui touche, que l'on entend agir, qui se rend visible en certain cas & qui a de l'odeur, peut-elle être autre chose qu'une matière en mouvement.' Nollet and his animals lodged in the Louvre; Nollet to Dutour, 27 April 1745, and Luynes, *Mémoires*, VII, 238.

rial, exists in all bodies, and flies from them whenever their parts are sufficiently agitated to expel it. Friction can agitate the electrical matter of electrics; only moving effluvia can do the trick in metals. The expelled matter, or 'effluent stream,' does not move vortically; it is often seen diffusing quietly from pointed objects as a luminous cone. Nollet argues from a wealth of examples and analogies that even when nonluminous the electrical matter issues from the pores of an electrified body in divergent jets. One demonstration will suffice. Spread fine powder on the prime conductor, and electrify it. The dust leaps up from many points on the conductor's surface, pouring forth like so many water-jets (fig. 11.3), whose streamlines, Nollet supposes, are those of the effluent flux.[24]

The divergent jets were the foundation, and very likely the immediate inspiration, of the système Nollet.[25] Here the German influence was decisive. Before he had learned of Bose's work Nollet had taught, following Dufay, that electrical matter was distinct from that of light.[26] Ignition by sparks, however,

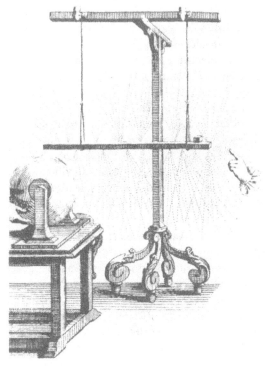

11.3 The stream lines of the hypothetical effluent flow. From Nollet, Essai *(1750²).*

24. *MAS* (1745), 124–36; *Essai* (1750²), 65–79, 84–93, 157–64.

25. The connection of ideas is perhaps less evident in the 'Conjectures' and the *Essai* than in the letter to Dutour, where the first six of the twenty-four experiments reported relate to discharges.

26. Nollet, *Programme* (1738), 99–104: 'On apprend que cette lumière est indépendante de l'électricité, que l'une subsiste quelquefois sans l'autre.' Writing Dutour on 13 (?) Aug. 1744 (Burndy), Nollet remarks that though electricity 'tient de tout près à la lumière' the glow in the

and the continuous luminosity produced by the electrical machine made the identification of the matters of fire, light and electricity irresistible. Premising their identity, Nollet could infer the trajectory of the effluent matter from the appearance of the brush discharge. He ignored the difficulties Dufay had raised. In an important shift in emphasis, Nollet preferred to identify the agents of light and electricity and leave unexplained instances where luminous and electrical effects do not occur together than to admit two distinct matters, and leave the German experiments unintelligible.

Let us return to the dusty prime conductor, where most of the powder still sat, trapped, we should say, by moisture or grease, or by irregularities in the metallic surface. But to Nollet, who believed with his mentors that 'mechanical explanations are the only ones capable of advancing experimental physics,'[27] the powder's apparent adherence showed that an incoming or 'affluent' flow of matter pressed down upon it, and that no effluent jet disturbed it. He inferred that effluences did not emanate from every pore of an electric and that, simultaneously with the outward flow, currents of electrical matter set in towards the electrified body from all substances, including the air, in its vicinity. Nollet found further evidence for this inward flow in sparks that dart from the finger towards electrified objects, and in the consideration that, without such compensation, a body could be exhausted of its electrical matter. He had never experienced even a dimunition, although he had worked repeatedly with the same glass globes.[28]

Nollet's system was much superior to earlier theories of double flux, such as Molières' or Hauksbee's. A chief difficulty with them (as with the theory of pulsating atmospheres) had been to explain why attraction occurs first, followed by contact and then repulsion. Nollet answered that the effluent and affluent flows differ not only in direction, but in velocity and spatial distribution as well. The effluent stream consists of many jets, each of which spreads out into a cone with apex at one of the scattered exit pores. The affluent current, being a hodgepodge of differently directed effluent streams, is approximately isotropic.

Consider an electrified body of cylindrical shape surrounded by small objects EFG immersed in a moderate affluent flow, well removed from the strong divergent effluent currents (fig. 11.4): nothing impedes the 'attraction,' the drift of these mites towards the excited electric. When the small body touches or nears the electric, it is stimulated to emit its own effluent jets, to which Nollet assigned properties similar to those of Hauksbee's stiff effluvia; the charging of the small object proceeds like the unfolding of the tentacles of an octopus, giving purchase to effluent spurts that 'repel' or drive it away (as at H, fig.

mercurial phosphorus is probably not accompanied by electricity; adventitious air currents, he thinks, explain why the German experiments apparently confirmed the contrary.

27. Nollet to Bergman, 20 Sept. 1766, in *Bergman's For. Corr.* (1965), 285.

28. *MAS* (1745), 135–7; *Essai* (1750²), 75–89. Of course the usual attraction also confirms the affluent's existence.

*11.4 The système Nollet, showing the divergent effluents and isotropic affluent; below
is the first depiction of the Leyden experiment. From Nollet,* Essai *(1750²).*

11.4). It can remain suspended at a height determined by its weight and the
local strengths of the opposed fluxes.[29]

Why does a suspended leaf fly to a finger presented to it? The explanation
requires the proposition, then granted by everyone, that the electrical matter
flows less readily through the air than it does through denser bodies, excepting
sulphurous, waxy, and resinous substances. The effluent stream from the floating leaf converges toward the finger, as does the luminous matter of a brush
discharge (cf. the extreme right of fig. 11.3); the affluent matter bombarding
the leaf has less purchase on its near side—that facing the finger—than on its
far side, and so drives it toward the hand. If instead of the finger a wax rod is
presented to the leaf no attraction occurs, as the wax, being less permeable to

29. *MAS* (1745), 139–42; *Essai* (1750²), 148–52. Cf. the reification of stream lines in Kepler's
anima motrix.

the electrical matter than the air, does not allow the necessary convergence of the effluent flow.[30]

Repeat the last experiment, having first rubbed the rod, and the affair turns out differently: the leaf goes to the wax, precisely the phenomenon that had led Dufay to recognize the two electricities. But there is no room in Nollet's scheme for *qualitatively* different electricities; there are only differences in the strength and direction of current flow, and in the electrical permeability of bodies.[31] Now a resinous electric is less opaque to the electrical matter when excited than when not, for when electrified it freely passes an effluent, and admits an affluent flow. Observation also shows that the effluent from a resinous body is weaker than that from a vitreous one. The suspended leaf comes to the rubbed wax for two reasons: the permeability of the rod to the leaf's effluent activates the same mechanism as in the case of the finger, while the weak effluent from the rod offers little opposition to the motion caused by the strong affluent striking the far side of the leaf. Had the leaf been resinously electrified repulsion would occur as between two vitreously charged objects, though less strongly.[32]

Besides its plausible accounts of ACR and the two electricities, the system provided a natural explanation of what Nollet called 'the most celebrated experiments of the last forty years,' Hauksbee's pointing threads. The threads tend toward the globe under the influence of affluent currents. Those within are oriented radially by an afflux from the enclosed air toward the concave side; those without obey a flow from the environment towards the convex side. The threads inside move away from the finger because the efflux from it, passing through the excited glass and augmenting that from the inside surface, overpowers the internal afflux. The assumed penetrability of glass is not, as in the case of the wax rod, a consequence of its electrification. With all his colleagues, Nollet believed glass to be freely penetrable up to thicknesses of the order of bottle walls, as evidenced by attractions across glass plates.[33]

Sparking also falls easily into Nollet's system. Luminous effects originate in violent collisions between opposing fluxes, stripping gross matter from elementary fire, which produces heat and light in its accustomed manner. The spark that springs from a strongly electrified object to the finger bespeaks an encounter between a powerful effluent jet and an affluent from the hand. In the collision the air is squashed or distended, giving rise to the discharge noise; the affluent, driven violently backwards, strikes the finger with the force experienced as a shock; while the ricocheting effluent, by plugging the available outlets, destroys the electricity of the electrified object.[34]

That is the système Nollet. 'At first sight,' according to Sigaud, 'nothing is

30. *MAS* (1745), 143–5; *Essai* (1750²), 152–4; Nollet to Dutour, 27 April (Burndy).

31. Later Nollet wrote that he had earlier accepted Dufay's distinction only 'à regret & avec une sorte de répugnance.' *MAS* (1746), 8.

32. *MAS* (1745), 145–6; *Essai* (1750²), 154–7. The leaf when resinously charged excites a relatively weak affluent to its far side.

33. *MAS* (1745), 146–7; *Essai* (1750²), 93–115, 164–5.

34. *MAS* (1745), 147–51; *Essai* (1750²), 178–216.

simpler, nothing more ingenious than this hypothesis.'[35] Contemporaries admired the boldness and clarity of the scheme, its merciless reduction of all electrical phenomena, not sparing the two electricities, to quantitative differences in the directions and speeds of flow of one universal, fiery, electrical matter. 'In fact, however,' continues Sigaud, 'when one looks into it more closely, one finds that the system involves insurmountable difficulties.' Sigaud's main objection is that, if the quantities of the effluent and affluent flows are equal, suspension will not be possible. But if one admits, with Nollet, that electrification increases the effective volume of a gold leaf, it might well find a position where its weight, the weak homogeneous affluent, and the strong divergent effluent balance.

The trouble in Nollet's system lay elsewhere. First, its selectivity, its close dependence on the details of ACR, made it inflexible, almost inapplicable to novelties like the Leyden experiment. Second, its imprecision allowed it to explain its own failures too easily: if repulsion, or nothing, occurs occasionally where attraction was expected, one inferred a momentary, local, inconsequential reversal of the usual flow, and ignored the anomaly.[36] Intially the system's stock of easy explanations was an asset. When confronted by Franklin's overly precise distinctions, however, its serious methodological weakness became clear. Finally, we might notice among the fundamental problems of Nollet's system that, like all effluvial theories, it contained the seed of its own destruction in the ambivalent role it assigned to glass.

These incipient difficulties did not become acute until the 1750s. From 1745 to 1752 Nollet's system enjoyed the widest consensus any electrical theory had yet received. The French, apart from scribblers we shall notice momentarily, embraced it enthusiastically. 'Not only does it suffice to explain the facts on which it was founded,' wrote Mairan's successor, Grandjean de Fouchy, in the Academy's *Histoire* for 1745, 'but all others since discovered.' Réaumur was delighted with his protégé's theory, 'a more probable and natural explanation can scarcely be expected.'[37] The scheme did well internationally also. William Watson, England's leading electrician, emphasized the similarity between his views and those of 'that excellent philosopher,' our abbé.[38] Musschenbroek, the Dutch physicist most knowledgeable in the matter, became a firm disciple and remained so until his death; and the most intransigent of the mechanists of the age, the Genevan Lesage, had only praise for a theory that so beautifully exemplified the true method in physics.[39]

35. Sigaud de Lafond, *Précis* (1781), 112.

36. Cf. Nollet, *Essai* (1750²), 148–50.

37. Grandjean de Fouchy, *HAS* (1745), 4; Réaumur to J. F. Séguier, 25 May 1747, in Réaumur, *Lett. inéd.* (1886), 60. Cf. Bose, *Recherches* (1745), liv–lv; Nollet, *Lettres* (1767), 209, 224, for the system's success among provincials, and Dangerville, *Mem. hist. sci. beaux-arts* (Aug., 1762), 2027: 'Dieux! quelle magnifique scène prépare la main des Nollets . . .'

38. Watson, *PT*, 44:2 (1746–7), 748; *PT*, 46 (1749–50), 348–56; *PT*, 48 (1753), 201–16. Cf. *FN*, 449–50; and for late survivals of Nollet's influence in Britain, Schofield, *Mechanism* (1970), 128, 132–3.

39. Sigaud de Lafond, *Précis* (1781), 113; Prevost, *Notice* (1805), 254; *infra*, xiii.

Bose did Nollet the honor of claiming to be an independent discoverer of the theory of the double flux, which he worked out in a little book in French published in 1745. His theory is indeed much the same as Nollet's;[40] and he graciously accepted the abbé's version of the chief point of difference, the mechanism of afflux. Nollet saw nothing extraordinary in this coincidence: 'Mr. Bose of Wittenberg,' he wrote Dutour on December 10, 1745, 'has just sent us a dissertation on the causes of electricity; his system and mine coincide in many respects; I'm not surprised. I flatter myself that anyone acquainted with physics who examines electrical phenomena at his leisure will hit on the same ideas.'[41]

The opposition the system intially encountered was so little serious that it underscored the consensus. Four gladiators rose against Nollet between 1746 and 1748. All were newcomers to electrical experiments, often wrong on matters of fact and, less pardonable, unable to discern that their opponent, with his experience and resources, was beyond the reach of their feeble dilettantism. Each attempted, in a different way, to revive the theory of impelled air. One of them, Jean Baptiste Secondat, the son of Montesquieu, held that rubbing the globe draws the subtlest parts of the air across the glass and shoots them out again to agitate the atmosphere;[42] another critic, Jean Morin, professor of philosophy at the Collège Royal at Chartres, believed that the air's pressure both caused attraction and retained the electrical atmosphere, which caused repulsion;[43] while a professor of philosophy at the University of Naples, Niccolò Bammacaro, invented a redundant mechanism whereby a pulsating electrical matter drove the air, which carried mobile bodies.[44] These gentlemen also

40. Bose ascribes different velocities and spatial distributions to the two streams (*Recherches* [1745], xviii–xix), contents himself with explaining ACR, suspension, Hauksbee's threads (*ibid.*), sparks (xxix–xxxv), and the two electricities, which he refers, without details, to a difference in effluent strengths (xxxv–xxxvii). Of particular interest are his denial that ACR is absolutely regular (xviii), his discovery that the prime conductor does not charge if rubber and machine are insulated (xli–xlii, and *infra*, xiii.3), and his strong, old-fashioned Cartesianism: the electric matter, composed of Cartesian globules, shoots forth when warmed by friction, and returns when cooled by air (ix–xvi), the double motion constituting a 'vortex,' which, however, is not in rotation (!), except around the globe, where a tangential force exists (xxxix–xl). Bose's presentation is less forceful and detailed than Nollet's. As for his independence, Bose claims to have entertained the theory since 1738, but to have written nothing until June 1745, when he sent his ideas to Nollet and Réaumur for comment. Both replied in July, informing him of Nollet's 'Conjectures' (xliii–lv).

41. 'Mr. Bose de Wittenberg vient de nous envoyer une dissertation sur les causes de l'électricité; son système et le mien se réunissent dans bien des choses; je ne suis pas surpris. je me flatte que les mêmes idées viendront à tout homme un peu initié en physique, qui aura vu à loisir tous les phénomènes électriques.' (Burndy.)

42. Secondat, *Mémoire* (1746), 17–33. Nollet, *Recherches* (1749), 5–32, wrote of an anonymous work with the same title, date and substance as Secondat's; in *Essai* (1750²), 242, he identified the author as a M. Boulanger.

43. Morin, *Nouv. diss.* (1748), 56–66, a pompous and obscure essay.

44. Bammacaro, *Tentamen* (1748); cf. Nollet, *Recherches* (1753³), 56–75. Bammacaro's work is literary and derivative; '[il] n'a commencé que depuis deux ans à se mesler de physique'; Nollet, 'Journal' (1749), f. 169. The fourth opponent, Louis, a surgeon at the Salpêtriére, directed himself

argued against Nollet, respectively, that the rigorous regularity ACR would not obtain on the dual-current system; that the affluent flow is opposed to sound physics and good sense; and that it is both useless and hypothetical.[45]

Nollet's prolix response ought to have cleared the field. Undiminished attraction across glass barriers and *in vacuo* shows, he says, that ordinary air cannot play the part his opponents required of it. As for the affluent stream, the backbone of his system, its existence is proved by a thousand experiments: attraction, the increased flow from water-jets placed near an electrified body, the sparks seen to issue from a finger advanced toward the prime conductor, etc. As to Secondat's objection, it falls to the ground; for Nollet does not hold that ACR is absolutely valid, and rehearses cases wherein, he says, repulsion preceded attraction. These clarifications did not warn off three of his critics, whose replies are too captious to interest us. The fourth, Bammacaro, Nollet personally converted when he visited Naples in 1749, on a mission that was itself a mark of his international stature as an electrician. He crossed the Alps in order to examine claims of electrical cures puffed by Italian physicians. His expert, tactful, decisive debunking of these claims further consolidated his position, and even won him a kind word from Benjamin Franklin.[46]

---

to medical applications; cf. Nollet, *Recherches* (1753³), 32–56, and [Mangin], *Hist.* (1752), Pt. ii, 46–103.

45. Secondat, *Mémoire* (1746), 15–16; Morin, *Nouv. diss.* (1748), 62, 101–2, 179–86; Nollet, *Recherches* (1753³), 56–75; [Mangin], *loc. cit.*

46. Nollet, *Recherches* (1753³), 12–20, and *Essai* (1750²), 148–51, 219–63; Quignon, *Nollet* (1905); *EO*, 115–16; Bose, *Recherches* (1745), xviii.

CHAPTER XII

Electricity in England after Gray

1. DESAGULIERS' RULES

During the 'interregnum' that, according to Wheler, stretched from 1737, when he last attended to electricity, until 'the Germans took up the Sphere in Hauksbee's Method,'[1] the Reverend J. T. Desaguliers generated most of England's electrical power. The son of a prominent Huguenot pastor and schoolmaster, he had helped to educate his father's charges before going to Oxford, where he matriculated unusually late, at the age of twenty-two. Five years later, in 1710, having become a graduate and a deacon, he succeeded John Keill as lecturer in experimental philosophy at Hart Hall (Hertford College).[2]

According to Desaguliers, Keill was the first who 'publicly taught *Natural Philosophy* by *Experiments* in a mathematical manner,' by which he meant order and system, not geometry: 'The experiments made at the first lecture prove the precept given at the second, and so on.'[3] The Oxford performances, which began in 1704 or 1705, were more sophisticated, if less professional, than those Hauksbee inaugurated about the same time in London. 'As they [Hauksbee's] were only shewn and explain'd as so many curious *Phaenomena,* and not made use of as *Mediums* to prove a series of philosophical Propositions in a mathematical Order, they laid no such Foundation for true Philosophy as *Dr Keill's Experiments*; tho' perhaps perform'd more dexterously and with a finer Apparatus: they were *Courses of Experiments,* and his a *Course of Experimental Philosophy.*' Desaguliers chose to follow Keill, whose relative excellence he probably exaggerated, to try to render Newton's principles persuasive and their connections evident to persons with little or no mathematics.[4]

In the year of Hauksbee's death, 1713, Desaguliers carried the Oxford method to London, where he was received with the same approbation that his Cartesian contemporary, Polinière, enjoyed in Paris. The public cared not a fig for system; as Desaguliers' technique acquired the slickness of Hauksbee's, his demonstrations were appropriated with equal facility by Newtonians like 'sGravesande and by Cartesian fellow-travelers like Nollet. In 1714 Desaguliers

1. Wheler to B. Wilson, 7 Sept. 1748, Add. Mss., 30094, f. 71 (BL).
2. Torlais, *Rochelais* (1937), 1–13; *DNB,* V, 850–1. Keill was deputizing for the Sedleian professor, Sir Thomas Millington, when lecturing at Hart Hall; he did not become Savilian Professor of Astronomy until 1712. Musson and Robinson, *Science* (1969), 33.
3. Desaguliers, *Course* (1763³), I, ix; Hans, *New Trends* (1951), 139.
4. Desaguliers, *Course* (1763³), I, ix; cf. *FN,* 243–61.

harvested what remained of Hauksbee's inheritance, succeeding him in the confidence of Newton and the service of the Royal Society. His intimacy with the scientific establishment (Newton acted as godfather to his third son), his resourcefulness in experiments, and his Oxford learning, made him no mere expositor, but a philosopher, or even a prophet: a Newtonian savant, a minister of the established church and, ever more actively, the leader of English Freemasonry.[5]

As savant Desaguliers considered it his task to supply some of the missing secondary causes and operative laws of gravity, cohesion, electricity, magnetism and elasticity, the five 'powers' or 'principles' to which, he believed, Newton had reduced the physical universe.[6] He carried this program furthest in connection with electricity, which may have figured in his lectures, as it did in the rival course established by Hauksbee's nephew, from the beginning. He did not begin to contribute significantly to the subject, however, until after Gray's death, being unwilling, he said, to interfere with the old man, who 'had wholly turn'd his Thoughts that Way, but was of a Temper to give it intirely over, if he imagin'd that anything was done in Opposition to him.'[7] The study of electricity may have suffered from Desaguliers' delicacy.

In a manuscript dating from June, 1731, Desaguliers offers a guess, 'by way of Query, at the first Cause, or the Use' of electricity, which, following 'sGravesande, he understands as the property to 'attract and repel small Bodies alternately.'[8] Three 'observations' and an 'experiment' precede the query: the observations report the phenomena of the hunted feather, Gray's discovery of communication, and the existence of electrics other than glass; the experiment shows that the leaves of a lopped tree branch will attract and repel when the tube is held to its butt. 'Query. May not this Property be of very great Use to direct the Farina faecundans [pollen] in plants to the proper Utriculi or Flowers that are to receive it for producing Fruit? . . . If a small grain of the Farina darts into the little Flower or Utriculus design'd for it, the Clammy Substance in the Flower overcoming the successive Repulsion keeps it in its Place; but if the Farina strikes against any other Part (by observ: 1) it remains there but a little while, but flies back again to the next Branch to it, and so backwards and forwards till it jumps into, and is retain'd by, some other Utriculus. If it shou'd

---

5. Torlais, *Rochelais* (1937), 14, 35–6; *supra*, ii.4. As Grand Master, Desaguliers as usual ordered and systematized, codified rules, specified decorum, designed uniforms. Lee, *Desaguliers* (1932), 20–3; Rowbottom, Hug. Soc. Lond., *Proc.*, 21 (1968), 205–9.

6. Desaguliers to Sloane, 4 March 1730–1 (Sloane Mss., 4051, ff. 200–1): 'Whereas now our Principles are Four or Five at least, whose Causes we do not know, nor all the Laws of some of them, viz., Gravity, Attraction of Cohesion, Electricity, Magnetism and Elasticity: we only know that there are such Powers in Nature . . .'

7. *PT*, 41:1 (1739–40), 187. Desaguliers had done nothing notable in electricity before Gray's discovery either, as appears from JB, XIII (1726–31), 301, 316 (20 Feb. and 13 March 1728–9), and from his *Course*, I (1734), 17–21, 41–4. For the material of the younger Hauksbee see his *Course* (1714), under 'Pneumatics.'

8. RS, Class. Pap., XVIII:2, 34. CF. J. B., XI, 180–3, 342–3.

be found by Experiment, or Observation, that some material Power (as the Wind, or anything else), rubbing against the Trunk of the Tree, shou'd render its small Branches and leaves electrical, it would strongly support my conjecture.'

Note that Desaguliers makes repulsion 'successive' to attraction. He had glimpsed the regularity ACR in the course of refuting 'Mr. Friewald [Mårten Triewald] of Stockholm,' who in December 1728 had favored the Royal Society with the hypothesis that repulsion arises from outgoing affluvia, attraction from reflected ones. Triewald's theory, which reminds one of Gray's, had a similar fate, for neither it nor its refutation was published.[9] Had Desaguliers persisted, he might have identified ACR before Dufay.

When Desaguliers next turned his attention to electricity he had the advantage of Dufay's example. He showed the Society the two electricities in May 1737. Early in 1738 he appeared in his own right, with experiments designed for 'Making a discovery of the true Laws of Electricity.'[10] These 'true Laws,' or emphatic rephrasings of Dufay's phenomenological rules, relate to the arrangement of figure 12.1, where G is the receiver, H the 'thread of trial,' EF a packthread, and AB a 'supporter' made of dry catgut. Here are the laws, which Desaguliers expressed as queries: any body G does (or does not) receive via EF the virtue of a tube approached to E according as the supporter CD is (or is not) an electric.[11] The reason? Electric supporters 'soon become saturated' with the 'electric Stream,' admit no more, and allow the rest to pass down the 'conductor;'[12] nonelectrics freely transmit it if they touch the ground, and so cut off communication with the receiver. Desaguliers seems to hold Gray's view of the process of communication: stimulated emission via the transmission and collection of effluvia.

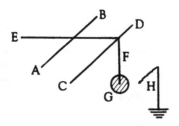

12.1 Desaguliers' scheme for testing for conductivity.

9. JB, XIII. 301, 306–7 (20 and 27 Feb. 1728-9); for Triewald, *supra*, ii.4.

10. JB, XVI (1736–9), 92 (5 May 1737), 211 (9 March 1737-8). Desaguliers had entertained 'foreigners' with electricity in the summer of 1736 (RS, Corr., D:2, 71–5).

11. *PT*, 41:1 (1739–40), 186–93. Desaguliers prefers 'such Bodies as are capable of having Electricity excited in them by Friction, Heating, Beating or Patting' to Gilbert's serviceable word. Later (*ibid.*, 203), he introduced 'electrical [and then 'electric'] *per se*' for Dufay's 'électrique par lui-même.'

12. Originally 'conductor' meant not a class of substances, but a component of an experimental arrangement; the modern meaning grew from shortening phrases like 'conductors that communicate electricity' or 'conductors that pass the virtue.' Desaguliers himself uses 'conductor' to mean the 'non-electrick Body' providing the communication (*Course* [1763³], II, 323), and Watson (*PT*, 43

Having grabbed Gray's scepter the interrex relaxed until challenged by the Bordeaux Academy's prize competition of 1742.[13] Another vacuum, another conquest: only two others entered, and Desaguliers won. Much of his dissertation is given over to stale novelties like Hauksbee's threads and glows, the dangling boy, the tube-and-feather, the two electricities, and the electrification of water.[14] It also offers an inconsistent yet briefly influential theory, in which deposits from several Newtonian strata can be identified. Recall that Desaguliers had been much concerned to legitimize repulsive force in general, and that, in the late 1730s, he had sketched a theory of Boscovich's type of the relation between attraction and repulsion. The occasion for his speculations then was Hales' work on air; and about 1740 he returned to the central question of pneumatics, the cause of the air's elasticity, which Hales' discoveries had aggravated. He now saw that air is a vitreous electric, naturally somewhat excited, and springy owing to electrical repulsion between its particles.[15] The picture is clear and advanced: air has been identified as an electric *per se*, and electricity reduced to distance forces. But Desaguliers must also have effluvia; everyone else, including Newton, admitted them. He therefore compromised, as follows.

Rubbing the tube as usual causes it to emit effluvia; the air returns them, again as usual, but not because of its coldness, pressure, or unwelcoming pores, but because of its electricity. Is the air electric in the same way as the tube? If so, its particles must pelt one another with effluvia; if not, what justifies ascribing electricity to it? Desaguliers says that the effluvia, driven back to the tube, 'dart out anew, and so move backward and forward with a vibratory motion, which continues their electricity.'[16] It appears that electricity has become the property of its agent and that 'repulsion' is nothing but impact. There are other curiosities. Touch the tube to an insulated conductor. 'The Electricity receiv'd by the Conductor advances from one end of it to the other in a kind of cylindrick Vortex.' The atmosphere now rotates as well as pulsates; but how does a vortex run down a string?[17] Now repeat everything with a wax

---

[1744–5], 486) speaks of 'non-electricks or Conductors of Electricity.' Desaguliers seems also to have introduced 'insulator,' or at least 'insulated,' meaning 'terminated by Electricks' (e.g., *Course* [1763³], II, 331). Cf. *FN*, 377.

13. Barrière, *Académie* (1951), 136. Nollet's lecture series at Bordeaux (*supra*, xi.2) probably influenced the selection of the prize question. In the interim Desaguliers had become better acquainted with the two electricities and had extended Dufay's results on the despoiling of the donor's electricity during sparking. *PT*, 41:2 (1740–1), 634–40.

14. Desaguliers, *Diss.* (1742), translated in *Course*, (1763³), II, 316–36.

15. *Supra*, i.6; Desaguliers, *PT*, 42 (1742–3), 14–18, 140–3; *FN*, 381–2. Desaguliers developed a theory of evaporation by electricity.

16. *PT*, 42 (1742–3), 140–3.

17. The problem was taken seriously by Musschenbroek, *Elements* (1744), I, 196–7. Desaguliers' alternative formulation, 'Electricity which a Non-electric of great Length . . . receives, runs from one End to the other in a Sphere of electrical Effluvia' (*PT*, 41:2 [1740–1], 634–7), does not help much. Home, *Effl. Th.* (1967), 72–7, analyzes Desaguliers' inconsistencies acutely.

rod: a vibrating atmosphere again sets up, and everything happens as before. But ought not the vitreously charged air attract the resinous effluvia?

That this farrago of fuzzy thinking and tired experiments won a prize has perplexed many people, including the fair-minded Gralath. The Bordeaux Academy would have done better to have withheld the award, as it had in about half its competitions; but perhaps the judges judged Desaguliers the better physicist for his masonry, which counted several adherents among their colleagues. Despite his weakness, which even Priestley, who made the most of his countrymen's contributions, did not attempt to conceal, Desaguliers played an important part in our story.[18] For it was he, the French-speaking intermediary, the terminologist and systematizer, who brought Dufay's results before anglophone philosophers.

2. FREKE AND MARTIN

The German sparking experiments aroused as much interest in London as they had in Leipzig, and soon were enjoyed in the countryside.[19] They made a fine impression. 'Who can comprehend how . . . flame issues out of my finger [it is John Wesley's], real flame, such as sets fire to spirits of wine? How these, and many more as strange phenomena, arise from turning round a glass globe? It is all mystery; if haply by any means God may hide pride from man.'[20] But many are the prideful, and the number of fools is legion. Among those who rushed to bare the mystery of the electrical fire were John Freke, M.D., F.R.S., and Benjamin Martin.

In 1746 Freke, ophthalmological surgeon at St. Bartholomew's Hospital, friend of Hogarth, Fielding and Watson, a kindly man but no physicist, revealed the cause of electricity. He did so, he said, because no one else had, although the business was plain as a pikestaff.[21] One need only consider that the electrical matter is the universal, common fire, pressed *out* of the air and onto the tube by the agent providing the friction. The fire particles have a great *attraction* to one another and so form a halo about an electrified body. That they come from the atmosphere and not from the body appears from the fact that lightning, which is 'entirely the same with Electricity,' is just fire collected from the air in the normal course of nature.[22] Freke has made some lucky strikes: his universal electrical matter, collected on but not from the electric,

18. Gralath, Nat. Ges., Danz., *Vers. Abh.*, 1 (1747), 262; Barrière, *Académie* (1951), 129–30, 142, 184; Priestley, *Hist.* (1775³), I, 82. Cf. Torlais, *Rochelais* (1937), 31–2.

19. See for example the doings of the Northampton Philosophical Society, in Musson and Robinson, *Science* (1969), 379–82, and *GM*, 16 (1745), 475–6; and the lectures of the Society's electrician, the engineer Thomas Yeoman, in Birmingham (E. Robinson, *Ann. Sci.*, 18 [1962], 205). Instructions for the German games were provided for those who wished to play *en famille;* Neale, *Directions* (1747), 3, 26–32.

20. Wesley, *Journal* (1906–16), III, 320–1.

21. Freke, *Essay* (1746), iii; Moore, *Hist.* (1918), II, 633–6. Freke evidently knew nothing of Dufay, Bose, Waitz or Winkler.

22. Freke, *Essay* (1746), 5–6, 28.

and identical with lightning, has characteristics in common with Franklin's electrical fire.

Some questions remain. Why do electrics refuse electrification by communication? Because fire is intimately connected with life—Freke prefers the term 'vivacity' to 'electricity'—and therefore cannot be transmitted easily to wax or silk, which are, respectively, the excrement of bees and worms. How do the fire particles effect attraction? The halo is actually a vortex; bits of leaf-brass, responding to the 'Invitation they receive from the curling Effluvia to a closer Contact,' slide gratefully into the whirlpool. Freke gives a beguiling account of repulsion, conduction, and the electrical wind, which ill accord with his idea of attraction; he knows no physics; he admits that he cannot understand the 'Feats attributed to the mighty Weight of our Atmosphere, in causing Siphons and Pumps, *etc.*, to operate.'[23]

The sitting duck was soon smelt out by that bumbling Newtonian watchdog Benjamin Martin, a former assistant of Desaguliers, a provincial lecturer become a London instrument maker, the author of an *Experimental Philosophy* so 'plain and easy [as] to be understood by all Capacities in general, and in particular by the Fair Sex.'[24] Martin barked at Freke for knowing nothing of electricity and less of modern philosophy. Freke wrote as if Newton had never existed, a sin worse than Cartesian. 'Forsaking Sir Isaac,' says Martin, 'is as bad, if not worse, than opposing him.' 'The Nature, Cause, Properties and Effects of electric Virtue are explicable only on the Principles of the *Newtonian Doctrine of Light and Fire.*'[25]

According to these principles, a body sufficiently agitated throws off subtle particles, which, if the agitation be regular, include the effluvia of fire, light or electricity. So far Martin stumbles after Newton. He then lays down as a 'phenomenon' that the 'electric Virtue consists in a fine subtle Matter,' thus forsaking both Sir Isaac and sound philosophy; for in Newtonian parlance this statement is an hypothesis, not a phenomenon, and virtue is not matter.[26] The emitted subtle matter brings about attraction and repulsion in the manner of 'sGravesande's atmospheres. Martin does better than Freke by giving clear if unoriginal accounts of classic experiments. For example, Hauksbee's threads show 'the astonishing subtlety' of electrical matter, which 'passes through glass as if nothing had been interposed between the thread and the hand.'[27]

Freke dismissed his 'unmannerly' adversary as a 'country showman,' a man who took money for philosophy. To him physics was a hobby; if his conjec-

---

23. *Ibid.*, 15–17, 34, 47; *Essay* (1746²), 59; Benjamin, *Hist.* (1898), 571.

24. *Course* (1743), Pref.; cf. *FN*, 145–7. 'But the most elegant account of the matter [attraction] is by that hominiform animal, Mr Benjamin Martin, who, having attended Dr Desaguliers' fine, raree, gallanty shew for some years in the capacity of a turnspit, has, it seems, taken it into his head to set up for a philosopher.' De Morgan, *Budget* (1872), 91; *DNB*, XII, 1156–7.

25. Martin, *Essay* (1746), Pref.

26. Cohen (*FN*, 414) emphasizes Martin's perversion of the Newtonian categories.

27. Martin, *Essay* (1746), *ad fin.*; cf. Martin, *Course* (1743), 8–9.

tures had missed the mark, it was no doubt because he had not had time to do a single experiment. Martin, piqued by Freke's sneers at his calling, returned a scurrilous attack on 'the *worst* Writer on Electricity that ever was;' the establishment, taking Freke's part, condemned Martin's essay as 'ungenteel' and 'pompous.'[28] The tiff then subsided, to the advantage of both parties. Martin's dexterity, inventiveness and pushiness allowed him to prosper in the highly competitive instrument business, until he ended in bankruptcy. Freke did better, for he may still be found, in the pages of *Tom Jones,* where his friend Fielding gently satirized his brief encounter with electricity.[29]

3. WATSON

William Watson was no dilettante or occasional philosopher. Never 'indolent in the slightest degree,' according to his friend and memorialist, 'he was a most exact economist of his time, and throughout life a very early riser, being up usually in summer at six o'clock, and frequently sooner; thus securing to himself daily two or three uninterrupted hours for study.'[31] He mastered his first subject, botany, during his apprenticeship to an apothecary, which he entered upon in London in 1731, and earned a prize for his competence from the Apothecaries Company. In 1738 he bought his freedom, married, and set up his own business, which won him the acquaintance of members of the Royal Society. He first attended a meeting on March 16, 1737/8, as guest of John Martyn, professor of botany at Cambridge in absentia, abridger of the *Philosophical Transactions,* and translator of Boerhaave.[32] Three years later Watson became a Fellow. During the next decade his diligence, energy and learning made him one of the Society's most active and productive members.[33]

Watson labored to keep his countrymen informed of advances made on the Continent. He was the first to report Peyssonnel's view that corals are animal remains, not vegetables, and he very early recognized and advertised the advantages of the Linnean system.[34] He took up electricity in the same way, by repeating the firing of spirits by sparks, and for many years he remained the chief link between English and Continental electricians.

Watson reported his success in kindling the German flare on March 27, 1745, in a letter to the Society's president, Martin Folkes.[35] A month later a

28. Freke, *Essay* (1746²), 63–4; Martin, *Supplement* (1746), 28–9, 33; *GM*, 16 (1746), 51–2, 567–70; Millburn, *Martin* (1976), 54–7.

29. Daumas, *Instruments* (1953), 314; *Tom Jones* (1749¹), Bk. IV, chap. 9; Bernoulli, *Lettres* (1771), 73; G. Turner, RS, *Not. Rec.*, 29 (1974), 64; Millburn, *Martin* (1976), chaps. 6–8.

31. Pulteney, *Hist. Biog. Sket.* (1790), II, 334.

32. JB, XVI, 212.

33. Pulteney, *Hist. Biog. Sket.* (1790), II, 295–340; *DNB*, XX, 956–8; JB, XVI, 455, 485. Watson was knighted in 1786, having made a meritorious career in medicine: he bought himself free of the apothecaries in 1756, became a licentiate of the Royal College of Physicians (1759), and honorary MD of the Universities of Halle and Wittenberg. *DSB*, XIV, 193–6.

34. Pulteney, *Hist. Biog. Sket.* (1790), II, 295, 304–5.

35. *PT*, 43 (1744–5), 481–9; the Society first learned of the electrical flare in November 1744, from a review of Winkler's *Gedanken* (*ibid.*, 166–9). Cf. Hollmann, *ibid.*, 239–49.

second letter announced firing of the volatile inflammables he had in stock, and explained, more clearly than Desaguliers had done, that atmospheric moisture destroyed electricity by conduction.[36] It also recorded what Watson mistakenly regarded as a new effect, a 'repulsive' firing of spirits, in which the igniting spark is drawn across the liquor in an electrified spoon, rather than towards that in a grounded one. Thus encouraged, he began to study electricity systematically.[37]

His first results included one pretty demonstration and a very drab one, whose significance Watson missed and Franklin later caught. First the drab one. Two men, call them A and B, stand on wax cakes. Electrify A. Let him touch B; he loses 'almost all his electricity,' which B receives in one snap. Electrify A again and proceed as before; the 'snapping' is less, and it diminishes with each electrification of A, becoming insensible when B is 'impregnated with electricity.' Watson draws no conclusion from these suggestive observations.

Here is the pretty demonstration. Put threads of glass, bits of metal, and a few cork balls in a pewter plate held by an insulated man. Electrify him and, while standing on the ground, bring your hand over the plate. The glass threads dart to your fingers and stick without snapping; the wire bits fly upwards, snap as they near you, 'discharge their Fire' and fall to the plate; the cork balls strike your hand and drop directly. The circus elegantly expresses Dufay's results on the sparking of conductors and insulators of diverse shapes and materials.[38]

Watson's next results, made public on February 6, 1746, brought him deep into the problems of electrical theory. He was forced to ascribe to insulators properties incipiently paradoxical. He first supposed that, 'in common with Light, Electricity pervades Glass, but suffers no Refraction therefrom;' its power is exercised in right lines, proceeds radially, not vortically, from a rubbed tube, and thence across barriers of glass, porcelain, wax, etc., to agitate threads or leaf-brass lying beyond. Refraction occurs in the case of conducting barriers or screens: the direction of the 'Electricity' changes 'as soon as it touches the Surface of a Non-electric, and is propagated with a Degree of Swiftness scarcely to be measured in all possible Directions to impregnate the whole Non-electric Mass in Contact with it.' Only if the mass is terminated by electrics *per se* will the 'Electricity' show itself, and then at a point most remote from the exciting tube.

A different picture emerged from another experiment. Watson placed a few books atop a warm glass jar turned bottom up and covering some leaf-brass. He brought the tube above the jar: 'The Motions of the light Bodies underneath correspond with the Motions of the Glass Tube held over them, the Electricity

36. The leak derives primarily from surface conductivity promoted by the moisture, not from loss to the air, a point not understood until the end of the 19th century. G. P. and J. J. Thomson, *Conduction*, I (1928), 1–9.

37. *PT*, 43 (1744–5), 484–591; *PT*, 44:1 (1746), 41–50. 'Repulsive' firing occurs in Hollmann's paper and in Gordon's bad joke (*supra*, x.3); that Watson laid 'great stress' (JB, VIII, 453–5) on the difference between attractive and repulsive firing suggests that he had not yet mastered Dufay's work.

38. *FN*, 398; Watson, *PT*, 43 (1744–5), 494; *PT*, 44:1 (1746), 41–50.

seeming instantaneously to pass through the Books and the Glass.' But the passage in fact is not instantaneous. No penetration of the glass occurs 'till the Electricity has fully impregnated the Non-electrics, which lie upon the Glass; which received Electricity is stopped by the Glass; and then these Non-electrics dart their Power directly through the upper Part of the Glass.' Now read 'electrical effluvia' for 'electricity' and consider the whimsical role of the glass: why, after stopping the effluvia, does it suddenly grant them passage?[39] If the effluvia 'dart directly' across the glass after 'impregnating' the books, how does a charged insulated conductor maintain its electricity?

Watson appears to have relied on the doctrine of the semitransparency of glass and the assumption that a flow of effluvia, either direct from the tube or from an 'impregnated' nonelectric, must attain a certain strength before penetration can occur. This only sharpened the incipient paradox, as appears from measurements he made to determine the thickness that cakes of wax and resin must have to function effectively as insulators. He found that 'when the Electricity is very powerful' it could penetrate at least 2.4 inches of resin wax! What he measured was not insulation in the modern sense, but the depth of wax across which the tube could attract; since the theory did not distinguish the range of attraction from the penetration of the effluvia, he recommended using cakes at least 2.8 inches thick.[40] And here the impending difficulty appears in full force: how could the effluvialists explain a clear case of successful insulation of a large charge by a thin piece of wax or glass? They were helpless when the discovery of the condenser supplied just such a case.

WATSON'S FLOWING AETHER

Everyone liked Watson's demonstrations. The Prince of Wales and lesser dignitaries watched them; the President of the Royal Society assisted him; and Sir Hans Sloane appointed him the Copley medalist for 1745.[41] His papers were printed privately and distributed as a pamphlet to his importunate admirers.[42] The public was only titillated. The pamphlet was again demanded, and thrice reprinted. Watson pursued his enquiries; and on October 20, 1746, he sent the Royal Society a 'Sequel,' which consolidated his position as the dean of English electricians.[43]

39. That this reading is correct appears from Watson's elucidation of the advantages of warming the jar used in the experiment: 'First, warm Glass does not condense the Water from the Air, which makes Glass . . . a Conductor of Electricity [i.e., of effluvia, as the next sentence shows]. Secondly, as heat enlarges the Dimensions of all known Bodies . . ., the electrical Effluvia, passing in straight lines, find probably a more ready Passage through their Pores.' Watson's inexact usage of 'electricity' perhaps reflects an unconscious realization of the ambiguity of his account of insulation.

40. Cohen (*FN*, 399) mistakenly accepts Watson's numbers as referring to insulation in the modern sense, and so credits him with an important 'improvement' in the usual experimental arrangement. Whatever thickness of resin-wax would support the weight of a man without cracking would insulate him.

41. Pulteney, *Hist. Biog. Sket.* (1790), II, 314–15.   42. *Experiments* (1746).

43. *PT*, 44:2 (1747), 704–49, issued as a pamphlet in 1746.

The 'Sequel' gives an improved account of electrical fire or effluvia based upon inferences Watson drew from an unexpected effect. It had occurred to him, as to several others, that a greater electricity could be excited in the tube if the operator stood upon wax as he rubbed it, and so prevented loss of effluvia to ground. Most of the independent discoverers of this effect—Bose, a colleague of Musschenbroek's, J.N.S. Allamand, Watson, Wilson (whom we shall meet momentarily), and Franklin—were astounded to find the event the reverse of their expectation.[44] The electricity of the tube or machine was diminished, not enhanced, by insulating the operator and his instrument. Bose allowed that the oddity 'tormented him terribly;' no illumination presenting itself, he invented a 'plausible subterfuge,' or piece of mystification, based on an unintelligible relation between action and reaction.[45] Allamand, less tormented, left the oddity an oddity. Watson, Wilson and Franklin all drew the same, insightful, paradoxical conclusion: the effluvia projected by an excited electric came not from itself, but from the body rubbing it and, ultimately, from the ground.[46] In a word, they saw that tubes and globes are *pumps* serving to raise effluvia from the earth onto insulated nonelectrics.

Watson devised a plausible demonstration of his novel hypothesis. With the machine and its operator insulated, no sparks would spring from the prime conductor; but if the operator touched his foot to the floor or held a grounded nonelectric, the gun barrel electrified as usual. Watson varied the arrangement several ways, notably by setting spark gaps to make the course of the effluvia visible, a technique later exploited by Franklin's disciple Giambattista Beccaria. Among the variations was inversion: an operator insulated together with his machine would electrify strongly if the prime conductor were grounded. Watson rightly regarded this inversion as a fact of great theoretical importance. His interpretation of it proved misleading, however, for he took the direct and inverted cases to be symmetric, whereas the prime conductor charges vitreously and the man resinously.[47] It was in this natural presumption of symmetry that

44. With half the philosophers of Europe at work with glass tubes, it is not surprising that several independently hit on this obvious application of the theory of effluvia. Only Wilson cried plagiarism, against Watson, who persuasively defended his innocence before the Royal Society (JB, XIX, 149–51). Wilson also was the only discoverer who claimed to have foreseen the effect.

45. Bose, *Recherches* (1745), xli: 'Quelle bizarrerie! Nous plaçons les corps sur des matières électriques per se, anfin qu'ils le deviennent par communication. Cet homme placé dans les circonstances les plus favorables, résiste invinciblement.'

46. Franklin, *infra*, xv.1; Allamand, *Bibl. brit.*, 24 (1746), 406–37. Rackstrow, the mender of broken vases, intimated that he, too, independently discovered the effect of Watson et al., which he interpreted in the manner of Freke: the globe pumps electrical fire from the air (where Rackstrow, following Desaguliers, located a great reservoir), and not from the ground, else why should the electricity of the globe depend upon the state of the atmosphere? Otherwise Rackstrow sticks close to Watson. Rackstrow, *Misc. Obs.* (1748), 11–18; *supra*, ii.4.

47. In the direct case the globe draws positive (vitreous) charge from the man and conveys it to the prime conductor (PC); the man's supply is replenished from ground and the process continues until the force of the accumulated charge in the PC equals that with which the globe presses new increments upon it. If the man is insulated, the negative charge he builds up quickly reduces the amount of positive he can rub onto the globe, and the reduced charge is unable to force much of itself onto the

Watson's system differed from the one Franklin was soon to construct. Just as Franklin's *qualitative distinction* between positive and negative electrification created a system with a logic of its own, so Watson's tacit identification of the two states led him to a theory with the characteristics and weaknesses of Nollet's.

Watson likens the agent of electricity, the effluvia, the fire driven round by the pump, to an 'aether' whose particles act upon one another over microscopic distances, and upon ponderable bodies by impact. The extreme subtlety of the electrical aether appears from its ability to penetrate glass. Its elasticity reveals itself in its swelling into an atmosphere around an electrified body, wherein electricity consists;[48] and in its tendency to seek equilibrium, to return to all bodies in proper measure, as when, accompanied by sparks and a cold wind, it rushes to a grounded conductor thrust into an electrical atmosphere. In fine, it behaves much like Boerhaave's elementary fire, with which Watson inclined to identify it.

The equilibrating aether flow is the agent of attraction and repulsion. To save the phenomena, Watson must make the current of effluvia run initially in a direction opposite to that demanded by the theory.[49] It is to flow *towards* electrified objects, towards objects already clothed with atmospheres, from the naked conductors about them: 'The blast of electrical aether constantly sets in from the nearest unexcited non-electrics toward those excited, and carries with it whatever light bodies lie in its course.' That is attraction. On approaching an excited object the light bodies encounter the equilibrating aether blast the original theory promised, and are repelled. As exemplar of the process of equilibration, Watson picked Wilson's 'fish,' a sliver of silver suspended by the force of electricity between two parallel metal plates, the lower grounded and the upper in contact with an electrical machine.[50] The fish takes up a position where the oppositely directed aether blasts compensate.

Watson recognized that his view paralleled Nollet's, which he thought did that 'excellent philosopher' great credit; for Nollet had been able to design the system of afflux and efflux without knowledge of the critical experiments with which Watson had begun. In fact, the theories differ significantly. In Nollet's, conduction, or communication, is the propagation of a state, the stimulation of a flow of effluvia; in Watson's, the aether itself sweeps through conducting bodies. Their divergence on this point instances a graver difference: whereas

---

PC. If, however, the PC is grounded and the opposing force of its electricity removed, charge will be conveyed through it until the operator becomes so strongly electrified that he can no longer force his positive charge onto the globe.

48. The stability of atmospheres implies a force opposing the aether's elasticity, such as resistance of the air or attraction to the electrified body. Watson does not specify which; either solution would raise difficulties for him.

49. Cf. *FN*, 407.

50. The 'silver fish' is Franklin's phrase (*infra*, xiv.4). Although Wilson's arrangement is to us an 'air condenser,' neither he, Watson nor Franklin recognized that it had anything in common with the Leyden jar. Cf. *FN*, 409n.

French electricians attended to the attractive, repulsive and luminous properties of bodies already electrified, the successors of Gray emphasized the conduction of electrical matter, the change of electrical state. In this sense, as well as by nationality, Franklin was a British electrician.

4. WILSON

Watson's chief rival among the British electricians was Benjamin Wilson, the fourteenth child of a wealthy clothier who named him Benjamin in the hope that he would be the last. Like Watson, Wilson made his career at a respectable trade that required manual dexterity and a knowledge of the properties of bodies; he became a painter, and in time acquired clientele and patronage from the higher aristocracy. He was taught to paint while still a child. That instruction, together with a brief attendance at the Grammar School at Leeds, where his friendship with his long-time adviser John Smeaton probably began, constituted his education. His father failed financially, withdrew him from school, and encouraged his emigration to London and opportunity. There Wilson strove like a Horatio Alger hero; of the weekly salary of three half-crowns he earned in his first job, a clerkship, he claims to have saved two. His assiduity recommended him to a better job, another clerkship at double the salary, in the registry of the Charterhouse, where memory of Gray's triumphs was still fresh.[51] The new position offered greater leisure as well, which Wilson used to improve his painting, with such success that established artists, like Hogarth, encouraged him to make it his profession.

As preparation he resolved to repair his culture, 'to know something of everything,' 'to render himself agreeable in all societies,' to insure 'that none should reproach him with ignorance.'[52] To accomplish this modest purpose he made himself learn something new each day for a year. He devoured whole *Spectators, Tatlers* and *Guardians*; took *Don Quixote* and the chief English writers at one gulp each; chewed over Locke, whom he found tougher; and ended with a light supper of Euclid, Bacon, Boyle, Newton and Desaguliers. He liked the taste of experimental philosophy, and adopted a more sensible diet. He probably attended the lectures of Desaguliers, and about 1745, when he became acquainted with Watson, he fixed on electricity as his philosophical specialty.

Wilson perceived the advantages of the electrical machine more quickly than Watson did. The Charterhouse, where he had rooms, again became an electrical laboratory. He wrote Smeaton for advice and found a collaborabor in John Ellicott, an accomplished clockmaker.[53] He did not need their help to penetrate to the true cause of electricity. It took only the sight of sparks passing between a

51. *DNB*, XXI, 553–5; Randolph, *Wilson* (1862), I, 3–42.

52. *Ibid.*, I, 7, quoting an autobiography Benjamin Wilson composed for his children.

53. *Ibid.*, 8. Ellicott probably began studying electricity before Wilson did; cf. *PT*, 44:1 (1746), 96–9.

grounded metal plate and one continuously electrified. It amazed Wilson that the sparks appeared to originate from *both* plates, since by the usual hypothesis only the electrified one possessed a surplus of effluvia. He inferred that an excited body could differ only in degree from one unelectrified; and he guessed that the difference related to the condition of the atmospheres with which Newton's optical theory clothes all bodies.[54] This happy thought occurred to him in June 1746. He discussed it with Watson and with Martin Folkes, who gave him two good pieces of advice: he should design experiments in support of his theory, and he should practice his painting in Ireland, where botches he might make would not haunt his career in England.[55]

On his journey Wilson hit on a way to confirm his views. He wrote Ellicott, through Watson, asking him to climb upon wax and try how much electricity he could excite from the tube. This is nothing other than the experiment of Bose and Watson; but whereas they expected an enhancement, Wilson (we have his own word for it) predicted a diminution. He may have reasoned as follows: the electrical effluvia come from previously existing aether atmospheres; aether is omnipresent and highly elastic; hence, if the tube or the rubber is not replenished, not much can be removed from them before the pressure of the ambient aether stops the process. Although the reasoning seems faulty and the experiment irrelevant,[56] the fulfillment of his prediction, which Wilson himself tested at Trinity College, Dublin, confirmed him in the theory from which he had derived it. He now needed a closer acquaintance with Newton's aether. Fortunately, the College still boasted Bryan Robinson, the author of two essays on the physical properties of that rare medium. Wilson profited so quickly from Robinson's books and conversation that in October 1746 he was able to send the Royal Society a full account of his new system.[57]

The electrical matter, he says, is Newton's gravitational-optical aether with an admixture of luminous and sulphurous particles. The aether will account for charging, conduction and the electrical motions; the light and sulphur, for the usual sparks, glows and smells. Now heavy objects act as if they repel this aether, for it stands rarest in the densest bodies. This principle unfortunately

---

54. *Supra*, i.6. Perhaps the earliest such association was made by Worster, *Comp. Meth. Acct.* (1730²), 29–30; but no one before Wilson applied it systematically to the discoveries of Gray and Dufay. Cf. J. Rowning, *Compendious System* (1737–43), quoted by Schofield, *Mechanism* (1970), 37: although without recourse to a subtle medium Rowning thought electricity inexplicable, he dismissed Newtonian aether as 'too much clogged with suppositions' to be useful.

55. Randolph, *Wilson* (1862), I, 8–9; B. Wilson, *Essay* (1746), viii–ix, 2.

56. From the same premises one might more readily infer that the proposed experiment would augment the tube's electricity, since Newton's optical aether is rarer the denser the object it surrounds. Because of its elasticity the denser aether mixed with the air should rush in to replenish the tube's atmosphere.

57. Wilson, who did not become FRS until 1751, sent three letters, which were read during November 1746 and printed with additions in Wilson's *Essay*, but not in *PT;* Wilson, *Essay*, Pref., and Randolph, *Wilson* (1862), I, 9–10. For Robinson see *DNB*, XVII, 4–5; *FN*, 418–19; Smeaton to Wilson, 28 Aug. 1749, Add. Mss., 30094, 74 (BL).

gives no purchase for distinguishing between insulators and conductors. Wilson allows the repulsion between aether and matter to be anomalously strong in the case of electrics *per se*.

The machine works by 'vibrating' the aether naturally in and around the hand or cushion; the vibration projects electrical matter onto the glass, which, being an electric substance and highly 'elastic' besides, drives it onto the prime conductor, the closest object more apt to receive it. The process can continue if a reservoir of electrical matter, like the ground, replenishes the particles lost by the rubber. When the PC is replete, excess electrical matter will run from it, most copiously from its extremities. The streaming causes the usual electrical motions. It is not the only cause, however; two bodies whose natural envelopes are electrified will recede from one another owing to the elastic power of the aether.[58]

So far had Wilson progressed before he ran across Robinson, who taught him that the aether could explain all the accidents of attraction and repulsion. Robinson's scheme, which violates common sense and most of the laws of motion, is illustrated in fig. 12.2, where Y is electrified and X unexcited and mobile. Both objects carry Newtonian atmospheres whose density increases with distance outwards. If X and Y are close enough the rarefactions at A and C will extend to B, and that, according to Robinson and Wilson, will enable the aether pressure at D to push X to Y. Even when X electrifies, that is, increases

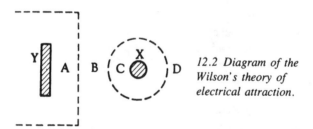

12.2 Diagram of the Wilson's theory of electrical attraction.

the density of its atmosphere, attraction can result if the pressure at B falls below that at D. Repulsion takes place if X is also excited and the aether at B becomes dense enough to overcome that at D. Wilson wisely avoided mentioning the phenomena which had led Dufay to assume two electricities. As for sparks, the snaps come from rarefactions of the air caused by the heat and light, which are themselves the work of the sulphurous and luminous particles expelled from rubbed bodies along with the electro-optical aether.[59]

The best demonstration of his system, Wilson thought, was the suspension of the silver fish, which he supposed to illustrate an equilibrium between its atmosphere and those of the plates.[60] It did indeed imply a balance of force, but it did

58. B. Wilson, *Essay* (1746), 6–7, 13–26.
59. *Ibid.*, 57–61. Much of this mechanism follows Newton's letter to Boyle (*supra*, i.6).
60. *Ibid.*, 63–77.

not show Newton's aether to be the responsible agent, any more than it confirmed Watson's system or Franklin's, as each in turn believed. The fish was to swim about tantalizingly until physicists came to see its tank—the parallel plates—for what it was, a Leyden jar without the glass.

The insufficiency of Wilson's system, the most literal of several attempts to exploit Newton's *Opticks* in the interest of electrical theory,[61] leaps to the eye. So the plain-spoken Smeaton unsparingly told its author. The fundamental hypothesis, that electricity and gravity have the same agent, cannot stand scrutiny: there are two kinds of electricity, he said, and but one gravity; and neither electricity appears to act, as does gravity, in proportion to the quantity of *matter* in the body it qualifies. Smeaton further observed that Wilson's theory obliged two *unelectrified* objects to tend towards one another more strongly than either should approach an electrified body. That was but one absurdity of a mechanism itself absurd, and in violation of all the laws of motion. Smeaton insisted that his friend pay attention to the balance of forces; he explained that rarefaction is not synonymous with low pressure; and he doubted the utility of microscopic models. What advantage is there to referring observed attractions and repulsions to a hypothetical medium the undetectable particles of which act upon one another at a distance? It is no less unsatisfactory, according to Smeaton, and far less artificial, to assign the necessary forces to the particles of matter.[62]

No more did Wilson's collaborator Ellicott fancy the Dublin fashion in electricity. He thought it patent that material bodies *attract* electrical effluvia; that the effluvia repel one another; and that forces between effluvia and material particles on the one hand, and between atmospheres on the other, determine the magnitude and direction of electrical motions. The density gradient in Ellicott's electrical medium necessarily ran counter to that of Newton's aether. Wilson, who was contemplating publishing refinements of the Dublin theory, grew alarmed at Ellicott's reasoning.[63] In uncharacteristic doubt he asked Smeaton whether he should print anything further. The engineer opted for Ellicott. 'I can't help being so far of Mr E's opinion that the electrical fluid is in a state of repulsion with regard to its own particles and is attracted by all non electrical Bodies, that I have long observed to be apparently the case in all the electrical

61. For other schemes invoking aether see Rackstrow, *Misc. Obs.* (1748), 8–10; Cooper, *Phil. Enq.* (1746), 22–6.

62. Smeaton to Wilson, 24 Sept. 1746 and 4 Feb. 1746/7, Add. Mss., 30094, ff. 22, 29–30 (BL). Wilson never learned elementary mechanics, as witness this proposition from his *Treatise* (1750), 174–5: a body immersed in a liquid will move only if the pressure gradient acting on it exceeds its 'vis inertiae'! Cf. Nollet to Dutour, 25 Dec. 1761 (Burndy): 'Soit dit entre nous, Mr Wilson ne paraît pas un homme fort initié en physique.'

63. Ellicott, *PT*, 45 (1748), 195–224. Although the first of these essays was not read until 25 Feb. 1747/8, Ellicott had communicated their principles 'several months before'; the date of composition is of interest, for Ellicott's principles resemble Franklin's, whose views Watson summarized before the Society on 21 Jan. 1747/8 (*PT*, 45 [1748], 93–120). Ellicott discusses the flow of syphons (combatting the opinions of Nollet), the appearance of glows at points, and the silver fish.

experiments that I have seen.'[64] Smeaton advised publication for possible methodological value, perhaps as a *reductio ad absurdum*.

Wilson needed no more encouragement to bring out a *Treatise on Electricity* (1750). The weakness of the theory was then so plain that he openly admitted its impotence.[65] In the shock of the Leyden jar the Dublin aether dissipated more easily than the baroque mechanisms of the Germans or the dual currents of Nollet.

64. Smeaton to Wilson, 28 Aug. 1749, Add. Mss., 30094, f. 74 (BL).
65. B. Wilson, *Treatise* (1752²), 34–47, 97–157.

PART FOUR
The Age of Franklin

What has led most People into Errors, is an immoder-
ate Desire of Knowledge, and the Shame of confessing
our Ignorance.

—'sGRAVESANDE, *Elements* (1731), viii.

Non plus sapere quam oportet sapere.
—PAUL, *Romans,* xii:3.

CHAPTER XIII

The Invention of the Condenser

The circumstances surrounding the invention of the condenser should interest the philosopher as well as the historian of science. The apparatus required—an electrical machine, a wire and a glass vessel containing water—was everyday equipment for the electrician. His constant manipulation of these elements might have created condensers wholesale once the spinning globe came into common use. But the theories received in 1745 would suggest the wrong arrangement to anyone intent upon constructing what the condenser proved to be, an instrument for concentrating and strengthening the force of electricity. Its invention, made independently in Germany and Holland, was a true discovery, the finding of something opposed to expectation, a piece of serendipity on the part of two unpracticed operators. The importance of this discovery, on which the development of modern electrostatics hinged, was recognized immediately. The power and unexpectedness of the condenser, the distress, even fear, it initially provoked, guaranteed that it did not become one of those shy 'anomalies' that are supposed gradually to prepare the ground for scientific revolution.[1]

1. KLEIST'S PHIAL

The first man to construct a condenser was Ewald Jürgen von Kleist, the son of a Prussian official and himself the dean of the cathedral chapter of Kammin, in Pomerania. He perhaps acquired his taste for experimental physics from 'sGravesande's demonstrations, for he studied at the University of Leyden, probably in the early 1720s.[2] Caught up in the enthusiasm engendered by Bose's fireworks, he built himself an electrical machine and repeated the experiments retailed by Krüger and Winkler. The remoteness of his situation and the press of business, however, prevented him from ever quite mastering the subject; and so it was that, initially, he doubted the novelty and overlooked the importance of his great discovery.[3]

Among Kleist's earliest experiments were efforts to increase the strength and reliability of the electrical flare. He persuaded himself of the truth of the false proposition that sparks increase in proportion to the mass of electrified material

---

1. Kuhn, *Structure* (1962), 52–65.
2. *DSB*, VII, 403.
3. Gralath, Nat. Ges., Danz., *Vers. Abh.*, 2 (1754), 406; Kleist to Krüger, 19 Dec. 1746, in Krüger, *Gesch.* (1746), 177–81. The letter's date, not given by Krüger, comes from Gralath, Nat. Ges., Danz., Vers. Abh., 2 (1754), 411.

from which they spring: the bigger the prime conductor and its attachments, the greater the *Schlag*.[4] In a typical experiment, he ran a wire from the prime conductor into an insulated water-filled glass vessel, which was to him a convenient way of increasing the nonelectric mass of the PC. He needed only to ground the vessel to make a condenser. But that was precisely what he could *not* do: the effluvia he aimed to collect would run out through the semipermeable glass. To strengthen the spark, to concentrate effluvia in the prime conductor and the water, he had to insulate the vessel, in accordance with the Rule of Dufay.

Continuing his play with the flare, Kleist thought to show that 'sparks and glows can break spontaneously from an electrified piece of wood.' He succeeded, by replacing the water-filled vessel with a wooden spool, and he varied the demonstration by setting a nail in such a way as to draw sparks from it and the spool alternately.[5] After describing these insipid doings to Krüger, Kleist went on, without pause or fanfare, to announce the following news:[6]

> If a nail, a strong wire, etc., is introduced into a narrow-necked little medicine bottle and electrified, especially powerful effects follow. The glass must be very dry and warm. Everything works better if a little mercury or alcohol is placed inside. The flare appears on the little bottle as soon as it is removed from the machine, and I have been able to take over sixty paces around the room by the light of this little burning instrument.

(The instrument appears to have been a condenser charged through the nail's point; after removal from the machine, the free electricity on the inside of the glass ran out the nail by corona discharge, while that on the outside flowed to ground.)

What was the rationale for bringing a nail stuck in a narrow bottle held in the hand up to the spinning globe? It was not to concentrate electricity, for the strength of the spark, as Kleist had found, increased with the size and insulation of the nonelectric that delivered it. What then? Kleist's discovery seems to have originated in the diverting notion of a portable sparking machine, an idea that may have come to him as he watched the sparks play between his spool and nail. The alcohol, whose presence was found to promote the effect, may have been added as fuel.

The first application Kleist made of his phial supports this reconstruction of its invention: 'If I electrify the nail strongly, so that the light within the glass

---

4. Gralath, *ibid.*, 402; Winkler, *Eigenschaften* (1745), 31–57, 77–83.

5. 'Auf diese Rolle wird ein eiserner Nagel schreg gestecket, so strömen die Flammen bald aus dem Holze, bald aus dem Metall hervor.' Krüger, *Gesch.* (1746), 178.

6. *Ibid.*: 'Wenn ein Nagel, starker Drath etc. in ein engälsiges Medicingläschen gestecket und electrisiret wird, so erfolgen besonders starke Wirkungen; das Gläschen muss recht trocken und auch warm sein. Thut man etwas Mercur, oder Spir. Vin hinein, so geht alles desto besser von statten. Sobald das Gläschen von der electrischen Machine weggenommen wird, so äussert sich an demselben der flammende penicillus, und habe ich mit dieser kleinen brennenden Machine über 60. Schritt in dem Gemach hell gehen können.'

and the sparks are visible, I can take it into another room and ingite spirit of wine or of terpentine.'[7] The portability did not surprise him: it was the effect he had sought. 'What really surprises me in all this,' he told Krüger, 'is that the powerful effect occurs only in the hand. No spirit can be ignited if it [the instrument] rests on the table. No matter how strongly I electrify the phial, if I set it on the table and approach my finger to it, there is no spark, only a fiery hissing. If I grasp it again, without electrifying it anew, it displays its former strength.' Kleist also found—and this completes his work—that if he touched the nail while electrifying it either at the machine or at the prime conductor, he received a smart blow: and that a phial four inches in diameter, very dry outside and partly filled with water,[8] had more punch than the medicine jar. 'I'm certain that with such sparks Herr Bose would give up kissing his charming Venus.'

Kleist's first experiments with the nail and the medicine bottle took place on October 11, 1745.[9] He apparently did not recognize their significance immediately, since he waited until November 4 before reporting them. He chose to confide in J. N. Lieberkühn, a distinguished physicist and physician, a member of the Berlin Academy of Sciences, and learned in reply that the experiments were 'novel and remarkable.' On November 28 Kleist initiated Paul Swietlicke of Danzig into the secret, and in December he wrote Krüger.[10]

Kleist is known to have communicated news of his discovery to two other persons. Not one of the five was able to repeat the experiment successfully. Each asked Kleist for further particulars, each received an answer, and each failed anew. The only man in Germany who managed to reproduce the result before the announcement of the Leyden experiment was Gralath, who worked from Kleist's letter to Swietlicke. Gralath also floundered initially and wrote Kleist for additional information. He learned that the experiment worked best when a wire was used instead of the nail and when a thermometer or barometer tube filled with alcohol replaced the medicine glass. Under such circumstances, according to the kindly dean, children of eight or nine could be knocked off their feet by the shock. These revelations, which Gralath received toward the end of February, did not immediately bring success. Not until March 5, 1746 were his efforts rewarded; and then not he but his assistant hit upon the successful variation.[11]

Kleist's correspondents were unable to reproduce his experiment because he

---

7. Cf. Watson's repulsive firing.

8. Kleist's description of this instrument is typically obscure: 'Noch habe ich eine 4 Zoll im diam. haltende mit etwas Feuchtigkeit gefüllte gläserne Kugel genommen, und das drein gefasste metallene Instrument [a nail, wire, or the original medicine bottle and nail?], welches wie eine kleine Cammer war, auf vorbeschriebene Art electrisirt . . .' Krüger *Gesch.* (1746), 179.

9. Gralath, Nat. Ges., Danz., *Vers. Abh.,* 2 (1754), 407–8.

10. *Ibid.*; Hoppe, *Gesch.* (1884), 18–19; Priestley, *Hist.* (1775³), I, 103–4, who gives an English translation of Gralath's version of Kleist's letter to Lieberkühn.

11. Kleist to Krüger, n.d., in Krüger, *Gesch.* (1746), 181–4; Gralath, Nat. Ges., Danz., *Vers. Abh.,* 2 (1754), 409–11, and 1 (1747), 512–13.

had failed to mention the counter-intuitive step that made a condenser from a nail in a bottle: he had neglected to say that one must grasp the external surface of the bottle and stand upon the floor while electrifying the nail. He did not specify that the bottle's exterior must be *grounded*. Without this direction, his readers, endeavoring to communicate a strong electricity to the nail and recognizing the theoretical quasi-permeability of glass, would tend to use a thick bottle, or to hold the jar loosely, or to insulate themselves during the electrification, variations that would diminish the charge on the condenser. And even if they chanced to electrify the bottle as Kleist intended, they might fail to discharge it effectively. The electrified jar delivers its greatest blow when the person who grasps its exterior draws its spark, say by advancing his finger or a brandy-filled spoon to the nail. But the standard ignition employed two people, one grounded to present the spoon and one insulated, in communication with the prime conductor, to deliver the flash. Kleist's directions specified only that the bottle must be held when it fired; in analogy to the usual practice, a second man would offer his finger to the spoon. This was Gralath's most persistent mistake;[12] success came only when his associate, Gottfried Reyger, inadvertently played both parts. We can imagine the scene. Gralath, fatigued and exasperated, calls to Reyger, who has held the nail to the machine for the thousandth time: 'Come, Reyger, it is useless; I will go to Kammin and shake the secret out of him. Put the apparatus away.' Reyger reaches up to disengage the nail from the electrified bottle . . .

2. THE LEYDEN JAR

As Kleist was tantalizing his correspondents with inadequate accounts of his inventions, Musschenbroek, who had succeeded to 'sGravesande's professorship of experimental physics in 1742,[13] was independently pursuing an experiment that was to culminate in a similar find. Its purpose, an idea of Bose's, was to draw 'electrical fire' from water electrified in a glass vessel.[14] Although Bose did not say how he electrified the water, he and Musschenbroek doubtless followed the method described at length by Andreas Gordon. One places a jar filled with water on an *insulating* stand and runs a wire into the water from the prime conductor: the effluvia flow into the water, where they are retained by the semipermeable glass and the insulating support.[15] When the water is 'filled,'

12. *Ibid.* In the unsuccessful experiments something stronger than usual must have been felt; but, owing to the small condenser and the poor grounding, nothing worthy of remark. Also, if two men discharge the jar, the shock passes through the arm, side, leg and semi-insulating shoes, whereas if one man acts alone it runs through both arms and across the breast, a much more sensitive path.

13. Musschenbroek taught mathematics and philosophy at Utrecht, 1723–40, and then at Leyden for two years before 'sGravesande's death; he turned down several offers in order to remain in the Netherlands. Savérien, *Hist.*, VI (1768), 348–50.

14. *Supra*, x.2.

15. Gordon, *Versuch* (1745), 32. That Musschenbroek accepted the semi-permeability of glass appears from his *Elements* (1744), I, 190: 'The electrical effluvia pass through glass and sealing wax . . . [but] do not extend themselves far along silk, glass or sealing wax.'

one stops the machine, approaches a finger to the PC, and elicits the desired spark. Following Gordon's procedure, Musschenbroek would have done as Kleist did when investigating the strength of sparks; again the received theory would not authorize the small change necessary to convert the water-filled jar into a condenser.

Here chance intervened in the person of Andreas Cunaeus, a lawyer who amused himself by visiting Musschenbroek's laboratory. Cunaeus was intrigued by Bose's experiment, and tried to reproduce it with what instruments he had at home. He was alone; he was not an adept; he knew nothing of the Rule of Dufay. He electrified the water-filled jar in the manner most natural to him, holding it in his hand; lacking assistance, he drew the spark himself, and transformed Bose's modest fireworks into the terrible Leyden jar.[16]

Cunaeus reported his discovery to Musschenbroek and his colleague Allamand, the assistant, disciple, and biographer of 'sGravesande. Allamand repeated the experiment, which succeeded: 'it knocked the wind out of me for several minutes.' Two days later Musschenbroek, 'naivety itself, who love[d] truth with the openness of a child,' substituted a globe for Cunaeus' jar. 'He was so painfully affected [Allamand reported] that when he came to my house a few hours later he was still shaken, and he told me that nothing in the world could make him try the experiment again.'[17] Musschenbroek reported his distressing experience to his appointed correspondent at the Paris Academy, Réamur:

> As I see that this sheet [containing meteorological observations] is not completely filled, I would like to tell you about a new but terrible experiment, which I advise you never to try yourself, nor would I, who have experienced it and survived by the grace of God, do it again for all the kingdom of France. I was engaged in displaying the powers of electricity. An iron tube AB was suspended from blue-silk lines; a globe, rapidly spun and rubbed, was located near A, and communicated its electrical power to AB [see fig. 11.4, where the bar is AB and the jar Musschenbroek's globe D]. From a point near the other end B a brass wire hung; in my right hand I held the globe D, partly filled with water, into which the wire dipped; with my left hand E I tried to draw the snapping sparks that jump from the iron tube to the finger; thereupon my right hand F was struck with such force that my whole body quivered just like someone hit by lightning. Generally the blow does not break the glass, no matter how thin it is, nor does it knock the hand away [from the phial]; but the arm and the entire body are affected so terribly I can't describe it. I thought I was done for. But here are some peculiarities. When the globe D is made of English glass there is no effect, or almost none; German glass must be used, Dutch doesn't work either; D does not have to be a globe, a drinking glass will do; nor does it matter if it is

16. Allamand, *Bibl. brit.*, 24 (1746), 432; Allamand to Nollet, n.d., in *MAS* (1746), 3–4; Musschenbroek to Bose, 20 April 1746, in Bose, *Tentamina*, II:ii (1747), 36–7: 'Plurima quae instituisti pericula electricitatis repetivi, & voto potitus fui. Experimentum a te optime inchoatum [the water-spark experiment] expolivi parumper, politura mihi fere fuit lethalis.' For Cunaeus see F. A. Meyer, *Duisb. Forsch.*, 5 (1961), 31–4.

17. *MAS* (1746), 3–4; Voltaire to Mairan, 11 Sept. 1738, in Voltaire, *Corresp.*, VII, 364; Musschenbroek, 'qui est la naiveté même et qui aime la Vérité avec une candeur d'enfant.'

large or small, thick or thin, tall or short, or of any particular shape; but it must be made of German or Bohemian glass. The globe D that almost killed me was of very thin white glass, five inches in diameter. Most other note-worthy phenomena I here omit. Suffice it that the man should stand directly on the ground; that the same one who holds the globe should draw the spark; the effect is small if two men participate, one grasping the globe and the other pulling the sparks. If the globe D rests on metal lying on a wooden table, and someone touches the metal with one hand and elicits sparks with the other, he also will be struck with an immense force. I've found out so much about electricity that I've reached the point where I understand nothing and can explain nothing. Well, I've filled this sheet up pretty well.[18]

Musschenbroek's clear professional account, very different from the murky letters of von Kleist, enabled any informed and courageous reader to reproduce Cunaeus' experiment. Nollet, who had Musschenbroek's report from Réaumur, immediately confirmed the effect, not waiting to obtain the German glass the Dutch physicists believed indispensable. Plain French glass bent him double and knocked out his wind. Fully persuaded, he read Musschenbroek's letter at the public meeting of the Paris Academy in April, 1746.[19] Meanwhile the English learned about the marvelous jar from Trembley, who happened to be in Holland about the time of Cunaeus' invention.[20]

Others tried the bottle. They experienced, or so they said, nosebleedings, fevers, temporary paralysis, concussions, convulsions, and prolonged dizziness. Winkler warned that his wife was unable to walk after he had gallantly used her to short-circuit a Leyden jar.[21] It was said that J. G. Doppelmayr, the man who had 'taught electricity to speak the language of the fatherland,' suffered a progressive paralysis for his curiosity.[22] Such exaggerations annoyed

18. Musschenbroek to Réaumur, 20 Jan. 1746, in AS, Proc. verb., LXV (1746), 6. The tone of the letter has not unnaturally given the impression that Musschenbroek was the author of the Leyden experiment. But here he describes the results of many trials, of which his own was the most memorable; he does not, and never did, claim the invention. The reference to his initial experiments 'displaying' or 'uncovering' the power of electricity was intended to provide a rationale for the arrangement of apparatus. Cf. Heilbron, *Isis*, 57 (1966), 264–7, and *RHS*, 19 (1966), 133–42.

19. *HAS* (1746), 10; Nollet, *MAS* (1746), 2–3. Nollet's French translation of Musschenbroek's Latin letter, though perfectly satisfactory for his purposes, omits particulars of interest to the historian and obscures the connection of events, e.g., by rendering 'Occupabar in detegendis electricitatis viribus. Tubus ferreus . . . suspendebatur' as 'Je faisois quelques recherches sur la force de l'électricité et pour cet effet j'avois suspendu . . .'

20. Trembley, *PT*, 44:1 (1746), 56–80, where Allamand is made the author of the Leyden experiment. Since Trembley's report is dated 'February 4, 1745 NS,' Dorsman and Crommelin (*Janus*, 46 [1957], 275–80) have suggested that the Leyden jar existed for a year before Musschenbroek announced its discovery, and thus antedated Kleist's phial. The silence would be incomprehensible. Apparently Trembley, the editor or the compositor misdated the letter, or failed to translate properly into the new style (NS). (Until 1752 the English began their legal year on March 25 so that, roughly speaking, their dates were a year behind continental ones for the first quarter of every continental year.)

21. Gralath, Nat. Ges., Danz., *Vers. Abh.*, 1 (1747), 513–14; Sigaud de Lafond, *Précis* (1781), 260–4: Priestley, *Hist.* (1775³), I, 85–6; Winkler, *PT*, 44:1 (1746), 211–12.

22. Gralath, Nat. Ges., Danz., *Vers. Abh.*, 3 (1757), 544, reporting newspaper stories, which, with his usual good sense, he doubts.

the early historians of electricity, Gralath, Priestley and Sigaud. In a well-known passage Priestley scolds Musschenbroek, 'the cowardly professor,' and praises 'the magnanimous' Bose, who, in a characteristic conceit, had wished that he might die of the electric spark, so as to do something worth noting in the Paris Academy's annual *Histoire*.[23]

These exaggerations should rather delight than annoy the historian: they confirm that the condenser flagrantly violated received principles of electricity. The electricians were no more able to explain the jar than was the general public who came to witness its power. Their theories had ceased to predict the outcome of events; their stories were the natural consequence of their uncertainty—even fear—in the face of a manifestation of electrical force far stronger than any they had yet experienced. If, by a slight simplification of Bose's little game, one could create a blow of unprecedented power, might not an equally trivial alteration of the Leyden experiment transport an unlucky electrician into the next world? 'Would it not be a fatal surprize to the first experimenter who found a way to intensify electricity to an artificial lightning, and fell a martyr to his curiosity?'[24]

Frank admissions that the jar had shattered accepted theory appeared on every side. Musschenbroek ended his letter to Réaumur with a confession of ignorance: he now 'understood nothing and could explain nothing' about electricity. Gordon at first disbelieved reports about the jar: 'What seemed particularly doubtful to me was that the man who held the bottle in one hand could draw a spark with the other, since we used to think that by touching the bottom one would hinder the electricity.' 'No one,' says Gralath, 'ever thought that the Rule [of Dufay] must allow an exception in the case of glass.' Before he learned about the jar Winkler believed that nothing of any consequence remained to be discovered about electricity. The genii in Cunaeus' bottle dissipated his complacency: 'One could not understand how the shock could be so powerful, as the unimpressive experimental arrangement . . . , with the phial resting on a non-electric, should have produced a weaker electrification than in the case of a silk support.' Neophytes could identify the difficulty. The defender of a dissertation at Uppsala, Olaus Södergren, recognized that the Leyden experiment was 'different in kind' from those in the classical repertoire, all of which 'willingly' obeyed Dufay's rules.[25]

23. The humorless English historian did not appreciate that Bose's remark was meant to be funny. Bose had written about an experiment with the jar, which, he said, had almost killed him. Réaumur answered that, though most unfortunate, such a death would have the advantage of furnishing 'matière à une page des plus intéressantes de l'histoire de l'Académie.' To which the lyrical physicist replied: 'Oui, Très-Illustre de Réaumur, ç'auroit été un monument qui défie le Mausolée d'Artimise, toutes les piramides de l'antiquité . . . Ni Caesar, ni Alexandre auroient eu un panégiriste si noble, et le hérault d'Achilles m'auroit cédé ces lauriers.' Bose, *Tentamina*, II:ii (1747), 39.

24. [Squario], *Dell'elett.* (1746), 379. Cf. Nollet, *MAS* (1746), 23: 'On ne peut traiter avec trop de prudence un élément qui nous est plus intime que l'air même que nous respirons, & qui peut s'animer & s'irriter par des moyens que nous connaissons si peu.'

25. Gordon, *Versuch* (1746²), 111; Gralath, Nat. Ges., Danz., *Vers. Abh.*, 2 (1754), 447;

An instructive expression of the fundamental theoretical problem raised by the jar occurs in a report of French electrical experiments that John Turberville Needham—English Catholic priest, professional travelling tutor, itinerant intelligencer, accomplished naturalist, first director of the Brussels Academy of Science—sent the Royal Society late in 1746.[26] Needham pointed out that the jar will not charge while resting upon a glass salver unless someone presents his finger to the outside of the jar, and that then the 'electrical Fire' can be seen to flow across the glass and into the water.[27] He continued:

> Here we see an Example, where an Electric *per se* [the jar] is so far from terminating or excluding the Power of Electricity, that it is even made a *Medium* of Communication in Circumstances where the Wire, which is a Non electric *per se*, refuses to perform its expected Office. When I speak of the Power of Electricity in this Case, I would not be understood of the Power of attracting light Bodies, which is well known to be scarce sensibly interrupted by a glass *Medium*, as appears in the common Experiment of an electrified Tube, acting upon Leaf-Gold, in a crystal Bottle: Though even this, if duly considered, might create some Difficulty; but I would only be understood of that communicated Virtue, which renders Non-electrics *per se* electrical. In one Word, the Singularity of this Experiment is, that, by the Addition of the glass Salver, the Wire and the Water, both of them Non-electrics *per se*, should not be in the least affected without the Approach of a Hand, and should then receive the electrical Fire from it through a glass *Medium*; notwithstanding they are in the very same Circumstances, that a Man is in, or any other Non-electric *per se*, placed upon a Cake of Wax and in Contact with the electrifying Spheroid.

The invention of the condenser actualized the potential inconsistency inherent in the ambiguous role that effluvial theory assigned to glass. It threw doubt on the identity of communicated and frictional electricity. It made a mockery of the confidence of the system-builders of 1745. It created a crisis.

3. EARLY EXPERIMENTS WITH THE LEYDEN JAR

No one was killed or even injured by the Leyden flask. Reports of experiments became soberer. The established electricians, Nollet, Watson, Wilson and Winkler, explored the properties of the new instrument, aided by two able newcomers: Daniel Gralath, a former judge, the mayor of Danzig, and the founder, director and leading spirit of its scientific society; and Louis-Guillaume Le Monnier, M.D., son and brother of Parisian academicians, Professor of Botany at the Jardin du Roi, a cautious, inventive, and well-trained scientist.[28] By the end of 1746 these men had reached a consensus on several

---

Winkler, *Stärke* (1746), Vor., 72; Södergren, *De rec. quib.* [1746], 12; Le Monnier, *PT*, 44:1 (1746/7), 292, 293: 'Strong Exceptions to the Rule laid down by Monsieur du Fay,' 'an effect so contrary to M. du Fay's rule.'

26. *DSB*, X, 9–11; Needham, *PT*, 44:1 (1746), 259, concerning the experiments of L. G. Le Monnier, for which see *infra*, xiii.3.

27. Evidently Needham did not quite grasp the experimental setup; cf. Neale, *Directions* (1747), 35–6.

28. Hoppe, *Gesch.* (1884), 17–18; Cuvier, *HAS*, 3 (an IX), 101–17.

facts and errors, which may be surveyed under four heads: the jar, the circuit, propagation of the shock, and nonviolent demonstrations.

THE JAR

According to Musschenbroek, the jar would not work unless its outside were perfectly clean and dry. Nollet found that only the inside above the water need be dry, and he guessed that superfluous moisture had caused the Dutch to fail in their attempts with French and English glass. He found further that metal or any non-oily liquid could replace water, but that no substance other than porcelain could act the part of glass.[29] He tried wax, wood, and metals, and sulphur excited by a sulphur globe, in case the two electricities somehow figured; but, because his dielectrics were thick or perforated,[30] he could not build a condenser of sulphur or wax. The consequent emphasis on the supposed uniqueness of glass proved unfortunate. Not so Nollet's discovery that the shape of the vessel did not count. Soon the English were making parallel plate condensers, later known as 'Franklin squares.'[31]

From the fact that the jar had to be grounded during electrification, several physicists guessed that the shock would grow stronger in proportion to the external surface in immediate contact with nonelectrics. Watson, following a suggestion made to him by John Bevis, M.D., an enthusiastic Newtonian natural philosopher (it is said that he went nowhere without his copy of the *Opticks*[32]), armed the bottom of his bottle with thin lead sheets; while Winkler, Dutour, Allamand and Wilson set their jars in water-filled dishes. ('All Mr W[ilson]'s kitchen utensils are [said to be] converted into Leyden bottles. If so, I should not much care to dine with him.'[33]) Ultimately the jars were armed with metal within and without, which removed a source of unwanted moisture and unnecessary weight.[34] The fearless physicists found three ways besides external arming to increase their shocks. Gralath connected jars in parallel, in what he called an electrical battery; a stroke from two such phials, taken to the forehead, felt like a blow from a bludgeon. Nollet, using a huge prime conductor, produced sparks he feared to take and a shock that killed a sparrow. The insides of this unlucky creature, which was opened by a surgeon friend of its executioner, resembled those of a man slain by lightning. The obvious analogy

29. Nollet, *MAS* (1746), 5–10. According to Mangin, *Hist.* (1752), Pt. i, 31, Manteuffel also used porcelain. Gralath, Nat. Ges., Danz., *Vers. Abh.*, 1 (1747), 513–14, and Watson, *Sequel* (1746) and *PT*, 44:2 (1747), 704–49, §14, emphasize harm caused by moisture.

30. Gralath, Nat. Ges., Danz., *Vers. Abh.*, 1 (1747), 513–14, noticed the deleterious effects of tiny cracks in the insulator.

31. Nollet, *MAS* (1746), 7. The invention of the parallel-plate condenser was the work of Wilson's circle, perhaps of Smeaton. Sigaud de Lafond, *Præecis* (1781), 279.

32. *DNB*, II, 451–2. Bevis was an indefatigable astronomer, a clever mechanic (Henderson, *Ferguson* [1867], xxiii), a physicist admired even by the French (Lalande to Boscovich, 15 Feb. 1766, in Varićak, *RAD*, 193 [1912], 229–31).

33. W. Henly to J. Canton, n.d., Canton Papers, II, 86 (RS).

34. Watson, *Sequel* (1746), §28; Nollet, *MAS* (1747), 161; Hoppe, *Gesch.* (1884), 20; Priestley, *Hist.* (1775³), I, 119; Sigaud de Lafond, *Précis* (1781), 226–7.

to lightning suggested by the sparks and snappings thereby gathered unexpected strength.[35] And, lastly, Gralath, Wilson, and Watson recognized the important fact that the shock was greater the thinner the glass.[36]

The power of the externally armed battery inspired Watson to invent a tasteless parlor trick he called the 'electrical mine.' Hide a battery in a corner of the room; attach a line to its upper wire and another to its lower, soldered to the lead armor of the last jar; run the upper line to, and the lower beneath, the prime conductor, concealing the second wire under a rug. Electrify the hidden battery. Now invite a friend (whom you can spare) to draw a spark from the prime conductor. Nothing happens to him unless he stands upon the rug directly over the concealed wire. It is most amusing to see his surprise when, by moving his toe but an inch, he is pulverized by the gun-barrel he had earlier grasped with impunity. 'The first time I experienced it,' wrote Watson, 'it seemed to me, used as I am to such trials, as though my arm were struck off at my shoulder, elbow and wrist, and both my legs, at the knees, and behind the ankles.'[37]

THE CIRCUIT

By another easy inference electricians concluded that if one man, say A, holds the jar and a second, Z, touches the conductor, both will feel the shock when they bring their hands together. How many others, B, C, D, etc., might be inserted between the first pair? Le Monnier tried 140 courtiers, in the presence of the King; Nollet shocked 180 gendarmes before the same fastidious company, and over 200 Carthusians at their monastery in Paris (figs. 13.1 and 13.2). 'It is singular to see the multitude of different gestures, and to hear the instantaneous exclamation of those surprised by the shock.'[38]

The chains were effective detectors. Gralath and Watson used them to show that only those who stood in a direct line joining the jar's wire to its outer coating passed the discharge.[39] Add a side chain a,b,c, . . . inserting a between A and B. The shock runs through AaB . . . Z; b,c . . . feel nothing. Disengage the branch line. Let C hold by the middle a metal bar whose extremities are grasped by B and D; ABD . . . Z are struck, but not C. If the bar is replaced by a dry stick or a glass rod, no discharge occurs. If consecutive members in the chain do not quite touch, sparks span the gaps when the bottle explodes. No clearer idea of the course of the electrical matter could be desired.

35. Gralath, Nat. Ges., Danz., *Vers. Abh.*, 1 (1747), 506–41; Hoppe, *Gesch.* (1884), 19–20; Watson, *Sequel* (1746), §26; Nollet, *MAS* (1746), 20–2. Other bird-slayers: Gralath, *loc. cit.*, 521–5, and Bose, *Tentamina*, II:ii (1747), 38. Winkler, *Stärke* (1746), 134–64, and Allamand, *Bibl. brit.*, 24 (1746), 436, also emphasized the analogy to lightning.

36. Gralath, *loc. cit.*, 513–14; Watson, *Sequel* (1746), §13; Wilson, *Essay* (1746), 89.

37. Watson, *Sequel* (1746), §§ 29–30.

38. Nollet, *MAS* (1746), 18; Priestley, *Hist.* (1775³), I, 125–6; Sigaud de Lafond, *Précis* (1781), 293; *HAS* (1746), 8.

39. Gralath, *loc. cit.*, 518; Watson, *Sequel* (1746), §§25, 31, 34–41; Le Monnier, *MAS* (1746), 452; Priestley, *Hist.* (1775³), I, 109–14, 130.

13.1 (a) A one-person and (b) a two-person discharge train. From Nollet, Leçons, VI (1748).

13.2 A Japanese version of Nollet's many-person discharge train. From a drawing of 1813 reproduced by Prinz, Schweiz. Elektr. Ver., Bull., 61 (1970), 8.

These experiments succeeded almost as well when the participants stood on the ground as when they were insulated. Le Monnier, who thought the degree of insulation made no difference, passed the shock through a mile of long-suffering Carthusians joined together by grounded iron wires. In fact moist ground may offer a discharge path as good as a human chain. Sigaud once tried to send a shock through a large company of grounded men, but only those at the extremes, A, B, and Y, Z, felt it. Now C was suspected, as Sigaud delicately put it, 'of not possessing everything that constitutes the distinctive character of a man.' Having learned from playful colleagues that it was impossible to electrify a eunuch, Sigaud brought the puzzle to the public. The Duc de Chartres demanded proof; three of the King's musicians, 'whose state was not equivocal,' were inserted in the discharge. The *castrati* jumped as the rest. Sigaud repeated his experiments. The appropriate explanation occurred to him when a man at whom the discharge appeared to stop complained that the shock struck his leg as well as his arm.[40]

PROPAGATION OF THE SHOCK

As the discharge trains lengthened, the range and speed of the shock became subject to study. Le Monnier made the first important estimate of velocity by trying to discern a delay between seeing the flash and feeling the shock when an electrified jar was brought into a mile-long circuit. He found none. Estimating that he could distinguish an interval of a quarter-second, Le Monnier concluded that the discharge propagated at at least thirty times the speed of sound. He also found that the shock traveled through bodies of water, like the basin in the Tuilleries Gardens, without losing power.[41] Watson and several Fellows of the Royal Society undertook to combine and improve these results by determining the velocity with which the shock ran down a mile or two of river. They found the discharge too swift to measure.[42]

NONVIOLENCE

Physicists soon learned that an electrified phial need not explode to be intriguing: when doing nothing at all, innocently insulated, it unaccountably preserved its punch for hours or days.[43] Even when grounded it remained potent, provided that its top wire was not touched. And if, with its bottom insulated, one

40. Sigaud de Lafond, *Précis* (1781), 284–91; cf. *VO*, III, 222–5, and Lichtenberg, *Briefe*, II, 144–5.

41. *MAS* (1746), 451, 456–7. Le Monnier believed that the entire basin was electrified during discharge, as did the reviewer in *Bibl. rais.*, 38:2 (1747), 112–13; Watson (*PT*, 44:2 [1747], 388–95), Nollet (*MAS* [1747], 196), and Gralath (Nat. Ges., Danz., *Vers. Abh.*, 3 [1757], 550–1) insisted that the electrical matter passed through only the filament of water lying in the circuit; in fact each coating of the jar discharges separately to ground, as with Sigaud's eunuchs.

42. Watson, *PT*, 45 (1748), 49–92, reviewed at length by Priestley, *Hist.* (1775³), I, 130–7.

43. Nollet, *MAS* (1746), 12, gives 36 hours, an observation Priestley, *Hist.* (1775³), I, 110, credits to Le Monnier; Musschenbroek gives one day (Bose, *Tentamina*, II:ii [1747], 36–7); and Watson, *Sequel* (1746), §23, gives 'many hours.'

attempted to discharge it by grounding its top, it yielded but a tiny spark; after which it could be 'revivified,' or made capable of the Leyden shock, merely by holding it in the hand. This behavior, which had perplexed Kleist, intrigued Le Monnier, who simulated it as follows.[44] He placed an electrified jar on a glass support and near a little ball hung by a silver thread. Everything remained quiet. He touched the wire; the ball flew to the jar's bottom, sparked, and fell back. He repeated the sequence twenty times or more. Apparently Le Monnier did not recognize that the play of the ball gradually discharged the phial. This important fact was noticed by John Canton, who detected the progressive enervation of an insulated jar as he drew sparks from its top and bottom alternately.[45]

4. FIRST THEORIES OF THE JAR

Nollet admitted that the condenser did not act in accordance with the accepted rules of electricity. Instead of using its violations as a guide to a new general theory, however, he preferred to treat the Leyden jar as an exception, a phenomenon *sui generis*. Did the bottle electrify and stay electrified when perched on nonelectrics, defying the Rule of Dufay? That is because highly electrified glass has the power to retain its electricity under such conditions. Did the experiment succeed with vitreous substances only? Glass and porcelain are unique in combining ready acceptance of, and great tenacity for, communicated electricity.[46]

Translated into Nollet's mechanical theory, the glass in the Leyden experiment had the power simultaneously to prevent loss of effluvia from the prime conductor and to acquire a strong electrification itself, to become a vigorous emitter and receiver of electrical matter, as long as its external surface was grounded. Now Nollet's electrical matter flowed more easily through metals than through air, wax, or resin, whence the need for grounding to keep up the jar's vitality: the charged glass maintained its electricity through a resistanceless exchange of electrical matter with its nonelectrical support. Le Monnier's jars lost their virtue when placed on wax or thick glass because their effluent streams could not penetrate them to excite the answering affluent.[47] The upper surface of the bottle acted like an insulator, the volume of the glass like a conductor, and the bottom surface like either a highly electrified metal or an unexcited body according as it was grounded or insulated, respectively. Nollet did not explain how the jar's effluent, extinguished by an insulating support, could be 'revivified' by grounding its bottom.

A person holding an electrified jar is the host of a slow double flux, an

44. *MAS* (1746), 454–5.

45. Watson, *Sequel*, (1746), §54; Priestley, *Hist.* (1775³), I, 122–3; *infra*, xv.1.

46. Nollet, *MAS* (1746), 11–14; *Essai* (1750²), 199–201, 204–6; *MAS* (1747), 195–6.

47. *MAS* (1746), 12–13; *Essai*, 202–4. Nollet's use of Le Monnier's result gave an opening to later critics, who interpreted him to say that an isolated jar lost its virtue, not, as he intended, that it became dormant. Cf. D. Colden in *EO*, 282–92; Beccaria, *Dell'elett.* (1753), 151.

affluent from the bottle's exterior through himself to ground, and a counter-current from himself into the jar. When he touches the prime conductor with his free hand, another and more rapid dual flow sets up, from the metal to the hand and vice versa. He is suddenly struck by affluents from the jar and the conductor, which drive back his body's effluents, painfully compressing the electrical matter naturally present in the arms and breast. The course of the commotion may be made visible by passing it through a string of monks connected by water-filled glass tubes. As the shock passes the water flashes, as would our bodies, says Nollet, 'were we as transparent as glass and water.'[48]

Winkler proceeds in the same spirit as Nollet. It is plain, he says, that the jar defies Dufay's rule owing to special properties of glass, which enable it simultaneously to retain and to transmit communicated electricity. Take an empty jar, electrify it as usual, and add water: the glass electrifies the liquid. In the Leyden experiment, therefore, the water is doubly excited, once by the wire and again by the glass; it thereby acquires an unusual electrification, especially when the jar is grounded while charging. It appears that the excited glass electrifies the conductor used to ground it; that the conductor's agitated atmosphere further disturbs the electrical matter within the glass; that the electricity of the *inner* surface of the bottle thereby increases; and that it, in turn excites the contained water to a still higher pitch. The bottle strikes oppressively because it attacks the body simultaneously with the feverish atmospheres of its water and its glass. Winkler's glass is more potent even than Nollet's, for it confers more electricity than it has, and receives back more than it gave. No wonder the Leyden experiment 'agitated' Winkler's blood and obliged him to use 're-frigerating medicines.'[49]

Watson, like everyone else, had initially been baffled by the bottle: 'The difficulties thereof I confess seemed insurmountable,' he wrote, 'until I had made the following discoveries.'[50] He referred to the need for an external discharge circuit; to his earlier finding that the machine is a pump; to his mistaken belief that an insulated person could not explode the jar; and to his important observation that the strength of the shock grows as the extent of the surface armed. This mixture of fact and fancy premised, Watson represented that the Leyden experiment followed directly from his own version of Nollet's theory. The electrical aether pumped onto an insulated nonelectric strives to dissipate, creating an efflux whose loss the environment compensates; the double flux preserves the accumulation, or disequilibrium, of the aether for a time after the pump ceases to operate.

While the bottle charges, electrical aether accumulates in the prime conductor and the water and, by seeping through the glass, in the bottom of the jar.

---

48. *MAS* (1746), 16–18; *Essai* (1750²), 194–9. The *Bibl. rais.*, 38:2 (1747), 113, allows that, although something like Nollet's double flux exists, it is premature to theorize about it.

49. Winkler, *Stärke* (1746), Vor., 73–9, 103 ff.; *PT*, 44:1 (1746), 211; *Anfangsgründe* (1754), 317–18.

50. *Sequel* (1746), §64. Watson later *(PT,* 44:2 [1747], 389) denied that the jar violated the Rule of Dufay on the quibble that it rests upon itself, i.e., upon glass, an electric *per se.*

These accumulations set up local, peaceful double fluxes. When the man holding the jar touches the prime conductor, the uneasy equilibrium breaks; he tries to reconstitute the accumulations of the bottle and the prime conductor, while the ground resupplies him. 'The man instantaneously parts with as much of the fire from his body, as was accumulated in the water and gun-barrel; and he feels the effects in both arms, from the fire of his body rushing through one arm to the gun-barrel, and from the other to the phial. . . . As much fire as the man then parted with, is instantaneously replaced from the floor of the room, and that with a violence equal to the manner in which he lost it.' If the man were cut off from ground, the mechanism would fail. Watson adduced the supposed incapacity of an insulated circuit to discharge the jar as the chief corroboration for his theory of its action.[51]

The feebleness, imprecision, and incompleteness of the theories of Nollet, Winkler and Watson, their failure to explain or even to mention characteristic features of the condenser, require no comment. They offered no guidance or stimulation. Many of the prominent electricians of the mid-forties, like Bose, Winkler, the Berlin essayists, Gordon, Watson, Le Monnier and Gralath, stopped working on electricity before 1750. Nollet engaged in long studies, which he thought unexciting,[52] of the effect of changes in atmospheric conditions, and of alterations of geometry and materials, on the standard phenomena of electricity. A new approach was needed.

51. Watson, *Sequel* (1746), §64. Watson later acknowledged that this experimental finding was spurious and that the condenser did not agree with his principles; *PT*, 45 (1748), 102.

52. Nollet, *PT*, 45 (1748), 187, regarding his papers in *MAS* (1747), 102–31, 149–99, 207–42.

CHAPTER XIV

Benjamin Franklin

1. BEGINNINGS

In January 1746, when Benjamin Franklin attained the age of forty and the leisure toward which he had long been working, he was busily engaged in advancing the study of natural philosophy in America. Already in 1743 he had joined in an effort to establish a society dedicated to 'promoting useful knowledge among the British Plantations in America.' The society at first grew briskly, but as the novelty wore off the 'very idle Gentlemen' who were its members found they preferred 'the Club, Chess and Coffee House' to the 'Curious amusements of natural observations.' By the fall of 1745 Franklin alone among the original projectors still argued for the scheme. And so, by coincidence, he became eager to produce, and free to attempt an exemplary philosophical study at precisely the moment when news of the German novelties in electricity first reached the new world.[1]

Franklin had been introduced to the subject by a Dr. Spencer of Edinburgh, an itinerant lecturer whose show he had seen in Boston in 1743 and whom he sponsored in Philadelphia the following year.[2] Spencer's bag of tricks included a demonstration that 'Fire is diffus'd through all Space, and may be produc'd from all Bodies:' he strung up a little boy and, according to an eyewitness, caused 'Sparks of Fire' to proceed from his face and hands 'by only rubbing a Glass Tube at his Feet.' This was but to repeat Gray's stale diversion; no doubt the competent Dr. Spencer exploited the boy's electricity as well as his fire.[3] An attentive auditor could gather that the tube put in motion or redistributed something much like fire, which might agitate small bodies but manifested itself primarily in sparks and glows. Franklin's first writings on electricity reflect these views.

It was not Spencer, however, who made Franklin an electrician, but a report of the German fireworks that reached Philadelphia in the fall of 1745, in the

---

1. Van Doren, *Franklin* (1938), 154; Hindle, *Pursuit* (1951), 66–72; Stearns, *Science* (1970), 670–4. Franklin's researches did not save the project in the early 1740s, but helped ensure its success, as the American Philosophical Society, when it was revived in 1766.

2. Cohen, *EO*, 49–54, and Lemay, *Maryl. Hist. Mag.*, 59 (1964), 199–216, identified Spencer, whom Franklin remembered as 'Dr. Spence.'

3. Cohen, *J. Frank. Inst.*, 235 (1943), 1–25. Cf. Heathcote, *Isis*, 46 (1955), 29–35, and Cohen, *FN*, 435, who argue that Spencer might not have known that the boy worked electrically. This is most unlikely, since Spencer owned a full electrical apparatus (*EO*, 57).

company of a glass tube suitable for producing them. This coincidence was the work of Peter Collinson, F.R.S., a Quaker merchant in London with colonial business connections, a good amateur botanist who encouraged Americans with philosophical predilections.[4] Collinson served as agent of one of Franklin's favorite projects, the Library Company of Philadelphia, which collected natural curiosities and scientific apparatus as well as books and journals. At his suggestion the Company had subscribed to the *Gentleman's Magazine,* a review of current happenings, fashionable, political, and scientific.[5] An 'historical account' of the 'wonderful' German electrical experiments, which appeared in the *Gentleman's* for April 1745, caught Collinson's attention, and, thinking to encourage the Philadelphians to repeat them, he added a glass tube to the shipment containing the journal.[6] One sees the influence of the *Gentleman's* résumé throughout Franklin's first paper on electricity,[7] not only in demonstrations and amusements, but also, and more significantly, in experimental procedures and arrangements that led to fundamental Franklinist doctrines.

The account turns out to be an imperfect translation of a report of the work of Bose, Hausen and Winkler published in the *Bibliothèque raisonnée* for the first quarter of 1745. The author, the tireless polymath and compulsive reviewer, Albrecht von Haller, was a professor at the University of Göttingen. The *Bibliothèque,* despite its language, was conducted by a group of Dutch professors that once included 'sGravesande. Franklin's first steps in electricity were not guided by Watson or Wilson, as is often said, or by his untutored imagination, but by an artifact of a continental professoriate then recently become enthusiastic about experimental physics: the work of Leipzig academicians as reported by a Swiss at Göttingen in a review run by professors in Holland.[8]

Haller's account begins with the discoveries of Hauksbee, Gray and Dufay. One initiated by Spencer would find familiar items, Gray's hanging boy for

4. Hindle, *Pursuit* (1951), 65 and *passim;* Brett-James, *Collinson* (1926); Stearns, *Science* (1970), 515–16 and *passim.*

5. Library Company, Philadelphia, Minute Book, f. 115, under Nov. 1741. An issue took about six months to come, and might be delivered by Franklin himself (*ibid.,* 12 Dec. 1743 and 11 April 1745); Brett-James, *Collinson* (1926), 161–6.

6. *GM,* 15 (1745), 193–7. That this might be the account mentioned by Franklin in letters to Collinson of 28 March 1747 (*EO,* 169) and to Collinson's son Michael, *c.* 1765 (*EO,* 19), was suggested by Lemay, *Kinnersley* (1964), 54; cf. Cohen, *FN,* 431. That it is the account is proved in Heilbron, *Isis,* 68 (1977), 539–44.

7. BF to Collinson, 25 May 1747 (*EO,* 171–8). The first installment of Franklin's *Experiments and Observations on Electricity* (1751) consisted of this letter, three later ones and a paper, the last additions dated 29 July 1750 (*EO,* 158–9, 171, 179). The note to Collinson of 28 March 1747, mentioned in the preceding note, initially appeared in *EO*⁴ (1769), the first revised edition of the letters. For their complicated printing history see Cohen's admirably clear account in *EO,* 139–61.

8. [Haller], *Bibl. rais.,* 34 (1745), 3–20; Haller to Gesner, Jan. 1745, in Sigerist, Ak. Wiss., Gött., *Abh.,* 9:2 (1923), 158. A striking indication of Franklin's dependence on Haller's review is Franklin's use of 'electrise,' an anglicization of 'électriser' and 'electrisiren.' There are only three significant occurrences of 'electrise' in *PT* before 1748, once in a snippet from Bose, and twice in passages from Franklin quoted by Watson, who always wrote 'electrify'; the poor translation of Haller in the *Gentleman's* employs 'electrise' throughout.

instance, and the assurance that 'what proceeds from the electrised body is really the production of fire.' And anyone with imagination would respond to Haller's view that 'lightening has pretty much the same qualities' as electricity; not only, in his opinion, because both produce light and flame, but also because lightning, like electricity, can be conducted by metals. 'It generally runs over the whole length of the solid bodies which it strikes, and it has been seen to descend along the wire of steeple clocks [*clochers*!] from top to bottom.' A broad hint indeed to the future inventor of the lightning rod.

The first experiment Haller described closely enough to give a neophyte repeating it a chance of success requires the electrification of a man standing on pitch. 'Whoever shall approach his finger to the body of the person thus electrised will cause a spark to issue from the surface, accompanied with a crackling noise and a sudden pain of which both parties are but too sensible.' 'If any other person not electrised puts his finger near one who is so, no matter where it be to his naked skin or his clothes, there issues thence a fire, with a painful sensation, which both persons feel at the same time. . . . If instead of a finger you hold a key or a potsherd near the electrised person, there proceeds a fire, even from his buttons, or the hoop of a lady's petticoat, how distant so ever from the body.' We shall meet almost identical experiments in Franklin's first letter on electricity.

From these glimmers Haller proceeds to the heart of his subject. 'To be a true fire,' he observes, the electric spark 'wants nothing but the power of kindling combustible bodies or liquors; in that case the electric matter would be fair to be either inseparable from that of fire, or to be one and the same thing.' He tells then of the great progress of electricity in Germany, of the applications of Hauksbee's whirling machine to the material of Gray and Dufay, of the impression made there upon the fashionable world, where 'electricity took [the] place of quadrille,' and of the success of the German electricians in igniting spirits with sparks. He reports Bose's division of sparks into the '*male* fire, which is attended with crackling, and has a considerable force, and the *female* fire, which is a luminous emanation, without violence or percussion.' And he describes several teutonic games, particularly *Venus electrificata,* whose caress was so painful, he says, that the gallant Bose 'durst not renew his kisses more than three times.' The hardier Philadelphians were to make this lady's endearments still more disagreeable.

As for theory, Haller takes it for granted that all electrical effects derive from the motions and displacements of an electrical matter, which he imagines to extend around 'electrised' bodies in what he calls 'atmospheres.' He does not believe, however, that the time had yet come for a physics of the electrical matter in the style of Winkler, Waitz or Nollet. Much remained to be done. 'These new experiments being made public, it is to be hoped that our philosophers will soon add some improvements by new ratiocinations, and perhaps by new discoveries.'

2. FIRST FRANKLINIST PRINCIPLES

We shall assume that Franklin and his associates—Philip Syng, silversmith, Thomas Hopkinson, lawyer, and the teacher-preacher Ebenezer Kinnersley— began to respond to this challenge soon after the arrival of Collinson's gift.[9] Although no unambiguous account of their progress during the winter of 1745/6 survives, two characteristic Franklinist principles, the power of points and the doctrine of electricity plus and minus, appear to date from that time. In the winter of 1746/7 they concentrated on the Leyden jar. By then Franklin was a full-time electrician. 'I never before was engaged in any study that so totally engrossed my attention and my time as this has lately done,' he wrote Collinson in March, 1747. 'What with making experiments when I can be alone, and repeating them to my Friends and Acquaintances, who, from the novelty of the thing, come continually in crowds to see them, I have, during some months past, had little leisure for anything else.' One of these friends, James Logan, merchant and book collector, the most knowledgeable scientific amateur in the middle colonies, advised Franklin to publish his results. In March, 1747, Franklin notified Collinson of his intention to communicate 'some particular phenomena that we look upon to be new.'[10]

He redeemed his promise on May 25 in a brisk, unaffected, yet careful letter intended for circulation, if not for publication. It takes up four topics. The first concerns Hopkinson's discovery of the readiness of pointed conductors to draw and emit 'the electrical fire;' the second follows the fire, 'diffus'd among other matter,' through a series of electrifications; the third suggests new diversions and improvements of continental amusements; and the fourth describes Syng's cheap colonial modification of the European electrical machine, a sphere turned with a crank attached directly to its axis, 'like a common grindstone.'[11]

The 'power of points' is their aptness in *'drawing off* and *throwing off* the electrical fire.' In the mouth of a clean dry bottle Franklin places a three-inch iron shot, against which rests a cork ball suspended by a long silk thread. He

9. The April number of the *Gentleman's* was in hand by October; Libr. Co., Phil., Minute Book, f. 146. For Franklin's collaborators, among whom the surveyor Lewis Evans should perhaps be included, see *EO*, 60n, 62n; Lemay, *Kinnersley* (1964), 62–87; *BFP*, III, 112.

10. *EO*, 169; *BFP*, III, 110–11. For Logan see Hindle, *Pursuit* (1951) 21–3, and Brasch, Am. Phil. Soc., *Proc.*, 86 (1942), 8–12; Stearns, *Science* (1970), 535–9.

11. *EO*, 177–8. Immediately before describing this crude but serviceable apparatus, Franklin advised that one should avoid sullying the *tube* with one's hand and, for best results, should store it in a flannel-lined case. 'This I mention [he says] because European papers on Electricity frequently speak of rubbing the tube as a fatiguing exercise.' Cohen (*EO*, 60) has inferred that the Philadelphians conceived their machine to be 'a great and significant innovation, an original contribution to ease the "tired" European electricians.' Franklin intended no connection between his advice respecting the tube and his specification of the local substitute for the expensive European machine, several of which, together with Hauksbee's prototype, were described by Haller. (The Library Company, according to Franklin's catalogue of 1741, owned the second edition of Hauksbee's *Physico-Mechanical Experiments*.) Cf. Benjamin, *Hist.* (1898), 542; van Doren, *Franklin* (1938), 156–7; Cohen, *FN*, 440.

electrifies the shot with the tube, and the ball flies off four or five inches. Then he advances a pointed metal tool towards the bottle; suddenly, at a distance of eight or ten inches, the shot loses its 'repellency' and the ball returns to it.[12] At night the effect is more striking. 'If you present the point in the dark, you will see, sometimes at a foot distance and more, a light gather upon it, like that of a fire-fly or a glow-worm; the less sharp the point, the nearer you must bring it to observe the light; and at whatever distance you see the light, you may draw off the electrical fire, and destroy the repellency.'

The report of these facts may invert the order of their discovery. The soft glow about the point is Bose's feminine spark, and we may suppose that the Philadelphians soon followed Haller's hints about male and female fires. They would have discovered quickly how best to produce each type, aided, perhaps, by reference to the papers of Gray and Dufay. The important observation that the shape of a conductor affected the distance over which it could solicit the fire would have occurred during the course of these researches. Naturally Franklin, initiated by Spencer and tutored by Haller, understood the glow to be the electrical fire itself, that is, the agent of attraction and repulsion; he then demonstrated the relationship with the elegant experiment of the shot and ball.[13] The discovery of the fire power of points followed that of their drawing power. Hopkinson guessed that a point affixed to a gun-barrel might focus the electrical fire, and so provide a stronger discharge. He found precisely the reverse: his bayonet refused to accumulate the fire, which ran out its point as often as the tube was applied to its butt. The unexpected symmetry between the drawing and firing power of points was to furnish Nollet with a forceful objection to the Franklinist system.

The doctrine of positive and negative electricity appears to have derived from the rediscovery of the effect of Bose, Watson and Wilson, which suggested to Franklin, as it had to the English electricians, that glass acted as a pump, pulling electrical fire from the ground, through the rubber and into adjacent conductors. Let two persons insulate themselves by standing on wax. Let the first, A, rub the tube, while the second, B, 'draws the fire' by extending his finger towards it. Both will 'appear to be electrised' to a person standing on the floor; that is, C will perceive a spark on approaching either of them with his knuckle. If A and B touch during the rubbing, neither will appear 'electrised'; if they touch after excitation, they will experience a spark stronger than that exchanged by either with C; and after the shock neither exhibits electricity. Gentleman A, says Franklin in explanation, the one who first collects the fire from himself into the tube, suffers a deficit in his usual stock of fire, and is thus 'electrised' minus; B, who draws the fire from the tube, receives a superabun-

12. *EO*, 171–3. As Nollet rightly objected, *Lettres* (1753), 134, the shot does not lose all its electricity.

13. The Philadelphians also destroyed the 'repellency' of the shot by breathing on it, by covering it with sand, and by exposing it to wood smoke or open flame. In connection with these demonstrations Franklin refers to the 'atmospheres' about shot and ball formed by smoke from burning resin. Cf. *infra*, xiv.4.

dance, and is 'electrised' plus; while C, who stands on the ground, retains his just and proper share. Any two, brought into contact, will experience a shock in proportion to their disparity of fire, that democratic element forever striving to attach itself to each equally.[14]

The question arises whether literary influences other than Haller's played a part in the genesis of Franklin's system. The date of his first communication (the spring of 1747), his references there to 'European papers of Electricity,' and the superficial similarity between his ideas and those of Watson and Wilson suggest that he might have derived something from their pamphlets of 1746.[15] But Franklin's letters show that he did not know these pamphlets when he intimated to Collinson his intention of communicating 'some particular phenomena that we look upon to be new.' And when he read them he discovered no key or guide to the scheme of electricity plus and minus; on the contrary, he learned that the Philadelphians had been anticipated on a number of points, especially the critically important effect of Bose et al.[16] He also found that the theories his group had evolved contained much both new and significant.

The approach and style of Franklin's first letters on electricity also demonstrate his ignorance of the European literature of the mid-forties. By 1746 the search for phenomenological regularities, as conducted by Dufay, had every-

14. Cf. Watson's similar manipulations (*supra*, xii.3), which led him nowhere. Gliozzi, *Elettrologia* (1937), I, 188–9, observes that Franklin's discussion of the sparks interchanged between A, B, and C implicitly invokes something besides the quantity of electrical fire transmitted. For if $A = -1$, $B = +1$, $C = 0$ and grounded, then between all pairs the same amount of fire must pass, namely one unit. Why then should a stronger spark jump from B to A than between either and C? Gliozzi suggests that Franklin here sought the idea of potential. The *form* of Franklin's analysis of the passage of sparks appears to have been habitual with him. Consider his *Dissertation on Liberty and Necessity, Pleasure and Pain*, written in 1725, which compares an animate object A that successively experiences ten units of pleasure and an equal amount of pain with a rock C, insensible to both (*BFP*, I, 55–71). There is a striking parallel between pleasure, pain and insensitivity on the one hand and positive electricity, negative electricity, and grounding on the other. Cf. BF to Priestley, 19 Sept. 1772, in *BFP* (Smyth), V, 437–8, on 'prudential algebra'; and Heilbron, *Phys. Today*, 29:7 (1976), 33–4.

15. Thus Cohen (*FN*, 390) takes Watson as Franklin's 'master and guide,' following Benjamin, *Hist.* (1898), 539–42, and anticipating Finn, *Isis*, 60 (1969), 367–8. Only Finn, who argues primarily from terminological similarities, e.g., 'vial' for 'Leyden jar,' justifies his position. But Finn's parallels not only do not relate to Franklin's *innovations*, they also require, to have any weight, that Franklin worked out his theory between 28 March and 25 May 1747. This assumes in turn that the theory of the contrary electricities owed its origin to an analysis of the Leyden jar, which the Philadelphians could have met no earlier than the autumn of 1746 (e.g., in *GM* for March). Finn, like most historians, therefore dates Franklin's earliest experiments in late 1746 or early 1747; but this is to overlook the evidence about the 'historical account' and to make unintelligible the grounding of the system in manipulations of the tube. The same objections apply to Yamazaki, *Jap. Stud. Hist. Sci.*, 15 (1976), 37–8.

16. BF to Collinson, 28 March 1747 (*EO*, 169); to Colden, 5 June 1747 (*BFP*, III, 142). Franklin acknowledged receipt of Watson's two pamphlets of 1746, the *Experiments and Observations* and the *Sequel* (*supra*, xii.3), in May 1747. Collinson shipped them the preceding March; BF to Collinson, 25 May 1747 (*EO*, 178). Wilson's *Essay* had arrived by July (same to same, 28 July 1747 [*EO*, 179]. Likewise Cohen's identification (*FN*, 440) of Needham, *PT*, 44:1 (1746), 247–63, as the 'European papers' fails before the dates, as that number of the *PT* was not printed until 1747.

where given way to elaborating aether mechanics. Franklin remained ignorant of this change until his principles had jelled. That proved fortunate, for the originality of his contributions decreased in proportion to his acquaintance with European writings, the style of which he began to affect in 1749.[17] Not that the Philadelphians operated *in vacuo*. They probably knew the texts of 'sGravesande and Musschenbroek,[18] and the papers on electricity published in the *Philosophical Transactions* up through Desaguliers'. Their main debt, however, they owed to Haller, who beguiled and challenged without overwhelming them.

Historians have further confused the contributions of the Philadelphians by ascribing to them the discovery of the conservation of electric charge. At the fountainhead of this recurrent myth stands Franklin himself, who wrote Colden in June 1747 that 'it is now discovered and demonstrated, both here and in Europe, that the Electrical Fire is a real Element, or Species of Matter, not *created* by the Friction, but *collected* only.'[19] Although the view that friction creates electrical matter may have been compatible with the teachings of Digby or Browne, no electrician appears to have held it explicitly; and since the time of Dufay, most authorities had regarded the electrical matter as either common fire or an element *sui generis*, both beyond man's power to create or destroy. Conservation of the electric matter enters easily and, in the case of Nollet's, explicitly into the grand effluvial systems.[20] Although Franklin did not 'discover' conservation, he was unquestionably the first to exploit the concept fruitfully. Its full utility appeared in his classic analysis of the condenser.

3. THE LEYDEN JAR

To analyze 'Musschenbroek's wonderful bottle,' the Philadelphians required more than the concepts of conservation, electricity plus and minus, and the machine as pump. They had also to make assumptions about the interaction of electric matter and electrics *per se*. Franklin characteristically adopted a simple and serviceable hypothesis, which happened to be novel besides: he held that glass is absolutely impermeable to the electrical fire, a doctrine recognized by his followers and opponents alike as the central principle of his system.[21]

Give an uncharged jar to a man standing on the floor. Rub the tube and touch it to the bottle's internal wire or hook, thereby throwing, say, one unit of electrical fire into the jar. None crosses the bottle's impermeable bottom: all accumulates within. Where, then, is the minus electrification necessarily associated with this unit of positive electricity? It resides, Franklin says, in the

17. Cf. Musschenbroek to BF, 15 April 1759 (*EO*, 71); Finn, *Isis*, 60 (1969), 363–9; *infra*, xiv. 4.

18. And perhaps also Boerhaave, whose 'fire' is much like Franklin's electrical matter. Schofield, *Mechanism* (1970), 172; *FN*, 230, 234; *supra*, i.6.

19. *BFP*, III, 142.

20. Cf. Gliozzi, *Arch.*, 15 (1933), 202–15; Home, *Effl. Th.* (1967), 180, 183.

21. *EO* (Wilcke), Vor.; Nollet, *Lettres* (1753), 50 ff.; Beccaria, *Dell'elett.* (1753), 144. Cf. Whittaker, *Hist.*, I (1951²), 49.

outer coating of the jar: 'At the same time that the wire at the top of the bottle, etc., is electrised *positively* or *plus,* the bottom of the bottle is electrised *negatively* or *minus,* in exact proportion.'[22] The unit of electrical fire within the bottle drives an equal unit from the other side, through the man holding the coating and into the ground; the ejected unit compensates for the fire pumped *from* the ground, via the tube, into the bottle. Rub the tube again and present it to the hook: a second unit enters the bottle, and another exits from it. The process can be repeated until the external surface of the jar has surrendered all its fire. Then the operation ceases: 'No more can be thrown into the upper part when no more can be driven out of the lower.' Apply the tube further and fire spews back from the hook.

Since glass is electrically opaque, a charged jar can regain its equilibrium only through an external contact between its inside and outside surfaces. The restoration may be effected instantaneously or gradually, according as one brings a nonelectric into contact with the two surfaces simultaneously or alternately. In neither case does charge remain in the bottle or on the nonelectric when the process is completed. Just as no fire can enter the top of the bottle if none can leave the bottom, the hook will keep its fire if the exterior cannot imbibe. A fresh phial cannot be charged, nor a prepared one discharged while its external surface rests solely upon electrics *per se.* Franklin thereby resolved the great enigma of the condenser, the necessity of grounding its outer surface during electrification, in the same terms he used to explain the completion of its charging. Nothing enters if nothing can leave; nothing leaves if nothing can enter.

Franklin gave eleven elegant experiments in support of his views, nine of which played variations on the following theme.[23] Solder a 'tail wire' to the external coating of the jar and bend it upwards as in fig. 14.1a. Place the charged bottle on wax so that a cork *f,* suspended from a silk thread, can play back and forth between *e* and *e,* fetching fire from the hook and delivering it to the tail, until neither the inside nor the outside of the bottle retains any electricity.[24] The same end is achieved if, instead of the cork, one uses an insulated wire to draw sparks alternately from the top and bottom of the phial; or if, giving an insulated man the bottle to hold, one approaches one's knuckle alter-

---

22. *EO,* 180. Franklin intends 'in equal amount' for 'in exact proportion.' Later he decided that the electricity resided in the glass, not the coatings.

23. The other two show the discharge path by including the gold filigree of a book (across which the fire leapt 'like the sharpest lightning'), and prove that a shorted condenser will not charge. The last proposition is not literally correct, since it electrifies like any (insulated) nonelectric; Nollet, *Lettres* (1753), 108–9, exploited this imprecision. Cf. E. Gray, *PT,* 78 (1788), 121–4.

24. In his first letter to Collinson, Franklin suggested the following diversion: Place a wire upright in a grounded table; between the wire and the hook of a charged, grounded jar suspend a weighted piece of burnt cork, with linen strips attached to suggest legs; this 'counterfeit spider' will dance back and forth for an hour or more in dry weather, 'appearing perfectly alive to persons unacquainted.' Does Franklin's powerful demonstration of his theory of the bottle derive from the counterfeit spider? Nothing would better illustrate the tie between parlor tricks and the early study of electricity.

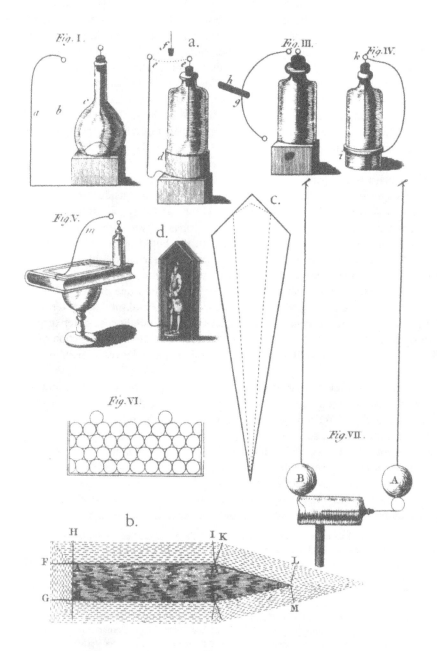

14.1 (a) Franklin's demonstration of the opposite electrifications on the coatings of a Leyden jar (b) his depiction of the statics of electrical atmospheres (c) his form of Wilson's fish (d) the sentry box for collecting lightning. From EO[5].

nately to the hook and to the tail. Or one can restore the bottle's equilibrium at a blow, by connecting together its external and internal surfaces. Again, the discharged phial discovers no electricity, nor do the bodies connecting its surfaces, even with the discharge circuit insulated. These demonstrations, showing the equal and opposite character of positive and negative electricity, were rightly regarded as the strongest confirmation of Franklin's system.[25]

Two other results concerning the condenser must be noticed.[26] Kinnersley, who had an eye for demonstrations,[27] showed that a jar could be charged backwards, by connecting its exterior to the tube or globe. The jar conformed perfectly to Franklin's principles, its outer coating now being positive and its inner negative. The Philadelphians exploited and demonstrated this fact in many ways, for example by charging an 'electrical battery,' a line of jars each having its hook attached to the tail of the one preceding it, as if the line were a single jar.

The second result, that the jar retained its power to shock though its inner coating be removed, was associated with the principle of equal and opposite charges. The jar was supposed to contain no more electrical fire when charged than when not, and to preserve its disequilibrium by the action of the glass. Franklin guessed that 'the whole force of the bottle' must be *in* the glass, the coatings and wires serving only to fetch the fire. The Philadelphians took a charged water-filled jar, set it upon wax, removed its own hook and inserted another: the phial gave the usual shock. They charged the jar again and removed both hook and water, taking care to decant into an insulated bottle. The recipient gave no shock; the donor, refilled with fresh water and furnished with a new hook, struck as before. To obviate the possibility that the shape of the glass played a role, they made a condenser of a glass pane, armed on both sides with lead plates, and electrified it as the bottle. They then removed the pane, drew sparks from it to show its electricity, reinserted it between the plates, and took the usual shock.

Further evidence that the force resided in the glass and not in the coatings came from the discharge of a window pane gilded on both sides and electrified as the bottle. After the explosion, holes appeared where the connecting wire touched the gilding, the upper hole testifying to the passage of the electrical fire *from* the glass to the wire, the lower to its passage from the wire into the glass.[28] Similar holes appeared in sheets of paper placed over the gilding.

25. E.g., *EO* (Wilcke), 237–8; there is, of course, some residual charge. Franklin later (1749 or 1750) illustrated the equal and opposite electrification of the two surfaces by insulating the machine and connecting it to the outer coating of the jar. Without the connection the bottle cannot be charged; with it the fire is pumped, via the machine, from the outer to the inner surface, the former gaining what the latter loses (*EO*, 237–8).

26. BF to Collinson, 29 April 1749 (*EO*, 187), a paper dated 1748 and reporting work of that year.

27. Kinnersley later made a living as an itinerant lecturer; his audience at one time included a Seneca chief. Lemay, *Kinnersley* (1964), 62–87; Overfield, *Emp. St. Res. Ser.*, 16:3 (1968), 46.

28. These experiments, though often repeated (cf. *infra*, xv.3–4, xviii), in fact reveal nothing about the direction of the discharge.

Franklin's solution to the problem, mainly of his own devising, of the location of a condenser's electrical fire illustrates both the strength of his system and the limitations of his own approach. His absolute abstractions—perfect insulators supplied by perfect conductors—however valuable as first approximations, fail to suggest the complex distribution of electricity on plate condensers, particularly on the poorly armed and imperfectly insulated ones he employed. But his unambiguous statement and solution of the problem pushed others, temperamentally better equipped than himself to investigate irregular small effects, to important discoveries. The dismembering of parallel-plate condensers, or Franklin squares,[29] resulted in the recognition of dielectric polarization and residual charge, and the invention of the electrophore.

The season of 1748 would not have been complete without elaborate new diversions. Here is one any number can play. Take a large picture of the king (may God preserve him!), cut off a two-inch border all around, and glue the remainder to the *inside* of the picture's glass cover. Now gild the empty glass inside, an exactly corresponding area outside, and the internal edge of the back of the frame, excepting a small part at the top. Connect the gilding on the edge to that inside the glass. Paste the remaining portion of the portrait outside the glass over the gilding so that the picture appears of a piece, and place a movable gilt crown on the king's forehead. Electrify the picture. Holding it by the ungilded upper edge, offer it to a friend and invite him to remove the crown. His left hand grasps the gilded edge, connected to the inner gilt surface of the glass; his right approaches the crown, connected to the outer gilt surface beneath the two-inch border; he receives a sharp blow for his treason. If several people, joining hands, take the shock together, the game is called 'The Conspirators.'

The conspirators planned a picnic on the banks of the Schuylkill. It was to begin with the firing of spirits by a spark sent across the river.[30] The dinner, a turkey humanely electrocuted, was to roast on an electrical jack, before a fire kindled by the electrified bottle. As a climax to the party, 'the healths of all the famous electricians in *England, Holland, France* and *Germany* are to be drunk in *electrified* bumpers, under the discharge of guns from the *electrified battery.*'[31]

4. MECHANICS OF THE ELECTRICAL FIRE

The principles of the contrary electricities, of the impenetrability of glass, and of the equality of electrification on the opposite surfaces of a Leyden jar are closer to phenomenological rules, or descriptions of experiments, than to a physics of the electrical fire. Though Wilcke goes too far in claiming Franklin's

29. Though Bevis and Smeaton had used glass panes before Franklin did (*supra*, xiii.3).

30. The experiments of Watson and Le Monnier (*supra*, xiii.3) doubtless inspired this extravagance.

31. *EO*, 200. An electrified bumper, a small thin tumbler nearly filled with wine and electrified as the bottle, gives an exquisite electrical kiss.

principles to be mere *Erfahrungssätze*,[32] he does thereby convey the instrumentalist spirit from which, in his judgment, the Philadelphia system drew its strength and allure. When Franklin became better acquainted with European theory, he began to concoct his own mechanics, and as he proceeded his innovative power weakened, like an Antaeus separated from the earth.

The phenomena of electricity, he says, depend upon the interactions of two types of matter, the common and the electrical.[33] Elements of the former attract, those of the latter repel one another; while between a particle of the common, and one of the electrical matter a strong attraction obtains. Because of these forces, and the great 'subtlety' of electrical fire, neutral 'matter' is a 'kind of sponge' crammed with as much of the fire as it can hold. Between two neutral bodies there is no net electrical interaction. For 'electrical signs' to appear—attractions, repulsions, sparks, shocks—electrical matter must be accumulated. The accumulation, or excess beyond the amount required to saturate a body, 'lies without upon the surface, and forms what we call an electrical atmosphere.' We know that, in general, bodies contain nearly their just quantities of electrical matter, for a small increment usually confers an atmosphere; and we know that, since bodies with atmospheres repel one another, common matter does not normally contain more of the electrical matter than it can absorb.

Generalizing his earlier observation that an electrified shot can maintain about itself a concentric halo of smoke, Franklin inferred that atmospheres always take the form of the bodies they surround. Consider the electrified spike of fig. 14.1b, which retains its envelope through the postulated attraction between common and electrical matter. The portions HABI and KLCB are held by the large areas AB and BC, respectively, while HAF, IBK and LCM rest on much smaller surfaces. Therefore, according to Franklin, the atmosphere may easily be drawn off from a corner or a point, whence the difficulty the Philadelphians had met in charging a bayonet. Some moisture always swims in the air; the slightest trace will pull the atmosphere contiguous to a point; and that portion once removed, the rest, moving in to replace it, dissipates step by step.

Why then do pointed conductors draw electrical fire as readily as they throw it? Consider two grounded iron probes, one rounded and one pointed, confronting a charged insulated plate. Although both instruments, being 'connected with the common mass of unelectrified matter,' pull at the plate's envelope with equal force, the spike draws more effectively, because it can act upon a smaller portion of atmosphere: 'And as in plucking the hairs from a horse's tail, a degree of strength not sufficient to pull away a handful at once, could yet easily

32. *EO* (Wilcke), Vor.

33. BF to Collinson, 29 July 1750 (*EO*, 213–38); cf. Ellicott, *PT*, 45 (1748), 195–224. Franklin here uses the phrase 'electrical matter,' the common terminology of the European electricians, for the first time; he had previously employed 'electrical fire,' the phrase he picked up from Spencer and the *Gentleman's*.

strip it hair by hair; so a blunt body presented cannot draw off a number of particles at once, but a pointed one, with no greater force, takes them away easily, particle by particle.'[34]

This reasoning would not earn its author high marks in physics even in the eighteenth century. Franklin has confused force and pressure in his analogy to the horse's tail; and he unwittingly supposes two inconsistent sets of forces, one to establish the conformal atmospheres, the other to preserve them. (If the forces that maintained the atmosphere determined its shape, it would be shallow opposite points and deep opposite plains.) Franklin did come to doubt his explanation of the power of points and published it, he said, only in the hope of inspiring a better one. 'A bad solution read, and its faults discovered, has often given rise to a good one, in the mind of an ingenious reader.'

In the event it took many ingenious readers and further hints from Franklin to clean up his atmospheres, which suffered from outright contradiction as well as from bad physics. Consider the double role Franklin forces on the air. First it must readily pass the electrical matter, for the atmospheres subsist in it and, moreover, as Franklin took the trouble to demonstrate by whirling a charged cork about his head, they move freely through it when the bodies they surround are transported. But again, the air must resist the electrical matter, for two reasons: as Franklin's 'pneumatic engine' misleadingly disclosed, atmospheres cannot be maintained *in vacuo*; and *in pleno* they dissipate the more quickly the moister or smokier—that is, the less pure—the air.[35] If one tries to put the burden of maintaining the atmosphere on attractive forces between the electrical and common matter the same conundrum obtrudes. First, metal draws electrical matter more strongly than air does, for a charged shot holds its atmosphere. *Sed contra*, air draws more powerfully than metal since, as Franklin affirmed, insulators differ from conductors only in their greater power to 'attract and retain' the electrical matter.[36] No solution to these puzzles appears to exist within the Philadelphia system.

A man of Franklin's insight and intelligence does not, however, give serious and prolonged attention to a subject without improving it. Fruitful tendencies may be discerned in his impossible atmospherics. Perhaps the most obvious is the impossibility itself: once recognized, it menaced all theories of effluvia. Second, Franklin's representation emphasized that electrical matter must be accumulated for signs to appear; his atmospheres differ from 'sGravesande's, which derive from the electrified body itself, and from the coursing effluvia of Nollet and Watson, which work by streaming. Franklin's representation leads naturally to the concept, and to the term, 'electrical charge,' literally an extra quantity or load of electrical matter resting on or in an electrified body. Third, the static character of the atmospheres implied a step away from mechanism;

34. *EO*, 219.

35. It is worth repeating that the belief of eighteenth-century electricians that moist air conducts better than dry is mistaken.

36. *EO*, 214, 248; Home, *Br. J. Hist. Sci.*, 6 (1972), 131–8.

only effluvia in motion can act by impact. It is noteworthy that the inventer of the other contemporary scheme that relied upon static atmospheres, Wilson, explicitly invoked the microscopic distance forces of Newton's aether. Fourth, Franklin covertly uses the notion of attraction over macroscopic distance; for the electrical matter at the confines of the atmosphere, which extends as far as the electrified body acts, is held by the attraction between electrical and common matter. Finally, Franklin's intimation that extent of surface supporting unit quantity of electrical matter measures its stability may be generously interpreted as a very rough anticipation of the results of Biot and Poisson: the depth of the atmosphere, measured perpendicularly to the electrified object, is proportional to electrical force.[37] But these were only tendencies, some very remote; in trying to express his principles in the language and mechanisms of European electrical theory, Franklin let his intuition outrun his physics.

ATTRACTION AND REPULSION

Like most Newtonian experimental philosophers, Franklin implicitly ascribed action at microscopic distances to the particles of his electrical and common matter, but balked at admitting forces over sensible intervals, even though he supposed them implicitly in his mechanics of atmospheres and explicitly in his theory of the Leyden jar. 'I agree with you,' he once said to Cadwallader Colden, 'that it seems absurd to suppose that a body can act where it is not. I have no idea of bodies at a distance attracting or repelling one another without the assistance of some medium, though I know not what the medium is, nor how it operates.' Like most electricians, therefore, Franklin assumed without question that electrical atmospheres mediated attractions and repulsions. He remained true enough to the spirit of his system, however, to omit circumstantial accounts of these interactions. Somehow an atmosphere causes its host and neighboring neutral bodies to come together, and two atmospheres, without mixing, force their (positively) charged possessors to move apart. Whether the atmospheres rotate, gyrate, pulsate or vegetate does not appear; compared to earlier models they seem lifeless, even vestigial.[38] Nonetheless, they remained vital in Franklin's thought. To take a notorious example, the repulsion between negatively charged bodies looked to him a contradiction in terms, an interaction between bodies that lacked *by definition* the mechanism responsible for electrical motions. The Philadelphia system grew up in ignorance of this awkward phenomenon; it always bothered Franklin, and was to torment his followers.[39]

Although Franklin could not bridge the gap between the atmospheres and the motions they induced, he could reduce complicated cases, like Wilson's fish and the feather-in-the-bottle, to the edge of the divide. Attach one of two paral-

37. *Infra*, xx.2.

38. BF re Cadwallader Colden, 4 Nov. 1756, in Franklin, *Works* (1840), VI, 180. Cf. Polvani, *Volta* (1942), 46–7; Finn, *Isis*, 60 (1969), 363–4.

39. BF to Collinson, 29 April 1749, *EO*, 199; to Kinnersley, 20 Feb. 1762, *EO*, 365; *infra*, xvi.3, xvii.3, xviii.2. The attraction between a neutral and a negative body would seem no less perplexing.

lel plates to the prime conductor of a working machine and launch the fish (fig. 14.1c) in the upper half of its tank. It swims toward the upper plate, 'with a brisk but wavering motion, like that of an eel through water,' obeying the attraction between bodies with and those without atmospheres. As it approaches, its point draws an envelope; it thereupon sinks under the usual repulsion between atmospheres until it can dispose of its surplus fire, via its pointed tail, to ground. In equilibrium it has no atmosphere, the attraction of the upper plate equalizing its weight at a distance where the net flow of fire through it vanishes. Note the presumed dissymmetry in the action; had Franklin tried the experiment with the plates interchanged, he would have been nonplussed to see the fish initially rise toward the grounded one. He did not see that the plates play similar roles, that they constitute the coatings of a parallel-plate condenser separated by air instead of glass. By the very nature of his conquest of the Leyden jar he could not discern the condenser in this arrangement; for a condenser, he thought, demanded an electrically opaque medium, and the electrical matter flowed visibly through the fish and its tank of air.

Franklin referred the impermeability of glass to the shrinking during cooling of its internal pores. The constriction, he said, lies only at the center; otherwise each surface, or rather half-depth, of the glass absorbs electrical fluid until saturated, until the mutual repulsions among the particles of the fluid just balance its attraction for the glass. The fluid in one half-depth acts upon that in the other *across* the impassable, macroscopic, intestine barrier.[40] Now add an element of fluid to one surface (A); that in the other (B) feels a greater repulsion from A than attraction from the glass, and part will exit into any available nonelectric. The process can be repeated until B has lost its native supply and the entire attractive force of the glass goes to balance the repulsions among the particles of the superabundant fluid. An external nonelectric connection between A and B restores the primitive equilibrium precisely and violently.

Franklin extended this model beyond the Leyden jar to solve Hauksbee's old puzzle, why a tube wet within or lined with a nonelectric cannot be charged by friction: the fluid extracted from the hand in the down-stroke enters the glass's pores and drives an equal charge from the inner surface of the tube into the nonelectric coating. What this laconic formulation implies may be gathered from Wilcke, who idealized the experiment as in fig. 14.2a. The tube AC is lined with gold foil inside.[41] The external surface AB is free and the left hand grasps the portion BC. The right hand, in rubbing AB, delivers to it a positive charge, which induces a negative one in the grounded hand cd (fig. 14.2b); and these charges act across the glass to segregate the electrical matter of the inner foil ab. Negative charge accumulates under AB and positive under BC; the tube exhibits very little external electricity because the opposite electricities almost annul one another outside the glass.

40. Franklin abandoned this theory on learning that a piece of glass could serve in the Leyden experiment after more than half its thickness had been ground away. BF to Ingenhousz, 14 June 1784, in *BFP* (Smyth), IX, 52.

41. *EO* (Wilcke), 273–6.

a.

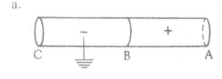

b.

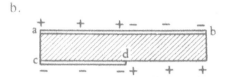

14.2 Elucidation of the dormancy of tubes with metallic inner linings. (a) The tube electrified plus outside on AB by rubbing, and negative on CB (covered by the grounded hand) by induction (b) the charges outside and inside the glass wall lined with gold foil within.

And now we come to the most compromising of phenomena, the demonstration that had been regarded since Hauksbee's time as decisive evidence for the passage of electrical matter through glass. Franklin suspends a feather within a sealed bottle and places the tube so that its atmosphere bathes the vessel's exterior. Just as the surplus thrust by the machine into the Leyden jar expels electrical matter from the tail wire, so the tube's atmosphere forms another from fluid it drives from the internal surface of the bottle. It is this hypothetical internal atmosphere that draws the feather via the unspecified mechanics of the Philadelphia system.

Many aspects of the system, and these ultimately the most fruitful, did not depend on the models Franklin devised to illustrate them. The doctrines of the impenetrability of glass, of electricity plus and minus, and their embodiment, the theory of the Leyden jar, had a logic and an appeal of their own. Without doubt the advantages of his explanation of the condenser would eventually have recommended his system to the Europeans. As it happened, however, his ideas rapidly gained a hearing owing to the success of experiments having no logical connection with the most important and characteristic of his theories.

5. LIGHTNING

A comparison to lightning inevitably presents itself to writers trying to describe the electric spark. From the metaphor linking the phenomena to the conjecture of their identity is a small step; electricians as early as Gray and Wall suggested that lightning was nothing other than laboratory glows on a cosmic scale.[42] In 1748 Nollet inventoried the appearances that testified to the electrical nature of lightning, and his friends at the Bordeaux Academy set the analogy as the subject of its essay competition of 1749. One Barbaret, a physician from Dijon, won the prize. 'L'électricité est entre nos mains,' sums up the laureate, stealing *verbatim* a conceit of Nollet's, 'ce que le tonnerre est entre les mains de la nature.'[43]

42. Brunet, *Lychnos,* 10 (1946–7), 117–48; Elvius, *AKSA,* 9 (1747), 184–5.

43. Nollet, *Leçons,* IV (1748), 314–15; Brunet, *Lychnos,* 10 (1946–7), 121–2. Another competitor at Bordeaux, B. Rackstrow, who printed his losing essay in his *Misc. Obs.* (1748), 59–72,

There was thus nothing fresh about the idea when, in 1749, Franklin began
to identify lightning and electrical fluid. His conjecture, like his predecessors',
rested on similarities in appearance and action; and his analogies, though more
extensive than theirs, contained nothing new in principle. Compare the bill of
particulars he drew up on 7 November 1749 with the list Nollet had published
the preceding year.[44]

'Electrical fluid agrees with lightning in these particulars:

1. Giving light   [son inflammabilité]
2. Colour of the light
3. Crooked direction
4. Swift motion   [promptitude de son action]
5. Being conducted by metals
6. Crack or noise in exploding
7. Subsisting in water or ice
8. Rending bodies it passes through   [la propriété . . . de frapper les corps
9. Destroying animals   extérieurement et intérieurement,
10. Melting metals   jusque dans leurs moindres parties]
11. Firing inflammable substances   [son activité à enflammer]
12. Sulphurous smell.'

Franklin's originality, to say nothing of his optimism, expressed itself in his
proposal to confirm the analogy of enticing lightning into the laboratory.

Franklin appreciated that thunderbolts are not toys. He thought, however,
that he commanded an agent that could tame lightning, a virtue discovered in
America, the power of points. Just as a sharp bodkin, presented to a charged
shot, silently drew off electrical fire a foot or more away, so a pointed metal
rod, projecting into the atmosphere, should noiselessly despoil a passing thunder cloud of its lightning. If the rod were insulated, the lightning would remain
upon it, and be available for electrical experiments. For example, as Franklin
proposed in 1750, let a sentry box large enough to hold a man and an insulating
stand A surmount a tower or steeple, as in fig. 14.1d.[45] The rod, of iron,
projects 20 to 30 feet above the box. When a thunder cloud passes, the observer
tries to draw a spark of lightning from the rod. 'If any danger to the man be
apprehended (though I think there would be none) let him stand on the floor of

---

emphasizes the similarity between lightning and electricity in working destruction, harming animals, igniting flammables, following the shortest path, etc., and concludes that 'Lightening and
Electricity are the same Fire.'

44. BF to 'Dr. L.,' 18 March 1755 (*EO* 334); Nollet, *Leçons*, IV (1748), 314–15; cf. Bertholon, *JP*, 20 (1782), 227, crediting Nollet with first setting out the analogies. Franklin had called
attention to some of them in a letter on 'thunder gusts' sent John Mitchell on 29 April 1749 (*EO*,
201–11).

45. *EO*, 222.

his box, and now and then bring near to the rod the loop of a wire that has one end fastened to the leads,[46] he holding it by a wax handle; so the sparks, if the rod is electrified, will strike from the rod to the wire, and not affect him.'

Franklin's easy dismissal of his agent's peril might appear disingenuous. The apparatus proposed would occupy much the same situation as a weather vane or gargoyle. Such ornaments were often damaged by lightning. In fact an experiment akin to Franklin's had been performed many times. The standard early-modern antidote to thunderbolts was a vigorous ringing of bells designed to break up menacing clouds. In the ordinary lengthy rite of consecration of bells the priest prayed that their sound might 'temper the destruction of hail and cyclones and the force of tempests and lightning; check hostile thunders and great winds; and cast down the spirits of storms and the powers of the air.'[47] He ought rather to have prayed for the bell ringer. The church tower, usually the highest structure in village or town, was often hit by lightning; according to a study published in 1721, about one church a year in Sweden suffered serious damage from thunderbolts, and of these almost 20 percent were totally destroyed.[48] Peter Ahlwardts, the author of 'reasonable and theological considerations about thunder and lightning,' accordingly advised his readers to seek refuge from storms anywhere than in or around a church. Had not lightning struck only the churches ringing bells during the terrific storm in lower Brittany on Good Friday, 1718?[49]

Thunder nonetheless continued to start the bells, and lightning to electrocute the bell ringers. In explanation of the failure of 1718 the Breton peasants pointed to the ritual prohibiting bell ringing on Good Friday. Diderot's *Encyclopédie* still teaches that 'one can break up and turn aside thunder by the sound of several large bells, or by firing a cannon.' The philosophical faculty at Prague earnestly debated the utility of the old method in 1753.[50] Still it was employed, long after the invention of lightning rods; several bell ringers met their end in France and Belgium during the spring storms of 1781.[51] By then authorities at last recognized the uselessness of the sacrifice, and in several places bell ringing during thunder storms was prohibited; not so much to save

---

46. The lead roof of the tower or steeple supporting the box.

47. Rocca, *De camp. comm.* (1612), 136–9; for the rite, *ibid.*, 31–42.

48. A. O. Rhyzelius, *Brontologia Theologico-Historica* (Stockholm, 1721), excerpted by Müller-Hillebrand, *Tids. Pastoratsf.*, 14:7 (1960), 3–16. It is noteworthy that the incidence of heavy damage by lightning to Swedish churches has not changed much since the eighteenth century.

49. Ahlwardts, *Bronto-Th.* (1745), 298–300; *HAS* (1719), 21–2.

50. *Encyclopédie*, art. 'Tonnerre,' quoted by Cohen, *Isis*, 46 (1955), 270n; Smolka, *Acta hist. rer. natur. techn.*, 1 (1965), 163. The canard that Nollet recommended the bells (Cohen in *EO*, 133n; Hujer, *Isis*, 43 [1952], 355) derives from a misreading of Franklin, *EO*, 393–5; in fact Nollet strongly advised against them (*MAS* [1764], 408–51).

51. Needham, Acad. roy. sci. Belg., *Mém.*, 4 (1783), 59–72. According to J. N. Fischer, *Beweis* (1784), 386 hits to bell towers over a period of 33 years resulted in the death of 103 bell ringers.

lives, it must be confessed, as to abate noise.[52] It need hardly be said that the unfortunate bell ringer stood in the same relation to the church steeple as Franklin's observer to his insulated pole.

In fact Franklin tacitly admitted the danger of the experiment in his momentous proposal to employ the rod as protection. He required only one alteration in the apparatus: rather than terminating the rod on an insulator, he advised running a wire along the protected building from the point directly to ground. He therefore recognized that a rod unable easily to free itself of the lightning it quietly drew from the clouds might accumulate enough electricity to be dangerous. Apparently, however, he did not consider sufficiently the relation between the insulated experimental rod and the grounded protective device, nor did he inform himself fully about the damage caused by lightning strokes. Perhaps he did not expect anyone to try the experiment; he never did himself. It is of course very dangerous. The first man to perform it 'successfully,' the first man to catch a lightning stroke on his insulated probe, was instantaneously electrocuted.[53]

Franklin's optimism derived from his faith in the power of points, whose effectiveness against clouds he had demonstrated on a laboratory model. Take a pair of brass scales hanging by silk threads from a two-foot beam; suspend the whole from the ceiling by a twisted cord attached to the center of the beam; set a small blunt instrument, like a leather punch, upright on the ground. Now electrify one pan and let the cord unwind; the charged pan inclines slightly each time it passes over the punch until the relaxation of the cord places it close enough for a spark to jump between them. If, however, you mount a pin, point uppermost, atop the punch, the pan silently loses fire to the point at each pass and, no matter how close it approaches, never throws a spark into the punch. Interpreting the punch as the steeple, the point as the rod and the pan as the cloud, Franklin could predict the fate of his sentry, *si parva licet componere magnis*. In this case the extrapolation fails.

It remains to understand how clouds charge themselves with lightning. In the sea the rubbing of particles of salt (an electric *per se*) against those of water (a conductor) collects the fire from the ocean depths onto its surface, as the electrical machine pumps fire from the ground to the prime conductor. Vapor obtained from electrified water must itself be charged;[54] this fire, added to the common, promotes the rise of water-laden air, and forms it into durable clouds. These eventually sail over land, meet a mountain or an uncharged, sweet-water cloud, discharge their electrical fire, deposit their water, and disappear. Any projecting object, like a church steeple, a tree, or the mast of a ship, entices the

52. Paris, Parl., *Arrest* (1784); Herbert, *Th. phen.* (1778[2]), 207ff. Cf. Nollet, *MAS* (1764), 408–51; Cohen, *J. Frank. Inst.*, 253 (1952), 393–440.

53. *EO*, 221; *infra*, xv.2.

54. The falsification of this plausible proposition played some part in the prehistory of the electron. Schuster, RS; *Proc.*, 37 (1884), 317–39.

fire, or lightning. Do not shelter under a tree during a thunder storm, but remain in open country, where no prominent projections menace and your soaked clothes will prove an advantage. Should you be struck, rest assured that the electrical fire will run into the ground over your dripping garments. For, as Franklin himself had shown in unsavory experiments with a Leyden jar, it is much more difficult to electrocute a wet rat than a dry one.[55]

55. Nollet, *Lettres* (1753), 10–11, had good sport with this prescription: 'Some people having assured us that a traveller in a barren countryside could defend himself [from lightning] by drawing his sword against the clouds, the clergy, who carry none, began to complain; whereupon they were shown in Mr. Franklin's book . . . ' That Nollet had not set up a straw man appears from John Wesley's *Journal*, IV, 53, under date 17 Feb. 1753: 'From Dr. Franklin's *Letters* I learned . . . that the electrical fire, discharged on a rat or fowl, will kill it instantly, but discharged on one dipped in water, will slide off, and do it no hurt at all. In like manner the lightning which will kill a man in a moment will not hurt him if he be thoroughly wet. What an amazing scene is here opened for after ages to improve upon!' If caught in open country the best measure *is* to stand away from trees, with feet pressed tightly together so that ground currents created by the lightning stroke will not run up one leg and down the other. Wiesinger, *Blitzforschung* (1972), 47–8.

CHAPTER XV

The Reception of Franklin's Views in Europe

The response of Europe's electricians to Franklin's subversive system provides the historian with instructive material. In England, where Franklin's letters to Collinson circulated before their publication in 1751, the American system was either ignored, as by Wilson and his group, or minimized, as by Watson, who thought it similar to his own. In France and Italy the reverse occurred; Franklin's innovations were unduly, often unfairly, emphasized to improve their utility in academic squabbles.

1. ENGLAND

On January 21, 1747/8, Watson read to the Royal Society from Franklin's first letter on electricity. He emphasized the analogy between tube and pump and the idea of electrification plus and minus. Several Fellows must already have seen the letter, as Watson had it not from Collinson but from the Society's Vice-President, Lord Charles Cavendish, who had collaborated in Watson's experiments to determine the velocity of electricity. The purpose of Watson's reading was to call attention to a gratifying coincidence, the simultaneous enunciation, on both sides of the Atlantic, of ideas he had introduced into his theory in the spring of 1747.[1]

Watson's ideas were suggested by an experiment of Bevis'. An insulated man (A) extended his finger towards a rotating globe, the machine and its operator (B) also being insulated. Both A and B became electrified, as evidenced by the sparks they exchanged with a grounded man, C. The experiment appeared to violate Watson's principle that the globe pumped electrical aether *from* the ground; since the machine and both men were insulated, no electricity should have developed. Bevis, nonplussed, conceived that the globe drew electrical matter from the air. Watson, conscious that this solution conflicted with well-known phenomena,[2] reasoned as follows: the machine conveyed electrical matter from B to A; in A's absence, as in the arrangement of Bose et al., B does not electrify, for the revolving globe restores to him as much electrical matter as it extracts. Both A and B appear electrified to C and to each other because they possess electrical matter at different *densities*.

1. Watson, *PT*, 45 (1748), 97–100. Though Watson did not publish these ideas until the following January, he had communicated them, so he says, before he saw Franklin's letter.
2. For example, the feeble electrification the operator obtains when he and the machine are completely isolated.

Watson's new theory was that electrical effects occur only when bodies with dissimilar electrical densities interact. In their natural states, all objects contain electrical matter at the same degree of compression; rubbing increases its density in one body, such as A, at the expense of another, like B. As one would expect, the strength of the spark that re-establishes equilibrium between bodies unequally electrified is proportional to the *difference* between the densities of their electrical matter. The spark exchanged between A, with condensed matter, and *B,* with it rarefied, exceeds that exchanged between either and *C,* who has the natural density.[3]

Watson pointed out the similarity between his elucidation of Bevis' experiment and Franklin's discussion of the three gentlemen and the tube. The major difference, the importance of which can easily be missed, is that Watson took density and Franklin quantity of electrical matter as the measure of electrification. (Neither measure alone will do; for these experiments, Watson's, which resembles potential, perhaps more nearly suffices.[4]) To Watson the difference did not look significant, and he, an electrician with an international reputation, graciously identified his ideas with those of the unknown American.

The Royal Society continued to hear Franklin's letters as they arrived. As one might expect, Watson thought the later ones less agreeable than the earlier; and he offered an experiment to disprove the proposition that the electrification of the condenser lay entirely within the glass.[5] When Franklin's letters were published in the early summer of 1751, Watson supplied an accurate review of the portions the Society had not yet heard, praised their 'great variety of curious and well-adapted experiments,' and recommended them highly despite 'some few opinions' with which he could not 'perfectly agree.'[6] (The book was seen through the press by a friend and coreligionist of Collinson's, the physician John Fothergill, who distributed copies and the announcement that Franklin 'has said more sensible things on the subject, and let us see more into the nature of this delicate affair, than all the other writers put together.'[7])

Since Bevis, Smeaton, Lord Charles Cavendish and other collaborators of Watson probably shared his opinions, it is likely that several competent English electricians knew and approved the theory of positive and negative electricity for well over three years before its publication. Nonetheless it affected their work very little, much less, for example, than Nollet's *Essai* had done. The same benign indifference prevailed during the year following publication, until reports of the French experiments on lightning reached England. Watson, hav-

3. *PT,* 45 (1748), 94–100.

4. Cf. *supra,* xiv.4, and *FN,* 441–9.

5. Decant water from a charged phial into a basin supported by an insulated man, who holds his other hand over a spoon filled with warm alcohol; as the water flows the spirit ignites because, according to Watson, the electrical matter accumulated in it (the water) runs into the man and thence to the spoon (JB, XX [1748–51], 215–16, partly published by Cohen, *FN,* 464–5). Franklin countered (BF to Collinson, 27 July 1750, in *EO,* 241–2), that the water merely conducts the electricity.

6. Watson, *PT,* 47 (1751–2), 203, 210.

7. Fothergill, letter of 18 May 1751, in Corner and Booth, *Chain* (1971), 143.

ing concluded experiments on electricity *in vacuo,*[8] turned to other subjects. Wilson brought out a *Treatise on Electricity,* first in 1750 and again in 1752, each time reaffirming his aether, representing the Leyden jar as an unanalyzable violation of the Rule of Dufay,[9] and ignoring the existence of Mr. Franklin of Philadelphia. Why did the English fail to exploit the promising system of their colonial electrician?[10] Obstinacy, obtuseness, complacency, poor judgment, bad luck? A look at the activities of the first Franklinists will suggest an answer.

2. BUFFON'S CLIQUE

As Franklin's *Experiments and Observations on Electricity* issued from the press in London, another set of letters, in five volumes duodecimo, appeared at Hamburg. Though they contain not a word about electricity, these *Lettres à un amériquain* perhaps occasioned, and certainly affected, the promotion of Franklin's views in France. They are the work of a close friend of Réaumur's, Joseph Adrien Lelarge de Lignac, onetime Oratorian and Malebranchiste, ultimately abbé and Newtonian, who, however, in 1750 still adhered to the sort of Cartesianism congenial to Réaumur and Nollet.[11] The *Lettres* mount a vigorous attack on the first volumes of the famous *Histoire naturelle* of Buffon, 'whose way of reasoning,' says Lelarge, 'is even more revolting than his hypotheses.' He has a 'taste for paradox and obscurity;' he encourages atheism, materialism, hedonism, epicureanism, and bad physics; whenever his reasoning fails, which is often, he simple-mindedly invokes that wonderful word 'attraction.'[12] Réaumur was delighted with Lelarge's *Lettres,* which he had inspired,[13] thereby returning the compliment paid him by Buffon, who had abused, insulted and ridiculed him from the first page of the *Histoire naturelle.* Their

8. *Ibid.*, 362–76, where an 'experimentum crucis' proves that glass is only a pump: the light appearing in a vacuum arranged between ground and the machine's operator shows conclusively, according to Watson, that the machine pulls its supply of electrical matter from the earth.

9. 'The accumulation of electric matter in a non-electric body, in *some* circumstances [where Dufay's rule holds], seems to be directly proportional, and in *other* circumstances [the charging of the condenser] reciprocally proportional to the resistance it meets with as it tends to expand or dissipate.' B. Wilson, *Treatise* (1752²), 34. The *Encycl. Brit.*³, VI, 442, unaccountably praises the consistency of Wilson's views.

10. Franklin, *BFP*, IV, 125, emphasized the indifference of the English reaction, and later (*Autob. Writ.* [1945], 750) wrote that his paper on lightning had been 'laughed at by the connoisseurs' of the Royal Society. Although Franklin was correct in remembering the English response as insipid in comparison to the French, he erred in thinking that he had been ignored (his letters were read at the Society and reviewed in *GM*, 20 [1750], 34–5) or ridiculed (a misconception repeated in much secondary literature, e.g., Tabarroni, *Coelum*, 34 [1966], 127–8). Cf. Heilbron, *Isis*, 68 (1977), 546, and Stearns, *Science* (1970), 628–9.

11. Bouillier, *Hist.* (1854), II, 616–21, who quotes the affecting announcement of Lelarge's tergiversation (1760): 'Ce n'est pas sans peine ni sans répugnance que je m'y suis déterminé. J'ai été si longtemps cartésien! J'ai été si longtemps touché de voir notre nation s'assujétir à penser à l'anglaise.'

12. Lelarge de Lignac, *Lettres,* (1751²), I, 7, 11, 96–9.

13. Letters of Réaumur to P. Mazzolini, 30 June, 3 Sept., 11 Dec. 1751, in dossier 'Réaumur' (AS); Caullery, *Ency. ent.*, XXXII (1955), Suppl., 41–8.

antipathy, by then long-standing, derived partly from personal dislike and partly from disagreement about the purpose and methods of science. Réaumur, cautious, thorough, painstaking, Cartesian, was the author of seven quarto volumes of *Mémoires pour servir à l'histoire des insectes*. Buffon, quick, superficial, self-appointed Newtonian panjandrum,[14] a polished popularizer and propagandist, affected to despise such narrow, meticulous labor; 'after all,' he said, 'a fly ought not to occupy a greater place in the head of a naturalist than it does in nature.'

There was a further ground of difference. Buffon was a great friend of Diderot, whom Réaumur disliked both for his person and for his ideas. It is said that Réaumur assisted Diderot's entry into prison in 1749 as retribution for sneers in Diderot's *Lettre sur les aveugles* at him and his friend Mme Dupré de Saint Maur; and that the sneers were repayment for Réaumur's refusal to grant Diderot permission to witness the removal of bandages from the eyes of a blind girl couched for cataract.[15] The activity of the *philosophes* and the enterprise of the *Encyclopédie,* on which Buffon had promised to collaborate, exacerbated the intellectual opposition between him and Réaumur. As for their personal antagonism, it needed no outside fuel; it probably dated from the late 1730s, when a youthful Buffon, untried and pushy, used ministerial influence to secure the Intendancy of the Jardin des Plantes in succession to Nollet's master, Dufay.[16]

The link between this gossip and the history of electricity was Nollet, Réaumur's most successful protégé and, by 1751, a close friend of his patron.[17] Nollet made it no secret that he regarded Réaumur as the model of a scientist, a man infinitely superior to a certain person of 'brilliant reputation,' who could 'risk many things because no one dared to contradict him.'[18] While still smarting from Lelarge's attack, Buffon chanced on Franklin's *Observations,* either in the original or in a poor French translation.[19] It was an irresistible opportunity. The book evidently had merit; it spoke English, which was fashionable; it

14. Cf. Brunet, Acad. sci., Dijon, *Mém.* (1936), 85–91; *supra,* i.5.

15. Roger, *Diderot St.,* 4 (1963), 221–3; A. M. Wilson, *Diderot* (1957), 97, 99, 110, 128, 197; Diderot, *Corresp.,* I (1955), 150, 152.

16. Torlais, *Réaumur* (1933), 29–40; *Esprit* (1961²), 237–47; *Presse méd.,* 66:2 (1958), 1057–8; *RHS,* 11 (1958), 27–8; Caullery, *Ency. ent.,* XXXII (1955), suppl., 47.

17. Réaumur, *Lettres* (1886), 60, 82; Torlais, *Esprit* (1961²), 382–3.

18. Torlais, *Réaumur* (1933), 42. That this was not empty rhetoric appears from the case of M. J. Brisson, another protégé of Réaumur's, who had helped coach Lelarge. After Réaumur's death, Brisson was denied access to natural history collections controlled by Buffon's clique; Brisson thereupon left ornithology for experimental physics and the welcome of Nollet. Birembaut, *RHS,* 11 (1958), 167–9; Torlais, *Esprit* (1961²), 343–5; *Physicien* (1954), 234–6; Caullery, *Ency. ent.,* XXXII (1955), suppl., 48.

19. Dalibard, *EO* (Dal), 4, claims that Franklin instructed Collinson to send 'one of the first copies' of *EO*¹ to Buffon; cf. *EO,* 101. Both Nollet (*Lettres* [1753], 5) and Franklin (*Autob. Writ.* [1945], 750), emphasize that Buffon encountered the book by chance. Possibly Collinson sent it on his own initiative since, according to Buffon, the two men had been on friendly terms since about 1745. Hanks, *Buffon* (1966), 129n.

plainly opposed Nollet's doctrine, which Buffon had never liked; and, if handled correctly, it could be used to embarrass the abbé, to annoy Réaumur, and to confound others among his 'great many enemies.'[20]

Buffon asked an old school chum and collaborator, the botanist T. F. Dalibard, to translate Franklin accurately into French. Dalibard's only qualification for the job was a knowledge of languages. To repair his ignorance of electricity, he applied to one Delor, a public demonstrator in experimental philosophy.[21] With the help of Delor, who owned and could operate the necessary apparatus, Dalibard repeated Franklin's experiments. By December 22, 1751, he had become adept enough to show games with Franklin squares to the Paris Academy, into which, not being a member, he was doubtless introduced by Buffon. Word of the novelties reached the King, who desired to see them. On February 3, 1752, MM. Buffon, Dalibard and Delor, poaching on Nollet's royal preserve, entertained Louis XV with the inventions of Philadelphia. The monarch's appreciation was so warm that it 'excited in Messieurs de Buffon, D'Alibard and De Lor, a desire of verifying the conjectures of Mr Franklin, upon the analogy of thunder and electricity.'[22]

Dalibard's translation appeared during March, prefaced with an 'Avertissement' and a 'Histoire abrégrée de l'électricité' intended to insult Nollet. The 'Avertissement' jibes at systematizers and 'hack physicists,' meaning Nollet, who 'make inductions neither just nor natural, [and] deduce consequences founded only on vague suppositions entirely foreign to the subject;' it rhapsodizes in Buffon's style about experimental method and represents that Franklin alone among electricians had followed it; and it insinuates that nothing was done in electricity before the advent of the Philadelphians. The 'Histoire abrégée' passes over Nollet in silence, though it notices several lesser electricians.[23] Contemporaries did not miss the intended slap. Nollet was not far wrong when, on first learning of the Philadelphia system, he conceived Franklin to be an invention of his (and Réaumur's) enemies.[24] He soon discov-

---

20. 'M. de Buffon a, parmi les savants de ce pays-ci, un très-grand nombre d'ennemis; et la voix prépondérante des savants emportera, à ce que je crois, la balance pour bien du temps.' Montesquieu to Cerrati, 11 Nov. 1749, in Montesquieu, *Oeuvres*, VII (1879), 328–9. For Buffon's early views on electricity see Needham, *PT*, 44:1 (1746), 247–63.

21. For Dalibard see Delaunay, *Prov. Maine*, 6 (1926), 53–9, 141–8, 192–9, 251–61, 292–8. Delor, apparently a competent and shrewd practitioner, was about 34 when he lectured Dalibard, a man some 14 years his senior. Nollet to Dutour, 22 Dec. 1754 (Burndy).

22. Dalibard, *EO* (Dal), 4, 20–1; Torlais, *RHS*, 9 (1956), 339–42; l'abbé Mazéas, *PT*, 47 (1752), 534–52.

23. Nollet to Dutour, 30 March 1752 (Burndy); Dalibard, *EO* (Dal), 11. The rhapsody resembles that in Buffon's preface to his translation of Stephen Hales' *Vegetable Staticks* (Paris, 1735). Bose, who had supported Nollet, is shabbily treated in the 'Histoire,' *EO* (Dal), lxv–lxix.

24. Torlais, *Physicien* (1954), 129–45; *J. de Trévoux*, June 1752, 1208, as cited in Torlais, *RHS*, 9 (1956), 340; Nollet to Dutour, 30 March 1752 (Burndy): 'Mr Dalibard, traducteur et éditeur de l'ouvrage sous les ordres de Mr de Buffon, y a joint de son chef une hist. de l'électricité qu'il a intitulée *abbregée* apparamment pour être en droit de ne me pas nommer; cette affectation dont je ne me plains point [!], a été remarquée de tous ceux qui ont vu cet ouvrage.'

ered the truth: 'Mr de Buffon is the promoter of the whole business. He does not appear openly himself, because he knows too little about the subject; he has two tradesmen in his service [Dalibard and Delor] who take care of everything.'[25]

Early in May, 1752, Dalibard set up a metal pole, pointed and insulated, at Marly-la-ville, a village six leagues from Paris, where he had earlier geologized at Buffon's request.[26] On May 10 it thundered in Marly. An old dragoon named Coiffier, previously instructed by Dalibard, who had returned to Paris, ran to the insulated pole, presented to it a brass wire stuck into a glass handle, and saw the first spark man ever intentionally drew from the sky. Coiffier alerted the curé, Raulet, who fearlessly repeated the experiment, 'at least six times within four minutes,' 'each trial lasting as long as a *pater* and an *ave*' (figs. 15.1 and 15.2). Dalibard, informed by Raulet, reported the Marly test to the Paris Academy on May 13.[27] He said that the test confirmed not only the kinship of lightning and electricity, but also Franklin's conjecture that an elevated, *grounded,* pointed rod could protect a building from thunderbolts. 'Perhaps only a hundred or so iron rods, so arranged and deployed in different quarters and in the highest places, would suffice to preserve the entire city of Paris from thunder storms.'[28]

Nollet objected to these sanguine extrapolations and to the suppression of his part in bringing out the analogy between electricity and lightning. He requested acknowledgment and urged the Academy not to publish Dalibard's paper until further evidence was in hand. Soon all the electricians of Paris were repeating the experiment of Marly. Delor succeeded on May 18, Buffon and Bouguer on May 19, and Le Monnier on June 7, with such frightful sparks that the ladies present begged him to desist.[29] Nollet also succeeded, 'dying of chagrin from it all,' as Buffon put it.[30] Indeed Réaumur's poor friend then had good reason to feel miserable. He had had to request a committee of academicians to verify some experiments of his impugned by the *nouveau arrivé,* Delor; he felt his subject, electricity, on which he had labored for a decade, snatched from him

25. Nollet to Dutour, 30 June 1752 (Burndy): 'Mr de Buffon est le promoteur de tout ce vacarme, mais il n'ose paraître luy même sur la scène, il est trop peu initié pour cela, il a deux ouvriers à son service qui se chargent de tout. . . .'

26. Delaunay, *Prov. Maine,* 6 (1926), 194. Delaunay, who treats his subject kindly, observes that Dalibard was assigned to geology after botching experiments on the physical properties of wood (*ibid.,* 192–4).

27. Dalibard's paper, including Raulet's report, was not published by the Academy, for reasons shortly to appear; Dalibard added it to *EO* (Dal²), II, 99–125, and most of it was printed in *EO*⁵ (London, 1774), 108–13 = *EO,* 257–62. The full paper is given in *BFP,* IV, 303–10.

28. *BFP,* IV, 310. This passage, significantly, does not appear in Dalibard's paper as reprinted by Franklin in *EO*⁵.

29. Mazéas, *PT,* 47 (1752), 535, 537; Le Monnier, *MAS* (1752), 233–43; Lamontagne *RHS,* 19 (1966), 231–2 (Bouguer). Ludolf and others soon confirmed Marly in Germany; Mylius, *PT,* 47 (1752), 559.

30. Nollet, *PT,* 47 (1752), 553–4, 557–8; Buffon, letter of 22 July 1752, in *Corresp.* (1885²), I, 84; Torlais, *RHS,* 9 (1956), 342–4.

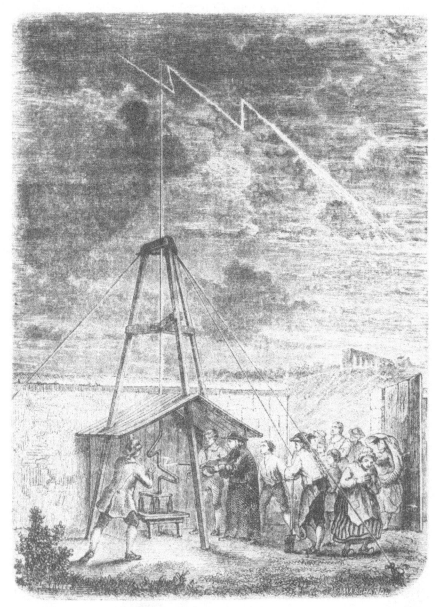

15.1 The experiment at Marly according to a 19th-century artist. From Figuier, Les merveilles de la science *(Paris, n.d.), I, 521.*

15.2 A Japanese rendition of the Marly experiment. From Tuge, Development (1961), 80–1.

by his unpracticed enemies; and he saw Franklin extolled for an idea to which he, and several others, could rightly lay claim.[31] Even the proposals to play with and protect against lightning were not Franklin's, free and clear: a French kite flier, a lawyer of Bordeaux named Jacques de Romas, claimed priority in experimenting with electricity from thunderclouds,[32] while a Czech priest, Prokop Diviš, has appeared as an independent inventor of the lightning rod.[33]

The Philadelphia system had proved more successful than Buffon's clique could have anticipated. It also proved more than they could handle. After the

31. On the dispute with Delor see Torlais, *Physicien* (1954), 342–4; on earlier anticipations of the connection between lightning and electricity, Brunet, *Lychnos*, 10 (1946–7), 117–24, and *supra*, xiv.5.

32. De Romas, who independently of Franklin hit on the electrical kite, seems to have proposed an experiment much like that of Marly in 1750 or 1751. He fiercely defended his independence against Priestley's insinuation (*Hist.* [1775³], I, 411) that he merely improved upon Franklin's demonstrations (Fouveaux de Courmelles, *Électricien* [1909], 14–16). In fact de Romas drew his inspiration from the Bordeaux prize question of 1749 which, in turn, derived from Nollet's speculations. De Romas, AS, *Mém. par div. sav.*, 2 (1755), 393–407; *Oeuvres* (1911), 28–36.

33. The case for Diviš rests upon a lightning rod erected in June, 1754, and upon a fine practical joke perpetrated, according to Diviš' eighteenth-century biographer, in 1750. The joke: Diviš went to see the electricity of Joseph Franz, S. J., professor of mathematics at the University of Vienna. No matter how hard Franz labored he was unable to electrify his prime conductor. Diviš 'had stuck over 20 very sharp iron wires in the forepart of his wig . . .; if he wished to discharge a body, and to make the effort of the electrician fruitless, he merely inclined his head'; Smolka, *Acta hist. rer. natur. techn.*, I (1965), 155. The story is to show that by 1750 Diviš knew and could exploit the power of

first excitement following Marly, they rapidly fell behind, and soon retired completely, leaving Franklin's theory precisely as they found it.[34] They had played an essential part. The experiment at Marly brought the Philadelphia system to the attention of every electrician in Europe. For a year men studied the fire they carelessly brought from the sky and learned that the atmosphere was often electrical without thunderclouds, that points were not necessary, and that captured electricity, contrary to Franklin's expectation, was more frequently negative than positive.[35] After a year of these amusements a German physicist working at St. Petersburg managed to do the experiment Franklin had designed. Instead of picking up the usual minor electrical disturbances associated with storms, G. W. Richmann caught a thunderbolt on his insulated pole and died instantly.[36] As he had said, 'in these times even the physicist has an opportunity to display his fortitude.'[37] That ended the craze begun at Marly. The Franklinists turned their attention to protecting buildings and to developing theory, under the continuous sallies of Nollet.

3. NOLLET

In the years between the shocks from Leyden and Philadelphia, Nollet had pursued two series of investigations that did him credit. The first set, completed in the summer of 1747, examined the accidents that affect electrical appearances: the size and shape of conductors, the nature of supporting stands, the presence of neighboring bodies, air pressure, atmospheric moisture. The merit of the investigation lay not in the general character, but in the detail of the results, especially those concerning the influence of shape—including the power

---

points. Smolka, *ibid.*, 153, concludes after careful investigation that Diviš probably began to think about protection against lightning after reading of Richmann's death, and hence was not independent of Franklin; Cohen and Schofield, *Isis*, 43 (1952), 358–64, agree; the contrary position, argued by Hujer, *Isis*, 43 (1952), 351–7, requires special pleading and neglect of dates. Diviš was in any case not a practical electrician but, as his editor styled him, a 'theologus et magus electricus,' who invoked electricity to explain the light of the first day (before the creation of the sun) and the mechanism of the Last Judgment (the wicked shall be literally discharged, deprived of their animating electrical force). Benz, *Theologie* (1971), 46, 56–7, 82–7.

34. Cf. Dalibard to BF, Feb. 1762, *BFP*, X, 61–2, excusing himself for having done nothing since 1756 because of opposition, lack of instruments, and palsy induced by shocks; so sensitive has he become, he says, that he can hardly handle sealing wax and a single spark prevents him from signing his name for a day.

35. For these improvements, owed largely to Le Monnier, Beccaria and Franklin, who set up a rod in September 1752, see Brunet, *Lychnos*, 10 (1946–7), 117–48; *infra*, xv.4; BF to Collinson, Sept. 1753 (*EO*, 269). Also Bouguer to Euler, 14 July 1752, in Lamontagne, *RHS*, 19 (1966), 231–2, suggesting that the electrical matter runs from the rod to the cloud up the channel provided by the lightning. Nollet obtained some satisfaction from these discoveries: 'Les franklinistes ont le visage bien allongé de cette affaire là.' Nollet to Dutour, 8 July 1752 (Burndy); cf. Nollet, *Lettres* (1753), 10–18, 162–7.

36. *PT*, 44 (1755), 61–9; Sigaud de Lafond, *Précis* (1781), 355–7; Nollet, *Lettres* (1753), 136; *DSB*, XI, 432–3. Richmann had about one chance in 100,000 of receiving such a stroke; Müller-Hillebrand, *Daed.* (1963), 49–50.

37. Richmann, *NCAS*, 4 (1752–3), 335.

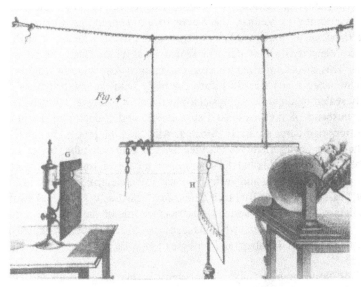

15.3 Nollet's electroscope: the lamp at G images the threads from the prime conductor on the screen H. From Nollet, MAS (1747), 102–31.

of points—on electrical action.[38] Nollet also recorded data suggestive of the concepts of induction and capacitance. His exemplary patience and skill may be illustrated by the electroscope he designed for the work. The instrument threw the image of the usual threads upon a calibrated scale some distance from them, so that neither the scale nor the observer would disturb the measurement (fig. 15.3).

The second set of investigations dealt with medical electricity. Nollet had tried the effect of the Leyden jar on paralytics in 1746 in the hope that the dual flux responsible for the shock would penetrate deeply enough to open or untangle malfunctioning nerves. He did not encourage vain hopes. Although he accepted well-attested instances like Jallabert's cure of the locksmith Noguès, he came to have little faith in the medical efficacy of shocks.[39] Another application of electricity appeared to him more hopeful. Having proved once more that electrification accelerates flow from a capillary siphon, he inferred that electricity might promote evaporation from animal and vegetable matter. Elaborate experiments confirmed the conjecture. Might not faster circulation similarly purge the body of the agents of disease?[40]

38. Nollet, *MAS* (1747), 102–31, 149–99, 207–42 (esp. 225–7), reprinted in his *Recherches* (1753³), 103–341. Jean Jallabert, professor of physics at the University of Geneva and a friend of Nollet's, had investigated the electricity of points and knobs; Nollet, *Recherches* (1753³), 229, and Jallabert, *Expériences* (1749), 40–1, 276–8.

39. Nollet, *Recherches* (1753³), 405–15; Jallabert, *Expériences* (1749), 143–73; cf. BF to John Pringle, 21 Dec. 1757, *BFP*, VII, 298–300.

40. Nollet, *MAS* (1747), 230–42; *PT*, 45 (1748), 187–94.

This reasoned expectation was then conservative. In 1747 Gianfrancesco Pivati, an attorney in Venice, the author of a *Scientific and Curious Sacred-Profane Dictionary*,[41] claimed to have saturated a friend with vapor of Peruvian balsam by electrifying him using a sealed glass globe lined *inside* with the perfume. This sweet-smelling success encouraged less frivolous applications: Pivati announced that he could infuse the body with the subtlest parts of hermetically sealed medicines. A physicist-physician in Bologna, Giovan Giuseppe Veratti, husband of the physicist Laura Bassi, and a doctor in Turin named Bianchi, reported cures similarly effected. All found the process beneficial even when the patient was merely electrified by the common tube, as he held the prescribed medicine in his hand. Mount upon a cake of wax, grasp some aloes or other strong purgative, and suffer yourself to be electrified: you will find 'on the spot,' as one whimsically put it, 'the same effect as if [you] had swallowed the Medicine.'[42] The only serious electrician north of the Alps to accept the Italian claims was Winkler, who managed to corroborate the transfusion of balsam; but Watson, Jallabert and Nollet, though they spared no effort, never succeeded.[43]

In 1749 Nollet toured Italy, demolishing in his progress all the electrical quacks and fakers bold enough to see him. They invented very ingenious excuses. For example, Veratti electrified a young woman for Nollet and his companion, a Cardinal. She did not perform as advertised. Veratti's explanation of the failure must have piqued their curiosity. 'Monsignore,' he said to the Cardinal, 'it is because we could not electrify her in your presence as she should be electrified.' Exposing other reputed Italian marvels he casually encountered, the 'wise' and 'kindly' abbé cut a swath through miracle mongers from Turin to Naples. Except by his victims he was royally received. Reporting the results of his travels occupied him until 1751, and matters other than electricity took up the following year.[44] Then, in the winter of 1751/2, while still aglow from his successful Italian campaign, the 'chief of the electrifying physicists' of Europe[45] found himself the quarry of Buffon.

Nollet realized that Franklin menaced more than his amour-propre. He saw, probably more clearly than Buffon, that the new system struck at the heart of his own. Perhaps he perceived as well that Franklin's exact but disconnected accounts of individual effects menaced all merely plausible, unified, qualitative

41. Pivati's *Nuovo dizionario scientifico e curioso sacro profano*, III (Venice, 1746), has a section on electricity, including a fine electrical scene presided over by Otto von Guericke, 'virium electricarum antesignanus.' It is reproduced in Prinz, Schw. Elek. Ver., *Bull.*, 65 (1974), 4.

42. H. Baker, *PT*, 45 (1748), 270–5. For details of these experiments see Nollet, *PT*, 46 (1749–50), 368–97.

43. Winkler, *PT*, 45 (1748), 262–70; Nollet, *MAS* (1747), 238; Watson, *PT*, 46 (1749–50), 3˙. 56.

44. Nollet, 'Journal' (1749), ff. 16, 93–4, 98–9; *MAS* (1749), 444–57; *MAS* (1750), 54; *PT*, 46 (1749–50), 368–97; Quignon, *Nollet* (1905). Nollet, *MAS* (1753), 429, says he had given up the study of electricity after his Italian trip and returned to it only in response to Franklin. 'Wise' and 'kindly' are Boscovich's words; Hill in Whyte, ed., *Boscovich* (1961), 61.

45. Paulian, *Electricité* (1768), xvii.

and comprehensive descriptions of electricity. Most physicists of Nollet's generation were effluvialists, and he expected them to remain firm;[46] the younger generation, as usual, needed guidance.

Nollet began his counteroffensive before Marly, at the public meeting of the Academy at Easter, 1752. Between protests of Franklin's exaggeration of the power of points and Dalibard's claims of the novelty of the Philadelphia experiments, Nollet paraded several new demonstrations of the permeability of glass. Here is the prototypical proof. A phial is sealed with its neck just emerging from the receiver of an air pump; one pours water into the phial and evacuates the receiver. If now a wire is run into the water from the prime conductor, sparks play within the receiver; if one touches its exterior and the wire, one feels the Leyden shock. The fiery electrical matter gleams as it squeezes through the phial's bottom into the vacuum beyond. Its accumulation in the receiver is evident enough to whoever takes the shock. 'If, after seeing these things, anyone still maintains that the electric fluid does not penetrate glass, his bias must be invincible.'[47] *Ad pleniorem scientiam,* one can fix a long-necked flask, evacuated and hermetically sealed, directly to the prime conductor, as in fig. 15.4. The sparks gambol as before, and can be drawn from the flask's exterior, making visible electrical matter that has crossed glass twice, once in entering and once in leaving the bottle. The flask can again serve for a Leyden experiment; and that, according to Nollet, conflicts with another Franklinist principle, the need for an external, nonelectric discharge circuit. (The vacuum, which, in Nollet's view, played the role of the inner coating, was separated from the external circuit by the glass seal across the neck of the flask.)

These sallies quickened after Marly. Nollet wrote to enlist Dutour's aid against the Franklinists; his ally returned three more demonstrations of the penetrability of glass. For example, he sealed a thin metal window into the wall of a glass cylinder adapted for the air pump and attached a nail to the outside of the window and another to the glass wall directly opposite. The electrical fire appeared within the exhausted cylinder whichever nail touched the conductor, though in one case the effluvia entered through the metal window and in the other, presumably, across the glass. It is sheer perversity, according to Dutour and Nollet, to believe with Franklin that the mechanisms of identical effects produced by virtually the same apparatus are in fact dissimilar.[48]

Franklin's explanation of the Leyden jar required more heresies than the im-

---

46. Nollet to Dutour, 30 June 1752 (Burndy): 'Je crois qu'à la fin les vieux ne seront pas de son [BF's] cotté.'

47. Nollet, *MAS* (1753), 435–40; *MAS* (1747), 188–9; AS, Proc. verb., LXXI (1752), 172–82. While experimenting with electricity *in vacuo*, a part of his program of 1747, Nollet had made a Leyden jar with 'vacuum' as internal coating (Nollet, *MAS* [1747], 189); these experiments probably suggested his first line of counterattack.

48. Dutour, AS, *Mém. par div. sav.*, 2 (1755), 541–2, a paper Nollet had by June 1752; Nollet to Dutour, 19 May and 30 June 1752 (Burndy). The other two demonstrations invoke sparks elicited by the finger through thin glass covering the end of an electrified prime conductor, and communication (by induction) of electricity across a pane of glass inserted between two metal bars.

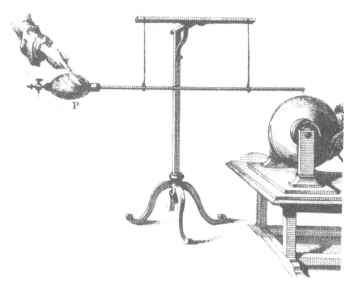

15.4 Nollet's apparatus for showing the penetrability of glass to electrical fluid. From Nollet, MAS (1753), 429–46.

permeability of glass. Recall Nollet's theory of the jar: electrical matter from the prime conductor seeps through the glass to excite the bottle's exterior; the water, the glass and, particularly, its outer coating all emit strong effluents; the shock comes from oppositely directed currents flowing simultaneously from the water and from the external surface of the glass into the man connecting them.[49] Franklin taught that the electricity lay entirely in the glass, and that the discharge went in *one direction,* from positive surface to negative; he put the outer charged surface in a state *qualitatively* different from what Nollet supposed to obtain there. Six of eight Franklinist 'heresies' Nollet mentioned to Dutour in May relate to negative electricity.[50] These, together with impenetra-

49. Nollet believed more firmly in his condenser theory in 1752 than he had when he proposed it in 1746. Nollet, *MAS* (1753), 434; *Lettres* (1753), 85–6, 95–6; AS, Proc. verb., LXXI (1752), 175.

50. Nollet to Dutour, 19 May 1752 (Burndy): 'Je vous dénonce comme autant d'hérésies en électricité les propositions suivantes, qui sont de Mr Franklin: . . . 1. La bouteille qu'on électrise pour l'exp. de Leyde perd autant par sa surface extérieure qu'elle acquiert par sa surface intérieure. 2. On ne peut rétablir l'équilibre entre les deux surfaces qu'en établissant une communication de l'une à l'autre par des corps non électr. 3. Le feu ne passe pas du doit qui touche au fil d'archal mais du fil d'archal au doit. 4. Un conducteur d'électricité ne peut *absolument* point s'électriser s'il est surpassé par une pointe fine. 5. La bouteille chargée par le crochet ne peut être déchargée que par le crochet; et celle qui a été chargée par le ventre, ne peut l'être que par le ventre. 6. Une fiole suspendue au conducteur ne peut *absolument* point se charger qu'elle n'ait communication avec le plancher, par des corps non électr. 7. Le rétablissement du feu électrique ne peut se faire à travers la substance du verre; le verre est imperméable à la matière électrique. 8. L'eau de la bouteille ne s'électrise pas, il n'y a absolument que le verre [qui s'électrise]; l'eau transvasée de cette bouteille n'a aucune électricité.' Numbers (4) and (6), which do not relate directly to negative electricity or the seat of the jar's electrification, protest Franklin's exaggerations.

bility, made up the chief targets of the polished polemic Nollet brought out early in 1753, a set of unmailed *Lettres sur l'électricité* addressed primarily to Benjamin Franklin.[51]

He begins by introducing himself, a necessity forced upon him, he says, by his omission from Dalibard's history. 'Philosophical frankness' obliges him to observe that most of the reputed electrical inventions of the New World had long before been made in the Old. Dufay, for example, discovered the universality of the electrical matter, and not the Philadelphians, as Franklin's editors claimed.[52] Without doubt, however, some features of the American work are original; for example, the supposed impermeability of glass, which is also wrong. Here Nollet rehearses the vacuum experiments he had shown the Academy. And he makes much of Franklin's story about the feather-in-the-bottle. Why, he asks, understandably missing that Franklin had not intended to give the mechanics of attraction, why is the feather not repelled by the derived atmosphere, which issues *from* the wall?[53]

He returns to the Leyden jar. He attempts to show first that the water shares in the bottle's electricity. Into a clean, dry phial, externally grounded, he decanted a charged jar, which he held as he poured; the fresh bottle struck as usual. Evidently the electricity of the second phial came *in and with* the water from the first, a plausible conclusion from a misleading experiment.[54] He moves next to negative electrification, trying to show by the path of the discharge and by the behavior of the bottom of a normally charged jar that no qualitative difference exists between the electricities of hook and tail. Holes blasted in cards by the discharge looked the same regardless whether they had rested on its lower or upper surface; apparently in both cases the stroke sprang from the condenser outwards; the lower surface, like the upper, *delivers* electrical matter in the Leyden shock.[55] The behavior of the so-called negative side of a charged bottle confirmed this conclusion: it yields sparks, and attracts and repels light objects.[56] How can it do so if it has lost all its electrical matter? If

51. Nollet, *Lettres* (1753). The first letter is addressed to Marie-Ange Ardinghelli, who, as Nollet mentions, had rendered Hales's *Haemostaticks* into Italian in 1749, at the age of 16. Nollet had met her in Naples, and was impressed (Journal [1749], f. 170v, 191v); cf. Lalande, *Voyage* [1786²], VII, 228–9); he chose her, perhaps, as a slap at Buffon, who prided himself on *his* translation of Hales. Letters 2–7 are to Franklin, number 8 to Jallabert and number 9, the last, to Bose.

52. Nollet, *Lettres* (1753), 25–7, 36, 40–1; *EO*, 165–6; *EO* (Dal), Pref.

53. Nollet, *Lettres* (1753), 61–73; *supra*, xiv. 4.

54. *Ibid.*, 94–5. When in June 1752 Nollet reported this experiment and inference to the Royal Society, Watson remarked that he had himself reasoned similarly some years before (Nollet, *MAS* [1752], 554–6). See *infra*, xv.4, for Beccaria's elegant counterargument.

55. Nollet, *Lettres* (1753), 126–7; cf. Nollet, *MAS* (1752), 554–6; AS, Proc. verb., LXXI (1752), 178; Nollet, *MAS* (1753), 440. The holes arise from attractions between the discharge wires and charges they induce in the card's surfaces, which cause them to explode or irrupt *outwards*. Hence the identical appearance of the perforations, irrespective of the sign of the coating on which the card rests. Cf. Atkinson, *Elements* (1887), 81–3; *VO* III, 222–5; and *infra*, xviii.2.

56. Nollet, *Lettres* (1735), 100–1. Nollet seems to say that a positively electrified object is repelled by the tail. Although one can sometimes obtain the effect (*infra*, n. 100), it is not necessary

it cannot emit or retain effluvia, it can neither possess nor confer an atmosphere; and without atmospheres, whether static like Franklin's or dynamic like Nollet's, there is no electrical action.

The principal Philadelphia demonstrations therefore require reinterpretation. The play of Franklin's cork between hook and tail wires, for example, the *experimentum crucis* of the system, depends upon different *degrees* of electrification in the inner and outer surfaces of the jar. The cork oscillates through the usual mechanism of ACR; it discharges the jar not by conveying electrical matter from one surface to another, but by decreasing the activity of the effluvia of either surface at each contact, through the collisions of its effluent with those of the hook and tail wires. A second important demonstration, that no discharge occurs if two electrified phials are joined hook to hook and tail to tail, made less difficulty for Nollet. Apparently using dissimilar jars unequally electrified, he received an agreeable shock, and contested Franklin's elegant experiment.[57]

Nollet's objections were of unequal force. The contours of card perforations, for example, do not reveal the direction of discharge. The sparks that appear to pass through glass can be interpreted without violence on Franklin's principles. The feather-in-the-bottle, however, and the repulsive 'atmospheres' of supposedly negative surfaces told heavily: they touched the new system at vulnerable points, where it relied on older ideas incompatible with it. Only gradually, and partly through Nollet's prodding, did continental physicists come to identify and to excise these holdovers.[58] The first Franklinists, however, answered him by trying vainly to reinterpret all his objections within the original Philadelphia system.

Friends spared Franklin the trouble of replying to Nollet. An early defender was Colden's son David, an able young man who unsystematically countered several of Nollet's objections to the theory of the condenser. Colden observes, for example, that decanting jars by Nollet's method proves nothing, for the water serves not as subject of the effluvia, but as their vehicle; had Nollet poured into an insulated, not a grounded bottle the second jar would not have struck.[59] Against the vacuum experiments supportive of permeability young Colden says nothing, lacking the apparatus necessary to repeat them.[60] Nor

---

to Nollet's argument. To secure repulsion from the negative surface insulate the charged jar, touch its hook, freeing some of the 'condensed' negative electricity, and present a resinously charged leaf, or touch a neutral one, to the tail. Nollet might have disingenuously omitted details about the preparation of the leaf.

57. *Ibid.*, 107–8, 123–4; Beccaria, *Dell'elett.* (1753), 157.

58. Cf. Jacquet, *Précis* (1775), 8, praising Nollet 'qui, même en combattant jusqu'à sa mort la théorie de Mr. Franklin, n'a pas peu contribué à étendre les connoissances électriques.'

59. Collinson to BF, 7 March 1753 (*BFP*, IV, 454); BF to C. Colden, 12 April 1753 (*ibid.*, 463–4); D. Colden to BF, 3 Dec. 1753 (*ibid.*, 282–92). Colden concedes that, when freshly charged, the water in an insulated jar has some electricity, which, however, can be removed as a small spark by touching its hook (*ibid.*, 289–90).

60. Franklin did, and found them to 'answer exactly as they should do on my Principles.' BF to C. Colden, 1 Jan. 1754 (*BFP*, V, 186).

does he discuss the feather-in-the-bottle, or the mutual repellency of negative bodies, except to remark that Nollet's experiment will not succeed as described.

Colden is quite effective, however, in defending the qualitative distinction between positive and negative electricity. Franklin's cork, he says, does not oscillate between *hooks* of jars electrified to different degrees, but stands between them, a telling criticism of Nollet's theory of the play between hook and tail. Colden further observes that an insulated man, though he can extract a spark from the hook of a jar presented him by a grounded colleague, cannot discharge it entirely; whereas if he grasps the hook of one and the tail of another, he can detonate both. (The hook of the first gives to the tail of the second through Colden's body, while the tail of the first receives, and the hook of the second empties, through his two grounded assistants.) Though one admires the disciple's defense, one laments the master's silence.

In France the original Franklinists, Buffon's clique, could not manage so much as a Colden. Dalibard applied to his author for advice, in terms that made Franklin doubt his competence.[61] Delor's tactic, hinting at faults in Nollet's experiments, backfired; Nollet requested an academic commission of enquiry that, after seeing them, unanimously attested to their veracity.[62] One of the younger commissioners, however, Jean-Baptiste Le Roy, looked beyond the experiments to the theory, and, at the close of 1753, presented three memoirs supporting negative electricity. Le Roy did not directly contradict Nollet, who was his neighbor, his superior in the Academy, and his mentor in electricity.[63] He affected a dispassionate examination into the existence of a 'rarefied electricity,' qualitatively different from the common, positive, 'compressed' variety.

The strongest evidence for rarefied electricity, according to Le Roy, is the spark exchanged between an object electrified by the prime conductor and one charged by contact with the cushion of an insulated machine. The spark is more vivid, and jumps further, than that which either object would exchange with ground. (Similarly, the attraction between a body electrified by the conductor and one charged by the cushion exceeded that of either for a neutral object.) Before sparking, both objects were electrified, as each attracted chaff; had they been similarly charged, say at the conductor, only a flicker could have crossed between them. The electrifications of conductor and cushion, concluded Le Roy, seemed truly contrary, and qualitatively different. Yet he hesitated to assert it: 'We can deceive ourselves in so many ways,' he said, 'and it is so important in physics to take as principles only facts confirmed by a great

---

61. *Ibid.* Dalibard misunderstood Nollet's statement about the enervation of insulated charged Leyden jars (*supra*, xiii.3).

62. The experiments in question concerned the perforation of cards, the decanting of jars, sparks across glass, repulsion from negative surfaces, shock between hooks, etc. The commissioners attested to more than they saw, for the report, drawn up by Nollet, is not a neutral description of the phenomena (*Lettres* [1753], 241–62). Herbert, *Th. phaen.* (1778²), 31–2, remarks on a similar swindle by Nollet in 1760.

63. Nollet, *MAS* (1747), 232; Nollet to Dutour, 18 Jan. and 22 Dec. 1754 (Burndy).

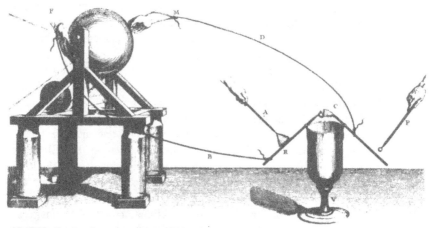

15.5 Le Roy's demonstration of the luminous signs: a conical discharge from the grounded point A and the positive point C facing the negative rod R; a dull glow on the grounded point P and the negative point R facing the positive rod C. From Le Roy, MAS (1753), 468–74.

number of experiments, that I will wait to draw the general conclusion until further, even more decisive evidence makes it absolutely certain.'[64]

Le Roy hesitated less from deference to Nollet than from sensitivity to a nice problem in electricity that had not bothered Franklin. The problem: how can one know that the cushion and not the conductor is the rarefied, or negative, body? Toward the end of November, 1753, the answer came.[65] According to Franklin's theory, a point mounted on a positively charged object should throw, and one stuck on a negative surface should draw the electrical fire. It occurred to Le Roy that the contrary flows might give rise to different luminous effects. The reality exceeded his expectations. From points affixed to bodies positive in Franklin's sense, or from grounded objects opposed to negative ones, issued the hissing, conical, feathery discharge on which Nollet had based his system; while on negative points, or on grounded ones exposed to positive bodies, a limited luminous glow appeared (fig. 15.5). Tacitly accepting the usual identification of sparks and electrical matter, Le Roy inferred that positively charged points do throw, and negative ones receive; Franklin guessed correctly, and luckily, in taking the conductor as positive and the cushion as negative.

Nollet was distressed by Le Roy's apostasy and by his results, which if allowed to stand would establish the unidirectional Franklinist flow against the dual fluxes. Nollet countered with a characteristic argument: Le Roy's distinction comes to nothing because the appearances vary with the size and shape of the points and the sparking distance.[66] The observation is certainly correct: the quality of discharges from points depends on the geometry of the electrode, its

64. Le Roy, *MAS* (1753), 447–59.
65. *Ibid.*, 459–68.
66. Nollet, *MAS* (1753), 503–14, 498–500; Le Roy, *MAS* (1753), 468–9.

potential, air pressure, and impurities in the atmosphere, as well as on the sign of the electricity. But Le Roy's phenomena *dominate,* and while, as Nollet perceived, they do not prove the direction of circulation of Franklin's fluid, they dependably indicate the conventional signs. Nollet came to rely more and more on deceptive exceptions to Franklin's rules, a tactic that succeeded the better because the weak French electrical machines did not always demonstrate the contrary electricities as neatly as the English.[67]

After this skirmish over the points, which opened a feud that divided the Academy until Nollet died in 1770, he shifted his fire from the novelties of the Philadelphia system to its lacunae, particularly to its failure to 'explain' attraction and repulsion, the oldest, most reliable, and most challenging of electrical phenomena.[68] Le Roy had confessed not to know how attraction and repulsion operate. 'That,' snapped Nollet, 'is almost like saying to a man who points to a weathervane as proof of the wind: Ho! we don't know what turns the vane.'[69] And how did the Franklinists account for opposite, *simultaneous* electrostatic effects, like the idealized Hauksbee demonstration of fig. 15.6? What retained the elastic electrical matter about positively charged conductors? Certainly not the air, as Franklinists claimed, for attraction, the first effect of electricity, shows the permeability of the atmosphere. And how could this elastic, compressible material propagate shocks, or itself, in an instant, as Le Monnier's experiments demanded? What, in Franklin's system, is the mechanism of

15.6 *Nollet's double-Hauksbee experiment. From Nollet, MAS (1753), 503–14.*

67. The French continued to use globes or cylinders after the English had switched to plates, and straggled in replacing the hand (as rubber) with the more effective amalgam-coated cushion (*infra,* xviii.2). When Le Roy demonstrated the Franklinist regularities with English instruments in 1772, 'beaucoup de gens ont été.étonnés de voir avec quelle évidence ils établissent la distinction de ces deux électricités.' BF to Priestley, 4 May 1772, in Schofield, *Sci. Autob.* (1966), 101–2. The primary cause of the difference in appearance between positively and negatively electrified points is the character of the space charge each establishes.

68. Nollet, *MAS* (1755), 293–317, responding to Le Roy, *MAS* (1755), 264–83, over the two electricities (*infra,* xviii); Nollet to Dutour, 4 July 1762 (Burndy); Nollet, *MAS* (1753 [read 1755]), 481–3, 486–97.

69. Nollet, *Lettres* (1760), 63. Cf. Yamazaki, *Jap. Stud. Hist. Sci.,* 15 (1976), 45, 50, 59.

sparks? How can one understand their sudden appearance, at fixed distances of the discharging bodies, on the assumption of a unidirectional flow? Franklin either fudged or disregarded these questions. Nollet had designed his system to answer them.

Many of the older electricians, like Bose, Gordon, Musschenbroek, Dutour, Paulian, de Romas and, perhaps, Watson, who thought the traditional problems more significant than the challenge of the Leyden jar, remained with Nollet, as he had expected.[70] The younger men, as he had feared and despite his efforts, tended to confuse the subject and drifted towards Franklin. The Paris Academy, paralyzed by the feud, could give no direction to the study of electricity. Nollet's influence continued strong enough to prevent widespread knowledge of improvements in the Franklinist system.[71] France, which had led the way since the time of Dufay, lost its position in the 1750s and remained behind for a generation. As it declined Italy rose.

4. BECCARIA

Giambattista Beccaria came to the study of electricity intending to do battle. He too found the Philadelphia system a convenient weapon in academic warfare; but, unlike Buffon, whose position had not been menaced, Beccaria fought for his professional life. The rivalry of clerical orders, nationalism, projects for university reform, personal jealousies, and disputes over the nature of physics, played a part in the struggle that turned the testy and undistinguished Beccaria into one of Europe's leading electricians.

In 1748, at the retirement of the incumbent professor, Francesco Garro, Beccaria had accepted the chair of physics at the University of Turin. Garro had held the chair since 1732, when he succeeded his teacher and fellow Minim, Giuseppe Roma, whose physics, so far as he had any, was Cartesian. Garro learned to be a competent instrument maker and demonstrator in the tradition of Rohault, and he collaborated fruitfully with Nollet in pursuit of Pivati. Nonetheless, his own predilections, reinforced by the traditions of their com-

---

70. Nollet, *Lettres* (1753), 31–2; Gordon, *Phys. exper.* (1751–3), I, 284; Dutour, *Recherches* (1760), vii–viii; Marie, 'Phil. quarta pars,' (1763/4), ff. 492–3 (Am. Phil. Soc.); Paulian, *Dictionnaire* (1773²), 156–97. Watson, *PT*, 48 (1753), 201–16 (cf. *FN*, 509–10), reviewed the first volume of Nollet's *Lettres* favorably. De Romas supplied arguments against the impenetrability of glass in an unpublished memoir of 1756 or 1757 (*Oeuvres* [1911], 116–38); in a typical demonstration, a variant of the feather-in-the-bottle, a Leyden jar, weakly electrified without grounding its outer surface, shows electrical signs at its exterior, caused by matter that sifted through its bottom.

71. Nollet to Dutour, 30 June 1752, 22 Dec. 1754, 4 Dec. 1755 (Burndy). Had Le Monnier taken up Franklin's theory Nollet might not have been able to stymie the Academy. But Le Monnier, temperamentally unfit for such a contest (Cuvier, *HAS*, 3 [an IX], 101–17), devoted himself entirely to medicine and botany after 1752, his last original paper on electricity being the account of his lightning experiments in *MAS* (1752). His article 'Electricité' in the *Encyclopédie* was out of date when written: dynamical atmospheres, 7–8 inch insulating layers, no mention of Franklin or the contrary electricities. Cf. Mercet, *Rev. hist. Vers. Seine-et-Oise*, 28 (1926), 179–84; and Home, *Cong. int. hist. sci.*, XIVᵉ, 1974, *Actes*, 2 (1975), 270–2.

mon order, insured that Garro would continue to march in the direction Roma had established.[72]

Garro's retirement in 1748 provided the reforming party the opportunity to free the chair of physics from the influence of the Cartesians and the grasp of the Minims. The king, Charles Emmanuel III, who wished his university to assume a more national flavor, supported the reformers, led by the Marchese Giuseppe Morozzo. Beccaria satisfied all Morozzo's conditions. He belonged not to the Minims but to the Piarists or Scolopians, a modernizing teaching order that had just begun to move into the Italian universities.[73] Though as yet only thirty-two, he had taught for a decade, and lastly at Rome, where the ability of his students had earned him a reputation. Furthermore, he knew and admired English physics, both the mathematics of the *Principia* and the disciplined empiricism of the *Opticks*.[74] Besides these institutional and intellectual qualifications, Beccaria had the necessary nationality, and the ideal character for an academic gladiator, truculent, resourceful, opinionated and ambitious.[75]

Garro's party had also found an excellent candidate, François Jacquier, a French Minim but no Cartesian, an excellent mathematician, coeditor of the so-called 'Jesuits' edition' of Newton's *Principia,* and an established writer and teacher.[76] Jacquier, five years Beccaria's senior, had been the yardstick against whom the younger man and his superiors measured his promise. Jacquier was known in Turin, having once taught there with such effect that the king had begged him to stay; but he had preferred Rome and the company of his collaborator Le Seur. During the negotiations over Garro's succession Jacquier was professor of experimental physics at the Sapientia. No doubt he was the more distinguished candidate.[77] After Beccaria's appointment Garro's clique intimated that the new professor's sole qualification had been his nationality, a charge that gained plausibility as years went by and Beccaria did nothing to win academic glory. Morozzo became concerned to justify his choice. One day, while looking through a French journal, he ran across a report of the experi-

72. Vallauri, *Storia* (1845–6), III, 10, 72–3, 82–5, 102–3; Pace, *Franklin* (1958), 50–1, and *DBI*, VII, 469–71; Calcaterra, *Nost. imm. Ris.* (1935), 377–8; Nollet, *MAS* (1749), 446n. Garro seems to have enjoyed baiting Newtonians (Nollet, 'Journal,' ff. 168v, 176v).

73. For the Piarists see art. 'Scolopi' in *Enciclopedia italiana* and Heimbucher, *Orden,* III (1908[2]), 287–96.

74. [Eandi], *Mem. ist.* (1783), 9–10. Beccaria's library included a first edition of the *Principia*, the *Optice*, Boyle's works, and the texts of Desaguliers, 'sGravesande and Musschenbroek. Piacenza, *Boll. stor.-bibl. subalp.*, 9 (1904), 209–28, 340–54, gives a catalogue.

75. A former student describes Beccaria as 'talvolta alquanto intollerante, pungente, sospettoso ed anche meticuloso' (Vassalli-Eandi, *Lo spett.*, Milan, 5, parte italiana [1816], 101–5, 117–22); Piacenza, *loc. cit.*, illustrates how opinionated and nasty he could be in small matters; and even his apologist remarks on his ambition and jealousy (Tana, *Elogio* [1781], 24–5).

76. For an appreciation of the 'perpetual commentary' that Jacquier and his collaborator, T. Le Seur, added to their edition, see Mme du Châtelet to Jacquier, 12 Nov. 1745 and 13 Nov. 1747, in du Châtelet-Lomont, *Lettres* (1958), I, 143–4, 155–6. Jacquier was also highly regarded by mathematicians; Clairaut to Cramer, 13 April 1744, in Speziali, *RHS,* 8 (1955), 217.

77. Bonnard, *Hist.* (1933), 178–83.

ment at Marly. He thought it might do. He summoned the beleaguered professor: 'Here is a new branch of physical science for you,' he said; 'spare no expense, but cultivate it so as to make yourself famous.'[78]

Electricity was not quite a 'new branch of science' in Italy in 1752. Like everyone else in Europe, the Italians had played with sparks during the mid-1740s. A Fleming, one Boissart, formerly of the Spanish Navy, showed his electricity around the peninsula, to great applause: 'In all my life,' wrote the historian L. A. Muratori, 'I've seen nothing that surprised me more; in other times people would have yelled about magic and some satanic pact.'[79] Several physicians then took up the subject: Pivati, who copied Boissart's equipment, and other medical-electrical quacks; C. A. Guadagni, M.D., Ph.D., the first professor of experimental physics at the University of Pisa, who added electrical demonstrations to his course of experimental physics in Florence;[80] and above all Eusebio Squario, if he is the author of the anonymous *Dell'elettricismo* published in Venice in 1746.[81]

Squario gives a full account of electrical phenomena known up to the invention of the Leyden jar, including the two electricities, the Hauksbee threads, and all the German sparks and glows; he knows intimately the work of Winkler, the Berlin essayists, Bose, Hausen and Nollet, and is himself, so he claims, an independent discoverer of a system of effluence and affluence.[82] The system has several progressive features. Squario's electrical matter is fundamentally the same as that of fire and light, expansive, constituted of particles mutually repulsive; it flows from the hand into the globe during the operation of the machine; its force depends upon its quantity and the rapidity of its motion.[83] And it is found everywhere, even in the heavens: 'Who can frankly deny that lightning is anything but a subtle electrical matter driven to the highest degree of violence?'[84]

The news of the Marly experiment therefore did not catch Italians unprepared. Not so the revelation of its author. 'Who,' wondered Veratti, 'who would have believed that electricity had learned students in North America? In Philadelphia, a city in Canada?' He confirmed the identity of electricity and

78. 'Eccovi un nuovo ramo di scienza fisica; non guardate a spesa, ma coltivatelo in modo da rendervi celebre.' Vassalli-Eandi (*Lo spett., loc. cit.*, and Acc. sci., Turin, *Mem.*, 26 [1821], xxvii), who supplied and probably invented this quotation, knew Beccaria personally and was the nephew of his biographer; there seems no reason to doubt the sense of the words, as they fit the known circumstances of the squabble.

79. Tabarroni, *Coelum*, 34 (1966), 105–6.

80. Guadagni, *Indice* (1745), a list of experiments designed for the instruction of his nephews; 13 of 116 concern electricity. Occhialini, *Notizie* (1914), 3, 10–11.

81. The identification was made in *Novelle della repubblica letteraria* (Venice, 1747); Gliozzi in Volta, *Op. scelt.* (1967), 10; Volpati, *Volta* (1927), 67–8.

82. [Squario], *Dell'elett.* (1746), 56–7, 105–6, 114; the afflux derives from the efflux, returned by the resistance of the air and by collisions among the effluent streams. Squario succeeded in obtaining electricity *in vacuo* (*ibid.*, 91, 104), and hence rejected Cabeo's theory, shortly to be revived by Bammacaro, *Tentamen* (1748), 122–8 (*supra*, xi.3).

83. [Squario], *Dell'elett.* (1746), 75, 94–100, 352.   84. *Ibid.*, 379.

lightning from the tower of the University of Bologna. The public feared that he might attract a lightning bolt and forced him to remove the rod.[85] Thereupon a colleague, Tommaso Marini, began systematic observations from his home, and learned independently of Le Monnier that atmospheric electricity occurred in tranquil weather and that clouds came negatively electrified.[86] This was precisely where Beccaria himself started. His lightning rod had begun to spark by July, 1752, and soon brought him abreast of Le Monnier and the Bologna group.[87] But he did not stop, as they did, at the outworks of Franklin's system. He meticulously examined the power of points, the impenetrability of glass, the pumping of the machine, and the action of the condenser, and wherever his hot glance fell it uncovered something new, at least to him. Most useful and agreeable was the discovery, independent of Le Roy's, of the different appearances at positive and negative points.

These finds did not suffice. Beccaria realized that his best chance for glory lay in producing the first European Franklinist treatise, and that priority was essential. He began to write at tremendous speed. His enemies discovered his plan and resolved to suffocate his book. They brought copies of Nollet's *Letters* from Paris to prove the frivolity of Beccaria's project. They purloined the sheets of his book as they were printed, and published a refutation of one half of it before Beccaria had corrected the proofs of the other. They failed to discourage him. He drew up an answer to Nollet, which he inserted in the middle of his book, and a blast against his adversaries, which he placed at the end. The work, *Dell'elettricismo artificiale e naturale libri due,* appeared at Turin in the late summer of 1753, to 'the applause of the learned and the despair of the spiteful.'[88]

The hasty work is a minor masterpiece. The first of its two parts, or 'books,' concerns 'artificial' or laboratory electricity, the second 'natural' or atmospheric electricity. Throughout reigns the spirit of system; where Franklin was disorganized, parochial, unassertive and open, his disciple ordered, developed, polished and generalized.[89] To establish the critical distinction between positive and negative electricity, for example, Beccaria eschewed the tube. He obtained

---

85. Anon., Acc. sci. ist. Bol., *Comm.*, 3 (1755), 96–7; Veratti, *ibid.*, 200–4; Tabarroni, *Coelum*, 34 (1966), 127–30. Cf. the report in the *Virginia Gazette*, 17 Nov. 1752, in R. Overfield, *Emp. State Res. Ser.*, 16:3 (1968), 45.

86. Marini, Acc. sci. ist. Bol., *Comm.*, 3 (1755), 209, 212–16.

87. Beccaria, *Dell'elett.* (1753), 160–7.

88. Vallauri, *Storia*, III (1846), 141–3; Vassalli-Eandi, *Lo spett.*, 5 (1816), 101–5, 117–22; Beccaria, *Dell'elett.* (1753), 235. The last sheets had not been printed in April; Le Roy, *MAS* (1753), 464, knew of the work, but had not seen a copy, in November; hence the book probably appeared in the summer or autumn.

89. Gliozzi, *Arch.*, 17 (1935), 20–1, 32, makes Book II 'the first treatise on atmospheric electricity' and Book I a novelty for its coverage and for its 'geometrical' form of presentation; cf. Gliozzi, *Elettrologia* (1937), I, 209, 246, and *Fisici* (1962), 5. Both statements are exaggerated: there is little in Book II beyond information in Le Monnier, *MAS* (1752), 233–43, except speculations about the electrical nature of waterspouts, auroras, and earthquakes; and earlier electricians, particularly Winkler and Nollet, had tried to exhaust the facts, and to present their theories

his contrary charges from the prime conductor, and / or the frame of the machine, by insulating one or the other or both during the rubbing; he described the sparks drawn between them, and from either to ground, varying the combinations in every conceivable way; and he reduced all the appearances to one universal fallacious rule, namely, that electrical signs occur only when electrical fire *passes* between two bodies containing unequal amounts, and then with a vivacity proportional to the quantity of fire transferred.[90]

Beccaria no doubt believed that this rule expressed everything useful in Franklin's representation of atmospheres. In fact it implied a distinctly different picture, a return to dynamical theories opposed to Franklin's nonmechanical atmospheres.[91] Beccaria later revived Cabeo's account of attraction. The rationale for the fallacious rule is not far to seek: Beccaria accepted Franklin's emphasis on sparks, as opposed to attractions, as indicators of electricity, and agreed with most electricians that sparks occur only during the passage of electrical matter. His discovery of the different characteristics of the discharges at positive and negative points, and at grounded points opposed to negatively and positively charged objects, enabled him to fix the direction of passage and to improve the exposition of the theory. Like Le Roy, he thought it evident that the 'fiocco elettrico,' or divergent luminous pencil, at the positive point betrayed an outward flow of electrical fire, while the 'stelletta,' or little star, at negative ones indicated an influx.[92]

These matters, which occupied the sheets Beccaria's enemies acquired from his printer, furnished the fuel for their ferocious attack. They exhibit Beccaria as an inept experimenter and imbecile plagiarist, futilely defending Franklin's demolished system with false observations. They observe that stars sometimes occur at positive, and pencils at negative points; where stars appear one cannot infer an influx of electrical matter, since an electrical wind blows from them; negative points repel as well as attract small objects.

Beccaria answers these effluvialist objections as best he can. He concedes the exceptions, but insists that they do not invalidate his discovery nor lessen its importance. As for the wind and the repulsion, he can only assert, against two

'geometrically.' Beccaria claimed to be the first to arrange the phenomena systematically, under general rules, a pretention that drew a vigorous denial from Nollet, who quite rightly pointed to Dufay. This from marginalia in Nollet's copy of Beccaria's *Dell'elett.* (1753), 5 (Burndy).

90. *Dell'elett.* (1753), 17. Gliozzi, *Arch.*, 17 (1935), 23, and *Elettrologia* (1937), I, 217, insists that, as in Franklin's formulation (*supra*, xiv.2), Beccaria's implicitly involves more than the quantity of charge exchanged. More to the point, it denies the reality of repulsion if 'electrical signs' include 'electrical motions.'

91. Cf. Home, *Br. J. Hist. Sci.*, 6 (1972), 146–9. Home, *ibid.*, 148, and *J. Hist. Biol.*, 3 (1970), 235–51, points out that Boscovich, Haller, Felice Fontana and M. A. Caldani accepted Beccaria's rule.

92. *Dell'elett.* (1753), 9–12, using the rendering of 'fiocco' in Priestley, *Hist.* (1775³), I, 362. Nollet would not accept the distinction: 'Votre stelletta est toujours une aigrette de rayons divergens . . . il suit sensiblement effluence, quand la pointe est assez grande.' Nollet, marginalia in *Dell'elett.* (1753), 9 (Burndy).

generations of electricians, that they do not bear on the matter. These defensive responses point towards the fruitful compromise the dogmatic effluvialists were to force on the Franklinists: relaxation of the connection between electrical appearances and the motion of electrical matter. Neither the repulsion, the wind, the star nor the pencil directly testifies to its course, and none immediately determines whether or where negative electricity exists. The best supports of Franklin's fundamental distinction available in 1753 were, as Beccaria told his critics in another connection, the effects that had inspired it: while nothing passes between bodies electrified at the conductor, or between those charged by the frame of the machine, an object from the first class exchanges a stronger spark with one of the second than either would with ground.[93]

The later Franklinists also differed from their opponents in eschewing mechanical analogies to attraction and repulsion. Here again Beccaria occupied a tentative middle ground. He tried to adopt an instrumentalism that he understood to be Newtonian: 'Although,' he says, 'it would complete this work if the mechanics of electrical motions could be specified, it is enough to do as Newton, and reduce the phenomena to a manifest principle.' But Beccaria's principle, that bodies unequally electrified attract one another, is neither manifest nor true: it is but a rider on his earlier erroneous generalization about electrical signs, and cannot be applied without recourse to mechanics.

What causes a cork suspended from a silk thread to play between the prime conductor and a grounded bar? Franklin would say that the (unexplained) action between (static) atmospheres and neutral bodies draws the pendulum to the conductor, where it acquires an atmosphere that pushes it away and drags it to the bar. Beccaria wishes a further reduction—the elimination of repulsion—primarily because Franklin's account left unintelligible the spread of threads attached to the negative frame of the machine. According to him, therefore, the pendulum initially flies to the conductor, or to the machine, via the unexplained dynamical action of electrical matter flowing to re-establish equilibrium; when the cork attains the same degree of electricity as its solicitor, it drops *under its own weight,* bombards (or receives from) the bar, which it then approaches in obedience to Beccaria's 'manifest principle.' As for the spread of a tassel of charged threads, the air draws each aside separately, by the usual attraction between neutral and electrified bodies.[94] Still claiming to eschew mechanism, Beccaria follows Franklin closely regarding the power of points and the charging of the Leyden jar, offers a new 'fiction' in place of the American model of the structure of glass, and adapts Franklin's theory of the feather-in-the-bottle.

For some years Beccaria balanced between the advanced positivism of his

93. 'Risposta alle obiezioni fatte contro il primo capo del primo libro,' *Dell' elett.* (1753), 235–45.

94. *Ibid.*, 17, 23–30, 40, 48–9, 55; Beccaria's model leaves unexplained Hauksbee's observation of the large velocity of recoil and 'repulsion' *in vacuo,* a point that did trouble him (*ibid.*, 34–9). Cf. Beccaria's response to Nollet's challenge to explain minus-minus repulsion (*ibid.*, 147, 153).

manifest principle and the effluvialist goal of mechanical explanations. Having a weaker stomach for inconsistencies than Franklin, he felt forced to choose, opted for tradition, and fell from the Franklinist forefront.

Among the experiments reported in *Dell' elettricismo,* those on the relative resistances of air, water, and metals deserve notice. Beccaria experienced neither difficulty nor surprise in showing the superior conductivity of metals.[95] To investigate the 'fluids,' he cemented two wires into a narrow glass tube, leaving a small gap between them, and exploded a jar through the instrument. The shock passed without incident through an air-filled tube and shattered one containing water. Beccaria inferred the novel doctrine that water offers greater resistance than air to the passage of electrical matter, an observation that, if vigorously pursued, could have embarrassed the effluvialists.[96]

In responding to Nollet, Beccaria first tackles the question of glass's impenetrability, which he recognizes as the essence of Franklinism. Nollet's experiments with sparks in evacuated receivers, he says, in no way disconfirm the doctrine, as the sparks only *appear* to pass through the glass walls separating the charged conductor from the void. In the prototypical case, a Leyden jar sealed into the neck of a vessel then evacuated, electric fire from the prime conductor accumulates in the upper surface of the glass, whence it expels fire from the bottom into the void, where it becomes visible. The evacuated vessel sealed directly onto the PC behaves similarly. If the outside of the vessel is grounded, it becomes the negative side of a condenser, denuded of its electrical matter by the force of fire expelled from the wall nearest the PC. Nollet's striking experiments amount to so many visible demonstrations of Franklin's condensing mechanism![97]

Beccaria added a new proof of impenetrability. Place a thin small pane of glass on the upper armature of a charged Franklin square, leaving a border all around. Bend a wire attached to the lower armature until it touches the center of the pane. The square remains quiet. Now gradually move the wire across the glass; the condenser explodes when, but not before, the probe reaches the edge of the pane. We are to infer that, because glass will not transmit the shock, it does not pass the electric matter in the Leyden experiment. To make the infer-

95. *Dell' elett.* (1753), 57–8, 67–9, 97–103, 111. Gliozzi, *Arch.,* 17 (1935), 31, exaggerates the novelty of the resistance experiments and of their results concerning metals; Nollet, *MAS* (1747), 149–99, for example, had carefully studied the relative resistance of air and 'vacuum,' and no one in 1753 doubted that metals conducted better than water. 'L'autheur [Beccaria] se fait icy un phantome pour avoir le plaisir de le combattre. Je ne connais aucun autheur qui ait dit que l'eau s'électrifiait mieux que le métal.' Nollet, marginalia in *Dell' elett.* (1953), 111 (Burndy).

96. *Dell' elett.* (1753), 113–15. Priestley, *Hist.* (1775³), I, 248–9, misses the novelty in these experiments. 'What astonishes us most,' he says, 'in Signior [!] Beccaria's experiments with water, is his making the spark visible in it.' Drawing sparks in water had been one of Bose's favorite pastimes and the proximate cause of the original Leyden experiment (*supra,* xiii.2).

97. *Dell' elett.* (1753), 146–9; similarly Nollet's argument (*supra,* xv.3) against the external discharge circuit falls to the ground (*ibid.,* 156).

ence more plausible, Beccaria announced that a sliver of talc, which prevented the explosion, could also be used in a Franklin square.[98]

Beccaria shows best when combatting Nollet's objections to the theory of the Leyden jar. He dismisses the card perforations as inconclusive. He shows clearly that Nollet's decanting failed its purpose: by grounding both vessels while pouring, he freed the excess electrical matter of the first to flow into the second via the water momentarily joining them. Here is an elegant proof of the neutrality of the liquid in a charged phial. Insert the short arm of a narrow-gauge siphon into a prepared, insulated flask. Water drips slowly from the longer arm, exactly as it would from an uncharged phial. Touch the bottle's exterior and the siphon flows freely, in a hydrodynamical version of Kleist's portable flame thrower.[99] Against Nollet's claim that the electrifications of the inner and outer surfaces differ in degree and not in kind, Beccaria reasserts fundamental Franklinist facts. A leaf attracted and repelled by the hook immediately flies to the tail, and vice versa, showing that the electrifications differ in kind.[100] That they are also equal follows from the total discharge of both surfaces in the Leyden explosion: 'Two quantities cannot destroy themselves exactly unless one is negative, the other positive, and both are equal.'[101]

Dell'elettricismo secured Beccaria's position at Turin, earned him a pension and silenced his enemies.[102] Not only did it wipe out Nollet's following in Turin, it also created international Franklinism. Or so said a friendly reviewer. 'Beccaria got Franklin's system so unformed, so devoid of proof, that rather than a true system of nature it scarcely merited being called an hypothesis . . . His singular modesty [!] cannot hide from us that he owes Franklin little more than Newton owed Pythagoras.'[103] The Parisian clique was delighted with the book. They considered translating it into French, but appar-

---

98. *Ibid.*, 144–5. Later Beccaria succeeded with wax (*Dell'elett.* [1758], 54; Gliozzi, *Elettrologia* [1937], I, 239), and others with various solids including ice (Achard, *Chem-phys. Schr.* [1780], 19).

99. The siphon acts like a point, through which a portion of the electrical fluid exits (via the water) when the jar is removed from the machine. Touching the external surface reduces its charge and frees more fluid to electrify the water and accelerate it through the 'point.'

100. *Dell'elett.* (1753), 149, 151, 154, 158. Beccaria observes that a leaf electrified at the hook may *initially* be repelled by the bottom of an unarmed jar, near places where, according to him, electrical matter not removed during the uneven charging has leaked into the air to form a local 'positive atmosphere.' (The phenomenon occurs opposite unarmed edges where the field of the positive charge within the bottle, or that of the polarized dielectric, can overpower the force of the negative coating.)

101. Shortly after finishing *Dell'elettricismo* Beccaria invented another demonstration of this property: a jar electrified at the conductor may be discharged by touching its (positive) hook to the frame of the machine and spinning the globe. Beccaria to P. Frisi (whom Beccaria wished to convert from Nolletist views), 1753, in Gliozzi, *Cong. int. hist. sci.*, XI[e], 1965, *Actes*, 3 (1968), 210–15. Cf. Kinnersley to Franklin, 3 Feb. 1752, *EO*, 250–1.

102. [Eandi], *Mem. ist.* (1783), 22–3; Pace, *Franklin* (1958), 52–3, 64–8.

103. Anon., *Nuovo gior. de' lett. d'Ital.*, 2 (1773), 356.

ently shrank before the labor. They contented themselves with rendering Beccaria's answer to Nollet, to free themselves from the obligation of composing one themselves.[104] The little book extirpated the Nolletists, according to Dalibard, and fell dead from the press, according to Nollet, who thought his Parisian front secure and feared only for his reputation in Italy.[105]

Franklin was pleased and flattered by the defense his apparently disinterested European supporters mounted. Beccaria, he said, was a 'Master of Method,' and *Dell'elettricismo* 'one of the best pieces on the subject . . . in any language,' though one may doubt that he understood it perfectly.[106] The dour professor, highly gratified in his turn by the applause of the growing Franklinist confederacy, continued to labor on the subject that, at the bidding of his academic superior, he had made his own.

Beccaria's ongoing studies dealt mainly with the electricity of the atmosphere, which he and his students probed perseveringly with poles, kites, and rockets. Their valuable observations made up two-thirds of Beccaria's second work on electricity, letters addressed to G. B. Beccari, F. R. S., professor of chemistry at the University of Bologna. These letters, published in 1758, improve on *Dell'elettricismo* in factual detail, but they lack its verve and retrograde in theory in trying to solve the 'oldest question in electricity,' the mechanics of electrical motions.[107]

As before, Beccaria rigidly distinguished the mechanisms of repulsion and attraction, for which he now proposed the theory of Cabeo. Against the usual demonstrations of electrical motions *in vacuo*, he offered two bits of evidence. First, take a tube arranged to test the resistance of air to the spark, leaving one end open. Immerse the tube in water to just beneath the spark-gap, as in fig. 15.7a. When the instrument passes the Leyden shock, the water falls, depressed, according to Beccaria, by air driven aside by electric fire rushing through it.[108]

104. Beccaria, *Lettere* (1754), v–vi. The proposed translation of the whole is mentioned in Dalibard to BF, 31 March 1754 (*BFP*, V, 253) and in Nollet to Dutour, 18 Jan. 1754 (Burndy).

105. Dalibard to BF, 31 March 1754 (*BFP*, V, 234). Nollet to Dutour, 26 May 1754 (Burndy): 'La lettre de P. B. traduite par Mr de Lor ne fait point grand bruit à Paris. Mr Freron dans les feuilles périodiques y a répondu pour moy de manière que je m'en contenterois, si ce n'étoit que ce religieux a écrit à Thurin où je suis fort connu et où il se donne bien des mouvements pour faire croire au grand nombre que le ciel l'a envoyé tout exprès pour relever mes erreurs et rétablir la doctrine de l'électricité.' Cf. same to same, 14 July 1754 (Burndy).

106. BF to Collinson, 29 July 1754, and to Colden, 30 Aug. 1754 (*BFP*, V, 395, 428), and BF to Dalibard, 29 June 1755 (*BFP*, VI, 98–9); Franklin particularly liked Beccaria's discovery, confirming his own, of the preponderantly negative electrification of clouds. For Franklin's proficiency in Italian see Pace, *Franklin* (1958), 1–2, 54.

107. Priestley, *Hist.* (1775³), I, 245–50, rates *Dell'elett.* (1758) very highly because of its new facts, missing entirely the profound opposition between its theory and Franklin's, which was not lost on Nollet (*infra*, xviii.1).

108. *Dell'elett.* (1758), 36; *PT*, 51:2 (1760), 515; *Dell'elett.* (1753), 113–15. Gliozzi, *Arch.*, 17 (1935), 25–7, argues that this experiment, first described in *Dell'elett.* (1753), marks the invention of the 'electrical thermometer' usually ascribed to Kinnersley (Kinnersley to BF, 12 March 1761, *EO*, 351–5), but Beccaria seems not to have realized before seeing the American work that the effect depended on the dilation of air by the heat of the spark.

a.

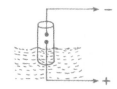

15.7 Diagrams of Beccaria's experiments in support of Cabeo's theory. (a) Sparking (b) motions in vacuo.

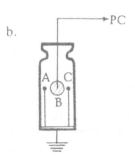

b.

The possibility of Cabeo's mechanism thus established, Beccaria tried to show its necessity. He arranged an evacuated, grounded receiver as in fig. 15.7b. When the prime conductor charged, the gilded paper bob B oscillated slightly and a tranquil glow filled most of the vessel. As air returned both the light and the motion altered: the diffuse glow disappeared and the pendulum vibrated between A and C, sparking at each contact, with a vigor proportional to the pressure of the admitted air.[109] The experiment seemed unambiguous. Beccaria next tried the standard demonstrations of electrical motions *in vacuo*, the dance of the paper bits, and found them equivocally favorable. In his dynamic Franklinism, therefore, electrical attraction occurs when and only when electrical fire from a body relatively plus leaps to one relatively minus *across an air space,* the irruption setting in train the choreography of Cabeo.[110]

Attractions also account for repulsion. One of Franklin's principles denies electric signs to a system of bodies (including the atmosphere) all of which have their natural content of electrical fire. Beccaria generalized this principle to include the case that the natural contents of the bodies all alter in the same proportion, so that, for example, neither sparks nor relative motions should occur between the ends of two threads suspended from an electrified conductor.

109. *Dell'elett.* (1758), 38–40. The residual gas, quickly ionized, carries the current from A through B and C, whence it flows to ground.

110. *Ibid.*, 43–9. Because of the standard vacuum experiment, Beccaria advanced the revived air theory reservedly in *Dell'elett.* (1758). He resolved his doubts before the book was published, however, and continued to perfect his own counter demonstrations. Beccaria to BF, 24 Dec. 1757, in *PT*, 51:2 (1760), 514–25, a résumé of the theory of electrical motions in *Dell'elett.* (1758), 36–49, 85–97; Ferrner, *Resa* (1956), 543–5.

Yet nothing is more evident than their separation. Beccaria's principle is not violated, however, for repulsion occurs whenever bodies equally electrified stand in air charged, as in this case, to a different degree.

The theory puts a great burden on the air, which must produce the *same* effect no matter whether it or the threads have more electrical fire. The solution had little vogue among the Franklinists, despite their eagerness to accommodate minus-minus repulsion. Franklin himself explicitly rejected it when Kinnersley proposed it, perhaps independently, in 1762. Seven years later Beccaria abandoned it for a proto-field theory that struck many electricians as the best available compromise.[111]

Meanwhile Beccaria supported his universal law of repulsion by characteristically intricate manipulations with the spreading threads. Charge the prime conductor. The threads spring to their maximum elongation, x_0, where x represents the distance between their ends. Then [1] as the machine continues, x decreases to about $x_0 / 2$ and remains constant; [2] when the rubbing ceases, x gradually diminishes to zero and [3] thereafter increases a little. According to Beccaria, all three appearances derive from leak of electrical fire into the air, which reduces the disparity causing the repulsion; equilibrium can occur only in [1], where the machine makes good the continual loss. The same processes are manifest, he says, if [4] one touches the conductor when $x = x_0$, or [5] when x has assumed its constant value. In [4], x immediately falls to zero, because the air has not yet imbibed any electrical fire from the conductor. The result varies in [5] according to the surroundings: if performed out of doors, the threads collapse, as drafts prevent the establishment of aerial electricity; if performed in a still room, the strings drop, vibrate and rise again without touching, acted upon only by the electrical fire accumulated in the air. One might also wait until [6] the threads hang parallel or [7] have rerisen before touching the conductor. In [6] a spark passes, removing the fire in the conductor balancing that in the air, which then draws the threads apart; while in [7] nothing happens, as the conductor has already returned to its natural state.

The last effect, as Beccaria emphasized, shows distinctly that 'aerial electricity,' the signs produced in *neutral* bodies immersed in electrified air, differs from 'proper electricity,' the signs between *charged* bodies surrounded by neutral air. In proper electricity, for example, bodies draw apart only when insulated and equally electrified, whereas in aerial electricity they separate when grounded.[112] (Both 'electricities' ultimately derive from mechanical action of the air.) The usual laboratory experiments, being mixtures of the two cases, do not yield easily to analysis. Should we expect attraction or repulsion when two freshly charged objects, both positive but to different degrees, come together in neutral, unelectrified air?

111. *Ibid.*, 93–4; *EO*, 349–50, 365; *infra*, xvii.3.
112. *Dell' elett.* (1758), 85–91, 94–6. Most of these appearances, except [3] and [7], are readily explained by reference to the space charge build up around the corners of the prime conductor.

CHAPTER XVI

The Atmospheres Attacked

1. CANTON'S INDUCTIONS

The leading English Franklinist of the 1750s was a London schoolmaster, John Canton, a native of Stroud in Gloucestershire. Canton received little formal education himself, for just as he reached the rudiments of geometry his father removed him from school to set him to the family trade of broadcloth weaving. He continued his studies on his own. Dr. Henry Miles of Stroud, a nonconformist minister and sometime physicist,[1] noticed the clandestine scholar. Through Miles, whose pulpit was in Surrey, Canton obtained a situation more suitable than weaving. He was articled to a London schoolmaster, whose business he acquired in 1745.[2]

By then he had met Watson and Ellicott, perhaps through Miles, who became F.R.S. in 1743. All of them no doubt belonged, or were to belong, to that Society for Mathematical and Electrical Studies, 'whose members consisted principally of tradesmen and artisans,' mentioned in the history of the Royal Society.[3] These new associates shifted Canton's interest to experimental philosophy, which he promptly enriched with a new, secret method for making artificial magnets. It brought him a small income, election to the Royal Society, and a charge of plagiarism, doubtless unfounded, preferred by a rival magnetizer, John Michell.[4]

Canton's entry into his special field, electricity, was also ill-omened. In 1747, in schoolmasterly fashion, he asked the readers of the *Gentleman's Magazine* to contrive a means whereby the prime conductor will charge strongly by turning the globe clockwise and not at all by turning it counterclockwise. John Smeaton supplied an answer, which his friend Benjamin Wilson forwarded to the *Gentleman's* with a patronizing sneer: 'We learn little more from these experiments than if they had never been made.'[5] Canton con-

---

1. Miles was one of the first to succeed with the electrical flare (*PT*, 43 [1744–5], 290–3); he kept a meteorological journal (*PT*, 44 [1746], 613); and he planned to write a history of electricity, apparently never completed (Miles to B. Wilson, 6 Oct. 1747, Ms. Add. 30094, f. 48 [BL]). Cf. *DNB*, XIII, 378, and Canton Papers I, liii–lvi (RS).

2. W. Canton in Kippis, ed., *Biog. brit.*, III (1784²), 215–22; *DSB*, III, 51–2.

3. Weld, *Hist.* (1848), I, 467n; Singer, *Ann. Sci.*, 6 (1949), 166n.

4. Canton Papers, I, xxiii–xxxi (RS); Hutton, *Diar. Misc.* (1775), II, 96, 116; Canton, *PT*, 47 (1751–2), 31–8. The regrettable dispute with Michell, which brought two of England's best physicists to daggers' points, is judiciously treated by De Morgan, *Athen.* (1849), 5–7, 162–4, 375.

5. *GM*, 17 (1747), 16, 183–4. Smeaton's solution: let X be the rubber, Y and Z two insulated metallic points placed opposite the globe's equator on either side of the chain hanging from the prime conductor (PC); and let the order, proceeding clockwise, be XYchainZ. The PC will charge when the

tinued to experiment and to show the usual tricks to the fair sex ('I shall not bring any ladies with me,' wrote a visitor, 'so there will be no occasion for your having the trouble of making Electrical Experiments'), but he published nothing further on the subject until 1752, when he had the good fortune to be the first in England to confirm the spectacle of Marly.[6]

As Canton mined this unexpected vein, he made the important discovery, independent of Beccaria, Le Monnier and Franklin, that clouds came electrified both positively and negatively. The response from Philadelphia bore no trace of Wilson's condescension: 'It is a great pleasure to me [Franklin wrote] that [Canton's] observations evince the various States of the clouds (as to positive and negative Electricity) as well as mine. I was afraid of being thought out of my Senses.'[7] On Franklin's theory of evaporation, Heaven is positive and lightning probes should charge plus. The discovery that a probe could be negative and, worse, might reverse its sign without receiving a stroke, weakened Franklin's system at its best advertised point. Canton made these problems his own and, while trying to understand the processes clouds stimulate in poles, invented his important experiments on induction.[8]

The apparatus required is meager: two cork balls, batons of glass and wax, an insulated tin cylinder. [1] Bring the glass rod (a positive cloud) under the corks hung from the ceiling by conducting threads: the balls diverge as the rod approaches, collapse as it recedes. The same phenomena occur with the wax stick (a negative cloud). [2] Now allow for the transfer of electrical matter, as assumed in the theory of the Marly experiment, by suspending the corks from the cylinder (fig. 16.1). Electrify the cylinder with a spark from the glass rod (a lightning stroke); the corks spread to a distance x of two inches. Place the rod at A, a distance y under the corks: x decreases as y diminishes, falls to zero and grows again as y continues to shorten; the sequence reverses as the rod is withdrawn, and the entire experiment succeeds as well with the stick of wax. If the cylinder receives a spark from the glass and the wax is moved under the balls (or vice versa), x increases monotonically as y diminishes. [3] Discharge the cylinder and bring the glass rod to its center at B: x will increase as z diminishes. If z falls to six inches the cylinder retains a negative charge when the tube withdraws.

Excepting the last, these effects are cases of induction that any modern schoolboy can reduce to rule. For Canton, however, who attempted to under-

---

globe spins clockwise or counterclockwise if X or Z, respectively, remains insulated. Cf. Beccaria's determinations of the course of the electric fluid (*supra*, xv.3).

6. Canton Papers, II, 19, 28 (RS); Watson, *PT*, 47 (1751–2), 567–70; Wilson, *Treatise* (1750), 63–4, reports Canton's experiments on the alternate discharging of a Leyden jar.

7. Canton, *PT*, 48:1 (1753), 350–8; BF to Collinson, 26 June 1755, *BFP*, VI, 88. Cf. Franklin's praise of Beccaria's first book (*supra*, xv.3 and *BFP*, V, 266, 394–5).

8. The rationale of Canton's investigations, which Priestley (*Hist.* [1775³], I, 287–8) missed, appears in Canton, *PT*, 48:1 (1753), 350–8, under 'Experiment IX.' Franklin did not insist on his theory of cloud formation (e.g., letter to 'J. B.,' 24 Jan. 1752, *EO*, 318), which he strove to improve (e.g., *EO*, 334–5, 348, 359–60); cf. *infra*, xix.1.

stand them in terms of atmospheres, they presented severe difficulties; the hints from Franklin not sufficing, he added mechanisms quarried helter-skelter from Watson, Wilson, and, perhaps, Beccaria. In [1], for example, Canton understood the balls to spread owing to Franklin's static atmospheres, but subject to Beccaria's rider that the strength of their repulsion is proportional to the difference in the density of electrical matter within and outside them.[9] Contrary to Franklin, Canton pondered over the physics of these density gradients and atmospheres. He was brought to invoke an implausible dynamic equilibrium.[10] Electrical matter runs from the cylinder and out through the corks, or vice versa, according to whether the wax or glass rod is used; in the first case the resistance of the air, and in the second that of the corks, opposes the flow, and heaps up atmospheres around the balls. Franklin's force of attraction between common and electrical matter played no explicit part in the retention of the atmospheres.

To explain why the corks spread further when hung by hemp than when hung by silk, Canton supposed that the air adjoining the cork's surface acts like the glass of the Leyden jar: the balls deck themselves with the tube's atmosphere, which drives their own electrical matter into the cylinder. The same *ad hoc* but potentially fruitful analogy helps explain the second set of experiments [2]. If the cylinder is positive the balls have atmospheres and diverge; the approaching tube drives back these envelopes, and *x* declines; when the density inside and outside the balls equalizes, they collapse, to spread again as the advance of the tube strengthens the external atmosphere. If the cylinder is negative the balls acquire their atmospheres from the air; the approaching tube supplies electrical matter more plentifully than the air, 'whence the distance between the balls will be increased, as the fluid surrounding them is augmented.' Finally, in [3], 'the common stock of electrical matter in the tin tube is supposed to be attenuated about the middle, and to be condensed at the ends, by the repelling power of the atmosphere of the excited glass tube.' The residual charge, according to Canton, probably arose from leakage of electrical matter from the ends of the cylinder.

One can raise objections. Where do the atmospheres come from when the wax stick operates on the negatively charged corks? How can the air-cork boundary play the part of the bottom of a Leyden jar? If effluvia leak from the cylinder to the air, ought not *a fortiori* the atmosphere of the tube attach itself to the conductors it bathes? These difficulties are but new instances of the old,

9. This principle was later explicitly condemned as un-Franklinst; cf. T. Ronayne to BF, 22 Oct. 1766 (*BFP*, XIII, 469–70), anent a similar proposition of Wilson.

10. Such phrases in Canton's account as 'according to Mr. Franklin, excited glass *emits* the electric fluid, but excited wax *receives* it' might suggest that Canton located the cause of electric motion in the streaming of the fluid, and not in its accumulation (Home, *Br. J. Hist. Sci.*, 6 [1972], 142–4). These phrases point rather to the direction of flow during the usual process of excitation; thereafter the accumulations leak away in the direction that leads most quickly to re-establishment of equilibrium. Since there is always some leak (insulation being imperfect) there is always streaming where there is electrical motion; but there would still be motion in the ideal case of perfect insulation.

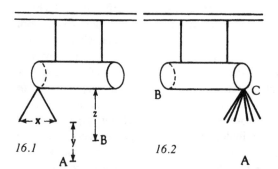

16.1 Canton's induction experiments.

16.2 Franklin's improvements of Canton's inductions.

fatal flaw of effluvialists, the burdening of the atmospheres with the twin roles of the electric matter and its field. Nonetheless Canton's expression of the difficulty proved more suggestive than earlier ones, especially to those who viewed the disparity between his elegant experiments and awkward explanations in the light from Philadelphia. Franklin himself took the first step towards a solution in a paper sent the Royal Society on March 14, 1755.[11]

Franklin's contribution is clarification. First he insists, more emphatically than Canton, on the distinction between induction effects, in which the tube withdraws its atmosphere entire, and those involving communication of electrical matter. For simplicity he uses only the glass tube, with which [1] he electrifies the cylinder, causing the tassel (fig. 16.2) to spread. He rubs the tube again and brings it to B; the tassel diverges further, but closes as much when the tube is withdrawn. The threads fall further with the tube at A, and again assume their original separation at its removal. [2] He discharges the cylinder and places the tube at B: the threads spread. He takes a spark from C: the threads collapse. He withdraws the tube, the threads open; he moves it to A and they diverge further. [3] He again discharges the cylinder and holds the tube in his right hand at A. The tassel spreads and flees the touch of his left hand.

To elucidate these striking effects Franklin introduced three new principles. The first was implicit in Canton's operations: 'Electric atmospheres, which flow round [positively charged] non-electric bodies, being brought near each other, do not readily mix and unite into one atmosphere, but remain separate, and repel one another.' The innovation here is the explicit inviolability of the atmospheres, which gives them a unitary, metaphorical character; although Franklin still tacitly supposes that they cause repulsion in the old manner, by pushing apart the bodies they surround, they have in fact become unmechanical, interacting, touching, perhaps interpenetrating, but not mixing. The second principle, also adumbrated by Canton, extends the analysis of the Leyden jar to the general case of induction: 'An electric atmosphere not only repels another electric atmosphere, but will also repel the electric matter contained in the sub-

11. *EO*, 302–6. Like Canton, Franklin took up these experiments primarily as models of the effect of charged clouds on experimental apparatus; cf. BF to Thomas Ronayne, 20 April 1766, *BFP*, XIII, 248, and *FN*, 516.

stance of a body approaching it; and, without joining or mixing with it, force it to other parts of the body containing it.' Now the 'atmosphere' plays an ambiguous role: it appears to act where it is not. It might by immediate contact drive the native electrical matter from the near to the far end of a long insulated conductor, but another mechanism is needed to confine the springy stuff there. In the first principle, 'atmosphere' signifies accumulated electrical matter; in the second, it signifies a field of force.[12]

The third principle is a *pis aller*: 'Bodies electrified negatively, or deprived of their natural quantity of electricity, repel each other, (or at least appear to do so, by a mutual receding) as well as those electrified positively, or which have electrified atmospheres.' Here, where the atmosphere's first role creates an enigma—the apparent repulsion of bodies lacking the repulsive mechanism—Franklin characteristically capitulates to the facts. He did not, like Canton, insist upon having a mechanical explanation of minus-minus repulsion; he elevated the perplexing behavior into a 'principle' and used it to 'explain' other phenomena.[13]

In [1] the threads spread initially in virtue of the atmosphere (effluvia) of the charged cylinder. They diverge further with the tube at B because the 'atmosphere [effluvia] of the prime conductor is pressed by the atmosphere [field!] of the excited tube, and driven towards the end where the threads are, by which each thread acquires more atmosphere [effluvia].' They close again when, in accordance with the first two principles, the tube withdraws its atmosphere (field) entire. They also close with the tube at A 'because the atmosphere [field] of the glass tube repels their atmospheres. . . .' In [2] the tube's atmosphere (field) 'drives away' the cylinder's native electrical matter, which is 'forced out' upon the threads, endowing them with repulsive atmospheres (effluvia). 'They close on withdrawing the tube, because the tube takes with it *all its own atmosphere,* and the electric matter, which had been drawn out of the substance of the prime conductor, and formed atmospheres round the threads, is thereby permitted to return to its place.' The spark causes the tassel to close by removing a portion of the threads' atmospheres; the cylinder therefore stands negative on the withdrawal of the tube. The tassel spreads in accordance with the inexplicable principle of minus-minus repulsion, which also accounts for the particulars of [3].

Franklin's development of Canton's ideas was the outcome of a struggle to extract clear notions from traditional concepts grown tangled and confused. It so sharpened the problem that Priestley could not conceal his impatience with Franklin for having missed the obvious solution, which 'foreigners' were quick to find. 'Doctor Franklin,' writes his usual admirer, 'put not forth . . . all his strength on this occasion;' otherwise he must have seen that the electrical matter of any body, charged or not, 'is held in close contact with it' and 'acts by

12. Cohen, *FN,* 527–31, goes too quickly in assimilating Franklin's 'atmospheres' to the modern concept of charge. Cf. Hoppe, *Gesch.* (1884), 46–7.

13. Franklin was not unaware of his move; cf. *FN,* 531–3, and *EO,* 365–9.

repulsion upon the electric fluid belonging to other bodies.'[14] The shearing of the atmospheres required a radical change of perspective.

Wilson's flailing with the problem of induction will instance the impasse of the effluvialists. In 1752, the year of Marly, Wilson updated his demonstration of the identity of Newton's aether and the electrical fluid. The novelty is a provision for increasing the amount of aether in the aether-poor regions surrounding ponderable bodies: heat (friction) expands an electric; the expansion draws aether; the accumulation is preserved by pressure from the denser aether of the air. Wilson says nothing about minus-minus repulsion and misses the contrariety of the electrifications on either side of the condenser.[15] After discovering Franklin, he disavowed his interpretation of the Leyden jar and accepted the opposite natures of plus and minus electricity. But he cleaved as stubbornly as Nollet, with whom he sympathized over Franklin's meteoric rise, to his special system, which he attempted to refurbish in 1756 with the help of Benjamin Hoadley, a physician and sometime comic playwright.[16]

Again the guiding idea is the resistance of the aether, which now, however, must preserve two kinds of atmospheres. In one (the positive), the elasticity of the air's aether retards the dissipation of surplus matter; in the other (negative), electrical matter piles up around the body (as in the positive case) because of resistance to the influx of aether striving to supply the body's deficiency. Two positive corks repel each other via the expansiveness of their atmospheres; negative ones do likewise because, according to Wilson, the 'condensed electrical fluid in the air, in order to force itself in at the surfaces of the balls between their two centers, crowds in, and forces them asunder;' oppositely charged corks attract, the outer-directed atmosphere of the one assisting the inner-directed atmosphere of the other.

The poverty of the theory appears when Wilson turns to Canton's experiments. For example, if the cylinder (fig. 16.1) is minus and a glass tube approaches the balls, their separation increases because its extroverted envelope, pressing upon their introverted ones, augments the aether density between them. Now electrify the cylinder plus: the tube causes the balls to converge by driving their extroverted atmospheres back upon them. And all the while an unelectrified object approaches a positively charged body through the push of the aether, supposed denser *beyond* the attracted object than in the atmosphere surrounding the excited body.[17] Despite its absurdity, the system had a following; its allegiance to Newton, which brought it to grief, recommended it to those more concerned to follow authority than to save phenomena.[18]

14. Priestley, *Hist.* (1775³), I, 286–7, 293–4.

15. Wilson, *Treatise* (1752²), 34, 42, 100–6, 135–6, 174–7.

16. Wilson, *PT*, 49 (1755–6), 682–3. Hoadley, a graduate of Corpus Christi, Cambridge, was the son of the combative Bishop of Winchester (*DNB*, IX, 910), and the author of a hit called 'The Suspicious Husband' (1747).

17. Hoadley and Wilson, *Observations* (1756), §§75–93, 271–4.

18. Among them Richard Lovett, 'of the Cathedral Church of Worcester,' *Sir I. Newton's Aether* (1759), panned in *Monthly Review*, 22 (1760), 341–2, which, however, was not unsympathetic to Wilson (*ibid.*, 20 [1759], 302, and 23 [1760], 91–3). A German version of Hoadley and

Franklin, having settled in London as the agent of the Pennsylvania Assembly, was able to deliver his opinion of Wilson and Hoadley directly to its senior author. He objected to the untenable mechanics, the inconsistent models, and the confounding of positive and negative electrification, whose contrariety Wilson's atmospheres fail to express.[19] As he proceeded with his critique,[20] Franklin reduced still further his dependence upon the mechanics of effluvia. He now insisted that the atmospheres of positively charged balls are closely confined to them, 'not only by the surrounding Air, but by a mutual Attraction between each Atmosphere and its Ball.' He spoke freely of negative atmospheres, meaning not material envelopes but *states* in the neighborhood of negatively charged bodies. He could not transfer this insight to the positive case; he had pictured the repulsion of the shot-and-ball in his earliest experiments in terms of atmospheres, and he could not break his habit of thought. Neither could Wilson, who ignored most of Franklin's suggestions in a new edition of his *Observations*, published in 1759.[21]

With this shot the Wilsonians and Franklinists joined battle. Wilson himself attacked the citadel of the Philadelphia system, the impenetrability of glass: he mounted his corks on an insulating conducting stand (as in figure 16.3), rubbed the pane with his finger at A, and determined that the balls rested in a positive atmosphere. *Quid clarius?* The atmosphere aroused at A squeezed through the glass, across the air gap and around the corks![22] This flummery was politely answered by Aepinus, who reminded Wilson that glass does not conduct, and by Franklin himself, in a note in Priestley's *History*, which reaffirmed the original, now obsolescent, theory of the Leyden jar: both sides of the glass are positive, Franklin and Priestley said, because the 'Electric Fluid' rubbed on one side of the glass 'repels an equal quantity of that contain'd in the other Side of the Glass, and drives it out on that Side, where it stands [!] as an Atmosphere.'[23]

Simultaneously Edward Delaval, Fellow of Pembroke College, Cambridge,

---

Wilson appeared in 1763 in the hope, according to the translator, that it would inspire an improvement.

19. Similar objections were raised in unpublished notes on Hoadley and Wilson by the statistician Thomas Bayes; Home, RS, *Not. Rec.*, 29 (1974), 86–9.

20. The marginalia in Franklin's copy of Hoadley and Wilson's *Observations* are given in *BFP*, VIII, 241–63; see esp. pp. 245, 253–4.

21. In 1780 (*Short View*, 22–37) Wilson, still trying to connect aether and electricity, gave out a new theory suggested by electrical experiments on Newton's rings.

22. Wilson, *PT*, 51:1 (1759), 327–30, approved in *Monthly Review*, 23 (1760), 93; cf. Wilson, *PT*, 51:2 (1760), 896–8; 53 (1763), 436–43. Wilson varied this experiment many ways, sometimes vitiating it by mixing up conduction and induction, but occasionally hitting the mark with effects depending upon dielectric polarization. A similar experiment was urged to the same purpose by the ill-informed Irish electrician Henry Eeles (*Phil. Ess.* [1771], xv). In November, 1752, Eeles sent the Royal Society, as a novelty, his conjecture of the identity of lightning and electricity (JB, XXI, 179–80).

23. Aepinus, *Recueil* (1762), 189; Priestly, *Hist.* (1767¹), 424; *Hist.* (1775³), I, 490–1; *BFP*, XIII, 543. Cf. de Herbert, *Th. phen.* (1778²), 122–3, who explains that the far side of the glass shows positive not because the electrical matter has crossed, but because its force has.

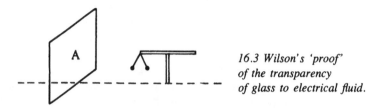

16.3 Wilson's 'proof'
of the transparency
of glass to electrical fluid.

tried to prove Wilson's tenet that heat counted for much in altering the aether density, and hence the electrical 'resistance,' of bodies. He discovered the 'very extraordinary' fact that Portland stone, clay, charred wood and tobacco pipe conducted well when cold or hot, but could act in a Leyden jar when heated moderately. These claims, sent the Royal Society with a note from Wilson testifying to their accuracy, brought an answer from Canton, with a similar note from Franklin. Delaval's experiments, Canton said, contain nothing new: when cold, the tobacco pipes pass electricity along condensed surface moisture and, when very hot, through air made conducting by the heat.[24]

These skirmishes were aggravated by social and political differences between the two groups, Franklin's gravitating about the Club of Honest Whigs, Wilson's levitating towards aristocratic partrons of the arts, including the Delaval family and the Duke of York.[25] The opposition ended in the comical battle of the knobs and points, which began in 1772, when the Board of Ordnance asked the Royal Society how best to protect its powder magazines at Purfleet from lightning. The Society appointed a committee chaired by Henry Cavendish and including Franklin, Watson and Wilson. The majority recommended pointed rods projecting above the roofs of the buildings; Wilson dissented, arguing that tall spikes, which pull electricity at a great distance, could bring down lightning from clouds that might pass harmlessly over short blunt rods.[26] He thereby affirmed the advice he had given in 1764, that the best lightning conductor is a thick iron rod ending in a copper ball placed within the building a foot or two beneath the roof. His reasoning: most clouds will ignore the ball, and should it be hit, it would protect just as well as Franklin's conductors. To be sure, the roof might be damaged. We ought not, however, to expect 'anything like absolute security' in these matters. In this pessimism he agreed with Nollet, who had gone even farther. Common sense and the analogy of nature, he said, warn that *all* conductors are dangerous during thunder storms; for maximum protection run from them, squat on a pedestal of wax, cover yourself with resin, or hang yourself in a glass box. But why make yourself ridiculous? Full protection

24. Delaval, *PT*, 51:1 (1759), 83–8, and 52 (1761), 353–6; Canton, *PT*, 52 (1762), 457–61. Delaval almost certainly did not discover that substances besides glass and porcelain could serve in a Leyden jar. For another dimension to the dispute see Priestley, *Hist.* (1775[3]), I, 277–86.

25. *DNB*, V, 766–7; Crane, *Will. Mary Q.*, 23 (1966), 210–33; Priestley to BF, 25 March and 13 April 1766, *BFP*, XIII, 201, 245.

26. Cavendish *et al.*, *PT*, 63:1 (1773), 42–7; Wilson, *ibid.*, 48, 57–8; Tunbridge, RS, *Not. Rec.*, 28 (1974), 207–19; Brunet, *RHS*, 1 (1948), 233–40.

16.4 Wilson's demonstration of the superiority of blunt lightning rods: the model of the Purfleet arsenal is at the right, the electrostatic generator in the center background. From Wilson, PT, 68:1 (1778), 245–313.

will always elude you. 'A strong electricity breaks through everything we can oppose to it, and, unfortunately, thunder is the greatest of electricities.'[27]

The Board of Ordnance armed Purfleet with points. In 1777, to Wilson's delight, the magazine suffered damage from lightning. Another committee, another dissent: the majority blamed poor grounding, Wilson blamed the points. The matter was serious, not only for the protection of powder kegs but even more for the preservation of 'a house, which is of the first consequence in this kingdom, that hath pointed conductors also fixed upon it: I mean the KING'S.'[28] The king, his vital interests doubly engaged, provided the means for Wilson to set up an 'artificial cloud,' or prime conductor, 155 feet long and 16 inches in diameter, in the great hall of the London Pantheon (fig. 16.4). An exact model of the Purfleet arsenal, armed with sharp or blunt conductors raised to diverse heights, was drawn on rails beneath the prime conductor to simulate the relative motion of clouds and buildings. As expected, the high points so-

27. Wilson, *PT,* 54 (1764), 249–51; *PT,* 63:1 (1773), 50, 55; Nollet, *MAS* (1764), 445; Cohen, *J. Frank. Inst.,* 253 (1952), 397–8, 439. Wilson's reasoning of 1764 was attacked by the orthodox Franklinist William Henley (*infra,* xviii.2), *PT,* 64:1 (1774), 133–52, who offered experiments to show that 'a sharp point will draw off a charge of electricity silently, at a much greater distance than a knob . . .; and this, I imagine, completely decides the question' (*ibid.,* 152).

28. Henley *et al.,* *PT,* 68:1 (1778), 236–8; Wilson, *ibid.,* 239–42, 247.

licited the electrical matter from a greater distance than the low knobs. The Franklinists replied that Wilson had misinterpreted his results, which showed rather that pointed rods can silently despoil clouds of their lightning; should a pregnant cloud come close enough to discharge into a knob, it would do so violently. In the case of Purfleet, according to the Franklinist instrument maker Edward Nairne, very little damage was done because the point had almost denuded the cloud before it struck. Had the rod been blunt, it—or the building it was designed to protect—would have received the entire destructive blast when the cloud came into range.[29]

The king saw Wilson's experiments and understood. 'They are so plain,' he is said to have said, 'they would convince the apple-women in the street.' He ordered Franklin's points torn down from Buckingham-House and Wilson's knobs installed instead. When Franklin learned about the charge he was in Paris, promoting revolution. 'The king's changing his pointed conductors for blunt ones is a matter of small importance to me. If I had a wish about it, it would be that he had rejected them altogether.' The idea got abroad that the shape of lightning rods was a matter of politics.[30] It was said that Franklin's crony Pringle resigned the presidency of the Royal Society owing to disputes about the rods. The rumor was embroidered into a fine story. As it reads in Cuvier's éloge of Pringle's successor, Joseph Banks, the king told Pringle to instruct the Fellows that lightning rods would henceforth end in knobs. 'The prerogatives of the president of the Royal Society do not extend to altering the laws of nature,' Pringle replied, and forthwith resigned.[31] 'In any other country, or at any other time [Cuvier added], Wilson would have been a laughing stock, the business was so clear.'

When Cuvier wrote, in 1821, all electricians favored points, the more the better. Their conviction found its finest expression in the 426 spikes, spears and bayonets planted on the Hôtel de Ville in Brussels in 1865 (fig. 16.5).[32] And yet it appears that sharp points have very little advantage over blunt ones raised to the same height. On nature's scale both appear to be points. As high-speed cameras have shown, lightning strikes after a tentative 'leader,' advancing step by step, has established the discharge path; and the leader, in picking the path, cannot distinguish between a point and a knob until so close an approach that it usually does not matter. A sharp conductor may afford a slightly greater radius

29. *Ibid.*, 245–310; cf. Wilson's ally Musgrave, *PT,* 68:2 (1778), 801–22.

30. Pringle *et al.*, *PT,* 68:1 (1778), 313–17; Nairne, *PT,* 68:2 (1778), 825, 835, 859–60.

31. R. T. Wilson, *Life* (1862), I, 35–8; BF to ?, 14 Oct. 1777, in Franklin, *Works* (1840), VII, 227–8; D. Garrick to Countess Spencer, 14 Sept. 1777, in Garrick, *Letters* (1963), III, 1189. Cf. Turgot's famous epigram (of 1778?), 'Eripuit caelo fulmen mox sceptrum tyrannis'; van Doren, *Franklin* (1938), 606; Cuvier, *HAS,* 5 (1821–2), 220–1. Kippis in Pringle, *Six Discourses* (1738), lvi–lvii, says that Pringle resigned because of ill-health, not because of the rods. Cuvier's tale has often been retold, always as if true, e.g., by Viemeister, *Light. Book* (1972), 196; R. Anderson, *Light. Cond.*, (1885³), 37–40; Weld, *Hist.* (1848), II, 64, 101–2; and Stearns, *Science* (1970), 634–5.

32. Melsens, *Des parat.* (1877), 114–42; Wiesinger, *Blitzforschung* (1972), 8–10; R. Anderson, *Light. Cond.* (1885³), 151–7, 178–80.

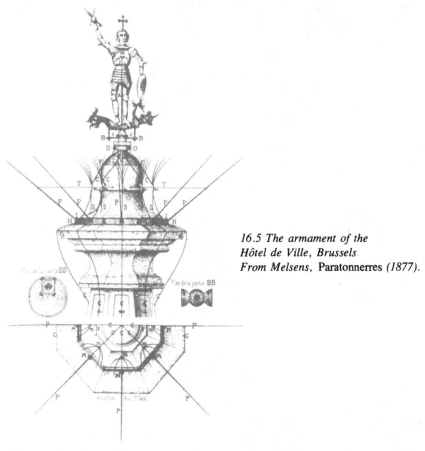

16.5 *The armament of the*
Hôtel de Ville, Brussels
From Melsens, Paratonnerres *(1877).*

of protection than a blunt one at the same elevation because the corona discharge at its point extends it effective height somewhat. Lightning specialists usually do not exploit this marginal advantage in their protective installations.[33] Wilson was correct in asserting that elevated pointed rods do not discharge distant clouds, and his adversaries were right in insisting that protection increases with the height of the conductor. Both parties erred in believing that the paltry discharges arranged in their laboratories aped the grand processes of nature.

To return to the 1760s. After some excellent work on the tourmaline, Wilson retired to his paints and the defense of his obtuse rods.[34] Franklin, surrendered to political and business affairs, promoted a translation of Beccaria's retrograde

33. Viemeister, *Light. Book* (1972), 110–23, 197–8; Schonland, RS, *Proc.*, 235A (1956), 440–2; Wiesinger, *Blitzforschung* (1972), 13–40.

34. Wilson thought, quite rightly, that his last papers on the tourmaline (*PT*, 51:1 [1759], 308–39, and *PT*, 53 [1763], 436–66) represented his best work; it won him a medal from the Royal Society and the reputation, according to a knowledgeable Swedish visitor, of being the best English electrician active in 1760. Cf. T. Bergman, *PT*, 51:2 (1760), 907–9; Wilson to Bergman, 23 March 1761,

ideas; Canton turned to other matters; and the field was left to the instrument makers, the textbook writers, and the historians.[35] 'Could but Dr. Franklin, Dr. Watson and yourself [so Priestley wrote Canton in 1766] be once more seriously engaged in these pursuits, something very considerable, I am confident, might yet be done in electricity.' Two years later, in a letter to Tobern Bergman, Wilson also admitted collapse: 'I wish you would spirit us up a little, for at present we are rather languid in our electrical researches.'[36] The initiative had returned to the Continent.

2. WILCKE AND AEPINUS

The effluvial model, reeling from Franklin's blows, received the coup de grâce from an unexpected quarter. The executioners—Johan Carl Wilcke and Franz Ulrich Theodor Aepinus—came from Baltic lands as remote as Philadelphia from the capitals of Europe. Free from the partisanship afflicting the study of electricity in France, Italy and England, and in full possession of the polyglot literature, they dispassionately examined the claims of Nollet and Franklin, expecting neither system to be altogether satisfactory. They were not disappointed. Although they found Franklin's scheme unquestionably superior, they recognized the occasional pertinence of Nollet's objections, which they accommodated in a manner that pleased neither party.

THE CONTRARY ELECTRICITIES

Like many Swedish savants of the eighteenth century, including his professors at the University of Uppsala, Samuel Klingenstierna and Mårten Strömer, Wilcke came from a clerical family. His father, Samuel Wilcke, the son of a shoemaker, had trained for the ministry at the University of Rostock. He supported himself by tutoring the children of his professor of theology, Franz Albert Aepinus. In 1739 he was called to minister to the German-speaking community in Stockholm, where he spent the remainder of his life.[37]

Wilcke received his secondary education at the German school associated with his father's church. In 1750 he entered the University of Uppsala to prepare for the ministry. It was not theology, however, but mathematics and physics that aroused his interest; he was led astray, it seems, by the rationalistic theology and mathematical methods of Wolff. These dangerous tendencies were then

---

in *Bergman's For. Corr.* (1965), I, 413–16; Fermer, *Resa* (1956), xliii; Home, *Isis*, 67 (1976), 21–30.

35. E.g., to Henley and Cavallo (*infra*, xvii.3 and xviii.2) and to Priestley, whose original work in electricity consisted of exploding wires by the Leyden discharge (Priestley, *Hist.* [1775³], II, 215–357, 362–4) and in determining relative conductivities (*ibid.*, 201–14, 368–71). Cf. R. E. Schofield, *AIHS*, 16 (1963), 277–86.

36. Priestley to Canton, 18 April 1766, Canton Papers, II, 66 (RS); Wilson to Bergman, 6 Sept. 1768, *Bergman's For. Corr.* (1965), I, 428. Cf. Priestley to BF, 25 March 1766, *BFP*, XIII, 200: 'That [Franklin's return to the study of electricity] seems to be universally acknowledged to be the great desideratum to further discoveries.'

37. Oseen, *Wilcke* (1939), 3–8.

represented by the physicists Klingenstierna and Strömer, both men of ability and experience.[38] Klingenstierna, the professor of physics, had travelled widely in Europe, and was esteemed a good mathematician by competent judges; in Uppsala he labored to collect an apparatus for experimental physics, which he taught according to the books of 'sGravesande and Musschenbroek. Strömer, himself a student of Klingenstierna's, succeeded Anders Celsius in 1745 as professor of astronomy; an excellent teacher, he was 'father and adviser [according to Wilcke] of all the young geniuses' of Uppsala.[39]

Klingenstierna was one of the first in Northern Europe to appreciate the work of Gray and Dufay.[40] Later the German flare caught his fancy. He applied to Euler, then at the Berlin Academy, for particulars. 'If you would be kind enough to tell me something about this business of electricity, which I hear has begun to make great strides with you, namely in the invention of a way of igniting various combustibles, you would do me a very great favor.' After consulting Bose too, Klingenstierna designed some impressive fireworks.[41] Then, in collaboration with Strömer, he made an important discovery: jars charged through the tail wire did not behave like those charged by the hook; insulated jars connected in parallel discharged if dissimilarly, but not if similarly, electrified. 'We consider it very remarkable [they wrote in 1746] that if two phials are filled with electricity in the two ways we have described, their electricities are of different types . . .; Dufay has found that glass and resin give unequal electricities; but no one yet has noticed that glass alone gives different kinds as well.' Other duties and Klingenstierna's constitutional inability to complete a project slowed their investigation of this duplicity, which still occupied them during Wilcke's student days. And so, at the very beginning of his college career, he became acquainted with the problem of the 'contrary electricities.'[42]

Hoping, perhaps, to save his son from science, Samuel Wilcke agreed to Johan Carl's wish to study at the University of Rostock. Samuel counted on his wholesome former pupil, A. I. D. Aepinus, now risen to the Rostock chair of oratory, to urge the merits of the ministerial life. The scheme backfired. At the home of Aepinus, where Wilcke boarded, also lived the rhetor's younger brother Franz, who, having rejected the family's plan to make him a physi-

38. Hildebrandsson, *Klingenstierna* (1919), 12, 17; Frängsmyr, *Hist. Sci.*, 12 (1974), 39, and *Wolff. genemb.* (1972), 82–7, 188–204, 221–5.

39. Hildebrandsson, *Klingenstierna* (1919), 9–16, 27–32; Cramer to Stirling, 20 June 1729, in Tweedie, *Stirling* (1922), 107; Klingenstierna to Stirling, 19 Sept. 1738, *ibid.*, 164–5, asking for help in procuring optical instruments, 'nullibi terrarum meliora quam Londini conficiantur'; Nordenmark, *Strömer* (1944), 7, 10, 72–3.

40. *Supra*, x. 1.

41. Klingenstierna to Euler, 31 Jan. 1745 (OS), in Zubov in Winter, ed., *Die deut.-russ. Beg.* (1958), 36; Ferrner, *Resa* (1956), xxxix.

42. Oseen, *Wilcke* (1939), 9–15; B. Hildebrand, *K. sven. vet. ak.* (1939), 283, 498–500; Strömer, *AKSA*, 9 (1747), 154–7. Klingenstierna later brought Franklin's work to the attention of the Stockholm Academy, and endorsed its chief conclusions, including the impenetrability of glass and the explanation of the feather in the bottle. *Tal*, (1755), 4, 9, 12, 15, 17.

cian,[43] now taught mathematics at the University for the pitiful salary of 100 RT a year. No more than his brother's oratory did Franz's poverty—a timely warning of the difficulties of a mathematical career—stop Wilcke's drift from theology. In October, 1753, when he matriculated at Göttingen to follow the lectures of J. A. Segner and Tobias Mayer, he no longer inscribed himself 'theologus,' as at Rostock, but 'mathematicus.' Two years later A. I. D. Aepinus brought Samuel Wilcke to acquiesce in the *fait accompli,* made more palatable, perhaps, by the growing success of Franz, who in the spring of 1755 became the astronomer of the Berlin Academy of Sciences at a salary of 400 or 500 RT.[44]

A few months later Wilcke joined Aepinus in Berlin, where both were drawn into a circle of savants centered on Leonhard Euler and his son Johann Albrecht. Several of them, including the Eulers, were then studying electricity.[45] Wilcke and Aepinus directed their efforts at the old problem of the contrary electricities. Were Dufay's two electrifications different only in degree, as Nollet insisted, or in kind, as the Franklinists contended? Aepinus initially inclined towards Nollet, while Wilcke remained uncommitted, predisposed towards Franklin by the old experiments of Klingenstierna and Strömer, but arrested by Nollet's 'apparently unanswerable' demonstrations of the permeability of glass.[46]

To resolve these uncertainties Wilcke repeated all the experiments urged on either side. He filled his rooms, chosen for their height, with the apparatus of atmospheric electricity; he prepared a translation of Franklin's letters, enriched with notes adjudicating the claims of Paris and Philadelphia; and he found that Nollet's objections usually rested on misinterpretations of obscure, imprecise or abbreviated passages in Franklin's work. The theory of the double flux proved useless or misleading; that of the contrary electricities easily covered new situations. As for Aepinus, he became a Franklinist when an experiment he designed to confirm Nollet failed. As he interpreted the theory, resinous electricity was merely a less powerful form of vitreous; hence diverging threads attached to an insulated conductor weakly charged by a glass tube should spread still further were a wax rod touched to them. But in the trial, which Aepinus made with

43. Koppe, *Jetzl. gel. Meckl.* (1783–4), I, 9–15. Franz Aepinus had studied physics and medicine with G. E. Hamberger at Jena in 1744/5.

44. Oseen, *Wilcke* (1939), 18–35; Pupke, *Naturw.,* 37 (1950), 49–52; Dorfman in Aepinus, *Opyt* (1951), 461–2; L. Euler, *Berl. Petersb. Ak.,* I (1959), 119–20. Although Aepinus had a local reputation as an astronomer (Home, *Slav. East Eur. Rev.,* 51 [1973], 77n), he was appointed at Berlin faute de mieux, after Maupertuis had failed to capture Mayer or an astronomer from France, 'where the best are' (Brunet, *Maupertuis* [1929], I, 164; *supra,* ii.2).

45. Home, *Isis,* 67 (1976), 22–3. The Eulers had sent off an essay on electricity for a prize offered by the Petersburg Academy (*infra,* xvi.3) shortly before Wilcke and Aepinus came to Berlin. L. Euler to G. F. Müller, 20/31 Dec. 1754 and 26 Sept. 1755, in *Berl. Petersb. Ak.,* I (1959), 72, 92.

46. Aepinus, *Recueil* (1762), 134; Wilcke, *EO* (Wilcke), Vor., 388–9. The demonstrations that bothered Wilcke involved 'streams of electric matter' (sparks) apparently forced into an evacuated bottle hermetically sealed to the prime conductor (*supra,* xv.2).

Wilcke's help, the threads fell, and did so whatever the order of application of the glass and wax. Who could doubt the radical opposition of positive and negative electricity, and the appropriateness of Franklin's terminology?[47]

Not that Franklin's theory was unexceptionable. Wilcke recognized that it contained both fundamental flaws and unimportant errors, distinctions too rigidly drawn or expressed. He showed that absolute insulation did not exist, that any electric could serve for the Leyden experiment, that the charges on the two coatings of the jar are not quite equal, and that substances do not come innately vitreous or resinous.[48] This last point Wilcke owed to Canton, who had found that roughened (unpolished) glass might be made minus or plus at will, by rubbing with flannel and oiled silk, respectively. Canton probably came to this discovery by following up problems of atmospheric electricity: since clouds charge either way by friction, the manner of rubbing as well as the nature of the body rubbed must determine the sign of the electricity conferred.[49] Recognizing that friction set up competition for electrical matter, Wilcke drew up a winners' list, the competitors being so placed that a given one becomes positive (or negative) when rubbed by those placed beneath (or above) it. His sequence, the first triboelectric series, was smooth glass, wool, quills, wood, paper, sealing wax, white wax, rough glass, lead, sulphur, and metals other than lead.[50]

Regarding fundamental Franklinist flaws, Wilcke recognized the old difficulty of minus-minus repulsion and a new one he thought weighty, sparking between unequally negatively electrified bodies. He was unable fully to identify the culprit, the quasi-literal atmospheres; but the information he supplied his friend enabled Aepinus to condemn them.

THE AIR CONDENSER

Sometime in 1756 a fellow academician, the mineralogist Johann Gottlob Lehmann, told Aepinus about the power of a heated tourmaline to attract and repel light bodies. Aepinus decided that the effect was electrical. He found that

47. *EO* (Wilcke), 280–6, 348; Aepinus, *Recueil* (1762), 134–7; Oseen, *Wilcke* (1939), 46, 53–5; Aepinus, *MAS*/Ber (1756), 109. Similar doubts were similarly resolved at Geneva under Jallabert (Deluc, *Traité* [1804], I, 19–20, 36–7) and De Saussure (Senebier, *Mémoire* [an IX], 38–9, and *infra*, xviii).

48. *EO* (Wilcke), 219–21, 271–2, 290, 308–9; Wilcke, *Disputatio* (1757), 59–60, 81–93. Cf. Wilcke, *AKSA*, 20 (1758), 241–68 (on non-glass condensers) and Oseen, *Wilcke* (1939), 57–9 (unpublished memoir on relative conductivities). Franklin too had come to doubt the uniqueness of glass—he suggested making a Franklin square of a thin, dry board in a letter to Kinnersley of 28 July 1759 (*BFP*, VIII, 417)—and he had long since given up his theory of the structure of glass (*BFP*, V, 522–3 [*EO*, 333], XI, 254, and XIII, 249–50).

49. *EO*, 298; Canton, *PT*, 48:2 (1754), 780–4; Priestley, *Hist.* (1775³), I, 259–63. Canton gives no rationale for trying stripped glass in which, however, we might see a further analogy to the formative stages of clouds. It was later found by de Herbert, *Th. phen.* (1778²), 163–4, that pieces of the same material but of different temperatures would charge when rubbed together.

50. Wilcke, *Disputatio* (1757), 44–64; Oseen, *Wilcke* (1939), 48; Priestley, *Hist.* (1775³), I, 273–6. Later Wilson, perhaps independently, made a very short series (diamond, tourmaline, glass,

the stone had 'poles,' which became oppositely charged by heat much as soft iron is magnetized by a lodestone; and that, when held in the hand and rubbed, it charged positively on one side and negatively on the other. He concluded that he dealt with a miniature Leyden jar: the positive charge rubbed on one face of the thin stone drove electrical matter from the opposite face through the hand to ground. An obvious corollary was that any gem and, indeed, any electric whatsoever, including air, could serve in a condenser.[51]

At this point Wilcke consulted Aepinus about problems he had encountered in interpreting Franklin's version of Canton's induction experiments. Apparently the difficulty related to the action of the atmospheres. In one brilliant, enduring insight Aepinus saw that the whole business—the insulated tin cylinder, the approaching glass tube, the spreading tassel—was nothing but an imperfect Leyden jar. The tube corresponded to the interior coating, the cylinder to the external, and—here is the breakthrough—the air gap to the glass; were the cylinder grounded the analogy would be complete. Wilcke could not at first bring himself to accept so novel a doctrine. To test it he and Aepinus connected an iron tube to the prime conductor, placed a grounded tin cylinder below and parallel to it, and started their machine. Aepinus grasped the iron in one hand and expectantly extended the other towards the cylinder. The blast he courted did not come. They tried again using foil-covered boards a foot and a half square: only the weakest titillation. Finally, from two metal-covered frames each 56 feet square, Aepinus received a stroke comparable to that of a well-charged bottle.[52]

The air condenser made it impossible to maintain the theory of literal atmospheres. While the repulsive force of the upper plate certainly reached the lower, its redundant electrical matter as certainly did not: for in that case the condenser, being short-circuited internally, could not have charged. To the true Franklinist the experiment posed the additional conundrum of the location of charge; for if, as Franklin taught, the charge of the Leyden jar lies in the glass, where should one look for the electricity of the air condenser?[53]

The only recourse, according to Aepinus, is to banish the effluvia, admit that similarly charged bodies repel, and dissimilarly charged bodies attract one another, and not bother about the mechanism of the forces. Indeed, he continued, Franklinist theory must end in agnosticism. For it taught that ordinary

amber) and misguessed that the triboelectric order always followed that of hardness (*PT*, 51:1 [1759], 330–7). Cf. Canton, *PT*, 52 (1761–2), 457–61; Bergman, *PT*, 54 (1764), 86–7; Beccaria, *PT*, 56 (1766), 105–18; Henley, *PT*, 67 (1777), 122–7; Hoppe, *Gesch.* (1884), 58–61.

51. Aepinus, *MAS*/Ber (1756), 106–7, 117–18; *Recueil* (1762), 50–1; *Tentamen* (1759), 2; *GM*, 28 (1758), 617–19. Wilcke, *AKSA*, 28 (1766), 99–100, says that Lehmann recommended the stone to the attention of electricians.

52. Aepinus, *MAS*/Ber (1756), 120, and *Tentamen* (1756), 75–83; *EO* (Wilcke), 306–9; Oseen, *Wilcke* (1939), 51–2. Wilcke, *Disputatio* (1757), 93, suggests that his comparative experiments on the triboelectric series helped Aepinus come to the generalization that included the air condenser. As will appear (*infra*, xvii.1) Volta later reached similar conclusions, perhaps independently.

53. See the 'problème sur l'électricité,' *JP*, 9 (1777), 6.

matter attracts the particles of the fluid, which arè mutually repulsive; whence it followed, in order to explain minus-minus replusion and even the possibility of normal, force-free distributions, that the particles of 'matter' must repel one another too. Physicists might well be 'horrified' at this last proposition, because of its apparent violation of Newton's teachings; it troubled Aepinus, so he says, until he considered that, since forces are merely descriptions, nothing prevented him from ascribing both gravitational attraction and electrical repulsion to the same particles.

Agnosticism is not ignorance. Although Aepinus declined to speculate about the causes of his forces, he knew enough to exclude one possibility: they could not be mediated by the electrical fluid itself, supposed extended out from charged bodies in the form of atmospheres. No such things exist, he insisted, and at best denote merely the 'sphere of activity' of a charged body, or the space in which it exercises attraction and repulsion. Only in the case of sparks does electrical matter jump from one body to another across air gaps. As for the phenomena used to establish the effluvia—the smell, the spider webs, the electric wind—they arise from the action of electric forces on our bodies and on the air.[54]

Wilcke was too much an experimental philosopher to embrace his friend's teachings altogether. The compromise he reached deserves attention. On the one hand, he tentatively accepted the most bizarre of the new postulates, the mutual repulsiveness of matter particles, in order to conquer the engima of minus-minus repulsion. On the other hand, he ascribed plus-plus repulsion to the 'pressure' of atmospheres, usually meant literally: contiguous atmospheres cause repulsion only if they have 'room to expand;' they can be 'driven' from bodies and 'conducted' away; they 'fill' tubes positively electrified, preventing further accumulation by their bodily presence.

The disparity between the positive and negative cases, as elucidated by Wilcke, appears clearly in his formulation of the general law of induction with which, after consulting Aepinus, he undertook to explain the experiments of Canton and Franklin. 'The portions of an object immersed in the atmosphere of a positive body became negative; in a negative atmosphere they become positive.' Positive envelopes, according to Wilcke, belong to bodies, surround bodies, are body, while negative atmospheres are free of body, spheres of activity, spaces distorted by the presence of a deficient object.[55] To jettison the effluvia required a mind that prized consistency above the evidence of sense and formal rules above the traditions of seven generations of electricians. Wilcke was not such a mind, despite his mathematics. Once free from the direct influence of Aepinus, he retrogressed towards effluvia and a literalism that brought him to advocate the anti-Franklinist theory of two electrical fluids.[56]

54. Aepinus, *MAS*/Ber (1756), 108; *Tentamen* (1759), 5–10, 35–40, 257–9.
55. *EO* (Wilcke), 221–4, 233–6, 262–3, 270–1, 307, 340–1.
56. Wilcke, *AKSA*, 25 (1763), 207–26; *infra*, xviii.2.

In June of 1756 G. F. Müller of the Petersburg Academy wrote Euler enquiring whether Aepinus might consider an appointment there. 'Mr Aepinus is not only a good astronomer . . . but also a very good physicist; however, since he boards with me and consequently hears me complain every day about the Academy's treatment of me [it owed Euler money], Your Excellency can guess the sort of impression your proposition made on him. True, he gets only 400 reichsthaler here, but he lives well enough and his fate does not depend on a chancery secretary.' But Aepinus decided to leave Berlin, perhaps in part because of the rapidly deteriorating political situation, which soon brought the Seven Years' War. After several months and Euler's intervention he secured permission to leave the Academy and Prussia. In May, 1757, he arrived in St. Petersburg, where he remained until retirement.[57]

Wilcke also received a Russian offer. He decided first to try his luck in Sweden. Only one of the four Swedish universities, Uppsala, had a chair of physics; and for that Wilcke had no hope, for a successor to Klingenstierna had only just been appointed. Nor did the Swedish Academy of Sciences in Stockholm seem more promising. 'What will [young Wilcke] do here, since he can't preach?' A clever maneuver of Klingenstierna and the Academy's secretary, P. W. Wargentin, kept Wilcke from St. Petersburg. A wealthy merchant, Sebastian Tham, had left a large sum in 1727 for the establishment of a 'collegium illustre' for the sons of the nobility. The college had not materialized and the Academy secured part of the income to support its secretary. In 1759 Wargentin suggested applying the remainder to maintain a lecturer in experimental physics; Klingenstierna nominated Wilcke, who accepted, at the starvation wage of 2000 Dkmt (700#) a year. With room and board provided by a fellow academician, whose children he tutored, he had just enough to live on. Not until 1777, when his salary rose to 2000#, did he believe that he could support a wife, who gave him no children to increase his expenses. Not until 1784, when he became secretary of the Academy, did his financial difficulties end.[58] Some of Samuel Wilcke's misgivings about his son's career had been well taken.

3. ELECTRICITY IN ST. PETERSBURG

Aepinus' material success exceeded his friend's. He came to a well-paid academic job, added other lucrative teaching chores, instructed the imperial corps of cadets, tutored the crown prince, became a knight, an educational reformer, a diplomat, a courtier and a privy councillor. By the late 1760s his multiplying benefices had effectively ended his scientific career. That was doubly unfortunate. For one, his preferment exacerbated tensions between the native Russian savants, like Lomonosov, and the Western academicians, with

57. L. Euler, *Berl. Petersb. Ak.*, I (1959), 116–20, 128, 140, 143; Winter, ed., *Registres* (1957), 227.

58. Oseen, *Wilcke* (1939), 56, 62–6, 165, 181, 192–3, 258; *DSB*, XIV, 352–6; Hildebrand, *K. sven. vet. ak.* (1939), 169–71, 449–50, 596–7. Cf. Lindroth, *Nord. tids.*, 43 (1967), 24–48.

whom Aepinus associated. For another, his apostasy ended the promising advance towards a quantified experimental physics begun by his predecessor, G. W. Richmann, and continued briefly and brilliantly by himself.[59]

After study at the universities of Halle and Jena, Richmann went to St. Petersburg in the early 1730s as tutor to the children of the then powerful Count Osterman. He entered the new imperial academy as 'adjunct' in 1735; ten years later he succeeded G. W. Krafft as its professor of physics and director of its physical cabinet. At about the same time, 1745, he began to experiment with electricity, with the encouragement of the Empress Elizabeth and the guidance of Doppelmayr's *Neu-entdeckte Phaenomena*.

Richmann, whose forte was accurate measurement, immediately felt the need for an electrometer. He first took as measure of electrical force the weight required to equilibrate a balance one pan of which hung over the electrified object. But the device was unworkable, and he changed to the angle formed between a vertical rule attached to an electrified body and a thread suspended from the top of the rule (cf. fig. 16.6). With this 'index,' as he called it, he measured the rate of loss of charge of an insulated prime conductor as a function of humidity. He also determined the magnitude of its loss when it gave sparks to nonelectrics of various sizes, placed at different distances; the concept of electrical capacity emerges clearly from these experiments, which appear to be the earliest of their kind.[60] The same sorts of measurements, and the same set of concepts, recur in Richmann's tests of Newton's law of cooling and his establishment of the calorimetric mixing formulae. Indeed, he became a martyr to his mania. Perceiving the danger of the study of lightning, he none the less persevered, 'it being my business to enquire into nature so far as I am able, and to neglect no occasion not only for observing, but also for measuring the phenomena of natural electricity.'[61] It was while bending over to read the electrometer he had fixed to his insulated atmospheric probe that he received the stroke that ended his life.

Six months after Richmann's death, and perhaps as an honor to him, the Petersburg Academy announced 'the true cause of electricity' as the subject of its prize competition for 1755.[62] The announcement called attention to similarities in the phenomena of heat and electricity, and cautioned competitors to distinguish among 'progressive, rotational and vibratory' motions of the electrical matter. The peculiar phrasing of the question immediately identifies its author: Mikhail Vasil'evich Lomonosov, onetime student of Wolff's, corre-

59. Morozov, *Lomonossow* (1954), 434–6, 441–3; Home, *Slav. East Eur. Rev.*, 51 (1973), 76–94. Müller began to complain to Euler of Aepinus' neglect of his duties as early as June, 1760; L. Euler, *Berl. Petersb. Ak.*, I (1959), 151. Cf. the exchange during the summer of 1761 (*ibid.*, 175–7), when Euler heard that his protégé had taken on the cadets.

60. *ADB*, XXVIII, 442–4; *DSB*, XI, 432–4; Boss, *Newton* (1972), 154; Richmann, *CAS*, 14 (1744–6), 299–324; Sotin, Ak. nauk, Inst. ist. est. i tekh., *Trudy*, 44 (1962), 1–42.

61. Richmann, *NCAS*, 4 (1752–3), 335; cf. *supra*, i.7.

62. Dorfman in Aepinus, *Opyt* (1951), 472–83; Zubov in Winter, ed., *Deut.-russ. Beg.* (1958), 32–43; 'Programma in conventu solemni academiae' in J. A. Euler *et al.*, *Dissertationes* (1757).

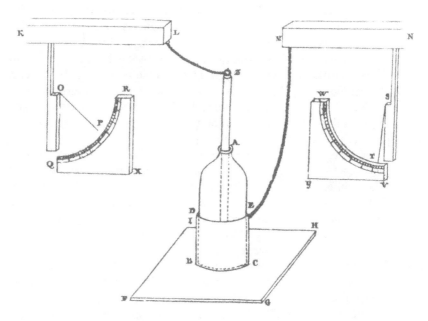

16.6 Richmann's indices QPR and WTV arranged to measure the electricity of the coatings of a Leyden jar ('Richmann's experiment'). From Richmann, NCAS, 4 (1752–3), *301–40.*

spondent of Euler's, and enemy of almost all the foreign-born Petersburg academicians. An exception was Richmann, whose death Lomonosov witnessed, and whose family he tried to help.[63] The two men worked together and had much the same understanding of the nature of electricity. They planned a joint lecture to the Academy, in which Richmann would 'bring forward his work and I [Lomonosov] would set out its theory and practical application.'[64]

This theory was not advanced for 1753. Richmann took electricity to be an 'agitation' of a 'certain electrical matter that surrounds electrified bodies to a certain distance;' he could not accept qualitatively contrary electricities, as the only distinction available on his theory was direction of motion. No more could Lomonosov, who refused to use distance forces of any kind,[65] handle or even represent Franklin's theory. According to notes he made in 1756, Lomonosov associated electricity with the motion of the aether, the same aether that carries light and heat; and according to a theory that he held as early as 1745, heat

63. Cf. Lomonosov to I. I. Shuvalov, 26 July 1753, asking that Richmann's salary of 860 rubles be continued to his widow; Lomonosov, *Ausg. Schr.* (1961), II, 194–5.

64. Lomonosov to Shuvalov, 31 May 1753, *ibid.*, II, 190–1. Cf. Kusnetzov, Ak. nauk, Inst. ist. est. i tekh., *Trudy,* 19 (1957), 333–44; Ak. nauk, *Protokoly,* II, 283.

65. Richmann, *NCAS,* 4 (1752–3), 305, 323–4; Lomonosov to Euler, 5 July 1748, *Ausg. Schr.* (1961), I, 140–1, reworked into a paper in 1758. Lomonosov, *Poln. sobr. soch.*, III, 360–2; Leicester, *Lomonosov* (1970), 228–9.

arises from a rotation and light from a vibration of the aether particles.[66] Since but one other motion is possible, progressive or translational, one might expect that Lomonosov assigned it to electricity; he delighted in such deductions, which earlier had brought him the knowledge that aether particles are spherical, contiguous and rough.[67] The competitors for the prize failed him. Despite their intention, they apparently convinced him that the motion of electricity is not progressive, and there left him.[68]

The winner of the Petersburg prize was J. A. Euler, who developed ideas of his father, who admired the electrical theories of Lomonosov.[69] The Eulers, who rejected the usual emission theory of light, placed the seat of electricity in their luminiferous aether. According to them, electrification consists in a local deficiency in this medium; bodies remain charged as long as a compensating aether current sets in towards them; the narrowness of a body's pores fully determines its electrical properties; only the water in the Leyden jar is electrified; etc., etc. They do not notice Franklin or the contrary electricities. The Euler genius appears only in the application of Bernoulli's principle, associating moving fluids with a drop in pressure, to the elucidation of ACR.[70] Rubbing the tube deprives it of electrical matter; aether streams in from a nearby bit; the aether pressure drops between them, and that in the air beyond effects the union. During contact the bit becomes the entry port for aether flowing from the air, the pressure falls beyond the bit, the mechanism inverts and repulsion ensues.

In the year that Johann Albrecht won his prize, Aepinus and Wilcke joined the Eulers' group in Berlin. They soon learned about the contrary electricities, and patched their hypothesis to suit.[71] They now allowed electricity to consist of any imbalance in the aether, accumulation corresponding to one, and deficiency to the other, of Franklin's opposed states.[72] The approach of oppositely

---

66. Lomonosov, *Poln. sobr. soch.*, III, 284, 302; *Ausg. Schr.* (1961), I, 318, 325; Leicester, *Lomonosov* (1970), 22. By associating the electrification of clouds with motions in the aether Lomonosov constructed an argument in favor of bell ringing in thunderstorms. *Poln. sobr. soch.*, III, 78; *Ausg. Schr.* (1961), I, 275.

67. *Poln. sobr. soch.*, III, 304–8; *Ausg. Schr.* (1961), I, 326–7.

68. *Poln. sobr. soch.*, III, 240–2; *Ausg. Schr.* (1961), I, 305–7.

69. E.g., L. Euler to Petersburg Academy, 1754: 'The considerations of the acute Lomonosov about the motion of the fine [electrical] matter in clouds is a great help to those who try their strength on the investigation of these phenomena' (Lomonosov, *Ausg. Schr.* [1961], II, 266–7); 'Geniuses such as he are very rare' (Boss, *Newton* [1972], 80, quoting Euler on Lomonosov). Cf. Kusnetzov, Ak. nauk, Inst. ist. est. i tekh., *Trudy*, 19 (1957), 348–62; Home, *AIHS*, 25 (1975), 3–7.

70. J. A. Euler in *Dissertationes* (1757), esp. 18–20, 27–35. Cf. Minchenko, Ak. nauk, Inst. ist. est. i tekh., *Trudy*, 28 (1959), 188–200; Stäckel, Naturf. Ges., Zürich, *Viertelj.*, 55 (1910), 63–90, a sympathetic sketch of Johann Albrecht; and W. Stieda, Ak. Wiss., Leipzig, Phil.-Hist. Kl., *Ber. Abh.*, 84:1 (1932), 1–43.

71. J. A. Euler, *MAS*/Ber (1757), 125, says that Aepinus convinced him of the importance of Franklin's distinction between positive and negative electricity.

72. The Eulers at first thought that their model required the inversion of Franklin's signs, making smooth glass charge *negatively* by friction (J. A. Euler, *MAS*/Ber [1757], 156–8), but they later accepted the usual arguments from the appearance of sparks (L. Euler, *Lettres* [1787–9], II, 310).

charged bodies, or of an electrified and a neutral one, still occurs through Bernoulli's principle, the pressure drop being independent of the sense of the aether flow. The mechanism of repulsion between like-charged bodies is less plausible: one must suppose the aether everywhere and always agitated, and that the region between the bodies, as the site of oppositely directed currents, has a lower net flow than elsewhere. The pressure must therefore be greater between than beyond the bodies. The Eulers are less persuasive still in their attempts to explain inductive effects. Consider the action of a glass tube brought a few inches from one end, say b, of an insulated iron bar. The Eulers say that the other end, a, charges positively because the 'impetuosity' of the flow of aether from the tube into the bar at b carries off some of the iron's natural supply; 'it can even happen that the electricity in b is negative, the rapidity of the motion having snatched from b more aether than it requires for its natural state.'[73] One need only consider that, on this theory, the bar must eventually become positively charged throughout even in the presence of the tube, to see the insufficiency of the model.

These amplifications, which omitted ACR and the Leyden jar, appeared in 1757 under the protection of heady and irrelevant mathematics. Here is a sample. In the aether-filled pore AB (fig. 16.7), $y^2(x)$ signifies cross-section at x, $\phi(x,t)$ and $v(x,t)$ the aether's density and velocity. The fluid within PQ at time t is $\int_{x_0}^{x_1} \phi(x,t)y^2(x)dx$, and the fluid gained in time dt is

$$dt \int_{x_0}^{x_1} \frac{\partial \phi}{\partial t} y^2(x)\, dx.$$

The amount that entered PQ in time dt may also be computed from the velocities at the boundaries P and Q, and the two expressions set equal:

$$\int_{x_0}^{x_1} \frac{\partial \phi}{\partial t} y^2(x)\, dx = y^2(x)v(x,\,t) \cdot \phi(x,\,t) \Big|_{x_1}^{x_0}.$$

The Eulers eliminated the velocities by assuming that the aether pressure or elasticity is n times its density. The accelerating pressure at x becomes $n[\phi(x,t) - \phi(x + dx,t)] = n(\partial\phi/\partial x)dx$; and the acceleration, or force per unit mass, is $-n(\partial\phi/\partial x)dx \cdot y^2/y^2\phi dx$. Since the acceleration equals the total rate of change of velocity,

$$\frac{\partial v}{\partial t} + v\frac{\partial v}{\partial x} = -\frac{n}{\phi}\frac{\partial \phi}{\partial x}.$$

73. J. A. Euler, *MAS / Ber* (1757), 126–8, 139–47. The model is changed slightly, and for the worse, in L. Euler's *Lettres à une princesse d'allemagne*, written in 1761. Repulsion after AC occurs either because the drawn object drops of its own weight (*Lettres*, II, 292), or because it is pushed off by the mechanism detailed in the prize essay (*ibid.*, 296–7). Euler also tries the Leyden jar: glass is impervious, so that the accumulated aether remains in the water until the explosion; the aether, rushing to the outside of the bottle, succeeds in driving *through* the glass, where it further agitates the remaining electrical matter. 'Consequently it escapes with renewed effort; and, as all happens in an instant, it will enter the finger with increased power, and penetrate the body' (*ibid.*, 317–18). This reminds one of Winkler's fumblings, as does Euler's uncritical acceptance (*ibid.*, 306, 314) of Bose's long-exploded 'beatification' (*supra*, x.2).

The last two equations 'contain all the conditions for determining the motion of the aether.'[74] Unfortunately they cannot be solved. One must simplify to a case that never occurs, to a steady state in time; the last equation integrates immediately, revealing that $\phi(x) = $ const. $\exp(-v^2/2n)$. The Eulers ended with a special case of the Bernoulli principle—the faster the streaming, the lower the pressure—and make no further use of their mathematical apparatus.

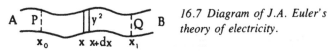

16.7 Diagram of J.A. Euler's theory of electricity.

The quantitative weakness and unregenerate effluvialism of the Eulers' effort provide a useful measure of the achievement of Aepinus. So does the feeble attempt of another geometer, the Barnabite Paolo Frisi, who placed second in the Petersburg contest.[75] Frisi labors to show that a small conducting sphere is not moved by the aether rushing from an extended charged plane as long as the globe of the electrical machine that feeds the plane spins equably. The aether, according to Frisi, leaves the plane perpendicularly; that between A and B (fig. 16.8) enters the sphere, pushing it with force F_1; that left of A and right of B, as at G or H, proceeds beyond the sphere where it expands laterally; a portion enters the sphere at right angles to its original motion, as at L', and drives the sphere downwards with a force F_2. Let the constant aether velocity be v, its constant (!) density β: the 'force' directed radially is $\beta v \cos\phi$, whose vertical component is $\beta v \cos^2\phi$; making the total contribution to F_1 of the infinitesimal band CC' $2\pi\beta v \cos^2\phi \sin\phi \cdot OCd\phi$. Similarly, the contribution of the band LL' to F_2 is $2\pi\beta v \cos\theta \sin^2\theta \cdot OLd\theta$, assuming (what seems scarcely plausible) that the flux along LM is βv. If $\theta = \pi/2 - \phi$, the contributions, being equal and opposite, cancel; summing over all angles, $F_1 + F_2 = 0$. The sphere rests immobile. Which is precisely what does not happen. Frisi explains that, in real cases, the direct flow between A and B, being more easily admitted to the sphere than the flanking flow to the air, diminishes (even with the globe turning!), F_2 prevails over F_1, and the sphere leaps to the plane! No more than the Eulers can he bring his mathematics to bear on the most elementary electrical phenomenon.

74. J. A. Euler, *MAS*/Ber (1757), 136–40; the equations are special cases of the generalized hydrodynamics L. Euler worked out in 1755 (*MAS*/Ber [1755], 316–61); cf. L. Euler, *Opera*, ser. 2, XII (1954), 100–8, and Truesdell, *ibid.*, xxxvi, xcii–xciii.

75. Frisi (who was professor of philosophy at the University of Pisa, and, from 1764, of applied mathematics at the Scuole palatine at Milan) also rejects the emission theory of light, takes the luminiferous aether to be the electrical matter and the redress of aether imbalance to be electricity: 'Ut sicuti quod aetheri est lux, illus aëri est sonus, ita quod aëri est ventus, illus aetheri est electricitas' (J. A. Euler *et al.*, *Dissertationes* [1757], 55). But he admits the contrary states and tries his hand at the Leyden jar (*ibid.*, 111–12). Verri, *Memorie* (1787), 23–4, says that the judges preferred Frisi but refused him the prize because he had written his name on his essay. In fact Frisi's essay was part of a larger work, not entered in the contest (J. A. Euler *et al.*, *Dissertationes* [1757], 13); it so impressed the judges that they gave it second place anyway (Müller to Euler, 22 June 1756, in L. Euler, *Berl. Petersb. Ak.*, I [1959], 115).

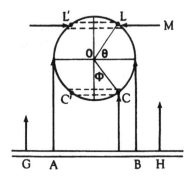

16.8 *Diagram of Paolo Frisi's theory of electricity.*

AEPINUS' *TENTAMEN*

Aepinus owed his success to strict adherence to the program of the *Principia*, which he read as an applied mathematician. Two consequences of his orthodoxy have already been noticed: the elimination of the atmospheres and the admission, in addition to the usual elasticity of the electrical fluid and its attraction to ponderable bodies, of a repulsive force between the particles of 'ordinary matter.' This last force, he observed, is implied by the other two and the mutual indifference of uncharged objects. For if the neutral body A, containing an amount of matter a and of fluid α, exerts no net force upon the fluid β in the neutral body B, the attraction of a for β must cancel the repulsion between α and β; there remains an unbalanced force on B, the attraction between its matter b and the fluid α, which Aepinus neutralizes by supposing a push between a and b. In the formal presentation of his theory, he endows matter particles with a repulsive force not, as is often said, to save the embarrassing phenomenon of minus-minus repulsion, but as an inescapable consequence of the forces assumed by the Franklinists.

This consequence was by no means obvious to contemporaries even after Aepinus had pointed it out. J. H. van Swinden, for example, thinking that the 'total force' between two neutral bodies is, in the notation just used, $a\beta - \alpha\beta + \alpha b - \beta a = 0$, inferred that the distasteful repulsion ab did not exist. Van Swinden erred in counting the repulsion $\alpha\beta$ twice, to take account, he said, of the 'reaction' Aepinus had ignored; his superfluous $\alpha\beta$ played the part of the true ab, and the forces balanced in pairs. A similar confusion confounded the reverend George Miller. A sinner by omission, he allowed the 'total' repulsion $\alpha\beta$ between the fluids to balance the 'net' attraction, αb and βa, between the 'bodies.'[76] It is in the rigorous application of the Newtonian concept of force that Aepinus' chief contribution to electricity lies.

76. Aepinus, *Tentamen* (1759), 16–37; van Swinden, *Recueil* (1784), III, 220–4; Miller, *J. Nat. Phil.*, 4 (1800/1), 465. Van Swinden says that Aepinus' attempt to treat electricity and magnetism on the same terms made him suspicious: 'J'ai donc cru devoir examiner les fondemens de ces calculs, en montrer . . . l'illusion de l'accord qui se trouve entre ces calculs et l'expérience.' Van Swinden to Deluc, 23 April 1784, Deluc Papers (Yale).

Consider first the action of a body A containing a charge δ in excess of its proper amount Q on an uncharged object B of normal content q. The force on B consists of (1) attractions between $Q + δ$ and B's matter b, and between q and A's matter a (2) repulsions between $Q + δ$ and q, and between a and b. Since normal bodies do not act upon one another at any distance, the dependence of force on distance must be the same for any pair of similar and dissimilar particles. Although Aepinus inclined to believe that the law of distance was inverse-square, 'from the analogy of nature,' he left the matter open.[77] All his calculations incorporate an undetermined function, $f(r)$, representing the force between the native electrical fluids Q and q of bodies implicitly considered very small in comparison to the distance r between them.[78] (We would prefer to write $qQ \cdot g(r)$, where $g = f/Q$.) Taking the forces proportional to the amounts of fluid (or common matter) interacting and applying the rule of three, we can sum the forces on B:[79]

$$F = \frac{1}{qQ}[(Q + δ)b + qa - ab - (Q + δ)q] \cdot f(r).$$

This formidable expression is zero. The portion $F_1 = (Qb + qa - Qq - ab) \cdot (f(r)/Qq)$ must vanish, as it represents the force between unelectrified bodies. Rewriting, $F_1 = (Q - a)(b - q) \cdot f(r)/Qq$; from symmetry we conclude that $a = Q$, $b = q$; in our measure the amount of matter in a body equals its normal content of fluid. Now $F - F_1$ is $(b - q) \cdot f(r)/Qq$, whence the total force exerted by an electrified on an unelectrified body is precisely nothing!

Has Aepinus proved as impotent as Frisi? Not in the least. An essential precondition for electrical interaction is the possibility of charge displacement in the attracted body; just as a magnet must first polarize the iron it draws, so a glass tube must induce charges before it can solicit unelectrified chaff.[80] The ease of induction determines the alacrity of response: a perfect insulator would remain indifferent, a piece of metal fly off at once. Differences of resistance to the flow of the fluid explain the diverse gambols of corks, pins and hairs with which electricians amused their visitors.

Assume that, under the influence of A, an amount of fluid $β$ accumulates on the far side of B, as indicated in figure 16.9. The net attractive force on B is $F = (δβ/Qq) \cdot (f(r) - f(r'))$. If now B is also given a positive charge, say $α$,

77. *Tentamen* (1759), 39, 114–17. Later he tried to determine the law by finding the number of axial points m outside a parallel-plate condenser at which the total force vanishes. Supposing, what is not the case, that all the roots of the force equation have physical significance, he expected $m = n$ where $f(r) \sim r^{-n}$ is the force from each plate. Aepinus, *NCAS*, 12 (1766–7), 340–2, where a slip identifies $m = 2$ (which Aepinus expected but could not prove) with an inverse proportion ($n = 1$).

78. Aepinus writes r for $f(r)$, which is employed here to avoid confusion with a direct-distance force. Otherwise the notation is Aepinus'.

79. The first quantity in the equation, e.g., follows from $Qq:f(r)::(Q+δ)b:?$ Cf. Robison, *System* (1822), IV, 9.

80. *Tentamen* (1759), 127–8. Aepinus stresses this point, as had Wilcke, *EO* (Wilcke), 310; cf. Wilcke, *AKSA*, 24 (1762), 220–2, and Miller, *J. Nat. Phil.*, 4 (1800/1), 465.

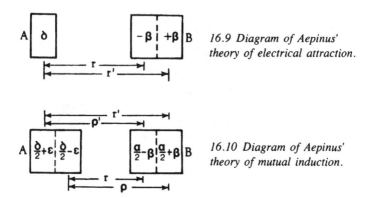

*16.9 Diagram of Aepinus'
theory of electrical attraction.*

*16.10 Diagram of Aepinus'
theory of mutual induction.*

and we suppose that induction on A displaces an amount of fluid ε, the situation of figure 16.10 arises. The force on B becomes

$$F = \frac{1}{2Qq}[(\delta - 2\varepsilon)\{(\alpha - 2\beta)f(r) + (\alpha + 2\beta)f(\rho)\}$$

$$+ (\delta + 2\varepsilon)\{(\alpha - 2\beta)f(\rho') + (\alpha + 2\beta)f(r')\}].   (16.1)$$

All quantities on the right of eq. (16.1) are positive, and, for large distances, the induced charges β and ε are much less than δ and α, respectively. Hence at these distances F too is positive (repulsive). But since β and ε depend upon r, there may be a region where F is negative (attractive). Assume for simplicity that $\varepsilon \sim 0$. Then F will vanish, and the force change sign, where

$$(\alpha - 2\beta)(f(r) + f(\rho')) + (\alpha + 2\beta)(f(\rho) + f(r')) = 0.$$

The same can happen with A and B both negative. No system but his own, according to Aepinus, could predict such effects.[81] Having confirmed them by experiment, he proclaimed that the rule requiring repulsion between bodies similarly electrified did not hold generally. At most he would allow the converse: if electrified bodies repel one another, they carry similar charges.[82]

An equally curious prediction emerged from Aepinus' inspired discussion of the condenser. Guided, as usual, by careful consideration of the forces, he saw that a Leyden jar ceases charging not, as Franklin affirmed, when all the electrical matter has been driven from its outer surface, but when the repulsion at the tail wire drops to zero. If the charge within is α, that without β, the repul-

81. *Tentamen* (1759), 132–6. The Eulers had predicted, though not demonstrated, that plus-plus attraction could occur if one body were so weakly electrified that it admitted the aether flow from another; *MAS*/Ber (1757), 144.

82. Referring to eq. (16.1), taking α negative and again neglecting ϵ, F can become positive (and therefore denote a repulsion between oppositely charged bodies) if

$$(2\beta - \alpha)(f(\rho) + f(r')) - (2\beta + \alpha)(f(r) + f(\rho')) > 0.$$

This can occur only if $f(\rho) - f(\rho')$ can exceed $f(r) - f(r')$, i.e., if df/dr changes sign between r and r'. Aepinus considered such an inflection point most unlikely (*Tentamen* [1759], 137–43).

sive force on a unit positive charge at B (fig. 16.11) is $(\alpha f(r') - \beta f(r))/Q$. At cessation, $\alpha = \beta f(r)/f(r')$, which implies $\alpha > \beta$: Franklin erred again in believing the charges on a condenser to be equal and opposite. The famous experiments in which an insulated circuit used to explode the jar showed no residual electricity were misleading, for the thinness of the jar made $f(r')$ little different from $f(r)$. With an air condenser with a plate separation of one or two inches, however, the circuit ends with a sensible proportion of the charge initially put into the upper plate.[83]

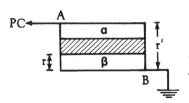

16.11 Diagram of Aepinus' theory of the charging of a condenser.

The masterpiece of the *Tentamen*'s treatment of electricity is a theory of delicate observations inspired by an experiment of Richmann's (fig. 16.6). Having charged the insulated jar ADE between the prime conductor KL and the large metal bar MN, Richmann found that he could destroy the electricity of KL with a touch, then restore it by grasping the bar MN. The sequence might be repeated ten or more times, the threads PO and TS alternately rising and falling, behaving as if his finger pushed electrical matter from one side of the glass to the other.[84] Aepinus refined the procedure by grounding the external surface and attaching the threads directly to the coatings (fig. 16.12).[85] When charging ceases, the index attached to the inner coating A stands at CP, that of the outer at Iρ. Detach the wires from the machine and ground: A's thread

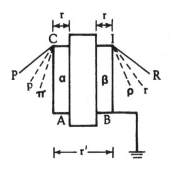

16.12 Aepinus' version of Richmann's experiment.

83. *Tentamen* (1759), 82–7. Richmann had detected the charge left by the explosion of a Leyden jar, whose operation baffled him. Aepinus erred in taking the residual charge proportional to the thickness of the condenser; glass refuses to part with all its accumulated electricity in an explosion (*infra*, xvii.2–3).

84. Richmann, *NCAS* (1750–1), 323; cf. the experiments of Le Monnier, *supra*, xiv.3.

85. Aepinus profited from the availability of Richmann's excellent apparatus; cf. Chenakal, Ak. nauk, Inst. ist. est. i tekh., *Trudy*, 43 (1961), 54–69.

descends about halfway, to Cp, while B's rises to IR. Left to themselves, both wires would slowly fall; but if either coating is touched, its wire drops completely while the other's ascends to its full height. All answers fully to Aepinus' theory.

Disconnection from the machine may result in a small leakage of charge ε from A into the air. The force exerted on a unit positive charge at I then diminishes from zero to $-\varepsilon f(r')/Q$, and a positive charge γ enters B. The forces at C and I accordingly become:

$$\left.\begin{aligned} f_C &= (\alpha/Qf)(f^2 - f'^2) + (1/Q)(\gamma f' - \varepsilon f) \\ f_I &= (1/Q)(\gamma f - \varepsilon f') \end{aligned}\right\}, \qquad (16.2)$$

where $\alpha f'/f$ has been written for β, and explicit dependence on r has been omitted.[86] To make these equations useful, Aepinus must eliminate γ. He does so via the reasonable assumption that the rates of leak of positive charge from A and B, $d\varepsilon/dt$ and $-d\gamma/dt$, are proportional to the forces f_C and f_I, respectively. On this assumption, which amounts to setting $d\varepsilon f_I = d\gamma f_C$, the first of Richmann's effects may be explained. Note that according to eq. (16.2), f_I is a maximum when $d\varepsilon f(r') = d\gamma f(r)$. Using the last two relations to eliminate γ, one obtains for the forces at time t_0, when f_I is maximum,

$$\left.\begin{aligned} f_C(t_0) &= (\alpha - \varepsilon_0)f(f^2 - f'^2)/Q(f^2 + f'^2) \\ f_I(t_0) &= -(\alpha - \varepsilon_0)f'(f^2 - f'^2)/Q(f^2 + f'^2) \end{aligned}\right\}. \qquad (16.3)$$

Remember that at disconnection A's thread fell to half its original height. Aepinus notes that since $r \sim r'$ and ε and γ are initially very small, $f_C(t_0)$ is about one half the initial value of f_C.[87]

What happens if either coating is earthed when the jar rests in this maximum state? Consider plate A first. It will lose an amount of positive charge μ, the magnitude of which Aepinus can calculate. The free charge μ is the difference between the actual charge on A, $\alpha - \varepsilon_0$, and that required to make $f_C(t_0)$ zero. Hence $(\alpha - \varepsilon_0 - \mu)f/Q = \lambda f'/Q$, where $-\lambda$ is the charge on plate B at t_0. We can find λ from eq. (16.3). Since $f_C(t_0) = ((\alpha - \varepsilon_0)f - \lambda f')/Q$,

$$\left.\begin{aligned} \lambda &= 2(\alpha - \varepsilon_0)ff'/(f^2 + f'^2) \\ \mu &= (\alpha - \varepsilon_0)(f^2 - f'^2)/(f^2 + f'^2) \end{aligned}\right\}. \qquad (16.4)$$

---

86. When the jar is freshly charged, $f_C = (\alpha f - \beta f')/Q$ and $f_I = (\alpha f' - \beta f)/Q = 0$.

87. *Tentamen* (1759), 355–70. The agreement is contrived, since Aepinus' analysis holds strictly only for the point charges Q_1 and Q_2 arranged as in the figure:

Touch A; μ runs to ground and the force at I becomes

$$f_I = ((\alpha - \varepsilon_0 - \mu)f' - \lambda f)/Q$$
$$= -2(\alpha - \varepsilon_0)f'(f^2 - f'^2)/Q(f^2 + f'^2). \qquad (16.5)$$

Comparing eqs. (16.5) and (16.3), one sees that when A is touched, B's thread should (and does!) double its displacement. The analysis may be inverted. Let ν be the free charge on B, whence $(\lambda - \nu)f(r) = (\alpha - \varepsilon_0)f(r')$ and

$$\left. \begin{array}{l} \nu = (\alpha - \varepsilon_0)f(f^2 - f'^2)/f'(f^2 + f'^2) \\ f_C = 2(\dot\alpha - \varepsilon_0)f(f^2 - f'^2)/Q(f^2 + f'^2) \end{array} \right\}. \qquad (16.6)$$

Ground B, and A's thread ascends to its initial position, CP.[88]

'Some, who can barely tolerate such things in physics, will perhaps be displeased that I have mixed in so much mathematics. But I judge quite differently, and believe that this brief dissertation on Richmann's experiment contributes a great deal towards proving the truth of my principles of electricity. For an hypothesis grows greatly in plausibility if it can explain complicated phenomena, which can only be deduced from it by a long chain of reasoning.'[89] Who will disagree? It is precisely the method of the gravitational theory, introduced for the first time, and with remarkable resourcefulness, into the study of electricity.

Aepinus' *Tentamen* appeared in 1759, rushed into print earlier than its author intended by the importunate Petersburg academicians, who were eager to show the world the splendid achievements of their new colleague. They wished it to appear before they announced the winner of their prize contest on artificial magnets, since one of Aepinus' most important contributions was a new method of magnetization.[90] The book did not at first have much influence. It had a small circulation. The Italian electricians, Beccaria, Barletti and Volta, did not know it, or, in the case of Beccaria, did not read it attentively until the late 1770s. Few French physicists knew it before R. J. Haüy's epitome appeared in 1787. In Germany it was also scarce; Bohnenberger complained as late as 1793 that he had only just succeeded in obtaining a copy.[91]

In England several important electricians knew the *Tentamen*, or at least fondled it, during the 1760s. Apart from Cavendish, they did not understand it. Franklin complimented Aepinus several times for attempting to apply 'my Principles of Electricity' to magnetism, but never recognized the transformation Aepinus had worked.[92] Priestly, while greatly admiring Aepinus, missed his

88. *Ibid.*, 371–6. Aepinus calculates f_c for f_1 for two values of $f(r)/f(r')$), 100/99 and 5/4, which fix ϵ. He gets a drop (and rise) to 1/2 for the first case (100/99), to less for the second, and confirms (by controlling ϵ) by experiment.

89. *Ibid.*, 375–6.

90. Aepinus, *Tentamen*, 'Pref.'; Ak. nauk, *Protokoly*, II, 428.

91. For Beccaria, Pace, *Franklin* (1958), 59, 376–8; for Barletti and Volta, *infra*, xvii.3; Haüy, *Exposition* (1787), xxvii; Bohnenberger, *Beyträge* (1793), I, 11.

92. BFP, X, 204, 266; XI, 254; XIII, 245, 274; Home, *Isis*, 63 (1972), 192, 198–201.

importance altogether. He praised him for his experiments—the air condenser, the tourmaline, the detection of plus-plus attraction—but warned that 'many of his reasonings and mathematical calculations cannot be depended upon, because he supposes the repulsion or elasticity of the electric fluid to be in proportion to its condensation.'

It appears that Priestley confused the proportionality factor $\Delta Q/Q$ in Aepinus' equations with the density of a traditional electrical atmosphere; only if Aepinus assumed that the electrical matter extended through space does Priestley's disproof of the suppositious proportionality of density and elasticity make sense. Priestley 'refuted' Aepinus by appealing to Newton, or rather to Newton's theory of the air, a fluid satisfying the proportionality of density and elasticity in so far as it obeys Boyle's law. Newton had obtained Boyle's law using a special and scarcely plausible model of a gas, a static array of 'particles' repelling one another with a force reaching only to nearest neighbors and decreasing reciprocally as the distance. Priestley mistakenly inferred that Newton had demonstrated in general that the particles of fluids of which the elasticity is proportional to the density 'repel one another in the simple reciprocal ratios of their distances.' And Priestley had already convinced himself, by an argument we shall examine later, that electrical repulsion diminishes as the distance squared.[93]

It is said that the poor reception of Aepinus' theory was owing to the 'very slight and almost unintelligible account, which Dr Priestley has given of it.'[94] That is at best a partial explanation. Priestley himself managed to explain clearly Aepinus' rejection of literal atmospheres (without observing, to be sure, the contradiction with his criticism of Aepinus' mathematics); indeed, he endorsed the rejection, and even sniped at Canton and Franklin for failing to go so far. The *Tentamen* itself probably put off many of those able to find it: it had too much mathematics, magnetism, Latin, and detail for most electricians of the eighteenth century. Moreover, Aepinus' program did not please the old guard. Euler, for example, thought the *Tentamen* strong on matter, that is experiments, but weak in spirit; his good old friend Aepinus could not expect physicists to 'agree with his theory and with the arbitrary attractive and repulsive forces.'[95] Especially physicists like Wilson. 'The introducing of algebra in experimental philosophy is very much laid aside with us,' he told Aepinus, 'as few people understand it.'[96] The acceptance of Aepinus' approach awaited the work of younger men who independently entered into the same path.

93. Priestley, *Hist.* (1775³), I, 298–303, 369–72; II, 34; Newton, *Math. Princ.* (1934), Bk. II, prop. xxiii, pp. 300–2.

94. Robison, *System* (1822), IV, 2–3; Priestley is also severely treated by Bohnenberger, *Beyträge* (1793), II, 111–20. Perhaps the misunderstandings of the author of the article on electricity in the third edition of the *Britannica*, VI, 458–9, and of van Marum, *Ann. Phil.*, 16 (1820), 449, should be laid at Priestley's door.

95. Priestley, *Hist.* (1775³), I, 303–4; II, 7; Euler to Müller, 30 Dec./10 Jan. 1761, in *Berl. Petersb. Ak.* (1959), 166. Cf. Home, *Isis*, 63 (1972), 196, 202–3, and Cong. int. hist. sci., XIIIᵉ, 1971, *Actes*, 6 (1974), 287–94.

96. Wilson to Aepinus, n.d., Add. Mss. 30094, f. 91 (BL).

PART FIVE
Quantification

Explicantur phaenomena gravitatis universalis, elas-
ticitatis, duritiei corporum, etsi causa earum mysterium
sit, neque melius faciliusque explicarentur vel ipsa ver-
issima causa patefacta. Quin ob id ipsum, quod non
valde utilis sit causarum remotiorum cognitio, natura
eas abscondidisse videtur, adeo non possimus non in
hoc veritatem agnoscere illius dicti; Natura nihil frustra
facere.
   —MAYER, 'Theoria magnetis', cap. 2, §12.

I confess that I am even now less acquainted with the
principle of [electrical] action than I thought I was
twenty years ago.
   —WILSON, *PT*, 63:1 (1773), 50.

CHAPTER XVII

The Atmospheres Destroyed

1. THE JESUITS' COMPASS BOX

In 1750 Jesuit missionaries in Peking received an electrical machine from a well-wisher in St. Petersburg. Richmann made them a present of his papers. These opportunities were not lost on Joseph Amiot, S. J., who knew all the sciences of Europe and most of the languages of Asia. He and his confrères began to amuse themselves with electricity; and in 1755 they were able to send the Petersburg Academy an account of experiments both new and important.[1]

They had placed a glass pane, rubbed side down, on top of a compass case. The needle rose to the top, stayed, and suddenly retired to its normal position. Removal of the pane caused a second ascent, replacement, another fall; the Jesuits repeated the sequence for an hour or more, without rerubbing the glass. 'These experiments are certainly remarkable,' wrote Aepinus, the first theoretician to notice them; 'it seems quite paradoxical that electricity, become almost extinct, after some time, as if instantaneously, without further rubbing, can be resuscitated.' 'Nonetheless,' he continued, 'I saw at a glance that these effects followed, as if spontaneously, from the Franklinist theory of electricity.'[2]

In fact a new postulate was required, which Aepinus collected from the analogy that underlay the *Tentamen*. Just as iron exposed to a magnet acquires poles, so electric force polarizes glass.[3] The cover of the compass box develops a negative charge above and a positive one below; the needle rises under the net force F between the charges in the plates and the negative electricity induced in its point. Because of the tiny conductivity of glass, the charge on the lower surface of the cover slowly leaks off through the needle. F vanishes; the needle drops. On removal of the upper plate it reascends under the pull of the negative charge remaining on the cover. It falls on replacing the plate, which again drives F to zero. This beautiful explanation, which demonstrated, without

1. Gaubil to Delisle, 25 Oct. 1750, and to Kratzenstein and Richmann, 30 April 1755, in Gaubil, *Corresp.* (1970), 617, 810–11. The electrical experiments were not done openly for fear they would alarm or irritate the Chinese. The Peking Jesuits frequently reported curiosities to the Petersburg Academy of which their leader, Gaubil, was a member; Richmann, among others, encouraged the connection (Ak. nauk, *Protokoly*, II, 282).

2. Aepinus, *NCAS*, 7 (1758–9), 277–302.

3. Owing to the small conductivity of glass, positive charge rubbed on one end of a rod migrates a bit toward the other (Aepinus, *Tentamen* [1759], 183–5). It pushes out neighboring electrical matter, which travels a short distance down the rod, stops, and repeats the process, so that the rod contains alternate positive and negative layers, whose positions Aepinus tried to measure.

mathematics, the power of Aepinus' approach,[4] did not appeal to everyone. Beccaria thought he could do better.

In 1764, having spent six years in measuring a degree of latitude in Piedmont, Beccaria returned to electricity, to protect Franklinism from his nephew, Gianfrancesco Cigna.[5] Cigna, Professor of Anatomy at the University of Turin, had been Beccaria's student and assistant in the early fifties, along with G. L. Lagrange. Both young men had been too independent long to suffer their mentor's domineering. They opposed one of his ideas about gases, wrongly as it happened; Beccaria cut them off completely and, to his own disadvantage, refused to participate in the scientific society they founded in 1759, the forerunner of the Turin Academy.[6] When news of this estrangement reached Nollet he perceived in Cigna, who occasionally wrote on electricity,[7] a likely leader of his beleaguered Italian garrison.

Nollet forwarded a French translation, commissioned by himself, of papers by Robert Symmer, who had invented two sets of experiments much like those of the Peking Jesuits. In one set, as idealized by Nollet, a silk ribbon was charged by rubbing as it rested on a glass plate; the combination gave few signs when united, although each member acted vigorously when apart. In the other set, a second pane replaced the ribbon and the pair, armed externally and charged like the jar, again showed no external signs but cohered very strongly. Symmer used these experiments to argue for the existence of two electrical fluids, and Nollet twisted them in support of his double flux. After varying them several ways, Cigna concluded that, rather than deciding between Franklin and Symmer, between one electricity and two, they raised a problem for both. 'We are left with one desideratum,' he wrote, 'namely an explanation of how the contrary electricities destroy each other when mixed and how, when they cannot combine [as in the experiments], they draw mutually, restrain one another, and act as if their reciprocal attraction were obstructed (intercederet).'[8]

Cigna improved the first pair of experiments with an abstraction closer to the

4. Aepinus slipped in assuming, against the principles of the *Tentamen*, that $F = 0$ when the induced positive charge q on the lower surface S of the case is neutralized; since the coercing charge $Q > q$, S must be negative for F to be zero.

5. [Eandi], *Memorie* (1783), 39–42; Beccaria, *PT*, 56 (1766), 105–18, and *A sua alt. sper. e osser.* (1764), a brief notice of devices for measuring spark lengths.

6. Vassalli-Eandi, Acc. sci., Turin, *Mem.*, 26 (1821), xiv–xx, xxix; Calcaterra, *Nostro imm. Ris.* (1935), 343, 347–9; Grassi, *Saluzzo* (1813), 11–13, 20–6. Gliozzi, *Fisici* (1962), 8–10, says that Saluzzo and Lagrange opposed Beccaria over whether combustion and breathing can take place in a vacuum. Toward the end of his life Beccaria relented, took Cigna as his physician and followed up Cigna's experiments (Beccaria, *Op. Mil.*, 3:3 [1780], 3–19).

7. Cigna had included electrical experiments, supplied by Beccaria while relations were cordial, in his doctoral thesis of 1757; after the break he studied analogies between electricity and magnetism (*Misc. taur.*, 1 [1759–60], 43–67), inventing an experiment important to Volta (*infra*, xvii.2), and electrical motions (*Misc. taur.*, 2 [1760–1], 77–9), arguing that, as attractions occur between charged bodies immersed in oil, Beccaria's air theory (*supra*, v.4) could not suffice.

8. Symmer, *PT*, 51:1 (1759), 340–89; Nollet, *MAS* (1761), 244–58, and *Lettres* (1767), iii–v, 1–151; Cigna, *Misc. taur.*, 3 (1762–5), 31, 72.

original, which employed a pair of stockings, than Nollet's: he employed two superposed ribbons of dry white silk, rubbed with an ivory stick while resting on glass. When removed together the ribbons cohered, appearing negative *on both sides*; when separated the upper showed strongly minus, the lower weakly plus; on recombining with the glass all signs disappeared. If, after excitation, the ribbons were stripped *seriatim,* both exhibited negative electricity on both surfaces.[9] That these effects imply conundrums for Franklinists appears from Cigna's version of Symmer's double glass: two panes now replace the ribbons and a lead plate the glass of the original experiment. The panes behave exactly like the ribbons, except for a change in sign, the lead being minus, the top glass plus.

Here is one conundrum. Consider the arrangement an incomplete Leyden jar, the lead plate being an armature: the sparks, the cohesion and the residual electrification of the lead suggest that the charge of the jar does not reside in the glass, as Franklin taught, but only passes to the dielectric as the coating is removed. Cigna confirms this conjecture by rubbing the topmost of a pile of ribbons laid on a glass plate. He removes the pile which, like the original pair, shows negative on both sides. He now strips the ribbons and notes that the intermediate ones become either plus or minus according as he begins at the top or bottom of the pile. At each disjunction a spark passes. The same flash occurs, he observes, when an armature is removed, via an insulating handle, from a charged Franklin square; since not all the coating's charge can pass in the spark, the denuded surface possesses less electricity than the side still armed, which, therefore, as experience shows, exhibits its virtue on both sides of the glass.[10]

Here is another conundrum. Place two superposed panes on an insulated lead plate; rub the upper glass and draw a spark from the metal; repeat; arm the upper glass with another lead plate, complete the circuit, and obtain the usual shock. 'What seems singular [according to Cigna] is that, although discharged and exercising no electrical force on external objects, yet the glasses strongly cohere, and on separation display opposite electricities on their unarmed faces.' Franklinist theory called for annihilation of the supposedly equal and contrary electricities of the armed surfaces. Cigna thinks to save the phenomena by introducing a second, or 'Symmerian,' mode of electrification, which, in distinction to the 'Franklinian,' dissipates slowly, like the aerial electricity of Bec-

9. Cigna, *ibid.,* 31–6. In the friction, the glass becomes plus; the lower ribbon electrifies weakly either way, depending upon the manner of rubbing; and the top ribbon charges minus, according as the ribbons are removed together (when it receives a spark from the glass) or separately (when it collects from its fellow). Cigna varied this experiment with ribbons of different colors rubbed on different substances (*ibid.,* 37–43).

10. *Ibid.,* 52–3, 60–8. On removing the pile, a spark passes between the glass and the lowest ribbon, which becomes plus, but not enough to overpower the influence of the negative top; if the stripping begins at the top, the negative charge of the uppermost jumps from one ribbon to the next; if at the bottom, the weaker positive charge of the lowest is similarly passed along. All the ribbons are polarized, which assists the sparking.

caria. Cigna hints that the new mode depends upon the small conductivity of glass (another Franklinist heresy!), but declines to give a mechanical theory.[11]

He offers further examples, however, and one deserves attention. Approach a ribbon charged with Symmerian electricity to a lead plate; draw a spark from the lead, endowing it with an equal dose of Franklinian electricity of the opposite sign; lay the ribbon on the plate and all signs vanish. Remove the ribbon, a small spark passes, ribbon and plate regain their electricities; the plate's residue may be drawn off entirely by grounding, but the ribbon's cannot. Indeed, it can still charge the metal many times by induction, as in the first two steps of the preceding experiment; 'sicque modum inveniemus facilem electricitatem absque frictu multiplicandi.' With this embellishment Cigna found the principle of the electrophore.[12]

VINDICATING ELECTRICITY

These notable demonstrations suffered, according to Beccaria, from Cigna's regrettable 'inclination to innovate and contradict.'[13] To begin to repair the damage, he worked up variations on the Peking experiment for the Royal Society and then settled down to elucidate the strange dormancy of the united glasses on principles as Franklinist as possible.

The effort came to print in 1769 in a difficult and ingenious pamphlet whose title, *Electricitas vindex*, betrays confusion between object and observer.[14] Here Beccaria subsumed all previous Symmerian and Peking phenomena under one experiment. Insulate a Franklin square whose upper surface A is charged to an amount represented by FO (fig. 17.1), as measured by the excursion of a thread held near A. Remove the upper coating, α, by silk strings; A loses an amount of electricity Fu. Replace α, touch the lower coating β; A's electricity increases: the net effect of the operation is to reduce A's original electrification to GP. Strip A again; it loses less then before (GX), and recovers to HQ when α returns and β is earthed. After a few repetitions, α comes away unelectrified (H); thereafter A *gains* a little electricity at each disjunction (YI, KS, LZ) and

11. *Ibid.*, 54–9; *supra*, xv.4. The effect depends both on polarization and on the 'absorption' or conductivity of glass. Charged slowly, by successive sparks, the double condenser soaks positive charge into its upper plate, negative into its lower, while the naked faces bear the charges resulting from the polarization, negative on the upper plate, positive on the lower. The explosion removes only the 'Franklinian' electricity on or near the coatings; the 'Symmerian' or absorbed electricity leaks back slowly, largely neutralized on the armed surfaces by the remaining polarization, which, on the unarmed surfaces, provides the force of cohesion. Cf. Wiedemann, *Lehre*, II (1883), 83.

12. Cigna, *Misc. taur.*, 3 (1762–5), 43–51; *infra*, xvii.2.

13. Beccaria to BF, 11 Oct. 1766, *BFP*, XIII, 453. Beccaria claimed the field of Symmerian electricity for himself because, in *Dell'elettricismo* (1753), 197, he had referred the sparks drawn on removing a shirt to electricity! Cf. Gliozzi, *Arch.*, 17 (1935), 38.

14. *Experimenta* (1769), not to be confused with Beccaria's brief interim report, *De el. vind.* (1767), reprinted *BFP*, XIV, 42–9. The phrase signifies 'electricity the vindicator' not 'vindicating electricity' (*electricitas vindicans*), which Beccaria intended and which Cavallo, *Comp. Treat.* (1795³), II, 196–206, employs in his excellent abstract of the theory. Cf. Beccaria, *Treatise* (1776), 395–413, and Polvani, *Volta* (1942), 64, who suggests *electricitatis vindicatio*.

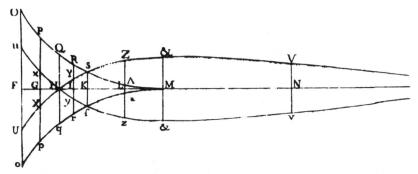

17.1 The course of vindicating electricity. From Beccaria, Elettricismo *(1772).*

loses at each restoration, until, at M and beyond, it shows no charge when armed. At the removal following M the free surface shows its maximum gain (M&); thereafter, when denuded, A's electricity slowly declines along the curve &V. These are the doings of vindicating electricity (VE). At first A suffers from negative VE, which causes it to share its charge with α at each separation; after spending its force, the affection changes sign and A recovers somewhat until the positive VE has also lost its virtue. Note that VE is a state, not a new electrical matter, and that it serves to save the Franklinist doctrine that the charge of a condenser lies entirely in the glass.[15]

To obtain a theory of the Peking experiment Beccaria had to combine VE with his theory of attraction. The link consisted of aerial electricity and the Franklnist conclusion, now raised to an axiom, that the two sides of glass charged for the Leyden experiment have equal and opposite electricities.

Consider first a charged double pane. Separate the laminae as in fig. 17.2: the top (AA') appears plus on both sides, the bottom (BB') similarly nega-

15. *Experimenta* (1769), 1–3. Let the original free charge on A be Q, that on α, q, and the bound or polarization charge of A, $-x(Q+q)$; FO (fig. 17.1) represents $(Q+q)(1-x)$. Remove α: the effective charge near A drops by $q(1-x)$, or Fu, and α's potential increases enough to leak a fraction of its charge, yq say, into the air. Replace α and touch β: the net charge at A becomes PG $= (Q + (1-y)q) \cdot (1-x)$. Remove α once more, decreasing the effective charge near A by XG $= q(1-y)(1-x)$; the coating, having again lost to the air, returns with approximately $q(1-y)^2$; and, after touching β, the net charge at A is QH $= (Q + (1-y)^2q)(1-x)$. At QH the charge which α brings away has become insensible. Beccaria now changes tactics, touching α and β alternately when the condenser is assembled, and neither when it is stripped. Grounding α removes the last of its original charge q and endows it with a *negative* one (by induction) proportional to QH. Remove α: A appears to 'revindicate positively' (IY) with the withdrawal of the counterbalancing negative coat. Return. Touch the coatings alternately: α obtains a larger induced charge than before (say LZ) because of the depolarization, which lags behind the removal of the coercing charge; at M, where the maximum revindication occurs at denudation (M&), the depolarization associated with q is complete. Thereafter A shows no force when covered and gradually loses its power of 'positive vindication' as the free charge on its surface (approximately $Q(1-x)$, ignoring leakage during the preceding manipulations) dissipates. From M on, α behaves like the shield of an electrophore; cf. *infra*, xvii.2, and Zeleny, *Am. J. Phys.*, 12 (1944), 329–39.

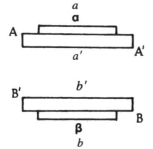

17.2 *Diagram of Beccaria's elucidation of the Peking experiment.*

tive.[16] This violation of the new axiom is only apparent, according to Beccaria. For suppose each pane the coating of the other: if in the charging A receives a quantity of fluid Q, A′ must deposit an equal amount in B′, and B must lose the same quantity Q to ground. At separation the internal surfaces vindicate negatively, B′ returns some positive charge to A′. Now, says Beccaria, A′ cannot take all of B″'s offering in their moment of parting, so a portion remains in the air $a′$, whence the region around A′ shows positive. A′ itself is negative, though not sufficiently so to balance the charge in A, which confers its surplus on the air at a. Thus the lamina AA′ appears positive on either side. Similarly B′, because of its gift to A′, cannot balance the original minus charge on B; compensating electrical matter flows out of the air at both b and $b′$, rendering BB′ negative on, or rather near, each surface. Hence the working of the Peking experiment: the flow of electrical matter into and out of the air whenever charged laminae separate or join provides the current 'attracting' the needle; the current ceases and the needle drops when the charges on the panes balance.[17]

The complex theory can deal with many variations of the Peking experiment. For single panes and partially dressed condensers alone Beccaria distinguished nineteen different cases, which he displayed in a cumbersome chart simplified in Table 17.1. The results of Group VI, which elaborate an effect he had mentioned in his paper of 1766, are noteworthy. Even Aepinus had missed the possibility of the reversal of the polarity of the coatings by the explosion, a phenomenon arising from induction from residual or absorbed charge. To Bec-

16. *Experimenta* (1769), 11–23. The phenomenon presents no difficulty for Aepinus' theory: the interior surfaces have bound or polarization charges much weaker than the free charges on the coatings and the armed surfaces, whose fields therefore dominate on both sides of the plates. Cf. Henley, *PT*, 66 (1776), 518. Beccaria has been complimented for discovering that the sign of the force near a surface of a condenser need not be that of the charge upon that surface; Cantoni, Ist. lomb. sci. lett., Milan, *Rend.*, 6 (1873), 112–21. The proposition was, however, already known to Aepinus; Beccaria contributed experimental demonstrations involving complicated maneuvers with plates of unequal capacities. *Experimenta* (1769), 23–35.

17. Beccaria lets A′ (the charged plate's bottom) share its negative charge with B′ (the top of the compass's lid), a poor but serviceable expedient. For equilibrium, B must receive a flow from the needle: on removing the plate, A′ revindicates negatively, B′ grows less minus, fluid flows from B into the air, and the mechanism of Cabeo works again. *Experimenta* (1769), 44–54; *PT*, 57 (1767), 298–9.

TABLE 17.1

Vindicating Electricity of a Single Pane

Experimental situation		Resultant electrification of			
		α	β	a	b
I. The pane, freshly charged, is stripped by the fingers	1. of α	0		*	−
	2. of β		0	+	**
II. The freshly charged pane is stripped by silk threads	3. of α	+		*	
	4. of β		−	+	
III. Both α and β are removed while	5. negative VE operates	as A freshly charged	as B freshly charged	as α in case 5	as β in case 5
	6. neither VE operates	0	0	0	0
	7. positive VE operates	as A after explosion	as B after explosion	as α in case 7	as β in case 7
IV. The pane, now exploded, is stripped by the fingers	8. of α	0		+	+
	9. of β		0	−	−
V. The exploded pane is stripped by silk threads	10. of α	−		+	+
	11. of β		+	−	−

*Slightly −, 0, or slightly +, according as the impressed electricity was weak, moderate, or strong.
**Slightly +, 0, or slightly −, according as the impressed electricity was weak, moderate, or strong.

caria it showed electricity vindicating. In experiment 8 or 10, for example, A revindicates positively, B (zero just after the explosion, according to Franklin) loses some of its electrical matter to the air in an unsuccessful attempt at equilibration, which comes when A jettisons its unbalanced excess. The air at *a* and *b* therefore behaves positively.[18]

The reversal of polarity, or 'oscillation of the electricities,' as Beccaria called it, brings out another purpose of VE: an explanation of the disappearance of electrical signs when equally and contrarily charged bodies are intimately united. In the case of an exploded condenser or of Cigna's ribbons, outward signs vanish because the contrary electricities have indeed annihilated one another; only on separation, through the action of VE, do the partners reacquire a portion of their former vigor. Beccaria's answer to Cigna's query about the mechanism of neutralization saves Franklinist algebra—plus + minus = zero—at the price of an intelligible account of the cohesions in Symmerian operations. Beccaria's clumsy physics traduced his beautiful experiments.

2. THE ELECTROPHORE

Alessandro Volta was born at Como, then a part of the Austrian Archduchy of Milan, in 1745. He came from an old Lombard family almost extinguished through its service to the Church. One of his three paternal uncles was a canon, another an archdeacon, the third a Dominican; his father had once been a Jesuit; and his three brothers followed precisely the careers of their uncles. Alessandro himself narrowly escaped the Jesuits, with whom he studied humanities; in 1761, as he began his second year of philosophy, his uncle the canon, who had become his guardian on the death of his father (1753), took alarm at the zealous recruiting of the philosophy professor, Girolamo Bonensi, and removed Alessandro to the school of a local seminary. He had reserved his nephew for the propagation of the Voltaic line, and hoped to make him a lawyer.[19]

Volta declined to train for law or even to attend a university. His eccentric older friend, Giulio Cesare Gattoni, who had acted as intermediary in the Bonensi affair, set up a museum and laboratory in a disused tower rented from the city of Como. On the tower he mounted a lightning rod that brought down electricity for him and Volta to experiment upon.[20] Encouraged by Gattoni and obeying what he called his 'genius,' Volta studied the works of Nollet and Beccaria. When questions or suggestions occurred to him, he wrote directly to his authorities; before leaving school in 1765 he had opened correspondence

18. *Experimenta* (1769), 6–10, 36–44; *PT*, 57 (1767), 298, 310. If the absorbed charge is large enough it can flow back into the coatings, overwhelm the induced charge, and give a second explosion. Friends of Cigna brought such a case to his attention in 1770. He doubted it initially, but then had no difficulty relating it to the slow march of 'Symmerian electricity.' Cigna, *Misc. taur.*, 5 (1770–3), 98–9; Barletti, *Physica* (1772), 41.

19. *VE*, I, 1–3; Z. Volta, *Volta* (1875), 36–68; *DSB*, XIV, 69–82; Bonensi's letters (*VE*, I, 6–33) make curious reading.

20. Gattoni, *Volt.*, 1 (1926), 493–7; Scolari, *ibid.*, 431–6.

with them and with Carlo Barletti, another energetic Piarist who was to be Volta's colleague at the University of Pavia.[21]

The letters to Paris and Turin were ill-calculated to please their recipients. Having embraced the method of the *Principia* and the program of Boscovich, Volta announced to Nollet and to Beccaria that all electrical phenomena arose from an attractive force operating, or appearing to operate, at a distance. Nollet, following his policy of divide and conquer, observed that it would be 'glorious,' though most difficult, to build a Newtonian theory of electricity, which 'no one yet has dared to do.' Beccaria replied a year later, after Volta had apologized for his 'very frivolous chatter.' As a cure for frivolity Beccaria recommended reading Beccaria and doing experiments.[22] Volta did so, without access to the usual apparatus; forced to invent cheap substitutes, he began to develop the genius for inexpensive, effective instrumentation that determined his career.

His earliest results, communicated to Beccaria in April, 1765, derived from the discovery, which Volta fancied new, that silk rubbed by hand became plus, and rubbed by glass, minus. He designed a machine to capitalize on the electrical properties of silk, and drew up a schematic triboelectric series, doubtless independent of Wilcke's. Again silence from Beccaria and apologies from Volta, now mixed with worry that he had alienated what a friend, Marsilio Landriani, called the 'electrical oracle,' and had injured his chances for a career in physics in Italy. Beccaria allowed himself to be cajoled and for four years corresponded fruitfully with Volta, sending his publications, urging vindicating electricity, and criticizing the principle of attraction that Volta refused to abandon. The exchanges ceased abruptly in 1769, when Volta published a dissertation in the form of a letter to Beccaria, in which the oracle's latest experiments, including those on VE, were explained in terms of Volta's *idée fixe*.[23]

Volta observed that Franklin's electrical matter cannot itself be the cause of electrical motions because it courses unidirectionally, from excess to defect, while the most common of experiments, as Nollet emphasized, shows that the same electrified body *simultaneously* imposes both attractions and repulsions. In fact the effluvia do not stream at all: the famous electrical wind, as Volta had earlier told Nollet, is nothing but a current of electrified air. Nor can the effluvia operate indirectly, by impelling the air, for electrical attraction takes place between objects immersed in oil, an experiment Volta lifted without acknowledgment from Cigna. We must therefore admit short-range attractive

21. *VE*, I, 4, where Franklin and Priestley are incorrectly included among Volta's earliest correspondents; no relevant letters appear in *BFP* and Priestley did not become known in Italy until after 1771, when the French translation of his *History* appeared. For Barletti see *DBI*, VI, 401–5, and *infra*, xviii.2.

22. *VE*, I, 33–6; *VO*, III, 23. Aepinus' *Tentamen* had played no part in Volta's thinking; he did not see the book until 1778. Polvani, *Volta* (1942), 58–9; *infra*, xvii.2.

23. *VE*, I, 36–43, 64–5, 91; *VO*, III, 6–7, 10–11, 15–16, 19–20, 23–4; Polvani, *Volta* (1942), 13–19. The pamphlet that roused Volta was *De atm. el.* (1769), which previewed VE; cf. *VO*, III, 23; Polvani, *Volta* (1942), 64; *infra*, xvii.3.

forces. To the usual objection that multiplying such forces clutters matter with special, nonmechanical powers, Volta countered that, since only 'mixed bodies' are electric, one need imagine no special virtue of electricity, but merely a net macroscopic force compounded from the different microscopic forces possessed by the particles of pure substances, or from the universal, elemental, multi-purpose force of Boscovich. Nor should one falter at the great range of electrical attraction: we have, on the one hand, the patent example of magnetism and, on the other, the existence of electrical atmospheres. These, according to Volta's even-handed compromise, consist of redundant electrical matter the attractive force of which extends somewhat beyond their physical limits. 'However that might be, for present purposes it need only be granted that attractive forces really exist in bodies; following the example of the most distinguished men, I will then seek to explain certain natural phenomena that do not arise from impulse or from known laws.'[24]

Volta's fundamental concept is that there exists for each body a state of saturation in which the integrated attractions of its particles for electric fluid are precisely satisfied. This integrated attraction may be altered by any process, mechanical or chemical, that displaces the particles relative to one another; friction, pressure, and perhaps evaporation, electrify bodies by destroying existing patterns of saturated forces and redistributing electrical fluid.[25] One sees in this proposition the seeds of the experiment of Volta, Lavoisier and Laplace on electrification by evaporation, and, perhaps, of Volta's consequential concept of contact charge. As for the notion of saturation, it vaguely foreshadows the concept of tension, Volta's qualitative equivalent of potential: the condition of electrical equilibrium between two bodies is not equality of quantity of electric fluid, but of degree of departure from saturation.[26]

For the rest, *De vi attractiva* is an exercise in reducing the standard phenomena—attraction, 'repulsion' (really attraction away from the 'repelling' body[27]), the Leyden experiment, the effects of VE—to the single attractive force. The operation of the Leyden jar, which Franklinist theory ascribed to macroscopic *repulsion* between particles of the electrical matter, taxed Volta's ingenuity. He decided that normal glass, although saturated with fire, can still admit a small increment, say to its upper surface, and that this increment *weakens* the sum of the attractions exercised by the glass on the fire in the lower surface, a part of which is drawn away to ground. To make this physics

24. Volta, *De vi attr.* (1769), in *VO*, III, 23–52, on pp. 25–7, 29; Volta to Landriani, 3 June 1775, *ibid.*, 85; cf. *VE*, II, 510–11. Boscovich refers electricity to his scheme in *Theory* (1763), §§ 511–12.

25. *VO*, III, 30–4.

26. Cf. *VO*, I, 459–90. In discerning redeeming qualities in *De vi attr.* we follow the Italian commentators, e.g., Massardi, Ist. lomb. sci. lett., *Rend.*, 59 (1926), 373–81, and oppose the French, who dislike the book: 'a very imperfect hypothetical explanation' (Biot in *Biographie* [1880²], XLIV, 77–81), which 'added little to Franklin' (Arago, *Oeuvres*, I [1865²], 190).

27. *VO*, III, 28–9, 85; cf. Polvani, *Volta* (1942), 68, 77, and *VO*, IV, 358. Volta located the attractions responsible for repulsions partly in the air, and believed that electrical motions cease *in vacuo*.

plausible, Volta invoked an analogy, again tacitly lifted from Cigna: suspend from a lodestone (the glass) the greatest weight of iron filings it can support (its natural charge); if more filings (the increment) fall onto the upper surface of the magnet, some of those attached to the lower drop off. The same effect occurs in the usual induction experiments where the surplus (or deficient) fire of an electrified body, 'applied' to a neutral one, creates a condenser effect by acting at a distance.[28]

The fruitful tendencies of Volta's approach appear from his analysis of experiments illustrating vindicating electricity: he assumes, but does not apply, his machinery, and analyses in the style of Aepinus. Place a metal plate on rubbed glass; induction occurs because the surplus fire cannot cross; touch the plate, rendering it negative and destroying all signs. Remove the plate. The charges reappear not, as Beccaria would have it, because they are revindicated anew, but because each can now express itself free from the *overlapping action* of the other. The 'oscillation of the electricities' on exploding the condenser, an effect that took Volta's fancy, derives, he says, from incomplete discharge and induction in the coatings. No signs appear until separation, as in the preceding case. With the top coating removed, the condenser shows plus on both sides not because of a flow of electrical matter to or from the air, but because the force of the free positive charge on the naked surface dominates even beyond the armed one. One analyzes the double panes similarly, by considering each the coating of the other. As to the switch from negative to positive VE, Volta keeps silent.[29]

Despite the hopelessness of the program and the residue of the atmospheres, Volta's first theory of electricity marked substantial progress, the first important independent affirmation of the approach of Aepinus. And, although Volta did not abandon his program before the mid-1780s[30]—reluctance to change or discard a once-fruitful theory was characteristic of him—he soon realized that he had unnecessarily sacrificed coverage and convenience in restricting himself to attractions. The difference between insulators and conductors is difficult to understand if both allure the electrical matter. This difficulty obtruded insistently when Volta, following an observation of Nollet's, discovered that many nonelectrics, like wood, become excellent insulators when dried and oiled. He turned to Boscovich to save both the phenomena and the simplicity of the single force. Oil, like water and metals, attracts electrical matter at short range, but unlike them repels it at greater distances. Hence, Volta conceded, his explanation of the Leyden jar was faulty: electrical matter adheres to the upper surface of the glass via the short-range attraction, but is driven from the underside by the long-range repulsion, which the surplus positive charge 'intends.'[31] Meanwhile the underside strives to drive away the excess, 'whence,' as Volta

---

28. *VO*, III, 36–42; Cigna, *Misc. taur.*, 1 (1759–60), 56.

29. *VO*, III, 46–50. In general, Volta limits himself to easy cases and claims that the rest follow; although assertive, he is respectful to Beccaria.

30. *VO*, III, 56–9, 62–71, 85; *VO*, IV, 410–13; cf. Polvani, *Volta* (1942), 69–74.

31. Volta, *Nov. simp. app.* (1771), in *VO*, III, 55–80, esp. pp. 56–9, 62–71.

quaintly put it to Priestley, 'the force that makes [charged] glass vomit electrical fire into the bosom of the conductor.'[32]

To concoct the electrophore, the most intriguing electrical device since the Leyden jar, 'the most surprising machine hitherto invented,'[33] Volta had only to combine the insight that resin retained its electricity longer than glass with the fact, emphasized by Cigna and Beccaria, that a metal plate and a charged insulator properly maneuvered can produce many flashes without enervating the electric. Beccaria inspired the combination. In 1772, despite a debilitating illness, he published a lengthy, difficult, updated *Elettricismo artificiale* that emphasized more strongly than before his odd view that the contrary electricities destroy one another in the union of a charged insulator with a momentarily grounded conductor, only to reappear 'revindicated' in subsequent separations. He also criticized the theory of the unique attractive force, without deigning to mention its author. Swearing vengeance on the vindicator, Volta undertook to topple Beccaria's cumbersome theory, which had found advocates in Barletti and Saussure.[34]

Volta conceived that, if he could greatly increase the duration of the effects ascribed to VE, the implausible idea of alternate destructions and incomplete recuperations would fall to the ground. After lengthy trials he discovered that an insulator made of three parts turpentine, two parts resin and one of wax answered perfectly; and on June 10, 1775, he announced to Priestley the invention of an inexhaustible purveyor of electricity, an *elettroforo perpetuo*, which 'electrified but once, briefly and moderately, never loses its electricity, and although repeatedly touched, obstinately preserves the strength of its signs.'[35]

The device consisted of a metal dish B (fig. 17.3) containing the dielectric cake, a wooden shield CC covered with tin foil rounded to remove all corners and points, and an insulating handle E. One can charge it readily by rubbing the dielectric while grounding the plate, a procedure that became standard; but as that would have obscured the connection with VE, Volta told Priestley to electrify the instrument like a Leyden jar, and to take sparks alternately from the

32. Volta to Priestley, May 1772, as quoted in Volta to Klinkosch, May 1776, *VO*, III, 140–1n. Volta began to announce his discoveries through Priestley in 1772 after learning from the French translation of the *History* that several of the fancied finds in *Nov. simp. app.* (1771) had been anticipated (Volta to Spallanzani, 8 Dec. 1771, *VO*, III, 77–9). Priestley allowed that he had written to save people the time of rediscovery (letter of 14 March 1772, *VE*, I, 59–60); and indeed, especially in Italy, his *History* guided the research of many electricians less original than Volta (Barletti, *Phys. sp.* [1772], 24–5).

33. *Encycl. Brit.*[3], VI, 424.

34. Barletti, *Nuove esp.* (1771); Frisi to Volta, 9 July 1771, *VE*, I, 52: '[De Saussure] could tell you other great things about vindicating electricity, to which word it seems he attaches some meaning'; Volta to Landriani, 3 June 1775, *VO*, III, 83–4; Polvani, *Volta* (1942), 94, 99; Beccaria, *Elettricismo* (1772), 407 ff. Cf. J. A. Deluc to Volta, 16 April and 22 June 1784, *VE*, II, 207: 'Je n'ai pu me dispenser de lire tout Beccaria *dell'elettr. artificiale*, et combien ne m'en a-t-il pas causé de tourment.'

35. *VO*, III, 96. Volta published this letter, with plates and supplementary instructions, in *Scelta di opuscoli interessanti* (Milan) for 1775.

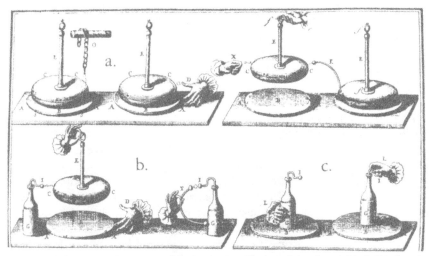

17.3 Volta's electrophore: (a) charging by 'oscillation of the electricities' (b) charging a bottle by an electrophore (c) charging an electrophore by a bottle. From VO, *III, 101.*

positive shield and the negative disk until the 'oscillation of the electricities' took place (fig. 17.3a). Here, where Beccaria supposed positive VE to set in, the shield could be removed and its *negative* charge given, say, to the hook of a Leyden jar; then replaced, touched, and again brought to the hook; and so on until the condenser was moderately charged (fig. 17.3b). Any number of jars and electrophores might be electrified without regenerating the original; and if it should decline, it might be reinvigorated by lightly rubbing its cake with the coating of a Leyden jar previously charged by it (fig. 17.3c). Volta set great store by this last property, which did seem to vouchsafe eternal life to the electrophore, and to justify the term *elettricità vindice indeficiente,* with which he proposed to celebrate his victory over Beccaria.[36]

The triumph was not cloudless. 'The perpetuity that D. Alessandro has attached to his electrophore is only a greater duration of electricity impressed on resin,' thundered Beccaria, asserting priority for himself and even for Cigna, who pressed his own claims.[37] A follower of Beccaria in Vienna, a physician named Klinkosch, emphasized the connection with vindicating electricity, to Volta's disadvantage; others credited Stephen Gray, Aepinus, or the Jesuits of Peking.[38] The inventor responded in a letter to Klinkosch, published in the

36. *VO,* III, 98–105. Volta knew of course that the cake would charge by friction, always minus when rubbed with cloth, plain paper or the hand, usually plus when gilded paper was used (*ibid.,* 102). Charging the electrophore by friction drove out Volta's initial procedure so effectively that a century later F. V. Neyreneuf brought it out as something new (*JP,* 1 [1872], 62–3).

37. Cigna, *Scelte,* 9 (1775), 83.

38. [Eandi], *Memorie* (1783), 132; Vassalli-Eandi, Acc. sci., Turin, *Mem.,* 26 (1821), xxx–xxxi; *VO,* III, 135–6, 159; *VE,* I, 81, 90, 116; J[acquet], *JP,* 7 (1776), 501–8; Henley, *PT,* 66

Opuscoli di Milan for 1776, acknowledging the role of Cigna, claiming no credit for the principle of VE, but insisting that he alone had made a usable instrument, had developed the dielectric, the armatures, and the play with the bottle. He recognized his tactical error, and dropped 'vindice' from his vocabulary: 'I now say *elettricità permanente,* and no longer *vindice,*' he wrote in November, 1775, 'since the idea conveyed by the latter term is less appropriate . . . , not to say altogether erroneous.'[39] The matter had thus been composed when a more imposing claimant appeared beyond the Alps.

After settling in as Thamisk lecturer in Stockholm, Wilcke had returned to a problem reopened by the air condenser, the location of the charge in the Leyden jar. He fashioned a multiple dissectible Franklin square (fig. 17.4), consisting of the glass ABCD, the coatings b, B, and the leads L, C (each furnished with detecting threads), the metal parts being mounted on insulating feet m which slid along a grooved bar RR. Wilcke assembled the square, charged it, disconnected C from the prime conductor and L from ground: C showed positive and L negative, as in the experiment of Richmann analyzed by Aepinus. [1] When he removed B and C *seriatim,* both proved positive, the latter more so (as measured by the excursions of electroscope strings), while the threads attached to L spread further. [2] Discharged, replaced and removed again, C was plus, but B minus, and strongly enough to make C negative when joined to it. 'Without further electrification by the globe,' Wilcke observed, 'at each removal C will be positive, B negative, as often as the trial is made.'

In another demonstration Wilcke came even closer to the manipulations of the electrophore as Volta initially described them. [3] He electrified the square as before, exploded it, and removed B and C. The lead L now appeared positive, B negative, giving the phenomenon Beccaria was to christen the 'oscillation of the electricities.' Wilcke took a spark from B and C, replaced them, briefly joined C and L, removed B and C, took another spark, and so on: 'In this way the glass can keep electrifying the coatings for many days or weeks, as often as the experiment is repeated.' An account of these experiments was published thirteen years before Volta invented the electrophore.[40]

(1776), 514–15; Barletti, *Dubbî* (1776), 47–50; Cf. Polvani, *Volta* (1942), 104–9. In the relevant experiments of Gray (*PT*, 37 [1731–2], 285–8), the electricity of sulphur lasted several months when the specimen stayed in a tight box; in those of Aepinus (and Wilcke), melted resin and sulphur charged while cooling in wooden frames or metal cups (Priestley, *Hist.* [1775³], I, 274; Wilcke, *Disputatio* [1757], 44–5; Aepinus, *Recueil* [1762], 69) but remained dormant until removed from their contrarily electrified receptacles. The same odd effect occurred in the manufacture of chocolate (Henley, *PT*, 67:1 [1777], 95; Liphardt, *JP*, 30 [1787], 431–3); it seems to arise primarily from friction during the pouring of the molten matter (van Marum and van Troostwijk, *JP*, 33 [1788], 248–53).

39. *VO*, III, 120, 137–43; *JP*, 8 (1776), 23. Cf. Barletti to Volta, 2 Jan. 1776, *VE*, I, 107, recommending 'elettricità spontanea indeficiente.'

40. Wilcke, *AKSA*, 24 (1762), 213–35, 253–74, esp. pp. 260–71. Cf. the 'electrical problem' in *JP*, 9 (1777), 6.

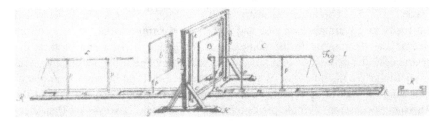

17.4 Wilcke's dissectible condenser. From Wilcke, AKSA, *24 (1762), 213–35.*

There is more. Volta gave no theory of 'eternal vindicating electricity.' Wilcke, however, had analyzed his own experiments to the last dissection, by distinguishing, as he had in his notes to Franklin's *Letters*, between the *communication* of electrical matter and the *segregation* of the normal supply of one body by the action of the atmosphere of another. Segregation occurs even in glass, as Wilcke showed by electrifying B as it stood an inch or so from the surface ABCD. The results of his dissections he explained as follows.

[1] When removed immediately after charging, B and C showed their communicated positive charge: L became more negative because the removal of the positive armature required a new segregation and a new equilibrium. [2] When B and C were grounded and returned, the positive electricity in the glass, retained there by the poor conductivity and the negative atmosphere of b, made B negative and C positive by segregation, as Wilcke demonstrated by separation *seriatim*; and B overcame C after joining because charge had leaked from C's far corners during segregation. Induction likewise accounted for the continuous re-electrification of B and C when, following segregation, they were grounded and returned. [3] When explosion preceded removal, b became positive and B negative entirely by segregation; after stripping B and C, L showed positive because the atmosphere of ABCD, unbalanced by that of its coating, drove electrical matter out of b.[41] When friends later drew Wilcke's attention to the electrophore, he had no trouble giving its theory or explanations of Symmer's and Beccaria's experiments, including the cohesion of charged double panes.[42]

The German translation of Wilcke's theory of the electrophore did not appear until 1782. Meanwhile the ubiquitous Dutch doctor Jan Ingenhousz advanced an approximation to it. After extensive studies that included physics with

41. Wilcke, *AKSA*, 24 (1762), 226, 231–3, 253–4, 272. Cf. the far less elaborate explorations of Herschel in 1780, *Sci. Pap.* (1912), I, lxxix–lxxx.

42. Wilcke acknowledged Volta's merit in designing a useful machine, but rightly asserted priority in discovering its principle, a claim supported by most German-speaking electricians, and ignored by Volta. Wilcke, *AKSA*, 39 (1777), 54–78, 116–30, 200–16, esp. 57–60 (relation to Volta), 71–8 (explanation of Beccaria's experiments), 119–21 (explanation of the electrophore); Lichtenberg, Ak. Wiss., Gött., *Novi Comm.*, 8 (1777), 168; Lichtenberg to Wolff, 10 Feb. 1785, in Lichtenberg, *Briefe*, III (1904), 203; Oseen, *Wilcke* (1939), 199, 215–17. Cf. Hoppe, *Gesch.* (1884), 72–3; Verrecchia, *Sudh. Arch.*, 51:4 (1967), 349–60; Gliozzi, *Cult. sc.*, 5 (1966), 235–9.

Musschenbroek at Leyden, Ingenhousz had set up practice in his home town, Breda, in 1757. The situation restricted his ambition. As a Catholic he found the roads to government service and academic position closed; as a nocturnal electrician—'sitting up in the company of the Father of Evil and plotting the destruction of mankind'[43]—he earned himself a reputation for necromancy that did not improve his business. When his father died in 1764 he went to England, following the advice of Pringle, who had met the family during the War of the Austrian Succession. Pringle presented his protégé, who 'always [kept] an electrical machine, ready for action, fix'd upon a table in my room,' to Franklin. The young doctor admired the Philadelphia system as a 'masterpiece of the human mind;' a strong friendship developed between himself and its inventor; and after the American Revolution Franklin invited Ingenhousz to join him in retirement. 'I have instruments if the Enemy did not destroy them all, and we will make Plenty of Experiments together.'[44]

Ingenhousz met the electrophore shortly after its invention. In 1767, in answer to a request from Maria Theresa and at the suggestion of Pringle, who was also royal physician, George III sent Ingenhousz to inoculate the House of Hapsburg. After dry runs on a hundred Viennese commoners, Ingenhousz scratched the imperial children, succeeded, became the Empress's physician, and won a pension that supported his studies and his travels. Now Volta, who did not neglect such things, had sent one of his first electrophores to the Archduke Ferdinand, who showed it to the family doctor. 'I can not express how much I was pleased with the first sight of it,' Ingenhousz told the Royal Society of London to which, as fourth Bakerian lecturer, he explained its operation in June, 1778.[45]

He required a new but not alien principle: 'The electric fluid that is accumulated upon a body, and finds obstruction in its free passage to another neighboring body, by the interposition of a non-conducting body (such as dry air, glass, etc.) forces by its repulsive power the electric fluid naturally contained in all bodies to the furthermost extremity of the neighboring body.' This inelegantly premised, Ingenhousz discussed the electrophore in the manner of Aepinus, and, having explained the oscillation of the electricities, showed the equivalence of Volta's machine and an exploded dissectible condenser.

Ingenhousz did not free himself entirely from Franklinist terms and models; he allowed himself phrases like 'the atmosphere of electrical fluid surrounding

---

43. Fitzmaurice, *Life* (1875), I, 447–8. The electrician Kratzenstein also had trouble with neighbors in the 1750s: believing that his apparatus would attract lightning and blow up the town, the citizens of Copenhagen sensibly declined to rent him a house. Snorrason, *Kratzenstein* (1967), 46.

44. J. Wiener, *Ingen-Housz* (1905), 13–21, 214–15; Ingenhousz, *PT*, 68:2 (1778), 1023, and *Nouv. exp.* (1785), x, xiii; *BFP* (Smyth), IX, 321.

45. J. Wiener, *Ingen-Housz* (1905), 23–9, 35; *DNB*, X, 433–4; Ingenhousz, *PT*, 68:2 (1778), 1031, and *Nouv. exp.* (1785), xiv–xv. For Baker, microscopist and would-be poet who had made a fortune teaching deaf-mutes, see G. Turner, *RS, Not. Rec.*, 29 (1974), 53–79; for the lectureship *supra*, ii.2.

the excited body' and 'the pressing action of the atmosphere of the cake.' Yet
the weight of explanation in his theory is carried by the charge 'accumulated' or
'crouded' upon the 'extremities' of conductors, the 'repulsive force' of which
acts at a distance to segregate the natural supply of neutral bodies.[46] The theory
met little opposition. Some, indeed, thought Ingenhousz belabored the obvious.
'As far as I understand it from your Description, it is only another form of the
Leyden Phial,' Franklin told his friend. 'Mr Ingenhousz has exhausted himself
in the last volume of the Philosophical Transactions,' wrote one of Volta's
correspondents, 'explaining or rather repeating the experiments and explanations given of your electrophore, which he likens to a Leyden jar.'[47]

3. LOCALIZATION OF CHARGE

Many had greeted the electrophore, 'which daunted the most celebrated [elec-
tricians] of Germany and Italy,' as a profound mystery, 'a riddle as great as the
Leyden jar,' 'a menace to the Franklinist system.'[48] In Vienna it drove the
local scientific gossip, the abbé Jacquet, 'to the end of my rope.' In Paris M.
Rouland, Sigaud's nephew and successor, a Nolletist once removed, warned, or
rather hoped, that, 'if it does not directly contradict [accepted theories], it presents many phenomena reconcilable with them only with the greatest
difficulty.' In London William Henley, a linen draper by trade and momentarily
the Royal Society's most prolific Franklinist, hunted unsuccessfully for the
source of the shield's endless charge; 'it is hard to say how or where the electricity is deposited,' he fretted, 'there is so much of it.'[49] The theories of Ingenhousz and Wilcke ended this distress. Still, those who anticipated that the
electrophore would affect electrical theory profoundly were not wrong.

In the considered judgment of the encyclopedist Gehler, 'the instrument has
been as valuable to theory as to practice.' At first as perplexing as the Leyden
jar, it enabled German-speaking electricians to bring forward the true interpretation of electrical phenomena. The same opinion appears in the physics text of
F. K. Achard, chief electrician of the Berlin Academy in the 1770s and 1780s:
'The announcement of the electrophore was the spark that drew the favorable

46. Ingenhousz, *PT*, 68:2 (1778), 1037–48; *PT*, 69 (1779), 661; *BFP*, XIV, 4, 165; *BFP*
(Smyth), VIII, 189–94. The experiments of Volta and Henley had meanwhile been confirmed by
Cavallo, *PT*, 67:2 (1777), 388–98, who gave no theory.

47. Ingenhousz, *JP*, 16 (1780), 117–26; Franklin to Ingenhousz, 26 April 1777, *BFP* (Smyth),
VII, 49–50; J. Senebier to Volta, 29 July 1780, *VE*, I, 416. Cf. Henley, *PT*, 68:2 (1778), 1049–
55, and the independent work of A. Socin (*infra*, xviii.2), *Anfangsgründe* (1778²), 95, 105–8,
and postscript, note 4, for concurrence.

48. De Herbert, *Th. phen.* (1778²), Praef.; Achard, *Vorlesungen* (1791), III, 60; Baldi, *De
igne* (1790), 59.

49. J[acquet], *JP*, 7 (1776), 501–8; Rouland, *ibid.*, 438–42; Henley, *PT*, 66 (1776), 514–16, 521.
Henley first became known, through Priestley, as the inventor of an electrometer (*PT*, 62 [1772],
359–64, and *infra*, xix.1), which helped make him F.R.S. in 1773; he died by his own hand in 1779.
Cf. *DNB*, IX, 421; Canton Papers, II, 85–6, 91–108 (RS); *BFP* (Smyth), VII, 222, 413; and Lichtenberg, letter of 2 Sept. 1779, *Briefe* (1901), I, 321.

attention of electricians to the long neglected approach [of Aepinus].'[50] Achard wrote from personal experience. He had given a rough theory of the electrophore in 1776, couched in Eulerian terms, electricity being a special motion or state of an electrical matter.[51] He converted under the guidance of Ingenhousz and Wilcke, after discovering for himself that evaporation and vegetable growth were promoted both by positive and by negative electricity. Hydrodynamical mechanisms or atmospheres could not account for artificial or animal electricity; what is left but appeal to forces acting at a distance?[52]

THE MARCH OF AEPINUS

The importance of Aepinus' views came clear to Europe's electricians at about the same time that they puzzled out the workings of the electrophore. No doubt some of them learned the new doctrine of atmospheres directly from the *Tentamen*, but many took it from qualitative theories of the electrophore and the dissectible condenser. Perhaps the most influential of those who spread Aepinus' views in this manner was Volta himself, who met with a copy of the 'incomparably profound' *Tentamen* in the late 1770s. At the same time he appears to have digested Cavendish's now famous memoir of 1771, then neglected, which develops a Newtonian theory of electricity.[53] The combination of this reading, of the natural development of his own views, and, perhaps, of the writings of Barletti, who first acquainted him with Aepinus, worked a change in Volta's approach to electrical theory. The first public expression of his new style, or 'second manner,'[54] came in 1778, in the form of an open letter to Saussure on the capacity of condensers.

Volta now did not seek to justify or even to apply his old microscopic model, but admitted what macroscopic forces he needed: repulsions between accumulations of homologous charges, attractions between those of contrary charges. Although he still spoke of electrical atmospheres, he meant thereby nothing more than 'spheres of activity,' the range over which the charge accumulations (or deficiencies) *confined* to electrified bodies could exercise their virtues. As appears from a manuscript of 1778, he supposed electricity to derive 'not from mechanical impulse, but from the principle of mutual forces,' whose dominion, 'already extensive in physics and chemistry, is becoming daily more evident in the phenomena of electricity.' 'Nothing real,' he says, passes between bodies interacting electrically beyond sparking range; the surplus electrical fluid of a positively charged body does not sit in the air around it. Spheres of activity do the work in his paper on the condensing electroscope, published in the *Philosophical Transactions* for 1782; and the following year, in the *Journal de*

50. Gehler, *Phys. Wört.*, I (1787), 817; Achard, *Vorlesungen* (1791), III, 60; cf. Kühn, *Die neu. Ent.* (1796), I, 174–5.

51. *MAS*/Ber (1776), 130–1; *Chem.-phys. Schr.* (1780), 226–33.

52. *JP*, 21 (1782), 199; *JP*, 25 (1784), 433; *Sammlung* (1784), 290. Cf. *DSB*, I, 44–5, and Stieda, Ak. Wiss., Leipzig, Phil.-Hist. Kl., *Abh.*, 39:3 (1928), 11–13.

53. *VO*, III, 210n, 236; Gliozzi, *Physis*, 11 (1969), 231–48; Barletti to Volta, 24 March 1776, *VE*, I, 121; Barletti, *Dubbi* (1776), 61–3, 103–19.

54. Gliozzi, *Cult. scuola*, 5 (1966), 235–9.

physique, 'rending [as he says] the veil that prevents little electricians from seeing clearly into the laws of electrical atmospheres,' he warns that there is no escape from *actio in distans,* 'for that is the action of electricity.'[55]

Although Volta never published a full account of his theory of electrical atmospheres, learned Europe did not languish in ignorance of it. From the chair of experimental physics at the University of Pavia, to which he was called in 1778 from the rectorship of the state-supported secondary school in Como,[56] he kept up an extensive international correspondence. On trips beyond the Alps he carried his message directly to foreign physicists. A visit to Paris in 1781/2 was especially fruitful. The Academy invited him to demonstrate electricity; Buffon, Le Roy, Lavoisier and Franklin himself, now the American agent in France, desired his company at dinner; and Laplace, then collaborating with Lavoisier,[57] and the expatriate Genevan geologist J. A. Deluc, Reader in Natural Philosophy to the Queen of England, confessed themselves smitten with his system.[58]

Deluc, who had been introduced to electricity by Jallabert in the 1740s, had hesitated between Nollet and Franklin until rescued by Volta. He then recognized Franklin as the Kepler (Volta being the Newton) of electricity, and promised to publish a faithful account of the new doctrine of atmospheres, 'a project which,' he assured Volta, 'is looked forward to here [England], in France, in Holland and in Switzerland.'[59] Deluc never epitomized Volta, but instead invented an idiosyncratic theory employing action at a distance and an electrical fluid compounded of 'electrical matter' and a 'deferent,' which turned out to be plain water.[60] A more useful agent was Tiberio Cavallo, a Neapolitan settled in

55. *VO,* III, 206, 236, 373; *VO,* IV, 65–8, 71–4; Volta, *PT,* 72:1 (1782), App., xii–xxxiii; cf. the unpublished notes on the electrophore in *VO,* III, 166–7, 182, 239–40; Volta to Landriani, 8 July 1775 and 11 Oct. 1778, *ibid.,* 155, 160, and Volta to Mascherini, 20 May 1784, *VE,* II, 213. Polvani, *Volta* (1942), 113–17, and Gliozzi, *Cult. scuola,* 5 (1966), 235–9, following earlier writers like Biot, *Biographie* (1880²), XLIV, 77–81, and J. D. Forbes, *Review* (1858), 164, err in taking Volta's atmospheres in a mechanical sense.

56. Volta's discovery of methane in 1776 had consolidated his reputation, and he was given the professorship at the expense of another. Pavia then had two chairs of physics, a 'general' held by Francesco Luini, a former Jesuit, and an 'experimental,' held by Barletti. Volta's patron, Count Carlo di Firmian, Austria's plenipotentiary in Lombardy, gave Barletti Luini's chair, Volta Barletti's, and moved Luini to Milan. Vaccari, *Storia* (1957²), 177–8; von Wurzbach, *Biog. Lex.,* XVI, 154, says that Luini's philosophical views had made him unpopular; Bernoulli, *Lettres* (1777–9), III, 59, calls him a promising young mathematician.

57. Volta joined them in an effort to detect electricity produced by evaporation; they succeeded (although the effect was spurious) probably by measuring charge generated by the friction of bubbles against the evaporating pan. The sensitivity of Volta's condensatore made the investigation possible. *VO,* III, 33–4, 301–5; *VO,* V, 173–87, 196–7; *VE,* II, 104–5. The French might also have obtained from Volta an important clue towards the discovery of the composition of water (*VO,* VI, 410–11).

58. *VE,* II, 51–141, esp. 84–5, 96; Laplace and Lavoisier, *MAS* (1781), 292–4.

59. Deluc to Volta, 29 June 1783, *VE,* II, 163–5; Deluc, *Idées,* I (1786), 3–4. Cf. Deluc to Volta, April–June, 1784, *VE,* II, 203–9; *DNB,* V, 778–9.

60. Deluc, *Traité* (1804), I, 36–9, 74ff; *Idées* (1786), I, 232–8; Lichtenberg, *Briefe,* I, 267–8, 282, 300, 384–5, II, 287, 300, 332; Miller, *J. Nat. Phil.,* 4 (1800/1), 462.

London where Volta met him in the spring of 1782. Cavallo had come to the metropolis in 1771 to study commerce, and remained to become the leading English electrician of the 1780s and a prolific writer of authoritative textbooks on natural philosophy, particularly electricity.[61]

In 1784 Volta invaded Germany. His chief conquest was the Göttingen professor G. C. Lichtenberg, a pastor's son now admired for his trenchant aphorisms but best known to his contemporaries as a mathematician, physicist and pitiless critic of second-rate work. Lichtenberg came to electricity from curiosity about the electrophore. He built himself a huge apparatus with a fifty-one-pound cake, whose copious shavings filled every cranny of his laboratory. The clutter caused a splendid discovery. Having lifted the ponderous shield from the excited electrophore, Lichtenberg found that dust scattered by a passing draft had settled on the great cake in patterns resembling suns and stars. The patterns, or 'Lichtenberg figures,' depend on the nature of the dust and the degree and sign of the electrification of the surface on which it falls. Lichtenberg supposed that they might map electrical atmospheres in the same way that iron filings traced magnetic ones.[62] He presently stopped seeing them as crosssections of material atmospheres and inclined towards Volta's position, without, however, sharing his friend Deluc's high estimate of Don Alessandro.[63]

Volta's visit changed that. Lichtenberg could not resist his 'lusty' guest, a genius who swore and cackled over his experiments, guzzled and disputed over his dinner, electrified the ladies (a 'Reibzeug für die Damen'), and understood more of electricity than anyone else in Europe. After a few days with this 'raisonneur sans pareil,' Lichtenberg was prepared to grant him the Newtonship of electricity.[64] And, what was more important, Lichtenberg stressed the new interpretation of atmospheres in his influential editions of Erxleben's *Anfangsgründe*: the rule of the working of atmospheres, he wrote, and the recognition that they do not exist, provide the 'key to the secrets of electricity.'[65]

The connection between the electrophore and the adoption of ideas akin to Aepinus' also appears in the work of Joseph Weber, modernizing priest, friend of the Enlightenment, and professor of physics at the universities of Dillingen

61. *DSB*, III, 153–4; Cavallo to Volta, 7 Feb. 1791, *VE*, III, 118; Cavallo, *Comp. Treat.* (1795³), III, 282: 'electric atmospheres (by which we mean only that power of acting on other bodies).' Cf. Paffrath, *Arch. Ges. Math. Naturw. Techn.*, 5 (1914), 86–92; Mondini, *Mach.*, 2 (1969–70), 19–26.

62. *VE*, II, 225–73; Lichtenberg to Schernhagen, 17 Feb. 1777, and to Hollenberg, March 1777, *Briefe*, I, 276–9, 378; Lichtenberg, Ak. Wiss., Gött., *Novi Comm.*, 8 (1777), 169, 172; *GGA*, 1 (1778), 345–8; P. Hahn, *Lichtenberg* (1927), 5–9, 20–3, 36–43.

63. Lichtenberg, Ak. Wiss., Gött., *Novi Comm.*, 1 (1778), 77; Deluc to Volta, 29 June 1783, *VE*, II, 163. The causes of the figures are complex; cf. Cavallo, *PT*, 70 (1780), 15–29; Beccaria, *Op. scelte sci. arti*, 3 (1780), 242–7; Przibram, *Handbuch*, XIV (1927), 391–404; von Hippel and Merrill, *J. App. Phys.*, 10 (1939), 873–87.

64. Lichtenberg to Sömmering, 25 Oct. 1784, to Wolff [1784] and 10 Feb. 1785, *Briefe*, II, 150, 153–4, 203.

65. Erxleben, *Anfangsgründe* (1791⁵), 513–14. Erxleben had doubted the materiality of atmospheres.

and Ingolstadt. In the early 1770s, while still studying theology, Weber bought an electrical machine from an old-clothes peddler. The purchase showed its value in 1778 when, just after his ordination, he won a prize from the Bavarian Academy of Sciences for an 'air electrophore,' an air condenser with a movable coating, whose action he explained in the modern manner, using 'atmosphere' to mean 'sphere of activity.' In 1783, in a formal account of electrical theory drawn up for a new scientific society in Berlin, he observed that, until experiment prove the contrary, 'we must maintain that electrical matter does not leave an [electrified] body to form an atmosphere.'[66]

The same message, coupled always with explanations of the electrophore and / or the dissectible condenser, occurs in much of the respectable literature on electricity published in the Germanies in the late '70s and '80s. Joseph de Herbert, a onetime Jesuit who became professor of physics at the University of Vienna, teaches in his excellent *Theoriae phenomenorum electricorum*: 'Electrical actions [other than sparking] do not originate in the transition of fluid from a body with a surplus to one deficient [as in Euler's theory], but by . . . action at a distance.' Cölestin Steiglehner, a friend of Ingenhousz, a Benedictine, professor of mathematics at the former Jesuit university in Ingolstadt, received third place in the Bavarian Academy's competition of 1780 for a rephrase and extension of the *Tentamen*. 'Corpora electrizata vires suas ad aliquod intervallum exerunt,' an ex-Jesuit pedagogue, J. B. Horvath, writes in 1782. 'Dieser Raum,' says Achard, 'nennt man den Wirkungskreis oder die electrische Atmosphäre.' 'You must not take these atmospheres to consist of special, invisible emanations,' another professor warns; they are simply spheres of activity, *Wirkungskreise*, the region of detectable actions apparently exerted *in distans*.[67] In sum, according to the best physics dictionary of the age, atmosphere, Einflüsse, sphaerae activitatis electricae, influences électriques, Wirkungskreis, all mean the same thing, 'the space through which electrical action reaches.'[68]

Outside the lands of self-conscious textbook writers the connection between the electrophore and localization is less evident. Perhaps it underlies the tergiversation of Barletti, who had grown disenchanted with the orthodox identification of vitreous and positive electricity shortly after publishing a routine Franklinist text in 1772. Why, he asks in 1776, should one hold to Franklin's

66. J. Weber, Ak. Wiss., Munich, *Neue phil. Abh.*, 1 (1778), 171–214, esp. pp. 184, 208; Ges. nat. Freunde, Berlin, *Schr.*, 4 (1783), 347, 360–3, 368–73; *ADB*, XLI, 316–18. Weber converted to Naturphilosophie in 1803–4, after endorsing Kant and being censured for his trouble; his career would repay study. Cf. Trefzger, *Weber* (1933), 1–15; Schmid, *Erinnerungen* (1953), 93–5; Specht, *Geschichte* (1902), 575–8.

67. De Herbert, *Th. phen.* (1778²), 72–80; Steiglehner, Ak. Wiss., Munich, *Neue phil. Abh.*, 2 (1780), 229–350, esp. 276–7; *ADB*, XXXV, 593–5; Horvath, *Physica* (1782), 318–19; Achard, *Vorlesung* (1791), III, 32–3; Hube, *Voll. fass. Unter.* (1801²), I, 406, 420. Cf. Kühn, *Neu. Ent.* (1796), I, 148–50, 175–6; Saxtorph, *El.-laere*, I (1802), 268–78, 303.

68. Gehler, *Phys. Wört.* (1787–91), I, 737–9, IV, 800–1. Gehler's articles on electricity were influenced by Aepinus.

arbitrary guess? 'Especially since he had a false idea of spheres of activity, which he called and continues to call *electric atmospheres,* an idea corrected by Aepinus.'[69]

In England one often finds succinct formulations, without reference to atmospheres, of what amount to rules of operation of the *Wirkungskreis,* as in the booklets of the instrument makers Nairne and Cuthbertson or in the spare texts of Ferguson, Enfield and Nicholson.[70] But except for the eccentric Cavendish, the English made no connection with Aepinus until John Robison, a Scot who had lived in Russia and knew Aepinus personally, wrote an admirable account of the *Tentamen* for the supplement to the third edition of the *Encyclopedia Britannica,* published in 1801.[71]

In the low countries the leading authorities divided over the foundations of electricity. Needham, now the head of the Brussels Academy of Sciences, had no trouble accepting the notion of *Wirkungskreis;* indeed he took it for granted that the 'action' of the 'sphere of electricity' drops off with the square of the distance, as in the case of gravity. In Holland van Marum at first accepted Aepinus, including the explanation of minus-minus repulsion; but he was soon bamboozled by his countryman van Swinden's spurious disproof of Aepinus' principles, and returned to literal Franklinism.[72]

Toward the mid-'80s the French shook off their torpor and embraced modern electrical theory. We noted Volta's success with Laplace. In 1784 a translation of Nairne's manual appeared, furnished with an introduction asserting the localization of charge; in 1785 Coulomb began his celebrated experiments, which rest on a theory equivalent to Aepinus'; and two years later the abbé R. J. Haüy, crystallographer and physicist, member of the Paris Academy, incorporated Coulomb's results and other phenomena discovered since 1759 in the nonmathematical epitome of the *Tentamen* that he issued to the great applause of reviewers in France and Italy.[73]

THE STRESSED MEDIUM

The views just sketched represent the majority response to the question raised by the *Journal de physique* shortly after the popularization of the electrophore. 'Does electricity reside only on the charged conductor, without extending be-

69. Barletti, *Phys. sp.* (1772), 30–4; Soc. ital., *Mem. mat. fis.*, 1 (1782), 28, 43.

70. Nairne, *Directions* (1773), 42–3; Cuthbertson, *Abhandlung* (1786), 1–11, in *Pract. El.* (1807), 1–10; J. Ferguson, *Intro.* (1778[3]); Nicholson, *Intro.* (1782), II, 403–6; Enfield, *Institutes* (1785), 342–3, 347. Neither Nairne nor Ferguson owed anything to the electrophore; cf. Ferguson, *Intro.* (1770), 44–5.

71. *Infra,* xix.2–3. Even Wilson seems to have pulled in his atmospheres a bit after 1775, e.g., *PT,* 68:1 (1778), 295.

72. Needham, Ac. sci., Brussels, *Mém.*, 4 (1783), 63–4; Hackmann in Forbes, *Marum* (1971), III, 337, 344, 373, n.53; van Marum, *Ann. Phil.*, 16 (1820), 450.

73. Caullet de Veaumorel, tr., *Description* (1784), ii, xxxi–xxxii; Haüy, *Exposition* (1787), of which lengthy reviews in *JP,* 31 (1787), 401–17, and *Bibl. oltr.*, 2 (1788), 139–60, and 3 (1788), 221–38 (Calcaterra, *Nostro imm. Ris.* [1935], 395, n. 104). Cf. Berthollet to van Marum, 30 July 1787, in Sadoun-Goupil, *RHS,* 25 (1972), 234.

yond its surface . . . or does the electric matter reach beyond the conductor, in the form of an atmosphere, up to the point where electric signs disappear?'[74]

It must not be supposed that this consensus concerned modern elementary electrostatics. The agreement related exclusively to the *location* of the charge of electrified bodies, not to the manner of its action. To us, as to the puzzler of the *Journal de physique,* the localization is the cardinal point: by rigorously distinguishing the electrical matter from its field, the electrician of the 1770s and 1780s discarded the last seriously misleading tenet of the effluvialists. Moreover, in practice this radical separation brought with it analyses in the style of Aepinus, since no competent mechanical representation of the field existed. But the consensus on localization did not oblige one to accept literal actions at a distance. Several sketches of approaches to field representations duly appeared.

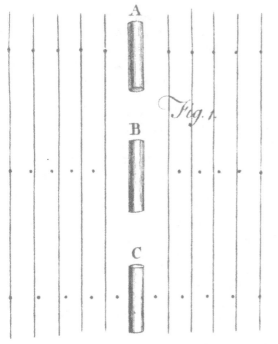

17.5 Canton's representation of the electric field. From Priestly, Hist., I, 305–6.

In 1766 Canton informed Priestley that electrical atmospheres 'are not made of Effluvia from excited or electrified Bodies, but are only Alterations of the State of the electrical Fluid contained in & belonging to the Air surrounding them to a certain Distance.' Canton's idea will be clear from figure 17.5, where A, B, and C are neutral, positive, and negative, respectively. Since B pushes and C pulls the surrounding electrical matter, the air near B has less and near C more than its normal complement. Because the stressed atmosphere forces the electrical matter in conductors immersed in it to assume the distribution appropriate to the air they displace, the figure also represents Canton's idea of a body suffering induction. Priestley, who already had reasons to 'suspect the existence of [electric] atmospheres,' published Canton's theory in his *History*, through which, perhaps, it came to the attention of Beccaria.[75] In February, 1769, just before bringing out the details of VE, which required literal atmospheres, Beccaria sent the Royal Society of London a pamphlet, *De atmosphaera*, which hinted at a model of Canton's type.

Details came three years later. 'The electricity of a body [Beccaria then wrote] resides within the superficial pores of it, and actuates the ambient air, not by diffusing itself into it, but by exciting either a tension or a relaxation in the natural fire inherent in it.' As indicated in fig. 17.6, the pressure gradients set up by positive bodies (e, E) oppose those surrounding negative ones (d, D). As in Canton's scheme, the tensed air induces charges by altering 'the equilibrium of the fire inherent in those bodies which are immersed in it.' Furthermore, it provided a new mechanism for electrical motions. Assume that bodies interacting electrically strive to move so as to reduce the tensions in the air. The usual rules of plus and minus electricity follow immediately.[76]

The approach òf Canton and Beccaria, which in effect assigned to the air some of the tasks Faraday later imposed upon the aether, seemed ratified by the perennial misleading failure to detect electrical motions in a vacuum. Many competent physicists therefore welcomed the approach, although none worked it out in detail. Wilcke followed his theory of the electrophore with a hint that induction, attraction and repulsion all arise from a state of the air analogous to the polarization of the glass of a Leyden jar. An admirer of Volta's, Barbier de Tinan, who rightly held that a correct theory of atmospheres was the desideratum in electrical theory, suggested a modification of Beccaria's scheme.[77]

The stressed-air model can be followed in Italy from Beccaria and Barletti to Avogadro, whose version has earned him credit, among Italian writers, for the discovery of dielectric polarization and specific inductive capacity.[78] No doubt a similar line connects Faraday with the early English proponents of the model, Canton, the authoritative Cavallo, and their echoes Thomas Milner, Abraham Bennet, and the instrument maker George Adams.[79] In France and Germany

75. Canton to Priestley, 5 April 1766, Canton Papers, II, 65 (RS); Priestley, *Hist.*, I (1775³), 305–6.

76. Beccaria, *De at. el.* (1769), in *PT*, 60 (1770), 277–301, §§ 71, 79–80; *Treatise* (1776), 186, 218, 220, 384–7. Cf. Gliozzi, *Arch.*, 17 (1935), 41–2. Beccaria did not free himself altogether from his earlier concept of atmospheres; see his *Treatise* (1776), 383.

77. Wilcke, *AKSA*, 39 (1777), 121–6; Barbier de Tinan to Volta, 10 Nov. 1778, *VE*, I, 296 (cf. *ibid.*, 326–7, 376, 411), and *JP*, 14 (1779), 20.

78. Barletti, *Dubbî* (1776), 17–24; Avogadro, *JP*, 63 (1806), 451, and 65 (1807), 130–2. Cf. Avogadro, *Opere* (1911), 375–7; Gliozzi, *Fisici* (1962), 14.

79. Cavallo, *Comp. Treat.* (1795³), III, 195–7; Milner, *Experiments* (1783), 91–9; Bennet, *New Exp.* (1789), x–xi; Adams, *Essay* (1785²), 56, 76–80. A possible link is Woods, *Phil. Mag.*, 17 (1803), 103.

the theory had an influential advocate in E. G. Fischer, a professor of mathematics, who followed Wilcke as far as he could in a textbook widely used on the Continent.[80] Weber flirted with the same program, especially in his later romantic physics, and even Haüy, the vulgarizer of Aepinus, attributed electrical motions to pressures in the air set up by localized charges acting *in distans.*[81]

Naturally there were those who did not share the consensus regarding localization. Loudest among them was the future revolutionary J. P. Marat, doctor to the grooms, or perhaps the horses, of the Duc d'Artois, a rabid anti-academician and bilious egotist, the world's only electrical authority: 'Everything written on electricity before me,' he said in 1783, 'is but a heap of isolated, complicated, ill-digested experiments, scattered in 500 volumes.' Science had to be extracted from this chaos. 'I shut myself in my darkened room, observe according to my method . . . ; everything becomes intuitive, science is born.'[82] *Parturiunt montes: nascetur ridiculus mus.* Marat delivered himself of a system of effluence and affluence that the Academy, which had outgrown such things, declined to discuss.[83] It was perhaps this system that Charles, who himself had not abandoned atmospheres, mocked in a public lecture in 1782; Marat repaid the compliment by assaulting him in his home. 'If one is to be beaten for that [criticizing Marat], all Europe must arm.'[84] Many of Marat's apologists think that the Paris Academy rebuffed him out of politics or

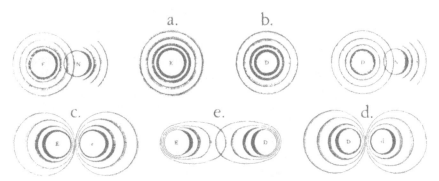

17.6 Beccaria's representation of the electric field. (a) About a positive body (b) about a negative one (c) between positive bodies (d) between negative ones (e) between unlike ones. From Beccaria, Elettricismo *(1772).*

80. E. G. Fischer, *Physique* (1806), 263–9; cf. Chappe, *JP*, 40 (1792), 329–32, and J. C. Fischer, *Anfangsgründe* (1797), 650–3.

81. Weber, Ges. nat. Freunde, Ber., *Schr.*, 4 (1783), 364–5, and *Dynam. Leben* (1816), 41–4; Haüy, *Traité* (1803), I, 362–5. Cf. Biot, *Traité* (1816), II, 317–26.

82. Marat, *Corresp.* (1908), 42; Cabanès, *Marat* (1891), 32–6, 54–5.

83. Marat, *Découvertes* (1779²), *Recherches* (1782), and *JP*, 21 (1782), 73–9.

84. Charles to ?, 17 March 1783, in France, *Elvire* (1893), 31; Fourier, *HAS* (1829), lxxxiv–lxxxv. Charles seems to have taught as late as 1795 that induction operates by 'pressure of atmospheres,' an 'improper and quite unsuitable expression' in the view of his auditor, T. Bugge; Bugge, *Science* (1969), 167–8.

jealousy.[85] That does not explain why competent electricians outside of Paris also rated him a crank.[86]

There were less extravagant revivals or survivals. Formey abridged Nollet and the author of the article on electricity in the third edition of the *Britannica* tried vainly to repair him.[87] An Italian priest, L. A. Ceppi, explained the electrophore, in 345 pages octavo, on lines simultaneously Nolletist and vorticist, while two of his countrymen made do with Beccaria's old Cabean theory and a third, G. F. Gardini, M.D., tried to update Euler.[88] Another revivifier, J. A. Donndorff, a lawyer in Quedlingberg, at once held atmospheres to be *Wirkungskreise* and envelopes of electrical matter, for 'no body can act immediately on another;' in his opinion, the atmospheres arrange themselves in the manner suggested by Canton and Beccaria, and endorsed by Abel Socin.[89] Meanwhile the youthful comte de Lacepède, uncertain whether to become a musician in the style of Gluck or a scientist à la Buffon, opted for science, attempted the long-missing Buffonian prose poem on electricity, and announced that the electric matter is a compound of fire and water, effective only when streaming. More fanciful, aristocratic and inexact was the comte de Gallitzin, a Russian diplomat by trade, whose experiments consisted primarily in 'sacrificing small animals to electricity.' He found it a relief from diplomacy and, when rightly interpreted, a confirmation of his belief that positive atmospheres came studded with spikes.[90]

In England, never bereft of eccentrics, the antiquarian John Lyon traced electricity to streams of polar effluvia, or electric lodestones, 'a distressing [idea], certainly,' said Lichtenberg, who, however, handled it relatively gently in review, 'for I've never been able to forget Marat.' And Lyon does compare favorably to one Peart, a country doctor who, having proved the 'absurdity' of Franklin's system, propounded a rococo replacement requiring two sets of electric atmospheres, one inside the other.[91] Few of these writers were respectable physicists, and none succeeded in explaining plausibly the behavior of novelties like the electrophore. The incontinent rejection of their atavistic schemes, which faithfully represent the dissenting minority, testifies abundantly to the strength of the consensus they declined to share.

85. Marat, *Corresp.* (1908), 5–87; Cabanès, *Marat* (1891), 150–62, 295–320; Dauben, *Arch. int. hist. sci.*, 22 (1969), 235–61.

86. Volta to van Marum, 8 Feb. 1786, *VO*, V, 66; Lichtenberg, *GGA* (1780:suppl.), 705–14.

87. Formey, *Abrégé*, I (1770), 276–82; Brisson, *Traité* (1800³), as reviewed by Deluc, *Traité* (1804), I, 9–35; *Encycl. Brit.*³, VI, 458–61.

88. Ceppi, *Dissertazione* (1784), 6, 38–42, 113–14; Adelasio, *De el.* (1781), 2–3; Follini, *Teoria* (1791), 56–62; Gardini, *De el. ignis nat.* (1792), 21.

89. Donndorff, *Lehre* (1784), 211, 217, 220–6; Socin, *Anfangsgründe* (1778²), 100, and *infra*, xix.1.

90. Lacepède, *Essai* (1781), esp. I, 53, 60–1, 80–1; Cuvier, *HAS*, 8 (1825), ccxii–ccxlviii; Roule, *Hist.* (1932), 35–6; Gallitzin, *Sendschreiben* (1780), 6–11, *JP*, 21 (1782), 73–9, and Ac. roy. sci. Belg., *Mém.*, 3 (1780), 2–11.

91. Lyon, *Experiments* (1780), 16–8, 85–6, 144–7; *DNB*, XII, 350; Lichtenberg, *GGA* (1780:suppl.), 709–10, and letter to Heyne, 20 Dec. 1780, *Briefe*, I, 369; Peart, *El. At.* (1793), 14–41. Cf. Read, *Summ. View* (1793), 99–102.

CHAPTER XVIII

Two Fluids or One?

1. SYMMER'S SOCKS

'I had for some time observed, that upon pulling off my stockings in an evening they frequently made a crackling or snapping noise; and in the dark I could perceive them to emit sparks of fire.' Pursuing this commonplace,[1] which began to intrigue him in November, 1758, Robert Symmer, an 'upstart,' as he put it, in electricity, was led to 'differ from all who had ever wrote upon the subject.' He proceeded from the discovery that a pair of stockings, one white, the other black, worn together on the same leg and then removed, showed almost no electricity *as long as they remained united*; but that, when separated, each eerily swelled as if it still contained a limb, and attracted small bodies five or six feet distant. And, conversely, the separated stockings again collapsed as they rushed together, until, lying flat upon one another, their enfeebled power extended no more than two or three inches.

'What appears most extraordinary is, that when they are separated, and removed of a sufficient distance from each other, their electricity does not appear to have been in the least impaired by the shock they had in meeting. They are again inflated, again attract and repel, and are as ready to rush together as before.' On cold crisp days Symmer could throw the expanded socks against the wall in his room, where they would stick, and pirouette under passing breezes, in a cancan that amused sophisticates like the Prince of Wales. The perfectly symmetrical and yet 'contradictory powers' of the black and white stockings—equally strong when separated, similarly impotent when united—showed Symmer that the contrary electricities arose from *two distinct positive principles,* perhaps materialized as two different, counterbalancing fluids. In this, as he rightly believed, he differed from all his predecessors.[2]

Symmer was not the clown his comical apparatus might suggest. He was educated at the University of Edinburgh, where he took a belated M.A. in 1735. Finding no local opportunities, he left Scotland as tutor and traveling companion to Lord Brooke, later Earl Warwick. He settled in London a few years later, rejoining old Scottish friends and fellow émigrés like the poet James Thomson, the mathematician Patrick Murdoch (F.R.S. 1745), and Sir

---

1. Cf. Miles, *PT*, 43 (1744–5), 443: 'There happened, about the month of November [1683], to one Mrs. Susanna Sewall . . . a Strange Flashing of Sparks . . . in all the wearing Apparell she put on.' Cooper, *Phil. Enq.* (1746), 16, reported much the same observations about doffing socks that Symmer did.

2. Symmer, *PT*, 51:1 (1759), 340, 354, 382; Symmer to Mitchell, 12 Feb. and 19 June 1760, Add. Mss., 6839, ff. 161–2, 182–3 (BL). Cf. Nollet, *Lettres* (1760), 3.

Andrew Mitchell (F.R.S. 1735), who became plenipotentiary to Frederick the Great during the Seven Years' War.[3] With the help of Warwick, Symmer obtained the post of paymaster to the Treasurer of the Chamber, the officer responsible for disbursements of the King's Household.[4] Connections forged in this position brought him the offer, which he declined, of the governorship of New Jersey.

Symmer's scientific interests were stimulated by Mitchell and by Pringle and, apparently, by mathematicians, since Abraham de Moivre, George Lewis Scott (perhaps de Moivre's most brilliant student), and John Gray signed his certificate for election to the Royal Society.[5] (Gray wrote on ballistics; he and Scott, like Symmer, made careers in the financial bureaucracy.) Symmer was welcomed to the Society on December 6, 1752, the same evening that John Canton read his important paper on electrostatic induction.[6] Six years later, while enjoying philosophical leisure forced by a change in ministry, Symmer delivered his first paper to the Society, an account of his astounding stockings.

Initially Symmer thought that the behavior of his socks owed much to their being animal substances excited by the same, to wit, his leg; but the evident analogy between their quiescent union and an insulated condenser gradually grew on him, and soon he was charging Leyden jars with his stockings. 'One frosty evening in [December, 1759], having thrown into a small phial filled with quicksilver, the electricity of one black stocking, I received from the explosion a smart blow upon my finger. With the electricity of two stockings, the blow reached both my elbows; and, by means of four, I kindled spirits of wine in a tea spoon . . . [and] felt the blow from my elbows to my breast.' The progressive, systematic penetration of the shock with increased electrification suggested a second argument in favor of Symmer's new view: apparently a real fluid sprang from each surface of the discharging jar. Nollet had claimed as much, and had tried to trace the double flux in holes struck through pads of paper placed in the discharge train. Symmer determined to do the same, although he recognized the deep division between his theory, which required two qualitatively different fluids, and Nollet's, which made do with one. Lacking the necessary apparatus, Symmer applied to Franklin who, 'with great civility,' assisted him in puncturing quires of paper and books with interleaved sheets of tin. The perforations did seem to confirm a double flux: as we know, the rend-

3. Heilbron, *Isis,* 67 (1976), 7–20; *DSB,* XIII, 224–5.

4. Symmer to Mitchell, 21 Jan. 1757 (Add. Mss., 6839, f. 33): 'Monday next I begin Payment of the Christmas Quarter, which, when it is finished, concludes my Administration in the Treasurer of the Chamber's Office. I understand Mr T [Charles Townshend, the fainéant treasurer] intends to have the business carried on by the Clerk and to put the Profits into his Brother Roger's Pockets.' Cf. Beatson, *Pol. Index* (1788²), II, 364–5, and *DNB,* XIX, 1043.

5. JB, XXI, 322; Pringle was Symmer's doctor (Add. Mss. 6839, f. 226). Symmer also was associated with the chemist William Lewis (*ibid.,* f. 171), with whom he shared an interest in platinum, and the botanist John Mitchell (*DNB,* XIII, 517), who assisted in some of his electrical experiments (*PT,* 51:1 [1759], 390–3).

6. JB, XXI, 410, 423.

ing proceeds from the covers toward the center, probably because of inductive forces between the electrodes and the pages.[7]

Inspired by the growing mania for measurement, Symmer found that his stockings could support up to ninety times their weight without separating. (The unexpected magnitude of this 'electrical cohesion'—to which Watson and John Mitchell testified—was to many the chief novelty of Symmer's researches.[8]) The feat suggested another analogy to a condenser: since each sock acts the part of one surface of a jar, a charged Franklin square should cohere even if cut longitudinally through its middle. Symmer armed one side only of each of two very thin glass panes, placed their bare sides together and electrified as usual: the panes stuck until he joined the coatings, solicited the explosion, and dissolved the cohesion. Two complete squares, superposed and electrified in series, gave the shock either separately or together, but declined to cohere. At their interface, according to Symmer, the independent, contrary electricities counterbalance, create a 'neutral state of electricity, and by that means prevent the two panes from acting on one another.'[9] He gave no further theory of these effects, which, however, follow readily from Aepinus' version of Franklin's theory.

Symmer had hoped to start a 'Revolution in the System of Electricity.' He failed at home. The Royal Society accorded his views little serious attention, partly, he thought, because of the unfortunate stockings. 'You may likewise be disgusted [he wrote Mitchell] with the frequent mention of pulling on, and putting off, of *stockings*: a circumstance, I confess, so little philosophical, and so apt to excite ludicrous Ideas, that I was not surprised to find it the Occasion of many a joke, among a sarcastical set of minute Philosophers, who do not love to have anything new forced upon them. With a certain physician (a friend of yours and mine [Pringle?]) I was *le philosophe déchaussé*, a bon mot I was much pleased with.'[10] Wilson thought to dispose of Symmer with a characteristic non sequitur: since, he argued, the type of electricity rubbed glass assumes can be reversed merely by roughening its surface, 'it follows, that the power of electrifying plus or minus arises from one and the same fluid.' Others objected, equally irrelevantly, that Dufay had shown the electrical powers of cloths to derive from their dyes, not from their colors. Symmer rightly rejoined that the nub of the matter was the *independence, constancy* and *contrariety* of the electricities of the stockings, not the mechanism of their production.[11]

7. Symmer, *PT*, 51:1 (1759), 358, 371–87; *supra*, xv.3.

8. Symmer, *PT*, 51:1 (1759), 359–70; Mitchell, *ibid.*, 390–3; Nollet, *MAS* (1761), 247; Priestley, *Hist.* (1775³), I, 311–12.

9. Symmer, *PT*, 51:1 (1759), 383–7.

10. Symmer to Mitchell, 19 June 1760, Add. Mss., 6839, ff. 182–3; cf. same to same, 7 April 1761, *ibid.*, 220–1. Symmer explained that he had chosen his uncouth presentation to protect himself from plagiarism: he reported his discoveries as he found them, before others could incorporate them into more elegant treatises.

11. Wilson, *PT*, 51:1 (1759), 329–30; Symmer, *ibid.*, 367–9; Wilson to Bergman, 26 Oct. 1761, in *Bergman's For. Corresp.* (1965), 418. Cf. Nollet, *MAS* (1761), 248–51.

As for the English Franklinists, they remained indifferent or antagonistic, despite the collaboration of their chief—whom Symmer thought an 'ingenious worthy sort of man'—in the supposed crucial experiment. 'The few that are in the way of these Things cabal together, and do what they can to stifle anything that is not of their own way of thinking,' Symmer complained. The Franklinist historians have joined the cabal. Priestley treats the dualist hypothesis fairly, but loses no opportunity to sneer at its restorer, our barefoot philosopher; Cohen adds that he cannot understand why Symmer's experiments aroused any interest at all.[12] 'If this little philosophical offspring of mine ever comes to make its fortune [Symmer wrote Mitchell], it must be abroad, I think, under your Protection; for it is starved, and brow-beaten at home, and will not be favored here, till it is honoured elsewhere.'[13]

Mitchell did what he could. He gave copies of his friend's papers to J. G. Sulzer, one of Frederick's academicians, who received it enthusiastically; to Frederick himself, who had no time for electricity; and to Winkler, who desired to know why Symmer retained the misleading Franklinist terminology. 'I confess [he replied through Mitchell] it was unlucky that I felt myself obliged to use, in some respect, the same terms that Mr Franklin and others, who follow his system, make use of, while there is an essential difference in the things meant by them and by me: By the terms *positive* and *negative,* they mean, as in algebra, simply *plus* and *minus*: By the same terms I mean two distinct Powers (both of them in reality positive) but acting in contrary Directions, or counteracting one another.'[14] Having secured a bridgehead in Germany, Symmer himself took charge of converting Nollet, to whom he sent his papers and a letter written 'with the greatest Deference possible.' Nollet returned a lengthy analysis, and his compliments for results 'not only new and surprising, but strictly true.' In his experience results both new and true were not characteristic of English science.[15]

Symmer's papers had arrived opportunely. Nollet's quiver stood empty. He had just shot off a second round of electrical letters, directed primarily against Beccaria, whom he transfixed with the inconsistency between the fervently Franklinist *Dell'elettricismo* and the Cabean *Lettere al Beccari.* Otherwise the new letters bring us no further than the old: we meet the same complaints about the arrogance of Franklinists, about their oversimplifications, about their refusal to give a plausible mechanical explanation of 'the first, the most infallible, the

---

12. Symmer to Mitchell, 19 June 1760 and 30 Jan. 1761, Add. Mss., 6839, ff. 183, 209; Priestley, *Hist.* (1775³), I, 308, 322–3, and II, 41, 47; *FN*, 543. The unfriendly French translator of Priestley's *History,* Brisson, caught the discrepancy in Priestley's treatment of Symmer and his hypothesis (*Histoire* [1771], II, 517, n. 121).

13. Symmer to Mitchell, 27 Feb. 1761, Add. Mss., 6839, f. 214.

14. Same to same, 27 Feb., 7 April and 10 July 1761, *ibid.,* ff. 214, 220–1, 229–30.

15. Same to same, 30 Jan. and 15 May 1761, *ibid.,* ff. 209, 225; *ibid.,* ff. 184–6, 231, giving excerpts from Symmer's correspondence with Nollet; Nollet to Dutour, 3 Oct. 1760 (Burndy): 'Je le vois avant de les avoir répétés que les résultats sont vrais, ils ne sont pas toujours tels quand ils viennent d'Angleterre.'

best known of all electrical phenomena,' simultaneous attraction and repulsion.[16] Beccaria's intimation that a minus body approaches a plus because the one wishes to give and the other to receive is not physics, says Nollet, but a theory of alms-giving; 'contrary electricities are phenomena, not causes.'[17]

Along with his *Letters* Nollet had seen through the press the baroque book of his friend Dutour, who had worked the contrary electricities into Nollet's system by supposing the electrical matter to 'vibrate' in two distinct manners, and by supposing a double set of pores in matter, one adjusted to each vibration.[18] Assume that everywhere there are more pores open to the fluid when vibrating in one way than when vibrating in the other. According to Dutour, in a body charged vitreously the effluent vibrates appropriately to the majority of pores, the affluent to the minority; in resinous electrification the association is reversed (fig. 18.1). Although Dutour could translate much of Franklin's system into his version of Nollet's, his cumbersome machinery could not grind out the crushing victory Nollet craved. Hence his enthusiasm over Symmer's paper, despite its 'pitiable reasoning.' 'I agree [he wrote Dutour, who had translated it for him] that here are some new proofs of simultaneous effluences and affluences; I be-

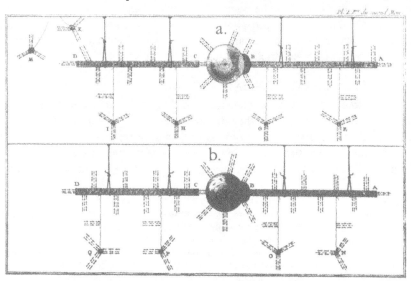

18.1 Dutour's updating of Nollet. (a) A glass globe gives positive electricity to the bar DC, whose effluent and affluent are represented by heavy and light arrows, respectively (b) a sulphur or resin globe conveys negative electricity to the bar, whose effluent is shown as a light arrow and affluent as a heavy one. From Dutour, Recherches *(1760).*

16. Nollet, *Lettres* (1760), esp. 56–7, 63, 155–62, 195–200; cf. Nollet to Dutour, 27 June 1759 (Burndy).

17. *Lettres* (1760), 155–62; *Lettres* (1767), 159.

18. Dutour, *Recherches* (1760); Nollet to Dutour, 17 May 1760 (Burndy).

lieve it will not be difficult to reduce the phenomena to them.'[19] All Nolletists will immediately understand his solution (fig. 18.2).

Being careful to keep the experiments secret from his archfoe Le Roy,[20] Nollet found the argument he sought in the behavior of the compound Franklin square. Symmer had implied that his opposite electrical powers, one on either coating, held the panes together; Nollet, who still proclaimed the permeability of glass, observed that under the circumstances contrary electricities would destroy one another, and therewith the cohesion. What experiment, he asks, could better prove the 'compatability' (and disprove the contrariety) of the electricities on either side of a Franklin square? Nollet simplified Symmer's favorite apparatus to a white silk ribbon and a glass plate, which showed the same cohesion as the stockings provided the ribbon were rubbed while resting on the glass. This suggested another argument against the contrary electricities. In Nollet's previous experience, white silk stockings always charged vitreously; in the present arrangement, however, they stuck strongly to glass. Should one infer that the ribbon charges vitreously when rubbed alone and resinously when rubbed on glass? It seemed more plausible to Nollet to admit an exception to the rule that bodies 'with the same electricity' always repel one another.[21] Then, as he observed, 'reciprocal attraction of two electrified bodies will not suffice to prove that their virtues are of different types;' indeed, one of the principles of the contrary electricities would fall to the ground.

2. A MATTER OF CONVENIENCE

We have already followed the elaboration of Nollet's Symmerian experiments by Cigna, who concluded that they did not determine whether electricity came in one fluid or two. By the late 1780s agnosticism, or forbearance, had won the day, complementing the newly-won consensus regarding the localization of charge; most physicists had stopped trying to demonstrate the exclusive truth of

19. Nollet to Dutour, 3 Oct. 1760 (Burndy): 'Je ne sçais pourquoy le bon Mr. Symmer est allé s'alambiquer dans les puissances distinctes, tous les raisonnemens qu'il met à ce sujet sont bien pitoyables . . . Je pense comme vous que voilà de nouvelles preuves pour les effluences et affluences simultanées, je crois qu'il ne me sera pas difficile d'y ramener les phénomènes dont il s'agit.'

20. Same to same, 3 Oct. 1760: 'Je garde le secret sur cet ouvrage jusqu'à ce que j'aye répété les expériences, et que je me sois mis en état d'y répondre; car s'il tombe entre les mains de le Roy et des Dalibards, ils ne manqueront pas de l'abuser pour embrouiller encore la matière.' The feud reached its climax in 1763, when Nollet succeeded in having the Academy censure Le Roy for errors of fact, including the claim that all Englishmen, and specifically Symmer, were Franklinists. Nollet to Dutour, 11 Oct. and 25 Dec. 1761, 14 July 1762, 15 Jan. 1763 (Burndy). Cf. Ferrner, *Resa* (1956), 384–5, 360, 366; Nollet to Wilson, 28 Feb. 1761, Add. Mss., 30094, ff. 45–7 (BL); Le Roy to Wilson, 5 Jan. and 9 May 1765, *ibid.*, 113–14, 117–18; and Wilson to Bergman, 29 June 1763, *Bergman's For. Corresp.* (1965), 423, where Wilson fatuously fancies that his advice will settle the vendetta.

21. Nollet, *MAS* (1761), 252–3, 257–8; Nollet to Dutour, 8 April 1761 (Burndy). The experiment in question differs from Wilcke's in that the silk is not intentionally moved against the glass; the finger alone is to rub the ribbon in both cases.

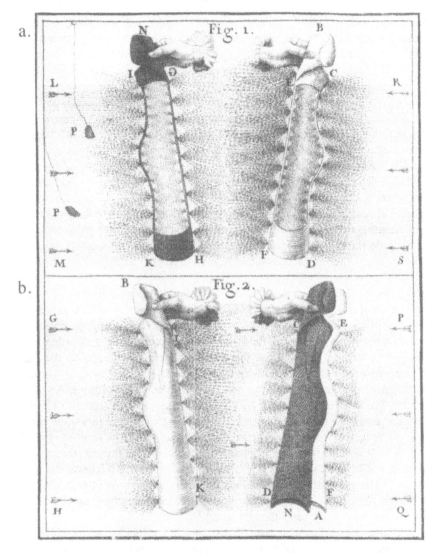

18.2 Nollet's elucidation of Symmer's socks. (a) When apart, the stockings swell owing to collisions between effluents from their internal surfaces, and 'attract' exterior bodies P by the usual affluent (b) they collapse when combined because of the interference between superposed resinous and vitreous jets. Note the short range of the jets from CD and EF. From Nollet, Lettres (1767).

either view, and affected to consult only their convenience in the choice of system. Before this positivistic paralysis could set in, however, Symmer's schism had to find leaders better placed and more powerfully armed than its founder. The earliest of these heresiarchs were Franklinist apostates like Wilcke and Barletti.

Wilcke's goal was to destroy the association, developed by Beccaria and Le Roy, between the character of point discharges and the direction of circulation of Franklin's hypothetical fluid. They had argued that the long, conical 'brush' consisted of electrical matter pouring freely *from* a positively charged conductor, while the 'star' showed it crowding to *enter* a negatively charged one. As Franklin's system developed, however, this association looked less plausible: electricians as opposed as Aepinus and Nollet argued that the air, being an insulator, should resist the outward passage of the electrical matter more than would crowding to enter permeable, if narrow, metallic points. Wilcke undertook to settle the matter by exploiting a phenomenon he had touched upon in his thesis, the scattering of phosphorescent powder from an electrified conductor.[22]

The powder, placed upon the spike B, shoots out in the narrow cloud C when the prime conductor A is electrified (fig. 18.3). C spreads from B irrespective of the sign of the charge given A. To us this proves nothing about the direction of circulation of the electrical fluid; the powder merely rides the electrical wind, or ion stream, driven from the point whatever the charge on the PC. Wilcke recognized that air currents carry the powder, but he held that the outstreaming electrical matter drove the wind. Those who accepted this effluvialist preconception inferred that negative as well as positive points spit forth electrical matter; and that, therefore, negative electricity consists not in a deficiency, but in an excess of some special material. This argument convinced one of Wilcke's colleagues, Torbern Bergman, then a junior faculty member at the University of Uppsala, where he had studied under Klingenstierna and Strömer.[23] The aggressive and ambitious Bergman lost no time in communicating Wilcke's results to the Royal Society of London, something the unpretentious Wilcke would not have thought to do.[24]

Franklinists less effluvialist than Wilcke had no difficulty answering him. The electrical wind, they said, neither consisted of, nor derived from, outstreaming electrical fluid; it was a current of air particles charged by communication at or near the points and driven away by the usual repulsion between

22. Wilcke, *Disputatio* (1757), 134.
23. Wilcke, *AKSA*, 25 (1763), 216–25; Bergman to Wilson, 18 Oct. 1763, *PT*, 54 (1764), 87. Cf. Wilcke to Bergman, 27 May 1765, in Oseen, *Wilcke* (1939), 121; Wilcke, *AKSA*, 28 (1766), 324; Bergman, *AKSA*, 27 (1765), 144–7; Priestley, *Hist.* (1775³), II, 44; E. G. Fischer, *Physique* (1806), 247. A similar argument was made by Kratzenstein; Snorrason, *Kratzenstein* (1967), 62–3.
24. Bergman was better known in England than Wilcke in 1780; Watt to Black, 1 March 1780, in Robinson and McKie, *Partners* (1970), 78. Cf. G. von Engestrom to Magellan, 24 Jan. 1783, in Carvalho, *Corresp.* (1952), 76 (Bergman is a man of great merit, but hasty, 'too eager to make discoveries'); *DSB*, II, 4–8; and Oseen, *Wilcke* (1939), 97–101, 120–32.

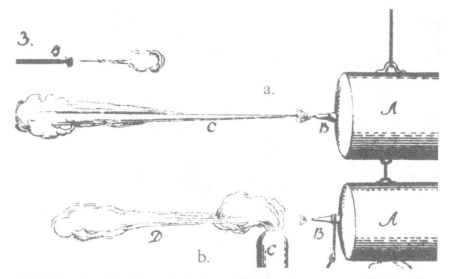

18.3 Wilcke's 'proof' of Symmer's theory. (a) The phosphorescent cloud C (b) its effect on the flame of the candle C. From Wilcke, AKSA, 25 (1763), 207–26.

bodies similarly electrified.[25] The wind and the phosphorescent clouds gave no clue to the circulation of the fluid(s), and in no way weakened the traditional arguments drawn from the star and the brush, which Beccaria reaffirmed with new demonstrations in 1766. The objections of Nollet and Aepinus, however, still stood, and forced Franklinists to improve upon Beccaria's signs.

William Henley undertook the project. He electrified a Leyden jar using a point sufficiently displaced from its coating to illustrate the passage of charge (as at B in fig. 18.4), and devised a new prime conductor expressly 'to ascertain the direction of the electric matter, as it passes through.' The 'conductor' (fig. 18.5) is the evacuated cylinder A, whose point F draws electrical matter from the globe of a machine not shown. The self-repellent electrical matter spreads out in the vacuum (which Henley, like most of his contemporaries, still supposed resistanceless), whence the diffuse light between E and D, and the intense glow around D, a spread-out star produced by whatever obstructs the entry of the charge into the conductor.[26] Modern physicists invert Henley's explanation, and ascribe his pretty picture to electrons flowing from D to E.

The improvement of the generating machines—the substitution of amalgam-covered cushions for the hand, of cylinders or disks for globes, of pointed collectors for wire tassels or chains, and of large-bore tin tubes for iron rods or gun barrels—eventually supplied the Franklinists with their strongest argument. Nairne's large cylinder machine (1773–4) and Wilson's generator for the Pantheon experiments of 1777 (fig. 16.4) could produce imposing sparks whose

25. E.g., Priestley, *Hist.* (1775³), II, 184–91, who says he discovered the wind from negative points, and Nicholson, *Intro.* (1782), II, 82–5.
26. Beccaria, *PT*, 56 (1766), 105–18; Henley, *PT*, 64 (1774), 400–1, 403–6.

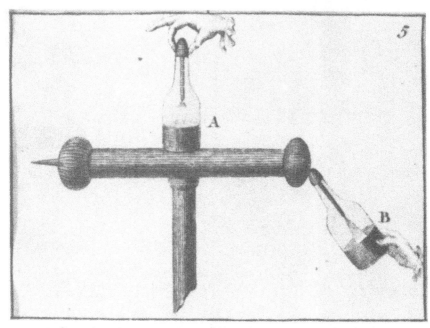

18.4 Henley's demonstration of the direction of flow of the electrical matter in the charging of a Leyden jar by the hook (as at B) or by the tail (A). From Henley, PT, 64 (1774), 389–431.

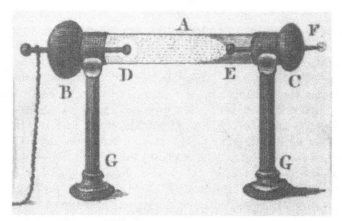

18.5 Henley's demonstration of the flow of electrical matter in a prime conductor. From Henley, PT, 64 (1774), 389–431.

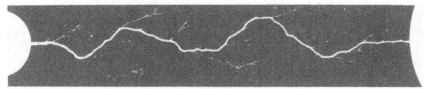

18.6 Van Marum's great spark. From van Marum, Teylers Tweede Genoot., Verh., *3 (1785).*

branching seemed to reveal the circulation more clearly than Beccaria's signs. Priestley, who began his experimental career with the 'confused notion that a person would stand the best chance for hitting upon some new discovery by applying a greater force than had hitherto been used,' was perhaps the first to capitalize on this effect, which he thought 'nearly decisive in favor of Dr Franklin's hypothesis.'[27]

The grandest of eighteenth-century sparks, and the most persuasive form of Priestley's argument, leapt from the machine that John Cuthbertson, an English instrument maker who moved to Holland in 1768, built for the Teyler Foundation of Haarlem in 1785. Cuthbertson had improved the mounting and rubbers of the plate machine, which had been introduced in the 1760s, to the point that its output exceeded that of the best cylinder machines.[28] In 1784 van Marum, who shared Priestley's faith in large installations, had added the directorship of the collections and library of the Teyler Foundation to his other activities — practicing physician, public lecturer, and curator of the cabinet of curiosities of the Hollandsche Maatschappij der Wetenschappen—on the understanding that he would have funds for building up the collection of physical instruments.[29] Ever a promoter, he commissioned Cuthbertson to make the largest electrostatic generator in the world: 'Reflecting on this [that 'electrical science has advanced in proportion to the increase in the machines'] I took it as certain that, if one could acquire a much greater electrical force than hitherto in use, it could lead to new discoveries.' It certainly produced a big spark, two feet long and as thick as a quill pen, not including prominent ramifications that always made acute angles with a line drawn from the prime conductor (plus) to the object struck (fig. 18.6).[30] The branches followed the main, unidirectional flow; there

27. Priestley, *Hist.* (1775³), II, 260; *supra*, i.7. Priestley's exploitation of the 'diverging long spark' appears from an undated letter (about 1766) from Henley to William Canton, in Canton Papers, II, f. 108 (RS). Henley also recommended the construction of vast machines (*PT*, 64 [1774], 430–1). For improvements in the generator see Priestley, *Hist.* (1775³), II, 86–118, and Hoppe, *Gesch.* (1884), 89–93.

28. Hackmann, *Cuthbertson* (1973), 14–15, 24–5; Cuthbertson found the Dutch very backward in electricity; Hackmann, RS, *Not. Rec.*, 26 (1971), 163. The early history of the plate or disk machine is obscure; Schimank, *Zs. für techn. Phys.*, 16 (1935), 250.

29. Muntendam in Forbes, *Marum*, I (1969), 15–21; Kurtz in *ibid.*, 87–8. The machine cost about 3000 florins, ten times van Marum's salary from the HMW.

30. Van Marum, Teyl. Tweede Genoot., *Verh.*, 3 (1785), x, 26–30, 92; *JP*, 27 (1785), 148–55; Dibner, *Nat. Phil.*, 2 (1963), 87–103. Van Marum's lightning should be compared to the

was no trace whatsoever of the contrary current required by the dualist hypothesis.

Many years later van Marum remembered that Franklin, Le Roy, and Lavoisier had accepted his big spark as conclusive when he discussed it with them in Paris in 1785. Volta certainly thought it decisive: 'The branches making acute angles with the direction in which the fluid supposedly moves show that the portions projected laterally preserve the motion common to the whole spark, and that consequently we are not deceived in a single electrical current coursing in the direction the Franklinian theory assumes. All *orthodox* electricians should be grateful to you, sir, for having delivered the final blow to the heresy of the *dualists,* the new partisans of the Dufays, Nollets [!] and Symmers.'[31]

The dualists defended themselves against the long spark by conceding what they had never denied, that positive electricity gave different optical effects than negative. The star and brush could be understood by assuming that the air resisted the passage of negative more than that of positive electricity. As for van Marum's 'proof,' one noted that he did not include a picture of the discharge from a negative prime conductor. This desideratum was supplied by William Nicholson, a solicitor's son, who had worked for the East India Company and the Wedgewood Potteries before setting up as a school teacher, inventor and scientific publicist in London.[32] Nicholson's mastery of instruments and concise, businesslike reporting compare favorably with the unnecessarily elaborate apparatus and diffuse ramblings of van Marum.

Using an improved silk rubber and a glass cylinder no more than twelve inches in diameter, Nicholson obtained sparks comparable to van Marum's at one-thirtieth the expense and showed that negatively charged balls gave off a characteristic, non-branching streamer (fig. 18.7 a,d,f).[33] His pictures of the simultaneous appearance of the plus and minus sparks (fig. 18.7 c,f) agreed nicely with the hypothesis of differential resistance, as did the observation that the negative discharge, like the positive, spread further and wider *in vacuo* than

---

13-inch spark of Nairne's machine of 1773, which had seemed promising to Franklin: 'From a greater Force used, perhaps more Discoveries may be made.' BF Ingenhousz, 30 Sept. 1773. *BFP* (Smyth), VI, 142.

31. Van Marum, *Ann. Phil.*, 16 (1820), 441–4; Volta to van Marum, *VO*, IV, 66; Levere in Forbes, *Marum*, I (1969) 176–7; Muntendam in *ibid.*, 22–3. Cf. Levere, RS, *Not. Rec.*, 25 (1970), 113–20, and Volta to Bellani, 2 Jan. 1804, *VO*, IV, 269–70. Van Marum did not mention that the big spark vindicated Franklinism in his paper in the Teyler proceedings of 1785.

32. *DNB*, XIV, 473–5. For a time Nicholson was secretary of the 'philosophical club' centered on that resourceful middleman of the instrument trade, J. H. Magellan (Magellan to Volta, 22 Feb. 1785, *VE*, II, 292–3). Cf. G. Turner, *Ant. Hor.*, 9:1 (1974), 74–7; Lilly, *Ann. Sci.*, 6 (1948), 82–3.

33. Lichtenberg, *GGA* (1785:2), 1562–3; Nicholson, *PT*, 79 (1789), 277–83, 288; and, regarding style, Nicholson, *Intro.* (1782), I, xi–xii: '[The author] has everywhere endeavored to preserve the solidarity of argument, and precision of expression, which so eminently distinguish the works of the best English philosophers.'

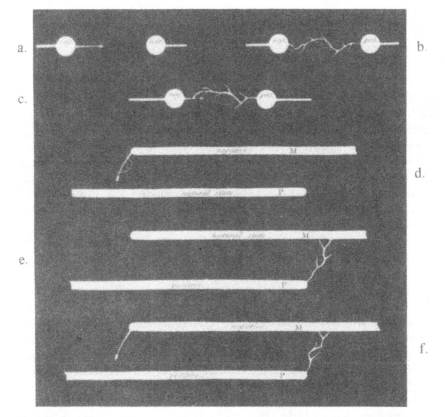

18.7 Nicholson's sparks between (a) − and 0 (b) 0 and + (c) − and + balls, and between (d) − and 0 (e) 0 and + (f) − and + rods. From Nicholson, PT, 79 (1789), 265–88.

in air. The upshot was that luminous signs could not decide the number of electrical fluids.[34]

A second dualist initiative emphasized perforations as performed by Symmer. The unitarians defended themselves in two ways. The old-fashioned, accepting their adversaries' terms, regarded the electrical matter as the immediate cause of the holes. They took great comfort in a demonstration described in a dissertation defended at Geneva in 1766. The dissertation, the work of Saussure, was ostensibly an impartial comparison of the systems of Franklin and Nollet. The demonstration, known as 'Lullin's experiment' after the respondent, Amadeus Lullin, consisted of exploding a jar through a card via parallel pointed

34. Among contributors to this agnosticism: Barletti, *Dubbî* (1776), 46; Lichtenberg to Wolff, 30 Dec. 1784, in Lichtenberg, *Briefe*, II, 176; Saxtorph, *El.-laere*, I (1802), 238, 254; Delambre, *HAS*, 7 (1806), 215; Trémery, *JP*, 48 (1799), 168–72, and *JP*, 54 (1802), 357–67; Remer, *Ann. Phys.*, 8 (1801), 323–41.

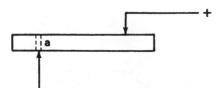

18.8 Diagram of Lullin's experiment; the hole is struck at a.

wires set a few inches apart (fig. 18.8). During the discharge a light appeared above the card only, and the hole was always struck directly over the point connected to the negative coating of the battery.[35] Those who found this demonstration persuasive saw the electrical matter rush along the top of the card, blast through the surface and jump into the negative wire. They were eventually answered by a mining engineer, Haüy's former student J. L. Trémery, who showed that the point of rupture could be displaced towards the positive wire by performing the discharge *in vacuo*.[36]

A more serious objection against Lullin's argument, and one perceived by the unitarians themselves, was its effluvialist presupposition: the electric matter *per se* could no more bore holes in paper than it could drive the air of an electrical wind. Beccaria and Kinnersley had demonstrated that sparks could heat the air they spanned enough to break small sealed vessels enclosing the spark gap; and in 1769 Priestley showed that such 'lateral' mechanical effects accompanied explosions even through a continued path, especially of poor conductors.[37] These experiments suggested that perforations were indirect consequences, and not footprints, of the electrical matter, and that nothing could be deduced from them about the direction of circulation.[38]

35. Lullin, *Dissertatio* (1766), 24, 38; Nollet, *Lettres* (1767); Priestley, *Hist.* (1775³), II, 374–5; Freshfield, *Saussure* (1920), 77–8. The thesis is identified as Saussure's by Senebier, *Mémoire* (an IX), 38–9.

36. Trémery, *JP*, 54 (1802), 357–67, and *Nich. J.*, 3 (1802), 223–4; Lacroix, Soc. fran. min., *Bull.*, 67 (1944), 49, 143. The effect can be reversed—the hole struck under the positive wire—by increasing the thickness of the paper above the negative electrode, or by coating the card with turpentine or glycerine. The 'Lullin experiment' interested several nineteenth-century physicists, notably Ernst Mach; it may not yet be understood. For the older literature see Wiedemann, *Lehre*, IV (1885), 775–8.

37. Priestley, *Hist.* (1775³), II, 336–42, and *PT*, 59 (1769), 57–62. Pursuing these experiments, suggested by side jumps of lightning strokes, Priestley found that an insulated conductor placed near the discharge wire received a spark during the explosion. But the spark did not seem to charge the conductor, a puzzle that Priestley, after a set of well-designed experiments (*PT*, 60 [1770], 192–210), referred to a double flux of electrical matter so rapid as to appear a single spark. Anti-Franklinists did not exploit this curious effect, of which Wilson had earlier noticed an example (*Treatise* [1750], Plate X). It has won Priestley the honor of discovering the oscillatory character of the discharge, although he spoke only of the flux to and from the side conductor and would have rejected the anti-Franklinist notion of the reversal of current in the discharge wire. Cf. Wiedemann, *Lehre*, IV (1885), 672–6, and Hartog, *Ann. Sci.*, 5 (1941), 11.

38. Franklin to Ingenhousz, 3 May 1780, *BFP* (Smyth), VIII, 191–2; Volta to Landriani, 27 Jan. 1776, *VO*, III, 157–8. Franklin had remarked earlier that punctures gave 'very equivocal signs of the Direction' of the discharge (letter to Collinson, 19 April 1754, *FN*, 502–3); he concluded that 'the holes are not made by the impulse of the fluid . . ., but by circumstances of explosion of parts of the matter.' BF to Ingenhousz, 16 May 1783, *BFP* (Smyth), IX, 45.

The strongest dualist argument exploited the Franklinist concept of negative electricity. Few liked Aepinus' desperate solution of the problem of minus-minus repulsion, a 'principle so inadmissable that his reasonings cannot be regarded as just explications of the phaenomena.' Even Haüy, in introducing his epitome of the *Tentamen*, suggested that the role of Aepinus' 'matter' might better be played by a second electrical fluid.[39] Another admirer of Aepinus, Barletti, who had published excerpts from the *Tentamen* in 1776, rejected the proposed intermolecular repulsions and insisted that negative states, defects, and emptiness (*il niente*)—the last being the electrical condition of the external coating of the jar according to orthodox Franklinists—are 'metaphysical ideas,' not positive causes: 'The great Torricelli has banished the *horror vacui*; Torricelli's great successor [by which Barletti apparently meant himself] banishes *il niente.*' Barletti's literal mind rebelled from endowing the Franklinist electrical void with the properties of 'real' substances.

Consider the annulment and resurrection of electrical signs in Symmer's socks or the electrophore: how can a negative thing, a void, resist the entry of a positive thing, the electrical fluid, and prevent the destruction of the contrary electricities of the stockings, or of the shield and the plate? The Franklinists supposed that, in charging a jar negatively, the electrical matter runs from the inner coating to the frame of the machine, voided of its natural content by the pull of the globe. But how can a vacuum attract? 'Although at first glance [the theory] seems plain enough, when well considered it has an air of magic and sympathy that smells of Gothic philosophy.'[40] One need only admit two positive real fluids to avoid all the difficulties of Franklin's 'moral theory,' says our monkish electrician, which requires a vacuum to resist, 'with heroic scrupulosity,' the entry of the matter it 'ardently desires.'[41]

The same point was made in a similar style by that combative crank, Major Henry Eeles of Linsmore, Ireland, who liked to write about electricity when laid up with the gout. He had vainly memorialized the Royal Society several times since 1751 about the electrification of clouds, about the errors of Franklin, and, in 1757, about what he called 'Eeles' new doctrine of electricity and magnetism,' which turned out to be a sketch of the theory that Symmer put forward two years later.[42] Not stopped by the Society's indifference, Eeles

39. Miller, *J. Nat. Phil.*, 4 (1800/1), 461; Haüy, *Exposition* (1787), Préf.; cf. Donndorff, *Lehre* (1784), 269, 273.

40. Barletti, *Dubbî* (1776), x–xi, 32–4, 45–50. Cf. Barletti, Soc. ital., *Mem. mat. fis.*, 1 (1782), 54, and Priestley, *Hist.* (1775³), II, 46, who agrees that a strong point of the dualists is avoidance of the Franklinist void.

41. Barletti, Soc. ital., *Mem. mat. fis.*, 7 (1794), 444–61. Barletti, who tended to aggrandize his contributions, affected to be strongly anti-Franklinist in teaching the partial penetrability of glass and the inequality of the charges on either side of the Leyden jar, doctrines he took directly from Aepinus; Barletti, *Dubbî* (1776), ix, 3–9, 17–18. That these points required emphasis, however, appears from E. Gray, *PT*, 78 (1788), 121–4. Cf. Volta to Landriani, 16 April 1785, *re* Barletti's *Fisica* (1785), 'ricchi di discorso e poveri di cose, almeno di cose nove, e neppure le cose degli altri sian presentate nel vero lume' (*VO*, VI, 414).

42. Eeles, *Phil. Essays* (1771), 3–7, 17–20, 45–6, 59, 68–9, 106. Eeles did not think that Symmer stole from him, but believed that Priestley's revisions of Symmer (*History* [1775³], II,

published his papers himself, and ridiculed Franklinist theory. '[It is] a perfect contradiction to say that two bodies shall repel each other the farther, the more they are divested of the repulsive power.' Eeles believed in atmospheres, but of two kinds; people who deny the existence of a real atmosphere answering to Franklin's negative electricity 'have removed themselves out of their senses.' 'I would ask these gentlemen a civil question, whether it is mere inanity which knocks down steeples and towers, rends trees, tears up the earth, kills men and cattle, sets places on fire, etc., or I might shorten the question by asking how mere inanity or nothing can act? but this would be a dispute about nothing.'[43]

The difficulty with minus-minus repulsion recommended Symmer's system to many who, in contrast to Eeles, shared the growing consensus about the localization of charge. Caullet de Veaumorel, the translator of Nairne, confessed himself dualist because negative electricity is 'void of sense.' 'I conceive more readily how a real $+E$ could draw, bind and hold another equally real $-E$ at a distance than I can picture [vorstellen] how a surplus or deficiency could.' Thus Gehler, who went on to question how a 'lack can show the same activity that our $-E$ so evidently displays in the phenomena of the electrophore.'[44] Moreover, no one knew which of the two, $+E$ or $-E$, is the lack; all attempts to identify the accumulation, the true positive charge, were circular, as the writer on electricity for the third edition of the *Britannica* observed.[45] These observations were not intended to be conclusive arguments for dualism; they pointed rather to its greater simplicity, or convenience, or plausibility. Haüy eventually endorsed Symmer's theory for its convenience, as did the other leading French electricians, Coulomb (whom van Marum considered the ringleader of the French dualists) and Biot. As the important text of Fischer put it: 'One can doubt that any of these hypotheses [Franklin's, Symmer's, and De Luc's] is entirely correct; but that of the two electricities, Symmer's, is incontestably preferable, because it explains electrical phenomena in the most convenient manner.' In particular, it avoids the forced Franklinist explanation of minus-minus repulsion.[46]

The unitarians could parry this thrust only by appealing to the same criterion, with the comical consequence that each side claimed to offer greater convenience and simplicity than the other. To many English minds, strengthened or immobilized by Franklin's eminence, the theory of a single fluid seemed the simpler. Two ways around the usual paradoxes of negative electricity presented themselves. Either one could swallow hard and follow Aepinus, as did Caven-

41–52) were a plagiarism (*Phil. Essays* [1771], xi–xiv). Cf. Schofield, 'Intro.,' in Priestley, *Hist.* (1775³), I, xxiv–xxvi; W. S. Harris, *Rud. El.* (1859⁵), 43–4.

43. *Phil. Essays* (1771), v–vii, xliv.

44. Caullet de Veaumorel, *Description* (1784), iv–vi, xxxi; Gehler, *Phys. Wört.*, I (1787), 766, 830–1. The notation $\pm E$ was introduced by Lichtenberg.

45. *Encycl. Brit.* (1797³), VI, 450; Gehler, *Phys. Wört.*, I, 764–5, and Saxtorph, *El.-laere*, I (1802), 28–30, agree.

46. Haüy, *Traité* (1803), I, 334; E. G. Fischer, *Physique* (1806), 257; van Marum, *Ann. Phil.*, 16 (1820), 445.

dish, Thomas Young and John Robison, Professor of Natural Philosophy at the University of Edinburgh.[47] 'I am unwilling to admit two electricities, since the redundance and deficiency of one does as well.'[48]

Or one could appeal vaguely to stresses in the medium, trusting that a satisfactory model someday would be found. Cavallo took this tack in his influential textbook, admitting that his particular representation, which had been severely and appropriately criticized, might not be fully competent. (He required the principle that no conductor could acquire a charge unless the insulating media around it, especially the air, had room to polarize; the pith balls of an electroscope move apart when charging in order to open a space for the accommodation of stressed air.) The utility of proto-field theories to the unitarians appears explicitly in an early paper by Avogadro, who recommended what he took to be a new form of the hypothesis of stressed air because it saved the 'simpler' single-fluid theory from the most telling objections of the dualists.[49] As Avogadro's remark suggests, the singlists more than the dualists tended to cling, or be driven, to mechanical representations. It is striking that Jacquet and Saxtorph, trying to be fair to both sides, describe Symmerian theory in instrumentalist terms and clutter Franklinist theory with the debris of atmospheres.[50]

Although by the late 1780s dualism dominated the continent and Franklinism England,[51] many Europeans remained orthodox and several Englishmen sided with Symmer.[52] The cardinal point for us was the recognition that, as Priestley put it, 'no fact can be shown to be absolutely inconsistent with either [theory].' *Valeat quantum valere potest.* One took one's choice or one remained agnostic. 'Oh monsieur, il faut être unitaire,' says Volta. 'I'm neither unitarian nor dualist,' replies Lichtenberg, 'but I'm ready to become one or the other as soon as I've seen a decisive experiment.' 'Let's not fight over the two fluids,'

47. Young, *Course* (1807), I, 670, where he erroneously reports that few dualists existed in his time; Robison, *System*, IV, 1–204 (a reprint of an article of 1801); *DNB*, XVII, 57–8.

48. Robison to Black, 20 Dec. 1803, in Robinson and McKie, *Partners* (1970), 382.

49. Cavallo, *Compl. Treat.* (1795⁴), III, 192–205; Miller, *J. Nat. Phil.*, 4 (1800/1), 463–4; *Encycl. Brit.* (1797³), VI, 444; Avogadro, *JP*, 65 (1807), 131–2.

50. Jacquet, *Précis* (1775), 30–3, 228–32; Saxtorph, *El.-laere*, I (1802), 311, 318, 419–21. Both think Franklin's theory simpler (fewer causes) but Symmer's more convenient and comprehensible. Neither doubts that either theory can save the phenomena; cf. J[acquet], *JP*, 7 (1776), 503.

51. For continental Symmerians see Gehler, *Phys. Wört.*, I, 766 ff. (Wilcke, Bergman, Kratzenstein. Karsten, J. R. Forster) and *Encyclopédie méthodique*, LXXVI, 58 ff.; to whom may be added J. C. Fischer, *Anfangsgründe* (1797), 650–3; J. Weber, Ges. naturf. Freunde, Berlin, *Schr.*, 4 (1789), 339–40; Horvath, *Physica* (1783), 309–36; and all the German Naturphilosophen, e.g., Galle, *Beyträge* (1813), ix, 1–2, and F. Hildebrand, *Anfangsgründe* (1807). In addition to the English singlists already mentioned, Enfield, *Institutes* (1785), 350; Read, *Summ. View*, (1793), 85–8.

52. Among traitors: de Herbert, *Th. phen.* (1778²), 26, singlist because Symmer's hypothesis is complex and obscure; Herschel, *Sci. Papers* (1912), I, lxxxiii–lxxxiv, dualist fellow traveller because, contrary to expectation, both positive and negative electrification accelerate the pulse. Adams, *Essay* (1784), 47–8, is unitarian without comment; the second edition, 1785, 51–2, quotes Eeles with approval; the fifth, 1799, 107 ff., is forthrightly dualist.

Lichtenberg writes another correspondent. 'I declare for neither; in speaking or writing, however, where the dispute is not the main subject, I am always a Franklinist, just as I write "deutsch" although I readily concede that "teutsch" is also possible, and perhaps even preferable.'[53] Lichtenberg had plumbed his subject. 'In physics,' he says, 'we have not yet reached puberty.' 'Hypotheses like [electric fluids] are in physics nothing more than convenient pictures. . . . The best method of representation is the simplist, however far it might be from the truth that we expect it to lead us to.'[54]

That brings us to the favorite criterion of both sides, the claim of convenience and simplicity. Lichtenberg again: 'The lofty simplicity of nature all too often rests upon the unlofty simplicity of the one who thinks he sees it.' After several days with Lichtenberg even Volta had to concede that everything could be explained either way. 'La nostra cara dottrina,' he later said, our dear singlist theory, although far preferable to the enemy's, is not the only one possible; 'I am far from regarding the dualist theory as absurd, only as very implausible, too complicated, and needing hypothesis on hypothesis,' that is, two hypothetical fluids rather than one.[55] We leave the last word in this matter to Mrs. Margaret Bryan, a beautiful and talented schoolmistress, who exploited the ambiguity of electrical theory to inculcate 'the love and practice of justice.' After discussing without prejudice the arguments of the unitarians and the dualists, she allows that the question cannot be decided. 'Cull the Sweets of religion as you rove through the flowery paths of Natural Philosophy,' she tells her charges, '[and don't] form hasty judgments.'[56]

53. Priestley, *Hist.* (1775³), II, 44, 52; Lichtenberg to Wolff, 30 Dec. 1784, in Lichtenberg, *Briefe*, II, 174–7.

54. Quoted by, respectively, Mautner and Miller, *Isis*, 43 (1952), 226, and Herrmann, *NTM*, 6:1 (1969), 81. Cf. Lichtenberg to Deluc, 10 Oct. 1779, Deluc Papers (Yale): 'I have done a great deal more [on electricity] than I have either leisure or inclination to describe. The reason is I cannot get hold of the trunk of which these are but very slender branches.'

55. Lichtenberg, *Briefe*, II, 174–7; *VO*, IV, 270, 359, 380; Volpati, *Volta* (1927), 75–7.

56. Bryan, *Lectures* (1806), Pref., 163, 168, 190; *DNB*, III, 154. Cf. G. C. Morgan, *Lectures* (1794), I, xlv–xlviii: 'Electricity possesses much of what is admirably adapted to discipline the mind.'

CHAPTER XIX

Quantifiable Concepts

The successful quantification of physical theory presupposes the existence of appropriate concepts expressible mathematically and amenable to 'transdiction,'[1] to test and refinement by measurement. The heart of the process lies in the selection of the concepts; and this, as the painful emergence of the notion of localized charge (Q) abundantly illustrates, may not be easy. Space and energy forbid our following the development of other central ideas of electrostatics, such as capacity (C) and potential (V), in the detail lavished on Q. But truth and harmony require an explicit recognition of the difficulties of selection and transdiction in their cases as well, and all the more because the usual histories take them for granted. As a compromise we shall follow the emergence of 'Volta's law,' $Q = CT$, connecting the charge localized in a conductor, its capacity, and the 'tension' of the contained electrical matter. T possessed many of the qualitative and a few of the quantitative properties of potential, for which it was often a serviceable substitute.

Further progress depended on successful reduction of the tensions and ponderomotive forces they represented to microscopic *forces* (italicized in what follows to distinguish them from macroscopic ones) assumed to act between *elements* of electric charge. Again a circumstantial account is not possible; the compromise this time is a suggestion of difficulties encountered in the formulation and demonstration of the law of squares, and a description of the successes of Robison and Coulomb. Coulomb's law by no means gained the immediate and unanimous assent of electricians, and even among those who accepted and employed it few appreciated that the exact concept 'potential,' introduced in all but name by Cavendish in 1771, was the appropriate Coulomb analogy to the imprecise notion 'tension.' The unique contribution of Cavendish demonstrates how the standard electrical theory of the 1780s, reworked unreservedly in the style of Newton and Aepinus, could be transformed into classical electrostatics.

1. CAPACITY AND TENSION

The business of measurement rapidly outpaced understanding of the quantities measured. In the 1740s Gray's simple 'thread of trial' gave way to a pair of strings whose angular separation α could be determined precisely. Nollet added a circular scale and an optical system (fig. 15.3); Beccaria, thinking that not α but its sine measured the force sustaining the threads against their gravity, de-

---

1. For a closer definition and history of the term see Mandelbaum, *Philosophy* (1964), 61–2.

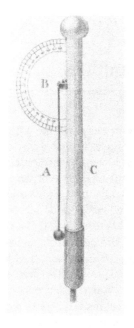

19.1 Henley's electrometer.
From Priestley, PT, 62 (1772), 359–64.

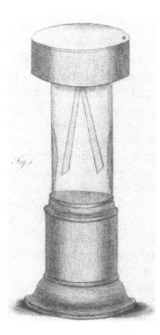

19.2 Bennet's electrometer;
note the earthed metal foil
on the interior walls to prevent
accumulation of charge that
otherwise might be brought
by the leaves to the glass.
From Bennet, PT, 76 (1786), 26–34.

termined the appropriate chord directly.[2] Others, like Richmann, substituted a fixed object for one of the threads and read $\alpha/2$ on a scaled quadrant (fig. 16.6). The chief inadequacies of the design—leak from the threads' ends and small charged surfaces—were lessened by Canton, who terminated the threads in small, light balls made from elder pith. The English used the pith balls until 1770, when Henley, inspired by Priestley's call for a good electrometer, invented a robust form of Richmann's instrument that quickly became standard (fig. 19.1).[3] With it and the later sensitive bottled electrometers of Bennet (fig. 19.2) and Volta (who employed, respectively, gold leaves and flat, thin straws in place of the original threads), electricians could determine α as reliably and exactly as they wished.[4]

But what did it measure? And how should it be associated with other measures, such as those proposed to rate the output or power of electrical machines? Here a commercial element entered to darken a situation already sufficiently obscure, for manufacturers naturally used the measure that showed their instruments to best advantage. Should one take the area of glass rubbed per turn of the wheel, or the length of the longest spark from the prime conductor, or the length of a standard fine wire fused by discharge of a Leyden jar or battery charged by the machine, or the number of sparks produced in unit time across a gap between the PC and ground? It was recognized that all these quantities measured something about amount of charge Q. But what? Two practiced electricians, Cuthbertson and van Marum, fell out over the meaning of w, the length of wire fused, which the former took proportional to Q^2 and the latter to Q; and they further squabbled over whether the maximum sparking distance or the rate of sparking was the better gauge of the performance of a machine.[5]

None of these quantities proved as useful as the measure most obvious to the laboring man, namely n, the number of times he must turn the wheel of the electrical machine to charge various objects to the same degree α.[6] Much more

2. Nollet, *MAS* (1747), 102–31; Beccaria, *PT*, 56 (1766), 105–18. The electrostatic force perpendicular to the string, $f \cdot \cos\alpha/2$, balances the gravitational, $W\sin\alpha/2$. Beccaria forgot or ignored the factor $\cos\alpha/2$.

3. Richmann, *CAS*, 14 (1744–6), 299–324, and *NCAS*, 4 (1752–3), 301–40; Canton, *PT*, 48:2 (1754), 780–5; Priestley (*re* Henley), *PT*, 62 (1772), 359–64. Canton's electroscope developed from cork balls used in his induction experiments, *PT*, 48:1 (1753), 350–8; cf. Cavallo, *PT*. 78 (1788), 1–22, and Walker, *Ann. Sci.*, 1 (1936), 70–5.

4. Bennet, *PT*, 76 (1786), 26–34; Volta to Lichtenberg, July and August, 1787, *VO*, V, 35–57. Cf. Walker, *Ann. Sci.*, 1 (1936), 84–8; Polvani, *Volta* (1942), 136–40; Nollet, *Leçons*, VI (1748), 323–4; Saussure, *Voyages* (1786), II, 205–9. See Kühn, *Neu. Ent.*, I (1796), 177–271, for an account of the electrometers in use in the 1780s and '90s.

5. Hackmann in Forbes, *Marum*, III, 347–57. Since w is proportional to the energy of discharge and the energy of a jar goes as Q^2, one might expect $w \sim Q^2$; but there are many complicating factors, for which see Mascart, *Traité* (1876), II, 326–32. The several measures mentioned do not always give the same relative ratings to the machines; varying the gap used in measuring the rate of sparking may change the order of performance (*ibid.*, 318).

6. Mascart's experiments showed the output of electrostatic machines to be proportional to the rate of rotation, 'other things being equal' (*ibid.*, 316). Many practical electricians, like Volta, *VO*, III,

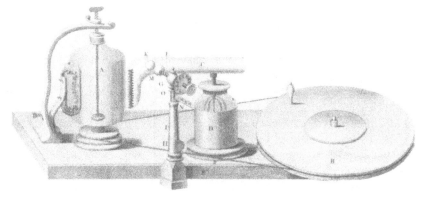

19.3 Lane's electrometer (the spark gap KM), as used with an inexpensive table electrical machine. From Lane, PT, 57 (1767), 451–60.

work goes into charging a Leyden jar than a dangling boy; and a much bigger spark, or more electrical fire, can be recovered from the one than from the other. Apparently α measured a property of electricity not readily visualized, its 'height,' to use Priestley's suggestive term, while n indicated the relative amounts of superfluous electric matter that could be stored on objects equally electrified according to the electrometer. These concepts found clear expression in the Lane electrometer, a device invented by an electrifying apothecary concerned to meter his patients' shocks. It consisted of the adjustable spark-gap KM (fig. 19.3), put in series with the patient and the jar(s) D. Lane found n to be proportional to the extent of the coated surface of D: the quantity of 'electric fluid' required to electrify the bottle to the 'degree' needed to force the gap was evidently inversely proportional to the effective size, or capacity, of the bottle.[7]

The notion of capacity also emerged in connection with simple conductors. Richmann, for example, used to charge pieces of metal to the same degree as measured by his index, then mount upon wax, draw a spark, and read his electrometer. The index fell the less the more 'capax' the conductor, and the smaller the object soliciting the spark. Le Monnier then found that not the mass but the surface of a conductor determined the quantity of charge it received by communication; and, further, that of bodies of equal surface, the longest enjoyed the largest capacity.[8]

217, and van Marum, *JP*, 38 (1791), 166, and *Ann. Phys.*, (1799), 81–4, took *n* as a good measure of *Q*; Beccaria objected to its inaccuracy. Eandi, *Memorie* (1783), 82.

7. Lane, *PT*, 57 (1767), 451–5; BF to Beccaria, 29 May 1766, praising Lane's 'elegant method' (*BFP*, XIII, 288); Walker, *Ann. Sci.*, 1 (1936), 72–3. Timothy Lane, perhaps an apprentice of Watson, to whose practice he succeeded, was Master of the Apothecaries' Company and the inventor of a widely-used set of weights and measures; *GM*, 77:2 (1807), 689. The proportionality to surface holds only if, as in the machine pictured, the capacity of the jar greatly exceeds that of the PC.

8. Richmann, *CAS*, 14 (1744–6), 307–8, 314–20; Le Monnier, *MAS* (1746), 461–3, and *PT*, 44 (1746/7), 295. Cf. Priestley, *Hist.* (1775³), I, 145, 155–6.

Many physicists extended these results, notably Nollet, Richmann, Beccaria, and Franklin, whose characteristically direct demonstration survives in courses of elementary electrostatics.[9] Place three yards of brass chain in an insulated silver can fitted with an electrometer. Charge the can with a spark from a Leyden jar and note the reading. Now draw up the chain with a silk line: as the chain rises the threads fall; the can is able to take another spark, which returns it to its original degree of electrification. In Franklin's old-fashioned terminology, the experiment shows that 'increase of surface makes a body capable of receiving a greater electric atmosphere.' In terms of the emerging quantifiable concepts, it suggests that, for the same charge, a body's 'degree of electrification' varies inversely as its capacity.[10]

THE CONDENSATORE

These ideas crystallized towards 1780, under the spell of Volta, who drew his inspiration from the inexhaustible electrophore and the newly discovered theories of Aepinus. Volta tried to relate the changing state of the electrical fluid of the shield to the progressive collapse of its electrometer threads during its approach to the plate. Assume the shield plus. It follows, almost as a corollary of the localization of charge, that when the shield stands outside the influence of the plate its surplus fluid must be compressed;[11] during the approach its capacity increases, and consequently the compression must diminish. According to Volta, the negative charge of the plate corresponds to a rarefaction, whose presence relieves the pressure of the 'constipated electricty' (to use Robison's quaint phrase) of the shield. The concept proved useful. It suggested that just as a charged body increases the capacities of neighboring conductors for electricity opposite to its own, it must decrease their capacities for homologous electricity; the presence of a plus body creates a 'tension' in others that in effect engages part of the room usually available for accumulating excess fluid. Volta considered further that the 'atmospheres,' or spheres of influence, of the various surface elements of a charged conductor might similarly inhibit one another, and consequently that, for a given surface, the longest conductor

9. Richmann, *NCAS*, 4 (1752–3), 307–9; Nollet, *MAS* (1747), 125 ff.; Beccaria, *PT*, 51 (1759/60), 517, 57 (1767), 309; BF to Collinson, Sept. 1753, *EO*, 272–5. Cf. Teichmann, *Entwicklung* (1974), 21–2.

10. Franklin tendered this experiment in support of a new theory of cloud formation designed to make the heavens negative: assume that a body, when rarefied, requires more electrical matter to establish its normal state than it did when condensed; a neutral volume of water will then make a negative cloud. The intended analogy does not hold, since the chain and can merely share a conferred atmosphere, while Franklin wants a normal body to electrify by expansion. Similar mechanisms occur in Wilson's (*supra*, xvi.1) and Volta's (*PT*, 72:1 [1782], §29) theories of triboelectrification.

11. Volta, *PT*, 72:1 (1782), x–xi, xix–xxi. This 'corollary' occurs implicitly in Franklin's theory of the Leyden jar and in all theories of stressed media (*supra*, xvii.3); Wilson (*PT*, 68:1 [1778], 305), Cavendish (*infra*, xix.3), Volta and many continentals (e.g., J. B. Le Roy, *MAS* [1772:1], 503) employ it explicitly.

would have the greatest capacity. In this way he rediscovered the result of Le Monnier, which he demonstrated with his usual flair in 1778.[12]

Take a set of twelve interconnected gilded cylinders, he told Saussure, in an open letter to the *Journal de physique,* each 8' long and ½" in diameter, hung in separated rows and charged like a prime conductor (fig. 19.4). You will find that, although the total surface does not much exceed that of a single 6' × 8" cylinder, it has six or seven times the capacity, as measured by wheel turns or by its 'intolerable shock, which shatters the whole body.' The concatenated cylinders struck like a small Leyden jar, as Volta had surmised: a condenser apparently differed from a common conductor only in the amount of fluid it could contain. He strengthened the analogy by showing that a Leyden jar exploded 'through' chains of people or 'across' rivers in fact discharged each surface separately to ground, precisely as his grand conductor relieved its 'constipation.'[13]

While Volta was hanging up his conductors in Pavia, Wilson was rigging his gigantic artificial cloud—3900 yards of wire plus a sectioned tin foil cylinder 155 feet long when entire—in a London dance hall. Like Volta, Wilson knew that for best effect the great conductor, including the wire, had to be placed so that the 'several atmospheres' of their various parts did not 'interfere too much with one another.' In general, the greater the length the greater the blast. The big cylinder alone struck like a Leyden jar when charged by only six turns of Wilson's cylinder machine. There the experiment stopped, for, says the prudent investigator, 'I could not prevail upon anyone present to take a higher charge.'[14]

In analyzing the action of his long conductor Volta departed from the micromechanical theory developed in his earlier work,[15] made do with the macroscopic concepts T, Q and C, and, what is more important, established a quantitative relationship among them. He called the degree or head of electricity of a conductor or jar its 'tension,' measured by α and representing 'the forces exerted by each point of an electrified body to free itself [disfarsi] of its electricity, and communicate it to other bodies.'[16] His experiments showed that, for a given T, the quantity of electricity Q localized on an isolated conduc-

12. Volta, *JP*, 12 (1779), 249–77 (*VO*, III, 201–29). Le Monnier's work had been forgotten; cf. Brook, *Misc. Exp.* (1797²), 102–5.

13. *VO*, III, 51, 203–6, 212–15, 224–5. Teichmann, *Entwicklung* (1974), 38–42, 54, 57, points out the relation between Volta's concept of capacity and Beccaria's earlier analogy between electrical and fluid reservoirs (cf. *VE* I, 62, *VO*, III, 106, 108).

14. Wilson, *PT*, 68:1 (1778), 252, 296–7; *supra*, xvi.1.

15. *Supra*, xvii.2. Again there is the possibility of influence from Cavendish, who had suggested that a conductor large enough could give the Leyden shock; yet Volta had already observed that a body takes charge from the PC 'in accordance with its capacity' in *De vi att.* (*VO*, III, 51). Cf. Cavendish, *PT*, 61 (1771), 666–70; *El. Res.* (1879), 56–8, 82–3, 98–9; Achard, *JP*, 26 (1785), 378–9; *VE*, I, 275, 280; Polvani, *Volta* (1942), 118–21.

16. *PT*, 78:1 (1788), 259. Cavallo translates (*ibid.*, xx), 'the endeavor by which the electricity of an electrified body tends to escape from it,' which, though unfaithful, corresponded better to the common notion (e.g., Détienne, *JP*, 10 [1777], 73).

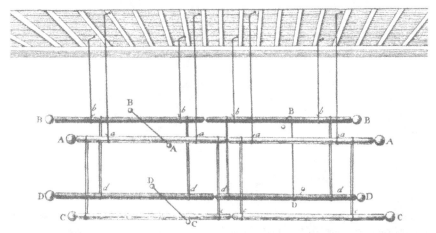

19.4 Volta's long conductor hung in eight sections in two parallel lines. From VO, III, 207.

tor increased with its capacity; and he guessed that the relationship was linear, that $Q = CT$.

'Guess' is used advisedly, for the relationship could not then be determined, or even confirmed, by comparing a measure of Q (say n, the number of wheel turns) with a measure of T (say $\alpha(n)$, the spread of leaves of an electrometer attached to the PC). The difficulty arose from the nonlinearity of the electrometers. Volta later devoted much time and ingenuity to calibrating his straw electrometer by subdivision: he would fix equal scale divisions by finding the spread for a charge Q upon a standard conductor, then for $Q/2$ (obtained by touching the conductor with an identical one), $Q/4$, and so on.[17] With such an instrument α or T is proportional to Q for a given isolated conductor; to show that the factor is capacity, Volta proposed to touch an insulated cylinder electrified to α degrees to a similar one n-1 times as long. If all went well, the electrometer read α/n.

Volta employed the relationship $Q = CT$ in a brilliant account, unfortunately unpublished in his time, of the charging of a pair of identical facing metal plates, which, with appropriate connections, might represent either a condenser or an electrophore. He begins with an insulated plate free from the influence of all other bodies and charged by the hook of a big Leyden jar. Let the jar's tension be unity, and the charge conferred (and therefore the plate's capacity) be a. (Note that, like potential, $T = 0$ for an uncharged conductor free from external electrical forces.) Volta confronts the plate, say A, fig. 19.5, with an insulated fellow, B, which thereby acquires a tension x (<1). He touches B, conferring a charge, say ΔQ, which he sets equal to ax. He thereby tacitly

17. *VO*, V, 37–42; Saussure, *Voyages* (1786), II, 205–9; Polvani, *Volta* (1942), 136–8; Cavendish, *El Res.* (1879), 279, 288–92.

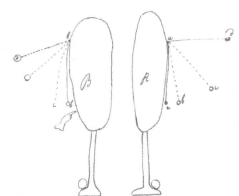

*19.5 Volta's analysis
of the charging of facing
or conjugate conductors.
From VO, III, 166.*

assumes that the tension $\Delta T = x$ lost in grounding a conductor produces a proportional change $\Delta Q = k\Delta T$ in its charge; since the plates are identical, $k = a$, and $Q = ax$. Now B's charge $Q = ax$ reduces the tension in A by an amount $-x^2$, as required by the rule of three $(a:x::-ax:?)$; and when Volta restores the initial potential with a kiss from the bottle, A increases its charge by ax^2. The process continues as follows:

	A		B	
	Tension	*Charge*	*Tension*	*Charge*
0. Initial State	1	a	x	0
1. Touch B to earth	$1-x^2$	"	0	$-ax$
2. Touch A to bottle	1	$a(1 + x^2)$	x^3	"
3. Touch B	"	"	0	$-ax(1 + x^2)$
4. Touch A	1	$a(1 + x^2 + x^4)$	x^5	"
5. Touch B	"	"	0	$-ax(1 + x^2 + x^4)$
6. Touch A	1	$a(1 + x^2 + x^4 + x^6)$	x^7	"

These tedious manipulations amount to the electrification of an air condenser in slow motion. Volta recognized that, if carried to an end, A and B would be charged $a/(1-x^2)$ and $-ax/(1-x^2)$, respectively. There are two noteworthy points about this calculation. First, Volta confirmed it by direct measurement, taking the angular separation of the trial threads as proportional to tension.[18] Second, it agrees precisely with modern electrostatic theory.[19]

Volta brought these ideas and the relation $Q = CT$ to the attention of electri-

18. *VO*, III, 248–58; Polvani, *Volta* (1942), 122–5.

19. The potentials ϕ and charges e of the identical conductors satisfy the simultaneous equations $\phi_1 = \alpha e_1 + \beta e_2$, $\phi_2 = \beta e_1 + \alpha e_2$. If initially $\phi_1 = 1$, $e_1 = a$, $e_2 = 0$, then $\alpha = 1/a$, $\beta = x/a$; the entries in the table are then easily obtained. Cf. Massardi, Ist. lomb. sci. lett., *Rend.*, 56 (1923), 293–308.

cians in 1782, in a description of a new instrument they could not ignore, a 'condensatore' for rendering sensible atmospheric electricity otherwise too weak for detection. This famous device is nothing but an electrophore with a layer of varnish as cake. One runs a wire to the shield from an apparently unelectrified atmospheric probe, waits, removes the wire and raises the shield, which can then affect an electroscope. The secret of the business, according to Volta, is the great capacity of the electrophore, which soaks up the electricity of the probe as often as it becomes charged, and the small capacity of the separated shield, which allows a weak electricity to manifest itself. Since the quantity of charge on a conductor increases as the product of its tension and capacity, for the same tension the condensatore will rob the probe of charge in proportion to their capacities. After separation the tension of the shield ascends in inverse proportion to the ratio of its own capacity to that of the electrophore.[20]

The condensatore had a vast progeny. Passing over the minor improvements of Volta and Cavallo, we may notice the Reverend Abraham Bennet's 'doubler,' (fig. 19.6), which multiplied electricity at the cost of mental as well as mechanical energy. ('I could not understand the operation from hearing it read, nor from enquiry.'[21]) The process was immediately mechanized, most successfully by Nicholson, whose ingenious doubler (fig. 19.7) anticipated the influence machines of the nineteenth century.[22]

These English adepts understood the notion of electrical capacity. Furthermore, they evidently disposed of a workable concept, 'tension,' related to charge via Volta's rule, $Q = CT$. The same knowledge appears in the texts of the time, usually in connection with the theory of facing or 'conjugate' conductors. It occurs in Cavallo, with 'degree of condensation' for 'tension'; in de Herbert, Achard and Saxtorph;[23] in Saussure, Landriani, and van Marum;[24] and in the excellent compendium of Kühn. 'The intensity of the shield becomes the less, and consequently its capacity the greater, the closer it comes to the cake.'[25]

20. Volta, *PT,* 72:1 (1782), vii–xxxiii, esp. §§3, 9, 33–4, 49. Volta's formulation is by no means perfect, for his enthusiasm for the rule $Q = CT$ causes him to associate T with the force between shield and plate, and to assume that it goes almost to zero when the electrophore is assembled. Cf. his hopeless theory of the action of lightning rods, *ibid.*, xxv.

21. Cavallo to Lind, 7 Aug. 1787, Add. Mss., 22897, f. 77 (BL). When he read a printed account Cavallo saw immediately that it was 'a very ingenious contrivance' based upon the principle of the electrophore.

22. Bennet, *PT,* 77 (1787), 280–96; Nicholson, *PT,* 78 (1788), 403–6; cf. Thompson, Soc. Tel. Eng., *Journal,* 17 (1888), 573–82. These doublers worked even without an initial charge, making do with electricity deposited by the air, or excited during handling. Cavallo and Bennet devised ways to reduce the resultant ambiguity; cf. Cavallo, *PT,* 78 (1788), 255–60, and Walker, *Ann. Sci.,* 1 (1936), 88–93.

23. Cavallo, *Comp. Treat.* (1795³), II, 219, 244–64, and *PT,* 78 (1788), 1–21; Herbert, *Th. phen.* (1778²), 96–7; Achard, *Vorlesungen* (1791), III, 79; Saxtorph, *El.-Laere* (1802), I, 217–18, 285–6.

24. According to Kühn, *Neu. Ent.,* I (1796), 154, and Hackmann in Forbes, *Marum,* III, 341–2; cf. van Marum to Volta, 31 Aug. 1788, *VE,* III, 6–7.

25. Kühn, *Neu. Ent.,* I (1796), 149–52.

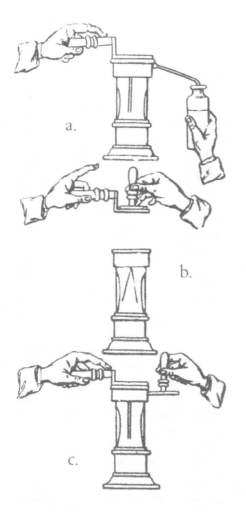

a.

b.

c.

19.6 Bennet's doubler, from PT, 77 (1787), 288–96. The doubler consists of the disks h (with a horizontal insulating handle) and v (with a vertical one.) (a) Give a weak charge Q from a debilitated bottle to the electrometer cap while h rests on it; momentarily ground h, giving it a change -Q. (b) Remove the bottle and raise h; if Q is large enough the electrometer leaves will spread; if not, place v on h and ground v, giving it a charge Q. (c) Return h, touch v to the underside of the cap, ground h, and remove both; the cap has a charge of nearly 2Q, h of -2Q, v of zero. The operation can be repeated as often as desired.

2. FORCE

The demonstration that gravity diminishes as a simple function of the distance encouraged even those who refused to accept Newtonian apologetics to look for similar rules for other apparent actions at a distance. The case of magnetism, examined earlier, illustrates both the appeal and the difficulty of the task. The earliest known attempt on electricity was made by Christian Gottlieb Kratzenstein, a clever man, able to discern interesting problems but unable to solve them, 'not a man of great insight,' according to Lichtenberg, but one who knew his way around.[26] In 1743–4, while yet a student at the University of Halle, he helped his professor, J. G. Krüger, in what were perhaps the earliest efforts to

26. Lichtenberg to Kästner, c. 1784, in Lichtenberg, *Briefe*, II, 123–4; *ibid.*, 294.

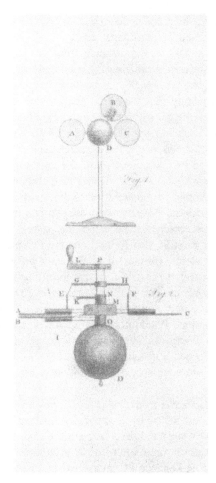

19.7 Nicholson's doubler, from Nicholson, PT, 78 (1788), 403–7. The doubler consists of two fixed metal disks A and C, a movable one B, and a metal ball D. (a) Give a small charge Q to A and bring B opposite; at that instant the pins E and F touch the protruding wires at G and H, connecting A and C, and B comes into contact with D via the wire at I. Owing to the great capacity of the conjugated plates A and B, the result of their confrontation is that most of Q remains on A and −Q is induced on B. (b) Bring B opposite C, breaking the first contacts and connecting C and D via the pin at K; C obtains a charge Q by induction. (c) When B returns to A, the connections between it and D, and between A and C are restored: A charges to almost 2Q at the expense of C and B charges to almost −2Q by induction. The charges may be doubled again at the next revolution.

apply electricity to medicine.[27] Simultaneously, under the influence of his professor of metaphysics and mathematics, the newly restored Wolff, he tried his hand at another novelty, quantitative physics, and with some success; for he won the Bordeaux Academy's physics prize in 1744 with an essay on evaporation that specified, among other things, the diameter of vapor particles.[28] In 1746 he combined his electricity and mathematics in an attempt to measure electrical action, which he understood to occur in a vortex, in the manner imagined by Hausen.

27. Snorrason, *Kratzenstein* (1967), 16–20, and *Kratzenstein* (1974), 11–24. Following Krüger's example, Kratzenstein became both physician and physicist.
28. Snorrason, *Kratzenstein* (1974), 70–2. Cf.Lichtenberg to Kästner, c. 1784, in Lichtenberg, *Briefe*, II, 128–9: 'I'm also no special friend of hypotheses in physics, especially of Kratzenstein's, who even gives the diameter of an aether particle.' Cf. Rogers, *Ann. Sci.*, 29 (1972), 239–55.

Kratzenstein's technique was to find the weight necessary to balance the electrical force between the globe of the machine and a disk hanging from a steelyard. He found that within what he called 'experimental error'—245 percent in one case—the force diminished inversely as the distance from the surface of the glass.[29] The approach was considered promising by such connoisseurs as Lambert and even Euler, who helped arrange Kratzenstein's appointment as 'mechanicus' to the Petersburg Academy in 1748. Five years later, at the expiration of his contract, Kratzenstein happily accepted the chair of experimental physics at the University of Copenhagen. The death of Richmann had touched him deeply. So had the 'forced labor,' as Kratzenstein characterized his duties as mechanic in charge of instruments, and the animosity of Lomonosov.[30]

Perhaps following Kratzenstein's example, Richmann tried to determine electrical 'force' by the balance, as did Gralath, Bose, and an anonymous correspondent of Ellicott's.[31] No simple rule emerged. The technique passed from the literature until revived by Daniel Bernoulli, also formerly of the Petersburg Academy, now professor of anatomy at the University of Basle, a powerful mathematician and the author of the hydrodynamical theory that Euler perfected and applied to electricity. Bernoulli took up the study of electricity towards 1755, about the time the Eulers, Wilcke and Aepinus also began. He and his former student Abel Socin, M.D., examined Franklin's system and (if one can judge from notes on lectures given by Socin in 1761) developed an advanced view of the nature of atmospheres.[32] To find a distance rule for electrical force Bernoulli applied the balance technique to a form of d'Arcy's hydrostatic electrometer.[33]

At a distance x below a 4″ disk suspended from the prime conductor (fig. 19.8) Bernoulli placed a similar disk supported by a vertically floating calibrated glass tube. He found by trial the weight required beneath the float ab to counter the attraction of the upper disk for several values of x. It appears from his results, as published by Socin, that he kept the PC connected to an operat-

29. Kratzenstein, *Th. El.* (1746), 34–5.

30. Lambert to Kratzenstein, 12 Dec. 1776, in Lambert, *Deut. gel. Briefw.*, II, 327–8; Bouguer to Euler, 23 Nov. 1752, in Lamontagne, *RHS*, 19 (1966), 233; Snorrason, *Kratzenstein* (1974), 75, 85–6.

31. Richmann, *CAS*, 14 (1744–6), 323; Gralath, Nat. Ges., Danzig, *Vers. Abh.*, 1 (1747), 525–34; Bose, *Tentamina*, II:ii (1747), 42; Ellicott, *PT*, 44:1 (1746), 96, and JB, XIX, 60–1. Cf. Boyle's similar proposal in M. B. Hall, *Boyle* (1965), 253–4.

32. The counts Teleki, who studied electricity *privatim* with Socin, noted a 'pretty experiment on the action of the electric atmosphere on surrounding bodies.' Socin's explanation: 'The fluid of the positive electricity induces a negative, i.e., it draws out the bodies' normal electrical fire, only one must arrange things so that no spark jumps, otherwise positive electricity will be transferred.' Spiess, *Basel* (1936), 104; cf. *ibid.*, 95, 120. Gehler, *Phys. Wört.*, I (1787), 817, lists Socin with Ingenhousz and Wilcke as pioneers in the correct interpretation of the electrophore.

33. D'Arcy, *MAS* (1749), 63–74; Socin, *Anfangsgründe* (1777), Vor.; cf. Herbert, *Th. phen.* (1778²), 189, and Tab. V, fig. 5.

19.8 Bernoulli's techique for measuring the 'force' of electricity: weights are added to the hook beneath the float ab to counter the attraction for the disk by a similar disk attached to the prime conductor. From Socin, Acta Helv., *4 (1760), 224–5.*

ing machine during each measurement. This arrangement, which gives the upper disk a different charge at each setting, would immediately suggest itself to one endeavoring to determine the force of a given degree (or tension) of electricity. Owing to it Bernoulli obtained a force rule misleadingly similar to Coulomb's. If the lower disk was effectively earthed through the moist tube and its basin, the two disks constituted a condenser of fixed potential V. Since $V \sim sx$, s being the charge density of the plates, and since the ponderomotive force f on either goes as s^2, $f \sim 1/x^2$. Bernoulli's relation does not refer to *force* (italicized to mean force between elements of electrical matter) and, as presented by Socin, supports no transdiction. Nonetheless some historians have thought it an anticipation of the law of squares.[34]

Volta once measured in much the same way as Bernoulli, using the steelyard technique, touching the movable disk (the one attached to the beam) with a source of constant force (a capacious Leyden jar) before each measurement. His best results, obtained with five-inch disks at a potential difference of about 5000 volt, gave the inverse-square dependence exactly. But he quite properly did not regard the relation as universal, or even as fundamental. He knew that changes in geometry or in technique influenced the results, and he had found

34. Socin, *Acta helv.*, 4 (1760), 224–5. Cf. Whittaker, *Hist.* (1951²), I, 53; Roller and Roller, in *Harv. Case St.* (1957), II, 610–11.

that repulsion and induction obeyed different relations. Indeed, he denied the general applicability of the law of Coulomb.[35]

All the quantifiers of electrical 'force' from Kratzenstein to Volta failed because they had no theory that enabled them to move from microscopic interactions to macroscopic measures, from *force* to force. A final illustration of the difficulties of transdiction in this case is the attempt of the precocious Charles Stanhope, Lord Mahon, a Fellow of the Royal Society before he was twenty. This distinction was not bestowed entirely for his science, which consisted of mathematics and physics acquired in Geneva from the arch-mechanist Lesage. Mahon's father, a Fellow and, what was more, an Earl, had merely asked that 'my son Mahon, who is but nineteen years old, might be admitted a Fellow, as an encouragement to persist in improving his natural turn for the mathematical, and especially the mechanical parts of knowledge.'[36] Mahon made his debut at the Society in 1774, on his return from Geneva. He chose to join in the battle then raging over the proper shape for lightning conductors.[37] While opposing 'obtuse Wilson's obtuse rods,' as Lichtenberg put it, he made the important discovery of the 'return stroke,' a sometimes lethal current produced when lightning instantaneously removes the cause of charges previously induced in grounded men or animals.[38] This effect furnished much of the material for a prolix book, to which Mahon prefaced his investigation of electrical action.

As a disciple of Lesage, Stanhope, as we shall now call him, declined to share the consensus localizing charge, and equated *force* with the density of electrical fluid excreted into the air about charged bodies. Let this density diminish as some function $f(r)$ of the distance measured from the center of the ball B (fig. 19.9), attached to the positively charged cylinder AB. Imagine the unelectrified cylinder CD divided at X such that the amount of electrical atmosphere or charged air RR 'superinduced' on CX equals the amount SS superinduced on XD; and suppose further that along the line XY the cylinder has its normal content of electrical matter, to the left less and to the right more. (Stanhope has an arbitrary formula, relating the density of the atmosphere to the amount of electrical matter displaced, which insures that XY is the equilibrium line both inside and outside the cylinder.) One more assumption is needed: because of XY's 'neutrality,' electrometer threads hung at X will stay plumb, while those left of it will diverge with negative electricity, and those right of it with positive. Find the null point by experiment, take the quantity of 'superin-

35. Volta to Lichtenberg, 28 July 1787, *VO*, V, 78–9; Polvani, *Volta* (1942), 145, found by experiment that Volta's standard straw electrometer (*VO*, V, 80) indicated 13,350 volt when it stood at a separation equivalent to 35° of a Henley electrometer.

36. Newman, *Stanhopes* (1969), 171.

37. *DNB*, XVIII, 887–90; G. Stanhope and Gooch, *Life* (1914), 10, 15, 32–5, 165–86; Aronson, *Nat. Phil.*, 3 (1964), 53–74.

38. Lichtenberg, *Briefe*, I, 321, and *GGA* (1782:Suppl.), 342–3. Let the cloud be stationed above one end of a long wire fence, nailed to wooden posts. The far end of the fence becomes strongly charged by induction, as does any animal touching it.

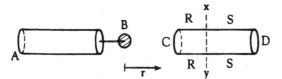

19.9 Diagram of Stanhope's apparatus for measuring f(r).

duced' electrical matter as the integral of $f(r)$, and f may be inferred from the definition of X:

$$\int_C^X f(r)dr = \int_X^D f(r)dr.$$

For the case that $f(r) \sim r^{-2}$, $2/BX = 1/BC - 1/BD$. Stanhope claimed to have found this equation precisely fulfilled in twelve trials, the measurements being accurate to 1%![39]

'I think that otherwise mathematical head sometimes behaves most unmathematically.' Thus Lichtenberg, faulting Stanhope's inept and arbitrary atmospherical physics, the obscure mechanics of segregation, and the passé assumption of electrified air.[40] Volta likewise thought Stanhope's theory had nothing to do with his experiments: the immediate action of the PC and cylinder on the electrometer threads determined their separation, and the cylinder must be in the same state, or at the same tension, throughout.[41] Technically Lichtenberg and Volta were correct. Yet Stanhope's work gained the assent and admiration of many, and not just of the algebraically illiterate like G. C. Morgan, who conceived that his noble author had 'discovered a most accurate agreement between the language of fact and the corollaries which he inferred from several abstract propositions.'[42]

Bennet accepted Stanhope's 'learned' principles, which he misunderstood as a stressed-medium theory; Needham translated Milord, 'one of the first mathematicians of Europe,' into French, and promoted him everywhere; and Thomas Young credited him with the earliest demonstration of Coulomb's law.[43] Two causes contributed to this favorable press. First, Stanhope's technique for inferring the law of *force* was ingenious and correct in principle, and

39. C. S. Stanhope, *Principles* (1779), 7–8, 34–61, 73–4.

40. Lichtenberg to Reimarus, 18 Oct. 1781, in Lichtenberg, *Briefe,* I, 386; *GGA* (1780:Suppl.), 338. Cf. Robison, *System* (1822), IV, 84.

41. Gehler, *Phys. Wört.*, IV (1791), 805–6. Deluc tried to mediate between Stanhope's theory and Volta's criticism (*Idées* [1786], I, §320).

42. G. C. Morgan, *Lectures* (1794), I, 15–16. Cf. Dorling, *Stud. Hist. Phil. Sci.*, 4 (1974), 340n.

43. Bennet, *New Exp.* (1789), ix; Needham in C. S. Stanhope, *Principes* (1781), Préf., and Ac. roy. sci. Belg. *Mém.*, 4 (1783), 76–7; Young, *Course* (1807), I, 664, repeated by Feather, *Electricity* (1968), 13–14. Cf. van Swinden, *Recueil* (1784), I, xxxiv–xxxvi, and *VE*, II, 12, reporting the favorable view of Le Roy. There is also a German translation.

few of those who read or professed to read him could discern that the essential step, the computation of the null point X, did not follow from his principles. Second, his result came as no surprise: most electricians of the time, as Volta observed to Lichtenberg, held the 'opinion that electrical action [!] is proportional to the inverse square of the distance.'[44]

This predisposition rested primarily on the evident enervation of electricity with distance and on an analogy to gravitation, which Priestley had strengthed in 1767 in an inspired account of a puzzling discovery of Franklin's: an insulated cork ball lowered into a charged metal cup does not go to its sides, nor, having touched bottom, does it show any electricity when withdrawn. 'The fact is singular,' Franklin had said. 'You require the reason; I do not know it.' Priestley used a pair of corks, and found the effect the same: the balls did not separate when suspended within the can and brought out little or no electricity even if they brushed its inner surface. 'May we not infer from this experiment,' he wrote, 'that the attraction of electricity is subject to the same laws with that of gravity, and is therefore according to the squares of the distances; since it is easily demonstrated that were the earth in the form of a shell, a body in the inside of it would not be attracted to one side more than another?'[45]

The argument is to the point. Yet, despite the friendly services of Franklinist historians, it is not a demonstration, or even a proper enunciation, of the law of squares. For one, Priestley gives us no warrant for equating the 'attraction of electricity' with the *force* of a particle of electrical matter. Apart from that, he requires not Newton's theorem—that a uniform gravitating shell exerts no force on a mass point anywhere inside it—but the converse, which he certainly could not prove. Again, a bucket is not a spherical shell. The force of gravity does not vanish within a uniform can; that of electricity does, except near the mouth, because the mobile electrical matter arranges itself on the external surface so as to nullify its action within. Cavallo noticed this difficulty, and rejected Priestley's law as an artifact of the can, in which opposing electrical forces fortuitously cancelled.[46] Finally, it appears from Priestley's explanation of the electrification of corks grounded while hanging in the charged can that he did not fully understand his parallel.[47] His sketchy argument did help solidify opin-

44. Volta to Lichtenberg, 1786, *VO*, V, 78. Earlier guessers include Aepinus, Needham, and D. Fordyce, Professor of Moral Philosophy at the University of Aberdeen, who inquired of Canton on 30 April 1748: '[Does Electricity] partake of the Nature of Gravitation & some other Qualities whose Force diminishes in the duplicate Ratio of the Distance?' Canton Papers, II, f. 14 (RS).

45. Franklin to Lining, 18 March 1755, *EO*, 336; Priestley, *Hist.* (1775[3]) II, 372–4.

46. Cavallo, *Comp. Treat.* (1782[2]), 199; cf. Milner, *Experiments* (1783), 99–103. Others who did not accept Priestley's 'law': Nicholson, *Intro.* (1782), II, 396, preferring relative agnosticism ($1/r^n$); van Marum, preferring the same, even after Coulomb (Hackmann in Forbes, *Marum*, III, 343); and Adelasio, *De el.* (1781), betting on $1/r$.

47. Priestley, *Hist.* (1775[3]), II, 374–6. The corks charge because of forces exerted on the electrical matter of the can owing to the connection to ground outside it. Priestley ascribed the effect to the new and false principle, related to Franklin's theory of the condenser, that 'no body can receive electricity in one place [the corks] unless an opportunity be given [the external ground] for parting

ion, and it pointed further. As a demonstration, however, it carried no more weight than Stanhope's, from which it differed chiefly in being susceptible to transdiction.[48]

ROBISON AND COULOMB

The first unexceptionable demonstration of the law of electric *force* was the work of John Robison, who came to physics in the manner of Wilcke. The son of a wealthy Scots merchant who wished to make him a minister, Robison contracted an unsuppressible interest in science at the University of Glasgow, where he studied under Joseph Black and listened to James Watt, then the school's mathematical instrument maker. After graduation he made his way to London and the characteristic Scottish position, tutor to the English establishment; his charge was the son of Admiral Charles Knowles, then about to go to Canada as a midshipman under Wolfe. Robison returned to England in 1761, sick of the navy and willing to study theology; but he could not resist science, in the shape of Benjamin Martin's lectures, or the offer of the Board of Longitude, which at Knowles' suggestion asked him to test the latest marine chronometer on a voyage to Jamaica. Seasickness and the Board's stinginess strengthened his clerical resolve, and in 1763 he was again in Glasgow preparing to tackle theology. Black and Watt once more intervened. Robison devoted himself seriously to science, with the consequence that six years later, in 1769, he had neither calling nor living.[49] His friends found him several possibilities, which testify to his capacity: the superintendency of pupils at the Royal Naval Academy; a government commission to survey North America; the professorship of Greek at Glasgow; the pursership of a man-of-war; and a prosperous cure, with the promise of rapid advancement, if he would only acquire holy orders.

He chose instead to go to Russia as private secretary to Knowles, who had been sent at Catherine's request to help update her navy for a war with the Turks. Robison flourished in St. Petersburg, mastered the language, caught the Empress' eye, and received a commission, similar to one he had rejected in England, to teach mathematics to imperial sea cadets. He went easily into society, 'untinged,' according to an admirer, 'with the slightest shade of that fastidious severity and supercilious peevishness to which some of the worthiest characters devoted to serious studies of scientific pursuits are occasionally prone.' He also established close connections with the foreign academicians,

---

with it in another.' Apparently he thought that the acquisition followed from a push from the can's interior.

48. Neither Franklin, *EO*, 336n, nor Beccaria, *De atm. el.* (1769), used the law of squares in elucidating experiments made within the electrified can. Lullin, *Dissertatio* (1766), 38, interpreted Franklin's effect in favor of Nollet: the can's restricted interior prevents the establishment of the contrary fluxes.

49. Playfair, RS, Edin., *Trans.*, 7 (1815), 495–507; *DNB*, XVII, 57–8; Robison to Watt, c. 1796, 1 Feb. and 11 Dec. 1799, in Robinson and McKie, eds., *Partners* (1970), 256, 300, 318.

particularly with Aepinus. In 1773 a suitable permanent post opened. The University of Edinburgh, at the suggestion of Black, who had moved there in 1766, offered Robison its vacant chair of natural philosophy. He did not hesitate to accept.[50]

The choice of science over Church or Navy meant relative poverty. In 1793, to augment his salary of £200, Robison began writing for the *Encyclopedia Britannica*; and in the supplement to the third edition (1801) he brought out the elegant account of Aepinus' theory noticed earlier. He also recorded there his investigation of the law of *force*, which, he said, he had unveiled 'in a public society in 1769.' Perhaps he mistook the date, for the inspiration of the investigation, as he acknowledged, derived 'from the reasonings of Mr. Aepinus,' whose personal acquaintance he had not yet made. Such errors come easily thirty years after the event, especially to brains addled by opium, which Robison habitually took to ease abdominal pains.[51]

In addition to the theory of the *Tentamen* Robison required Newton's proposition that a gravitating shell or sphere acts on external bodies as if its entire mass were located at its center. If electrical *force* also obeyed the law of squares, two similarly charged spheres should flee one another with a force diminishing as r^{-2}, provided that r, the separation of their centers, is enough larger than their radii to protect them from the effects of induction. Robison's apparatus appears in fig. 19.10, where A represents a brass ball, and B and D gilt cork balls, each ¼″ in diameter; BD, the electroscope, is a stiff, waxed silk thread free to turn in the amber cheek C; AFEL is a glass rod. NH, a freely turning index, measures the inclination α of LA to the horizon and the inclination β of BD to the vertical (fig. 19.11).

Electrify the balls A and B with AL horizontal (at A_0L), the charges being insufficient to cause separation. Turn I and therefore AL to the angle α at which the balls diverge. To balance moments about C, the electrical force f must satisfy

$$f = \frac{\sin \beta}{\sin \gamma}(W_B - W_D \cdot CD/BC) \qquad (19.1)$$

where W_B and W_D represent the weights of the balls and γ is angle ABC. Setting BC = AL = 1, CL = x, AC = y, one finds

$$\sin \gamma = \sin(\beta + \phi) \cdot y[1 + y^2 + 2y \cos(\beta + \phi)]^{-1/2}, \text{ where}$$
$$\sin \phi = \cos \alpha/y, \quad y^2 = 1 + x^2 - 2x \sin \alpha.$$

50. 'Particulars respecting Mr Robison,' Ms. Murray 503, Glasgow Univ. Lib.; Robison to Watt, Dec. 1796, in Robinson and McKie, eds., *Partners* (1970), 248.

51. Playfair, RS, Edin., *Trans.*, 7 (1815), 520–2; Robinson and McKie, eds., *Partners* (1970), 250–1, 299, 375–6, 385; Robison, *System* (1822), IV, 68. For Robison's money problems see Add. Mss. 37, 915, ff. 30, 85, 98–101, 104–5, 207–9 (BL).

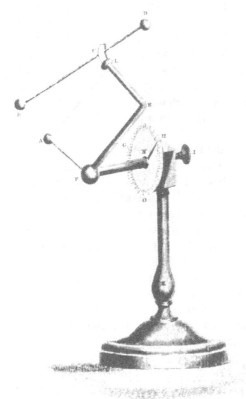

19.10 Robison's apparatus for measuring f(r) *between small charged spheres. From Robison,* System, *IV (1822).*

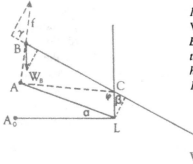

19.11 Schema of Robison's experiment. W_B *and* W_D *are the weights of the balls at B and D, and* f *is the repulsion between the balls at A and B; the other symbols have the same significance as in fig. 19.10.*

Since α and β are measurable and x fixed, $\sin\gamma$ may be found with a given charge on A and B for any number of positions of AL and BC. If the law of squares holds, f as computed from eq. (19.1) should be proportional to AB^2, or to $1+y^2+2y\cdot\cos(\beta+\phi)$. Robison says that he made 'many hundreds' of these measurements, which agreed 'far beyond his expectation,' and showed that the two spheres repelled one another inversely as the 2.06 power of the distance. Who is empiricist enough to fault him for concluding that 'the action between [the] spheres is exactly in the inverse duplicate ratio of the distance of their centers?'[52]

One might well wonder why Robison did not publish a contemporary account of his important experiments. Although Watt rated him the 'man of the clearest head and the most science of anybody I have known,' Robison periodically succumbed to self-doubt and despondency; 'I never possessed great talents,' he wrote Watt, 'and have often been much distressed by seeing more imputed to me than I deserve.'[53] 'I want vigour of mind and activity, and have jogged on in a sort of easy and decent obscurity which I might have escaped from had I been more active.'[54] His first publication, a calculation of the orbit of Uranus, was wrung from him as the Secretary's contribution to the opening volume of the *Transactions* of the new Royal Society of Edinburgh (1783); his other scientific work, the Encyclopedia articles, owe their existence to financial necessity. Had he had more 'fastidious severity and supercilious peevishness' electrostatics might rest on 'Robison's law' instead of on Coulomb's.

Coulomb was a military engineer well-trained in mathematics, to which he had been drawn by lectures of P. C. Le Monnier (the brother of the onetime electrician), a professor at the Collège Royal. At that time, c. 1750, Coulomb attended the Collège Mazarin. His father, a soldier turned tax-farmer, had hoped to make him a doctor; but he declined, was thereupon disinherited by his mother, and retired to Montpellier with his father, who meanwhile had speculated away his fortune. Coulomb made the best of his altered circumstances. He haunted the local scientific society, which appointed him 'adjoint,' with no salary, in 1757. More usefully, it provided him with introductions to d'Alembert, Le Roy, and P. C. Le Monnier when, the following year, he returned to Paris to prepare for the entrance examinations to the engineering school at Mézières. He entered in February, 1760. Perhaps he heard Nollet's summer lectures there, which began in 1761. He certainly followed the lectures and studied the books of the mathematician Charles Bossut, who may have alerted him to possibilities beyond an engineering career; for Bossut, who was already a correspondent of the Paris Academy, went in for prize contests, occasionally won, rose, and eventually obtained a post in Paris and a chair in the

---

52. Robinson, *System* (1822), IV, 68–74. He also measured the attraction between unlike charges by twisting AF below the horizon.

53. Robinson and McKie, eds., *Partners* (1970), 376, 389, 412–13.

54. Robinson to Watt, 7 April 1797, *ibid.*, 269.

Academy.[55] Coulomb was to follow the same steps, in the same order.

After nine grueling years in Martinique, where he directed the rebuilding of forts damaged by the English during the Seven Years' War, Coulomb returned to France in 1772, shaken in health but eager to communicate the results of his experience with earthworks, dikes, and masonry piers. He presented a memoir on these matters to the Paris Academy in 1773, which in return made him a correspondent of Bossut. The next step was to win a prize. Coulomb selected the Academy's competition for 1777, which required improved compass needles and an explanation of the diurnal variation of the earth's magnetic force. His sharing of the prize with the practiced magnetizer van Swinden was a double triumph, as the subject had not earlier concerned him and no one had been judged a winner in a similar contest held two years before. The victory almost brought him into the Academy in 1779. As it happened he had to win another prize and get himself to Paris, in the capacity of maintenance engineer of the Bastille, before he could enter the class of mechanics in December, 1781.[56] The memoir on magnetism did more than help realize his ambition. It set him on a path of research that led to the measurement of electrical *force*.

The investigation of the diurnal variation required a sensitive instrument. The standard arrangement, in which the needle sat upon a pointed pivot, developed frictional forces about as strong as the magnetic vagaries it was supposed to measure. Thread suspension had been tried, by Lana among others, but neither their reliability nor their theory had been established. Coulomb, who knew all about ropes, perceived that the torsion, or twisting force, of the suspension could be made a negligible fraction of the magnetic force by using a fine silk thread. One need only set the needle, magnetized by Aepinus' version of the double touch, in the average magnetic meridian, and the thread would offer no sensible resistance to the small forces of the variation. This demonstration rested upon the extremely important fact that, within the elastic limit of the thread, a freely hanging weight, twisted in its plane and released, vibrates isochronously. Having ascertained this truth, Coulomb inferred from the second law of motion that the force of torsion increases in direct proportion to the angle of twist; by measuring the torsional force for one angle, he could show its impotence at all angles reached in the variational measurements.[57]

Coulomb's threaded needle soon replaced the old pivoted model installed by Le Monnier at the Paris Observatory. It proved too good. It twitched when an assistant sneezed, or when the door opened, and trembled when carriages

55. *J. de l'Empire* (20 Sept. 1806), 3–4; Gillmor, *Coulomb* (1971), 1–22; *DSB*, II, 334–5. Bossut rose in the Academy perhaps more quickly than the rules allowed; Birembaut, *RHS*, 10 (1957), 151–2, 155–6.

56. Delambre, *HAS*, 7 (1806), 206–10; *HAS* (1775), 40; Gillmor, *Coulomb* (1971), 18–41, 175–81.

57. *Ibid.*, 140–5; Lana, *Acta erud.* (1686), 560–2; Coulomb, AS, *Mém. par div. sav.*, 9 (1780), §§26, 43–4, 48–50. Coulomb's papers will be cited by paragraph numbers (§) whenever possible to facilitate reference to Coulomb, *Mémoires* (1884).

passed in the street. Coulomb turned to the problem in 1782, shortly after settling into the Academy. Elaborate precautions against draughts and vibrations
failed to control the prize-winning instrument. He learned from the assistants
that the needle jumped whenever they touched an eyepiece fitted to its housing.
He suspected electricity. The silk thread insulated the needle from the frame,
whose wooden feet imperfectly insulated it from the floor. An assistant shuffled
over, charged himself, electrified the frame and disturbed the needle.[58]

It was not in the diagnosis but in the solution of this problem that Coulomb
showed his strength. The fact that the mariner's compass was sometimes disturbed by electricity inadvertently rubbed onto its glass housing had long been
known,[59] and many suspected that atmospheric electricity also could jostle the
needle. This last menace had been discussed in the *Journal de physique* for
several years before Coulomb began to worry about the caprices of his instrument. The comte de Milly, a soldier-chemist who liked to revivify calxes by
electricity, a noble free associate of the Paris Academy, recommended that
compass boxes be firmly insulated; 'utility,' he says, 'ought to be the only goal
of scientists [savans] and technologists [ceux qui cultivent les arts].' Milly's
useless solution, which could only have aggravated the problem, was endorsed
by the bumbling comte de Lacepède; both were set right by E. F. Gattey, who
advised grounding the needle, a procedure then promoted by Needham.[60]

Coulomb understood the merit of Gattey's solution to the problem of adventitious electricity. To employ it he would have to replace the silk thread of
suspension by a metal wire. But the aim of his new suspension had been the
reduction of resistance; he feared that wires might spoil the effect, owing to the
larger torque needed to twist them; and he had no proof that their torsional
vibrations were isochronous. He advised stopgap measures to correct the Observatory's compass and undertook a study of piano wires, which he found to
be more sensitive and elastic than he had supposed. Within certain limits, the
torque exerted by a twisted wire equalled $\mu(r^4/l)\theta$, r being its radius, l its
length, θ the angle of torsion, and μ a constant characteristic of the metal employed. Having satisfied himself of this strict proportionality, Coulomb inverted
his procedure, turned his wire into a balance, and set out to measure forces that
had eluded his predecessors. He first investigated the resistance of liquids to the
motion of solids by immersing a lead cylinder, hanging from a twisted wire,
in a pot of water, and noting the rate at which the amplitude of the torsional
motions decreased. The force measured, equivalent to a torque of $3.14 \cdot 10^{-2}$
dyne-cm, could not have been detected by any other instrument then in use.[61]
Coulomb proceeded to electricity.

58. Gillmor, *Coulomb* (1971), 146–9.

59. Anon., *PT*, 44:1 (1746), 242–5, puts forward the observation as something new.

60. Milly, *JP*, 13 (1779), 391–4; Lacepède, *JP*, 15 (1780), 140–3; Gattey, *JP*, 17 (1781), 296–
303; Needham, Ac. roy. soc. Belg., *Mém.*, 4 (1783), 81–6. Cf. the éloge of Milly, *HAS* (1784),
64–9.

61. Coulomb, *MAS* (1784), §§v, xiv, xviii, xx–xxvii. The law of torsion seems to have been
Coulomb's discovery; an earlier theory of Euler's (1764), e.g., had made the torque proportional to

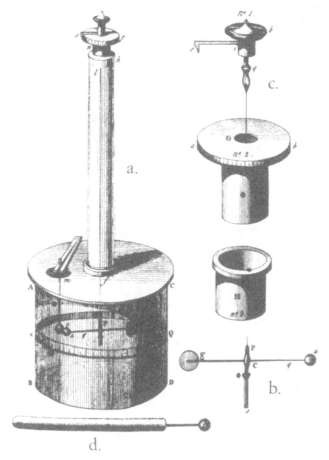

19.12 Coulomb's apparatus for measuring f(r). (a) The torsion balance (b) the electrometer stalk (c) detail of the knob by which the wire is twisted (d) the probe. From Coulomb, MAS (1785), 569–77.

Within the glass cylinder ABCD (fig. 19.12a) he suspended a wire carrying the electrometer stalk *q* (fig. 19.12b), which, like the similar part in Robison's instrument, was a waxed silk thread. One end of *q* bore an elder ball *a*, ¼″ in diameter, and the other a paper disk *g*, dipped in turpentine, which served to counterbalance *a* and to reduce oscillations. Another ball of elder, *t*, was rigidly fixed by a stick of gum-lac so that its center occupied the plane of rotation of the stalk *q*. Two scales completed the apparatus: the paper band ZOQ, pasted

sin θ; Truesdell in L. Euler, *Opera*, XI:2 (1960), 405. The notion of a torsion balance for measuring small forces occurred independently to John Michell, who, according to Cavendish (*PT*, 88 [1798], 49), expected the oscillations to be approximately isochronous. Cf. Geikie, *Michell* (1918), 84–91, and McCormmach, *Br. J. Hist. Sci.*, 4 (1968), 153–4.

on the outside of the cylinder, and the graduated disk ab (fig. 19.12c), which measured the rotation of the knob b, by which the wire could be twisted from outside the apparatus. To operate, Coulomb adjusted b until a touched t without twisting the wire. He then electrified the probe (fig. 19.12d) and inserted it through the hole m to touch t, which divided its charge with a. Ball a thereupon withdrew counterclockwise to a distance x, where the torque exerted by the electrical forces f balanced that of torsion, where $f l \cos\phi/2 = k\phi$ (fig. 19.13).

The wire can be twisted further by turning the knob b clockwise, which increases the angle of torsion to θ, say, and reduces the value of ϕ. Since the diameters of the balls may be disregarded in comparison to the half-length of the stalk, the separation of their centers is $2l\sin\phi/2$; if f diminishes as the inverse square, the balancing of torque requires $\theta = \text{const.}(\cos\phi/2)/(\sin^2\phi/2)$. For small angles the sine may be replaced by its argument and the cosine by unity, especially since the errors thus introduced tend to compensate; in the most unfavorable case Coulomb employed, $\phi = 36°$, the discrepancy amounts to less than 2%. He therefore expected to find the angle of torsion inversely proportional to the square of the angle of separation.[62]

At the first touch from the probe the ball a retired from t to $\phi_1 = \theta_1 = 36°$. Coulomb twisted the knob b through 126°, at which ϕ had dropped to half its initial value, making $\phi_2 = 18°$, $\theta_2 = 126° + 18° = 144°$. Further twisting produced $\phi_3 = 8°30'$, $\theta_3 = 575°30'$. Evidently $\theta_2/\theta_1 = \phi_1^2/\phi_2^2 = 4$, while the last measurement deviated from expectation by only 30'. 'It follows,' Coulomb wrote, 'that the mutual repulsive action between two balls charged with the same kind of electricity follows the inverse ratio of the square of the distances.' He did not doubt that attraction between unlike charges satisfied the same expression, but he found it more difficult to confirm, for the balls had an annoying habit of rushing together when he tried to twist them apart. He therefore turned to another, less exact procedure based upon a common technique for determining the strength of magnets.

An insulated copper globe G, one foot in diameter, stands with its center in the plane occupied by a gum-lac rod lg, suspended by a *silk* thread sc from the movable block V (fig. 19.14). Give G a positive charge and touch the gilded paper l, electrifying it negatively by induction. Since l had a diameter of only 7/12 inch, and never stood less than 9 inches from G's center, Coulomb supposed that the disk, like the globe, acted as if its charge were at its center. Displace lg from equilibrium by a small angle ϕ; the silk exerts a negligible torsion and the disk feels a restoring torque inversely proportional (by hypothesis) to y^2, y being the distance between the centers. As appears from fig. 19.15,

$$\ddot{\phi} \sim - (\phi/y^2)(1 + cl/y),$$

approximating (since y is sensibly constant and lc/y never exceeds 1/15) a simple harmonic motion with a period proportional to y. Coulomb found that the

62. Coulomb, *MAS* (1785), 569–77; *Mémoires* (1884), 107–15.

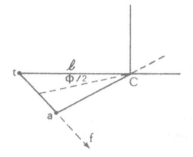

19.13 Schema of Coulomb's experiment: t, a, and C have the same significance as in figs. 19.12a and 19.12b.

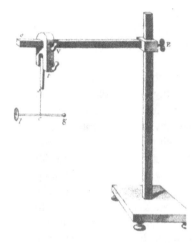

19.14 Coulomb's apparatus for obtaining f(r) *between bodies with unlike charges. From Coulomb, MAS (1785), 578–611.*

disk required 20, 41, and 60 seconds to execute 15 vibrations at 9, 18, and 24 inches, respectively. The law of proportionality required 20, 40, and 54 seconds. Ascribing half the large error of the third determination (some 10%) to leakage during the four minutes the experiment lasted,[63] Coulomb declared that he had 'arrived, by a method entirely different from the earlier, at a similar result: we can conclude that the reciprocal attraction of the electrical fluid called *positive,* on the electrical fluid ordinarily called *negative,* is in the inverse proportion of the square of the distances.'[64]

THE RECEPTION OF COULOMB'S LAW

Nowhere in these first memoirs does Coulomb make explicit the theory behind the experiments, and not until the very end, in announcing the law of attraction, does he reveal himself a dualist. On a parochial level his silence both reflected and rewarded the consensus regarding the localization of charge; more gener-

63. Several electricians in the 1780s were concerned to estimate error in electrical experiment, especially losses owing to leak; *infra,* xix.3.
64. *MAS* (1785), 578–85; *Mémoires* (1884), 116–23.

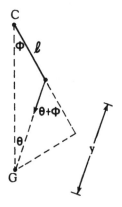

19.15 Schema of Coulomb's experiment: C and G have the same significance as in fig. 19.14.

ally, it illustrates the shift that was to produce the classical physics of the nineteenth century.

Compare Coulomb to the leader of the older school of French electricians. Nollet had little formal training and rose not by organized competition, but by assisting established practitioners and by making his knowledge accessible to the public. He thought it the business of the physicist to explore carefully, describe accurately, and explain qualitatively, using images drawn from common experience with macroscopic mechanical systems. Coulomb learned physics together with mathematics, and in his engineering practice sought to compute dependably, without fretting about mathematical rigor or mechanical orthodoxy. He entered the Academy as a victorious gladiator; he wrote for the knowledgeable, never for the fashionable; and he understood his primary task to be the quantification of physical relationships. In his winning essay on magnetic needles he had observed that, since the force of the earth's field upon a compass is independent of the orientation of the needle, terrestrial magnetism could not derive from a flux of effluvia confined to the magnetic meridian. 'Consequently, it seems that it is not vortices that produce the various magnetic phenomena, and that, to explain them, we must have recourse to attractive and repulsive forces [meaning *forces*] of the kind one must employ to explain the weight of bodies and the physics of the heavens.'[65] Nollet would have decried this non sequitur, and railed at its sentiment. But Nollet was dead; the astronomers and mathematicians in the Academy did not demur; and Coulomb's instrumentalism brought him such conquests as we have seen.

His colleagues did not doubt that his 'law' and the Newtonian propositions that, as he made plain in a memoir of 1788, had provided the rationale for his experiments, constituted the backbone of the theory of electricity. The law figures thus in Haüy's epitome of Aepinus (1787), which probably owed its inspiration to Coulomb's example. With a few exceptions, like the old Franklinist Le Roy, the academicians became both dualists and instrumen-

65. Coulomb, AS, *Mém. par div. sav.*, 9 (1780), §5.

talists, agreeing with Coulomb that, 'whatever be the cause of electricity, one can explain all the phenomena, and calculation will be found agreeable to experiment, if one supposes two electrical fluids, whose particles attract and repel one another inversely as the square of the distance.'[66] As for Le Roy, after succeeding his enemy Nollet as pensionary in 1770, he adapted the English plate electrostatic generator, then new to France, recapitulated the Franklinist side of the battle between knobs and points, and then retired from the field, ran away from his wife, and left the vacuum that Coulomb was to fill.[67]

Outside France applause came slower. Neither Cavallo, in the fourth edition of his text (1795), nor the *Encyclopedia Britannica*, in the lengthy article on electricity in its third edition (1797), mentions Coulomb's work. Similar silence reigned in Germany. Among foreigners who did notice it, two leading authorities thought it failed its purpose. The cosmopolitan Deluc, who, from his room in Windsor, issued bulletins on all aspects of meteorology and electricity, rejected the law of squares as an artifact of the torsion balance; still holding, like Cavallo, to the old theory of Beccaria (that electrical motions arise from disparities between the electricities of the interacting bodies and that of the air), Deluc insisted that Coulomb had no right to refer the forces to the centers of the repelling spheres! Had he taken an appropriate origin of coordinates—an undisturbed point in the surrounding medium—then, according to Deluc, he would have found it 'impossible' to obtain a precise law of *force*.[68] No doubt.

Volta placed Coulomb in the same category as Stanhope, a shortsighted pair who used procedures he had already rejected 'to follow a more direct path, which leads further.' It led him directly into a morass of particular cases, in only one of which the law of squares obtained. Repulsion not only obeyed a different law, but its relative weakness confirmed Volta's belief that it did not exist, that it arose from differential attractions as taught by Beccaria.[69] Moreover, the relation he obtained depended critically on the manner of charging the plates, and on their sizes and shapes: 'Various combinations have led me to discover diverse other laws, as curious as they are novel.' For want of connections between his chief theoretical entity—the universal primary attractive *force* of his first essay—and his measurements, Volta did not perceive that his macroscopic relations might all be consequences of Coulomb's rules for the particles of electrical fluid(s).

66. Coulomb, *MAS* (1786), 67–77, *Mémoires* (1884), 173–82, and *MAS* (1788), §xl; cf. Barruel, *Physique* (1799), Tab. 34.

67. Le Roy, *MAS* (1772:1), 499–512, and *MAS* (1773), 671–86; *BFP* (Smyth), X, 455. Le Roy permitted himself a public slap at Nollet, in the course of urging the construction of lightning rods: 'Tant il est vrai que la vérité, quoique lente dans sa marche, triomphe toujours, & ne manque jamais de renverser les vains obstacles que l'orgueil & l'envie veulent lui opposer'; *MAS* (1773), 673.

68. Deluc, *JP*, 36 (1790), 456.

69. *VO*, V, 78–9, 81–3; cf. Gliozzi in Volta, *Opere* (1967), 17, 101n. Volta argued that because the air—which, like any neutral body, must suffer induction before acting electrically—polarizes imperfectly, 'repulsion' must be relatively weak. Why his experiments appeared to show this weakness is a nice puzzle.

Coulomb's foreign fortunes began to improve with Robison's treatise in the *Britannica*, the first modern exposition of electrostatic theory in any language. Thomas Young, in his lectures at the Royal Institution in 1807, reduced and adapted Robison to a level above the heads of his fashionable auditors, and reached a larger public through the printed version of his course. The better German textbooks gave accurate accounts of the torsion balance, its use and its results; Fischer, in particular, emphasized the resultant reduction of Symmer's ideas to 'an exact theory,' from which the phenomena follow by 'rigorous calculation.' No respectable physicist would settle for less, he wrote, embracing the French theory while warning his readers not to believe it. 'It is but a convenient way to explain the facts, and one can only conclude that the phenomena occur as if they were produced by two fluids endowed with the preceding properties [the law of action], for the true nature of electricity lies hidden.'[70]

The growing enthusiasm for Coulomb must be interpreted as both cause and consequence of the penetration of the new French physics. Taken alone the results of the torsion experiments were not compelling. Coulomb reported very few measurements and some of them deviated considerably from expected values. To obtain even this agreement he had had to take many precautions, particularly with the wire, which had an annoying habit of twisting about even when unstressed. In a typical run the zero point would migrate two or three degrees. Many who tried to perform on his 'all too unsteady twisting machine,' as the Berlin physicist P. L. Simon called it, had trouble. Professor G. F. Parrot of Dorpat, for example, got errors of 12 percent or more, and Simon, emboldened by news of Volta's disbelief in Coulomb's law, announced that measurements made on his new gravity balance consistently gave $1/r$, not $1/r^2$. As late as the 1830s a respectable physicist, William Snow Harris, F.R.S., could call the law of squares into question. He bamboozled Whewell, who allowed that it was not yet supported by 'that complete evidence . . . which the precedents of other permanent sciences have led us to look for,' and stirred up William Thomson, who pointed out that the geometry of Harris' experiments had not been appropriate.[71]

There are other important sources of error. If the measurement lasts more than two minutes, leak can cause discrepancies comparable to those introduced by imperfect geometries or unsteady balances. To be able to compensate for leak Coulomb investigated its rate as a function of humidity; but the results only halved the earlier errors, and his calculations, which did not include the effects of charges induced on the damp walls of the balance's glass housing, were not free from criticism. In fact they do not agree with subsequent studies. As a preliminary to these investigations Coulomb showed that the repulsive

70. Robison, *System* (1822), IV, 1–204; Young, *Course* (1807), I, 662–4; E. G. Fischer, *Physique* (1806), 257–61.

71. Coulomb, *MAS* (1785), 574–5; *Mémoires* (1884), 113–15; Simon, *Ann. Phys.*, 27 (1807), 325–7, 28 (1808), 277–98; Parrot, *ibid.*, 60 (1818), 22; Harris, *PT*, 124:1 (1834), 239–41, 126:1 (1836), 431–7; Whewell, *History* (1858³), II, 210; Thomson, *J. math. pures appl*, 10 (1845), 209–16. Cf. Hoppe, *Gesch.* (1884), 106.

force between the balls varied directly as the product of the 'densities' of the electricity, without, however, estimating the consequences of charge segregation caused by mutual induction.[72] This effect probably infected the measurements from which he had deduced the law of attraction between unlike charges. Later, in a masterful study, he determined by experiment the distribution of charge on interacting spheres and cylinders, but he did not use the approximate theory he deduced to refine the earlier measurements of force.[73] That so few contemporaries thought those measurements inadequate confirms our earlier conclusion: they were quite prepared to believe before Coulomb announced the signs of the faith.

3. PRINCIPIA ELECTRICITATIS

Henry Cavendish was reputed to be 'a man so unsociable and cynical that he could stand honorably in the same tub with Diogenes.'[74] The son of Lord Charles Cavendish, himself an accurate and energetic physicist,[75] Henry had the unusual acquirement for an English scientist of a Cambridge education. One presumes that he found in college, which he left without a degree in 1753, a don with whom he could study Newton; he mastered the technique of fluxions and the content of the *Principia*, which became his model for the construction of electrical theory. After Cambridge came continental travel and a retired life in his father's London home, where he equipped or improved the laboratory, and followed a course of research suggested by such difficulties in Newton's theories of interparticulate *forces* as appear in the derivation of Boyle's law.

The *Principia* constitutes air of stationary particles, evenly spaced and repelling one another with a *force* diminishing inversely as their separations. Now the cardinal property of such a *force*, as of all *forces* decreasing more slowly than the third power of the distance, is that particles far away from a given one will, in the aggregate, exercise a greater force upon it than those close to hand.[76] Newton's air would stand close to the walls of rooms. To ease the

72. Coulomb, *MAS* (1785), 616–38 (*Mémoires* [1884], 147–72), discussed leakage, tried to divide the effects of imperfect insulation from those of humidity, and discovered the false law that the loss to the air increases with its moistness. The establishment of the law regarding densities—in which the charge of the fixed ball t would be halved ('by symmetry') by touching it with an identical insulated neutral ball—depended critically on estimates of leak (*ibid.*, §iv). Richmann, *CAS*, 14 (1744–6), was perhaps the first to publish measurements of leak; cf. Achard, *JP*, 19 (1782), 418.

73. Cf. Coulomb, *Mémoires* (1884), 123–4, ed. note; for Coulomb's studies of distribution (*MAS* [1787], 421–67, and *MAS* [1788], 617–705), which opened a beautiful realm of electrostatic theory, *infra*, xx.

74. Landriani to Volta, 9 Oct. 1788, *VE*, III, 10–11. Cf. G. Wilson, *Cavendish* (1851), 165–70.

75. BF to Kinnersley, 20 Feb. 1762, *EO*, 365. Lord Charles assisted in Watson's attempts to measure the velocity of electricity (*supra*, xiii.3), sought the temperature dependence of the conductivity of glass, and improved the usual demonstration of the electric light *in vacuo* (Watson, *PT*, 47 [1752], 370–1). Wilson, *PT*, 51:1 (1759), 310–12, adapted this demonstration, a spark in a Torricelli space, into an argument for the conventional direction of circulation of the electrical matter.

76. Consider the force f exerted by the particles within a narrow cone upon another at the apex. Taking the axis of the cone as x, $f \sim \int_\varepsilon^\infty x^2 dx/x^n$, which is infinite for $n<3$ and convergent for $n>3$. Cf. Cavendish, *PT*, 61 (1771), 586–8, 647–8; *El. Res.* (1879), 5, 43, 411; and Newton, *Math. Princ.* (1934), Bk. I, Prop. 86, pp. 214–15.

derivation of Boyle's law, and probably also to preserve uniform density, the *Principia* allows air particles to act only upon nearest neighbors. This arbitrary and implausible restriction, which did not bother Priestley, put off Cavendish, who directed his earliest experimental investigations, including the first study of hydrogen, at understanding the mechanics of aeriform fluids.[77] When he found he could not improve upon Newton, he turned to another elastic fluid, the electrical, expecting its law of force to be less intractable.[78]

The connection of ideas appears clearly from a comparison of Cavendish's 'Thoughts on Electricity' (c. 1770) with his long and difficult paper in the *Philosophical Transactions* for 1771. The 'Thoughts' are about a universal fluid 'consisting of particles mutually repelling one another.' Like Franklin's fire it can be accumulated, or decreased, beyond the amount normal to common matter, which opposes its elasticity; unlike Franklin's, it does not act through atmospheres, which Cavendish restricted to 'imperceptible distance[s],' but by the integrated effects of all electrical matter, acting immediately at a distance.[79] With a proper law of *force*, he hoped, the repulsion between the fluids in charged bodies, plus the repulsion arising from the uniform sea of surrounding fluid, would bring about observed electrical motions. Like Volta, Cavendish initially tried to reduce electricity to the action of a single elementary *force*, and tacitly assumed an interaction between the particles of electric and common matter. Again, like Volta, Cavendish found his theory unequal to its task.[80]

The paper of 1771 dispenses with aethers and effluvia to make do with localized charge, actions at a distance between elements of the electrical fluid, and even repulsions between the particles of common matter. In acknowledging Aepinus' priority in 'this way of accounting for the phenomena of electricity,' Cavendish pointed out that he had carried the theory much farther and had developed it 'in a different, and, I flatter myself, in a more accurate manner.' He also said that he had come to the theory independently of the *Tentamen*, which he saw only after he 'first wrote' his paper. But here perhaps he misremembered, for it appears that he had sought and found a copy of the book in 1766 or 1767, and it is certain that, about the same time, he read Priestley's garbled account of Aepinus' principles.[81] At the least, reference to the *Tenta-*

77. Cavendish, *PT*, 56 (1766), 141–84; G. Wilson, *Cavendish* (1851), 32–49; Wolf, *Hist.* (1952²), 362; *supra*, xvi.3.

78. G. Wilson, *Cavendish* (1851), 16–27; McCormmach, *Isis*, 60 (1969), 299–301, and *El. Res.* (1967), 68–9, 168–73, 241–6. Cf. Schofield, *Mechanism* (1970), 254–60.

79. Cavendish, 'Thoughts,' in *El. Res.* (1879), 94–103. Cavendish rejects extended atmospheres because experiment shows that, unless a spark passes, a body immersed in another's atmosphere acquires none of it.

80. Among other blemishes, Cavendish's theory seems incapable of handling minus-minus repulsion and assumes that the universal sea can contain charged bodies, maintain uniform density and retain its equilibrium (*El. Res.* [1879], 102–3). He must soon have realized that uniform density and equilibrium were incompatible, at least for the long-range forces he assumed (*ibid.*, 411–17). Cf. Maxwell's note, *ibid.*, 409–11, and McCormmach, *El. Res.* (1967), 108–13, 178–82.

81. Home, *Isis*, 63 (1972), 190–5, gives good reasons for assigning Cavendish's acquisition of the *Tentamen* to June, 1766. It certainly did not occur between the reading of his paper at the Royal

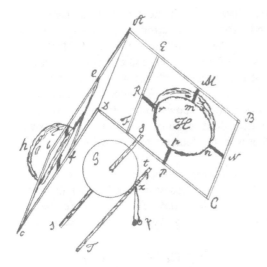

19.16 Cavendish's null experiment to show that electricity obeys the same law as gravity. From Cavendish, El. Res. (1879), 104.

men must have assisted his thinking when his aether theory of electricity proved more unsatisfactory than Newton's account of Boyle's law.

Whatever he may have owed Aepinus, Cavendish made good his claim to have 'carried the theory much farther.' In Aepinus' manner he explains Canton's experiments, predicts plus-plus attraction, emphasizes that segregation must precede electrical motions, etc; but he also introduces an advanced concept of potential and estimates the aggregate force of various distributions of electric fluids characterized by different laws of repulsion. We pass over his circles, disks, infinite cylinders and planes, and his shrewd guesses of the approximate behavior of fluids not regulated by the inverse square, to emphasize a result both exact and useful, namely that particles of redundant fluid on a conducting sphere uninfluenced by other charges will crowd into the shallowest possible depth beneath its surface.[82] Cavendish deduced this consequence from the Newtonian doctrine Priestley had crudely exploited,[83] and tested it in the following way, which he saw no occasion to publish.

A metal globe G mounted on waxed glass sticks SS stands within the frame BCADbc, which opens like a book (fig. 19.16).[84] The frame carries the paste-

---

Society (Dec. 1771 and Jan. 1772) and its printing, for the note recording the purchase, dated 'June 23,' is addressed to Canton, who died in March, 1772.

82. Cavendish, *PT*, 61 (1771), 584–98, 608–11; *El. Res.* (1879), on pp. 3–11, 18–19, 48–51.

83. A gravitating shell exerts no force on a mass point within it, while a gravitating sphere acts upon points outside it as if its entire mass were at its center. Suppose that electrical fluid, whose elements interact according to the law of squares, is frozen in a uniform distribution throughout a conducting sphere. Now imagine a thaw: assuming the fluid to be far from its maximum density, it will congregate at the surface, for any particle situated within the sphere will experience only a centrifugal force, arising from fluid closer to the center than itself. Hence—always supposing the law of squares—no fluid should stand on the internal surface of a charged shell of sensible thickness.

84. *El. Res.* (1879), 104–13. Cf. Dorling, *Stud. Hist. Phil. Sci.*, 4 (1974), 327–33.

board hemispheres H, h, which enclose G when the frame shuts, leaving a space of about ½ inch all around. Tt is the electrometer, a stick of glass wrapped with tin foil at x, from which the threads hang. Cavendish closes the frame, runs a short wire W (not shown) from H to G, and electrifies the pasteboard with a Leyden jar. Then, at one stroke, with the help of an elaborate system of strings, he disconnects the jar, removes W, opens the frame, and applies the electrometer. The interior globe shows no electricity. Cavendish infers the law of squares: the electrical fluid, contrary to the aerial, distributes itself in a manner entirely consistent with its elementary *force,* without the help of *ad hoc* restrictions of its range.

In a masterful analysis of experimental error, perhaps the earliest of its kind, Cavendish shows that, assuming the *force* to decrease as r^{-n}, his null result implies that n cannot differ from 2 by more than 1/50. The force exerted by the pasteboard sphere on a point inside it, say P (fig. 19.17), is by hypothesis

$$F = \int_{-b}^{b} \frac{s \cdot 2\pi y \cdot bd\theta}{QP^n} \cdot \cos\phi = \int_{-b}^{b} 2\pi sb \frac{a-x}{(b^2 + a^2 - 2ax)^{(n+1)/2}} \cdot dx$$

$$= \frac{2\pi sb}{(b^2 + a^2 - 2ax)^{(n-1)/2}} \cdot \frac{1}{a^2(n-1)(3-n)} \cdot$$

$$\left[a(3-n)(a-x) - (b^2 + a^2 - 2ax) \right]_{-b}^{b} \simeq \frac{Q}{ba} \cdot \frac{n-2}{(n-1)(3-n)} \cdot$$

Here s, the sphere's charge density, has been replaced by $Q/4\pi b^2$, and quantities raised to the $n-2$ power have been set equal to unity. Cavendish assumes that the globe G acquired a charge q too small to measure. Since the *force* differs little from the inverse square, q will act as if it were collected at O. Hence, a particle of fluid lying on the wire W would feel no force if

$$\frac{Q}{ba} \cdot \frac{n-2}{(n-1)(3-n)} = \frac{q}{a^2}, \quad \text{or if } \varepsilon \simeq \frac{q}{Q} \cdot \frac{b}{a},$$

where $\varepsilon = n-2$ measures the deviation from the square. Cavendish noticed that he could detect a charge on the globe equal to 1/60th that originally given the pasteboard sphere. If P lies midway between G and H (fig. 19.16), $\varepsilon \sim 1/57$; 1/50 is a generous estimate of the maximum deviation from Coulomb's law allowed by the experiments of Cavendish.

POTENTIAL

The high points of Cavendish's paper of 1771, indeed of the mathematical electrostatics of the eighteenth century, were considerations respecting narrow resistanceless tubes called canals, in which the compressible electrical matter behaved like an incompressible fluid. These considerations anchored a distinction that most electricians felt intuitively, but failed to recognize in the rigorous form and regrettable terminology introduced by Cavendish. 'Though the terms

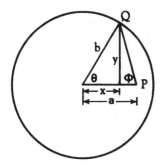

19.17 Diagram for Cavendish's estimate of error in the null experiment of fig. 19.16.

positively and negatively electrified are much used [he wrote], yet the precise sense in which they are to be understood seems not well ascertained.' No earlier electrician, not even Nollet, had doubted the operational significance of Franklin's words: a body was plus when it possessed the same sort of electricity as a glass rod or the hook of a Leyden jar charged by the globe. Cavendish proposed another operational definition, not equivalent to the original and much more difficult to apply. Procure yourself a standard test body B and an infinite canal C; place B an infinite distance from A (the body under investigation) and from all other electrified objects; join A and B *via* C; if A becomes plus ('overcharged' in Cavendish's terminology), A is 'positively electrified.'

The canals represent wires. The assumed incompressibility of the fluid within them is an artifice introduced to preclude accumulations or deficiencies Cavendish could not calculate; in the case of long, thin wires, as he expected, the idealization creates unimportant errors.[85] As for 'positive electrification,' it corresponds exactly to positive potential with respect to ground. Consider a system of conductors A, A', A'' . . . , joined by canals. Touch A with the tube: all bodies will electrify positively to the same *degree,* that is, each will confer the same amount of superfluous fluid on the test body B, although no two need contain the same charge. Cavendish observed that a body can be both positively electrified and undercharged or, to use modern terminology, to have a positive potential and a net negative charge. Bring an insulated positively charged body D up to one of the conductors, say A'. If D's electricity is sufficiently strong, it will drive all of A''s surplus, and some of its natural fluid, into its interconnected fellows A, A'', A''' . . . , increasing the 'electrification' (potential) of the system. Although now minus, A' will convey a larger plus charge to B than it did before. It was to secure this distinction that Cavendish proposed to alter the meaning of the old Franklinist terms.[86]

Cavendish exploited these ideas in computing or estimating the distribution of the fluid on systems of conductors. The case of two facing parallel planes led

85. Cf. Green, *Essay* (1828), v.

86. *PT*, 61 (1771), 628–9, 650–3; *El. Res.* (1879), 31, 45–6. Cf. the parallel distinction in the earlier 'Thoughts,' *ibid.*, 95–6, where the degree of electrification is associated with the condensation, or rarefaction, of the fluid.

without difficulty to a theory of condensers like Aepinus'. Greater instruction resulted from considering two charged disks or spheres connected by a canal so long 'that the repulsion of either body on the fluid in the canal shall not be sensibly less than if they were at an infinite distance.' Let the radii of the bodies be a_1, a_2, and the law of force be r^{-n}, $1<n<3$, and compute the pushes df_1, df_2, in the direction of the canal exerted by similar and similarly situated elements q_1, q_2 (fig. 19.18) on tiny cylinders of fluid dx_1, dx_2. Since *force* is proportional to the product of the amounts of interacting electrical matter,

$$df_1 = \frac{q_1\,dx_1\cos\phi_1}{z_1{}^n} = \frac{q_1\,dx_1\,(x_1 - y_1\cos\theta)}{z_1{}^{n+1}}.$$

A similar expression holds for df_2. Because of the symmetry of the elements q_1 and q_2, $y_1/y_2 = a_1/a_2$ and $q_1/q_2 = Q_1/Q_2$, the Q's representing the total charge on the bodies. Choose dx_1, dx_2, so that for *every* pair q_1, q_2, $df_1 = df_2$. Then

$$\frac{Q_1\,dx_1}{Q_2\,dx_2} = \left(\frac{x_1{}^2 + y_1{}^2 - 2x_1y_1\cos\theta}{x_2{}^2 + y_2{}^2 - 2x_2y_2\cos\theta}\right)^{(n+1)/2} \cdot \frac{x_2 - y_2\cos\theta}{x_1 - y_1\cos\theta}.$$

In order that the relation be independent of θ, $x_1/x_2 = y_1/y_2 = a_1/a_2$, and

$$\frac{Q_1}{Q_2} = \left(\frac{a_1}{a_2}\right)^{n-1}.$$

Now consider the points P_1 and P_2 so chosen that (1) df_1 at P_1 and df_2 at P_2 are insensible and, as usual, (2) the distances O_1P_1 and O_2P_2 are in the ratio of a_1 to a_2. The 'total force' or pressure in the canal at P_1 arising from the element q_1 is the integral of $df_1 = df(x_1)$ from a_1 to P_1; that arising from q_2 and acting at P_2, the integral of $df_2 = df(x_2)$ from a_2 to P_2. But P_1 and P_2 are by hypothesis far from O_1 and O_2. The upper limits in the integrals may therefore be replaced by infinity. It is easy to see that the canal is in equilibrium under the actions of q_1 and q_2. Since $x_1/x_2 = a_1/a_2$,

$$\int_{a_1}^{\infty} df(x_1) = \int_{a_2}^{\infty} df\left(\frac{a_2}{a_1}x_1\right) = \int_{a_2}^{\infty} df(x_2).$$

Consequently, in the case assumed, the two bodies divide up the available charge in the ratio of the $(n-1)$st power of their radii.[87] For the special case $n = 2$, $Q \sim a$: when two positively charged disks or spheres are connected by a long wire, each obtains an amount of fluid directly proportional to its radius.

87. *PT*, 61 (1771), 626–9; *El. Res.* (1879), 30–1. The integral of $df(\theta,x)$ from $x = a_1$ to ∞ is the negative of the potential of a unit positive charge at S_1 (fig. 19.18) under the influence of charge $q_1(\theta)$; to get the entire potential one must also integrate with respect to θ. Consider a spherical shell of radius a:

$$df(\theta,x) = q(\theta)dx(x - a\cos\theta) / (x^2 + a^2 - 2ax\cos\theta)^{(n+1)2},$$

$$\int_a^{\infty} df = \frac{1}{n-1}\left[\frac{1}{(2a^2)(1 - \cos\theta)}\right]^{\frac{n-1}{2}} \cdot q(\theta).$$

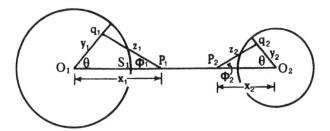

19.18 Diagram for Cavendish's deduction of the relative capacities of similar bodies.

This interesting proposition relates directly to that perennial problem, the power of points. Assume the law of squares. The force with which a sphere of charge Q and radius a drives a particle of fluid from its surface is Q/a^2; hence, by the result just obtained, the smaller of two spheres connected by a long canal repels at its surface more vigorously than the larger in the inverse ratio of their radii. Now represent a common experimental arrangement—a conical spike standing on a globe of large diameter—by two spheres, one much smaller than the other, communicating by a Cavendish canal. As far as the approximation holds, one understands immediately the apparent paradox that portions of the same surface, electrified to the same degree, can act very differently upon the electric fluid.[88] Cavendish was the first electrician to explain the power of points without recourse to *ad hoc* hypotheses.

The theory also allowed him to set Priestley straight about the electrical well. Recall that Priestley guessed the law of squares from Franklin's null experiment, but did not understand why corks hanging within the well charged when the strings suspending them were touched. Cavendish supposed a positively charged sphere ABC containing a body D, originally neutral, communicating with outside objects *via* the canal DG (fig. 19.19). Whatever the law of distance ($1<n<3$), there will be a point on the canal, say F, such that DF is uncharged and in equilibrium under the repulsion of ABC. (If $n = 2$, F coincides with B.) But the particles of fluid in the canal FG are driven towards G; and a current therefore flows from D until it and the objects at G come to the same potential. D accordingly will be minus. If, as in Priestley's case, it is

For equilibrium with another shell of radius b, charge distribution $r(\theta)$, $q/r = (a/b)^{n-1}$; 'equilibrium' means equal potential, brought about by the connecting canal (wire). Since $q(\theta) = (Q/4\pi)\sin\theta\, d\theta\, d\phi$, Q being the total charge and ϕ the azimuth of q,

$$f = -\text{potential} = \int_0^\pi d\theta \int_0^{2\pi} d\phi \int_a^\infty df = \frac{Q}{a^{n-1}} \cdot \frac{1}{(n-1)(n-3)} \cdot 2^{2-n}.$$

In the case $n = 2, f = Q/a$.

88. Cavendish, *PT*, 61 (1771), 660–4; *El. Res.* (1879), 52–4; cf. Young, *Course* (1807), I, 662–3.

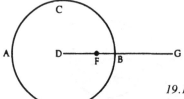

*19.19 Cavendish's elucidation of Priestley's exper-
iment in Franklin's well.*

separated from the canal before withdrawal, it will retain its charge (but not its potential!) on emerging from the well.[89]

The paper of 1771 established Cavendish as an authority on electricity; the Royal Society immediately made him chairman of its committee to consider the protection of the Purfleet arsenal from lightning. Nonetheless, few profited from Cavendish's great work. His difficult paper became a classic, and went unread; none but routine references to it have been found in any English work, including the *Encyclopedia Britannica,* before the beginning of the nineteenth century.[90] Cavendish's cold independence outmatched his colleagues' ignorant indifference. For a decade he continued to do electrical work of unprecedented scope and originality, most of which he did not publish. We have already noticed his demonstration of Coulomb's law. The rest of the work falls under three heads: measurements of the capacitance of (1) simple conductors and (2) condensers, and (3) painful determinations of the relative conductivities of metals and solutions.

The first set of experiments employed an extensible tin plate T to test the relative capacities of bodies B (fig. 19.20). Two equally capacious Leyden jars A and a charge B and T in opposite senses through rRSs and mMNn, respectively, while the wire dDδ is kept from A and B by the silk strings passing over the pullies L and *l.* Then, with another of his intricate arrangements, Cavendish lifts rR and mM, drops dδ, and measures the net charge by the pith ball electroscope at D. If the electroscope registers zero, the capacities of T and B are equal; if not, Cavendish adjusts T and tries again. One of his most striking results, which agrees well with later theory, made the capacity of a disk 18.5 inches in diameter about equal to that of a globe of 12.1 inches. These experiments are models of method; Cavendish not only tries to minimize or to correct for leak and the perturbations introduced by his own body considered as an electrical object, he also takes into account the contribution of the walls of the laboratory to the measured capacitance of the globe.[91] Although others, such as Barletti and Lichtenberg, had begun to appreciate these fine points,[92] no con-

89. *PT,* 61 (1771), 611–14; *El. Res.* (1879), 20–1.

90. Maxwell in Cavendish, *El. Res.* (1879), xxxii–xxxiv; McCormmach, *El. Res.* (1967), 256–7, 476–9.

91. *El. Res.* (1879), 114–37, 166–8.

92. Barletti, Soc. ital., *Mem. mat. fis.*, 4 (1788), 306–7, and *Dubbî* (1776), letter v (warning lest measuring instruments perturb the phenomena); Lichtenberg to Wolff, 3 Feb. 1785, in *Briefe,* II,

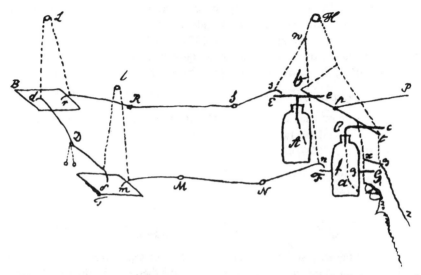

19.20 Cavendish's arrangement for measuring relative capacities. The purpose of the strings on the left is explained in the text; those on the right are to connect the jars to the prime conductor and to ground for charging, and to disconnect them for measuring. From Cavendish, El. Res. *(1879), 117.*

temporary, not even Coulomb, approached Cavendish in subtlety, complexity, and precision of electrical measurement.

The theory throws some light on the relative capacities obtained by experiment. Consider a positively charged disk and globe equal in diameter and connected by an infinite canal. The distribution of the redundant fluid of the globe is uniform; that of the disk, with which Cavendish says he is 'unacquainted,' he approximates by supposing a part of its excess to be spread evenly, and the remainder to reside in its circumference. If the distributed charge is p times the circumferential, the quantity of redundant fluid in the disk to that in the globe must be as $p + 1$ to $2p + 1$; and from Cavendish's measurement of the relative capacities, $p = 13/11$.[93]

194–5 (advising that the duration of electrical experiments be recorded to make possible corrections for leak).

93. *El. Res.* (1879), 30–6, 64. Cavendish's argument, which as usual exploits the equilibrium of the canal, amounts to computing the potential V of the disk. At an axial point a distance x from the center,

$$V(x) = (2pq/a^2)\int_0^a r(x^2 + r^2)^{-1/2}\,dr + q(x^2 + a^2)^{-1/2},$$

where q is the circumferential charge. At the disk's surface, $V = (q/a)(2p + 1)$. Since the total charge on the disk is $(p + 1)q$, its capacity, C_{disk}, is $a(p + 1)/(2p + 1)$; or, since the capacity of the sphere is a, $C_{disk} = C_{sphere} \cdot (p + 1)/(2p + 1)$. In Cavendish's case, $(2p + 1)\,(12.1) = (p + 1)\,(18.5)$, giving $p = 13/11$. Cavendish's value for the ratio of capacity, namely $1/1.545$, is extremely good; later theory gives $1/1.57$, an agreement that much impressed Lord Kelvin. Cf. Maxwell in Cavendish, *El. Res.* (1879), xxxix, 447–8; Mascart, *Traité* (1876), I, 121–2.

He tested this fraction ingeniously, by comparing the capacity of a disk of radius $2a$ with that of a *pair* of disks of radius a, hung parallel to one another a distance ka apart. Returning to his canals, Cavendish computed the capacity of each of the smaller disks by an analysis equivalent to finding its potential V, always assuming the distributed charge to be p times the circumferential:

$$V = p(2q/a)(\sqrt{k^2 + 1} + 1 - k) + (q/a)(1 + 1/\sqrt{1 + k^2}).^{94}$$

Since the charge on each disk is $(p + 1)q$, their combined capacitance, $2C(a)$, is $2q(p + 1)/V$; and the ratio r of their capacitance to that of a single disk of twice the radius, $C(2a)$, is

$$r = \frac{2C(a)}{C(2a)} = \frac{2q(p + 1)}{V} \cdot \frac{2p + 1}{2a(p + 1)}$$

$$= \frac{2p + 1}{2p(\sqrt{k^2 + 1} + 1 - k) + 1 + 1/\sqrt{k^2 + 1}} \cdot \qquad (19.2)$$

Cavendish obtained r separately for three different separations of the smaller plates, as shown in the table, which also gives the theoretical values of the ratio computed from the preceding equation and $p = 13/11$.

k	r computed	r measured
4	.867	.811
5.33	.897	.859
8	.932	.899

'[These discrepancies] can by no means be looked upon as a sign of any error in the theory, but, on the contrary, I think that the difference being so small is a strong sign that the theory is true.'[95]

Franklin squares proved more troublesome. Cavendish calculated the capacity of two facing circular plates of radius a, separated by an impenetrable barrier of thickness d, by assuming one plate connected *via* the usual canal to a plus body, and the other to a neutral one. Again this comes to computing the potential at the center of the positive plate:

$$V = \frac{2Q}{a^2}[a - (\sqrt{a^2 + d^2} - d)] \simeq \frac{2d}{a^2} \cdot Q,$$

if $d \ll a$ and the charges $\pm Q$ are equally spread upon the disks.[96]

---

94. This equation follows from that in n. 93 if one adds in the contribution of the other disk (for which x is replaced by $ka - x$) and sets $x = 0$.

95. *El. Res.* (1879), 64–6, 132–5, 142–3; Cavendish made similar comparisons with lengths of wire. (The computed values of r come from eq. (19.2); they differ slightly from those calculated by Cavendish.)

96. *El. Res.* (1879), 77–80; the contributions of circumferential charges cancel in this approximation. The equation in the text follows immediately from the integral in n. 93.

To test the result, Cavendish made several Franklin squares of the same glass, ground to different thicknesses and armed with circles of tin foil. Using a method similar to that depicted in figure 19.20, he obtained the relative capacities of the squares, which differed somewhat from the ratios of $a^2/2d$. The discrepancy, he found, could be almost removed by allowing for sparking from the foil to the glass, which effectively increased the radius of the coatings by 0.07 inch.[97] But when he compared the capacities of the squares to that of a simple conductor, the computed values, $a^2/2d$, turned out to be about eight times too small. The discrepancy, moreover, depended sensitively on the insulating material: it amounted to four for beeswax and nearly vanished for air. Cavendish suggested several explanations, none very plausible, and all directed towards reducing d to a smaller, 'effective' separation. These results, which point towards the concept of specific inductive capacity, greatly surprised and perhaps also disappointed Cavendish; for the discovery that each insulator had a characteristic electrical property besides its poor conductivity limited the generality of his theory, and may have checked his hope of preparing a *Principia electricitatis*.[98]

Estimating relative conductivities of metals came in with Priestley, whose earliest electrical experiments included comparisons of the lengths of wires of different bores and kinds that could be melted by a blast from his battery.[99] These crudities might have inspired but could not content Cavendish. In the winter of 1773, intrigued by accounts of an electrical fish, the torpedo, he began to measure the 'resistance' of solutions, placing himself in series with the test body and estimating the length x of each fluid required to diminish the explosion to a standard sensation.[100] The technique, which established that fresh water resisted 100 times better than salt, failed when applied to metals, for the resistance of Cavendish's body swamped that of any practicable length of wire. He therefore put himself in parallel with the test body and received the same stroke across 5.1 inches of saturated solution of salt—note the astounding accuracy—as across 2540 inches of a certain iron wire.

His reduction of these figures to comparative 'resistances' is most interesting. He tacitly assumed the total force F with which an object resists electricity to be proportional to $s \cdot p \cdot v^n$, s being its length, p its 'resistance,' and v the 'velocity of the electricity,' a substitute for current density. From data obtained with different tubes of salt water, Cavendish made $n = 1.08$; then, assuming F to be

97. *Ibid.*, 144–53, 161. Cavendish seems to have discovered this sparking. He also recognized, as had Franklin, that the fluid seeped into the glass from the coatings. He found that the process took time, and contributed nothing sensible to the capacity when he measured rapidly.

98. *Ibid.*, 172–82; McCormmach, *El. Res.* (1967), 469–76, 494–6.

99. Priestley, *PT*, 59 (1769), 65–70, wants to estimate the 'loss of force occasioned by the length of circuit,' 'force' being measured by the amount of fine test wire melted.

100. Cavendish, *El. Res.* (1879), 262; McCormmach, *El. Res.* (1967), 432–3, shows that Maxwell erred in dating these experiments in 1772. Cf. Maxwell in Cavendish, *El. Res.* (1879), xxxiv–xxxvii, and Walsh, *PT*, 63 (1773), 461–7.

the same for each branch of a parallel circuit, he computed the ratio of the ρ's of salt water and iron to be 560,000. It is in fact about 500,000.[101]

The little fish that inspired these investigations provided the occasion for Cavendish's second and last publication on electricity, an answer to those who refused to believe the torpedo electrical. And why not electrical? Because, as physicists had slowly and painfully learned, electrical matter manifests itself only after being collected upon an insulated body. 'When a Gentleman can so give up his reason as to believe in the possibility of an accumulation of electricity *among conductors* [i.e., the ocean] sufficient to produce the effects ascribed to the Torpedo, he need not hesitate a moment to embrace *as truths* the greatest contradictions that can be laid before him.'[102] Cavendish replied that one need not require the fish to maintain a charge and direct it towards a particular victim. It is only necessary for the torpedo to create electricity, and for the electrician to understand the nature of divided circuits.

When one discharges a Leyden jar through a wire held in the hands one opens two paths to the electricity, which, according to Cavendish, divides between the circuits inversely as their resistances. With a short, thick wire the experimenter feels no shock, not, as many electricians held, because no fluid traverses his body, but because the quantity that does cannot amount to a millionth of the whole.[103] When placed in parallel with a long, fine wire, however, one receives a sensible stroke, in exact analogy to the working of the torpedo. Here the parallel circuits are closed paths in the water, which carry less fluid the greater their lengths; if the experimenter, who conducts almost as well as the ocean, occupies a portion of one of the smaller circuits, he will pass most of its electricity.[104]

Another objection to ascribing the torpedo's prowess to electricity exploited its incompetence to throw sparks or attract chaff. How could it be electrical, lacking all the usual signs? Cavendish's answer returns us to our chief theme:

101. Cavendish, *El. Res.* (1879), 293–5, giving 607,000, apparently an arithmetical slip; cf. Maxwell, *ibid.*, 443–4. Several writers (Maxwell, *ibid.*, lix; Gliozzi, *Elettrologia* [1937], II, 23; and Süsskind, Frank. Inst., Phil., *Journal,* 249 [1950], 181–7), interpreting *F* as the e.m.f. of the circuit, credit Cavendish with Ohm's law. But Cavendish nowhere identifies *F* with the 'degree of electricity' of the Leyden jar, and employs 'resistance' in at least three different senses: for *F*, for ρ, and, as we shall see immediately, for something closer to the modern meaning. Cf. Teichmann, *Entwicklung* (1974), 26–38.

102. Henley to W. Canton, 21 May 1775, quoting Thomas Ronayne, Canton Papers, II (RS). For Ronayne, an Irish gentleman with an interest in the electricity of fogs and drizzles, see *BFP,* IX, 350.

103. Cavendish here revealed that iron wire conducts 400,000,000 times, saturated solution of salt 720 times, and sea water (one part of salt in 30 of water) 100 times better than rain or distilled water. He promised 'shortly' to give an account of these figures, which we recognize as those obtained in November, 1773. Cf. the measurements of the conductivity of water and ice by Vassalli and Zimmermann, Soc. ital, *Mem.*, 4 (1788), 264–77.

104. *PT,* 66 (1776), 197–9, 210–11; *El. Res.* (1879), 195–6, 205; cf. the general theory of divided circuits (*ibid.*, 301), where 'resistance' of a branch of a parallel circuit is inversely proportional to the quantity of fluid passed.

one must distinguish very carefully, he said, between quantity of electricity and its intensity, as measured by the spread of a Henley or the gap of a Lane electrometer, and recognize that the shock depends upon quantity and intensity conjointly. Although the stroke from a large battery greatly exceeds that of a single jar charged to the same degree, the battery will not discharge across a greater Lane gap than will any of its members taken separately. By multiplying the number of elements one can preserve the stroke while decreasing the electrification. Cavendish designed a battery of forty-nine jars that struck like the fish when so weakly electrified that he needed a microscope to discern the range of its spark.[105] To amuse himself and convince doubters, he built torpedos of wood and leather, placed them in baths of sea water, hooked them up to his battery, and invited some colleagues, including Lane and Priestley, to play with them. They got shocks to their complete satisfaction.[106] Cavendish's elegant paper on the torpedo was therefore assured of at least a few informed readers, whom it doubtless assisted in assimilating the electrophore, the condensatore, and the law of Volta.[107]

105. Volta, who knew of Cavendish's analysis of the torpedo by 1782 (*VO,* I, 10–11), later used the same arguments to show the electricity of the pile. In announcing his discovery (1800), he termed it an 'artificial electric organ,' an apparatus 'fundamentally the same' as the equipment of the torpedo (*VO,* I, 556, 580): neither pile nor torpedo gives electrostatic signs because they work at too low a tension.

106. *PT,* 66 (1776), 200–9, 216–22; *El. Res.* (1879), 197–206, 210–13; Maxwell, *ibid.,* xxxviii.

107. Cavendish never published the account of relative resistances he had promised (*supra,* n. 103). Many writers explain his silence either by his morbid shyness (e.g., Berry, *Cavendish* [1960], 21) or by his scorn for the opinions of others (e.g., Lépine and Nicolle, *Cavendish* [1964], 26). It appears, however, that in his scientific life he struggled against these and other disabilities, a stutter, a shrill voice, and social awkwardness; he would often force himself to attend the meetings of the Royal Society Club and the scientific soirées at the home of Joseph Banks. Sir Joseph's fashionable parties so terrified the poor man that he often hesitated on the landing; but once inside he enjoyed the conversation, provided no one addressed him directly (G. Wilson, *Cavendish* [1851], 165–9). He could collaborate with others, as witness the lightning committee, the torpedo, and electrometer calibrations done at Nairne's (*El. Res.* [1879], 297–8). He felt a need, and perhaps even an obligation to make public his results; not misanthropy, but the incomplete state of his researches, the want of occasions like the torpedo, and a growing interest in chemical problems kept him from publishing on electricity after 1776. Cf. McCormmach, Isis, 60 (1969), 293–306.

CHAPTER XX

Epilogue

The work of Volta, Cavendish and Coulomb brought electrostatics to the point that, within a few years, it could suffer its definitive quantification at the hands of the mathematical physicists of the Ecole Polytechnique. The step permanently removed higher electrical theories from the reach of the Gilberts, Franklins and Voltas who had prepared it. It did not, however, resolve all or perhaps even the chief qualitative difficulties that had worried the electricians of the Ancien Régime. The question of the number of electrical fluids remained unanswerable; proto-field theories continued their shadowy existence, in disagreement with one another and with many of the phenomena they were supposed to elucidate; and the recurrent problem of the electrical properties of the vacuum came no closer to resolution.

These classic problems claimed a smaller and smaller fraction of the attention of electricians from about 1790 until the revival of field theory by Faraday. A major shift of interest occurred in the late 1780s and early 1790s in favor of animal and medical electricity, and electrochemistry. 'Foreigners attend much more to [electro-]chemistry than electricity,' a traditionalist Briton told his students in 1794; 'therefore, you will discover nothing to reward the toil of reading [their journals], if you except Volta's papers.'[1] A few numbers will suggest the nature and magnitude of the change in interest. Table 20.1 gives the percentages of the papers concerned with various aspects of electricity reviewed or abstracted in the *Commentarii de rebus in scientia naturali et medicina gestis* from 1752 to 1797. 'Traditional electricity' signifies theories of electrical action, descriptions of apparatus, demonstrations of 'artificial' or laboratory electricity; 'medical electricity' includes accounts of therapy and experiments of electricity on animals; and 'natural electricity' means primarily the electricity of the atmosphere.

Perhaps the impression of some old-timers that (to quote the *Britannica* for 1797) 'the science of electricity seems to be at a standstill' arose from the diversion of electrical energy into new channels. Van Marum, unable as usual to design work for his big machine and now unguided by others, announced in 1795 that he was changing his line of research.[2] The machine remained idle for five years. Then, in 1800, Volta made public a discovery of the first impor-

---

1. Morgan, *Lectures* (1794), I, lxii; cf. Gehler, *Phys. Wört.*, I (1787), 767–70, and *Encyclopédie méthodique*, LXXVI, 68–9.

2. *Encycl. Brit.* (1797³), VI, 424; Hackmann in Forbes, *Marum*, III, 329–30. The perception of standstill did not arise from a decline in literature; cf. Kühn, *Neuste Ent.*, I (1796), 'Vorrede,' who complains about the huge increase in works on electricity since 1785.

TABLE 20.1

Distribution of Articles on Electricity by Field[a] (percent)

	1752/61	1762/74	1769[b]	1775/88	1789/97
Traditional Electricity	50	60	60	60	10
Medical Electricity	45	35	30	30	70
Natural Electricity	5	5	10	5	10
Electrochemistry	0	0	0	5	10

[a] All data from the *Commentarii* except those for 1769. The editors may have been biased towards medical electricity in the 90s.
[b] From Krünitz, *Verz.* (1769). Since Krünitz' bibliography is retrospective, the agreement of the figures with those extracted from the *Commentarii* is a striking confirmation of the validity of the concept 'traditional electricity.'

tance, which, by uniting work on animal electricity with traditional concerns and the new electrochemistry, not only reawakened the interest of van Marum, but marked an epoch in the history of electricity.

1. ELECTROSTATICS AND THE PILE

In 1791 Luigi Galvani, professor of anatomy at the University of Bologna and of obstetrics at the Institute of Sciences of the Bologna Academy, published his now famous study of the electrical excitation of disembodied frog legs. He explained the jerking of a leg on completing a circuit through the crural nerve and the leg muscle as the direct result of the discharge of a 'nerveo-electrical fluid' previously accumulated in the muscle, which he supposed to act like a Leyden jar. In Galvani's opinion the special structure of the muscle, like the peculiar anatomy of the torpedo or electrical eel, effected and retained the accumulation of nerveo-electrical fluid; as for the fluid, it was similar to but distinct from frictional electricity, an 'animal' electricity *sui generis*.

There were reasons to doubt Galvani's elucidation of the results of his odd studies. The well-informed electrician knew that the electricity of the eel differed in no detectable way from artificial electricity: the invocation of electrical fish as evidence for the existence of a special animal electricity could not have been persuasive to those who understood Cavendish's theories or had fondled his mock torpedo. As for the analogy between frog muscle and Leyden jar, it was undermined by Galvani's own experiments. In order to try whether atmospheric electricity could cause frog legs to jerk, he had hung the usual grisly preparations in the open air, and saw them dance without electrical disturbances. The observation that the legs had been fixed to an iron rail by brass hooks led to a capital discovery: prepared frog legs would jump when nerve and muscle were joined by a bimetallic arc. A bridge of a single metal would not do, or would not do as well.[3] But any metallic circuit will discharge a Leyden jar.

3. Galvani, *Commentary* (1953), 23–81, esp. p. 45; *DSB*, V, 267–9.

Perhaps for these reasons Volta at first dismissed Galvani's experiments as 'unbelievable' and 'miraculous.' He had a low opinion of physicians who dabbled in his subject, 'ignorant of the known laws of electricity.'[4] He tried the experiments, 'with little hope of success,' on the urging of his colleagues in pathology and anatomy at the University of Pavia. By April 1, 1792, the experiments had succeeded. Volta began the brilliantly planned and executed experiments that brought him to the invention of the pile.

His first instinct was to measure the minimum tension of artificial electricity that would cause the frog to jerk. Legs prepared in the style of Galvani proved to be the most sensitive electroscope yet discovered. When placed in a discharge train of a Leyden jar they responded to a tension of as little as five thousandths of a degree of Volta's straw electrometer, an amount he could only detect after manipulation by the condensatore. He also succeeded in inducing convulsions in a *live* frog by joining its leg and back externally by a bimetallic circuit.[5]

Volta's use of the whole frog, a move unnatural for an anatomist like Galvani, proved consequential. When the animal was intact it could be made to tremble only when struck by a discharge from a Leyden jar or when part of a bimetallic circuit. Volta inferred that the electricity put in action in the second case arose from the mere contact of dissimilar bodies, a property he had already identified in electrics, but was surprised to meet in metals.[6] The fact, however, was plain, as well as the conclusion that animal electricity played no part in spasms inspired by bimetallic arcs. No more did the Leyden jar supposedly fashioned from muscular tissue; as Volta showed, electricity excited the nerve, and the nerve the muscle, which could be omitted from the circuit. As for Galvani's classic case—a freshly killed and stripped frog, highly excitable, joined crural nerve to leg muscle by a single metal—Volta supposed that the electricity arose from the contact between the metal and unobserved impurities in it. Nothing remained of the theory of animal electricity, or so he told Galvani's nephew and defender, Giovanni Aldini, professor of physics at the University of Bologna, in an open letter published early in 1793.[7]

While the Galvanists pondered their response, Volta ranked the metals according to their electromotive power, and tried to determine the seat of the electromotive force. He recognized that an effective circuit contained, besides a bimetallic joint, at least one 'moist conductor,' the nerve to be excited; and he thought it more probable that the electrical imbalance occurred in the contact between the metals and the moist conductor than in the joint between the metals.[8] Volta was accordingly prepared to answer the Galvanists' counterattack. Exploiting their discovery that spasms could be induced in freshly prepared frogs using a Galvanist as arc, they insisted that convulsions could be excited without the metallic contact that Volta had just asserted to be necessary.[9] Volta

4. *VO*, I, 10–11, 21–3, 26; *VE*, III, 143–5. 5. *VO*, I, 15, 30–3.
6. *VO*, I, 55, 64–6, 73–4, 136. 7. *VO*, I, 147–59.
8. *VO*, I, 205, 212–14, 231–4, 304. 9. *VO*, I, 274, 279, 295n, 308.

observed that their key experiment worked best when the nerves and muscles were moistened with blood or saliva. He neutralized the Galvanist objection by supposing that any sequence of dissimilar conductors could generate an electrical current by contact forces without the intervention of metals.[10]

Volta's next, and characteristic step, was to determine what he called the 'electromotive force' of various combinations of conductors. The results, in order of decreasing power, expressed in Volta's notation (where capital and small letters signify dry and moist conductors, respectively): rABr (where r, the frog, is both a moist conductor and the electroscope); raBr; rabr; rAr and rar, both zero. What about ABCA? Volta thought that analogy favored the possibility of a weak finite current in such a circuit. But how to detect it when the only electroscope sensitive enough to register Galvanic electricity was itself a conductor of the second kind?[11] The difficulty instanced a much more serious one, which had long bothered Volta: his claim of the identity of galvanic and common electricity rested on experiments in which pieces of animals played an indispensable part.[12]

The contact of zinc and silver develops about 0.78 volt. Volta's most sensitive straw electrometer marked about 40 volt / degree.[13] By the summer of 1796 he had managed to multiply the charges developed by touching dissimilar metals together enough to stimulate his electrometer.[14] He first succeeded with a Nicholson doubler, and then with an unaided Bennet electroscope; he later rendered contact electricity easily sensible by a 'condensing electroscope,' a straightforward combination of the condensatore and the straw electrometer.[15] All these devices, including the doubler, came directly or indirectly from Volta's earlier work. They—or rather the need to use them to obtain multipliable contact charges—forced Volta to change his mind about the principal seat of the emf; he now (1797) located it in the junction of metals, and not in their union with moist conductors.[16]

It remained to find a way to multiply galvanic electricity directly. Volta discovered soon enough that piling metal disks on one another, say aABAB . . . a, did not help, and that a circuit made only of metals gave no emf. These results led to the useful rule that the emf of a pile of disks is equal to what its extreme disks would generate if put into immediate contact.[17] How or when Volta hit on the far from obvious artifice of repeating the apparently unimportant moist conductors in his generator is not known. The definitive pile, AZaAZaAZa . . . AZ, consisting of pairs of silver and zinc disks separated by

10. *VO*, I, 255–6, 295–7.

11. *VO*, I, 230, 371–2, 377–82, 396–7, 401–6, 411–13.   12. *VO*, I, 490, 540–5.

13. One degree of Volta's best straw electroscope equalled 0.1° of a Henley (*VO*, I, 486; V, 37, 52, 81; *supra*, xix.1), 35° of which marked about 13,350 volt (Polvani, *Volta* [1942], 145). Hence one degree of the straw electrometer indicated about 40 volt. Volta later estimated the tension between zinc and silver at one-sixtieth degree straw (*VO*, II, 39), or about 0.7 volt.

14. *VO*, I, 525; *VE*, III, 349, 359.

15. *VO*, I, 420–4, 435–6; *VE*, III, 438.   16. *VO*, I, 393–447, 472.

17. *VO*, II, 61; Volta had already glimpsed the rule in 1793 (*VO*, I, 226–7).

pieces of moist cardboard, was first made public in 1800, in a letter addressed to Joseph Banks.[18]

It is instructive that in explaining the operation of the pile Volta appealed to an analogy to the torpedo, using precisely those concepts of tension and quantity of electricity that had been worked out in the 1770s and 1780s. The analogy was not hard to grasp: a medium-size pile, one with 40 or 50 metallic pairs, gave anyone who touched its extremities about the same sensation he could enjoy holding an electric fish. In both cases, Volta said, a constant current running externally from top to toe of the electromotor passed through the arms and breast, and agitated the sense of touch. Were it directed at the senses of vision, taste or hearing, the current would cause light, taste or sound instead.[19] Neither the pile nor the torpedo give electrostatic signs because, as Cavendish had argued long before, they operate at too low a tension; their effects derive rather from the quantity of electrical matter they move. As for the cause and continuance of the electricity generated by the contact of dissimilar conductors, Volta feigned no hypothesis: 'This perpetual motion may appear paradoxical, perhaps inexplicable; but it is nonetheless true and real, and can be touched, as it were, with the hands.'[20]

The pile was the last great discovery made with the instruments, concepts, and methods of the eighteenth-century electricians. It opened up a limitless field. It was immediately applied to chemistry, notably to electrolysis, and soon brought forth the shy elements sodium and potassium from fused soda and potash.[21] Its steady current provided the long-sought means for establishing a relationship between electricity and magnetism. The consequent study of electromagnetism transformed our civilization.

2. THE QUANTIFICATION OF ELECTROSTATICS

As Galvani was examining the electricity of decapitated frogs, Coulomb was exploiting the newly demonstrated law of electrical force in researches of equal originality. After stressing, and confirming, that the charge on a conductor pressed itself within a layer of insensible thickness at its surface,[22] he set out to discover the distribution of the charge on the surfaces of bodies more interesting than isolated spheres. He represented the distribution as the varying surface density of one or the other electrical fluid, which he probably pictured as spread one molecule thick at the geometrical boundary of the conductor. The distribu-

---

18. *PT*, 90:2 (1800), 403–31; *VO*, I, 563–82.

19. *VO*, I, 556, 578–82.

20. *VO*, I, 576; cf. *ibid.*, 489, and the 'perpetuity of the electrophore' (*supra*, xvii. 2).

21. *DSB*, III, 601–2.

22. Coulomb gave an argument much like Cavendish's (*supra*, xix.3) to show that, if the force went as r^{-n}, $n < 3$, the charge must stand at the surface, and he confirmed the proposition by demonstrating (1) that a solid and a hollow wooden block of identical external dimensions have the same electrical capacity and (2) that no electricity was transferred to a probe touched to the bottom of a shallow hole drilled into the charged solid block. Coulomb, *MAS* (1786), §§vii–xi; *MAS* (1787), §xviii; *Mémoires* (1884), 178–82, 205.

tion can also be pictured as a variation in the depth of a fluid of uniform density, as will appear.

Coulomb's methods, simple in principle, required great manual dexterity and conceptual clarity. In one series of experiments he placed a charged sphere in his torsion balance, touched it with another of different radius, measured the decreased force of the first, compensated for leaks during the operation, and then calculated the uniform densities of the separated partners. In another series, he used an invention of his now known as a proof plane, a thin small disk of gilded paper attached to an insulating handle, which picks up an electricity proportional to the charge density of the place on a conductor to which it is touched. Coulomb measured the acquired electricity of the plane in his torsion balance. This method, 'the easiest, the simplest, and perhaps the most exact,' and, one might·add, the most general, way to compare charge densities, also allowed a convenient compensation for leak. To compare the densities at two places, A and B, Coulomb would at equal intervals of time measure that at A, at B, and again at A; assuming leak proportional to the time, he took the average of the measurements at A as the value of the density there at the time of the measurement at B.[23]

The sophistication of the technique, as compared to the manipulations of a Gray or a Franklin, speaks for itself. Our interest is in the approximate theory with which Coulomb accounted for his numerical results. Two examples will illustrate its strengths and limitations: the distribution of electricity among (1) three equal collinear spheres (2) two unequal ones. In the first case, assume the charge spread uniformly, with density ρ on the extreme spheres and ρ' on the middle one. (The approximation is not bad, as may be inferred from Table 20.2.) At either point of contact a unit charge would be pushed towards the center ball by the nearest extreme one with a force f_1, and away from the center by forces f_2 and f_3, arising from the middle and furthest extreme ball, respectively. For equilibrium, $f_1 = f_2 + f_3$, or $\rho = \rho' + 2\rho/9$. Hence $\rho/\rho' = 1.29$; measurement made it 1.34.[24]

Coulomb took the expressions for the forces from the gravitational theory without explanation or acknowledgment. That for f_3 is the Newtonian proposition that a uniformly gravitating spherical shell acts upon points outside it as if its mass were concentrated at its center. Within the shell the force vanishes. To find its value *on* the surface, as required for f_1 and f_2, one could either integrate directly or invoke an argument such as the following. Consider points P and Q, respectively inside and outside the surface and distant from it by a very small amount x; call s the portion of the force at Q arising from a very small bit of surface surrounding the point of intersection of the line PQ and the shell, and S the portion arising from the rest of the surface. At Q, $s + S = 4\pi a^2 \rho/(a+x)^2$; at P, $s = S$. Hence $S = 2\pi a^2 \rho/(a+x)^2$; and, since $x = 0$ on the surface, the

23. Coulomb, *MAS* (1787), §§ii–iii; *Mémoires* (1884), 187–90.
24. Coulomb, *MAS* (1787), §§xiii–xvi, xxii–xxiii; *Mémoires* (1884), 200–4, 210–11.

force there is $2\pi\rho$.[25] This pretty argument, as applied to a charged sphere, is credited by Poisson to Laplace.[26] But it is already implicit in Coulomb's writings.

To grasp the case of the two unequal spheres, Coulomb imagined the charge to be made up of 'an infinity of little electrified conducting globules.' A point P on one such globule m (fig. 20.1) suffers from the same sorts of forces already considered: an $f_1 = 2\pi\delta$ arising from the globule itself, assumed charged to a density δ; an $f_2 = 2\pi\rho'$ from the rest of the smaller sphere, assumed uniformly charged; and the normal component, $f_{3m} = (4\pi a^2/p^2)(x/b)\rho$, of a force f_3 from the larger sphere, also supposed uniformly charged. Since the globule must be in equilibrium,

$$\delta(\alpha) = \rho' - 2\rho \cdot \frac{a^2[(a + b)\cos\alpha - b]}{[(a + b)^2 + b^2 - 2b(a + b)\cos\alpha]^{3/2}} \cdot \quad (20.1)$$

This approximation is better than one might think when the spheres do not differ greatly in size, as may be gathered from Coulomb's results for the special case $a = b$, $\rho = \rho'$ (Table 20.2). 'There is a conformity here between the results of experience and of theory that could scarcely have been expected.'[27] could scarcely have been expected.'[27]

TABLE 20.2

Charge Density $\delta(\alpha)$ for Touching Equal Spheres

α°	δ/ρ observed	δ/ρ calc. by eq. (20.1)	δ/ρ calc. by Poisson
30	0.23	0.26	0.23
60	1.00	1.00	1.00
90	1.18	1.25	1.34
180	1.22	1.31	1.52

And the theory could be improved. Consider $\delta(180°)$, the electrical density of the smaller sphere at A', directly opposite the point of contact. Calculating according to eq. (20.1), and again using the approximation, no longer very good, that $\rho = \rho'$, Coulomb deduced the numbers designated $(\delta/\rho)_1$ in Table 20.3.

The numbers marked $(\delta/\rho)_2$ are Coulomb's second approximation, which he obtained by using the values of $\delta(\alpha)$ of eq. (20.1) to compute the condition of equilibrium at the surface of the smaller sphere. When a/b may be considered infinite, $\delta(\alpha) = \rho' - 2\rho\cos\alpha$. The force arising from the several infinitesimal ribbons of thickness $ad\alpha$, circumference $2\pi a\sin\alpha$, and density $\delta(\alpha)$, exerted by

25. Coulomb, *MAS* (1786), §xi; *MAS* (1787), §xix; *Mémoires* (1884), 181–2, 207.

26. Poisson, *MAS* (1811:1), 5–6.

27. Coulomb, *MAS* (1787), §§xxviii–xxix; *Mémoires* (1884), 216–18. Cf. Harris, *PT*, 126:1 (1836), 437–51.

TABLE 20.3

Charge Density δ(180) for the Smaller of Unequal Touching Spheres

a/b	δ/ρ observed	(δ/ρ)₁	(δ/ρ)₂	(δ/ρ) calculated by Poisson
1	1.27	1.22		1.32
2	1.55	1.50		1.83
4	2.35	1.89	2.31	2.48
8	3.18	2.28		3.09
∞	>4	3.00	4.33	4.27

the small sphere at the point of contact, must equal the action f_3 of the large sphere there. Analytically,

$$\int_0^\pi (\rho' - 2\rho \cos \alpha) \frac{\pi}{\sqrt{2}} \cdot \frac{\sin \alpha}{(1 - \cos \alpha)^{1/2}} \cdot d\alpha = 2\pi\rho. \qquad (20.2)$$

The integration yields $\rho_2' = 1.67\rho$. Coulomb improved his approximation by using the new value, instead of $\rho_1' = \rho$, to compute the equilibrium at the back point, $\alpha = 180°$; and he gilded the lily by taking into account the variation of $\delta(\alpha)$ by an integral of the form of eq. (20.2). Equilibrium requires

$$2\pi\delta_2(180) = \int_0^\pi (\rho_2' + 2\rho \cos \alpha) \frac{\pi}{\sqrt{2}} \cdot \frac{\sin \alpha}{(1 + \cos \alpha)^{1/2}} d\alpha + 4\pi\rho.$$

The outcome, $(\delta/\rho)_2 = 4.39$, came considerably closer than the first approximation to the experimental value of slightly over 4.[28]

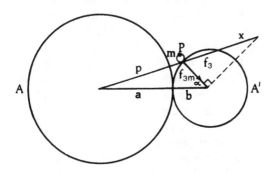

20.1 Diagram from Coulomb's approximation to the distribution of electricity on touching conducting spheres.

28. *MAS* (1787), §§xxx–xxxii; *Mémoires* (1884), 218–24. Eq. 20.3 contains three terms, one each from the imaginary globule m at A', the integrated δ(α) of the smaller sphere, and the total charge of the larger sphere; eq. 20.2 has but two terms because, since there is no charge about the point of contact, nothing there answers to the globule m.

Coulomb's approximations were those of a resourceful engineer: he worked directly with the quantity he could measure, force, and with the theoretical object he could handle, a uniformly charged sphere. The definitive quantification of electrostatics came when mathematicians not limited to forces or to homogeneous spheres took up his problem of the distribution of electricity on conducting bodies. These mathematicians, or mathematical physicists, succeeded by applying Laplace's reformulation of the theory of gravitating bodies to the case of electricity.

The reformulation centered on a certain function V, later christened 'potential,' introduced by Laplace in about 1785. The function is the integral of the quotients of the gravitational masses dm by their respective distances from the point P at which V is to be computed: if x, y, z are the coordinates of P, and x', y', z' those of P', $V(x, y, z) = \iiint dm(x', y', z') / PP'$. Laplace emphasized that the negative partial derivatives of V are proportional to the components of the gravitational force; for example, the force in the direction of x is -const.$\partial V/\partial x$.[29] V must be constant within an isolated spherical shell.

The function V simplified the calculations of the gravitational theory by allowing one to work with a scalar, additive quantity, rather than with force, whose direction must be taken into account. V was to prove even more useful in electrostatics. The electrical fluid, as opposed to ponderable mass, is free to distribute itself over the surface of a closed conductor in such a way that, irrespective of the presence of other bodies however electrified, there is no electrical force within or tangent to the surface of the conductor. This complicated distribution can be represented easily by imposing conditions on the potential: V must be constant within all conductors and $\partial V/\partial s$ must be zero, where s signifies any direction in any plane tangent to the conductor's surface.

Laplace began to encourage others to apply the reformulated gravitational theory to electricity about the time Volta visited Paris to promote the pile and to help cement relations between the Napoleonic regimes of France and Italy. Among the earliest to look where Laplace pointed was Jean-Baptiste Biot, one of the first graduates of the new, high-powered, Ecole Polytechnique. His brilliance, and his offer to read the proofs of the *Mécanique céleste,* brought him the attention and the patronage of Laplace. While enjoying one of the fruits of this association, a professorship of mathematical physics at the former Collège Royal, Biot solved a difficult problem in electrostatics by a simple method suggested by his patron.[30] The problem: the distribution of electricity on the surface of an ellipsoid of revolution. The answer: the fluid arranges itself so that its depth y (assumed to be insensible) beneath any point P on the surface is proportional to the distance from P to the corresponding point on the surface of a similar and similarly-placed ellipsoid within, and infinitesimally distant from, the original body.[31] The proof: Laplace's expression for the force or potential

---

29. Todhunter, *Figure* (1873), II, 25–7, 31.

30. Crosland, *Society* (1967), 254–6; *DSB*, II, 133–40; Picard, *Eloges* (1931), 226–8, 231–3.

31. Biot's representation, in which the depth of the charge varies rather than its surface density, grew from the analogy to concentric gravitating shells.

of a gravitating ellipsoid implied that no tangential force existed on a particle of electric fluid located anywhere between the similar ellipsoids.[32]

At this point all the ingredients of classical electrostatics had been identified: the law of squares, Coulomb's measurements of distribution, Laplace's machinery and its demonstrated utility in problems of electricity, the potential function. Siméon-Denis Poisson, a precocious mathematician, a graduate of the Ecole Polytechnique, tenant of professorships at the Ecole and the University, disciple of Lagrange and Laplace, put them together.[33] He showed in great generality that the depth y, proportional to the quantity of redundant vitreous or resinous fluid in unit volume just beneath the surface, is proportional to the force just outside it; and that, since this force must be perpendicular to the surface, $y \sim -\partial V/\partial n$, where n is the direction of the normal. Since the quantity of Poisson's electrostatic theory corresponding to the element of mass dm of the gravitational theory is $y\rho dS'$, where ρ is presumed constant and dS' is an element of surface, one has the integral equation

$$y \sim -\frac{\partial}{\partial n}\left[\int (y\rho/\text{PP}')dS'\right].$$

The function V must be constant within conductors, and its derivatives in their surfaces must vanish. Poisson managed to find V and y for the case of touching spheres by expressing the integrands as series, a trick at which he was a great master.[34] His numerical results, as compared with Coulomb's measurements, have been given in Tables 20.2 and 20.3.

Poisson's V is the analytic form of Cavendish's 'electrification' and Volta's 'tension.' It is more supple than either, for it permits the statement of the classic problems of electrostatics—finding the distribution of electricity and the resultant forces—in full generality. To be sure, Poisson had to approximate to obtain numerical results; but his approximations, which could be made as close as desired, were approximations in the evaluation of mathematical forms, not substitutions of counterfactual physical systems for the ones under consideration.

It need scarcely be said that Poisson shared the instrumentalism of the French school of mathematical physicists. Like them, he was a dualist for convenience. It made no difference to him whether electricity came in one fluid or two. He did not care whether these hypothetical fluids pressed themselves into a layer one molecule thick at the surface of the conductor to make a charge of varying surface density, as in Coulomb's representation, or whether, as in his and Biot's, they occupied an infinitesimal volume of constant surface density and variable depth. The problem of the mechanism of electrical action had no interest for him. As for the vexed question of the agency that prevented the escape

32. B[iot], Soc. phil., *Bull.*, 3 (1805), 21–3. The demonstration is not flawless; cf. Poisson, *MAS* (1811:1), 27–8n.

33. Anon., *Rev. deux mondes*, 23:3 (1840), 410–37; Crosland, *Society* (1967), 127–8, 212–13, 318–19; Arago, *Notices*, II (1854), 593–698.

34. Poisson, *MAS* (1811:1), 1–92.

of the fluids from the surface of conductors, a question by no means resolved by the elimination of atmospheres, Poisson was content to suppose with Biot that the 'pressure' of the air clamped electricity to its carrier. That was to ignore the evidence that vacuum insulates. But to Poisson the mechanism did not matter: it was enough that electricity remained on conductors, and distributed itself in accordance with his calculations.

The chief moral of this long *History* may be that, when confronted with a choice between a qualitative model deemed intelligible and an exact description lacking clear physical foundations, the leading physicists of the Enlightenment preferred exactness. It was not a choice peculiar to them. As the dean of contemporary theoretical physicists put it many years ago: 'The only object of theoretical physics is to calculate results that can be compared with experiment, and it is quite unnecessary that any satisfactory description of the whole course of the phenomena should be given.'[35]

35. Dirac, *The Principles of Quantum Mechanics* (1930), 7.

Bibliography

The Bibliography lists only works cited in the notes; other early writings about electricity may be found in the catalogs of Ekelöf, Frost, Gartrell, Mottelay, Rosetti and Cantoni, and Weaver. Information sufficient for retrieval is given with each entry except for articles in a collaborative volume cited more than once and for references to reprints in collected works. In such cases referral to a fuller entry is given in short form, e.g., editor's last name, brief title, date (for collected works, brief title and date only). Names of learned societies and citations to their publications conform as closely as possible to the usage of the Library of Congress.

Titles of some articles, particularly in *PT*, have been shortened by dropping irrelevant opening phrases; e.g., 'An Extract of a Letter from Samuel Pickwick, Esq., to Lord So & So, FRS, on the Theory of Tittlebats,' would appear as 'The Theory of Tittlebats.' Book titles always begin with their original word or phrase; where necessary they are shortened internally by ellipsis. Subtitles, and the final phrase(s) of lengthy ones, are usually omitted.

The Bibliography has two parts, manuscripts and printed works. No distinction is made between primary and secondary sources or between books and articles. Russian words are transliterated according to the system of the Library of Congress. Accents are omitted from capital letters in French titles.

A few special abbreviations besides those used in the notes (for which see *supra*, pp. xiii–xiv) appear in the Bibliography:

AIHS.	*Archives internationales d'histoire des sciences.*
Ann. Sci.	*Annals of Science.*
BJHS.	*British Journal for the History of Science.*
CIHS.	Congrès International d'Histoire des Sciences.
IIET.	Akademiia nauk S.S.S.R. Institut istorii estestvoznaniia i tekhniki.

PART I. MANUSCRIPTS

Académie des Sciences, Paris. Dossiers [of members]. AS.

———. Procès verbaux. AS.

Accademia del Cimento. 'Diario,' etc. Manoscritti Galileiani. Biblioteca Nazionale Centrale, Florence.

Anon. 'Particulars Respecting Mr. Robison.' Ms. Murray, 503. Glasgow University Library.

Boscovich, Roger Joseph. Papers. Bancroft Library, University of California, Berkeley.

Canton, John. Papers. RS.

Cavallo, Tiberio. Papers. Add. Mss. 22897/8. BL.

Deluc, Jean André. Papers. Yale University Library.

Dutour, Etienne-François. *See* Nollet.

Fabri, Honoré. 'Synopsis recens inventorum in re literaria et quibusdam aliis.' Biblioteca Nazionale Centrale, Florence. *See* Delandine. *Manuscrits* (1812), I, 393–4.

Flamsteed, John. Papers. Royal Observatory, Herstmonceux.

Formey, Jean Henri Samuel. Nachlass. Staatsbibliothek, Handshriften-Abteilung, Berlin.

Kircher, Athanasius. Carteggio. Università Gregoriana, Rome.

Library Company, Philadelphia. Minute Book.

Marie, Joseph François. 'Philosophiae quarta pars seu Physica.' [1763–64]. American Philosophical Society Library, Philadelphia.

Nollet, Jean Antoine. Correspondence with E. F. Dutour, 1742–1770. Burndy Library, Smithsonian Institution, Washington.

———. Annotations in Beccaria. *Dell'elettricismo* (1753). Burndy.

———. 'Journal du voyage de Piédmont et d'Italie en 1729.' Bibliothèque Municipale, Soissons.

Royal Society, London. Classified Papers. RS.

———. Correspondence and Miscellaneous Correspondence. RS.

———. Guard Book. RS.

———. Journal Book (JB). RS.

———. Scientific Papers. RS.

Sloane Mss. BL.

Symmer, Robert. Letters to Andrew Mitchell, 1757–63. Add. Mss. 6839. BL.

Wall, Samuel. Papers. Sloane Mss. 1731A. BL.

Wilson, Benjamin. Papers. Add. Mss. 30094. BL.

PART II. PRINTED WORKS

Abat, Bonaventure. *Amusemens philosophiques.* Amsterdam, 1763.

Abetti, Giorgio, ed. *Le opere dei discepoli di Galileo Galilei, edizione nazionale.* I. *L'Accademia del Cimento. Parte prima.* Florence, 1942.

———. 'L'Accademia del Cimento.' *Celebrazione dell'Accademia del Cimento nel trecentenario della fondazione.* Pisa, 1958. Pp. 3–10.

Académie des Sciences, Paris. *Histoire de l'Académie royale des sciences (1666–1699).* 11 vols. Paris, 1733.

———. *Index biographique des membres et correspondants de l'Académie des sciences du 22 décembre 1666 au 15 décembre 1967.* Paris, 1968.

Achard, Franz Karl. 'Abhandlung von der elektrischen Erscheinungen, die durch das Reiben des Quecksilbers auf verschiedene, vorzüglich harzige Körper, hervorgebracht werden.' Gesellschaft Naturforschender Freunde, Berlin. *Beschäftigungen,* 2 (1776), 12–19. (*Schriften,* 205–11.)

———. 'Expériences sur l'électrophore avec une théorie de cet instrument.' *MAS/Ber* (1776), 122–34.

———. 'Mémoire sur la force avec laquelle les corps solides adhèrent aux fluides.' *Ibid.,* 149–59. (*Schriften,* 354–67.)

————. 'Expériences sur la célérité avec laquelle des corps de différentes figures se chargent du fluide électrique.' *MAS*/Ber (1777), 25–35. (*Sammlung*, 20–45.)

————. 'Sur l'analogie qui se trouve entre la production et les effects de l'électricité et de la chaleur.' *MAS*/Ber (1779), 27–35. (*Sammlung*, 141–53.)

————. *Chemisch-physische Schriften.* Berlin, 1780.

————. 'Von der Elektricität des Eises.' *Ibid.*, 11–23.

————. 'Theorie des Elektricitätsträgers.' *Ibid.*, 226–33.

————. 'Von dem Unterschied zwischen ursprünglich elektrishen Körpern und Leitern der Elektricität.' *Ibid.*, 246–64.

————. 'Mémoire renfermant le récit de plusieurs expériences électriques.' *MAS*/Ber (1781), 9–19. (*Sammlung*, 283–96; *JP*, 25 [1784], 429–36.)

————. 'Notices des expériences électriques.' *JP*, 19 (1782), 417–24.

————. 'Mémoire sur la mesure de la force de l'électricité, et sa comparaison avec celle de la gravité.' *JP*, 21 (1782), 190–209.

————. *Sammlung physikalischer und chemischer Abhandlungen.* Berlin, 1784.

————. 'Expériences qui prouvent que des corps de même nature . . . se chargent de la matière électrique en raison de leur surface.' *MAS*/Ber (1780), 47–8. (*JP*, 26 [1785], 378–9.).

————. *Vorlesungen über die Experimentalphysik.* 4 parts. Berlin, 1791.

Adams, George. *An Essay on Electricity.* London, 1784; 1785[2].

Addison, William Innes. *A Roll of the Graduates of the University of Glasgow.* Glasgow, 1898.

Adelasio, Girolamo. *De electricitate disputatio physica.* Rome, 1781.

Aepinus, Franz Ulrich Theodor. *Opyt teorii elektrichestva i magnetizma.* IA. G. Dorfman, ed. Moscow, 1951.

————. 'Quelques nouvelles expériences électriques remarquables.' *MAS*/Ber (1756), 105–21.

————. 'Descriptio ac explicatio novorum quorundam experimentorum electricorum.' *NCAS*, 7 (1758–9), 277–302.

————. *Tentamen theoriae electricitatis et magnetismi.* St. Petersburg, 1759.

————. *Recueil de différents mémoires sur la tourmaline.* St. Petersburg, 1762.

————. 'De electricitate barometrorum disquisitio.' *NCAS*, 12 (1766–7), 303–24.

————. 'Examen theoriae magneticae a celeberr. Tob. Mayero propositae.' *Ibid.*, 325–50.

————. 'Descriptio novi phaenomeni electrici detecti in chrysolitho.' *Ibid.*, 351–5.

————. 'Two Letters on Electrical and other Phaenomena.' RS Edin. *Transactions*, 2 (1790), 234–44.

Agoston, George A. 'Voltaire on the Characteristics of Fire and its Propagation.' *Fire Research Abstracts and Reviews*, 12 (1970), 191–9.

Agrippa, Henry Cornelius. *The Vanity of Arts and Sciences* [1530]. London, 1676.

Ahlwardts, Peter. *Bronto-Theologia oder: Vernünftige und theologische Betrachtungen über den Blitz und Donner.* Greifswald and Leipzig, 1745.

Aiton, E. J. 'The Vortex Theory of Planetary Motions.' *Ann. Sci.*, 13 (1957), 249–64; 14 (1958), 132–47, 157–72.

————. *The Vortex Theory of Planetary Motions.* London and New York, 1972.

Akademiia Nauk, Leningrad. *Materialy dlia istorii imperatorskoi akademii nauk* [*1716–1750*]. 10 vols. St. Petersburg, 1885–1900.

————. *Procès verbaux* [*Protokoly zasiedanii*] *des séances de l'Académie impériale depuis sa fondation jusqu'à 1803*. 4 vols. St. Petersburg, 1897–1911.

Aldridge, A. O. 'Jacques Barbeu-Dubourg, A French Disciple of Benjamin Franklin.' *American Philosophical Society. Proceedings*, 95 (1951), 331–92.

Alembert, Jean Le Rond d'. *Oeuvres philosophiques, historiques et littéraires*. Jean François Bastien, ed. 18 vols. Paris, 1805.

————. *Oeuvres*. A. Belin, ed. 5 vols. Paris, 1821–2.

————. *Traité de l'équilibre et du mouvement des fluides*. Paris, 1744. (*Oeuvres* [1805], XIV, 173–202; *Oeuvres* [1821], I, 406–19.)

————. *Introduction et analyse des trois parties composant les réflexions sur la cause générale des vents*. Paris, 1747. (*Oeuvres* [1805], XIV, 7–44; *Oeuvres* [1821], I, 420–37.)

————. *Essai d'une nouvelle théorie de la résistance des fluides*. Paris, 1752.

————. *Recherches sur différens points importans du système du monde*. 3 vols. Paris, 1754–6.

————. 'Essai sur les élémens de philosophie, ou sur les principes des connoissances humaines.' *Mélanges de littérature, d'histoire, et de philosophie*. 5 vols. Amsterdam, 1759–67. IV (1760), 1–292. (*Oeuvres* [1805], II, 3–476; *Oeuvres* [1821], I, 115–348.)

————. 'Article du Comte de Clermont.' *Oeuvres* (1805), XI, 405–25.

————. *Preliminary Discourse to the Encyclopedia of Diderot*. Richard N. Schwab, tr. Indianapolis, 1963.

Alexander, H. G. *The Leibniz-Clarke Correspondence*. Manchester, 1956.

Allamand, Jean Nicolas Sébastien. 'Mémoire contenant diverses expériences d'électricité.' *Bibliothèque britannique*, 24 (1746), 406–37.

————. 'Histoire de la vie et des ouvrages de M. 'sGravesande.' Gravesande, *Oeuvres*, I, ix-lix.

Allen, Phyllis. 'Medical Education in 17th Century England.' *Journal of the History of Medicine*, 1 (1946), 115–43.

————. 'Scientific Studies in the English Universities of the 17th Century.' *JHI*, 10 (1949), 219–53.

Amodeo, Federico. *Vita matematica napoletana: studio storico*. 2 vols. Naples, 1905–24.

Anderson, P. J., ed. *Studies in the History and Development of the University of Aberdeen*. Aberdeen, 1906.

Anderson, Richard. *Lightning Conductors. Their History, Nature, and Mode of Application*. London and New York, 1885[3].

Anon. 'That the Electricity of Glass disturbs the Mariner's Compass, and also nice Balances.' *PT*, 44 (1746), 242–5.

————. 'De electricitate caelesti.' Accademia delle scienze dell'Istituto di Bologna. *Commentarii et opuscula*, 3 (1755), 94–100.

————. [Review of Beccaria's *Elettricismo artificiale*.] *Nuovo giornale de' letterati d'Italia*, 2 (1773), 355–412.

————. 'Mémoire historique.' Accademia delle scienze, Turin. *Mémoires*, 1 (1784–5), i-lvi.

————. [Obituary notice of Coulomb.] *Journal de l'Empire* (20 September 1806), 3–4.

————. 'Lettres à un Américain sur l'état des sciences en France. III. Poisson.' *Revue des deux mondes*, 23:3 (1840), 410–37.

————. 'Documents sur le Collège de France à la veille de la Révolution.' *Revue internationale de l'enseignement*, 5 (1883), 406–12.

————. 'Biographie de Christiaan Huygens.' Huygens. *Oeuvres complètes*, XXII, 385–771.

————. 'Aaresredogörelse.' Uppsala. University. *Aarsskrift* (1910), 30–50.

Anthiaume, Albert. *Le Collège du Havre*. 2 vols. Le Havre, 1905.

Antinori, Vincenzo. 'Notizie istoriche.' Accademia del Cimento. *Saggi dell'esperienze naturali fatte nell'Accademia del Cimento*. Florence, 1841³. Pp. 3–133.

Arago, Dominique François Jean. *Notices biographiques*. 3 vols. Paris, 1854–5.

————. 'Alexandre Volta.' *Oeuvres*. J. A. Barral, ed. 13 vols. Paris, 1865². I, 187–240.

Arcy, Patrick d' [and J. B. Le Roy]. 'Mémoire sur l'électricité, contenant la description d'un électromètre.' *MAS* (1749), 63–74.

Aronson, S. 'The Gravitational Theory of Georges-Louis Le Sage.' *Natural Philosopher*, 3 (1964), 53–74.

Arrighi, Gino. 'Ruggiero Giuseppe Boscovich e Giuseppe Antonio Slop.' *Studi trentini di scienze storiche*, 43 (1964), 209–42.

Atkinson, Philip. *Elements of Static Electricity*. New York, 1887.

Avogadro, Amedeo. 'Considérations sur l'état dans lequel doit se trouver une couche d'un corps non-conducteur de l'électricité, lorsqu'elle est interposée entre deux surfaces douées d'électricités de différente espèce.' *JP*, 63 (1806), 450–62; 65 (1807), 130–45.

————. *Opere scelte*. Turin, 1911.

Bacon, Francis. *Works*. J. Spedding, R. L. Ellis and D. D. Heath, eds. 15 vols. New York, 1863–9.

————. *The Advancement of Learning*. G. W. Kitchin, ed. London, 1915.

Bacon, Roger. *Opus maius*. J. H. Bridges, ed. 2 vols. London, 1900.

Baker, Henry. 'Medical Experiments of Electricity.' *PT*, 45 (1748), 270–5.

Baker, Keith. 'Les débuts de Condorcet au secrétariat de l'Académie royale des sciences (1773–1776).' *RHS*, 20 (1967), 229–80.

Balbo, Prospero. 'Vita del Signor Carlo Ludovico Morozzo.' Società italiana delle scienze, Rome. *Memorie di matematica e di fisica*, 15:1 (1811), lxv–lxxvii.

Baldi, Francesco. *De igne, luce et fluido electrico propositiones physicae*. Florence, 1790.

Bammacaro. Niccolò. *Tentamen de vi electrica eiusque phaenomenis*. Naples, 1748.

Barber, William Henry. *Leibniz in France from Arnauld to Voltaire*. Oxford, 1955.

Barbier de Tinan, ————. 'Considérations sur les conducteurs en général.' *JP*, 14 (1779), 17–46.

Barbosa Machado, Diogo. *Biblioteca lusitana*. 4 vols. Lisbon, 1930².

Barletti, Carlo. *Nuove esperienze elettriche secondo la teoria del Sig. Franklin e le produzioni del P. Beccaria*. Milan, 1771.

————. *Physica specimina*. Milan, 1772.

————. *Dubbî e pensieri sopra la teoria degli elettrici fenomeni*. Milan, 1776.

————. 'Introduzione a nuovi principi della teoria elettrica dedotti dall'analisi de' fenomeni delle elettriche punte.' Società italiana delle scienze, Rome. *Memorie di matematica e di fisica*, 1 (1782), 1–54.

——. 'Della supposta eguaglianza di contrarie elettricità nelle due opposte faccie del vetro.' *Ibid.*, 4 (1788), 304–9.

——. 'Della legge d'immutabile capacità e necessaria contrarietà di eccesso e difetto di elettricità sugli opposti lati del vetro.' *Ibid.*, 7 (1794), 444–61.

Barrett, William, and Theodore Besterman. *The Divining-rod.* London, 1926.

Barrière, Pierre. *L'Académie de Bordeaux . . . au XVIIIe siècle.* Bordeaux, 1951.

Barruel, Etienne. *La physique réduite en tableaux raisonnés ou Programme du cours de physique fait à l'Ecole polytechnique.* Paris, 1799.

Bartholin, Thomas. *De luce animalium libri III.* Leyden, 1647.

Bartholmess, Christian Jean Guillaume. *Histoire philosophique de l'Académie de Prusse depuis Leibniz jusqu'à Schelling.* 2 vols. Paris, 1850–1.

Bartoli, Daniello. *Lettere edite e inedite.* Bologna, 1865.

Bayle, François. *Institutiones physicae ad usum scholarum accomodatae.* 3 vols. Toulouse, 1700.

Bayle, Pierre. *Dictionnaire historique et critique.* 4 vols. Amsterdam, 1740⁵.

Beatson, Robert. *A Political Index to the Histories of Great Britain and Ireland.* 2 vols. London, 1788².

Beccaria, Giambatista. *Dell'elettricismo artificiale e naturale libri due.* Turin, 1753.

——. *Lettre sur l'électricité.* Delor, tr. Paris, 1754.

——. *Dell'elettricismo. Lettere . . . al chiarissimo sig. Giacomo Bartolomeo Beccari.* Turin, 1758.

——. 'Experiments in Electricity.' *PT*, 51 (1760), 514–25.

——. *A sua altezza reale il signor duca di York sperienze, ed osservazione.* Turin, 1764.

——. 'Novorum quorundam in re electrica experimentorum specimen.' *PT*, 56 (1766), 105–18; 57 (1767), 297–311.

——. *De electricitate vindice . . . epistola.* Turin, 1767. (*BFP*, XIV, 42–9.)

——. *Experimenta, atque observationes, quibus electricitas vindex late constituitur, atque explicatur.* Turin, 1769.

——. *De atmosphaera electrica.* Turin, 1769. (*PT*, 60 [1770], 277–301.)

——. *Elettricismo artificiale.* Turin, 1772.

——. *A Treatise upon Artificial Electricity . . . To which is added An Essay on the Mild and Slow Electricity which prevails in the Atmosphere during Serene Weather.* London, 1776.

——. 'Dei fiori elettrici.' *Opuscoli scelti sulle scienze e sulle arti,* 3 (Milan, 1780), 242–7.

Beckman, Anna. 'Två svenska experimentalfysiker på 1700 - talet: Mårten Triewald och Nils Wallerius.' *Lychnos* (1967–8), 186–214.

Becquerel, Henri. 'Notice sur Charles François de Cisternai du Fay.' Paris. Muséum d'histoire naturelle. *Centenaire de la fondation, 10 juin 1793–10 juin 1893.* Paris, 1893. Pp. 163–85.

Bedini, Silvio A. 'The Evolution of Science Museums.' *Technology and Culture,* 6 (1965), 1–29.

Bednarski, Stanislas. 'Déclin et renaissance de l'enseignement des Jésuites en Pologne.' *Archivum historicum Societatis Jesu,* 2 (1933), 199–223.

Ben-David, Joseph. *The Scientist's Rôle in Society. A Comparative Study.* Englewood Cliffs, N.J., 1971.

Benjamin, Park. *A History of Electricity (The Intellectual Rise in Electricity) from Antiquity to the Days of Benjamin Franklin.* New York, 1898.

Bennet, Abraham. 'Description of a new Electrometer.' *PT,* 76 (1786), 26–34.

——. 'An Account of a Doubler of Electricity.' *PT,* 77 (1787), 288–96.

——. *New Experiments on Electricity.* Derby, 1789.

Benz, Ernst. *Theologie der Elektrizität.* Mainz and Wiesbaden, 1971. (Akademie der Wissenschaften und der Literatur, Mainz. Geistes- und Sozialwissenschaftliche Klasse. *Abhandlungen* [1970], no. 12.)

Béraud, Laurent. *Dissertation sur le rapport qui se trouve entre la cause des effets de l'aiman et celle des phénomènes de l'électricité.* Bordeaux, 1748.

——. 'Theoria electricitatis.' J. A. Euler, et al. *Dissertationes,* 133–204.

Berger, Peter. 'Johann Heinrich Lamberts Bedeutung in der Naturwissenschaft des 18. Jahrhunderts.' *Centaurus,* 6 (1959), 157–254.

Bergman, Torbern Olof. 'A Letter to Mr. Benjamin Wilson, F.R.S. concerning Electricity.' *PT,* 51 (1760), 907–9.

——. 'Elektrische Versuche mit Seidenbande von unterschiedlicher Farbe.' *AKSA,* 25 (1763), 244–52.

——. 'Some Experiments in Electricity.' *PT,* 54 (1764), 84–8.

——. 'Elektrische Versuche mit an einander geriebnen Glasscheiben.' *AKSA,* 27 (1765), 132–47.

——. *Torbern Bergman's Foreign Correspondence.* G. Carlid and J. Nordstrom, eds. I. *Letters from Foreigners to Torbern Bergman.* Stockholm, 1965.

Bernal, J. D. 'Les rapports scientifiques entre la Grande-Bretagne et la France au XVIIIᵉ siècle.' *RHS,* 9 (1956), 289–300.

Bernoulli, Daniel, and Jean [II] Bernoulli: 'Nouveaux principes de mécanique et de physique, tendans à expliquer la nature et les propriétés de l'aiman.' *AS. Pièces qui ont remporté le prix* [en 1743 et 1744/6]. Paris, 1748. Pp. 117–44. (*AS. Recueil des pièces . . . ,* 5.)

Bernoulli, Jean [III]. *Lettres astronomiques.* Berlin, 1771.

——. *Lettres sur différens sujets écrites pendant le cours d'un voyage par l'Allemagne, la Suisse, la France méridionale et l'Italie, en 1774 et 1775.* 3 vols. Berlin, 1777–9.

Berry, Arthur John. *Henry Cavendish. His Life and Scientific Work.* London, 1960.

Berthé de Besaucèle, Louis. *Les Cartésiens d'Italie . . . aux XVIIᵉ et XVIIIᵉ siècles.* Paris, 1920.

Bertholon de Saint Lazare, Pierre. 'Sur l'identité de l'électricité aérienne et artificielle.' *JP,* 20 (1782), 224–8.

Bianco, Franz Josef von. *Die alte Universität Köln und die spätern Gelehrten-Schulen dieser Stadt.* 2 vols. Cologne, 1855.

Bidal, Anne-Marie. 'Inventaire des Archives du Muséum national d'histoire naturelle.' Paris. Muséum national d'histoire naturelle. *Archives,* 11 (1934), 175–230.

Bierens de Haan, J. A. *De Hollandsche Maatschappij der Wetenschappen, 1752–1952.* Haarlem, 1952.

Bigourdan, Guillaume. 'Sur diverses mesures d'arcs de méridiens faites dans la première moitié du XVIIIᵉ siècle.' *Bulletin astronomique,* 18 (1901), 320–36, 351–68, 389–400, 444–8; 19 (1902), 35–47, 80–91, 118–28, 166–76, 217–24, 252–6, 287–8, 471–80; 20 (1903), 30–46, 71–80, 122–8, 195–208, 284–8.

Bioche, Charles. 'Les savants et le gouvernement aux XVIIe et XVIIIe siècles.' *Revue des deux mondes,* 107:2 (1937), 174–89.

B[iot], Jean Baptiste. 'Sur un problème de physique, relatif à l'électricité.' Société philomathique, Paris. *Bulletin,* 3 (1805), 21–3.

———. 'Alexandre Volta.' *Biographie universelle.* 45 vols. Paris, 1880². XLIV, 77–81.

———. *Recherches expérimentales et mathématiques sur les mouvemens des molécules de la lumière autour de leur centre de gravité.* Paris, 1814.

———. *Traité de physique expérimentale et mathématique.* 4 vols. Paris, 1816.

Birch, Thomas. *History of the Royal Society of London.* 4 vols. London, 1756–7.

———. 'The Life of the Honourable Robert Boyle.' Boyle. *Works* (1772), I, v-cl.

Birembaut, Arthur. 'Sur les lettres du physicien Magellan conservées aux Archives Nationales.' *RHS,* 9 (1956), 150–61.

———. 'L'Académie royale des sciences en 1780 vue par l'astronome suédois Lexell (1740–1784).' *RHS,* 10 (1957), 148–66.

———. 'Les liens de famille entre Réaumur et Brisson, son dernier élève.' *RHS,* 11 (1958), 167–9. (Centre international de synthèse. *La vie,* 168–70.)

Biwald, Leopold. *Institutiones physicae in usum philosophiae auditorum.* 2 vols. Vienna, 1779².

Boas, Marie. *See* Hall, Marie Boas.

Böhm, Wilhelm. *Die Wiener Universität.* Vienna, 1952.

Boerhaave, Herman. *A New Method of Chemistry.* Peter Shaw, tr. 2 vols. London, 1741².

Börner, Friedrich. *Nachrichten von den vornehmsten Lebensumständen und Schriften jetztlebender berümter Aerzte und Naturforscher in und um Deutschland.* 3 vols. Wolfenbüttel, 1749–54.

Boffito, Giuseppe. *Il volo in Italia.* Florence, 1921.

———. 'Un *qui pro quo* degli storici galileiani,' *Bibliofilia,* 44 (1942), 176–84.

Bohnenberger, Gottlob Christian. *Beyträge zur theoretischen und praktischen Elektrizitätslehre.* 5 parts. Stuttgart, 1793–5.

Bonnard, Fourier. *Histoire du Couvent royal de la Trinité du Mont Pincio à Rome.* Rome, Paris, Thouars, 1933.

Bopp, Karl. 'Johann Heinrich Lamberts Monatsbuch mit dem zugehörigen Kommentaren.' Akademie der Wissenschaften, Munich. Math.-Phys. Kl. *Abhandlungen,* 27:6 (1915). 84 p.

———. 'Leonhard Eulers und Johann Heinrich Lamberts Briefwechsel.' Akademie der Wissenschaften, Berlin. Phys.-Math. Kl. *Abhandlungen* (1924:2). 45 p.

———. 'J. H. Lamberts und H. G. Kaestners Briefe.' Akademie der Wissenschaften, Heidelberg. *Sitzungsberichte* (1928:18). 34 p.

Borgeaud, Charles. *Histoire de l'Université de Genève.* I. *L'Académie de Calvin, 1559–1798.* Geneva, 1900.

Borsetti Ferranti Bolani, Ferrante. *Historia almi Ferrariae gymnasii.* 2 vols. Ferrara, 1735.

Bortolotti, Ettore. 'Origine e progressi della R. Accademia delle scienze dell'Istituto di Bologna.' Accademia delle scienze dell'Istituto di Bologna. *Memorie,* 1 (1923–4: suppl.), 1–20.

———. *La storia della matematica nella Università di Bologna.* Bologna, 1947.

Boscovich, Roger Joseph. *De viribus vivis.* Rome, 1745.

———. *A Theory of Natural Philosophy* [1763²]. J. M. Child, tr. Cambridge, Mass., 1961.

Bose, Georg Mathias. *Tentamina electrica.* Wittenberg, 1744. (This contains three essays, designated in the notes as follows:

    I. i: 'De attractione et electricitate,' 1738;

    I. ii: 'De electricitate,' 1742;

    I. iii: 'De electricitate et beatificatione,' 1744.)

———. *Recherches sur la cause et sur la véritable théorie de l'électricité.* Wittenberg, 1745.

———. *Tentamina electrica tandem aliquando hydraulicae chymiae et vegetabilibus utilia.* Wittenberg, 1747. (This contains two essays, as follows:

    II. i: 'De electricitate commentarius novus,' 1746;

    II. ii: 'De electricitate commentarius epistolaris,' n.d.)

———. *L'électricité son origine et ses progrès. Poème en deux livres.* Leipzig, 1754.

Bosmans, Henri. 'Grégoire de St. Vincent.' *Biographie nationale de Belgique,* XXI (1911), cols. 141–71.

Boss, Valentin. *Newton and Russia; the Early Influence, 1698–1796.* Cambridge, Mass., 1972.

Bossut, Charles. *Traité théorique et expérimental d'hydrodynamique.* 2 vols. Paris, an IV².

Bouguer, Pierre. *Entretiens sur la cause de l'inclinaison des orbites des planètes.* Paris, 1748². (*AS. Recueil des pièces qui ont remporté les prix.* 2:7 [Paris, 1752]. 140 p.)

Bouillier, François. *Histoire de la philosophie cartésienne.* 2 vols. Paris, 1854.

Boutroux, Pierre. 'L'enseignement de la mécanique en France au XVIIe siècle.' *Isis,* 4 (1921–2), 276–94.

Bowles, Geoffrey. 'John Harris and the Powers of Matter.' *Ambix,* 22 (1975), 21–38.

Boyle, Robert. *Certain Physiological Essays.* London, 1661; 1669². (*Works,* I, 298–457; Fulton, *Bibl.,* nos. 25, 26.)

———. 'A Diamond that shines in the Dark.' *Experiments and Considerations Touching Colours.* London, 1664. Pp. 413–23. (*Works,* I, 789–99; Fulton, *Bibl.,* no. 57.)

———. *Notes, etc., about the Atmospheres of Consistent Bodies here Below.* London, 1669. (*Works,* III, 277–87; Fulton, *Bibl.,* no. 16.)

———. *Essays of the Strange Subtlety, Great Efficacy, Determinate Nature of Effluviums.* London, 1673. (*Works,* III, 659–706; Fulton, *Bibl.,* no. 105.)

———. *The Excellency and Grounds of the Corpuscular or Mechanical Philosophy* [annexed to *The Excellency of Theology*]. London, 1674. (*Works,* IV, 67–78; Fulton, *Bibl.,* no. 116; M. B. Hall, *Boyle* [1965], 187–209.)

———. *The Excellency of Theology compar'd with Natural Philosophy.* London, 1674. (*Works,* IV, 1–66; Fulton, *Bibl.,* no. 116.)

———. *Experiments and Notes about the Mechanical Origine or Production of Electricity.* London, 1675. (*Works,* IV, 345–54; Fulton, *Bibl.,* no. 123; Old Ashmolean Reprints, no. 7 [Oxford, 1927].)

———. *Some Considerations about the Reconcileableness of Reason and Religion.* London, 1675. (*Works,* IV, 151–91; Fulton, *Bibl.,* no. 122.)

———. *Experiments, Notes, etc., about the Mechanical Origine or Production of Divers Particular Qualities.* London, 1675–6. (*Works,* IV, 230–354; Fulton, *Bibl.,* no. 123.)

————. *The Works . . . To which is prefixed The Life of the Author.* Thomas Birch, ed. 6 vols. London, 1772². (Fulton, *Bibl.,* no. 241.)

Brandes, Ernst. *Betrachtungen über den Zeitgeist in Deutschland in den letzten Decennien des vorigen Jahrhunderts.* Hannover, 1808.

Brasch, F. E. 'James Logan.' American Philosophical Society. *Proceedings,* 86 (1942), 8–12.

Braun, F. 'Das physikalische Institut.' Tübingen. University. *Festgabe zum 25. Regierungs-Jubiläum seiner Majestät des Königs Karl von Wittenberg.* Tübingen, 1889. Pp. 1–6.

Brett-James, Norman George. *The Life of Peter Collinson.* London, 1926.

Briggs, J. Morton. 'D'Alembert: Philosophy and Mechanics in the 18th Century.' Colorado. University. *Studies in History,* no. 3 (1964), 38–56.

Brischar, Karl. *P. Athanasius Kircher. Ein Lebensbild.* Würzburg, 1877.

Brisson, Mathurin-Jacques. *Pésanteur spécifique des corps.* Paris, 1787.

————. 'Manière de construire un aréomètre.' *MAS* (1788), 583–616.

Brook, Abraham. *Miscellaneous Experiments and Remarks on Electricity, the Air Pump and the Barometer* [1789]. London, 1797².

Brown, Harcourt. 'Buffon and the Royal Society of London.' *Studies and Essays in the History of Science and Learning Offered in Homage to George Sarton.* New York, 1946. Pp. 137–65.

Browne, Thomas. *Pseudodoxia epidemica: or, Enquiries into very many Received Tenets and Commonly Presumed Truths.* London, 1646.

————. *The Works of Sir Thomas Browne.* G. Keynes, ed. 4 vols. London,. 1964².

Browning, John. 'Fray Beneto Jerónimo de Feijóo and the Sciences in 18th-Century Spain.' Hughes and Williams, eds. *The Varied Pattern,* 353–71.

Brunet, Pierre. *Les physiciens hollandais et la méthode expérimentale en France au XVIIIᵉ siècle.* Paris, 1926.

————. *Maupertuis.* 2 vols. Paris, 1929.

————. *L'introduction des théories de Newton en France au XVIIIᵉ siècle.* Paris, 1931.

————. 'Un grand débat sur la physique de Malebranche au XVIIIᵉ siècle.' *Isis,* 20 (1934), 367–95.

————. 'Buffon mathématicien et disciple de Newton.' Académie des sciences, arts et belles-lettres, Dijon. *Mémoires* (1936), 85–91.

————. 'L'oeuvre scientifique de Charles François Du Fay (1698–1739).' *Petrus nonius,* 3:2 (1940), 1–19.

————. 'Les premières recherches expérimentales sur la foudre et l'électricité atmosphérique.' *Lychnos,* 10 (1946–7), 117–48.

————. 'Les origines du paratonnerre.' *RHS,* 1 (1948), 213–53.

————. *La vie et l'oeuvre de Clairaut (1713–1765).* Paris, 1952. (*RHS,* 4 [1951], 13–40, 109–53; 5 [1952], 334–49; 6 [1953], 1–17.)

Bryan, Margaret. *Lectures on Natural Philosophy.* London, 1806.

Bryden, D. J. *Scottish Scientific Instrument Makers 1600–1900.* Edinburgh, 1972.

Buchdahl, Gerd. 'History of Science and Criteria of Choice.' *Historical and Philosophical Perspectives of Science.* R. H. Stuewer, ed. Minneapolis, 1970. Pp. 204–30. (*Minnesota Studies in the Philosophy of Science, V.*)

Buffon, Georges-Louis Leclerc, Comte de. *Oeuvres complètes.* 6 vols. Paris, 1835–6. (*Oeuvres* [1954], *Bibl.,* no. 177.)

————. 'Traité de l'aimant.' *Oeuvres complètes.* III (1836), pp. 76–113. (*Histoire naturelle des minéraux,* V [1788]; *Oeuvres* [1954], *Bibl.,* no. 143.)

——. *Correspondance.* Nadault de Buffon, ed. 2 vols. Paris, 1885².

——. *Oeuvres philosophiques.* Jean Pivateau, ed. Paris, 1954. (*Corpus général des philosophes français,* XLI:1.)

——. 'Seconde vue de la nature.' *Oeuvres* (1954), 35–41. (*Histoire naturelle,* XIII [1765].)

Bugge, Thomas. *Science in France in the Revolutionary Era.* Maurice P. Crosland, ed. Cambridge, Mass., 1969.

Buonanni, Filippo. *Musaeum kircherianum.* Rome, 1709.

——. *Rerum naturalium historia nempe quadrupedum . . . existentium in Museo Kircheriano.* J. A. Battarra, ed. Rome, 1773.

Burtt, Edwin Arthur. *The Metaphysical Foundations of Modern Physical Science.* London, 1932².

Cabanès, Augustin. *Marat inconnu.* Paris, 1891.

Cabeo, Niccolò. *Philosophia magnetica.* Cologne, 1629.

——. *Meteorologicorum Aristotelis commentarii.* 4 vols. Rome, 1646.

Calcaterra, Carlo. *Il nostro imminente Risorgimento.* Turin, 1935.

Calinger, Ronald S. 'Frederick the Great and the Berlin Academy of Sciences (1740–1766).' *Ann. Sci.,* 24 (1968), 239–49.

Campo, Mariano. *Cristiano Wolff e il razionalismo precritico.* 2 vols. Milan, 1939. (Università cattolica del S. Cuore. *Pubblicazioni,* ser. 1, 30.)

[Canton, John.] 'Electrical Problems.' *GM,* 17 (1747), 16.

——. 'A Method of making Artificial Magnets without the Use of Natural Ones.' *PT,* 47 (1751–2), 31–8.

——, and John Michell. *Traité sur les aimans artificiels.* Antoine Rivoire, tr. Paris, 1752.

——. 'Electrical Experiments, with an Attempt to account for Their several Phaenomena, together with Observations on Thunder Clouds.' *PT,* 48 (1753), 350–8. (*BFP, V,* 149–54; *EO,* 293–9.)

——. 'New Electrical Experiments.' *PT,* 48 (1754), 780–5.

——. 'Some Remarks on Mr. Delaval's Electrical Experiments.' *PT,* 52 (1762), 457–61.

Canton, William. 'John Canton.' *Biographia britannica.* A. Kippis, ed. III, 215–22. London, 1784².

Cantoni, G. 'Importanti osservazioni di G. B. Beccaria sui condensatori elettrici.' Istituto lombardo di scienze e lettere, Milan. *Rendiconti,* 6 (1873), 112–21.

Caramuel, Juan. *Rationalis et realis philosophia.* Louvain, 1642.

Cardano, Girolamo. *De subtilitate libri XXI.* Lyon, 1550.

Carpenter, Nathanael. *Geographie Delineated Forth in Two Bookes.* Oxford, 1635².

Carré, Jean Raoul. *La philosophie de Fontenelle.* Paris, 1932.

Carteron, H. 'L'idée de force mécanique dans le système de Descartes.' *Revue philosophique de la France et de l'Etranger,* 47 (1922), 243–77, 483–511.

Carvalho, Joaquim de. *Correspondência científica dirigida a João Jacinto de Magalhães.* Coimbra, 1952. (Coimbra. University. Faculdade de ciências. *Revista,* 20 [1951], 93–283.)

Casati, Paolo, S. J. *Terra machinis mota.* Rome, 1658.

Casini, Paolo. *L'universo-macchina.* Bari, 1969.

Cassirer, Ernst. *The Philosophy of the Enlightenment.* Princeton, 1951.

Catania. University. *Storia della Università di Catania.* Catania, 1934.

Caullery, Maurice. 'Les papiers laissés par de Réaumur et le tome VII des *Mémoires pour servir à l'histoire des insectes.'* Encyclopédie entomologique, XXXII (1955), supplement. Pp. 1–63.

Caullet de Veaumorel, L., tr. *Description de la machine électrique négative et positive de M. Nairne.* Paris, 1784.

Caut, Ronald J. *The College of St. Salvator.* Edinburgh and London, 1950.

Cavallo, Tiberius. *A Complete Treatise on Electricity, in Theory and Practice; with Original Experiments.* London, 1777; 1782²; 3 vols., 1786–95.

——. 'New Electrical Experiments and Observations. With an Improvement of Mr. Canton's Electrometer.' *PT*, 67 (1777), 388–400.

——. 'Some new Experiments in Electricity, with the Description and Use of two new Electrical Instruments.' *PT*, 70 (1780), 15–29.

——. 'Of the Methods of manifesting the Presence, and ascertaining the Quality, of Small Quantities of Natural or Artificial Electricity.' *PT*, 78 (1788), 1–22.

——. 'A new Electrical Instrument capable of collecting together a diffused or little condensed Quantity of Electricity.' *PT*, 255–60.

Cavendish, Henry. 'Experiments on Factitious Air.' *PT*, 56 (1766), 141–84.

——. 'An Attempt to explain some of the Principal Phaenomena of Electricity by Means of an Elastic Fluid.' *PT*, 61 (1771), 584–677. (Cavendish. *Electrical Researches*, 3–63.)

——. et. al. 'A Method for securing the Powder Magazines at Purfleet.' *PT*, 63 (1773), 42–7.

——. 'Some Attempts to imitate the Effects of the *Torpedo* by Electricity.' *PT*, 66 (1776), 196–225. (Cavendish. *Electrical Researches*, 194–215.)

——, et al. 'The best Method of adjusting the fixed Points of Thermometers; and of the Precautions necessary to be used in making Experiments with those Instruments.' *PT*, 67 (1777), 816–57.

——. 'Experiments to determine the density of the Earth.' *PT*, 88 (1798), 469–526.

——. *The Electrical Researches of Henry Cavendish.* J. C. Maxwell, ed. Cambridge, 1879.

Caverni, Raffaello. *Storia del metodo sperimentale in Italia.* 5 vols. Florence, 1891–8.

Ceñal, Ramon. 'Emmanuel Maignan: su vida, su obra, su influencia.' *Revista de estudios políticos*, 46 (1952), 111–49.

——. 'Juan Caramuel. Su epistolario con Atanasi Kircher, S. J.' *Revista de filosofía*, 12 (1953), 101–47.

——. 'La filosofía de Emmanuel Maignan.' *Ibid.*, 13 (1954), 15–68.

Centre international de synthèse. Section d'histoire des sciences. *La vie et l'oeuvre de Réaumur (1683–1757).* Paris, 1962.

Ceppi, Luigi Antonio. *Dissertazione serio-giocosa sull'elettricità artificiale.* Vercelli, 1784.

Cesalpino, Andrea. *Quaestionum peripateticarum libri V.* Venice, 1573; 1593³.

Ceyssens, L. 'L'action antijanséniste de Brunon Neusser, O.F.M.' *Franziskanische Studien*, 35 (1953), 401–11.

Chaldecott, J.A. *Handbook of the King George III Collection of Scientific Instruments.* London, 1951.

Chalmers, Gordon Keith. 'Sir Thomas Browne, True Scientist.' *Osiris*, 2 (1936), 28–79.

Chapin, S. L. 'The Academy of Sciences During the 18th Century: An Astronomical Appraisal.' *French Historical Studies*, 5 (1968), 371–404.

Chappe, Claude. 'Exposition des principes d'où découle la propriété qu'ont les pointes pour recevoir & émettre à de grandes distances la matière électrique.' *JP*, 40 (1792), 329–32.

Charles le géomètre. 'Essai sur les moyens d'établir entre les thermomètres une comparabilité.' *MAS* (1787), 567–82.

Charleton, Walter. *Physiologia Epicuro-Gassendo-Charletoniana*. London, 1654.

Chenakal, V. L. 'Elektricheskie mashiny v rossii xviii veka.' IIET. *Trudy*, 43 (1961), 50–111.

Cheyne, George. *Philosophical Principles of Religion: Natural and Reveal'd*. 2 vols. in 1. London, 1715–6².

Chipman, R. A. 'An Unpublished Letter of Stephen Gray on Electrical Experiments, 1707–1708.' *Isis*, 45 (1954), 33–40.

———. 'The Manuscript Letters of Stephen Gray, F.R.S., 1666/7–1736.' *Isis*, 49 (1958), 414–33.

Chossat, Marcel. *Les Jésuites et leurs oeuvres à Avignon 1553–1768*. Avignon, 1896.

Christie, John R. R. 'The Origins and Development of the Scottish Scientific Community, 1680–1760.' *History of Science*, 12 (1974), 122–41.

Cigna, Gianfrancesco. 'De analogia magnetismi et electricitatis.' *Miscellanea taurinensia*, 1 (1759–60), 43–67.

———. 'De motibus electricis experimentum.' *Ibid.*, 2 (1760–1), 77–9.

———. 'De novis quibusdam experimentis electricis.' *Ibid.*, 3 (1762–5), 31–72.

———. 'De electricitate.' *Ibid.*, 5 (1770–3), 97–108.

———. 'Esperienze elettriche.' *Scelte di opuscoli interessanti tradotti di varie lingue*, 9 (1775), 68–83.

Clairaut, Alexis Claude. 'Sur les explications cartésienne et newtonienne de la réfraction de la lumière.' *MAS* (1739), 259–75.

Clark, Sir George Norman. *A History of the Royal College of Physicians of London*. 2 vols. Oxford, 1964.

Clavelin, Maurice. *La philosophie naturelle de Galilée*. Paris, 1968.

Cohen, Ernst, and W. A. T. Cohen-de Meester. 'Daniel Gabriel Fahrenheit.' *Chemisch Weekblad*, 33 (1936), 374–92; 34 (1937), 727–30.

Cohen, I. Bernard. 'Benjamin Franklin and the Mysterious Dr. Spence. The Date and Source of Franklin's Interest in Electricity.' *Journal of the Franklin Institute*, 235 (1943), 1–25.

———. *Some Early Tools of American Science*. Cambridge, 1950.

———. 'Guericke and Dufay.' *Ann. Sci.*, 7 (1951), 207–9.

———, and Robert Schofield. 'Did Diviš Erect the First European Protective Lightning Rod, and Was His Invention Independent?' *Isis*, 43 (1952), 358–64.

———. 'Prejudice against the Introduction of Lightning Rods.' *Journal of the Franklin Institute*, 253 (1952), 393–440.

———. 'Neglected Sources for the Life of Stephen Gray.' *Isis*, 45 (1954), 41–50.

———. 'A Note Concerning Diderot and Franklin.' *Isis*, 46 (1955), 268–72.

———. *Franklin and Newton*. Philadelphia, 1956.

———. 'Some Problems in Relation to the Dates of Benjamin Franklin's First Letters on Electricity, with Remarks by Francis S. Philbrick.' American Philosophical Society. Library. *Bulletin* (1956), 537–44.

Colangolo, Francesco. *Storia dei filosofi e dei matematici napolitani*. 3 vols. Naples, 1833–4.

Collegium conimbricense. *Commentarii . . . in octo libros Physicorum.* 2 vols. Lyon, 1602; Cologne and Frankfurt, 1609.

——. *Commentarii in quattuor libros de Coelo.* Lyon, 1603².

——. *Commentarii in duos libros de Generatione et Corruptione.* Lyon, 1606².

——. *Commentarii in universam dialecticam Aristotelis* [I]. Lyon, 1607.

Condorcet, Marie-Jean-Antoine-Nicolas Caritat, Marquis de. 'Eloge de Fouchy.' *HAS* (1788), 37–49.

Convegno internazionale celebrativo del 250° anniversario della nascità di R. G. Boscovich e del 200° anniversario della fondazione dell'Osservatorio di Brera. *Atti.* Milan, 1963.

Cooper, M. *Philosophical Enquiry into the Properties of Electricity.* London, 1746.

Copenhagen, Polytekniske Laereanstalt. *Den polytekniske laereanstalt.* Copenhagen, 1910.

Cordara, Giulio Cesare. *Historiae Societatis Jesu pars sexta.* 2 vols. Rome, 1750–1859.

Cornelio, Tommaso. *Epistola, qua motuum illorum, qui vulgo ob fugam vacui fieri dicuntur, vera causa per circumpulsionem ad mentem Platonis explicatur.* Rome, 1648.

——. *Progymnasmata physica.* Venice, 1663.

Corner, Betsy C., and Christopher C. Booth, eds. *Chain of Friendship: Selected Letters of Dr. John Fothergill of London. 1735–1780.* Cambridge, Mass., 1971.

Corson, D. W. 'Pierre Polinière, Francis Hauksbee and Electroluminescence: A Case of Simultaneous Discovery.' *Isis,* 59 (1968), 402–13.

Cosentino, Giuseppe. 'L'insegnamento delle matematiche nei collegi gesuitici nell' Italia settentrionale. Nota introduttiva.' *Physis,* 13 (1971), 205–17.

Costa, Gustavo. 'Il rapporto Frisi-Boscovich alla luce di lettere inedite di Frisi, Boscovich, Mozzi, Lalande e Pietro Verri.' *Rivista storica italiana,* 79 (1967), 819–76.

Costabel, Pierre. 'Le *De viribus vivis* de R. Boscovich ou De la vertu des querelles de mots.' *AIHS,* 14 (1961), 3–21.

——. 'La correspondance Le Sage-Boscovich.' Convegno Boscovich. *Atti,* 205–16.

——. 'L'Oratoire de France et ses collèges.' Taton, ed. *Enseignement,* 67–100.

Costello, William Thomas. *The Scholastic Curriculum at Early 17th Century Cambridge.* Harvard, 1958.

Cotes, Roger. *Hydrostatical and Pneumatical Lectures.* Robert Smith, ed. London, 1775³.

Coulomb, Charles-Augustin. 'Recherches sur la meilleure manière de fabriquer les aiguilles aimantées.' AS. *Mémoires de mathématique et de physique, présentés . . . par divers savans,* 9 (1780), 166–264.

——. 'Recherches théoriques et expérimentales sur la force de torsion et sur l'élasticité des fils de métal.' *MAS* (1784), 229–69.

——. 'Description d'une boussole dont l'aiguille est suspendue par un fil de soie.' *MAS* (1785), 560–8.

——. 'Sur l'électricité et le magnétisme.'

[I] 'Construction et usage d'une balance électrique.' *Ibid.,* 569–77.

[II] 'Où l'on détermine suivant quelles lois le fluide magnétique ainsi que le fluide électrique agissent.' *Ibid.,* 578–611.

[III] 'De la quantité d'électricité qu'un corps isolé perd dans un temps donné.' *Ibid.,* 616–38.

[IV] 'Où l'on démontre deux principales propriétés du fluide électrique.' *MAS* (1786), 67–77.

[V] 'Sur la manière dont le fluide électrique se partage entre deux corps conducteurs mis en contact.' *MAS* (1787), 421–67.

[VI] 'Suite des recherches sur la distribution du fluide électrique entre plusieurs corps conducteurs.' *MAS* (1788), 617–705.

——. *Mémoires.* Paris, 1884. (Société française de physique. *Collection de mémoires relatifs à la physique*, I.)

Courtney, W. P. 'Stephen Gray, F.R.S.' *Notes and Queries*, 6 (1906), 161–3, 354.

Cousin, Jean. 'L'Académie des sciences, belles lettres et arts de Besançon au XVIIIe siècle et son oeuvre scientifique.' *RHS*, 12 (1959), 327–44.

Cousin, Jules. *Le Comte de Clermont. Sa cour et ses maîtresses.* 2 vols. Paris, 1867.

Cousin, Victor. *Fragments philosophiques pour servir à l'histoire de la philosophie.* 5 vols. Paris, 1865–6^5.

Coutts, James. *A History of the University of Glasgow from its Foundation in 1451 to 1909.* Glasgow, 1909.

Cowper, J. M. *The Roll of the Freemen of the City of Canterbury.* Canterbury, 1903.

Crane, V. W. 'The Club of Honest Whigs: Friends of Science and Liberty.' *William and Mary Quarterly*, 23 (1966), 210–33.

Crino, Anna Maria. *Fatti e figure del seicento anglo-toscano.* Florence, 1957.

Cromaziano, Agatopisto. *Della restaurazione di ogni filosofia ne' secoli XVI, XVII e XVIII.* 3 vols. Venice, 1785–9.

Crombie, Alistair C. *Robert Grosseteste and the Origins of Experimental Science, 1100–1700.* Oxford, 1962^2.

Crommelin, Claude-August. 'Die holländische Physik im 18. Jahrhundert mit besonderer Berücksichtigung der Entwicklung der Feinmechanik.' *Sudhoffs Archiv*, 28 (1935), 129–42.

——. 'Die Elektrisiermachine des Dr. Deimans und deren Verfertiger John Cuthbertson.' *Zeitschrift für technische Physik*, 17 (1936), 105–8.

——. 'L'ascension du Mont-Blanc par H. B. de Saussure en 1787 et un relief du Mont-Blanc de 1790.' *Nederlands aardrijkskundig genootschap. Tijdschrift*, 66 (1949), 327–31.

——. *Descriptive Catalogue of the Physical Instruments of the 18th Century in the Rijksmuseum voor de Geschiedenis der Natuurwetenschappen.* Leyden, 1951. (Leyden. Rijksmuseum voor de Geschiedenis der Natuurwetenschappen. *Communications*, no. 81.)

Crosland, Maurice P. *Historical Studies in the Language of Chemistry.* London, 1962.

——. 'The Development of Chemistry in the 18th Century.' *Studies on Voltaire and the 18th Century*, 24 (1963), 369–441.

——. *The Society of Arcueil.* Cambridge, Mass., 1967.

Curtis, Mark H. *Oxford and Cambridge in Transition 1558–1642.* Oxford, 1959.

Cuthbertson, John. *Abhandlung von der Elektricität.* Leipzig, 1786.

——. *Practical Electricity.* London, 1807.

Cuvier, Georges-Frédéric. 'Notice sur la vie et les ouvrages du citoyen Lemonnier.' *HAS*, 3 (an IX), 101–17.

——. 'Eloge historique de M. Banks.' *HAS*, 5 (1821–2), 204–31.

——. 'Eloge historique de M. le comte de Lacépède.' *HAS*, 8 (1825), ccxii–ccxlviii.

Dähnert, Johann Carl, ed. *Sammlung gemeiner und besonderer Pommerscher und Rügischer Landes-Urkunden.* II. Stralsund, 1767.

Dainville, François de. *La géographie des humanistes.* Paris, 1940.

———. 'L'enseignement des mathématiques dans les collèges Jésuites de France du XVIᵉ au XVIIIᵉ siècle.' *RHS,* 7 (1954), 6–21, 109–23.

———. 'L'enseignement des mathématiques au XVII siècle.' *XVIIᵉ siècle,* no. 30 (1956), 62–8.

———. 'L'enseignement scientifique dans les collèges des Jésuites.' Taton, ed. *Enseignement,* 27–65.

Dalibard, Thomas François. 'Histoire abrégée de l'électricité.' *EO* (Dal)[1], i-lxx.

Dalzel, Andrew. *History of the University of Edinburgh from its Foundation.* 2 vols. Edinburgh, 1862.

Dangerville, l'abbé. 'Ode sur l'électricité.' *Mémoires pour l'histoire des sciences et beaux-arts* (August, 1762), 2027–34.

Danzel, T. W. *Gottsched und seine Zeit. Auszüge aus seinem Briefwechsel.* Leipzig, 1855.

Dauben, J. W. 'Marat: His Science and the French Revolution.' *AIHS,* 22 (1969), 235–61.

Daujat, Jean. *Origines et formation de la théorie des phénomènes électriques et magnétiques.* 3 vols. Paris, 1945.

Daumas, Maurice. *Les instruments scientifiques aux XVIIᵉ et XVIIIᵉ siècles.* Paris, 1953.

———. 'La vie scientifique au XVII siècle.' *XVIIᵉ siècle,* no. 30 (1956), 110–33.

———. 'Precision Mechanics.' Charles Singer et al., eds. *A History of Technology.* IV. *The Industrial Revolution.* Oxford, 1958. Pp. 379–416.

———. 'Precision of Measurement and Physical and Chemical Research in the 18th Century.' Alistair C. Crombie, ed. *Scientific Change.* New York, 1963. Pp. 418–30.

Debus, Allen G. 'Robert Fludd and the Use of Gilbert's *De magnete* in the Weapon-Salve Controversy.' *Journal of the History of Medicine,* 19 (1964), 389–417.

———. 'Mathematics and Nature in the Chemical Texts of the Renaissance.' *Ambix,* 15 (1968), 1–28.

———. *Science and Education in the 17th Century. The Webster-Ward Debate.* London and New York, 1970.

Dee, John. *Propaedeumata aphoristica.* London, 1568[2].

Dekker, T. 'De popularisering der natuurwetenschap in Nederland in de achttiende eeuw.' *Geloof en wetenschap,* 53 (1955), 173–88.

Delambre, Jean Baptiste Joseph. 'Eloge historique de M. Coulomb.' *HAS,* 7 (1806), 206–23.

———. 'Notice sur la vie et les oeuvres de M. Lagrange.' Lagrange. *Oeuvres,* I, ix-li.

Delandine, Antoine François. *Couronnes académiques.* 2 vols. Paris, 1787.

———. *Manuscrits de la bibliothèque de Lyon.* 3 vols. Paris and Lyon, 1812.

Delattre, Pierre. *Les établissements des Jésuites en France depuis quatre siècles.* 5 vols. Paris, 1939–57.

Delaunay, Paul. 'Th. F. Dalibard, botaniste et physicien.' *La province du Maine,* 6 (1926), 53–9, 141–8, 192–9, 251–61, 292–8.

Delaval, Edward. 'Some Electrical Experiments and Observations.' *PT,* 51 (1759), 83–8.

———. 'Several Experiments in Electricity.' *PT,* 52 (1761), 353–6.

Delfour, J. *Les Jésuites à Poitiers (1604–1702).* Paris, 1902.

Deluc, Jean André. *Recherches sur les modifications de l'atmosphère.* 2 vols. Geneva, 1772.

——. *Idées sur la météorologie.* 2 vols. London, 1786–7.

——. 'Cinquième lettre . . . sur le fluide électrique.' *JP,* 36 (1790), 450–69.

——. *Précis de la philosophie de Bacon et des progrès qu'ont fait les sciences naturelles par ses préceptes et son example.* 2 vols. Paris, 1802.

——. *Traité élémentaire sur le fluide électrogalvanique.* 2 vols. Paris and Milan, 1804.

De Morgan, Augustus. 'The Canton Papers.' *The Athenaeum* (1849), 5–7, 162–4, 375.

——. *A Budget of Paradoxes.* London, 1872.

Denis, J. B. 'Seconde conférence, présentée à monsieur le Dauphin.' *Journal des sçavans,* 3 (1672–4) [Amsterdam, 1682], 222–33.

Derham, William. 'Experimenta & observationes de soni motu.' *PT,* 26 (1708–9), 2–35.

Desaguliers, Jean Théophile. 'An Account of a Book entitul'd *Vegetable Staticks.*' *PT,* 34 (1728), 264–91; 35 (1729), 323–31. (*Course* [1763³], II, 403–11.)

——. 'An Attempt to Solve the Phenomenon of the Rise of Vapours, Formation of Clouds, and Descent of Rain.' *PT,* 36 (1729), 6–22.

——. *A Course of Experimental Philosophy.* 2 vols. London, 1734–44; 1745²; 1763³.

——. 'Some Thoughts and Conjectures Concerning the Cause of Electricity.' *PT,* 41 (1739–40), 175–85.

——. 'Some Thoughts and Experiments Concerning Electricity.' *Ibid.,* 186–93.

——. 'Experiments made before the Royal Society, Feb. 2, 1737–8.' *Ibid.,* 193–9, 200–8; *PT,* 41 (1740–1), 637–40, 666–7.

——. 'Some Things Concerning Electricity.' *Ibid.,* 634–7.

——. *A Dissertation concerning Electricity.* London, 1742. (*Course* [1763³], II, 316–36.)

——. *Dissertation sur l'électricité des corps.* Bordeaux, 1742. (*Recueil de dissertations de ceux qui ont remporté le prix à l'Académie royale des belles-lettres, sciences et arts de Bordeaux,* V.)

——. 'Some Conjectures Concerning Electricity and the Rise of Vapours.' *PT,* 42 (1742–3), 140–3.

——. 'Some Further Observations concerning Electricity.' *Ibid.,* 14–18.

——. 'Dissertation on the Cause of the Rise of Vapours.' *Course* (1763³), II, 336–50.

Descartes, René. *Oeuvres.* D. Adam and P. Tannery, eds. 13 vols. Paris, 1897–1913; 1964–74².

——. *Principes de la philosophie* [1644]. *Oeuvres,* IX:2.

——. *Correspondence of Descartes and Constantyn Huygens, 1635–1647.* L. Roth, ed. Oxford, 1926.

——. *Correspondance.* C. Adam and G. Milhaud, eds. 8 vols. Paris, 1936–63.

——. *Discourse on Method, Optics, Geometry and Meteorology.* Paul J. Olscamp, tr. Indianapolis, Ind., 1965.

Desmaze, Charles. *Infanterie française. Le régiment de Picardie.* Paris, 1888.

Desplat, Christian. *Un milieu socio-culturel provincial: L'Académie royale de Pau au XVIIIᵉ siècle.* Pau, 1971.

Détienne, ——. 'De l'électricité de pression.' *JP,* 10 (1777), 72–80.

Dibner, Bern. *Moving the Obelisks.* Norwalk, Conn., 1952.

——. *Early Electrical Machines.* Norwalk, Conn., 1957.

——. 'The Great van Marum Electrical Machine.' *Natural Philosopher,* 2 (1963), 89–103.

Dictionnaire de physique. Gaspard Monge et al., eds. 4 vols. Paris, 1793–1822. (*Encyclopédie méthodique, ou par ordre des matières.* Vols. 186–9.)

Diderot, Denis. *Correspondance.* I. *1713–1757.* Georges Roth, ed. Paris, 1955.

Diemer, Alwin, ed. *Der Wissenschaftsbegriff. Historische und systematische Untersuchungen.* Meisenheim am Glan, 1970.

Digby, Kenelm. *Two Treatises. In the One of which the Nature of Bodies; in the Other, the Nature of Man's Soul, is looked into.* London, 1644.

——. *A Late Discourse . . . touching the Cure of Wounds by the Powder of Sympathy.* R. White, tr. London, 1658.

Dijksterhuis, E. J. *The Mechanization of the World Picture.* C. Dikshoorn, tr. Oxford, 1961.

Dircks, Henry. *The Life, Times and Scientific Labours of the Second Marquis of Worcester.* London, 1865.

Dobbs, Betty Jo. 'Studies in the Natural Philosophy of Sir Kenelm Digby.' *Ambix,* 18 (1971), 1–25.

Doerberl, Michael. *Entwicklungsgeschichte Bayerns.* 3 vols. Munich, 1916–31[3].

Donndorff, Johann August. *Die Lehre von der Elektricität.* Erfurt, 1784.

Doppelmayr, Johann Gabriel. *Neu-entdeckte Phaenomena von bewunderswürdigen Würkungen der Natur.* Nuremburg, 1774.

Dorfman, IA. G. 'Epinus i ego traktat po elektrichestvu i magnetizmu.' *Aepinus. Opyt teorii,* 461–538.

——. 'B. Franklin i russkie elektriki XVII v.' IIET. *Trudy,* 19 (1957), 290–312.

Dorling, Jon. 'Henry Cavendish's Deduction of the Electrostatic Inverse Square Law from the Result of a Single Experiment.' *Studies in History and Philosophy of Science,* 4 (1974), 327–48.

Dorsman, C., and C. A. Crommelin. 'The Invention of the Leyden Jar.' *Janus,* 46 (1957), 275–80.

Doublet, E. 'L'abbé Bossut.' *Bulletin des sciences mathématiques,* 38:1 (1914), 93–6, 121–5, 158–9, 186–90, 220–4.

Droysen, H. 'Die Marquise du Châtelet, Voltaire und der Philosoph Christian Wolff.' *Zeitschrift für französiche Sprache und Literatur,* 35 (1910), 226–48.

Duboul, J. 'Dortous de Mairan. Etude sur sa vie et sur ses travaux.' L'Académie impériale des sciences, belles-lettres et arts, Bordeaux. *Actes,* 24 (1862), 163–97.

Du Châtelet-Lomont, Gabrielle-Emilie, Marquise. *Institutions de physique.* Paris, 1740.

——. *Les lettres de la marquise du Châtelet.* Theodore Besterman, ed. 2 vols. Geneva, 1958.

Dülmen, Richard van. 'Ein unbekannter Brief von Athanasius Kircher.' *Studia Leibnitiana,* 4 (1972), 141–5.

Dufay, Charles François de Cisternay. 'Mémoire sur les baromètres lumineux.' *MAS* (1723), 295–306.

——. 'Mémoire sur un grand nombre de phosphores nouveaux.' *MAS* (1730), 524–35.

——. 'A Letter from Mons. Du Fay . . . concerning Electricity.' *PT,* 38 (1733–4), 258–66.

————. 'Mémoires sur l'électricité. Ier. L'histoire de l'électricité.' *MAS* (1733), 23–35.

'2e. Quels sont les corps qui sont susceptibles d'électricité.' *Ibid.*, 73–84.

'3e. Des corps qui sont les plus vivement attirés par les matières électriques, et de ceux qui sont les plus propres à transmettre l'électricité.' *Ibid.*, 233–54.

'4e. L'attraction et la répulsion des corps électriques.' *Ibid.*, 457–76.

'5e. Des nouvelles découvertes sur cette matière, faites depuis peu par M. Gray; et où l'on examine quelles sont les circonstances qui peuvent apporter quelque changement à l'électricité pour l'augmentation ou la diminution de sa force, comme la température de l'air, la vide, l'air comprimé, etc.' *MAS* (1734), 341–61.

'6e. Quel rapport il y a entre l'électricité et la faculté de rendre de la lumière, qui est commune à la plupart des corps électriques, et ce qu'on peut inférer de ce rapport.' *Ibid.*, 503–26.

'7e. Quelques additions aux mémoires précédents.' *MAS* (1737), 86–100.

'8e.' *Ibid.*, 307–25.

————. 'Lettres de Dufay à Réaumur.' *La correspondance historique et archéologique*, 5 (1898), 306–9.

Du Fay, Charles Jérôme de Cisternay. *Bibliotheca Fayana*. Gabriele Martin, ed. Paris, 1725.

Duhamel, Jean Baptiste. *De corporum affectionibus cum manifestis tum occultis libri duo*. Paris, 1670.

Duhem, Pierre. 'L'optique de Malebranche.' *Revue de métaphysique et de morale*, 23 (1916), 37–91.

————. *To Save the Phenomena. An Essay on the Idea of Physical Theory from Plato to Galileo*. E. Doland and C. Maschler, trs. Chicago, 1969.

Duhr, Bernhard. *Geschichte der Jesuiten in den Ländern deutscher Zunge in der zweiten Hälfte des XVII. Jahrhunderts*. 4 vols. Munich and Regensberg, 1921.

Dukov, V. M. 'Issledovaniia Niutona v oblasti elektrichestva i magnetizma.' *IIET. Voprosy*, 7 (1959), 120–7.

Dulieu, Louis, 'La contribution montpelliéraine aux Recueils de l'Académie royale des sciences.' *RHS*, 11 (1958), 250–62.

————. 'Le mouvement scientifique montpelliérain au XVIIIe siècle.' *Ibid.*, 227–49.

Dunken, Gerhard. *Die Deutsche Akademie der Wissenschaften zu Berlin in Vergangenheit und Gegenwart*. Berlin, 1960.

Dupont-Ferrier, Gustave. *La vie quotidienne d'un collège parisien pendant plus de 350 ans: du collège de Clermont au lycée Louis-le-Grand, 1563–1920*. 3 vols. Paris, 1921–5.

Dutour, Etienne François. 'Essai sur l'aiman.' AS. *Pièces qui ont remporté le prix [en 1743 et 1744/6]*. Paris, 1748. [Prix de 1744/6], 51–114. (AS. *Recueil des pièces . . .* , 5.)

————. 'De la nécessité d'isoler les corps qu'on électrise par communication; et des avantages qu'un corps convenablement isolé retire du voisinage des corps non électriques.' AS. *Mémoires de mathématique et de physique présentés . . . par divers savans*, 2 (1755), 516–42.

————. *Recherches sur les différens mouvemens de la matière électrique*. Paris, 1760.

Dyment, S. A. 'Some 18th Century Ideas concerning Aqueous Vapor and Evaporation.' *Ann. Sci.*, 2 (1937), 465–73.

[Eandi, Giuseppi Antonio Francesco Girolamo]. *Memorie istoriche intorno gli studi del padre Giambatista Beccaria.* [Torino], 1783.

Eccher, A. 'La fisica sperimentale dopo Galileo.' *La vita italiana nel settecento.* Milan, 1903. Pp. 365–420.

Edleston, J. *Correspondence of Sir Isaac Newton and Professor Cotes.* London and Cambridge, 1850.

Eeles, Henry. *Philosophical Essays in Several Letters to the Royal Society.* London, 1771.

Ekelöf, Stig, ed. *Catalogue of Books and Papers relating to the History of Electricity in the Library of the Institute for Theoretical Electricity, Chalmers University of Technology.* 2 vols. Göteborg, 1964–6. (Chalmers Tekniska Högskola. Institution för elektricitetslära och elektrisk mätteknik. *Meddelande,* 11, 20.)

Ellicott, John. 'Of Weighing the Strength of Electrical Effluvia.' *PT,* 44 (1746), 96–9.

———. 'Several Essays toward Discovering the Laws of Electricity.' *PT* 45 (1748), 195–224.

Elvius, Pehr. 'Geschichte der Wissenschaften. Von der Electricität.' *AKSA,* 9 (1747), 179–85.

———. 'Geschichte der Wissenschaften, vom Ausdünsten des Wassers.' *AKSA,* 10 (1748), 3–10.

Encyclopédie méthodique. See *Dictionnaire de physique.*

Encyclopédie, ou dictionnaire raisonné des sciences, des arts, et des métiers [1751–65]. 36 vols. Geneva, 1778³.

Enfield, William. *Institutes of Natural Philosophy, Theoretical and Experimental.* London, 1785.

Erxleben, Johann Christian Polykarp. *Anfangsgründe der Naturlehre* [1772]. G. C. Lichtenberg, ed. Göttingen, 1784³; 1787⁴; 1791⁵; 1794⁶.

Eschenberg, Johann Joachim. *Lehrbuch der Wissenschaftskunde. Ein Grundriss encyclopädischer Vorlesungen.* Berlin and Stettin, 1792.

Eschweiler, K. 'Die Philosophie der spanischen Spätscholastik auf den deutschen Universitäten des siebzehnten Jahrhunderts.' Görresgesellschaft. *Spanische Forschungen,* 1 (1928), 251–324.

———. 'Roderigo de Arriaga, S. J. Ein Beitrag zur Geschichte der Barockscholastik.' *Ibid.,* 3 (1931), 253–85.

Espenschied, Lloyd. 'The Electrical Flare of the 1740s.' *Electrical Engineering,* 74 (1955), 392–7.

———. 'More on Franklin's Introduction to Electricity.' *Isis,* 46 (1955), 280–1.

Eulenburg, Franz. *Die Frequenz der deutschen Universitäten von ihrer Gründung bis zur Gegenwart.* Leipzig, 1904. (Akademie der Wissenschaften, Leipzig. Phil.-Hist. Kl. *Abhandlungen,* 24:2.)

Euler, Johann Albrecht, et al. *Dissertationes selectae quae ad imperialem scientiarum petropolitanam academiam an. 1755 missae sunt.* St. Petersburg and Lucca, 1757.

———. 'Disquisitio de causa physica electricitatis.' *Ibid.,* 1–40.

———. 'Recherches sur la cause physique de l'électricité.' *MAS/Ber* (1757), 125–59.

Euler, Leonhard. 'Dissertatio de magnete.' AS. *Pièces qui ont remporté le prix* [*en 1743 et 1744/6*]. Paris, 1748. [Prix de 1744/6], 3–47. (AS. *Recueil des pièces . . . ,* 5.)

———. *Lettres . . . sur différentes questions de physique et de philosophie.* M. J. A. Condorcet and S. F. de Lacroix, eds. 3 vols. Paris, 1787–9.

———. *Opera omnia*. Ferdinand Rudio et al., eds. Leipzig and Berlin, 1911+.

———. *Die Berliner und die Petersburger Akademie der Wissenschaften im Briefwechsel Leonhard Eulers. I. Der Briefwechsel L. Eulers mit G. F. Müller*. A. P. Juškević and E. Winter, eds. Berlin, 1959.

———. *Leonhard Euler und Christian Goldbach. Briefwechsel 1729–1764*. A. P. Juškević and E. Winter, eds. Berlin, 1965.

[Fabri, Honoré.] *Metaphysica demonstrativa sive Scientia rationum universalium*. Lyon, 1648. (Issued under pseudonym, Petrus Mousnerius.)

———. *Dialogi physici in quibus de motu terrae disputatur, marini aestus nova causa proponitur, necnon acquarum & mercurii supra libellam elevatio examinatur*. Lyon, 1655.

———. *Tractatus duo quorum prior de plantis et de generatione animalium, posterior de homine*. 2 vols. Paris, 1666.

———. *Synopsis geometrica*. Lyon, 1669.

———. *Physica id est Scientia rerum corporearum in decem tractatus distributa*. 5 vols. Lyon, 1669–71.

———. *Epistolae tres de sua hypothesi philosophica*. Mainz, 1674.

Fabroni, Angelo. *Lettere inedite di uomini illustri*. 2 vols. Florence, 1773–5.

Farrell, Allan P. *The Jesuit Code of Liberal Education*. Milwaukee, Wis., 1938.

Faucillon, Jean-Marcellin-Ferdinand. *Le Collège de Jésuites de Montpellier (1629–1762)*. Montpellier, 1857.

Favaro, Antonio. 'Per la storia della Accademia del Cimento.' Istituto veneto di scienze, lettere ed arti. *Atti*, 71 (1912), 1173–8.

———. *L'Università di Padova*. Venice, 1922.

Fawcett, Trevor. 'Popular Science in 18th-Century Norwich.' *History Today*, 22 (1972), 590–5.

Faÿ, Bernard. 'Learned Societies in Europe and America in the 18th Century.' *American Historical Review*, 37 (1932), 255–66.

Feather, Norman. *Electricity and Matter*. Edinburgh, 1968.

Fellmann, Emil A. 'Die mathematischen Werke von Honoratus Fabry (1607–1688).' *Physis*, 1 (1959), 6–29, 73–102.

Ferguson, Alan, ed. *Natural Philosophy through the 18th Century and Allied Topics*. London, 1948.

Ferguson, James. *An Introduction to Electricity in Six Sections*. London, 1770; 1775[2]; 1778[3].

Fermi, Stefano. *Lorenzo Magalotti scienziato e letterato (1637–1712)*. Piacenza, 1903.

Fernández Diéguez, D. 'Un matemático español del siglo XVIII: Juan Caramuel.' *Revista matemática hispano-americana*, 1 (1919), 121–7, 178–89, 203–12.

Ferretti-Torricelli, A. 'Padre Francesco Lana nel III centenario della nascità.' Ateneo di Brescia. *Commentari* (1931), 321–90.

Ferrner, Bengt. *Resa i Europa 1758–1762*. S. G. Lindberg, ed. Uppsala, 1956.

Fester, Richard. *'Der Universitäts-Bereiser.' Friedrich Gedike und sein Bericht an Friedrich Wilhelm II*. Berlin, 1905. (Archiv für Kulturgeschichte. *Ergänzungsheft*, 1.)

Fichter, Joseph Henry. *Man of Spain, Francis Suarez*. New York, 1940.

Finch, Jeremiah Stanton. *Sir Thomas Browne. A Doctor's Life of Science & Faith*. New York, 1950.

Finn, Bernard S. 'An Appraisal of the Origins of Franklin's Electrical Theory.' *Isis,* 60 (1969), 362–9.

———. 'The Influence of Experimental Apparatus on 18th Century Electrical Theory.' CIHS, 1968. *Actes,* X:A (1971), 51–5.

———. 'Output of 18th Century Electrical Machines.' *BJHS,* 5 (1971), 289–91.

Fisch, Max H. 'The Academy of the Investigators.' E. Ashworth Underwood, ed. *Science, Medicine and History. Essays* . . . [*for*] *Charles Singer.* 2 vols. Oxford, 1953. I, 521–63.

Fischer, Ernst Gottfried. *Physique mécanique.* J. B. Biot, tr. Paris, 1806.

Fischer, Johann Carl. *Anfangsgründe der Physik in ihren mathematischen und chemischen Theilen nach den neuesten Entdeckungen.* Jena, 1797.

———. *Geschichte der Physik seit der Wiederherstellung der Künste und Wissenschaften bis auf die neuesten Zeiten.* 8 vols. Göttingen, 1801–8.

Fischer, Johann Nepomuk. *Beweis dass das Glockenlaeuten bey Gewittern mehr schaedlich als nuetzlich sey.* Munich, 1784.

Fisher, J. 'The History of Electricity in Germany during the First Half of the 18th Century.' British Society for the History of Science. *Bulletin,* 2 (1956), 49–51.

Fitzmaurice, Edmond George Petty-Fitzmaurice. *Life of William, Earl of Shelbourne.* 3 vols. London, 1875–6.

Fitzpatrick, Edward Augustus, ed. *St. Ignatius and the Ratio Studiorum.* New York and London, 1933.

Fletcher, Harris Francis. *The Intellectual Development of John Milton.* 2 vols. Urbana, Ill., 1956–61.

Fletcher, John E. 'A Brief Survey of the Unpublished Correspondence of Athanasius Kircher, S. J. (1602–1680).' *Manuscripta,* 13 (1969), 150–60.

———. 'Medical Men and Medicine in the Correspondence of Athanasius Kircher.' *Janus,* 56 (1969), 259–77.

———. 'Astronomy in the Life and Correspondence of Athanasius Kircher.' *Isis,* 61 (1970), 52–67.

Flourens, Pierre. *Fontenelle ou De la philosophie moderne relativement aux sciences physiques.* Paris, 1847.

Förster, Johann Christian. *Übersicht der Geschichte der Universität zu Halle in ihrem ersten Jahrhunderte.* Halle, 1799.

Fogolari, Gino. 'Il museo Settala. Contributo per la storia della coltura in Milano nel secolo XVII.' *Archivio storico lombardo,* 14 (1900), 58–126.

Follini, Giorgio. *Teoria elettrica brevemente esposta ad uso della studiosa gioventù.* Ivrea, 1791.

Fontana, Felice. 'Description abrégée du cabinet de physique et d'histoire naturelle du grand-duc de Toscane, à Florence.' *JP,* 9 (1777), 41–8, 104–12, 194–203, 267–72, 377–9.

Fontenelle, Bernard le Bovier de. 'Doutes sur le système physique des causes occasionnelles' [1686]. *Oeuvres,* IX, 37–68.

———. 'Digression sur les anciens et les modernes' [1687]. *Oeuvres,* IV, 114–31.

———. 'Eloge de M. Hartsoeker.' *HAS* (1725), 137–53. (*Oeuvres,* VI, 135–54.)

———. *The Elogium of Isaac Newton.* London, 1728. (Newton. *Papers and Letters,* 444–74.)

———. 'Sur l'attraction newtonienne.' *HAS* (1732), 112–17.

———. 'Préface sur l'utilité des mathématiques et de la physique' [1733]. *Oeuvres,* V, 1–14.

——. 'Eloge de M. Du Fay.' *HAS* (1734), 73–83. (*Oeuvres*, VI, 372–83.)

——. 'Eloge de M. Saurin.' *HAS* (1737), 110–20. (*Oeuvres*, VI, 331–42.)

——. *Oeuvres*. 12 vols. Amsterdam, 1764.

Forbes, Eric, and I. K. Kopelevich. 'Neiznestnye pisma L. Eǐlera k T. Maǐery.' *Istorikoastronomicheskie issledovaniia*, 10 (1969), 285–310.

Forbes, Eric G. 'Tobias Mayer, Zur Wissenschaftsgeschichte des 18. Jahrhunderts.' *Jahrbuch für Geschichte der oberdeutschen Reichsstädte*, 16 (1970), 132–67.

Forbes, James David. *A Review of the Progress of Mathematical and Physical Science in More Recent Times and particularly between the Years 1775 and 1850*. Edinburgh, 1858.

Forbes, Robert J., ed. *Martinus van Marum, Life and Work*. Haarlem, 1969+.

——. 'A Member of the First Class.' Forbes, ed. *Martinus van Marum*. III, 1–21.

Formey, Jean Henri Samuel. *Abrégé de physique*. 2 vols. Berlin, 1770–2.

——. *Souvenirs d'un citoyen*. 2 vols. Paris, 1797².

Forster, J. R. ['Über die Natur des Feurs und der Elektricität.'] *Neueste Entdeckungen in der Chemie*, 12 (1784), 154–8.

Fourier, J. J. 'Eloge historique de M. Charles.' *MAS*, 8 (1829), lxxiii–lxxxviii.

Fouveaux de Courmelles, François Victor. *Un électricien oublié. Le magistrat de Romas*. Paris, 1909.

Fracastoro, Girolamo. *De sympathia & antipathia rerum*. Lyon, 1550.

France, Anatole. *L'Elvire de Lamartine. Notes sur M. et Mme Charles*. Paris, 1893.

Frängsmyr, Tore. *Wolffianismens genembrott i Uppsala. Frihetstida universitets-filosofi till 1700 talets mitt*. Uppsala, 1972. (Uppsala. University. *Acta*, ser. C, no. 26.)

——. 'The History of Swedish Science.' *History of Science*, 12 (1974), 29–42.

Frank, Robert G., Jr. 'Science, Medicine and the Universities of Early Modern England: Background and Sources.' *History of Science*, 11 (1973), 236–69.

Franklin, Alfred. *Les origines du Palais de l'Institut. Recherches historiques sur le Collège des Quatre-Nations*. Paris, 1862.

——. *Histoire de la Bibliothèque Mazarine et du Palais de l'Institut*. Paris, 1901².

——. *Dictionnaire historique des arts, métiers et professions*. Paris and Leipzig, 1906.

Franklin, Benjamin. *Experiments and Observations on Electricity*.

    *EO*¹ London, 1751–4.

    *EO*² London, 1754.

    *EO*³ London, 1760–5.

    *EO*⁴ London, 1769.

    *EO*⁵ London, 1774.

    EO (Dal)¹ Paris, 1752. (Tr. by T. Dalibard of the first part of *EO*¹.)

    EO (Dal)² 2 vols. Paris, 1756. (Tr. of the complete *EO*¹.)

    EO (Wilcke) Leipzig, 1758. (Tr. of *EO*¹ by J. C. Wilcke.)

——. *The Works of Benjamin Franklin*. Jared Sparkes, ed. 10 vols. Boston, 1840.

——. *The Writings of Benjamin Franklin*. A. H. Smyth, ed. 10 vols. New York, 1907. (Cited as *BFP* [Smyth].)

——. *Benjamin Franklin's Experiments. A New Edition of Franklin's Experiments and Observations on Electricity*. I. B. Cohen, ed. Cambridge, Mass., 1941. (Taken from *EO*⁵; cited as *EO*.)

——. *Benjamin Franklin's Autobiographical Writings*. C. van Doren, ed. New York, 1945.

——. *Papers*. L. W. Labaree, et al., eds. New Haven, 1959+. (Cited as *BFP*.)

Freind, John. *Praelectiones chymicae*. Amsterdam, 1710.

Freke, John. *An Essay to Shew the Cause of Electricity: and why Some Things are Non-Electricable*. London, 1746; 1746[2]; 1752[3].

French, Peter J. *John Dee. The World of an Elizabethan Magus*. London, 1972.

Freshfield, D. W. *The Life of Horace Benedict de Saussure*. London, 1920.

Freyberg, Bruno von. *Johann Gottlob Lehmann*. Erlangen, 1955.

Friedländer, Paul. 'Athanasius Kircher und Leibniz; ein Beitrag zur Geschichte der Polyhistorie im XVII. Jahrhundert.' *Pontificia accademia romana di archeologia. Rendiconti*, 13 (1937), 229–47.

Frisi, Paolo. 'De causa electricitatis dissertatio.' J. A. Euler et al. *Dissertationes*, 41–131.

Frost, A. J. *Catalogue of Books and Papers relating to Electricity, Magnetism, and the Electric Telegraph, etc., including the Ronalds Library*. London, 1880.

Fulton, John Farquhar. *A Bibliography of the Honourable Robert Boyle, Fellow of the Royal Society*. Oxford, 1961[2].

Gabler, Matthias. *Naturlehre zum Gebrauch öffentlicher Erklärungen*. Munich, 1778.

Gabrieli, Vittorio. *Sir Kenelm Digby*. Rome, 1957.

Galilei, Galileo. *Opere*. 20 vols. Ed. A. Favaro. Florence, 1890–1909.

Galle, Meingosus. *Beyträge zur Erweiterung und Vervollkommung der Elektricitätslehre in theoretischer und praktischer Hinsicht*. Salzburg, 1813.

Gallitzin, Demetrius Augustine. 'Lettre sur la forme des conducteurs électriques.' Académie royale des sciences, des lettres et des beaux-arts de Belgique. *Mémoires*, 3 (1780), 2–11.

———. *Sendschreiben an die kaiserliche Akademie der Wissenschaften zu St. Petersburg über einige Gegenstände der Electricität*. Münster, 1780.

———. 'Sur quelques objects d'électricité.' *JP*, 21 (1782), 73–9.

Galvani, Luigi. *Commentary on the Effect of Electricity on Muscular Motion*. Robert Montraville Greca, tr. Cambridge, Mass., 1953.

García Villoslada, Riccardo. *Storia del Collegio romano del suo inizio (1551) alla soppressione della Compagnia di Gesù (1773)*. Rome, 1954. (*Analecta Gregoriana*, 66 [1953]. Series Facultatis historiae ecclesiasticae, sec. A, no. 2.)

Gardini, Giuseppe. *De electrici ignis natura dissertatio*. Mantua, 1792.

Garin, Eugenio. 'Antonio Genovesi e la sua introduzione storica agli *Elementa physicae* di Pietro van Musschenbroek.' *Physis*, 11 (1969), 211–22.

Garrick, David. *Letters*. D. M. Little and G. M. Kahrl, eds. 3 vols. Cambridge, Mass., 1963.

Gartrell, Ellen G. *Electricity, Magnetism and Animal Electricity. A Checklist of Printed Sources, 1600–1850*. Wilmington, Del., 1975.

Gassendi, Pierre. *The Mirror of True Nobility and Gentility*. London, 1657.

———. *Opera omnia*. 6 vols. Lyon, 1658.

Gattey, Etienne François. 'Expériences sur les moyens de préserver les aiguilles des boussoles de l'influence de l'électricité atmosphérique.' *JP*, 17 (1781), 296–303.

Gattoni, Giulio Cesare. 'Notizie storiche della prima età di Alessandro Volta' [1806]. *Voltiana*, 1 (1926), 493–7.

Gaubil, Antoine. *Correspondance de Pékin, 1722–1759*. Renée Simon, ed. Geneva, 1970.

Gaullieur, Ernst. *Collège de Guyenne d'après un grand nombre de documents inédits*. Paris, 1874.

Gautruche, Pierre. *Mathematicae totius . . . institutio*. Cambridge, 1668.

Gehler, Johann Samuel Traugott. *Physikalisches Wörterbuch oder Versuch einer Erklärung der vornehmsten Begriffe und Kunstwörter der Naturlehre.* 4 vols. Leipzig, 1787–91; 11 vols. in 22, 1825–45².

Geike, Archibald. *Memoir of John Michell.* Cambridge, 1918.

Gelbart, Nina Rattner. 'The Intellectual Development of Walter Charleton.' *Ambix,* 18 (1971), 149–68.

Genoa. University. *L'Università di Genova.* Genoa, 1923.

Genovese, Antonio. 'Disputatio [later Dissertatio] physico-historica de rerum corporearum origine et constitutione.' Petrus van Musschenbroek. *Elementa physicae conscripta in usus academicos.* 2 vols. Naples, 1745. I, 1–79.

Gerhardt, C. I., ed. *Briefwechsel zwischen Leibniz und Christian Wolf.* Halle, 1860.

Gericke, Helmuth. *Zur Geschichte der Mathematik an der Universität Freiburg i. Br.* Freiburg, 1955. *(Beiträge zur Freiburger Wissenschafts- und Universitäts-geschichte, 7.)*

Gerini, S. B. 'I seguaci di Cartesio in Italia sul finire del secolo XVII ed in principio del XVIII.' *Il nuovo risorgimento,* 9 (1899), 426–43, 449–80.

Gerland, Ernst, and Friedrich Traumüller. *Geschichte der physikalischen Experimentier-kunst.* Leipzig, 1899.

Geymonat, Ludovico. *Galileo Galilei.* Stillman Drake, tr. New York and London, 1965.

Gibbs, F. W. 'Boerhaave's Chemical Writings.' *Ambix,* 6 (1958), 117–35.

————. 'Itinerant Lecturers in Natural Philosophy.' *Ambix,* 8 (1961), 111–17.

Gibson, James. *Locke's Theory of Knowledge and its Historical Relations.* Cambridge, 1917.

Gilbert, William. *De magnete, magneticisque corporibus, et de magno magnete tellure; physiologia nova.* London, 1600. P. F. Mottelay, tr. New York, 1893; S. P. Thompson, tr. London, 1900.

————. *De mundo nostro sublunari philosophia nova.* Amsterdam, 1651.

Gilen, Leonhard. 'Über die Beziehungen Descartes' zur zeitgenössichen Scholastik.' *Scholastik,* 32 (1957), 41–66.

Gillmor, C. Stewart. *Coulomb and the Evolution of Physics and Engineering in 18th-Century France.* Princeton, 1971.

Gilson, Etienne Henri. *Index scolastico-cartésien.* Paris, 1912.

Gimma, Giacinto. *Idea della storia dell'Italia letterata.* 2 vols. Naples, 1723.

Glick, Thomas F. 'On the Influence of Kircher in Spain.' *Isis,* 62 (1971), 379–81.

Gliozzi, Mario. 'L'elettrologia nel secolo XVII.' *Periodico di matematiche,* 13 (1933), 1–14.

————. 'Studio comparativo delle teorie elettriche del Nollet, del Watson e del Franklin.' *Archeion,* 15 (1933), 202–15.

————. 'Giambatista Beccaria nella storia dell'elettricità.' *Archeion,* 17 (1935), 15–47.

————. *L'elettrologia fino al Volta.* 2 vols. Naples, 1937.

————. *Fisici piemontesi del settecento nel movimento filosofico del tempo.* Turin, 1962. (Biblioteca filosofica di Torino. *Quaderni,* no. 2.)

————. 'La costituzione della materia nella concezione di Boscovich e di Faraday.' Convegno Boscovich. *Atti,* 115–19.

————. 'Lettres inédites de Giambattista Beccaria.' CIHS, 1965. *Actes,* III (1968), 210–15.

————. 'Consonanze e dissonanze tra l'elettrostatica di Cavendish e quella di Volta.' *Physis,* 11 (1969), 231–48.

————. 'Il Volta della seconda maniera.' *Cultura e scuola,* 5 (1966), 235–9.

——. 'Introduzione.' Volta. *Opere* (1967), 9–39.

Godley, Alfred Denis. *Oxford in the 18th Century*. London, 1908.

Gordon, Andreas. *Versuch einer Erklärung der Electricität*. Erfurt, 1745; 1746[2].

——. *Physicae experimentalis elementa in usus academicos conscripta*. 2 vols. Erfurt, 1751–3.

Gott, Samuel. 'Of Magnetical Virtue and Electricity.' *The Divine History of the Genesis of the World Explicated and Illustrated*. London, 1670. Pp. 274–86.

Gottsched, J. C. *Historische Lobschrift des weiland hoch- und wohlgeborenen Herrn Christians, des H.R.R. Freyherrn von Wolff*. Halle, 1755.

Grabmann, M. 'Die disputationes metaphysicae des Franz Suarez in ihrer methodischen Eigenart und Fortwirkung.' *Mittelalterliches Geistesleben*. 3 vols. Munich, 1926–56. I, 525–60.

Gralath, Daniel. 'Geschichte der Electricität.' [Erster Abschnitt.] Naturforschende Gesellschaft, Danzig. *Versuche und Abhandlungen*, 1 (1747), 175–304.
'Zweiter Abschnitt.' *Ibid.*, 2 (1754), 355–460.
'Dritter Abschnitt.' *Ibid.*, 3 (1757), 492–556.

——. 'Nachricht von einigen electrischen Versuchen.' *Ibid.*, 1 (1747), 506–41.

Grandjean de Fouchy, Jean-Paul. 'Eloge de Clairaut.' *HAS* (1765), 144–59.

——. 'Eloge de Nollet.' *HAS* (1770), 121–37.

Grant, Alexander. *The Story of the University of Edinburgh during its first Three Hundred Years*. 2 vols. London, 1884.

Grassi, Giuseppe. *Giuseppe Angelo Saluzzo di Menusiglio*. Turin, 1813.

Gravesande, Willem Jacob van 's. *Physices elementa mathematica, experimentis confirmata, sive Introductio ad philosophiam Newtonianam*. 2 vols. Leyden, 1720–1; 1725[2]; 1742[3].

——. *Mathematical Elements of Natural Philosophy confirmed by Experiments, or an Introduction to Isaac Newton's Philosophy*. J. T. Desaguliers, tr. 2 vols. London, 1721; 1731[4].

——. *Eléments de physique, ou Introduction à la philosophie de Newton*. 2 vols. Paris, 1747.

——. *Oeuvres philosophiques et mathématiques*. J. N. S. Allamand, ed. 2 vols. Amsterdam, 1774.

Gray, Edward Whitaker. 'Observations on the Manner in which Glass is charged with the Electric Fluid, and discharged.' *PT*, 78 (1788), 121–4.

Gray, Stephen. 'An Account of Some New Electrical Experiments.' *PT*, 31 (1720–1), 104–7.

——. 'Several Experiments concerning Electricity.' *PT*, 37 (1731–2), 18–44.

——. 'The Electricity of Water.' *Ibid.*, 227–30.

——. 'Farther Account of his Experiments concerning Electricity.' *Ibid.*, 285–91; 397–407.

——. 'Experiments and Observations upon the Light that is produced by communicating Electrical Attraction to Animate or Inanimate Bodies.' *PT*, 39 (1735–6), 16–24.

——. 'Some Experiments relating to Electricity.' *Ibid.*, 166–70.

Great Britain. Parliament. House of Commons. 'Evidence, Oral and Documentary, Taken and Received by the Commissioners for Visiting the Universities of Scotland.' *Sessional Papers*, 1837: Edinburgh (92), *35;* Glasgow (93), *36;* St. Andrews (94), *37;* Aberdeen (95), *38*. (Command no. in parentheses, vol. in italics.)

Greaves, Richard L. *The Puritan Revolution and Educational Thought*. New Brunswick, N. J., 1969.

Green, George. *An Essay on the Application of Mathematical Analysis to the Theories of Electricity and Magnetism*. Nottingham, 1828.

Greene, Robert. *The Principles of the Philosophy of the Expansive and Contractive Forces*. Cambridge, 1727.

Greenwood, Isaac. *Experimental Course of Mechanical Philosophy*. Boston, 1726.

Grimsley, Ronald. *Jean d'Alembert (1717–83)*. Oxford, 1963.

Grosier, Jean Baptiste Gabriel Alexandre. *Mémoires d'une société célèbre considérée comme corps littéraire . . . ou Mémoires des Jésuites sur les sciences, les belles-lettres et les arts*. 3 vols. Paris, 1792.

Gross, B. 'On the Experiment of the Dissectible Condenser.' *American Journal of Physics*, 12 (1944), 324–9.

Guadagni, Carlo Alfonso. *Indice di esperienze naturali*. Florence, 1745.

Guareschi, Icilio. 'Amedeo Avogadro e la sua opera scientifica. Discorso storico critico.' Avogadro. *Opere scelte*, i–cxl.

Günther, Siegmund. 'Note sur Jean-André de Segner. Fondateur de la météorologie mathématique.' *Bulletino di bibliografia e di storia delle scienze matematiche e fisiche*, 9 (1876), 217–28.

———. 'Die mathematischen und Naturwissenschaften an der Nürnbergischen Universität Altdorf.' Verein für Geschichte der Stadt Nürnberg. *Mitteilungen*, Heft 3 (1881), 1–36.

Guericke, Otto von. *Die Belagerung, Eroberung und Zerstörung der Stadt Magdeburg am 10.–20. Mai 1631*. F. W. Hoffmann, ed. Leipzig, 1912².

———. *Experimenta nova (ut vocantur) magdeburgica de vacuo spatio*. Amsterdam, 1672.

———. *Neue (sogenannte) magdeburger Versuche über den leeren Raum*. Hans Schimank et al., trs. and eds. Düsseldorf, 1968.

Guerlac, Henry. 'The Continental Reputation of Stephen Hales.' *AIHS*, 4 (1951), 393–404.

———. 'A Note on Lavoisier's Scientific Education.' *Isis*, 47 (1956), 211–16.

———. 'Newton's Changing Reputation in the 18th Century.' Raymond O. Rockwood, ed. *Carl Becker's Heavenly City Revisited*. Ithaca, 1958. Pp. 3–26.

———. 'Francis Hauksbee: expérimentateur au profit de Newton.' *AIHS*, 16 (1963), 113–28.

———. *Newton et Epicure*. Paris, 1963. (Paris. Palais de la Découverte. *Conférences*. Series D, no. 91.)

———. 'Sir Isaac and the Ingenious Mr. Hauksbee.' *Mélanges Alexandre Koyré*. I, 228–53.

———. 'Where the Statue Stood: Divergent Loyalties to Newton in the 18th Century.' Earl R. Wasserman, ed. *Aspects of the 18th Century*. Baltimore, 1965. Pp. 317–34.

———. 'Newton's Optical Aether: His Draft of a Proposed Addition to His *Opticks*.' RS. *Notes and Records*, 22 (1967), 45–57.

———. 'An Augustan Monument: The Opticks of Sir Isaac Newton.' Hughes and Williams, eds. *Varied Pattern*, 131–63.

———. 'Chemistry as a Branch of Physics: Laplace's Collaboration with Lavoisier.' *Historical Studies in the Physical Sciences*, 7 (1976), 193–276.

Guitton, Georges. *Les Jésuites à Lyon sous Louis XIV et Louis XV.* Lyon, 1954.

Gunther, Robert T. *Early Science at Oxford.* 14 vols. London, 1920–45.

Gurlt, Ernst Julius, and A. Hirsch. *Biographisches Lexikon der hervorragenden Ärzte aller Zeiten und Völker.* 6 vols. Vienna and Leipzig, 1884–8.

Gutmann, Joseph. *Athanasius Kircher, 1602–1680, und das Schöpfungs- und Entwicklungsproblem.* Fulda, 1938.

Haberzettl, Hermann. *Die Stellung der Exjesuiten in Politik und Kulturleben Österreichs zu Ende des 18. Jahrhunderts.* Vienna, 1973. (Vienna. University. *Dissertationen,* 94.)

Hackmann, Willem D. 'Electrical Researches.' Forbes, ed. *Martinus van Marum.* III, 329–78.

———. 'The Design of the Triboelectric Generators of Martinus van Marum, F.R.S. A Case History of the Interaction between England and Holland in the Field of Instrument Design in the 18th Century.' RS. *Notes and Records,* 26 (1971), 163–81.

———. *John and Jonathan Cuthbertson. The Invention and Development of the 18th Century Plate Electrical Machine.* [Leyden], 1973. (Leyden. Rijksmuseum voor de Geschiedenis der Natuurwetenschappen. *Communications,* no. 142.)

Hagelgans, Johann Georg. *Orbis literatus academicus germanico-europaeus.* Frankfurt a.M., 1737.

Hahn, Paul. *Georg Christoph Lichtenberg und die exakten Wissenschaften.* Göttingen, 1927.

Hahn, Roger. 'L'enseignement scientifique aux écoles militaires et d'artillerie.' Taton, ed. *Enseignement,* 513–35.

———. 'The Chair of Hydrodynamics in Paris, 1775–1791: A Creation of Turgot.' CIHS, 1962. *Actes,* II (1970), 751–4.

———. 'Determinism and Probability in Laplace's Philosophy.' CIHS, 1971. *Actes,* I (1974), 170–5.

———. *The Anatomy of a Scientific Institution. The Paris Academy of Sciences 1666– 1803.* Berkeley, 1971.

———. 'Scientific Careers in 18th Century France.' Maurice P. Crosland, ed. *The Emergence of Science in Western Europe.* London, 1975. Pp. 127–38. (*Minerva,* 13 [1975], 501–13.)

———. 'Scientific Lecturers in Paris, 1777–1789.' Unpublished ms.

Hales, Stephen. *Vegetable Staticks . . . Also, a Specimen of an Attempt to Analyse the Air.* London, 1727.

Hall, Alfred Rupert, and Marie Boas Hall. *Unpublished Scientific Papers of Isaac Newton.* Cambridge, 1962.

Hall, Marie Boas. 'Bacon and Gilbert.' *JHI,* 12 (1951), 466–7.

———. 'The Establishment of the Mechanical Philosophy.' *Osiris,* 10 (1952), 412–541.

———. *Robert Boyle and 17th Century Chemistry.* Cambridge, 1958.

———. *Robert Boyle on Natural Philosophy. An Essay with Selections from his Writings.* Bloomington, Ind., 1965.

———, and Alfred Rupert Hall. 'Newton's Electric Spirit: Four Oddities.' *Isis,* 50 (1959), 473–6.

[Haller, Albrecht von.] 'Histoire des nouvelles découvertes faites, depuis quelques années en Allemagne, sur l'électricité; mémoire sur les nouvelles découvertes qu'on a faites par rapport à l'électricité.' *Bibliothèque raisonnée des ouvrages des sa-vans de l'Europe*, 34 (1745), 3–20.

Hamberger, Georg Christoph, and Johan Georg Meusel. *Das gelehrte Teutschland*. 23 vols. Lemgo, 1796–1834.

Hamberger, Georg Erhard. *Elementa physices methodo mathematica in usum auditorii conscripta*. Jena, 1735²; 1741³.

Hammermayer, Ludwig. *Gründungs- und Frühgeschichte der bayerischen Akademie der Wissenschaften*. Kallmünz/Opf., 1959. (*Münchener historische Studien. Abteilung bayerische Geschichte, 4.*)

Hankins, Thomas L. 'The Influence of Malebranche on the Science of Mechanics during the 18th Century.' *JHI*, 28 (1967), 193–210.

———. *Jean d'Alembert. Science and the Enlightenment*. Oxford, 1970.

Hanks, Lesley. *Buffon avant 'l'Histoire Naturelle.'* Paris, 1966.

Hanna, Blake T. 'Polinière and the Teaching of Experimental Physics at Paris 1700–1730.' P. Gay, ed. *18th Century Studies Presented to Arthur M. Wilson*. Hanover, N. H., 1972. Pp. 15–39.

Hans, Nicholas A. *New Trends in Education in the 18th Century*. London, 1951.

Hansteen, Christopher. *Untersuchungen über den Magnetismus der Erde*. Christiania, 1819.

Hardin, Clyde L. 'The Scientific Work of the Reverend John Michell.' *Ann. Sci.*, 22 (1966), 27–47.

Harding, Diana. 'Mathematics and Science Education in 18th-Century Northamptonshire.' *History of Education*, 1 (1972), 139–59.

Harnack, Adolf von. *Geschichte der Königlich Preussischen Akademie der Wissenschaften*. 3 vols. in 4. Berlin, 1900.

Harper, Wallace Russell. *Contact and Frictional Electrification*. Oxford, 1967.

Harris, John. *Lexicon technicum*. 2 vols. London, 1704–10.

Harris, William Snow. 'On some Elementary Laws of Electricity.' *PT*, 124:1 (1834), 213–45.

———. 'Inquiries concerning the Elementary Laws of Electricity. Second series.' *PT*, 126:1 (1836), 417–52.

———. *Rudimentary Electricity*. London, 1859⁵.

———. *Rudimentary Magnetism*. Henry M. Noad, ed. London, 1872².

Harrison, John Anthony. 'Blind Henry Moyes. "An Excellent Lecturer in Philosophy."' *Ann. Sci.*, 13 (1957), 109–25.

Hartmann, P. J. 'Succincta succini prussici historia & demonstratio.' *PT*, 21 (1699), 5–40.

Hartog, Sir Philip J. 'Newer Views of Priestley and Lavoisier.' *Ann. Sci.*, 5 (1941), 1–56.

Hartsoeker, Nicolas. *Conjectures physiques*. 3 vols. Amsterdam, 1706–12.

———. *Cours de physique*. The Hague, 1730.

Harvey, Edmund Newton. *A History of Luminescence from the Earliest Times until 1900*. Philadelphia, 1957.

Hauch, A. W. 'Forsøg til et forbedret Udlade-Elektrometer.' Danske Videnskabernes Selskab. *Skrifter*, 5 (1793), 415–22.

Hauksbee, Francis. 'Experiments on the Production and Propagation of Light from the *Phosphorus in Vacuo*.' *PT*, 24 (1704–5), 1865–6.

———. 'Several Experiments on the Mercurial Phosphorus.' *Ibid*., 2129–35.

————. 'Several Experiments on the Attrition of Bodies in *Vacuo.' Ibid.*, 2165–75.

————. 'The Production of a Considerable Light upon a slight Attrition of the Hand on a Glass Globe Exhausted of its Air: with other Remarkable Occurrences.' *PT*, 25 (1706–7), 2277–82.

————. 'The Extraordinary Elistricity [!] of Glass, produceable on a smart Attrition of it; with a Continuation of Experiments on the same subject, and other *Phaenomena.' Ibid.*, 2327–31.

————. 'A Continuation of the Experiments on the Attrition of Glass.' *Ibid.*, 2332–5.

————. 'Several Experiments shewing the Strange Effects of Glass, produceable on the Motion and Attrition of it.' *Ibid.*, 2372–7.

————. 'The Production of Light, by the Effluvia of one Glass falling on another in Motion.' *Ibid.*, 2412–13.

————. 'An Experiment touching Motion given Bodies included in a Glass, by the Approach of a Finger near its outside: with other Experiments on the Effluvia of Glass.' *PT*, 26 (1708–9), 82–6.

————. 'The *Electricity* and *Light* produceable on the Attrition of Several Bodies.' *Ibid.*, 87–92.

————. *Physico-Mechanical Experiments on Various Subjects.* London, 1709; 1719².

————. 'An Account of Experiments concerning the Proportion of the Power of the Load-stone at different Distances.' *PT*, 27 (1710–2), 506–11.

Hauksbee, Francis, the Younger. *Course of Mechanical, Optical and Pneumatical Experiments, to be performed by F. H., and the Explanatory Lectures read by Wm. Whiston, M. A.* London, 1714.

Hausen, Christian August. *Novi profectus in historia electricitatis.* Leipzig, 1743.

Haüy, René Just. *Exposition raisonnée de la théorie de l'électricité et du magnétisme d'après les principes de M. Aepinus.* Paris, 1787.

————. *Traité élémentaire de physique.* 2 vols. Paris, 1803.

Hawes, J. L. 'Newton and the "Electrical Attraction Unexcited." ' *Ann. Sci.*, 24 (1968), 121–30.

————. 'Newton's Revival of the Aether Hypothesis and the Explanation of Gravitational Attraction.' RS. *Notes and Records,* 23 (1968), 200–12.

Heathcote, N. H. de V. 'Guericke's Sulphur Globe.' *Ann. Sci.*, 6 (1948–50), 293–305.

————. 'Franklin's Introduction to Electricity.' *Isis,* 46 (1955), 29–35.

————. 'The Early Meaning of Electricity: Some Pseudodoxia Epidemica.' *Ann. Sci.*, 23 (1967), 261–75.

Heilbron, J. L. 'G. M. Bose: The Prime Mover in the Invention of the Leyden Jar?' *Isis,* 57 (1966), 264–7.

————. 'A propos de l'invention de la bouteille de Leyde.' *RHS,* 19 (1966), 133–42.

————. 'Honoré Fabri and the Accademia del Cimento.' CIHS, 1968. *Actes,* III:B (1971), 45–49.

————. *H. G. J. Moseley. The Life and Letters of an English Physicist, 1887–1915.* Berkeley, 1974.

————. 'Robert Symmer and the Two Electricities.' *Isis,* 67 (1976), 7–20.

————. 'Franklin's Physics.' *Physics Today,* 29:7 (1976), 32–7.

————. 'Franklin, Haller, and Franklinist History.' *Isis,* 68 (1977), 539–49.

————. 'Volta's Path to the Battery.' Electrochemical Society. *Anniversary Volume* (1977), in press.

————. 'Introductory Essay.' Wayne Shumaker and J. L. Heilbron. *John Dee on Astronomy.* Berkeley, 1978. Pp. 1–99.

Heimann, P. M., and J. E. McGuire. 'Newtonian Forces and Lockean Powers: Concepts of Matter in 18th Century Thought.' *Historical Studies in the Physical Sciences*, 3 (1971), 233–306.

Heimbucher, A. *Die Orden und Kongregationen der katholischen Kirche*. 3 vols. Paderborn, 1907–8².

Heinrichs, Norbert. 'Scientia magica.' Diemer, ed. *Der Wissenschaftsbegriff*, 30–46.

Helbig, Herbert. *Universität Leipzig*. Frankfurt a.M., 1961.

Helden, Albert van. 'Eustachio Divini versus Christiaan Huygens. A Reappraisal.' *Physis*, 12 (1970), 36–50.

Hellman, C. D. 'John Bird (1709–1776), Mathematical Instrument Maker in the Strand.' *Isis*, 17 (1932), 127–53.

Hemmer, l'abbé. 'Sur l'électricité des métaux.' *JP*, 16 (1780), 50–2.

Henderson, Ebenezer. *Life of James Ferguson*. Edinburgh, 1867.

[Henley, William.] 'An Account of a New Electrometer, contrived by Mr. Henley.' *PT*, 62 (1772), 359–64.

———. 'Experiments concerning the different Efficacy of pointed and blunted Rods, in securing Buildings against the Stroke of Lightning.' *PT*, 64 (1774), 133–52.

———. 'An Account of some New Experiments in Electricity, containing . . . Some Experiments to ascertain the Direction of the Electric Matter in the Discharge of the Leyden Bottle.' *Ibid.*, 389–431.

———. 'Experiments and Observations on a New Apparatus, called, A Machine for exhibiting perpetual Electricity.' *PT*, 66 (1776), 513–22.

———. 'Experiments and Observations on Electricity.' *PT*, 67 (1777), 85–143.

———. 'Observations and Experiments tending to Confirm Dr. Ingenhousz's Theory of the Electrophorus; and to shew the Impermeability of Glass to Electric Fluid.' *Ibid.*, 1049–55.

Henley, W., et al. 'Report of the Committee appointed by the Royal Society, for examining the Effect of Lightning, May 15, 1777, on the Parapet-Wall of the House of the Board of Ordnance, at Purfleet.' *PT*, 68 (1778), 236–8.

Henry, Charles. 'Correspondance inédite de d'Alembert.' *Bulletino di bibliografia e di storia delle scienze*, 18 (1885), 507–70, 605–45.

Herbert, Josephus Nobilis de. *Theoriae phaenomenorum electricorum quae seu electricitatis ex redundante corpore in deficiens trajectu, seu sola atmosphaerae electricae actione gignuntur*. Vienna, 1772; 1778².

Hermann, Armin. 'Die Anfänge der "akademischen" Physik in München.' *Physikalische Blätter*, 22 (1966), 388–96.

Hermelink, Heinrich, and S. A. Kähler. *Die Philipps-Universität zu Marburg, 1527–1927*. Marburg, 1927.

Herrmann, Dieter B. 'Georg Christoph Lichtenberg als Herausgeber von Erxlebens Werk *Anfangsgründe der Naturlehre*.' *NTM*, 6:1 (1969), 68–81; 6:2 (1969), 1–12. (*Mitteilungen der Archenhold-Sternwarte Berlin-Treptow*, no. 102.)

Herschel, William. *The Scientific Papers of William Herschel*. 2 vols. J. L. E. Dreyer, ed. London, 1912.

Hesse, Mary B. 'Action at a Distance in Classical Physics.' *Isis*, 46 (1955), 337–53.

———. 'Gilbert and the Historians.' *British Journal for Philosophy of Science*, 11 (1960), 1–10, 130–42.

———. *Forces and Fields*. London, 1961.

Hesse, Werner. *Beiträge zur Geschichte der früheren Universität in Duisburg*. Duisburg, 1875.

Heyman, H. J. 'Samuel Klingenstierna och Fredrik Mallet.' *Symbola literaria*. Uppsala, 1927. Pp. 187–209. (*Hyllningsskrift till Uppsala Universitet vid Jubelfesten 1927.*)

Higenbottam, F. 'The Apparition of Mrs. Veal to Mrs. Bargrave at Canterbury, 8th Sept., 1705.' *Archaeologia Cantiana*, 73 (1959), 154–66.

Hildebrand, Bengt, *Kungl. Svenska Vetenskapsakademien. Förhistoria, grundläggning och första organisation*. 2 vols. Stockholm, 1939.

Hildebrand, Friedrich. *Anfangsgründe der dynamischen Naturlehre*. Erlangen, 1807.

Hildebrandsson, Hugo Hildebrand. *Samuel Klingenstiernas levnad och verk*. I. *Levnadsteckning*. Uppsala, 1919.

Hilgers, Joseph. *Der Index der verbotenen Bücher*. Freiburg, 1904.

Hill, Christopher. *Intellectual Origins of the English Revolution*. Oxford, 1965.

Hill, Elizabeth. 'Biographical Essay.' Whyte, ed. *Roger Joseph Boscovich*, 17–101.

Hindle, Brooke. *The Pursuit of Science in Revolutionary America, 1735–1789*. Chapel Hill, 1951.

Hine, W. L. 'The Interrelationship of Science and Religion in the Circle of Marin Mersenne.' Ph.D. thesis. University of Oklahoma, 1967.

Hippel, A. von, and F. W. Merrill. 'The Atomphysical Interpretation of Lichtenberg Figures.' *Journal of Applied Physics*, 10 (1939), 873–87.

Hoadley, Benjamin, and Benjamin Wilson. *Observations on a Series of Electrical Experiments*. London, 1756; 1759².

Hölscher, Uvo. 'Urkundliche Geschichte der Friedrichs-Universität zu Bützow.' Verein für Mecklenburgische Geschichte und Altertumskunde. *Jahrbuch*, 50 (1885), 1–110.

Hoff, H. E., and R. Guillemin. 'The First Experiments on Transfusion in France.' *Journal of the History of Medicine*, 18 (1963), 103–24.

Hoffmann, Friedrich Wilhelm. *Geschichte der Stadt Magdeburg*. 3 vols. Magdeburg, 1850.

Hofmann, Josef Ehrenfried. *Die Mathematik an altbayerischen Hochschulen*. Munich, 1954. (Akademie der Wissenschaft, Munich. Math.-Wiss. Kl. *Abhandlungen*, 62.)

Hollmann, Samuel Christian. 'Epistola . . . de igne electrico.' *PT*, 43 (1744–5), 239–49.

———. 'Succincta attractionis historia.' Akademie der Wissenschaften, Göttingen. *Commentarii*, 4 (1754), 215–44.

Holmberg, Arne. *Kungl. Vetenskapsakademiens äldre skrifter i utlandska oversättninger och referat*. Stockholm, 1939. (Svenska Vetenskapsakademien, Stockholm. Årsbok [1939], Bihang.)

Home, Roderick W. *The Effluvial Theory of Electricity*. Ph.D. thesis. University of Indiana, 1967.

———. 'Francis Hauksbee's Theory of Electricity.' *Archive for History of Exact Sciences*, 4 (1967–8), 203–17.

———. 'Electricity and the Nervous Fluid.' *Journal of the History of Biology*, 3 (1970), 235–51.

———. 'The Reception Accorded Aepinus's Theory of Electricity.' CIHS, 1971. *Actes*, VI (1974), 287–94.

———. 'Aepinus and the English Electricians: The Dissemination of a Scientific Theory.' *Isis*, 63 (1972), 190–204.

——. 'Franklin's Electrical Atmospheres.' *BJHS*, 6 (1972), 131–51.

——. 'Science as a Career in 18th-Century Russia: The Case of F. U. T. Aepinus.' *Slavonic and East European Review*, 51 (1973), 75–94.

——. 'Electricity in France in the Post-Franklin Era.' *CIHS*, 1974. *Actes*, II (1975), 269–72.

——. 'Some Manuscripts on Electrical and other Subjects attributed to Thomas Bayes, F.R.S.' *RS. Notes and Records*, 29 (1974), 81–90.

——. 'On Two Supposed Works by Leonhard Euler on Electricity.' *AIHS*, 25 (1975), 3–7.

——. 'Aepinus, the Tourmaline Crystal, and the Theory of Electricity and Magnetism.' *Isis*, 67 (1976), 21–30.

Hooke, Robert. *Micrographia*. London, 1665.

——. *Posthumous Works*. R. Waller, ed. London, 1705.

——. *Diary*. Henry W. Robinson and Walter Adams, eds. London, 1935.

Hoppe, Edmund. *Geschichte der Elektrizität*. Leipzig, 1884.

Horne, David Bayne. *A Short History of the University of Edinburgh 1556–1889*. Edinburgh, 1967.

Horvath, Joannes Baptist. *Physica particularis auditorum usibus accommodata*. Venice, 1782.

Hoskin, M. A. '"Mining all within": Clarke's notes to Rohault's *Traité de physique*.' *The Thomist*, 24 (1951), 353–63.

Houghton, W. E., Jr. 'The English Virtuoso in the 17th Century.' *JHI*, 3 (1943), 51–73, 190–219.

Hube, Michael. *Vollständiger und fasslicher Unterricht in der Naturlehre*. 3 vols. Leipzig, 1801².

Huber, Paulus. *Der Parnassus Boicus. Ein Beitrag zur Kulturgeschichte Baierns während der ersten Hälfte des 18. Jahrhunderts*. Munich, 1868. (Munich. Ludwigs-Gymnasium. *Program*, [1867–8]. 20 p.)

Hughes, Arthur. 'Science in English Encyclopedias, 1704–1875. I.' *Ann. Sci.*, 7 (1951), 340–70.

'II. Theories of the Elementary Composition of Matter.' *Ann. Sci.*, 8 (1952), 323–67.

Hughes, Peter, and David Williams, ed. *The Varied Pattern: Studies in the 18th Century*. Toronto, 1971. (MacMaster University. Association for 18th Century Studies. *Publications*, 1.)

Hujer, Karel. 'Father Procopius Diviš, The European Franklin.' *Isis*, 43 (1952), 351–7.

Hume, David. *A Treatise of Human Nature* [1739–40]. L. A. Selby-Bigge, ed. 3 vols. Oxford, 1888.

——. *An Enquiry concerning the Human Understanding* [1748]. L. A. Selby-Bigge, ed. Oxford, 1894.

Hunter, Michael. 'The Social Basis and Changing Fortunes of an Early Scientific Institution: an Analysis of the Membership of the Royal Society, 1660–1685.' *RS. Notes and Records*, 31(1976), 9–114.

Huter, Franz. *Die Fächer Mathematik, Physik und Chemie an der Philosophischen Fakultät zu Innsbruck bis 1945*. Innsbruck, 1971. (Innsbruck. University. *Veröffentlichungen*, 66; *Forschungen zur Innsbrucker Universitätsgeschichte*, 10.)

Hutton, Charles. *The Diarian Miscellany*. 5 vols. London, 1775.

——. *Philosophical and Mathematical Dictionary.* 2 vols. London, 1815².

Huygens, Christian. *Treatise on Light* [1690]. Sylvanus P. Thompson, tr. Chicago, 1945.

——. *Oeuvres complètes.* 22 vols. The Hague, 1888–1950.

Imhof, Maximus. *Grundriss der öffentlichen Vorlesungen über die Experimental-Naturlehre.* 2 vols. Munich, 1794–5.

Index librorum prohibitorum. Vatican City, 1948.

Ingenhousz, Jan. 'A Ready Way of Lighting a Candle, by a very moderate Electrical Spark.' *PT,* 68 (1778), 1022–6.

——. 'Electrical Experiments, to explain how far the Phaenomena of the Electrophorus may be accounted for by Dr. Franklin's Theory of positive and negative Electricity.' *Ibid.,* 1027–48.

——. 'Improvements in Electricity.' *PT,* 69 (1779), 661–73.

——. 'Exposition de plusieurs loix qui paroissent s'observer constamment dans les divers mouvemens du fluide électrique & auxquelles les physiciens n'avoient fait une suffisante attention.' *JP,* 16 (1780), 117–26.

——. *Nouvelles expériences et observations sur divers objets de physique.* Paris, 1785.

Irsay, Stephen d'. *Histoire des universités françaises et étrangères.* 2 vols. Paris, 1933–5.

Isnardi, Lorenzo. *Storia dell'Università di Genova.* 2 vols. Genoa, 1861–7.

Italy. Ministero della pubblica istruzione. *Monografie delle università e degli istituti superiori.* 2 vols. Rome, 1911–3.

Jacquet de Malzet, Louis Sébastien. *Précis de l'électricité ou Extrait expérimental & théorétique des phénomènes électriques.* Vienna, 1775.

——. *Schreiben eines Geistlichen zu Wien von dem immer währenden Elektrophor.* Vienna, 1776.

[——.] 'Sur l'électrophore perpétuel de M. Volta.' *JP,* 7 (1776), 501–8.

Jacquot, Jean. 'Sir Charles Cavendish and his Learned Friends. A Contribution to the History of Scientific Relations between England and the Continent in the Earlier Part of the 17th Century.' *Ann. Sci.* 8 (1952), 13–27, 175–91.

——. *Le naturaliste Sir Hans Sloane (1660–1753) et les échanges scientifiques entre la France et l'Angleterre.* Paris, 1953. (Paris. Palais de la Découverte. *Conférences,* series D, no. 25.)

——. 'Sir Hans Sloane and French Men of Science.' RS. *Notes and Records,* 10 (1953), 85–98.

Jallabert, Jean. *Expériences sur l'électricité avec quelques conjectures sur la cause de ses effets.* Paris, 1749.

Jansen, B. 'Die scholastische Philosophie des 17. Jahrhunderts.' Görres-Gesellschaft. *Philosophisches Jahrbuch,* 50 (1937), 401–44.

——. 'Die Pflege der Philosophie im Jesuitenorden während des 17./18. Jahrhunderts.' *Ibid.,* 51 (1938), 172–215, 344–66, 435–56.

Jarrell, Richard A. 'The Latest Date of Composition of Gilbert's *De mundo.*' *Isis,* 63 (1972), 94–5.

Joachim, Johannes. *Die Anfänge der königlichen Sozietät der Wissenschaften zu Göttingen.* Berlin, 1936. (Akademie der Wissenschaft, Göttingen. Math.-Phys. Kl. *Abhandlungen,* 19.)

Jöcher, Christian Gottlieb. *Allgemeines Gelehrten-Lexikon.* 4 vols. Leipzig, 1750–1.
———. *Fortsetzung und Ergänzungen.* Johann Christoph Adelung et al., eds. 7 vols. Leipzig, 1784–1897.
Johnson, Francis Rarick. *Astronomical Thought in Renaissance England.* Baltimore, 1937.
———. 'Gresham College: Precursor of the Royal Society.' *JHI,* 1 (1940), 413–38.
Jourdain, Charles. *Histoire de l'Université de Paris au XVIIe et XVIIIe siècle.* 2 vols. Paris, 1888.
Jurin, James. 'An Account of some Experiments . . . with an Enquiry into the Cause of the Ascent and Suspension of Water in Capillary Tubes.' *PT,* 30 (1717–9), 739–47.
———. 'Disquisitiones physicae de tubulis capillaribus.' *CAS,* 3 (1728), 281–92.
Jussieu, Antoine-Laurent de. 'Notices historiques sur le Muséum d'histoire naturelle.' Paris. Muséum national d'histoire naturelle. *Annales,* 1 (1802), 1–14; 2 (1803), 1–15; 3 (1804), 1–17; 4 (1804), 1–19; 6 (1805), 1–29; 11 (1808), 1–41.

Kästner, Abraham Gotthelf. 'Über die Verbindung der Mathematik und Naturlehre.' [1768]. *Vermischte Schriften,* II, 358–64.
———. *Vermischte Schriften.* 2 parts. Althenburg, 1783^3.
———. *Selbstbiographie und Verzeichnis seiner Schriften. Nebst Heynes Lobrede auf Kästner.* Rudolf Eckart, ed. Hanover, [1909?].
———. *Briefe aus sechs Jahrzehnten, 1745–1800.* Berlin, 1912.
[Kangro, Hans.] 'Zur Geschichte der Physik an der Universität Freiburg i. Br.' Zentgraf, ed. *Geschichte der Naturwissenschaften,* 9–22.
Kargon, Robert Hugh. *Atomism in England from Hariot to Newton.* Oxford, 1966.
Karsten, W. C. G. *Physisch-chemische Abhandlungen durch neuere Schriften von hermetischen Arbeiten und andre neuere Untersuchungen veranlasset.* 2 Hefte. Halle, 1786.
Kauffeldt, Alfons. 'Otto von Guericke on the Problem of Space.' CIHS, 1965. *Actes,* III (1968), 364–8.
———. *Otto von Guericke. Philosophisches über den leeren Raum.* Berlin, 1968.
Keill, John. 'Epistola . . . in qua leges attractionis aliaque physices principia traduntur.' *PT,* 26 (1708), 97–110.
———. *An Introduction to Natural Philosophy; or, Philosophical Lectures read in the University of Oxford, Ann. Dom. 1700.* London, 1720; 1745^4.
Kelly, Sister Suzanne. 'Gilbert's Influence on Bacon: a Revaluation.' *Physis,* 5 (1963), 249–58.
———. *The De Mundo of William Gilbert.* Amsterdam, 1965.
Kernkamp, G. W. *De Utrechtsche Universiteit, 1636–1936. I. De Utrechtsche Academie 1636–1815.* Utrecht, 1936.
Kestler, Johann Stephan. *Physiologia Kircheriana.* Amsterdam, 1680.
Keys, Alice Mapelsden. *Cadwallader Colden.* New York, 1906.
Khell von Khellburg, Joseph. *Physica ex recentiorum observationibus accomodata usibus academicis.* 2 vols. Vienna, 1751.
Kilian, Hermann Friedrich. *Die Universitäten Deutschlands in medicinisch-naturwissenschaftlicher Hinsicht.* Heidelberg and Leipzig, 1828.
King, W. J. 'The Quantification of the Concepts of Electric Charge and Electric Current.' *Natural Philosopher,* 2 (1963), 107–27.
Kink, Rudolf. *Geschichte der kaiserlichen Universität zu Wien.* 2 vols. Vienna, 1854.

Kippis, Andrew. 'Life of the Author.' John Pringle. *Six Discourses*. London, 1783. Pp. i–xcvii.

Kircher, Athanasius. *Magnes sive De arte magnetica libri tres*. Rome, 1641; Cologne, 1643².

———. *Mundus subterraneus in XII libros digestus*. 2 vols. Amsterdam, 1665.

Kirchvogel, Paul Adolf. 'Das astronomisch-physikalische Kabinett des hessischen Landesmuseums.' *Physikalische Blätter*, 9 (1953), 259–63.

Klee, Friedrich. *Die Geschichte der Physik an der Universität Altdorf bis zum Jahre 1650*. Erlangen, 1908.

Klingenstierna, Samuel. *Tal om de nyaste rön vid elektriciteten*. Stockholm, 1755.

Knappert, L. 'Les relations entre Voltaire et 'sGravesande.' *Janus*, 13 (1908), 249–57.

Knight, Gowin. *An Attempt to demonstrate that all the Phaenomena in Nature may be explained by Two Simple Active Principles, Attraction and Repulsion*. London, 1748.

Knight, Isabel F. *The Geometric Spirit. The abbé de Condillac and the French Enlightenment*. New Haven and London, 1968.

Körber, Hans-Günther. 'Die Berliner Instrumentenmacher im 18. Jahrhundert.' CIHS, 1971. Actes, VI (1974), 271–6.

Kohfeldt, Gustav. *Rostocker Professoren und Studenten im 18. Jahrhundert*. Rostock, 1919.

Kopelevich, I. K. 'Perepiska Leonarda Eĭlera i Tobiasa Maĭera.' *Istoriko-astronomicheskie issledovaniia*, 5 (1959), 271–444.

Koppe, Johann Christian. *Jetzlebendes gelehrtes Mecklenburg*. 3 vols. in 1. Rostock and Leipzig, 1783–4.

Koyré, Alexandre. 'An Experiment in Measurement.' American Philosophical Society. *Proceedings*, 97 (1953), 222–37. (Koyré. *Metaphysics and Measurement*, 89–117.)

———, and I. B. Cohen. 'Newton's "Electric and Elastic Spirit".' *Isis*, 51 (1960), 337.

———. 'Les Queries de l'optique.' *AIHS*, 13 (1960), 15–29.

———, and I. B. Cohen. 'Newton and the Leibnitz-Clarke Correspondence.' *AIHS*, 15 (1962), 63–136.

———. *Newtonian Studies*. London, 1965.

———. *Metaphysics and Measurement. Essays in Scientific Revolution*. Cambridge, Mass., 1968.

Krafft, Fritz. 'Experimenta nova. Untersuchungen zur Geschichte eines wissenschaftlichen Buches.' Eberhard Schmauderer, ed. *Buch und Wissenschaft*. Düsseldorf, 1969. Pp. 103–29. *(Technikgeschichte in Einzeldarstellungen, 17.)*

———. 'Sphaera activitatis — orbis virtutis.' *Sudhoffs Archiv*, 54 (1970), 113–40.

Krafft, Georg Wolfgang. 'De calore ac frigore experimenta varia.' *CAS*, 14 (1744–6), 218–39.

Kratzenstein, Christian Gottlieb. *Theoria electricitatis more geometrico explicata*. Halle, 1746.

Krones, Franz von. *Geschichte der Karl-Franzens Universität in Graz*. Graz, 1886.

Kronick, D. A. 'Scientific Journal Publication in the 18th century.' Bibliographical Society of America. *Papers*, 59 (1965), 28–44.

Krüger, Johann Gottlob. *Naturlehre*. 2 vols. Halle, 1740–3; 1771⁵.

————. *Zuschrift an seine Zuhörer worinnen er Ihnen seine Gedancken von der Electricität mittheilet und Ihnen zugleich seine künftige Lektionen bekannt macht.* Halle, 1743; 1745².

————. *Geschichte der Erde.* Halle, 1746.

Krünitz, Johann Georg. *Verzeichniss der vornehmsten Schriften von der Elektricität und den elektrischen Curen.* Leipzig, 1769.

Kubrin, David. 'Newton and the Cyclical Cosmos: Providence and the Mechanical Philosophy.' *JHI,* 28 (1967), 325–46.

Kühn, Karl Gottlob. *Die neuesten Entdeckungen in der physikalischen und medizinischen Elektrizität.* 2 vols. Leipzig, 1796–7.

Kuhn, Thomas S. *The Structure of Scientific Revolutions.* Chicago, 1962.

————. 'Historical Structure of Scientific Discovery.' *Science,* 136 (1962), 760–4.

————. 'Mathematical vs. Experimental Traditions in the Development of Physical Science.' *Journal of Interdisciplinary History,* 7 (1976), 1–31.

Kunik, Ernst-Edward. 'Einleitung.' Wolff. *Briefe* (1860), vii–xxxv.

Kurtz, Gurda H. 'Martinus van Marum, Citizen of Haarlem.' Forbes, ed. *Martinus van Marum.* I, 73–126.

Kusnetzov, G. B. 'Razvitie ucheniia ob elektrichestve v russkoi nauke XVIII v.' IIET. *Trudy,* 19 (1957), 313–85.

Lacepède, B. G. E. de la Ville-sur-Illon, Comte de. 'Mémoire sur les variations des aiguilles aimantées, & sur les boussoles.' *JP,* 15 (1780), 140–3.

————. *Essai sur l'électricité naturelle et artificielle.* 2 vols. Paris, 1781.

Lacoarret, M., and Mme Ter-Menassian. 'Les universités.' Taton, ed. *Enseignement,* 125–68.

Lacroix, Alfred. 'La vie et l'oeuvre de l'abbé René-Just Haüy.' Société française de minéralogie. *Bulletin,* 67 (1944), 15–226.

Laer, Pierre Henri van. *Philosophico-Scientific Problems.* Henry J. Koren, tr. Pittsburgh, 1953. (*Duquesne Studies, Philosophical Series,* 3.)

Lagrange, Joseph Louis. 'Sur la percussion des fluides.' Académie royale des sciences, Turin. *Mémoires,* 1 (1784–5), 95–108. (*Oeuvres,* II, 237–52.)

————. *Oeuvres.* J. A. Serret, ed. 14 vols. Paris, 1867–92.

Laissus, Yves. 'Le Jardin du roi.' Taton, ed. *Enseignement,* 287–341.

Lalande, Joseph Jérôme le Français de. *Voyage d'un français en Italie en 1765–66.* 9 vols. Paris, 1786².

Lallamand, P. J. *Histoire de l'éducation dans l'ancien Oratoire de France.* Paris, 1888.

Lambert, Johann Heinrich. *Photometrie (Photometria, sive De mensura et gradibus luminis, colorum et umbrae, 1760).* E. Anding, tr. 3 vols. Leipzig, 1892. (*Ostwalds Klassiker der exakten Wissenschaften,* nos. 31–3.)

————. 'Analyse de quelques expériences faites sur l'aiman.' *MAS/Ber* (1766), 22–48.

————. 'Sur la courbure du courant magnétique.' *Ibid.,* 49–77.

————. *Deutscher gelehrter Briefwechsel.* Johann Bernoulli, ed. 5 vols. Berlin, 1781–7.

Lamétherie, Jean-Claude de. 'Précis de quelques expériences électriques faites par M. Charles.' *JP,* 30 (1787), 433–6.

Lamontagne, Roland. 'Lettres de Bouguer à Euler.' *RHS,* 19 (1966), 225–46.

Lamprecht, S. P. 'The Role of Descartes in 17th Century England.' Columbia University. Department of Philosophy. *Studies in the History of Ideas.* 3 vols. New York, 1918–35. III, 178–240.

Lana Terzi, Francesco. *Prodromo; overo Saggio di alcune inventioni nuove premesso all'Arte maestra.* Brescia, 1670. (The 'Proemio' to the *Prodromo* is reprinted in M. L. A. Biagi, ed. *Scienziati del seicento.* Milan, 1969. Pp. 711–29.)

——. *Magisterium naturae et artis.* 3 vols. Brescia, 1684–92.

[——]. 'Excerpta actis Novae academiae brixiensis philo-exoticorum naturae et artis.' *Acta eruditorum* (1686), 556–65.

——. 'Nova methodus construendae pyxidis magneticae.' *Ibid.*, 560–2.

Lane, T. 'Description of an Electrometer.' *PT,* 57 (1767), 451–60. (*BFP,* XII, 459–65.)

Langdon-Brown, Sir Walter. *Some Chapters in Cambridge Medical History.* Cambridge, 1946.

Langenmantelius, H. A., ed. *Fasciculus epistolarum Adm. R. P. Athanasii Kircheri.* Augsburg, 1684.

Langevin, Paul. 'La physique au Collège de France.' Paris. Collège de France. *Le Collège de France,* 59–79.

Lankester, Ian. 'A Note on Itinerant Science Lecturers, 1790–1850.' *Ann. Sci.,* 28 (1972), 235–7.

Lantoine, H. E. *Histoire de l'enseignement secondaire en France au XVII⁰ siècle.* Paris, 1874⁴.

Lasswitz, Kurd. *Geschichte der Atomistik vom Mittelalter bis Newton.* 2 vols. Hamburg and Leipzig, 1890.

Laudan, Laurens, 'The Idea of a Physical Theory from Galileo to Newton: Studies in 17th-Century Methodology.' Ph.D. thesis. Princeton, 1966.

——. 'The Clock Metaphor and Probabilism: The Impact of Descartes on English Methodological Thought, 1650–65.' *Ann. Sci.,* 22 (1966), 73–104.

——. 'Comment [on a paper by G. Buchdahl].' Roger H. Stuewer, ed. *Historical and Philosophical Perspectives of Science.* Minneapolis, 1970. Pp. 230–8. (*Minnesota Studies in the Philosophy of Science,* 5.)

Lavèn, W. J., and J. G. van Cittert-Eymers. *Electrostatical Instruments in the Utrecht University Museum.* Utrecht, 1967.

Lavoisier, A. L., and P. S. Laplace. 'Mémoire sur la chaleur.' *MAS* (1780), 355–408.

——. 'Sur l'électricité qu'absorbent les corps qui se réduisent en vapeurs.' *MAS* (1781), 292–4.

Lecot, Victor Lucien Sulpice. *L'abbé Nollet de Pimprez.* Noyon, 1856.

Le Dru, Fils, ——. 'Lettre sur quelques expériences de M. Marat.' *JP,* 18 (1781), 402–4.

Lee, Duncan Campbell. *Desaguliers of No. 4 and his Services to Free-Masonry.* London, 1932.

Leeuwen, Henry G. van. *The Problem of Certainty in English Thought 1630–1690.* The Hague, 1970². (*International Archives of the History of Ideas,* 3.)

Lefranc, Abel. *Histoire du Collège de France.* Paris, 1893.

Leibniz, Gottfried Wilhelm, freiherr von. *Die philosophischen Schriften.* C. I. Gerhardt, ed. 7 vols. Berlin, 1875–90.

————. *Hypothesis physica nova*. Mainz, 1671. (*Philosophischen Schriften*, IV, 177–219.)

————. *Briefwechsel zwischen Leibniz und Christian Wolf*. C. I. Gerhardt, ed. Halle, 1860.

————. *Sämtliche Schriften und Briefe*. Ser. 2, I: *Philosophischer Briefwechsel*. Darmstadt, 1926.

Leicester, Henry M. *Mikhail Vasil'evich Lomonosov on the Corpuscular Theory*. Cambridge, Mass., 1970.

Leide, Arvid. *Fysiska institutionen vid Lunds universitet*. Lund, 1968. (Lund. University. *Acta*. Sectio 1: *Theologica, juridica, humaniora*, no. 8.)

Leipzig. University. 'Das physikalische und das theoretisch-physikalische Institut.' *Festschrift zur Feier des 500 jährigen Bestehens*. 4 vols. Leipzig, 1909. IV:2, 24–69.

Lelarge de Lignac, Joseph Adrien. *Lettres à un amériquain sur l'histoire naturelle générale et particulière de Mr. de Buffon*. 5 vols. in 3. Hamburg, 1751–6.

Lemay, Joseph A. Leo. 'Franklin and Kinnersley.' *Isis*, 52 (1961), 575–81.

————. 'Franklin's "Dr. Spence": The Reverend Archibald Spencer (169?-1760), M. D.' *Maryland Historical Magazine*, 59 (1964), 199–216.

————. *Ebenezer Kinnersley, Franklin's Friend*. Philadelphia, 1964.

Le Monnier, Louis-Guillaume. 'Recherches sur la communication de l'électricité.' *MAS* (1746), 447–64.

————. 'Extract of a Memoir concerning the Communication of Electricity.' *PT*, 44:1 (1746-7), 290–5.

————. 'Observations sur l'électricité.' *MAS* (1752), 233–43.

Lenoble, Robert. *Mersenne; ou, La naissance du mécanisme*. Paris, 1943.

Léotaud, Vincent. *Magnetologia in qua exponitur nova de magnetis philosophia*. Lyon, 1668.

Lépine, Pierre, and Jacques Nicolle. *Sir Henry Cavendish ou L'homme qui a pesé la terre*. Paris, 1964.

Le Roy, Georges. 'Introduction.' Etienne Bonnot de Condillac. *Lettres inédites à Gabriel Cramer*. Paris, 1953. Pp. 1–29.

Le Roy, Jean Baptiste. 'Mémoire sur l'électricité. [I] Où l'on montre . . . qu'il y a deux espèces d'électricités . . . & qu'elles ont chacune des phénomènes particuliers qui les caractérisent parfaitement.' *MAS* (1753), 447–59.
[II] 'Où l'on rapporte les expériences qui confirment l'existence des deux électricités *par condensation & par raréfaction*.' *Ibid.*, 459–68.
[III] 'Supplément au mémoire précédent, où l'on fait voir . . . que tous les feux que l'on observe aux extrémités des corps présentés à ceux qui sont électrisés *par condensation* . . . sont formés *par l'entrée du fluide ou feu électrique dans ces corps, & non par sa sortie.*' *Ibid.*, 468–74.

————. 'Mémoire sur l'électricité résineuse, où l'on montre qu'elle est réelement distincte de *l'électricité vitrée*.' *MAS* (1755), 264–83.

————. 'Mémoire . . . sur la différence des distances auxquelles partent les étincelles.' *MAS* (1766), 541–6.

————. 'Mémoire sur les verges ou barres métalliques destinées à garantir les édifices des effets de la foudre.' *MAS* (1770), 53–67.

———. 'Mémoire sur une machine à électriser d'une espèce nouvelle.' *MAS* (1772), 499–512.

———. 'Mémoire sur la forme des barres ou des conducteurs métalliques, destinés à préserver les édifices des effets de la foudre.' *MAS* (1773), 671–86.

[———]. 'Electromètre.' *Encyclopédie, ou dictionnaire raisonné.* Geneva, 1778³. XII, 52–60.

Leupold, W. *Die Aristotelische Lehre in Molières Werken.* Berlin, 1935. (*Romanische Studien.* E. Ebering, ed. No. 38.)

Levere, T. H. 'Martinus van Marum and the Introduction of Lavoisier's Chemistry in the Netherlands.' Forbes, ed. *Martinus van Marum.* I, 158–286.

———. 'Friendship and Influence. Martinus van Marum, F.R.S.' RS, *Notes and Records,* 25 (1970), 113–20.

Libanori, Antonio. *Ferrara d'oro imbrunito.* 3 parts. Ferrara, 1665–74.

Lichtenberg, Georg Christoph. 'De nova methodo naturam ac motum fluidi electrici investigandi.' Akademie der Wissenschaften, Göttingen. Phys.-Math. Kl. *Novi commentarii,* 8 (1777), 168–80. (German tr.: Lichtenberg. *Über eine neue Methode,* 17–28.)

———. 'Super nova methodo motum ac naturam fluidi electrici investigandi.' Akademie der Wissenschaften, Göttingen. Math. Kl. *Novi commentarii,* 1 (1778), 65–79. (German tr.: Lichtenberg. *Über eine neue Methode,* 31–45.)

———. *Briefe.* Albert Leitzmann and Carl Schüddekopf, eds. 3 vols. Leipzig, 1901–4.

———. *Aphorismen.* Albert Leitzmann, ed. 5 vols. Berlin and Leipzig, 1902–8.

———. *Über eine neue Methode, die Natur und die Bewegung der elektrischen Materie zu erforschen.* Herbert Pupke, ed. Leipzig, 1956. (*Ostwalds Klassiker der exakten Wissenschaften,* no. 246.)

———. 'Johann Heinrich Lambert.' Steck. *Bibliographia lambertiana,* vii–xiii.

Lilley, S. 'Nicholson's Journal, 1797–1813.' *Ann. Sci.,* 6 (1948), 78–101.

Lindborg, Rolf. *Descartes i Uppsala. Striderna om 'nya filosofien' 1663–1689.* Stockholm, 1965. (*Lychnos-Bibliotek,* 22.)

Lindroth, Sten. *Kungl. Svenska Vetenskapsakademiens historia 1739–1818.* 2 vols in 3. Stockholm, 1967.

———. 'Svensk-ryska vetenskapliga förbindelser under 1700-talet.' *Nordisk tidskrift för vetenskap, konst och industri,* 43 (1967), 24–48.

Liphardt, [J.Ch.L.?]. 'Mémoire sur l'électricité du chocolat et quelques objets relatifs.' *JP,* 30 (1787), 431–3.

Lipski, A. 'The Foundation of the Russian Academy.' *Isis,* 44 (1953), 349–54.

Lissa, Giuseppe. *Cartesianismo e anticartesianismo in Fontenelle.* Naples, 1971. (Università degli studi, Salerno. *Collana di studi e testi,* 6.)

Locke, John. *An Essay Concerning Human Understanding* [1690]. Alexander Campbell Fraser, ed. 2 vols. Oxford, 1894.

Loeb, Leonard Benedict. *Fundamental Processes of Electrical Discharges in Gases.* New York, 1939.

———. *Electrical Coronas.* Berkeley and Los Angeles, 1965.

Lohne, J. 'Newton's "Proof" of the Sine Law and his Mathematical Theory of Colours.' *Archive for History of Exact Sciences,* 1 (1961), 389–406.

Lomonosov, Mikhail Vasil'evich. 'Oratio de meteoris vi electrica ortis [et] Orationis de meteoris electricis explanationes' [1753]. *Poln. sobr. soch.* III, 15–133. *Ausg. Schr.* I, 238–304.

——. 'Theoria electricitatis methodo mathematica concinnata' [1756]. *Poln. sobr. soch.* III, 265–313. *Ausg. Schr.* I, 313–28.

——. '127 zametok k teorii sveta i elektrichestva' [1756]. *Poln. sobr. soch.* III, 237–63. *Ausg. Schr.* I, 305–12 (excerpted).

——. 'De ratione quantitatis materiae et ponderis' [1758]. *Poln. sobr. soch.* III, 349–71. (Leicester. *Lomonosov*, 224–32.)

——. *Polnoe sobranie sochinenii.* 10 vols. Moscow and Leningrad, 1951–9.

——. *Ausgewählte Schriften.* 2 vols. Berlin, 1961.

Lorey, Wilhelm. 'Die Physik an der Universität Giessen im 17. und 18. Jahrhundert.' Giessener Hochschulgesellschaft. *Nachrichten,* 14 (1940), 14–39.

——. 'Die Physik an der Universität Giessen im 19. Jahrhundert.' *Ibid.,* 15 (1941), 80–132.

Lorraine de Vallemont, Pierre. *La physique occulte, ou Traité de la baguette divinatoire.* Paris, 1693; 1709².

Lovett, Richard. *Sir Isaac Newton's Aether Realized; or the Second Part of the Subtil Medium Prov'd, and Electricity Rendered Useful.* London, 1759.

Louyat, Henri. 'Emmanuel Maignon (1601–1626), un religieux toulousain, mathématicien, astronome, philosophe.' 95. Congrès des Sociétes Savantes (1971). Section des Sciences. *Comptes rendus.* Paris, 1974. I, 15–29.

Lullin, Amadeus, respondent. *Dissertatio physica de electricitate.* Geneva, 1766. (H. B. de Saussure, praeses.)

Lungo, Carlo del. 'Per la storia dell'Accademia del Cimento: Una lettera del Card. Leopoldo al senese cav. Lodovico dei Vecchi.' *Archivio storico italiano,* 76 (1918), 109–19.

Luynes, Charles Philippe d'Albert, Duc de. *Mémoires du Duc de Luynes sur la cour de Louis XV (1735–1758).* L. Dussieux and E. Soulié, eds. 17 vols. Paris, 1860–5.

Lyon, John. *Experiments and Observations made with a View to point out the Errors of the present Received Theory of Electricity.* London, 1780.

McClellan, James Edward, III. *The International Organization of Science and Learned Societies in the 18th Century.* Ph.D. thesis. Princeton, 1975.

McCormmach, Russell K. *The Electrical Researches of Henry Cavendish.* Ph.D. dissertation. Case Institute of Technology, 1967.

——. 'John Michell and Henry Cavendish: Weighing the Stars.' *BJHS,* 4 (1968), 126–55.

——. 'Henry Cavendish: A Study of Rational Empiricism in 18th-Century Natural Philosophy.' *Isis,* 60 (1969), 293–306.

McDonald, John F. 'Properties and Causes: An Approach to the Problem of Hypotheses in the Scientific Methodology of Sir Isaac Newton.' *Ann. Sci.,* 28 (1972), 217–33.

McGuire, J. E. 'Body and Void and Newton's *De mundi systemate:* Some New Sources.' *Archive for History of Exact Sciences,* 3 (1966), 206–48.

——. 'Transmutation and Immutability: Newton's Doctrine of Physical Qualities.' *Ambix,* 14 (1967), 69–95.

——. 'Force, Active Principles and Newton's Invisible Realm.' *Ambix,* 15 (1968), 154–208.

——. 'The Origin of Newton's Doctrine of Essential Qualities.' *Centaurus,* 12 (1968), 233–60.

542   *Bibliography*

————. 'Boyle's Conception of Nature.' *JHI*, 33 (1972), 523–42.

McKie, Douglas. 'The Honorable Robert Boyle's *Essays of Effluviums* (1673).' *Science Progress*, 29 (1934), 253–65.

————, and N. H. de V. Heathcote. *The Discovery of Specific and Latent Heats*. London, 1935.

————. 'Mr. Warltire, a good Chymist.' *Endeavor*, 10 (1951), 46–9.

Mackie, J. D. *The University of Glasgow, 1451–1951*. Glasgow, 1954.

Maclaurin, Colin. *An Account of Sir Isaac Newton's Philosophical Discoveries*. London, 1748; 1750².

Madeira Arrais, Duarte. *Novae philosophiae et medicinae de qualitatibus occultis a nemine unquam excultae pars prima philosophis, et medicis, pernecessaria, theologis vero apprime utilis*. 2 vols. Lisbon, 1650.

[Magalotti, Lorenzo, Conte]. *Saggi dell'esperienze naturali fatte nell'Accademia del Cimento*. Florence, 1667; 1841³.

[————]. *Essayes of Natural Experiments Made in the Academie del Cimento Under the Protection of His Most Serene Prince Leopold of Tuscany*. R. Waller, tr. London, 1684. (Reprinted, New York, 1964, with introduction by A. R. Hall.)

————. *Lettere scientifiche ed erudite*. Venice, 1721; 1740².

Magrini, Silvio. 'Il *De magnete* del Gilbert e i primordi della magnetologia in Italia alla lotta intorno ai massimi sistemi.' *Archivio di storia della scienza*, 8 (1927), 17–39.

Maheu, Gilles. 'Introduction à la publication des lettres de Bouguer à Euler.' *RHS*, 19 (1966), 206–24.

Mahieu, l'abbé Léon. *François Suarez, sa philosophie et les rapports qu'elle a avec sa théologie*. Paris, 1921.

Maignan, Emanuel. *Perspectiva horaria*. Rome, 1648.

————. *Cursus philosophicus*. 4 vols. Toulouse, 1653; Lyon, 1673².

Maindron, Ernest. 'Les fondations de prix à l'Académie des sciences (1714–1880).' *Revue scientifique*, 18 (1880), 1107–17.

————. *L'Académie des sciences*. Paris, 1888.

Mairan, Jean-Baptiste Dortous de. 'Eloge de Molières.' *HAS* (1742), 195–205.

Malebranche, Nicolas. *Oeuvres complètes*. André Robinet, ed. 20 vols. in 18. Paris, 1958–68.

————. 'Reflexions sur la lumière et les couleurs et la generation du feu.' *MAS* (1699), 22–36. (Reprinted, with additions, in the 'XVI. Eclaircissement' to the *Recherche de la vérité*, in *Oeuvres*, III, 255–69.)

Mallet, Sir Charles Edward. *A History of the University of Oxford*. 3 vols. London, 1924–7.

Mandelbaum, Maurice. *Philosophy, Science and Sense Perception: Historical and Critical Studies*. Baltimore, 1964.

Manegold, K. H., ed. *Wissenschaft, Wirtschaft und Technik*. Munich, 1969.

[Mangin, l'abbé de]. *Histoire générale et particulière de l'électricité*. Paris, 1752.

Marais, Mathieu. *Journal et mémoires . . . sur la régence et le règne de Louis XV. (1715–1737.)* M. de Lescure, ed. 4 vols. Paris, 1863–8.

Marat, Jean Paul. *Découvertes sur le feu, l'électricité et la lumière, constatées par une suite d'expériences nouvelles*. Paris, 1779².

[————]. 'Détail des découvertes de M. Marat, sur l'électricité.' *JP*, 17 (1781), 317–20, 459–65.

——. *Recherches physiques sur l'électricité*. Paris, 1782.

——. *La correspondance*. Charles Vellay, ed. Paris, 1908.

Marcović, Zeljko. 'Boscovich's *Theoria*.' Whyte, ed. *Roger Joseph Boscovich*, 127–52.

Marek, J. 'Un physicien tchèque du XVIIe siècle: Joannes Marcus Marci de Kronland (1595–1667).' *RHS*, 21 (1968), 109–30.

——. 'Zu der Entwicklung der Physik im postrudolphinischen Prag.' *Bohemia*, 16 (1975), 98–109.

Marini, Tommaso. 'De electricitate caelesti, sive, ut alii vocant, naturali.' Accademia delle scienze dell'Istituto di Bologna. *Commentarii et opuscula*, 3 (1755), 205–16.

Marsak, Leonard M. 'Bernard de Fontenelle: The Idea of Science in the French Enlightenment.' American Philosophical Society. *Transactions*, 49:7 (1959). 64 p.

Mårtensson, Hans [Johannes Mortensson], respondent. *Dissertatio physica de electricitate*. Uppsala, 1740. (Samuel Klingenstierna, praeses.)

——. *Dissertatio physica gradualis de electricitate*. Uppsala, 1742.

Martin, Benjamin. *A Course of Lectures in Natural and Experimental Philosophy, Geography and Astronomy*. Reading, 1743.

——. *An Essay on Electricity: being an Enquiry into the Nature, Cause and Properties thereof, on the Principles of Sir Isaac Newton's Theory of Vibrating Motion, Light and Fire*. Bath, 1746.

——. *A Supplement: containing Remarks on a Rhapsody of Adventures of a Modern Knight-Errant in Philosophy*. Bath, 1746.

Martin, Geneviève. 'Documents de l'Académie de Rouen concernant l'enseignement des sciences au XVIIIe siècle.' *RHS*, 11 (1958), 207–26.

Martin, T. H. 'Observations et théories des anciens sur les attractions et les répulsions magnétiques et sur les attractions électriques.' Pontificia accademia delle scienze, Rome. *Atti*, 18 (1865), 17–32, 97–123.

Martine, George. *An Examination of the Newtonian Argument for the Emptiness of Space and of the Resistance of Subtle Fluids*. London, 1740.

Martini, Angelo. *Manuale di metrologia, ossia misure, pesi e monete*. Turin, 1883.

Mascart, Eleuthère-Elie-Nicolas. *Traité d'électricité statique*. 2 vols. Paris, 1876.

Massardi, F. 'Concordanza di risultati e formule emergenti da manoscritti inediti del Volta con quelli ricavati dalla fisico-matematica nella risoluzione del problema generale dell'elettrostatica.' Istituto lombardo di scienze e lettere, Milan. *Rendiconti*, 56 (1923), 293–308.

——. 'Sull'importanza dei concetti fondamentali esposti dal Volta nel 1769 nella sua prima memoria scientifica, *De vi attractiva ignis electrici*.' *Ibid.*, 59 (1926), 373–81.

Matignon, Camille. 'La chimie générale au Collège de France.' Paris. Collège de France. *Le Collège de France*, 81–103.

Maupertuis, Pierre Louis Moreau de. 'Sur les loix de l'attraction.' *MAS* (1732), 343–62.

——. *Oeuvres*. 4 vols. Lyon, 1756.

Mautner, Franz H., and Franklin Miller, Jr. 'Remarks on G. C. Lichtenberg, Humanist-Scientist.' *Isis*, 43 (1952), 223–31.

Mayer, Johann Tobias. *Anfangsgründe der Naturlehre zum Behulf der Vorlesungen über die Experimental-Physik*. Göttingen, 1801; 1804^2; 1812^3.

Mayer, Tobias. *The Unpublished Writings of Tobias Mayer*. III. *The Theory of the*

Magnet and its Application to Terrestrial Magnetism. Eric G. Forbes, ed. Göttingen, 1972. (Göttingen. Niedersächsische Staats- und Universitätsbibliothek. *Arbeiten*, 11.)

Mayor, J. E. B., ed. *Cambridge under Queen Anne.* Cambridge, 1911.

Mazéas, Guillaume. 'Concerning the Success of the Late Experiments in France.' *PT,* 47 (1752), 534–52.

Mazières, Jean Simon. *Traité des petits tourbillons de la matière subtile.* Paris, 1727.

Mazzuchelli, Giammaria. 'Notizie intorno alla vita ed agli scritti del padre Francesco Terzi Lana.' *Nuova raccolta d'opuscoli scientifici e filologici,* 40 (Venice, 1784), 1–132.

Medici, Michele. *Memorie storiche intorno le accademie scientifiche e letterarie della città di Bologna.* Bologna, 1852.

Meister, Richard. 'Geschichte des Doktorates der Philosophie an der Universität Wien.' Akademie der Wissenschaften, Vienna. Phil.-Hist. Kl. *Sitzungsberichte*, 232:2 (1958). 158 p.

Mélanges Alexandre Koyré, publiés à l'occasion de son soixante-dixième anniversaire. 2 vols. Paris, 1964.

Melsens, Louis Henri Frédéric. *Des paratonnerres.* Brussels, 1877.

Mendham, Joseph. *The Literary Policy of the Church of Rome Exhibited.* London, 1830².

Mercati, A. 'Il fisico tedesco Giorgio Matteo Bose e Benedetto XIV.' Pontificia accademia delle scienze, Rome. *Acta,* 15 (1952), 57–70.

Mercet, L. 'La maison des Italiens au grand Montreuil.' *Revue de l'histoire de Versailles et de Seine-et-Oise,* 28 (1926), 77–105, 168–90, 245–63.

Mercier, Barthélemy. *Notice raisonnée des ouvrages de Gaspar Schott.* Paris, 1785.

Merland, Constant. 'Mathurin-Jacques Brisson.' *Biographies vendéennes,* II. Nantes, 1883, Pp. 1–47.

Mersenne, Marin. *Correspondance du P. Marin Mersenne.* M. Tannery and C. de Waard, eds. 12 vols. Paris, 1932–72.

Metzger, Hélène. *Newton, Stahl, Boerhaave et la doctrine chimique.* Paris, 1930.

———. 'La théorie du feu d'après Boerhaave.' *Revue philosophique,* 109 (1930), 253–85.

Metzler, J. 'Der apostolische Vikar Nikolaus Steno und die Jesuiten.' *Archivum historicum Societatis Jesu,* 10 (1941), 93–152.

Meyer, Friedrich Albert. 'Petrus van Musschenbroek. Werden und Werk und seine Beziehungen zu Daniel Gabriel Fahrenheit.' *Duisburger Forschungen,* 5 (1961), 1–51.

Michell, John. *A Treatise of Artificial Magnets.* London, 1750.

———. *Traité sur les aimans artificiels.* A. Rivoire, tr. Paris, 1752.

Middleton, William Edgar Knowles. *The History of the Barometer.* Baltimore, 1964.

———. *A History of the Thermometer and its Use in Meteorology.* Baltimore, 1966.

———. *Invention of the Meteorological Instruments.* Baltimore, 1969.

———. 'The Title of the Saggi.' *BJHS,* 4 (1969), 283–6.

———. *The Experimenters. A Study of the Accademia del Cimento.* Baltimore, 1971.

———. 'Science in Rome, 1675–1700, and the Accademia Fisicomatematica of Giovanni Giustino Ciampini.' *BJHS,* 8 (1975), 138–54.

Middlington, E. C. 'Theories of Cohesion in the 17th Century.' *Ann. Sci.*, 5 (1945), 253–69.

——. 'Studies in Capillarity and Cohesion in the 18th Century.' *Ibid.* 7 (1947), 352–69.

Miles, Henry. 'Firing Phosphorus by Electricity.' *PT*, 43 (1744–5), 290–3.

——. 'Observations of Luminous Emanations from human Bodies and from Brutes; with some Remarks on Electricity.' *Ibid*, 441–6.

——. 'The Effects of a Cane of black Sealingwax, and a Cane of Brimstone, in Electrical Experiments.' *PT*, 44 (1746), 27–32.

——. 'Several Electrical Experiments.' *Ibid.*, 53–7.

——. 'Some Electrical Observations.' *Ibid.*, 158–62.

Millburn, John R. 'Benjamin Martin and the Royal Society.' RS. *Notes and Records*, 28 (1973), 15–23.

——. 'Benjamin Martin and the Development of the Orrery.' *BJHS*, 6 (1973), 378–99.

——. *Benjamin Martin, Author, Instrument-Maker, and 'Country Showman'*. Leyden, 1976.

Miller, George. 'Observations on the Theory of Electric Attractions and Repulsions.' *Journal of Natural Philosophy, Chemistry and the Arts*, 4 (1800–1), 461–6.

Milly, Nicolas-Christian de Thy, Comte de. 'Une aiguille de boussole indestructible par l'action des acides, & sur un moyen de diminuer la variation de l'aiguille aimantée.' *JP*, 13 (1779), 391–4.

Milner, Thomas. *Experiments and Observations in Electricity*. London, 1783.

Minchenko, L. S. 'Ob odnoï dissertatsii Leonarda Eïlera po elektrichestvu.' IIET. *Trudy*, 28 (1959), 188–200.

Mitchell, John. 'The Force of Electrical Cohesion.' *PT*, 51 (1759), 390–3.

Molières, Joseph Privat de. *Leçons de physique*. 4 vols. Paris, 1734–9.

Moll, Gerard. 'A Biographical Account of J.H. van Swinden.' *Edinburgh Journal of Science*, 1 (1824), 197–208.

Monchamp, Georges. *Histoire du cartésianisme en Belgique*. Brussels, 1886.

Monconys, Balthasar de. *Voyages*. Lyon, 1665–6; Paris, 1695².

Mondini, Alberto. 'Un precursore dimenticato. Tiberio Cavallo.' *Machine*, 2 (1969–70), 19–26.

Montandon, Cléopâtre. *Le développement de la science à Genève aux XVIIIe et XIXe siècles. Le cas d'un communauté scientifique*. Vevey, 1975.

Montesquieu, Charles Louis de Secondat, Baron de la Brède et de. *Oeuvres complètes*. E. Laboulaye, ed. 7 vols. Paris, 1875–9.

Montucla, Jean Etienne. *Histoire des mathématiques*. 4 vols. Paris, an VII².

Montzey, Charles de. *Histoire de la Flèche et de ses seigneurs*. 3 vols. Le Mans and Paris, 1877.

Moore, Norman. *The History of St. Bartholomew's Hospital*. 2 vols. London, 1918.

Mor, Carlo Guido. *Storia della Università di Modena*. Modena, 1953.

More, Louis Trenchard. *Isaac Newton*. New York, 1934.

Morey, George Washington. *The Properties of Glass*. New York, 1938. (American Chemical Society. *Monograph Series*, no. 77.)

Morgan, Alexander. *Scottish University Studies*. Oxford, 1933.

Morgan, George Cadogan. *Lectures on Electricity.* 2 vols. Norwich, 1794.

Morgan, William, 'Electrical Experiments made in order to ascertain the non-conducting Power of a perfect Vacuum.' *PT,* 75 (1785), 272–8.

Morhof, Daniel Georg. *Polyhistor literarius, philosophicus, et practicus.* 2 vols. Lübeck, 1747⁴.

Morin, Jean. *Nouvelle dissertation sur l'électricité des corps.* Chartres, 1748.

Mornet, Daniel. *Les origines intellectuelles de la révolution française.* Paris, 1947.

Morozov, Aleksandr Antonovich. *Michail Wassiljewitsch Lomonossow, 1711–1765.* Berlin, 1954.

Morrell, J. B. 'The University of Edinburgh in the Late 18th Century: Its Scientific Eminence and Academic Structure.' *Isis,* 62 (1971), 158–71.

Moscovici, Serge. *L'expérience du mouvement. Jean-Baptiste Baliani, disciple et critique de Galilée.* Paris, 1967. (*Histoire de la pensée,* 16.)

Mottelay, Paul Fleury. 'Biographical Memoir.' Gilbert. *De magnete* (1893), ix–xxvii.

———. *Bibliographical History of Electricity and Magnetism Chronologically Arranged.* London, 1922.

Mousnerius, Petrus, *pseud. See* Fabri, Honoré.

Mouy, Paul. *Le développement de la physique cartésienne, 1646–1712.* Paris, 1934.

Müller, Conrad Heinrich. 'Studien zur Geschichte der Mathematik an der Universität Göttingen.' *Abhandlungen zur Geschichte der mathematischen Wissenschaften,* 18 (1904), 51–143.

Müller-Hillebrand, D. 'Åskan och kyrkan.' *Tidskrift för Pastorats förvaltning,* 14:7 (1960), 3–16.

———. 'Torbern Bergman as a Lightning Scientist.' *Daedalus; Tekniska museets årsbok* (1963), 35–76.

Mumford, Alfred Alexander. *The Manchester Grammar School, 1515–1915.* London, 1919.

Muntendam, Alida M. 'Dr. Martinus van Marum (1750–1837).' Forbes, ed. *Martinus van Marum.* I, 1–72.

Murray, David. *Memories of the Old College of Glasgow.* Glasgow, 1927.

Musgrave, Samuel. 'Reasons for dissenting from the Report of the Committee to consider of Mr. Wilson's Experiments, including some remarks on some Experiments exhibited by Mr. Nairne.' *PT,* 68 (1778), 801–22.

Musschenbroek, Pieter van. 'De viribus magneticis.' *PT,* 33 (1724-5), 370–7.

———. *Epitome elementorum physico-mathematicorum conscripta in usus academicos.* Leyden, 1726.

———. *Physicae experimentales, et geometricae, de magnete, tuborum cappillarium vitreorumque speculorum attractione, magnitudine terrae, coherentia corporum firmorum dissertationes.* Leyden, 1729.

———. *Tentamina experimentorum naturalium captorum in Academia del Cimento . . . Quibus commentarios, nova experimenta, et orationem de methodo instituendi experimenta physica addidit.* 2 vols. Leyden, 1731. (French tr.: *Collection académique.* 1. Dijon, 1775.)

———. *Elementa physicae conscripta in usus academicos.* Leyden, 1734; 1741². (The second and third editions of *Epitome elementorum.*)

———. *Essai de physique . . . avec une Description de nouvelles sortes de machines pneumatiques et un recueil d'expériences par Mr. J. V. M.* Pierre Massuet, tr. 2 vols. Leyden, 1739; 1751².

———. *The Elements of Natural Philosophy.* John Colson, tr. 2 vols. London, 1744.

———. *Grundlehren der Naturwissenschaft.* Leipzig, 1747. (Tr. of *Elementa physicae,* 1734.)

———. *Institutiones physicae, conscriptae in usus academicos.* Leyden, 1748. (A later version of *Epitome elementorum.*)

———. *Dissertatio physica experimentalis de magnete.* Vienna, 1754².

———. *Introductio ad philosophiam naturalem.* 2 vols. Leyden, 1762.

———. *Cours de physique expérimentale et mathématique.* J. A. Sigaud de la Fond, tr. 3 vols. Paris, 1769. (Tr. of *Introductio* [1762].)

Musson, Albert Eduard, and Eric Robinson. *Science and Technology in the Industrial Revolution.* Toronto, 1969.

Mylius, Christlob. 'Extract of a Letter . . . to Mr. W. Watson.' *PT,* 47 (1752), 559.

Nairne, Edward. *Directions for using the Electrical Machine as made and sold by Edward Nairne.* [London], 1773.

———. 'Electrical Experiments made with a Machine of his own Workmanship; a Description of which is prefixed.' *PT,* 64 (1774), 79–89.

———. 'An Account of some Experiments made with an Air-pump on Mr. Smeaton's Principle; together with some Experiments with a common Air-pump.' *PT,* 67 (1777), 614–48.

———. 'Experiments on Electricity, being an Attempt to shew the Advantage of elevated pointed Conductors.' *PT,* 68 (1778), 823–60.

———. *Description and Use of Nairne's Patent Electrical Machine.* London, 1793⁴.

———. *See* Caullet de Veaumorel, L.

Nauck, E. T. 'Die Vertretung der Naturwissenschaften durch Freiburger Medizinprofessoren.' *Beiträge zur Freiburger Wissenschafts- und Universitätsgeschichte,* no. 4 (1954).

Neale, John. *Directions for Gentlemen, who have Electrical Machines, How to proceed in making their Experiments.* London, 1747.

Neave, E. W. J. 'Chemistry in Rozier's Journal.' *Ann. Sci.,* 6 (1950), 416–21; 7 (1951), 101–6, 144–8, 284–99, 393–400; 8 (1952), 28–45.

Nedeljković, Dusan. *La philosophie naturelle et relativiste de R. J. Boscovich.* Paris, 1922.

Needham, J. T. 'Some new Electrical Experiments lately made at Paris.' *PT,* 44 (1746), 247–63.

———. 'Recherches sur la question: si le son des cloches, pendant les orages, fait éclater la foudre en la faisant descendre sur le clocher.' Académie royale des sciences, des lettres et des beaux-arts de Belgique. *Mémoires,* 4 (1783), 59–72.

———. 'Recherches sur les moyens les plus efficaces d'empêcher le dérangement produit souvent dans la direction naturelle des aiguilles aimantées, par l'électricité de l'atmosphère.' *Ibid.,* 73–87.

Nelli, Giovambatista Clemente. *Saggio di storia letteraria fiorentina del secolo XVII.* Lucca, 1759.

Neumann, Caspar. *Lectiones publicae von vier subjectis pharmaceuticis, nehmlich von succino, caryophyllis, aromaticis und castoreo.* Berlin, 1730.

Newman, Aubrey. *The Stanhopes of Chevening: A Family Biography.* London and New York, 1969.

Newton, Isaac. *Philosophiae naturalis principia mathematica.* Thomas Le Seur and François Jacquier, eds. 3 vols. Geneva, 1739–42.

———. *Mathematical Principles of Natural Philosophy.* A. Motte, tr.; F. Cajori, ed. Berkeley, 1934.

———. *Opticks.* New York, 1952. (Reprint of London, 1730[4].)

———. *Isaac Newton's Papers and Letters on Natural Philosophy.* I. B. Cohen, ed. Cambridge, Mass., 1958.

———. *Correspondence.* H. W. Turnbull et al., eds. 6 vols. Cambridge, 1959–76.

Neyreneuf, F. V. 'Expériences sur le condensateur d'Aepinus.' *JP*, 1 (1872), 62–3.

Nicéron, Jean Pierre, ed. *Mémoires pour servir à l'histoire des hommes illustres dans la république des lettres, avec un catalogue raisonné de leurs ouvrages.* 43 vols. in 44. Paris, 1729–45.

Nicholson, William. *An Introduction to Natural Philosophy.* 2 vols. London, 1782; 1790[3]; 1797[4].

———. 'A Description of an Instrument which, by the turning of a Winch, produces the Two States of Electricity without Friction or Communication with the Earth.' *PT*, 78 (1788), 403–7.

———. 'Experiments and Observations on Electricity.' *PT*, 79 (1789), 265–88.

———. 'A Comparison between Electrical Machines with a Cylinder, and those which produce their Effect by means of a Circular Plate of Glass. With a Description of a Machine of great Simplicity and Power, invented by Dr. Martinus van Marum.' *Journal of Natural Philosophy, Chemistry and the Arts*, 1 (1797), 83–8.

Nollet, Jean Antoine. *Programme, ou, Idée générale d'un cours de physique expérimentale, avec un catalogue raisonné des instrumens qui servent aux expériences.* Paris, 1738.

———. *Leçons de physique expérimentale.* 6 vols. Paris, 1743–8. (Often reprinted.)

———. 'Discours sur les dispositions et sur les qualités qu'il faut avoir pour faire du progrès dans l'étude de la physique expérimentale.' *Leçons de physique expérimentale.* I, xlv–xciv.

———. 'Conjectures sur l'électricité des corps.' *MAS* (1745), 107–51.

———. 'Observations sur quelques nouveaux phénomènes d'électricité.' *MAS* (1746), 1–23.

———. *Essai sur l'électricité des corps.* Paris, 1746; 1750[2].

———. 'Eclaircissemens sur plusieurs faits concernant l'électricité.' *MAS* (1747), 102–31.

'Second Mémoire. Des circonstances favorables ou nuisibles à l'électricité.' *Ibid.*, 149–99.

'Troisième Mémoire, dans lequel on examine 1°. si l'électricité se communique en raison des masses, ou en raison des surfaces; 2°. si une certaine figure, ou certains dimensions du corps électrisé, peuvent contribuer à rendre sa vertue plus sensible; 3°. si l'électrification qui dure long-temps, ou qui est souvent répétée sur la même quantité de matière, peut en altérer les qualités ou en diminuer la masse.' *Ibid.*, 207–42.

'Quatrième Mémoire. Des effets de la vertu électrique sur les corps organisés.' *MAS* (1748), 164–99.

———. 'Concerning Electricity.' *PT*, 45 (1748), 187–94.

———. *Recherches sur les causes particulières des phénomènes électriques et sur les effets nuisibles ou avantageux qu'on peut en attendre.* Paris, 1749; 1753[3].

——. 'An Examination of certain Phaenomena in Electricity.' *PT*, 46 (1749–50), 368–97.

——. 'Expériences et observations faites en différens endroits d'Italie.' *MAS* (1749), 444–88.

——. 'Extracting Electricity from the Clouds.' *PT*, 47 (1752), 553–8.

——. 'Comparaison raisonnée des plus célèbres phénomènes de l'électricité.' *MAS* (1753), 429–6.

——. 'Examen de deux questions concernant l'électricité.' *Ibid.*, 475–502.

——. 'Réponse au supplément d'un mémoire lu à l'Académie par M. Le Roy.' *Ibid.*, 503–14.

——. 'Suite du mémoire dans lequel j'ai entrepris d'examiner si l'on est bien fondé à distinguer des électricités *en plus* & *en moins*, *résineuse* & *vitrée*, comme autant d'espèces différentes.' *MAS* (1755), 293–317.

——. 'Nouvelles expériences de l'électricité, faites à l'occasion d'un ouvrage publié depuis peu en Angleterre, par M. Robert Symmer.' *MAS* (1761), 244–58.

——. 'Mémoire sur les effets du tonnerre comparés à ceux de l'électricité.' *MAS* (1764), 408–51.

——. *Lettres sur l'électricité.* [I] *dans lesquelles on examine les dernières découvertes qui ont été faites sur cette matière, & les conséquences que l'on en peut tirer.* Paris, 1753.

[II] *dans lesquelles on soutient le principe des Effluences et Affluences simultanées contre la doctrine de M. Franklin, & contre les nouvelles prétensions de ses partisans.* Paris, 1760.

[III] *dans lesquelles on trouvera les principaux phénomènes qui ont été découverts depuis 1760, avec des discussions sur les conséquences qu'on en peut tirer.* Paris, 1767.

Nordenmark, N. V. E. *Mårten Strömer.* Uppsala, 1944. (Svenska Vetenskapsakademie. *Årsbok* [1944], Bilaga.)

Nordin-Pettersson, B. S. 'K. Svenska Vetenskapsakademiens äldre prisfrågor och belöningar 1739–1820.' Svenska Vetenskapsakademie. *Årsbok* (1959), 435–516.

Occhialini, Augusto. *Notizie sull'Istituto di fisica sperimentale dello studio pisano.* Pisa, 1914.

Oldenburg, Henry. *Correspondence.* A. R. and M. B. Hall, eds. and trs. 10 vols. Madison, 1965–75.

Olson, Richard. 'The Reception of Boscovich's Ideas in Scotland.' *Isis*, 60 (1969), 91–103.

Opel, J. O. 'Otto von Guerickes Bericht an den Magistrat von Magdeburg über seine Sendung nach Osnabrück und Münster 1646–47.' Thüringisch-Sächsicher Verein für Erforschung des Vaterländischen Altertums. *Neue Mitteilungen aus dem Gebeit historisch-antiquarischer Forschungen*, 11 (1867), 23–94.

Ornstein, Martha. *The Rôle of Scientific Societies in the 17th Century.* Chicago, 1928.

Oseen, Carl Wilhelm. *Johan Carl Wilcke, experimentalfysiker.* Uppsala, 1939.

Ostertag, Heinrich. *Der philosophische Gehalt des Wolff-Manteuffelschen Briefwechsels.* Leipzig, 1910. (*Abhandlungen zur Philosophie und ihrer Geschichte*, Heft 13.)

Overfield, Richard A. 'Science in the *Virginia Gazette*, 1736–1780.' *Emporia State Research Series*, 16:3 (1968), 1–53.

Pacaut, ——. 'Le physicien Jacques Rohault. 1620–1672.' Académie des sciences, des lettres et des arts, Amiens. *Mémoires*, 8 (1881), 1–26.

Pacchi, Arrigo. *Cartesio in Inghilterra da More a Boyle*. Bari, 1973.

Pace, Antonio. *Benjamin Franklin and Italy*. Philadelphia, 1958.

Pachtler, G. M. *Ratio studiorum et Institutiones scholasticae societatis Jesu per Germaniam olim vigentes*. 4 vols. Berlin, 1887–94. (*Monumenta germaniae paedegogica*, II, V, IX, XVI.)

Paffrath, J. 'Tiberius Cavallos Beiträge zur Lehre von der Elektrizität.' *Archiv für Geschichte der Mathematik, der Naturwissenschaften und der Technik, 5 (1914)*, 86–92.

Pahl, G. *Geschichte des naturwissenschaftlichen und mathematischen Unterrichts*. Leipzig, 1913. (*Handbuch des naturwissenschaftlichen und mathematischen Unterrichts*, 1.)

Palter, Robert. 'Early Measurements of Magnetic Force.' *Isis*, 63 (1972), 544–58.

Paris, Collège de France. *Le Collège de France 1530–1930. Livre jubilaire*. Paris, 1932.

Paris, Parlement. *Arrest de la cour de parlement qui fait défense à toutes personnes de sonner les cloches pendant le tems des orages*. Paris, 1784.

Parrot, Georg Friedrich. 'Über das Gesetz der elektrischen Wirkungen in der Entfernung.' *Annalen der Physik*, 60 (1818), 22–32.

——. 'Über die Sprache der Electricitäts-Messer.' *Ibid.*, 61 (1819), 263–93.

Paulian, Aimé Henri. *L'électricité soumise à un nouvel examen, dans différentes lettres addressées à M. l'abbé Nollet*. Paris, 1768.

——. *Dictionnaire de physique*. 3 vols. Paris, 1773².

Paulsen, Friedrich. 'Wesen und geschichtliche Entwicklung der deutschen Universitäten.' Wilhelm H. R. A. Lexis, ed. *Die deutschen Universitäten*. 2 vols. Berlin, 1896. I, 1–168.

——. *Geschichte des gelehrten Unterrichts*. 3 vols. Leipzig, 1896–7.

——. 'Professorengehalt und Kollegenhonorat in geschichtlicher Beleuchtung.' *Preussische Jahrbücher*, 87 (1897), 136–44.

Pavia. University. *Contributi alla storia dell'Università di Pavia*. Pavia, 1925.

Peart, Edward. *On Electric Atmospheres in which the Absurdity of the Doctrine of Positive and Negative Electricity is incontestibly proved*. Gainsborough, 1793.

Pemberton, Henry. *A View of Sir Isaac Newton's Philosophy*. London, 1728.

Penrose, Francis. *A Treatise on Electricity*. Oxford, 1752.

Pernetti, l'abbé Jacques. *Recherches pour servir à l'histoire de Lyon*. 2 vols. Lyon, 1757.

Petersen, Peter. *Geschichte der aristotelischen Philosophie im protestantischen Deutschland*. Leipzig, 1921.

Petersson, Robert Torsten. *Sir Kenelm Digby*. Cambridge, Mass., 1956.

Philip, J. R. 'Samuel Johnson as Antiscientist.' RS. *Notes and Records, 29* (1975), 193–203.

Phillips, Edward C. 'The Correspondence of Father Christopher Clavius, S. J., preserved in the Archives of the Pont. Gregorian University.' *Archivum historicum Societatis Jesu*, 8 (1939), 193–222.

Piacenza, M. 'Note biografiche e nuovi documenti su G. B. Beccaria.' *Bolletino storico-bibliografico subalpino*, 9 (1904), 209–28, 340–54.

Picard, Charles E. 'La vie et l'oeuvre de Jean Baptiste Biot.' *Eloges et discours académiques*. Paris, 1931. Pp. 221–87.

Piderit, Johann Rudolph Anton. *Dissertatio inauguralis philosophica de electricitate.* Marburg, 1745.

———. *Dissertatio philosophica [secunda] de electricitate.* Marburg, 1746.

Pietro, Pericle di. *Lo studio pubblico di S. Carlo in Modena (1682–1772). Novant' anni di storia della Università di Modena.* Modena, 1970.

Pighetti, Clelia. 'Boyle e il corpuscolarismo inglese del seicento.' *Cultura e scuola,* 10:40 (1971), 213–19.

Playfair, John. 'Biographical Account of the late John Robison.' RS Edin. *Transactions,* 7 (1815), 495–539.

Plot, R. 'A Catalogue of Electrical Bodies.' *PT,* 20 (1698), 384.

Poggendorff, Johann Christian. *Geschichte der Physik.* Leipzig, 1879.

Pohl, Joseph. *Tentamen physico-experimentale, in principiis peripateticis fundatum, super phaenomenis electricitatis . . . compositum.* Prague, 1747.

Poisson, Siméon D. 'Sur la distribution de l'électricité à la surface des corps conducteurs.' *MAS* (1811), 1–92. 163–274.

Polinière, Pierre. *Expériences de physique* [1709]. Paris, 1718².

Polvani, Giovanni. *Alessandro Volta.* Pisa, 1942.

Power, Henry. *Experimental Philosophy in Three Books . . . in Avouchment and Illustration of the now famous Atomical Hypothesis.* London, 1664.

Prandtl, Karl von. *Geschichte der Ludwig-Maximilians-Universität in Ingolstadt, Landshut, München.* 2 vols. Munich, 1872.

Prevost, Pierre. *Recherches physico-mécaniques sur la chaleur.* Geneva, 1792.

———. *Notice de la vie et des écrits de Georges-Louis Le Sage de Genève.* Geneva, 1805.

Price, Derek J. 'The Early Observatory Instruments of Trinity College, Cambridge.' *Ann. Sci.,* 8 (1952), 1–12.

Priestley, Joseph. *The History and Present State of Electricity, with Original Experiments.* 2 vols. London, 1767; 1775³, reprinted, New York, 1966.

———. 'Experiments on the Lateral Force of Electrical Explosions.' *PT,* 59 (1769), 57–62.

———. 'Various Experiments on the Force of Electrical Explosions.' *Ibid.,* 63–70.

———. 'An Investigation of the Lateral Explosion, and of the Electricity communicated to the electrical Circuit, in a Discharge.' *PT,* 60 (1770), 192–210.

———. *Histoire de l'électricité.* M.-J. Brisson, tr. and ed. 3 vols. Paris, 1771.

———. *Scientific Correspondence of Joseph Priestley.* H. C. Bolton, ed. New York, 1892.

———. *Memoirs of Dr. Joseph Priestley, Written by Himself (To the Year 1795). With a Continuation to the Time of his Decease by his Son, Joseph Priestley.* London, 1904.

———. *A Scientific Autobiography of Joseph Priestley, 1733–1804; Selected Scientific Correspondence.* R. E. Schofield, ed. Cambridge, Mass., 1966.

Pringle, John, *et al.* 'A Report of the Committee, appointed by the Royal Society, to consider of the most effectual method of securing the Powder Magazines at Purfleet against the Effects of Lightning.' *PT,* 68 (1778), 313–17.

Prinz, Hans. 'Nachdenkliches und Belustigendes über das Hochspannungsfeld.' Schweizerischer Elektrotechnischer Verein. *Bulletin,* 61:1 (1970), 8–18.

———. 'Erschütterndes und Faszinierendes über gespeicherte Elektrizität.' *Ibid.,* 62:2 (1971), 97–109.

———. 'Fulminantes über Wolkenelektrizität.' *Ibid.,* 64:1 (1973), 1–15.

——. 'Nachdenkenswertes über nützliche Elektrizität. Méditation sur l'électricité utile.' *Ibid.*, 65:1 (1974), 1–24.

——. 'Dalla "commozione elettrica" alle affascinanti prospettive dell'elettricità immagazzinata nel condensatore statico.' *Museoscienza*, 11:5 (1971), 3–27. (Tr. from Schw. Elekt. Ver. *Bull.*, 62:2 [1971].)

——. 'I mirabili artifici sperimentali dell'elettricità scintillante.' *Museoscienza*, 12:5 (1972), 3–26. (Tr. from Schw. Elekt. Ver. *Bull.*, 63:1 [1972].)

Privat de Molières, Joseph. *Leçons de physique, contenant les élémens de la physique déterminés par les seules loix des mécaniques.* 4 vols. Paris, 1734–9.

Przibram, Karl. 'Die elektrischen Figuren.' *Handbuch der Physik*, XIV. Berlin, 1927. Pp. 391–404.

Pütter, Johann Stephan. *Versuch einer academischen Gelehrten-Geschichte von der Georg-August Universität zu Göttingen.* 4 vols. Göttingen and Hanover, 1765–1838.

Pujoulx, Jean Baptiste. *Paris à la fin du 18ᵉ siècle.* Paris, 1801.

Pulteney, Richard. *Historical and Biographical Sketches of the Progress of Botany in England, from its Origin to the Introduction of the Linnaean System.* 2 vols. London, 1790.

Pupke, H. 'Franz Ulrich Theodosius Aepinus.' *Naturwissenschaften*, 37 (1950), 49–52.

Putnam, George Haven. *The Censorship of the Church of Rome and Its Influence upon the Production and Distribution of Literature.* 2 vols. New York, 1906.

Quignon, G. Hector. *L'abbé Nollet, physicien. Son voyage en Piémont et en Italie (1749).* Amiens, 1905. (Académie des sciences, des lettres et des arts, Amiens. *Mémoires,* 51 [1904], 473–539.)

Quinn, Arthur J. *Evaporation and Repulsion: A Study of English Corpuscular Philosophy from Newton to Franklin.* Ph.D. thesis. Princeton, 1970.

Rackstrow, B. *Miscellaneous Observations, together with a Collection of Experiments on Electricity.* London, 1748.

Rait, Robert Sangster. *The Universities of Aberdeen. A History.* Aberdeen, 1895.

Randolph, Herbert. *Life of General Sir Robert Wilson.* 2 vols. London, 1862.

Ranke, Leopold. *The Popes of Rome: Their Ecclesiastical and Political History during the 16th and 17th Centuries.* Sarah Austin, tr. 2 vols. London, 1847³.

Read, John. *A Summary View of the Spontaneous Electricity of the Earth and Atmosphere.* London, 1793.

Réaumur, René Antoine Ferchault de. *Lettres inédites de Réaumur.* G. Musset, ed. La Rochelle, 1886.

Régis, Pierre Sylvain. *Système de philosophie.* 7 vols. Lyon, 1691.

Regnault, Noël. *Les Entretiens physiques d'Ariste et d'Eudoxe; ou Physique nouvelle en dialogues.* 4 vols. Paris, 1732².

——. *L'origine ancienne de la physique nouvelle.* 3 vols. Paris, 1734.

Reicke, Emil. *Neues aus der Zopfzeit. Gottscheds Briefwechsel mit dem Nürnberger Naturforscher Martin Frobenius Ledermüller.* Leipzig, 1923.

Reif, Sister Patricia. 'The Textbook Tradition in Natural Philosophy, 1600–1650.' *JHI,* 30 (1969), 17–32.

Reilly, Conor. 'Father Athanasius Kircher, S. J.' *Studies* (Dublin), 44 (1955), 457–68.

(Cf. *Journal of Chemical Education*, 32 [1955], 253–8, and *Month*, 29]1963], 20–9.)

———. 'A Catalogue of Jesuitica in the *Philosophical Transactions* of the Royal Society of London (1665–1715).' *Archivum historicum Societatis Jesu*, 27 (1958), 339–62.

———. *Francis Line, S. J. An Exiled English Scientist, 1595–1675*. Rome, 1969. (*Bibliotheca instituti historici Societatis Jesu*, XXIX.)

Remer, Wilhelm. 'Beschreibung einiger electrischen Versuche.' *Annalen der Physik*, 8 (1801), 323–41.

Reusch, Franz Heinrich. *Der Index der verbotenen Bücher*, 2 vols. Bonn, 1883–5.

Riccioli, Giovanni Battista. *Almagestum novum*. Bologna, 1651.

Richmann, G. W. 'De electricitate in corporibus producenda nova tentamina.' *CAS*, 14 (1744–6), 299–324.

———. 'De quantitate caloris, quae post miscelam fluidorum certo gradu calidorum, oriri debet, cogitationes.' *NCAS*, 1 (1747–8), 152–67.

———. 'Formulae pro gradu excessus caloris supra gradum [O°F] . . . post miscelam duarum massarum acquarum, diverso gradu calidarum, confirmatio per experimenta.' *Ibid.*, 168–73.

———. 'De argento vivo calorem celerius recipiente et celerius perdente quam multa fluida leviora experimenta et cogitationes.' *NCAS*, 3 (1750–1), 309–39.

———. 'Inquisitio in decrementa et incrementa caloris solidorum in aëre.' *NCAS*, 4 (1752–3), 241–70.

———. 'Tentamen rationem calorum respectivorum lentibus et thermometris definiendi.' *Ibid.*, 277–300.

———. 'De indice electricitatis et de eius usu in definiendis artificialis et naturalis electricitatis phaenomenis dissertatio.' *Ibid.*, 301–40.

Riedl, C. C. 'Suarez and the Organization of Learning.' Gerard Smith, ed. *Jesuit Thinkers of the Renaissance*. Milwaukee, 1939. Pp. 1–62.

Riess, Peter Theophil. *Die Lehre von der Reibungselektricität*. 2 vols. Berlin, 1853.

Rigaud, Stephen Peter, ed. *Correspondence of Scientific Men of the 17th Century*. 2 vols. Oxford, 1846.

Ring, Walter. *Geschichte der Stadt Duisburg*. Ratingen, 1949².

Rivoire, Antoine. 'Préface.' Michell. *Traité sur les aimans artificiels*, i–xcii.

Rizza, Cecilia. *Peiresc e l'Italia*. Turin, 1965.

Robinet, André. 'La vocation académicienne de Malebranche.' *RHS*, 12 (1959), 1–18.

———. 'La philosophie malebranchiste des mathématiques.' *RHS*, 14 (1961), 205–54.

———. 'Malebranche dans la pensée de Fontenelle.' *Revue de synthèse*, 82 (1961), 79–86.

———. *Malebranche vivant. Documents biographiques et bibliographiques*. Paris, 1967. (Malebranche. *Oeuvres complètes*, XX.)

———. *Malebranche de l'Académie des sciences: L'oeuvre scientifique, 1674–1715*. Paris, 1970.

Robinson, Bryan. *A Dissertation on the Aether of Sir Isaac Newton*. Dublin, 1743.

———. *Sir Isaac Newton's Account of the Aether*. Dublin, 1745.

Robinson, Eric. 'The Profession of Civil Engineer in the 18th Century: A Portrait of Thomas Yeoman, F. R. S., 1704?–1781.' *Ann. Sci.*, 18 (1962), 195–215.

———. 'Benjamin Donn (1729–1798), Teacher of Mathematics and Navigation.' *Ann. Sci.*, 19 (1963), 27–36.

————, and Douglas McKie, eds. *Partners in Science. Letters of James Watt and Joseph Black.* London, 1970.

Robinson, Myron. 'Movement of Air in the Electric Wind of the Corona Discharge.' *Communication and Electronics,* no. 54 (1961), 143–50.

————. 'A History of the Electric Wind.' *American Journal of Physics,* 30 (1962), 366–72.

————. 'The Origins of Electrical Precipitation.' *Electrical Engineering,* 82 (1963), 559–64.

Robison, John. *A System of Mechanical Philosophy.* David Brewster, ed. 4 vols. Edinburgh, 1822.

Rocca, Angelo. *De campanis commentarius.* Rome, 1612.

Roche, Daniel. 'Milieux académiques provinciaux et société des lumières. Trois académies provinciales au 18ᵉ siècle: Bordeaux, Dijon, Châlons-sur-Marne.' *Livre et société dans la France du XVIIIᵉ siècle.* Paris, 1965. Pp. 93–184. (Paris. University. *Civilisations et société,* 1.)

Rochemonteix, Camille de. *Un collège de Jésuites aux XVIIᵉ et XVIIIᵉ siècles: le Collège Henri IV de la Flèche.* 4 vols. Paris, 1889.

Rochot, Bernard. *La correspondance scientifique du père Mersenne.* Paris, 1966. (Paris. Palais de la Découverte. *Conférences,* series D, no. 110.)

Rodis-Lewis, Geneviève. *Nicolas Malebranche.* Paris, 1963.

Roger, Jacques. 'Diderot et Buffon en 1749.' *Diderot Studies,* 4 (1963), 221–36.

Rogers, G. A. J. 'Descartes and the Method of English Science.' *Ann. Sci.,* 29 (1972), 237–55.

Rohault, Jacques. *Traité de physique.* 2 vols. Paris, 1671; 1692⁶.

————. *Physica.* S. Clarke, ed. London, 1718⁴.

————. *Rohault's System of Natural Philosophy, Illustrated with Dr. Samuel Clarke's Notes Taken Mostly out of Sir Isaac Newton's Philosophy.* 2 vols. London, 1735³.

Roller, Duane Henry Du Bose. 'Did Bacon Know Gilbert's *De magnete*?' *Isis,* 44 (1953), 10–13.

————. *The De magnete of William Gilbert.* Amsterdam, 1959.

Roller, D., and Duane Henry Du Bose Roller. 'The Development of the Concept of Electric Charge.' J. B. Conant, ed. *Harvard Case Histories in Experimental Science.* 2 vols. Cambridge, Mass., 1957. II, 543–639.

Romas, Jacques de. 'Mémoire où après avoir donné un moyen aisé pour élever fort haut, et à peu de frais, un corps électrisable isolé, on rapporte des observations frappantes, qui prouvent que plus le corps isolé est audessus de la terre, plus le feu d'électricité est abondant.' AS. *Mémoires de mathématique et de physique, présentés . . . par divers savans,* 2 (1755), 393–407.

————. *Oeuvres inédites de J. de Romas sur l'électricité . . . avec une notice biographique et bibliographique par Paul Courteault.* J. Bergognié, ed. Bordeaux, 1911.

Rooseboom, Maria. *Bijdrage tot de geschiedenis der instrumentmakerkunst in de norrdelijke Nederlanden tot omstreeks 1840.* Leyden, 1950. (Leyden. Rijksmuseum voor de Geschiedenis der Natuurwetenschappen. *Mededeeling,* no. 74.)

[Rosa, Enrico]. 'Un grande fisico e precursore: il gesuità bresciano P. Lana Terzi dopo il terzo centenario della sua nascità.' *La civiltà cattolica,* 1 (1932), 211–22.

[————]. 'Le invenzioni e i criterii scientifici del P. Francesco Lana Terzi.' *Ibid.*, 424–37.

Rosenberger, Ferdinand. *Die Geschichte der Physik.* 3 vols. Braunschweig, 1882–90.

————. 'Die erste Entwicklung der Elektrisiermaschine.' *Abhandlungen zur Geschichte der Mathematik*, no. 8 (1890), 69–88.

————. 'Die erste Beobachtungen über elektrische Entladung.' *Ibid.*, 89–112.

————. *Die moderne Entwicklung der elektrischen Prinzipien.* Leipzig, 1898.

Rosenfeld, Léon. 'Newton and the Law of Gravitation.' *Archive for History of Exact Sciences*, 2 (1965), 365–86.

Rosenfeld, Leonora Cohen. 'Peripatetic Adversaries of Cartesianism in 17th Century France.' *Review of Religion*, 22 (1957), 14–40.

Rosetti, Francesco, and G. Cantoni. *Bibliografia italiana di elettricità e magnetismo.* Padua, 1881.

Rossi, Paolo. 'Hermeticism, Rationality and the Scientific Revolution.' M. L. Righini-Bonelli and William R. Shea, eds. *Reason, Experiment and Mysticism in the Scientific Revolution.* New York, 1975. Pp. 247–73.

Rota Ghibaudi, Silvia. *Ricerche su Ludovico Settala. Biografia, bibliografia, iconografia e documenti.* Florence, 1959.

Rouland, ————. 'Lettre.' *JP*, 7 (1776), 438–42.

Roule, Louis. *L'histoire de la nature vivante. Lacépède et la sociologie humanitaire selon la nature.* Paris, 1932.

Rouse Ball, W. W. *A History of the Study of Mathematics at Cambridge.* Cambridge, 1889.

Rowbottom, Margaret E. 'The Teaching of Experimental Philosophy in England, 1700–1730.' CIHS, 1965. *Actes*, IV (1968), 46–53.

————. 'John Theophilus Desaguliers (1683–1744).' Hugenot Society of London. *Proceedings*, 21 (1968), 196–218.

Roy, William. 'Experiments and Observations made in Britain, in order to obtain a Rule for measuring Heights with the Barometer.' *PT*, 67 (1777), 653–787.

Royal Society of London. *The Record.* London, 1940⁴.

Ruestow, Edward G. *Physics at 17th and 18th Century Leiden.* The Hague, 1973. (*Archives internationales d'histoire des idées*, series minor, no. 11.)

Russo, François. 'L'hydrographie en France aux XVIIᵉ et XVIIIᵉ siècles: écoles et ouvrages d'enseignement.' Taton, ed. *Enseignement*, 419–40.

Sabra, A. I. *Theories of Light: From Descartes to Newton.* London, 1967.

Sadoun-Goupil, Michelle. 'La correspondance de Claude-Louis Berthollet et Martinus van Marum (1786–1805).' *RHS*, 25 (1972), 221–52.

Sailer, Johann Michael. *Leben und Briefe.* Hubert Schiel, ed. 2 vols. Regensburg, 1948–52.

St. Andrews. University. *Matriculation Roll, 1747–1897.* James Maitland Anderson, ed. Edinburgh and London, 1905.

Sanden, Heinrich von. *Dissertatio physico-experimentalis de succino electricorum principe.* Königsberg, 1714.

Sander, Franz. *Die Auffassungen des Raumes bei Emmanuel Maignan und Johannes Baptiste Morin.* Paderborn, 1934. (*Geschichtliche Forschungen zur Philosophie der Neuzeit*, 3.)

556 Bibliography

Sarton, George. 'Lagrange's Personality (1736–1813).' American Philosophical Society. *Proceedings*, 88 (1944), 456–96.

——, René Taton and Guy Beaujouan. 'Documents nouveaux concernant Lagrange.' *RHS*, 3 (1950), 110–32.

Saurin, Joseph. 'Examen d'une difficulté considérable proposée par M. Hughens contre le système cartésien sur la cause de la pésanteur.' *MAS* (1709), 131–48.

Saussure, Horace Bénédict de. *Voyages dans les alpes*. 4 vols. Geneva and Neuchâtel, 1779–96.

——. *Essais sur l'hygrométrie*. Neuchâtel, 1783.

Savérien, Alexandre. *Histoire des philosophes modernes avec leur portrait ou allégorie*. VI. *Histoire des physiciens*. Paris, 1768. Notices of Jacques Rohault, pp. 1–62; Pierre Polinière, pp. 165–216; Joseph Privat de Molières, pp. 217–48; Jean Théophile Desaguliers, pp. 249–88; Petrus van Musschenbroek, pp. 345–98.

Saxtorph, Friedrich. *Electricitets-Laere grundet paa Erfaring og Forsøg*. 2 vols. Copenhagen, 1802–3.

Scaduto, Mario. 'Il matematico Francesco Maurolico e i gesuiti.' *Archivum historicum Societatis Jesu*, 18 (1949), 126–41.

Schäfer, Ursula. *Physikalische Heilmethoden in den ersten medizinischen Schulen*. Vienna, 1967. (Akademie der Wissenschaften, Vienna. Phil.-Hist. Kl. *Sitzungsberichte*, 254:3.)

Schaff, Josef. *Geschichte der Physik an der Universität Ingolstadt*. Erlangen, 1912.

Schilling, Johann Jacob. 'Observationes et experimenta de vi electrica vitri aliorumque corporum.' *Miscellanea berolinensia*, 4 (1734), 334–43.

——. 'Continuatio.' *Ibid.*, 5 (1737), 109–12.

Schimank, Hans. *Otto von Guericke. Bürgermeister von Magdeburg. Ein deutscher Staatsman, Denker, und Forscher*. Magdeburg, n.d. *(Magdeburger Kultur- und Wirtschaftsleben, 6.)*

——. 'Otto von Guericke.' *Beiträge zur Geschichte der Technik und Industrie*, 19 (1929), 13–30.

——. 'Geschichte der Elektrisiermachine bis zum Beginn des 19. Jahrhunderts.' *Zeitschrift für technische Physik*, 16 (1935), 245–54.

——. 'Traits of Ancient Natural Philosophy in Otto von Guericke's World Outlook.' *Organon*, 4 (1967), 27–37.

——. 'Die Wandlung des Begriffs "Physik" während der ersten Hälfte des 18. Jahrhunderts.' Manegold, ed. *Wissenschaft*, 454–68.

——. 'Zur Geschichte der Physik an der Universität Göttingen vor Wilhelm Weber (1734–1830).' *Rete*, 2 (1974), 207–52.

Schimberg, André. *L'éducation morale dans les collèges de la Compagnie de Jésus en France sous l'ancien régime*. Paris, 1913.

Schmid, Christoph von. *Erinnerungen aus meinem Leben*. Hubert Schiel, ed. Freiburg, 1963.

Schmidt-Schönbeck, Charlotte. *300 Jahre Physik und Astronomie an der Kieler Universität*. Kiel, 1965.

Schofield, Robert E. 'Introduction.' Priestley. *History* (1966), I, ix–xlix.

——. 'Electrical Researches of Joseph Priestley.' *AIHS*, 16 (1963), 277–86.

——. 'Histories of Scientific Societies: Needs and Opportunities for Research.' *History of Science*, 2 (1963), 70–83.

———, ed. *See* Priestley, J. *Scientific Autobiography.*

———. *Mechanism and Materialism. British Natural Philosophy in an Age of Reason.* Princeton, 1970.

Schonland, Basil Ferdinand Jamieson. *The Flight of Thunderbolts.* Oxford, 1950.

———. 'Benjamin Franklin: Natural Philosopher.' RS. *Proceedings,* 235:A (1956), 433–44.

Schott, Gaspar. *Mechanica hydraulico-pneumatica.* Würzburg, 1657.

———. *Thaumaturgus physicus sive Magiae universalis naturae et artis pars IV.* Würzburg, 1659.

———. *Pantometrum kircherianum.* Würzburg, 1660.

———. *Physica curiosa sive Mirabilia naturae et artis.* 2 vols. Würzburg, 1662.

———. *Technica curiosa sive Mirabilia artis.* Würzburg, 1664.

———. *Organum mathematicum.* Würzburg, 1668.

———. *Magia optica, das ist, Geheime doch naturmässige Besicht- und Augen-Lehre.* Bamberg, 1671.

Schrader, Wilhelm. *Geschichte der Universität Halle.* 2 vols. Berlin, 1894.

Schreiber, Heinrich. *Geschichte der Stadt und Universität Freiburg im Breisgau.* 7 vols. in 2 Bände. Freiburg, 1857–60. (Band II: *Geschichte der Albert-Ludwigs-Universität.*)

Schur, Friedrich H. *Johann Heinrich Lambert als Geometer.* Karlsruhe, 1905.

Schuster, Arthur. 'On the Discharge of Electricity through Gases.' RS. *Proceedings,* 37 (1884), 317–39.

Schwarzburger, Maria. 'Die Mathematikpersönlichkeiten der Universität Leipzig, 1409–1945.' Leipzig. University. *Chronik der Karl-Marx-Universität, Leipzig, 1409–1959.* Leipzig, 1959. Pp. 350–73.

Scolari, Felice. 'Il canonico Giulio Cesare Gattoni (1741–1809).' *Voltiana,* 1 (1926), 431–6.

Scolopio, Everado. *Storia dell'Università di Pisa dal 1737 al 1859.* Pisa, 1877.

Scott, J. F. *The Scientific Work of René Descartes (1596–1650).* London, [1952].

Scriba, Christoph J. 'The Autobiography of John Wallis, F.R.S.' RS. *Notes and Records,* 25 (1970), 17–46.

Secondat, Jean Baptiste, Baron de. *Mémoire sur l'électricité.* Paris, 1746.

Sédillot, M. L. A. 'Les professeurs de mathématiques et de physique générale au Collège de France.' *Bulletino di bibliografia e di storia delle scienze matematiche e fisiche,* 2 (1869), 343–68, 387–448, 461–510; 3 (1870), 107–70.

Segner, Johann Andreas von. *De mutationibus aeris a luna pendentibus.* Jena, 1733.

Selle, Götz von. *Die Georg-August Universität zu Göttingen, 1737–1937.* Göttingen, 1937.

Senebier, Jean. *Mémoire historique sur la vie et les écrits de Horace Bénédict Desaussure.* Geneva, an IX.

Senguerdius, Wolferdus. *Philosophia naturalis.* Leyden, 1680; 1685².

Shapin, Steven. 'Property, Patronage, and the Politics of Science: The Founding of the Royal Society of Edinburgh.' *BJHS,* 7 (1974), 1–41.

Shapiro, Barbara. *John Wilkins, 1614–1672.* Berkeley and Los Angeles, 1969.

Shorr, Philip. *Science and Superstition in the 18th Century. A Study of the Treatment of Science in Two Encyclopedias of 1725–1750.* New York, 1932. (Columbia University. *Studies in History, Economics and Public Law,* no. 364.)

Shuckburgh-Evelyn, Sir George Augustus William, Bart. 'Observations made in Savoy, in order to ascertain the height of Mountains by means of the Barometer.' *PT*, 67 (1777), 513–97.

──. 'An Account of Some Endeavors to ascertain a standard of Weight and Measure.' *PT*, 88 (1798), 133–76.

Sigaud de la Fond, Joseph Aignan. *Précis historique et expérimental des phénomènes électriques, depuis l'origine de cette découverte jusqu'à ce jour.* Paris, 1781.

Sigerist, Henry E. 'Albrecht von Hallers Briefe an Johannes Gesner, 1728–1777.' Akademie der Wissenschaften, Göttingen. Math.-Phys. Kl. *Abhandlungen*, 9:2 (1923).

Silliman, Robert H. 'Fresnel and the Emergence of Physics as a Discipline.' *Historical Studies in the Physical Sciences*, 4 (1974), 137–62.

Simeoni, Luigi. *Storia della Università di Bologna.* II. *L'età moderna (1500–1888).* Bologna, 1940.

Simon, Joan. *Education and Society in Tudor England.* Cambridge, 1966.

Simon, Paul Ludwig. 'Auszug aus einem Schreiben . . . an den Professor Gilbert.' *Annalen der Physik*, 27 (1807), 325–7.

──. 'Über die Gesetze, welche dem elektrischen Abstossen zum Grunde liegen.' *Ibid.*, 28 (1808), 277–98.

Singer, Dorothea Waley. 'Sir John Pringle and his Circle. I. Life.' *Ann. Sci.*, 6 (1949), 127–80.

Skempton, A. W., and Joyce Brown. 'John and Edward Troughton, Mathematical Instrument Makers.' RS. *Notes and Records*, 27 (1973), 233–62.

Smeaton, John. 'Some Improvements . . . in the Air Pump.' *PT*, 47 (1751–2), 415–28.

Smend, Rudolph. 'Die Göttinger Gesellschaft der Wissenschaften.' Akademie der Wissenschaften, Göttingen. *Festschrift des zweihundertjährigen Bestehens.* Berlin, Göttingen and Heidelberg, 1951. Pp. x–xix.

Smolka, Josef. 'Prokop Diviš and his Place in the History of Atmospheric Electricity.' *Acta historiae rerum naturalium necnon technicarum*, 1 (1965), 149–69.

──. 'Otto Guericke et son rôle dans l'histoire de l'électricité.' *Ibid.*, special issue, 2 (1966), 43–56.

Snorrason, Egill. *Kratzenstein, en flittig Dansk professor fra det 18. århunderde.* Copenhagen, 1967.

──. *C. G. Kratzenstein, Professor Physices Experimentalis Petropol. et Havn., and his Studies on Electricity during the 18th Century.* Odense, 1974. (*Acta historica scientiarum naturalium et medicinalium*, XXIX.)

Socin, Abel. 'Tentamina electrica in diversis morborum generibus, quibus accedunt levis electrometri bernoulliani adumbratio, et quorundam experimentorum instituendorum ratio.' *Acta helvetica*, 4 (1760), 214–30.

──. *Anfangsgründe der Elektricität.* Hanau, 1777; 1778².

Södergren, Olaus. *De recentioribus quibusdam in electricitate detectis.* Uppsala, [1746].

Sommervogel, Carlos, ed. *Bibliothèque de la Compagnie de Jésus.* 11 vols. Paris, 1890–1932.

Sortais, Gaston. *Le cartésianisme chez les Jésuites français au XVIIᵉ et au XVIIIᵉ siècle.* Paris, 1929. (*Archives de philosophie*, IV:3.)

Sotin, B. S. 'Raboty G.-B. Rikhmana po elektrichestvu.' IIET. *Trudy,* 44 (1962), 1–42.

Spallanzani, Lazzaro. *Epistolario.* B. Biagi et al., eds. 5 vols. Florence, 1958–64.

Spano, Nicola. *L'Università di Roma.* Rome, 1935.

Specht, Thomas. *Geschichte der ehemaligen Universität Dillingen (1549–1804).* Freiburg i. Br., 1902.

Speziali, Pierre. 'Une correspondance inédite entre Clairaut et Cramer.' *RHS,* 8 (1955), 193–237.

Spiess, Otto. *Basel anno 1760. Nach den Tagebüchern der ungarischen Grafen Joseph und Samuel Teleki.* Basel, 1936.

Sprat, Thomas. *The History of the Royal Society of London* [1767]. J.I. Cope and H.W. Jones, eds. St. Louis and London, 1958.

[Squario, Eusebio]. *Dell' elettricismo: o sia delle forze elettriche dei corpi.* Venice, 1746.

Stäckel, Paul. 'J.A. Euler.' Naturforschende Gesellschaft, Zurich. *Vierteljahrsschrift,* 55 (1910), 63–90.

Stanhope, Charles Stanhope, 3rd Earl. *Principles of Electricity.* London, 1779.

———. *Principes d'électricité.* J. T. N[eedham], tr. London, 1781.

Stanhope, Ghita, and G. P. Gooch. *The Life of Charles, Third Earl Stanhope.* London, 1914.

Stearns, Raymond Phineas. *Science in the British Colonies of America.* Urbana, Ill., 1970.

Steck, Max. *Bibliographia lambertiana.* Hildesheim, 1970².

Steffens, Henry John. *The Development of Newtonian Optics in England.* New York, 1977.

Steiglehner, Cölestin. 'Beantwortung der Preisfrage über die Analogie der Elektricität und des Magnetismus.' Akademie der Wissenschaften, Munich. *Neue philosophische Abhandlungen,* 2 (1780), 229–350.

Steinmetz, Max, et al., eds. *Geschichte der Universität Jena 1548/58–1958.* 2 vols. Jena, 1958.

Stieda, Wilhelm. 'Die Anfänge der kaiserlichen Akademie der Wissenschaften in St. Petersburg.' *Jahrbücher für Kultur und Geschichte der Slaven,* 2 (1926), 133–68.

———. 'Franz Karl Achard und die Frühzeit der deutschen Zuckerindustrie.' Akademie der Wissenschaften, Leipzig. Phil.-Hist. Kl. *Abhandlungen,* 39:3 (1928).

———. 'Die Übersiedlung Leonhard Eulers von Berlin nach St. Petersburg.' Akademie der Wissenschaften, Leipzig. Phil.-Hist. Kl. *Abhandlungen,* 83:3 (1931), 62 p.

———. 'Johann Albrecht Euler in seinen Briefen 1766–1790. Ein Beitrag zur Geschichte der kaiserlichen Akademie der Wissenschaften in St. Petersburg.' Akademie der Wissenschaften, Leipzig. Phil.-Hist. Kl. *Bericht über die Verhandlungen,* 84:1 (1932), 1–43.

———. *Erfurter Universitätsreformpläne im 18. Jahrhundert.* Erfurt, 1934. (Akademie Gemeinnütziger Wissenschaften, Erfurt. *Sonderschriften,* Heft 5.)

Stone, Lawrence, ed. *The University in Society.* 2 vols. Princeton, 1974.

———. 'The Size and Composition of the Oxford Student Body, 1580–1909.' Stone, ed. *University* (1974), I, 3–110.

Strömer, Martin. 'Untersuchung von der Electricität.' *AKSA,* 9 (1747), 154–7.

Strong, Edward W. *Procedures and Metaphysics. A Study in the Philosophy of Mathematical-Physical Science in the 16th and 17th Centuries*. Berkeley, 1936.

――. 'Newtonian Explications of Natural Philosophy.' *JHI*, 18 (1957), 49–83.

Sturm, Johann Christoph. *Physica electiva sive hypothetica*. 2 vols. Nuremberg, 1697–1722.

――. *Collegium experimentale sive curiosorum*. 3 vols. in 1. Nuremberg, 1701–15.

――. *Kurzer Begriff der Physic*. Hamburg, 1713.

Susskind, Charles. 'Henry Cavendish, Electrician.' Franklin Institute, Philadelphia. *Journal*, 249 (1950), 181–7.

Swinden, Jan Hendrik van. *Dissertatio philosophica inauguralis de attractione*. Leyden, 1766.

――. *Oratio inauguralis de causis errorum in rebus philosophicis*. Freneker, 1767.

――. *Oratio de philosophia newtoniana*. Freneker, 1779.

――. 'Dissertatio de analogia electricitatis et magnetismi.' Akademie der Wissenschaften, Munich. *Neue philosophische Abhandlungen*, 2 (1780), 1–226.

――, ed. *Recueil de mémoires sur l'analogie de l'électricité et du magnétisme*. 3 vols. The Hague, 1784.

――. 'Remarques sur le principe employé par M. Aepinus.' *Ibid.*, II, 217–66.

Symmer, Robert. 'New Experiments and Observations Concerning Electricity. I. Of the Electricity of the Human Body, and the Animal Substances, Silk and Wool.' *PT*, 51 (1759), 340–7.

'II. Of the Electricity of Black and White Silk.' *Ibid.*, 348–58.

'III. Of Electrical Cohesion.' *Ibid.*, 359–70.

'IV. Of Two Distinct Powers in Electricity.' *Ibid.*, 371–89.

Tabarroni, G. 'La torre dell'Università di Bologna e l'elettricità atmosferica.' *Coelum*, 34 (1966), 102–6, 127–33.

Taglini, Carlo. *Lettere scientifiche sopra vari dilettevoli argomenti di fisica*. Florence, 1747.

Tana, Agostino Amedeo. *Elogio del padre Beccaria*. Turin, 1781.

Tandberg, J. G. 'Die Triewaldsche Sammlung am Physikal. Institut der Universität zu Lund und die Original-luftpumpe Guerickes.' Lund. University. *Årsskrift*, avd. 2, 16:9 (1920). 31 p.

Targe, Maxime. *Professeurs et régents des collèges dans l'ancienne Université de Paris*. Paris, 1902.

Targioni-Tozzetti, Giovanni. *Notizie degli aggrandimenti delle scienze fisiche accaduti in Toscana nel corso di anni LX del secolo XVII*. 4 vols. in 3. Florence, 1780.

Taton, René. 'A propos de l'oeuvre de Monge en physique.' *RHS*, 3 (1950), 174–9.

――, ed. *Enseignement et diffusion des sciences en France au XVIIIe siècle*. Paris, 1964.

――. 'L'école royale du génie de Mézières.' *Ibid.*, 559–615.

Taylor, Brook. 'An Account of an Experiment made . . . in order to discover the Law of Magnetical Attraction.' *PT*, 29 (1714–6), 294–5.

――. 'An Account of an Experiment, made to ascertain the Proportion of the Expansion of the Liquor in the Thermometer, with Regard to the Degrees of Heat.' *PT*, 32 (1723), 291.

――. 'An Account of some Experiments relating to Magnetism.' *PT*, 31 (1720–1), 204–8.

Taylor, Eva Germaine Rimington. *The Mathematical Practitioners of Tudor and Stuart England*. Cambridge, 1954.

——. *The Mathematical Practitioners of Hanoverian England, 1714-1840*. London, 1966.

Taylor, F. S. 'Science Teaching at the End of the 18th Century.' Ferguson, ed. *Natural Philosophy*, 144-64.

Tega, Walter. 'Le *Institutiones in physicam experimentalem* di Giambattista Beccaria.' *Rivista critica di storia della filosofia*, 24 (1969), 179-211.

Teichmann, Jürgen. *Zur Entwicklung von Grundbegriffen der Elektrizitätslehre, inbesondere des elektrischen Stromes bis 1820*. Hildesheim, 1974. (*Arbor scientiarum*, 4.)

Tenca, Luigi. 'Guido Grandi, matematico cremonese (1671-1742).' Istituto lombardo di scienze e lettere, Milan. Classe di scienze matematiche e naturali. *Rendiconti*, 83 (1950), 493-510.

——. 'Relationi fra Gerolamo Saccheri e il suo allievo Guido Grandi.' Pavia. Collegio Ghislieri. *Studi matematici-fisici*, 1 (1952), 19-46.

Thackray, Arnold. *Atoms and Powers*. Cambridge, 1970.

Thijssen-Schoute, C. Louise. 'Le cartésianisme aux Pays-Bas.' *Descartes et le cartésianisme hollandais*. Amsterdam, 1950. Pp. 182-260.

Thomas Aquinas, Saint. *Commentary on Aristotle's Physics*. R. J. Blackwell, *et al.*, trs. London, 1963.

Thompson, Silvanus P. 'The Influence Machine, from 1788 to 1888.' Society of Telegraph Engineers and Electricians. *Journal*, 17 (1888), 569-628.

——. *Notes on the De Magnete of Dr. William Gilbert*. London, 1901.

Thomson, George P., and J. J. Thomson. *Conduction of Electricity through Gases*. 2 vols. Cambridge, 1928-33.

Thomson, William. 'Sur les lois élémentaires de l'électricité statique.' *Journal de mathématiques pures et appliquées*, 10 (1845), 209-21.

Thorndike, Lynn. *A History of Magic and Experimental Science*. 8 vols. New York, 1923-58.

——. '*L'Encyclopédie* and the History of Science.' *Isis*, 6 (1924), 361-86.

Todhunter, Isaac. *A History of the Mathematical Theories of Attraction and the Figure of the Earth from the Time of Newton to that of Laplace*. 2 vols. London, 1873.

Tolnai, Gabriel, ed. *La cour de Louis XV; Journal de voyage du Comte Joseph Teleki*. Paris, 1943.

Tomek, Wenzel Waldiwoj. *Geschichte der Prager Universität. Zur Feier der 500 jährigen Gründung derselben*. Prague, 1849.

Tonelli, Giorgio. 'La nécessité des lois de la nature au XVIIIe siècle et chez Kant en 1762.' *RHS*, 12 (1959), 225-41.

Torlais, Jean. *Réaumur et sa société*. Bordeaux, 1933.

——. *Un esprit encyclopédique en dehors de "l'Encyclopédie". Réaumur, d'après les documents inédits*. Paris, 1935; 1961².

——. *Un Rochelais grand-maître de la franc-maçonnerie et physicien au XVIIIe siècle, le révérend J. T. Désaguliers*. La Rochelle, 1937.

——. *Un physicien au siècle des lumières, l'abbé Nollet*. Paris, 1954.

——. 'Un prestidigitateur célèbre, chef de service d'électrothérapie au XVIIIe siècle. Ledru dit Comus (1731-1807).' *Histoire de la médicine* (Feb. 1955). 13-25.

———. 'Une grande controverse scientifique au XVIIIe siècle. L'abbé Nollet et Benjamin Franklin.' *RHS*, 9 (1956), 339–49.

———. 'Réaumur philosophe.' *RHS*, 11 (1958), 13–33. (Centre international de synthèse. *La vie*, 144–65.)

———. 'Une rivalité célèbre. Réaumur et Buffon.' *Presse médicale*, 66:2 (1958), 1057–8.

———. 'L'Académie de la Rochelle et la diffusion des sciences au XVIIIe siècle.' *RHS*, 12 (1959), 111–25.

———. 'Qui a inventé la bouteille de Leyde?' *RHS*, 16 (1963), 211–19.

———. 'Le Collège royal.' Taton, ed. *Enseignement*, 261–86.

———. 'La physique expérimentale.' *Ibid.*, 619–46.

Trefzger, Hermann. *Joseph Weber als Philosoph der katholischen Romantik*. Freiburg i. Br., 1933.

Trembley, Abraham. 'The Light caused by Quicksilver shaken in a glass Tube, proceeding from Electricity.' *PT*, 44 (1746), 58–60.

Trémery, J. L. 'Observations sur les émissions du fluide électrique.' *JP*, 48 (1799), 168–72.

———. 'On certain Facts commonly urged against the Doctrine of two Electric Fluids.' *Journal of Natural Philosophy, Chemistry and the Arts*, 3 (1802), 223–4.

———. 'Examen des phénomènes électriques qui ne paroissent pas s'accorder avec la théorie des deux fluides.' *JP*, 54 (1802) 357–67.

Tressan, H. A. G. de La Vergne, Marquis de. *Souvenirs du Comte de Tressan, Louis-Elisabeth de la Vergne*. Versailles, 1897.

Truesdell, Clifford Ambrose. 'Rational Fluid Mechanics, 1687–1765.' Leonhard Euler. *Opera omnia*, XII (1954), ix–cxxv.

———. *The Rational Mechanics of Flexible or Elastic Bodies, 1638–1788*. Zürich, 1960. (Euler, *Opera omnia*, XI:2.)

Tuge, Hideomi. *Historical Development of Science and Technology in Japan*. Tokyo, 1961.

Tunbridge, Paul A. 'Franklin's Pointed Lightning Conductor.' RS. *Notes and Records*, 28 (1974), 207–19.

Turner, Dorothy Mabel. *History of Science Teaching in England*. London, 1927.

Turner, Gerard L'E. 'The Auction Sales of the Earl of Bute's Instruments, 1793.' *Ann. Sci.*, 23 (1967), 213–43.

———. 'Henry Baker, F.R.S., Founder of the Bakerian Lecture.' RS. *Notes and Records*, 29 (1974), 53–79.

———. 'The Portuguese Agent: J. H. de Magellan.' *Antiquarian Horology*, 9:1 (1974), 74–7.

Turner, R. Stephen. 'University Reform and Professorial Scholarship in Germany, 1760–1806.' Stone, ed. *University* (1974), II, 495–531.

Tweedie, Charles. *James Stirling. A Sketch of his Life and Works, along with his Scientific Correspondence*. Oxford, 1922.

Urbanitzky, Alfred von. *Elektrizität und Magnetismus in Alterthume*. Vienna, 1887.

Vaccari, Pietro. *Storia della Università di Pavia*. Pavia, 1957².

Valentinus, Benedictus Pererius (Benito Pereira). *Adversus fallaces et superstitiosas artes, id est, de magia, de observatione somniorum, & de divinatione astrologica, libri tres*. Ingolstadt, 1591.

Vallauri, Tommaso. *Storia delle Università degli Studi del Piemonte.* 3 vols. Turin, 1845–6.

Van Doren, Carl. *Benjamin Franklin.* New York, 1938.

Van Marum, Martinus. 'Description d'une très grande machine électrique placée dans le Muséum de Teyler, à Harlem, et des expériments faits par le moyen de cette machine.' Teyler's Tweede Genootschap. *Verhandelingen,* 3 (1785). (Excerpted in *JP,* 27 [1785], 145–55.)

———. 'Première continuation des expériences faites par le moyen de la machine teylerienne.' Teyler's Tweede Genootschap. *Verhandelingen,* 4 (1787). (Excerpted in *JP,* 31 [1787], 343–50.)

———, and A. Paets von Troostwijk. 'Sur la cause de l'électricité des substances fondues & refroidies.' *JP,* 33 (1788), 248–53.

———. 'Description de frottoirs électriques d'une nouvelle construction, dont l'effet surpasse de beaucoup celui des frottoirs ordinaires.' *JP,* 34 (1789), 274–96.

———. 'Seconde lettre . . . concernant la description des nouveaux frottoirs électriques adaptés à la machine Teylerienne.' *JP,* 38 (1791), 19–25.

[———. Excerpts and translations from *Seconde continuation des expériences faites par le moyen de la machine électrique Teylerienne.* Harlem, 1795.] *Annalen der Physik,* 1 (1799), 68–122, 145–57, 242–9.

———. 'On the Theory of Franklin, according to which Electrical Phenomena are explained by a single Fluid.' *Annals of Philosophy,* 16 (1820), 440–53.

Varićak, V. 'Drugi ulomak Boškovićeve Korrespondencije (sa 2 slike van teksta).' Jugoslavenska akademija znanosti i umjetnosti. *Rad,* 193 (1912), 163–338.

Vassalli, A. M., and C. A. G. Zimmermann. 'Sperienze elettriche sopra l'acqua e il ghiaccio.' Società italiana delle scienze, Rome. *Memorie di matematica e di fisica,* 4 (1788), 264–77.

Vassalli-Eandi, A. M. 'Notizia storica di Giambattista Beccaria.' *Lo spettatore,* 5, parte italiana (Milan, 1816), 101–5, 117–22.

———. 'Memorie istoriche intorno alla vita ed agli studi di Gianfrancesco Cigna.' Accademia delle scienze, Turin. *Memorie,* 26 (1821), xiii–xxxvi.

Veratti, Giovan Giuseppe. 'De electricitate caelesti.' Accademia dell scienze dell'Istituto di Bologna. *Commentarii et opuscula,* 3 (1755), 200–4.

Verrecchia, A. 'Lichtenberg und Volta.' *Sudhoffs Archiv,* 51 (1967), 349–60.

Verri, Pietro. *Memorie appartenenti alla vita ed agli studi del signor don Paolo Frisi.* Milan, 1787.

Vidari, Giovanni.'L'educazione in Italia dall'umanesimo al risorgimento. Rome, 1930.

Viemeister, Peter E. *The Lightning Book.* New York, 1961.

Visconti, Alessandro. *La storia dell'Università di Ferrara.* Bologna, 1950.

Vleeschauwer, H. J. de. 'La genèse de la méthode mathématique de Wolff. Contribution à l'histoire des idées au XVIII siècle.' *Revue belge de philologie et d'histoire,* 11 (1932), 651–77.

Volpati, Carlo. *Alessandro Volta nella gloria e nell'intimità.* Milan, 1927.

Volta, Alessandro. *De vi attractiva ignis electrici.* Como, 1769. (*VO,* III, 23–52; Italian tr. in *Opere scelte* [1967], 49–90.)

———. *Novus ac simplicissimus electricorum tentaminum apparatus: seu De corporibus heteroelectricis quae fiunt idioelectrica experimenta atque observationes.* Como, 1770. (*VO,* III, 55–80.)

———. 'Sur l'électrophore perpétuel.' *JP,* 8 (1776), 21–4.

———. 'Osservazioni sulla capacità dei conduttori elettrici.' *Opuscoli scelti sulle scien-*

ze e sulle arti, 1 (Milan, 1778), 273–80. (*VO,* III, 201–29; *JP,* 12 [1779], 249–77.)

———. 'Del modo di render sensibilissima la più debole elettricità sia naturale, sia artificiale.' *PT,* 72 (1782), 237–80. (*VO,* III, 271–300; English tr.: *PT,* 72 (1782), Appendix, xii–xxxiii.)

———. 'Sur la capacité des conducteurs coniugués.' *JP,* 22 (1783), 325–50; 23 (1783), 1–16, 81–99. (*VO,* III, 313–77.)

———. 'On the Electricity excited by the mere Contact of Conducting Substances of Different Kinds.' *PT,* 90 (1800), 403–31. (*VO,* I, 563–82.)

———. *Le opere.* 7 vols. Milan, 1918–29. (Cited as *VO.*)

———. *Epistolario.* 5 vols. Bologna, 1949–55. (Cited as *VE.*)

———. *Opere scelte.* Mario Gliozzi, ed. Turin, 1967. (*Classici della scienza,* 10.)

Volta, Zanino. *Alessandro Volta. Parte prima: Biografia. Libro primo: Della giovinezza.* Milan, 1875.

———. 'Il primo viaggio di Alessandro Volta a Parigi [1781/2].' Istituto lombardo di scienze e lettere, Milan. *Rendiconti,* 15 (1882), 29–37.

Voltaire, François Marie Arouet de. *Correspondance.* 107 vols. Theodore Besterman, ed. Geneva, 1953–65.

———. *Lettres philosophiques* [1734]. *Edition critique.* Gustave Lanson, ed.; rev. by André M. Rousseau. 2 vols. Paris, 1964.

Vregille, P. de. 'Un enfant du Bugey. Le père Honoré Fabri.' *Bulletin de la société Gorini,* 9 (1906), 5–15.

Vries, Joseph de. 'Zur aristotelisch-scholastischen Problematik von Materie und Form.' *Scholastik,* 32 (1957), 161–85.

Vucinich, Alexander S. *Science in Russian Culture. A History to 1860.* Stanford. 1963.

Waard, Cornelis de. *L'expérience barométrique.* Thouars, 1936.

Wahl, Richard. 'Professor Bilfingers Monadologie und prästabilierte Harmonie in ihrem Verhältniss zu Leibniz und Wolf.' *Zeitschrift für Philosophie und philosophische Kritik,* 85 (1884), 66–92, 202–23.

Waitz, Jacob Sigismund von, et al. *Abhandlungen von der Elektricität.* Berlin, 1745.

Walker, W. Cameron. 'The Detection and Estimation of Electric Charges in the 18th Century.' *Ann. Sci.,* 1 (1936), 66–99.

———. 'Animal Electricity before Galvani.' *Ann. Sci.,* 2 (1937), 84–113.

Wall, Samuel. 'Experiments of the Luminous Qualities of *Amber, Diamonds* and *Gum Lac.*' *PT,* 26 (1708–9), 69–76.

Waller, Samuel. 'Lorenzo Magalotti in England, 1668–9.' *Italian Studies,* 1 (1937), 49–66.

Wallerius, Niels. 'Versuche von Beschaffenheit der Dünste und dem Ursachen ihres Aufsteigens.' *AKSA,* 9 (1747), 272–81.

Walsh, John. 'Of the Electric Property of the Torpedo.' *PT,* 63 (1773), 461–77.

Watson, William. 'Experiments and Observations tending to illustrate the Nature and Properties of Electricity.' *PT,* 43 (1744–5), 481–501.

———. 'Further Experiments and Observations.' *PT,* 44 (1746), 41–50.

———. *Experiments and Observations tending to illustrate the Nature and Properties of Electricity.* London, 1745–6. (A reprint of the four letters in *PT,* 43 (1744–5), 481–501, and 44 (1746), 41–50; first issued by J. Ilive, 1745, and then by C. Davis, 1746, the latter being thrice reprinted. Confusion has resulted (cf., *FN,*

391–2n) as Davis called his successive reprints 'editions' and numbered them without counting Ilive's: Davis' 'editions,' collectively called the second in the text, are all identical, and differ from Ilive's, called the first, in possessing a preface thanking Folkes and describing the tubes employed.)

————. *A Sequel to the Experiments and Observations tending to illustrate the Nature and Properties of Electricity*. London, 1746. (Reprint [or preprint] of *PT*, 44 [1747], 704–49; cf., *FN*, 399n.)

————. 'Observations upon . . . Monsieur le Monnier the Younger's Memoir.' *PT*, 44 (1747), 388–95.

————. 'A Continuation of a paper Concerning Electricity.' *Ibid.*, 695–704. (Continues *PT*, 43 [1744–5], 481–501; read 6 Feb. 1745–6, but published out of date in the Appendix to *PT*, 44 [1747].)

————. 'A Sequel to the Experiments and Observations tending to Illustrate the Nature and Properties of Electricity.' *PT*, 44 (1747), 704–49. (Published as an appendix 'containing some Papers which were not ready to be inserted in the order of their Date'; read 30 Oct. 1746.)

————. 'A Collection of Electrical Experiments communicated to the Royal Society by Wm. Watson.' *PT*, 45 (1748), 49–92.

————. 'Some Further Inquiries into the Nature and Properties of Electricity.' *Ibid.*, 93–120.

————. 'A Letter [concerning Pivati's transmissions of odors and Bose's beatification].' *PT*, 46 (1749–50), 348–56.

————. 'An Account of Mr. Benjamin Franklin's Treatise, lately published, intituled, *Experiments and Observations on Electricity*.' *PT*, 47 (1751–2), 202–10.

————. 'An Account of the Phaenomena of *Electricity in vacuo*.' *Ibid.*, 362–76.

————. 'Concerning the electrical Experiments in England upon Thunder Clouds.' *Ibid.*, 567–70. (*BFP*, IV, 390–1; *EO*, 262–4.)

————. 'An Account of a Treatise, presented to the Royal Society, intituled, "Letters concerning Electricity".' *PT*, 48 (1753), 201–16.

————. 'An Answer to Dr. Lining's Query relating to the Death of Professor Richmann.' *Ibid.*, 765–77.

Weaver, William D. *Catalogue of the Wheeler Gift of Books, Pamphlets and Periodicals in the Library of the American Institute of Electrical Engineers*. 2 vols. New York, 1909.

Weber, Heinrich. *Geschichte der gelehrten Schulen im Hochstift Bamberg von 1007–1803*. Bamberg, 1879–82. (Historischer Verein, Bamberg. *Bericht über Bestand und Wirken*, 42–4.)

Weber, Joseph. 'Abhandlung vom Luftelektrophor.' Akademie der Wissenschaften, Munich. *Neue philosophische Abhandlungen*, 1 (1778), 171–214.

————. 'Die Theorie der Elektricität.' Gesellschaft Naturforschender Freunde, Berlin. *Schriften*, 4 (1783), 330–77.

————. *Vom dynamischen Leben der Natur überhaupt, und vom elektrischen Leben im Doppelelektrophor insbesondere*. Landshut, 1816.

Webster, Charles. 'Henry Power's Experimental Philosophy.' *Ambix*, 14 (1967), 150–78.

————. 'Henry More and Descartes: Some New Sources.' *BJHS*, 4 (1969), 359–77.

Webster, John. *Academiarum examen, or the Examination of Academies*. London, 1654. (Debus. *Science and Education*, 67–192.)

Weld, Charles Richard. *A History of the Royal Society, with Memoirs of Presidents.* 2 vols. London, 1848.

Werner, Karl. 'Die Cartesisch-Malebranche'sche Philosophie in Italien.' Akademie der Wissenschaften, Vienna. Phil.-Hist. Kl. *Sitzungsberichte*, 102 (1883), 75–141, 679–754.

———. *Franz Suarez und die Scholastik der letzten Jahrhunderte.* 2 vols. Regensburg, 1889[2].

Wesley, John. *The Journal of the Rev. John Wesley.* Nehemiah Curnock, ed. 8 vols. London, 1909–16.

Westenrieder, Lorenz von. *Geschichte der Bayerischen Akademie der Wissenschaften.* 2 vols. Munich, 1784–1807.

Westfall, Richard S. 'Unpublished Boyle Papers Relating to Scientific Method.' *Ann. Sci.*, 12 (1956), 63–73, 103–17.

Wheler, G. 'Some Electrical Experiments, chiefly regarding the Repulsive Force of Electrical Bodies.' *PT*, 41 (1739–40), 98–117.

Whewell, William. *History of the Inductive Sciences.* 2 vols. New York, 1858[3].

Whiston, William. *Memoirs.* London, 1749.

White, Thomas. *Institutionum peripateticarum ad mentem . . . K. Digbaei pars theorica.* London, 1646.

———. *Peripatetical Institutions. In the Way of that Eminent Person and Excellent Philosopher Sir Kenelm Digby.* London, 1656.

Whiteside, Derek T. 'Newton's Early Thoughts on Planetary Motion: A Fresh Look.' *BJHS*, 2 (1964), 117–37.

Whitmore, P. J. S. *The Order of Minims in 17th-Century France.* The Hague, 1967.

Whittaker, Sir Edmund Taylor. *A History of the Theories of Aether and Electricity.* I. *The Classical Theories.* London, 1951[2].

Whyte, Lancelot Law, ed. *Roger Joseph Boscovich, S.J., F.R.S., 1711–1787.* London, 1961.

Wiedemann, Gustav Heinrich. *Die Lehre von der Elektricität.* 4 vols. in 5. Braunschweig, 1882–5.

Wieleitner, E. 'Die *Elémens de géometrie* des Paters J. G. Pardies.' *Archiv für Geschichte der Naturwissenschaften und Technik*, 1 (1908–9), 436–42.

Wiener, Julius. *Jan Ingen-Housz. Sein Leben und Wirken als Naturforscher und Arzt.* Vienna, 1905.

Wiener, Philip Paul. 'The Experimental Philosophy of Robert Boyle.' *Philosophical Review*, 41 (1932), 594–609.

Wiesinger, Johannes. *Blitzforschung und Blitzschutz.* Munich, 1972. (Munich. Deutsches Museum. *Abhandlungen und Berichte*, 40, Heft 1–2.)

Wightman, William P. D. 'The Copley Medal and the Work of Some of the Early Recipients.' *Physis*, 3 (1961), 344–55.

Wilcke, Johan Carl. *Disputatio physica experimentalis de electricitatibus contrariis.* Rostock, 1757.

———. 'Electrische Versuche und Untersuchungen, wie die electrische Ladung und Schlag durch mehr körper, als Glas und Porzellan, erhalten werden können.' *AKSA*, 20 (1758), 241–68.

———. *Tal om methoden uti naturkunnigheten.* Stockholm, 1760.

———. 'Fernere Untersuchung von den entgegengesetzten Elektricitäten bei der Ladung und den dazu gehörenden Theilen.' *AKSA*, 24 (1762), 213–35, 253–74.

——. 'Elektrische Versuche mit Phosphorus.' *AKSA*, 25 (1763), 207–26.

——. 'Geschichte des Tourmalins.' *AKSA*, 28 (1766), 95–113.

——. 'Abhandlung von Erregung der magnetischen Kraft durch die Elektricität.' *Ibid.*, 306–27.

——. 'Neue Versuche vom Gefrieren des Wassers.' *AKSA*, 31 (1769), 87–108.

——. 'Von des Schnees Kälte beim Schmelzen.' *AKSA*, 34 (1772), 93–116.

——. 'Untersuchung der bey Herrn Voltas *neuen* Electrophoro perpetuo vorkommenden elektrischen Erscheinungen.' *AKSA*, 39 (1777), 54–78, 116–30, 200–16.

——. 'Über die specifische Menge des Feuers in festen Körpern, und derselben Abmessung.' *AKSA*, 2 (1781), 48–79.

——. Über der Wärme Ausdehnung und Vertheilung, nach Veranlassung des Aufstiegens der Dunste, und der Kälte in verdünnter Luft.' *Ibid.*, 146–64.

Wilkins, John. *Mathematicall Magic, or the Wonders that may be performed by Mechanicall Geometry, concerning Mechanicall Powers and Motions.* London, 1691[4].

——, and Seth Ward. *Vindiciae academicarum, containing Some Briefe Animadversions upon Mr. Webster's Book, stiled, The Examination of Academies.* Oxford, 1654. (Debus. *Science and Education*, 193–259.)

Will, Georg Andreas. *Nürnbergisches Gelehrten-Lexikon.* 4 vols. Nuremberg and Altdorf, 1755–8.

Williams, Anna. *Miscellanies in Prose and Verse.* London, 1766.

Wilson, Arthur M. *Diderot. The Testing Years, 1713–1759.* Oxford, 1957.

Wilson, Benjamin. *An Essay towards the Explication of Electricity deduced from the Aether of Sir Isaac Newton.* London, 1746.

[——], and J. Smeaton. 'Electrical Problems.' *GM*, 17 (1747), 183–4.

——. *A Treatise on Electricity.* London, 1750; 1752[2].

——. 'Retraction of his former Opinion concerning the Explication of the Leyden Experiment.' *PT*, 49 (1755–6), 682–3.

——. 'Experiments on the Tourmalin.' *PT*, 51 (1759), 308–39.

——. 'Further Experiments in Electricity.' *Ibid.*, 896–906.

——. 'A Letter . . . to Mr. Aepinus.' *PT*, 53 (1763), 436–66.

——. 'Some Observations on the Effects of Lightning.' *PT*, 54 (1764), 246–53.

——. 'Dissent to Part of the Preceding Report.' *PT*, 63 (1773), 48.

——. 'Observations upon Lightning, and the Method of securing Buildings from its Effects.' *Ibid.*, 49–65.

——. 'Dissent from the [Henley] Report.' *PT*, 68 (1778), 239–42.

——. 'New Experiments and Observations on the Nature and Use of Conductors.' *Ibid.*, 245–313.

——. *A Short View of Electricity.* London, 1780.

Wilson, George. *The Life of the Honorable Henry Cavendish.* London, 1851.

Wimsatt, W. K., Jr. 'Johnson on Electricity.' *Review of English Studies*, 23 (1947), 257–60.

Windgårdh, K. A. 'Otto von Guericke och hans fysikaliska arbeten.' *Fra fysikkens verden*, 12 (1950), 119–39.

Winkler, Johann Heinrich. *Gedanken von den Eigenschaften, Wirkungen und Ursachen der Electricität, nebst einer Beschreibung zwo neuer electrischen Maschinen.* Leipzig, 1744. ('Essai sur la nature, les effets et les causes de l'électricité, avec une description de deux nouvelles machines à l'électricité.' *Recueil de traités sur l'électricité traduits de l'allemand et de l'anglois.* Paris, 1748. 1ᵉ partie. References in the text are to this translation, as the more accessible version.)

———. *Die Eigenschaften der elektrischen Materie und des elektrischen Feuers aus verschiedenen neuen Versuchen erkläret und nebst etlichen neuen Maschinen zum Elektrisiren.* Leipzig, 1745.

———. 'Concerning the Effects of Electricity upon Himself and his Wife.' *PT*, 44 (1746), 211–12.

———. *Die Stärke der electrischen Kraft des Wassers in gläsernen Gefässen, welche durch den Musschenbrökischen Versuch bekannt geworden.* Leipzig, 1746.

———. 'Novum reique medicae utile electricitatis inventum.' *PT*, 45 (1748), 262–70.

———. *Anfangsgründe der Physick.* Leipzig, 1754.

———. *Elements of Natural Philosophy.* 2 vols. London, 1757. (Tr. from *Anfangsgründe* [1754²].)

Winstanley, Denys Arthur. *Unreformed Cambridge. A Study of Certain Aspects of the University in the 18th Century.* Cambridge, 1935.

Winter, Eduard. *Die Registres der Berliner Akademie der Wissenschaften, 1746–1766. Dokumente für das Wirken Leonhard Eulers in Berlin.* Berlin, 1957.

Wolf, Abraham. *A History of Science, Technology and Philosophy in the 16th and 17th Centuries.* New York, 1950².

———. *A History of Science, Technology and Philosophy in the 18th Century.* London, 1952².

[Wolff, Christian.] 'Responsio ad imputationes Johannis Freindii in Transactionibus anglicis.' *Acta eruditorum* (1713), 307–14.

———. *Gedanken über das ungewöhnliche Phaenomenon welches des 17 Martii 1716 . . . gesehen worden.* Halle, 1716.

———. *Cosmologia generalis, methodo scientifica pertracta.* Frankfurt and Leipzig, 1731; 1737².

———. *Ausführliche Nachricht von seinen eigenen Schrifften die er in deutscher Sprache . . . herausgegeben.* Frankfurt, 1733².

———. *Allerhand nützliche Versuche, dadurch zu genauer Erkäntniss der Natur und Kunst der Weg gebähnet wird* [1721–3]. 3 vols. Halle, 1745–7³.

———. *Kleine Schriften.* Halle, 1755.

———. 'Lebensbeschreibung.' Wuttke, ed. *Lebensbeschreibung*, 107–201.

———. *Briefe aus den Jahren 1719–1753. Ein Beitrag zur Geschichte der Kaiserlichen Akademie der Wissenschaften zu St. Petersburg.* St. Petersburg, 1860.

Wood, Alexander, and Frank Oldham. *Thomas Young, Natural Philosopher, 1773–1829.* Cambridge, 1954.

Woods, Samuel. 'Essay on the Franklinian Theory of Electricity.' *PM*, 17 (1803), 97–113.

Wordsworth, Christopher. *Social Life at the English Universities in the 18th Century.* Cambridge, 1874.

———. *Scholae academicae. Some Account of the Studies at the English Universities in the 18th Century.* Cambridge, 1877.

Worster, Benjamin. *A Compendious and Methodical Account of the Principles of Natural Philosophy.* London, 1730².

Wurzbach, Constantin von. *Biographisches Lexikon des Kaiserthums Österreich.* 60 vols. Vienna, 1856–91.

Wuttke, Heinrich. *Christian Wolffs eigene Lebensbeschreibung.* Leipzig, 1841.

Yamazaki, Elizio. 'L'abbé Nollet et Benjamin Franklin. Une phase finale de la physique cartésienne: la théorie de la conservation de l'électricité et de l'expérience de Leyde.' *Japanese Studies in the History of Science,* 15 (1976), 37–64.

Yates, Frances A. 'The Hermetic Tradition in Renaissance Science.' C. S. Singleton, ed. *Art, Science and History in the Renaissance.* Baltimore, 1968. Pp. 255–74.

Young, Thomas. *A Course of Lectures on Natural Philosophy and the Mechanical Arts.* 2 vols. London, 1807.

Zeleny, John 'Observations and Experiments on Condensers with Removable Coats.' *American Journal of Physics,* 12 (1944), 329–39.

Zentgraf, Eduard. *Aus der Geschichte der Naturwissenschaften an der Universität Freiburg i. Br.* Freiburg i. Br., 1957. (*Beiträge zur Freiburger Wissenschafts- und Universitätsgeschichte,* Heft 18.)

Ziggelaar, August. *Le physicien Ignace Gaston Pardies, S. J. (1636–1673).* Odense, 1971. (*Acta historica scientiarum et medicinalium,* XXVI.)

Zilsel, E. 'The Origins of Gilbert's Scientific Method.' *JHI,* 2 (1941), 1–32.

Zubov, Vasiliĭ Pavlovich. 'Kalorimetricheskaia formula Rikīmana i ee predistoriia.' IIET. *Trudy,* 5 (1955), 69–93.

———. *Istoriografie estestvennykh nauk v Rossii (xviii v. - pervaia polovina xix v.)* Moscow, 1956.

———. 'Die Begegnung der deutschen und der russischen Naturwissenschaft im 18. Jahrhundert und Euler.' E. Winter, ed. *Die deutsch-russische Begegnung und Leonhard Euler.* Berlin, 1958. Pp. 19–48.

———. 'La formule calorimétrique et ses origines.' *Mélanges Alexandre Koyré,* I, 654–61.

Zucchi, Nicola. *Nova de machinis philosophia . . . .* Rome, 1649.

Index

Besides the usual references to texts and notes, the Index refers to all early-modern authors mentioned in the notes and to modern authors whenever their opinions are quoted or paraphrased. Otherwise sources are not indexed. A given item may appear several times in the notes, or in both notes and text, on the same page. In these cases only one reference is given. Names of institutions will be found under their locations, e.g., Paris, Académie des Sciences; cross references are provided when the name does not indicate place. Long entries are usually arranged from the general to the particular. The Index supplements the text by giving full names of persons cited and birth and death dates, when available, of the early-modern physicists.

The following abbreviations are used:

AS. Académie des Sciences, Accademia delle Scienze. No distinction is made between an AS and a general academy, e.g., Académie des Sciences, Arts, et Belles-lettres.

AW. Akademie der Wissenschaften or its equivalent in Dutch or Swedish.

BF. Benjamin Franklin.

±E. Plus and Minus Electricity.

El. Electric, Electrical.

Expt(s). Experiment, Experiments.

LJ. Leyden Jar.

Px. Physique expérimentale, Experimental Physics.

SJ(s). Jesuit, Jesuits.

SJC. Jesuit College.

U(s). University, Universities.

Thermometer, *cont.*
improvement of, 81; electrical, 370n
Thirty Years' War: 183-4, 213
Thomas Anglicus. *See* White, Thomas
Thomas Aquinas, Saint: 21n, 25, 108; on
magnetic attraction, 23; on authority in
science, 109; praised by Gilbert, 172n
Thomson, George Paget: 297n
Thomson, James: 431
Thomson, Joseph John: 297n
Thomson, William: on Coulomb's law, 476; on
Cavendish, 485n
Thread electroscope: 418, 449-50
Thread experiments (Hanksbee's): 232-4, 238,
241, 264, 275, 295, 364; elucidated by
Dufay, 256; Nollet, 286, 361; Bose, 288n
Tides: cause of ocean, 33; atmospheric, 74
Tilly, Johannes Tserklaes, Count of: 213
Timaeus Locrensis. *See* Cornelio
Time: in propagation of light, 172; effect of, in
el. expts., 476, 485n, 487
Tobacco industry: 120
Tobacco pipe: el. properties of, 380
Torpedo: 209; analogy of, to magnet, 23; to
pile, 489n, 494; violates Dufay's rule, 488;
explained by Cavendish, 487-9; and
Galvanism, 491
Torricelli, Evangelista (1608-1647): 186, 195,
230, 445
Torricelli space: 477n; expts. in, 205-8; as
insulator, 207-8. *See also* Vacuum
Torsion: 95; vibrations of, 470
Torsion balance: 495; Michell's, 471n;
Coulomb's, 469-71; problems with, 476
Toulouse: 107, 210, 211
Toulouse, AS: 127
Toulouse, U.: 111, 223
Tourmaline: 387n, 402; Aepinus' work on,
387-8; Wilson's, 383
Townshend, Charles: 432n
Townshend, Roger: 432n
Traditional electricity: defined, 490, 491
Transdiction: 461, 465; defined, 449; difficulty
of, 87-97, 459-89
Transfusion of blood: 124
Transsubstantiation: 35
Traviginus, F.: 183n
Treasurer of the Chamber: 432
Trembley, Abraham (1710-1784): 314
Trémery, Jean Louis (1733-1851): 443n; and
Lullin's expt., 444
Tressan, Louis Elizabeth de la Vergne, *comte
de* (1705-1783): 164n
Tria prima: 111
Triboelectricity: and surface conditions, 4;
theories of, 453n; substances ordered by,
387, 388n, 417n
Triewald, Mårten (1691-1747): his cabinet, 80,

148; as Px lecturer, 163; and AW Stockholm,
125; on el. motions, 292
Trootswijk, Adriaan Paets van (1752-1837):
418n
Truth in physics: 61, 292, 403, 436, 438,
499-500; unseemly desire for, 307; and
mathematics, 76, 85, 104; according to
Fontenelle and Boyle, 41-3; Newton, 46-7;
Locke, 56; Boscovich, 67n; Mayer, 92;
Aepinus, 92, 401; Lana, 110-11;
Lichtenberg, 448; Fischer, 476
Tube: 232; instructions for use of, 242, 327n;
why ineffective, 338
Tübingen, U.: 44, 121
Tuilleries, gardens of the: 320
Turgot, Anne Robert Jacques: 382n
Turin: 38, 280, 354, 413; as Beccaria's
battleground, 363-4, 369-70
Turin, AS: 122, 406
Turin, U.: 138, 406; its cabinet, 151, 280n; Px
at, 146, 362-3
Turpentine: 416
Two electricities: 263, 272, 280, 317, 386; not
inherent, 257, 387; mechanical models of,
242, 272, 275-6, 285-6, 288n, 303, 435;
compared with contrary electricities, 385.
See also Electricity plus and minus
Tyrnau, SJC: 152
Tytler, James (1747?-1805). See *Encyclopedia
Britannica*

Undercharged: defined, 481
Unger, Johann Friedrich (1716-1781): 274-5
Unicorn: 185
Universities: and academies, 121-2, 128-31,
134, 137, 163; and Jesuit schools, 104; and
Px, 147-52; distinguished in physics, 138;
and Cartesiansim, 35-8; enrollments, 138-9
Unmoved mover: 20n
Unsociableness: 52
Uppsala, U.: 6, 86, 141, 390, 438; and
Cartesianism, 35; Px at, 148, 157, 384-5;
dissertations on electricity, 263, 315; and
AW Stockholm, 130n
Uranus: 468
Utility: in physical theory, 72
Utrecht, U.: 44, 142, 312n; and Cartesianism,
35; and Px, 138, 148-9; salaries, 154n

VE: defined, 416-18
Vacuum: 172; fear of, 197; as interplanetary
space, 214; resembles −E, 445; as coating of
LJ, 335, 357-8, 368; el. light in, 230-1,
477n; its conductivity, 82, 336, 364n, 371;
el. expts. in, 205-8, 233, 238, 346n, 370-1,
428, 439-40, 442-5, 490
Valentinus, Benedictus Pererius: 190n
Variation, magnetic. *See under* Magnetism

Designer: Dave Pauly
Compositor: Typesetting Services of California

Text: VIP Times Roman
Display: VIP Times Roman

CPSIA information can be obtained
at www.ICGtesting.com
Printed in the USA
JSHW060719020123
35616JS00001B/6